Lecture Notes in Computer Science 9986

Commenced Publication in 1973
Founding and Former Series Editors:
Gerhard Goos, Juris Hartmanis, and Jan van Leeuwen

More information about this series at http://www.springer.com/series/7410

Martin Hirt · Adam Smith (Eds.)

Theory of Cryptography

14th International Conference, TCC 2016-B
Beijing, China, October 31 – November 3, 2016
Proceedings, Part II

 Springer

Editors
Martin Hirt
Department of Computer Science
ETH Zurich
Zurich
Switzerland

Adam Smith
Pennsylvania State University
University Park, PA
USA

ISSN 0302-9743 ISSN 1611-3349 (electronic)
Lecture Notes in Computer Science
ISBN 978-3-662-53643-8 ISBN 978-3-662-53644-5 (eBook)
DOI 10.1007/978-3-662-53644-5

Library of Congress Control Number: 2016954934

LNCS Sublibrary: SL4 – Security and Cryptology

Printed on acid-free paper

This Springer imprint is published by Springer Nature
The registered company is Springer-Verlag GmbH Germany
The registered company address is: Heidelberger Platz 3, 14197 Berlin, Germany

Preface

The 14th Theory of Cryptography Conference (TCC 2016-B) was held October 31 to November 3, 2016, at the Beijing Friendship Hotel in Beijing, China. It was sponsored by the International Association for Cryptographic Research (IACR) and organized in cooperation with State Key Laboratory of Information Security at the Institute of Information Engineering of the Chinese Academy of Sciences. The general chair was Dongdai Lin, and the honorary chair was Andrew Chi-Chih Yao.

The conference received 113 submissions, of which the Program Committee (PC) selected 45 for presentation (with three pairs of papers sharing a single presentation slot per pair). Of these, there were four whose authors were all students at the time of submission. The committee selected "Simulating Auxiliary Inputs, Revisited" by Maciej Skórski for the Best Student Paper award. Each submission was reviewed by at least three PC members, often more. The 25 PC members, all top researchers in our field, were helped by 154 external reviewers, who were consulted when appropriate. These proceedings consist of the revised version of the 45 accepted papers. The revisions were not reviewed, and the authors bear full responsibility for the content of their papers.

As in previous years, we used Shai Halevi's excellent Web review software, and are extremely grateful to him for writing it and for providing fast and reliable technical support whenever we had any questions. Based on the experience from the last two years, we used the interaction feature supported by the review software, where PC members may directly and anonymously interact with authors. The feature allowed the PC to ask specific technical questions that arose during the review process, for example, about suspected bugs. Authors were prompt and extremely helpful in their replies. We hope that it will continue to be used in the future.

This was the third year where TCC presented the Test of Time Award to an outstanding paper that was published at TCC at least eight years ago, making a significant contribution to the theory of cryptography, preferably with influence also in other areas of cryptography, theory, and beyond. The Test of Time Award Committee consisted of Tal Rabin (chair), Yuval Ishai, Daniele Micciancio, and Jesper Nielsen. They selected "Indifferentiability, Impossibility Results on Reductions, and Applications to the Random Oracle Methodology" by Ueli Maurer, Renato Renner, and Clemens Holenstein—which appeared in TCC 2004, the first edition of the conference—for introducing indifferentiability, a security notion that had "significant impact on both the theory of cryptography and the design of practical cryptosystems." Sadly, Clemens Holenstein passed away in 2012. He is survived by his wife and two sons. Maurer and Renner accepted the award on his behalf. The authors delivered a talk in a special session at TCC 2016-B. An invited paper by them, which was not reviewed, is included in these proceedings.

The conference featured two other invited talks, by Allison Bishop and Srini Devadas. In addition to regular papers and invited events, there was a rump session featuring short talks by attendees.

We are greatly indebted to many people who were involved in making TCC 2016-B a success. First of all, our sincere thanks to the most important contributors: all the authors who submitted papers to the conference. There were many more good submissions than we had space to accept. We would like to thank the PC members for their hard work, dedication, and diligence in reviewing the papers, verifying their correctness, and discussing their merits in depth. We are also thankful to the external reviewers for their volunteered hard work in reviewing papers and providing valuable expert feedback in response to specific queries. For running the conference itself, we are very grateful to Dongdai and the rest of the local Organizing Committee. Finally, we are grateful to the TCC Steering Committee, and especially Shai Halevi, for guidance and advice, as well as to the entire thriving and vibrant theoretical cryptography community. TCC exists for and because of that community, and we are proud to be a part of it.

November 2016

Martin Hirt
Adam Smith

TCC 2016-B

Theory of Cryptography Conference

Beijing, China
October 31 – November 3, 2016

Sponsored by the International Association for Cryptologic Research and organized in cooperation with the State Key Laboratory of Information Security, Institute of Information Engineering, Chinese Academy of Sciences.

General Chair

Dongdai Lin   Chinese Academy of Sciences, China

Honorary Chair

Andrew Chi-Chih Yao   Tsinghua University, China

Program Committee

Masayuki Abe	NTT, Japan
Divesh Aggarwal	NUS, Singapore
Andrej Bogdanov	Chinese University of Hong Kong, Hong Kong
Elette Boyle	IDC Herzliya, Israel
Anne Broadbent	University of Ottawa, Canada
Chris Brzuska	TU Hamburg, Germany
David Cash	Rutgers University, USA
Alessandro Chiesa	University of California, Berkeley, USA
Kai-Min Chung	Academia Sinica, Taiwan
Nico Döttling	University of California, Berkeley, USA
Sergey Gorbunov	University of Waterloo, Canada
Martin Hirt (Co-chair)	ETH Zurich, Switzerland
Abhishek Jain	Johns Hopkins University, USA
Huijia Lin	University of California, Santa Barbara, USA
Hemanta K. Maji	Purdue University, USA
Adam O'Neill	Georgetown University, USA
Rafael Pass	Cornell University, USA
Krzysztof Pietrzak	IST Austria, Austria
Manoj Prabhakaran	IIT Bombay, India
Renato Renner	ETH Zurich, Switzerland
Alon Rosen	IDC Herzliya, Israel
abhi shelat	Northeastern University, USA
Adam Smith (Co-chair)	Pennsylvania State University, USA

John Steinberger   Tsinghua University, China
Jonathan Ullman   Northeastern University, USA
Vinod Vaikuntanathan   MIT, USA
Muthuramakrishnan   University of Rochester, USA
   Venkitasubramaniam

TCC Steering Committee

Mihir Bellare   UCSD, USA
Ivan Damgård   Aarhus University, Denmark
Shafi Goldwasser   MIT, USA
Shai Halevi (Chair)   IBM Research, USA
Russell Impagliazzo   UCSD, USA
Ueli Maurer   ETH, Switzerland
Silvio Micali   MIT, USA
Moni Naor   Weizmann Institute, Israel
Tatsuaki Okamoto   NTT, Japan

External Reviewers

Hamza Abusalah   Michele Ciampi   Carmit Hazay
Shashank Agrawal   Aloni Cohen   Brett Hemenway
Shweta Agrawal   Ran Cohen   Felix Heuer
Joël Alwen   Angelo Decaro   Ryo Hiromasa
Prabhanjan Ananth   Jean Paul Degabriele   Dennis Hofheinz
Saikrishna   Akshay Degwekar   Justin Holmgren
   Badrinarayanan   Itai Dinur   Pavel Hubáček
Marshall Ball   Léo Ducas   Tsung-Hsuan Hung
Raef Bassily   Tuyet Duong   Vincenzo Iovino
Carsten Baum   Andreas Enge   Aayush Jain
Amos Beimel   Antonio Faonio   Chethan Kamath
Fabrice Benhamouda   Oriol Farras   Tomasz Kazana
Itay Berman   Pooya Farshim   Raza Ali Kazmi
Nir Bitansky   Sebastian Faust   Carmen Kempka
Alexander R. Block   Omar Fawzi   Florian Kerschbaum
Tobias Boelter   Max Fillinger   Dakshita Khurana
Zvika Brakerski   Nils Fleischhacker   Fuyuki Kitagawa
Brandon Broadnax   Eiichiro Fujisaki   Susumu Kiyoshima
Ran Canetti   Peter Gaži   Saleet Klein
Andrea Caranti   Satrajit Ghosh   Ilan Komargodski
Nishanth Chandran   Alexander Golovnev   Venkata Koppula
Yi-Hsiu Chen   Siyao Guo   Stephan Krenn
Yilei Chen   Divya Gupta   Mukul Ramesh Kulkarni
Yu-Chi Chen   Venkatesan Guruswami   Tancrède Lepoint
Seung Geol Choi   Yongling Hao   Kevin Lewi

Contents – Part II

Attribute-Based Encryption

Functional Encryption

Secret Sharing

New Models

Contents – Part I

Foundations of Multi-Party Protocols

Round Complexity and Efficiency of Multi-party Computation

Differential Privacy

Delegation and IP

Delegating RAM Computations with Adaptive Soundness and Privacy

Prabhanjan Ananth[1]([✉]), Yu-Chi Chen[2], Kai-Min Chung[2], Huijia Lin[3], and Wei-Kai Lin[4]

[1] Center for Encrypted Functionalities,
University of California Los Angeles, Los Angeles, USA
prabhanjan@cs.ucla.edu
[2] Academia Sinica, Taipei, Taiwan
{wycchen,kmchung}@iis.sinica.edu.tw
[3] University of California, Santa Barbara, USA
rachel.lin@cs.ucsb.edu
[4] Cornell University, Ithaca, USA
wklin@cs.cornell.edu

Abstract. We consider the problem of delegating RAM computations over persistent databases. A user wishes to delegate a sequence of computations over a database to a server, where each computation may read and modify the database and the modifications persist between computations. Delegating RAM computations is important as it has the distinct feature that the run-time of computations maybe *sub-linear* in the size of the database.

We present the first RAM delegation scheme that provide both soundness and privacy guarantees in the *adaptive* setting, where the sequence of delegated RAM programs are chosen adaptively, depending potentially on the encodings of the database and previously chosen programs. Prior works either achieved only adaptive soundness without privacy [Kalai and Paneth, ePrint'15], or only security in the selective setting where all RAM programs are chosen statically [Chen et al. ITCS'16, Canetti and Holmgren ITCS'16].

Our scheme assumes the existence of indistinguishability obfuscation (iO) for circuits and the decisional Diffie-Hellman (DDH) assumption. However, our techniques are quite general and in particular, might be applicable even in settings where iO is not used. We provide a *"security lifting technique"* that "lifts" any proof of selective security satisfying certain special properties into a proof of adaptive security, for arbitrary cryptographic schemes. We then apply this technique to the delegation scheme of Chen et al. and its selective security proof, obtaining that their scheme is essentially already adaptively secure. Because of the general approach, we can also easily extend to delegating parallel RAM (PRAM) computations. We believe that the security lifting technique can potentially find other applications and is of independent interest.

This paper was presented jointly with "Adaptive Succinct Garbled RAM, or How To Delegate Your Database" by Ran Canetti, Yilei Chen, Justin Holmgren, and Mariana Raykova. The full version of this paper is available on ePrint [2]. Information about the grants supporting the authors can be found in "Acknowledgements" section.

M. Hirt and A. Smith (Eds.): TCC 2016-B, Part II, LNCS 9986, pp. 3–30, 2016.
DOI: 10.1007/978-3-662-53644-5_1

1   Introduction

In the era of cloud computing, it is of growing popularity for users to outsource both their databases and computations to the cloud. When the databases are large, it is important that the delegated computations are modeled as RAM programs for efficiency, *as computations maybe sub-linear*, and that the state of a database is kept persistently across multiple (sequential) computations to support continuous updates to the database. In such a paradigm, it is imperative to address two security concerns: *Soundness* (a.k.a., integrity) – ensuring that the cloud performs the computations correctly, and *Privacy* – information of users' private databases and programs is hidden from the cloud. In this work, we design *RAM delegation schemes* with both soundness and privacy.

Private RAM Delegation. Consider the following setting. Initially, to outsource her database DB, a user encodes the database using a secret key sk, and sends the encoding $\hat{DB}$ to the cloud. Later, whenever the user wishes to delegate a computation over the database, represented as a RAM program M, it encodes M using sk, producing an encoded program $\hat{M}$. Given $\hat{DB}$ and $\hat{M}$, the cloud runs an evaluation algorithm to obtain an encoded output $\hat{y}$, on the way updating the encoded database; for the user to verify the correctness of the output, the server additionally generates a proof π. Finally, upon receiving the tuple $(\hat{y}, \pi)$, the user verifies the proof and recovers the output y in the clear. The user can continue to delegate multiple computations.

In order to leverage the efficiency of RAM computations, it is important that RAM delegation schemes are *efficient*: The user runs in time only proportional to the size of the database, or to each program, while the cloud runs in time proportional to the run-time of each computation.

Adaptive vs. Selective Security. Two "levels" of security exist for delegation schemes: The, *weaker*, selective security provides guarantees only in the restricted setting where all delegated RAM programs and database are chosen statically, whereas, the, *stronger*, adaptive security allows these RAM programs to be chosen adaptively, each (potentially) depending on the encodings of the database and previously chosen programs. Clearly, adaptive security is more natural and desirable in the context of cloud computing, especially for these applications where a large database is processed and outsourced once and many computations over the database are delegated over time.

We present an adaptively secure RAM delegation scheme.

Theorem 1 (Informal Main Theorem). *Assuming DDH and $i\mathcal{O}$ for circuits, there is an efficient RAM delegation scheme, with adaptive privacy and adaptive soundness.*

Our result closes the gaps left open by previous two lines of research on RAM delegation. In one line, Chen et al. [20] and Canetti and Holmgren [16] constructed the first RAM delegation schemes that achieve *selective privacy* and *selective soundness*, assuming $i\mathcal{O}$ and one-way functions; their works, however, left open security in the adaptive setting. In another line, Kalai and Paneth [35], building upon the seminal result of [36], constructed a RAM delegation scheme with *adaptive soundness*, based on super-polynomial hardness of the LWE assumption, which, however, does not provide privacy at all.[1] Our RAM delegation scheme improves upon previous works — it simultaneously achieves adaptive soundness and privacy. Concurrent to our work, Canetti, Chen, Holmgren, and Raykova [15] also constructed such a RAM delegation scheme. Our construction and theirs are the first to achieve these properties.

1.1 Our Contributions in More Detail

Our RAM delegation scheme achieves the privacy guarantee that the encodings of a database and many RAM programs, chosen adaptively by a malicious server (i.e., the cloud), reveals nothing more than the outputs of the computations. This is captured via the simulation paradigm, where the encodings can be simulated by a simulator that receives only the outputs. On the other hand, soundness guarantees that no malicious server can convince an honest client (i.e., the user) to accept a wrong output of any delegated computation, even if the database and programs are chosen adaptively by the malicious server.

Efficiency. Our adaptively secure RAM delegation scheme achieves the same level of efficiency as previous selectively secure schemes [16, 20]. More specifically,

- CLIENT DELEGATION EFFICIENCY: To outsource a database DB of size n, the client encodes the database in time linear in the database size, $n \operatorname{poly}(\lambda)$ (where λ is the security parameter), and the server merely stores the encoded database. To delegate the computation of a RAM program M, with l-bit outputs and time and space complexity T and S, the client encodes the program in time linear in the output length and polynomial in the program description size $l \times \operatorname{poly}(|M|, \lambda)$, independent of the complexity of the RAM program.
- SERVER EVALUATION EFFICIENCY: The evaluation time and space complexity of the server, scales linearly with the complexity of the RAM programs, that is, $T \operatorname{poly}(\lambda)$ and $S \operatorname{poly}(\lambda)$ respectively.
- CLIENT VERIFICATION EFFICIENCY: Finally, the user verifies the proof from the server and recovers the output in time $l \times \operatorname{poly}(\lambda)$.

The above level of efficiency is comparable to that of an *insecure* scheme (where the user simply sends the database and programs in the clear, and does not verify the correctness of the server computation), up to a multiplicative $\operatorname{poly}(\lambda)$ overhead at the server, and a $\operatorname{poly}(|M|, \lambda)$ overhead at the user.[2] In particular, if the run-time of a delegated RAM program is sub-linear $o(n)$, the server evaluation time is also sub-linear $o(n) \operatorname{poly}(\lambda)$, which is crucial for server efficiency.

---

[1] Note that here, privacy cannot be achieved for free using Fully Homomorphic Encryption (FHE), as FHE does not directly support computation with RAM programs, unless they are first transformed into oblivious Turing machines or circuits.

[2] We believe that the polynomial dependency on the program description size can be further reduced to linear dependency, using techniques in the recent work of [5].

Technical Contributions. Though our RAM delegation scheme relies on the existence of $i\mathcal{O}$, the techniques that we introduce in this work are quite general and in particular, might be applicable in settings where $i\mathcal{O}$ is not used at all.

Our main theorem is established by showing that the selectively secure RAM delegation scheme of [20] (CCC+ scheme henceforth) is, in fact, also adaptively secure (up to some modifications). However, proving its adaptive security is challenging, especially considering the heavy machinery already in the selective security proof (inherited from the line of works on succinct randomized encoding of Turing machines and RAMs [10,17]). Ideally, we would like to have a proof of adaptive security that uses the selective security property in a black-box way. A recent elegant example is the work of [1] that constructed an adaptively secure functional encryption from any selectively secure functional encryption without any additional assumptions.[3] However, such cases are rare: In most cases, adaptive security is treated independently, achieved using completely new constructions and/or new proofs (see examples, the adaptively secure functional encryption scheme by Waters [44], the adaptively secure garbled circuits by [34], and many others). In the context of RAM delegation, coming up with a proof of adaptive security from scratch requires at least repeating or rephrasing the proof of selective security and adding more details (unless the techniques behind the entire line of research [16,20,37] can be significantly simplified).

Instead of taking this daunting path, we follow a more principled and general approach. We provide an abstract proof that "lifts" any selective security proof satisfying certain properties — called a "nice" proof — into an adaptive security proof, for arbitrary cryptographic schemes. With the abstract proof, the task of showing adaptive security boils down to a mechanic (though possibly tedious) check whether the original selective security proof is nice. We proceed to do so for the CCC+ scheme, and show that when the CCC+ scheme is plugged in with a special kind of positional accummulator [37], called *history-less accummulator*, all niceness properties are satisfied; then its adaptive security follows immediately. At a very high-level, history-less accummulators can statistically bind the value at a particular position q irrespect of the history of read/write accesses, whereas positional accumulators of [37] binds the value at q after a specific sequence of read/write accesses.

Highlights of techniques used in the abstract proof includes a stronger version of complexity leveraging—called small-loss complexity leveraging—that have much smaller security loss than classical complexity leveraging, when the security game and its selective security proof satisfy certain "niceness" properties, as well as a way to apply small-loss complexity leveraging locally inside an involved security proof. We provide an overview of our techniques in more detail in Sect. 2.

Parallel RAM (PRAM) Delegation. As a benefit of our general approach, we can easily handle delegation of PRAM computations as well. Roughly speaking, PRAM programs are RAM programs that additionally support parallel (random)

---

[3] More generally, they use a 1-query adaptively secure functional encryption, which can be constructed from one-way functions by [32].

accesses to the database. Chen et al. [20] presented a delegation scheme for PRAM computations, with selective soundness and privacy. By applying our general technique, we can also lift the selective security of their PRAM delegation scheme to adaptive security, obtaining an adaptively secure PRAM delegation scheme.

Theorem 2 (Informal — PRAM Delegation Scheme). *Assuming DDH and the existence of iO for circuits, there exists an efficient PRAM delegation scheme, with adaptive privacy and adaptive soundness.*

1.2 Applications

In the context of cloud computing and big data, designing ways for delegating computation privately and efficiently is important. Different cryptographic tools, such as Fully Homomorphic Encryption (FHE) and Functional Encryption (FE), provide different solutions. However, so far, none supports delegation of *sub-linear* computation (for example, binary search over a large ordered data set, and testing combinatorial properties, like k-connectivity and bipartited-ness, of a large graph in sub-linear time). It is known that FHE does not support RAM computation, for the evaluator cannot decrypt the locations in the memory to be accessed. FE schemes for Turing machines constructed in [7] cannot be extended to support RAM, as the evaluation complexity is at least linear in the size of the encrypted database. This is due to a refreshing mechanism crucially employed in their work that "refreshes" the entire encrypted database in each evaluation, in order to ensure privacy. To the best of our knowledge, RAM delegation schemes are the only solution that supports sub-linear computations.

Apart from the relevance of RAM delegation in practice, it has also been quite useful to obtain theoretical applications. Recently, RAM delegation was also used in the context of patchable obfuscation by [6]. In particular, they crucially required that the RAM delegation satisfies adaptive privacy and only our work (and concurrently [15]) achieves this property.

1.3 On the Existence of IO

Our RAM delegation scheme assumes the existence of IO for circuits. So far, in the literature, many candidate IO schemes have been proposed (e.g., [9,14,26]) building upon the so called graded encoding schemes [23–25,29]. While the security of these candidates have come under scrutiny in light of two recent attacks [22,42] on specific candidates, there are still several IO candidates on which the current cryptanalytic attacks don't apply. Moreover, current multilinear map attacks do not apply to IO schemes obtained after applying bootstrapping techniques to candidate IO schemes for NC^1 [8,10,18,26,33] or special subclass of constant degree computations [38], or functional encryption schemes for NC^1 [4,5,11] or NC^0 [39]. We refer the reader to [3] for an extensive discussion of the state-of-affairs of attacks.

1.4   Concurrent and Related Works

Concurrent and independent work: A concurrent and independent work achieving the same result of obtaining adaptively secure RAM delegation scheme is by Canetti et. al. [15]. Their scheme extends the selectively secure RAM delegation scheme of [16], and uses a new primitive called adaptive accumulators, which is interesting and potentially useful for other applications. They give a proof of adaptive security from scratch, extending the selective security proof of [16] in a non-black-box way. In contrast, our approach is semi-generic. We isolate our key ideas in an abstract proof framework, and then instantiate the existing selective security proof of [20] in this framework. The main difference from [20] is that we use historyless accumulators (instead of using positional accumulators). Our notion of historyless accumulators is seemingly different from adaptive accumulators; its not immediately clear how to get one from the other. One concrete benefit our approach has is that the usage of i$\mathcal{O}$ is falsifiable, whereas in their construction of adaptive accumulators, i$\mathcal{O}$ is used in a non-falsifiable way. More specifically, they rely on the i$\mathcal{O}$-to-differing-input obfuscation transformation of [13], which makes use of i$\mathcal{O}$ in a non-falsifiable way.

Previous works on non-succinct garbled RAM: The notion of (one-time, non-succinct) garbled RAM was introduced by the work of Lu and Ostrovsky [40], and since then, a sequence of works [28, 30] have led to a black-box construction based on one-way functions, due to Garg, Lu, and Ostrovsky [27]. A black-box construction for *parallel* garbled RAM was later proposed by Lu and Ostrovsky [41] following the works of [12, 19]. However, the garbled program size here is proportional to the worst-case time complexity of the RAM program, so this notion does not imply a RAM delegation scheme. The work of Gentry, Halevi, Raykova, and Wichs [31] showed how to make such garbled RAMs reusable based on various notions of obfuscations (with efficiency trade-offs), and constructed the first RAM delegation schemes in a (weaker) offline/online setting, where in the offline phase, the delegator still needs to run in time proportional to the worst case time complexity of the RAM program.

Previous works on succinct garbled RAM: Succinct garbled RAM was first studied by [10, 17], where in their solutions, the garbled program size depends on the space complexity of the RAM program, but does not depend on its time complexity. This implies delegation for space-bounded RAM computations. Finally, as mentioned, the works of [16, 20] (following [37], which gives a Turing machine delegation scheme) constructed fully succinct garbled RAM, and [20] additionally gives the first fully succinct garbled PRAM. However, their schemes only achieve selective security. Lifting to adaptive security while keeping succinctness is the contribution of this work.

1.5   Organization

We first give an overview of our approach in Sect. 2. In Sect. 3, we present our abstract proof framework. The formal definition of adaptive delegation for RAMs is then presented in Sect. 4. Instantiation of this definition using our abstract proof framework is presented in the full version.

2   Overview

We now provide an overview of our abstract proof for lifting "nice" selective security proofs into adaptive security proofs. To the best of our knowledge, so far, the only general method going from selective to adaptive security is *complexity leveraging*, which however has (1) exponential security loss and (2) cannot be applied in RAM delegation setting for two reasons: (i) this will restrict the number of programs an adversary can choose and, (ii) the security parameter has to be scaled proportional to the number of program queries. This means that all the parameters grow proportional to the number of program queries.

Small-loss complexity leveraging: Nevertheless, we overcome the first limitation by showing a stronger version of complexity leveraging that has much smaller security loss, when the original selectively secure scheme (including its security game and security reduction) satisfy certain properties—we refer to the properties as *niceness* properties and the technique as *small-loss complexity leveraging*.

Local application: Still, many selectively secure schemes may not be *nice*, in particular, the CCC+ scheme. We broaden the scope of application of small-loss complexity leveraging using another idea: Instead of applying small-loss complexity leveraging to the scheme directly, we dissect its proof of selective security, and apply it to "smaller units" in the proof. Most commonly, proofs involve hybrid arguments; now, if every pair of neighboring hybrids with indistinguishability is *nice*, small-loss complexity leveraging can be applied *locally* to lift the indistinguishability to be resilient to adaptive adversaries, which then "sum up" to the global adaptive security of the scheme.

We capture the niceness properties abstractly and prove the above two steps abstractly. Interestingly, a challenging point is finding the right "language" (i.e. formalization) for describing selective and adaptive security games in a general way; we solve this by introducing *generalized security games*. With this language, the abstract proof follows with *simplicity* (completely disentangled from the complexity of specific schemes and their proofs, such as, the CCC+ scheme).

2.1   Classical Complexity Leveraging

Complexity leveraging says if a selective security game is $\mathsf{negl}(\lambda)2^{-L}$-secure, where λ is the security parameter and $L = L(\lambda)$ is the length of the information that selective adversaries choose statically (mostly at the beginning of the game), then the corresponding adaptive security game is $\mathsf{negl}(\lambda)$-secure. For example, the selective security of a public key encryption (PKE) scheme considers adversaries that choose two challenge messages v_0, v_1 of length n statically, whereas

$$CH_s \xleftarrow{\quad v_0, v_1 \quad} A \quad CH_a \xleftarrow{\quad v_0, v_1 \quad} A$$

$$\xrightarrow{pk, \mathsf{Enc}(v_b)} \qquad\qquad \xrightarrow{\mathsf{Enc}(v_b)}$$

$$\xrightarrow{\quad pk \quad}$$

Fig. 1. Left: Selective security of PKE. Right: Adaptive security of PKE.

adaptive adversaries may choose v_0, v_1 adaptively depending on the public key. (See Fig. 1.) By complexity leveraging, any PKE that is $\mathsf{negl}(\lambda)2^{-2n}$-selectively secure is also adaptively secure.

The idea of complexity leveraging is extremely simple. However, to extend it, we need a general way to formalize it. This turns out to be non-trivial, as the selective and adaptive security games are defined separately (e.g., the selective and adaptive security games of PKE have different challengers CH_s and CH_a), and vary case by case for different primitives (e.g., in the security games of RAM delegation, the adversaries choose multiple programs over time, as opposed to in one shot). To overcome this, we introduce generalize security games.

2.2 Generalized Security Games

Generalized security games, like classical games, are between a challenger CH and an adversary A, but are meant to separate the information A chooses statically from its interaction with CH. More specifically, we model A as a non-uniform Turing machine with an additional write-only *special output tape*, which can be written to only at the beginning of the execution (See Fig. 2). The special output tape allows us to capture (fully) selective and (fully) adaptive adversaries naturally: The former write all messages to be sent in the interaction with CH on the tape (at the beginning of the execution), whereas the latter write arbitrary information. Now, selective and adaptive security are captured by running the same (generalized) security game, with different types of adversaries (e.g., see Fig. 2 for the generalized security games of PKE).

Now, complexity leveraging can be proven abstractly: If there is an adaptive adversary A that wins against CH with advantage $\mathsf{negl}(\lambda)$, there is a selective adversary A' that wins with advantage $\mathsf{negl}(\lambda)/2^L$, as A' simply writes on its tape a random guess ρ of A's messages, which is correct with probability $1/2^L$.

With this formalization, we can further generalize the security games in two aspects. First, we consider the natural class of semi-selective adversaries that choose only partial information statically, as opposed to its entire transcript of messages (e.g., in the selective security game of functional encryption in [26] only the challenge messages are chosen selectively, whereas all functions are chosen adaptively). More precisely, an adversary is F-*semi-selective* if the initial choice ρ it writes to the special output tape is always consistent with its messages $m_1, \cdots, m_k$ w.r.t. the output of F, $F(\rho) = F(m_1, \cdots, m_k)$. Clearly, complexity leveraging w.r.t. F-semi-selective adversaries incurs a 2^{L_F}-security loss, where $L_F = |F(\rho)|$.

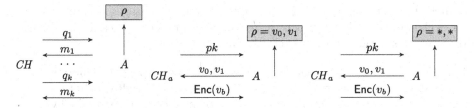

Fig. 2. Left: A generalized game. Middle and Right: Selective and adaptive security of PKE described using generalized games.

Second, we allow the challenger to depend on some partial information $G(\rho)$ of the adversary's initial choice ρ, by sending $G(\rho)$ to CH, after A writes to its special output tape (See Fig. 3)—we say such a game is *G-dependent*. At a first glance, this extension seems strange; few primitives have security games of this form, and it is unnatural to think of running such a game with a fully adaptive adversary (who does not commit to $G(\rho)$ at all). However, such games are prevalent *inside* selective security proofs, which leverage the fact that adversaries are selective (e.g., the selective security proof of the functional encryption of [26] considers an intermediate hybrid where the challenger uses the challenge messages v_0, v_1 from the adversary to program the public key). Hence, this extension is essential to our eventual goal of applying small-loss complexity leveraging to neighboring hybrids, inside selective security proofs.

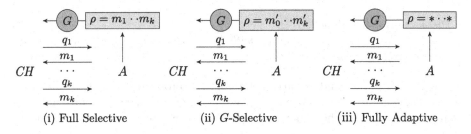

Fig. 3. Three levels of adaptivity. In (ii) G-selective means $G(m_1 \cdots m_k) = G(m_1' \cdots m_k')$.

2.3   Small-loss Complexity Leveraging

In a G-dependent generalized game CH, *ideally*, we want a statement that $\mathsf{negl}(\lambda) 2^{-L_G}$-selective security (i.e., against (fully) selective adversaries) implies $\mathsf{negl}(\lambda)$-adaptively security (i.e., against (fully) adaptive adversaries). We stress that the security loss we aim for is 2^{L_G}, related to the length of the information $L_G = G(\rho)$ that the challenger depends on,[4] as opposed to 2^L as in classical

---

[4] Because the challenger CH depends on L_G-bit of partial information $G(\rho)$ of the adversary's initial choice ρ, we do not expect to go below 2^{-L_G}-security loss unless requiring very strong properties to start with.

complexity leveraging (where L is the total length of messages selective adversaries choose statically). When $L \gg L_G$, the saving in security loss is significant. However, this ideal statement is clearly false in general.

1. For one, consider the special case where G always outputs the empty string, the statement means $\mathsf{negl}(\lambda)$-selective security implies $\mathsf{negl}(\lambda)$-adaptive security. We cannot hope to improve complexity leveraging unconditionally.
2. For two, even if the game is 2^{-L}-selectively secure, complexity leveraging does not apply to *generalized* security games. To see this, recall that complexity leveraging turns an adaptive adversary A with advantage δ, into a selective one B with advantage $\delta/2^L$, who guesses A's messages at the beginning. It relies on the fact that the challenger is oblivious of B's guess ρ to argue that messages to and from A are information theoretically independent of ρ, and hence ρ matches A's messages with probability $1/2^L$ (see Fig. 3 again). However, in generalized games, the challenger does depend on some partial information $G(\rho)$ of B's guess ρ, breaking this argument.

To circumvent the above issues, we strengthen the premise with two niceness properties (introduced shortly). Importantly, both niceness properties still only provide $\mathsf{negl}(\lambda)2^{-L_G}$-security guarantees, and hence the security loss remains 2^{L_G}.

Lemma 1 (Informal, Small Loss Complexity Leveraging). *Any G-dependent generalized security games with the following two properties for $\delta = \mathsf{negl}(\lambda)2^{-L_G}$ are adaptively secure.*

- *The game is δ-G-hiding.*
- *The game has a security reduction with δ-statistical emulation property to a δ-secure cryptographic assumption.*

We define δ-G-hiding and δ-statistical emulation properties shortly. We prove the above lemma in a modular way, by first showing the following semi-selective security property, and then adaptive security. In each step, we use one niceness property.

δ-**semi-selective security:** We say that a G-dependent generalized security game CH is δ-semi-selective secure, if the winning advantage of any G-semi-selective adversary is bounded by $\delta = \mathsf{negl}(\lambda)2^{-L_G}$. Recall that such an adversary writes ρ to the special output tape at the beginning, and later choose adaptively any messages $m_1, \cdots, m_k$ consistent with $G(\rho)$, that is, $G(m_1, \cdots, m_k) = G(\rho)$ or $\perp$ (i.e., the output of G is undefined for $m_1, \cdots, m_k$).

Step 1 – From Selective to G-semi-selective Security. This step encounters the same problem as in the first issue above: We cannot expect to go from $\mathsf{negl}(\lambda)2^{-L_G}$-selective to $\mathsf{negl}(\lambda)2^{-L_G}$-semi-selective security unconditionally, since the latter is dealing with much more adaptive adversaries. Rather, we consider only cases where the selective security of the game with CH is proven using a *black-box straight-line* security reduction R to a game-based intractability assumption with challenger CH' (c.f. falsifiable assumption [43]). We identify the following sufficient conditions on R and CH' under which semi-selective security follows.

Recall that a reduction R simultaneously interacts with an adversary A (on the right), and leverages A's winning advantage to win against the challenger CH' (on the left). It is convenient to think of R and CH' as a compound machine $CH' {\leftrightarrow} R$ that interacts with A, and outputs what CH' outputs. Our condition requires that $CH' {\leftrightarrow} R$ emulates statistically every next message and output of CH. More precisely,

δ-statistical emulation property: For every possible $G(\rho)$ and partial transcript $\tau = (q_1, m_1, \cdots, q_k, m_k)$ consistent with $G(\rho)$ (i.e., $G(m_1, \cdots, m_k) = G(\rho)$ or $\perp$), condition on them $(G(\rho), \tau)$ appearing in interactions with CH or $CH' {\leftrightarrow} R$, the distributions of the next message or output from CH or $CH' {\leftrightarrow} R$ are δ-statistically close.

We show that this condition implies that for any G-semi-selective adversary, its interactions with CH and $CH' {\leftrightarrow} R$ are $\text{poly}(\lambda)\delta$-statistically close (as the total number of messages is $\text{poly}(\lambda)$), as well as the output of CH and CH'. Hence, if the assumption CH' is $\text{negl}(\lambda)2^{-L_G}$-secure against arbitrary adversaries, so is CH against G-semi-selective adversaries.[5]

FURTHER DISCUSSION: We remark that the statistical emulation property is a strong condition that is sufficient but not necessary. A weaker requirement would be requiring the game to be G-semi-selective secure directly. However, we choose to formulate the statistical emulation property because it is a typical way how reductions are built, by emulating perfectly the messages and output of the challenger in the honest games. Furthermore, given R and CH', the statistical emulation property is easy to check, as from the description of R and CH', it is usually clear whether they emulate CH statistically close or not.

Step 2 – From G-semi-selective to adaptive security we would like to apply complexity leveraging to go from $\text{negl}(\lambda)2^{-L_G}$-semi-selective security to adaptive security. However, we encounter the same problem as in the second issue above. To overcome it, we require the security game to be G-hiding, that is, the challenger's messages computationally hides $G(\rho)$.

δ-G-hiding: For any ρ and ρ', interactions with CH after receiving $G(\rho)$ or $G(\rho')$ are indistinguishable to any polynomial-time adversaries, except from a δ distinguishing gap.

Let's see how complexity leveraging can be applied now. Consider again using an adaptive adversary A with advantage $1/\text{poly}(\lambda)$ to build a semi-selective adversary B with advantage $1/\text{poly}(\lambda)2^{L_G}$, who guesses A's choice of $G(m_1, \cdots, m_k)$

---

[5] Technically, we also require that CH and CH' have the same winning threshold, like both $1/2$ or 0.

later. As mentioned before, since the challenger in the generalized game depends on B's guess τ, classical complexity leveraging argument does not apply. However, by the δ-G-hiding property, B's advantage differ by at most δ, when moving to a hybrid game where the challenger generates its messages using $G(\rho)$, where ρ is what A writes to its special output tape at the beginning, instead of τ. In this hybrid, the challenger is oblivious of B's guess τ, and hence the classical complexity leveraging argument applies, giving that B's advantage is at least $1/\operatorname{poly}(\lambda)2^{L_G}$. Thus by G-hiding, B's advantage in the original generalized game is at least $1/\operatorname{poly}(\lambda)2^{L_G} - \delta = 1/\operatorname{poly}(\lambda)2^{L_G}$. This gives a contradiction, and concludes the adaptive security of the game.

Summarizing the above two steps, we obtain our informal lemma on small-loss complexity leveraging.

2.4   Local Application

In many cases, small-loss complexity leveraging may not directly apply, since either the security game is not G-hiding, or the selective security proof does not admit a reduction with the statistical emulation property. We can broaden the application of small-loss complexity leveraging by looking into the selective security proofs and apply small loss complexity leveraging on smaller "steps" inside the proof. For our purpose of getting adaptively secure RAM delegation, we focus on the following common proof paradigm for showing indistinguishability based security. But the same principle of local application could be applied to other types of proofs.

A common proof paradigm for showing the indistinguishability of two games $Real_0$ and $Real_1$ against selective adversaries is the following:

- First, construct a sequence of hybrid experiments $H_0, \cdots, H_\ell$, that starts from one real experiment (i.e., $H_0 = Real_0$), and gradually morphs through intermediate hybrids H_i's into the other (i.e., $H_\ell = Real_1$).
- Second, show that every pair of neighboring hybrids H_i, H_{i+1} is indistinguishable to selective adversaries.

Then, by standard hybrid arguments, the real games are selectively indistinguishable.

To lift such a selective security proof into an adaptive security proof, we first cast all real and hybrids games into our framework of generalized games, which can be run with both selective and adaptive adversaries. If we can obtain that neighboring hybrids games are also indistinguishable to adaptive adversaries, then the adaptive indistinguishability of the two real games follow simply from hybrid arguments. Towards this, we apply small-loss complexity leveraging on neighboring hybrids. More specifically, H_i and H_{i+1} are adaptively indistinguishable, if they satisfy the following properties:

- H_i and H_{i+1} are respectively G_i and G_{i+1}-dependent, as well as δ-$(G_i\|G_{i+1})$-hiding, where $G_i\|G_{i+1}$ outputs the concatenation of the outputs of G_i and G_{i+1} and $\delta = \operatorname{negl}(\lambda)2^{-L_{G_i}-L_{G_{i+1}}}$.

– The selective indistinguishability of H_i and H_{i+1} is shown via a reduction R to a δ-secure game-based assumption and the reduction has δ-statistical emulation property.

Thus, applying small-loss complexity leveraging on every neighboring hybrids, the maximum security loss is $2^{2^{L_{max}}}$, where $L_{max} = \max(L_{G_i})$. Crucially, if every hybrid H_i have small L_{G_i}, the maximum security loss is small. In particular, we say that a selective security proof is *"nice"* if it falls into the above framework and all G_i's have only *logarithmic length* outputs — such "nice" proofs can be lifted to proofs of adaptive indistinguishability with only polynomial security loss. This is exactly the case for the CCC+ scheme, which we explain next.

2.5 The CCC+ Scheme and Its Nice Proof

CCC+ proposed a selectively secure RAM delegation scheme in the persistent database setting. We now show how CCC+ scheme can be used to instantiate the abstract framework discussed earlier in this Section. We only provide with relevant details of CCC+ and refer the reader to the full version for a thorough discussion.

There are two main components in CCC+. The first component is *storage* that maintains information about the database, and the second component is the *machine* component that involves executing instructions of the delegated RAM. Both the storage and the machine components are built on heavy machinery. We highlight below two important building blocks relevant to our discussion. Additional tools such as iterators and splittable signatures are also employed in their construction.

– *Positional Accumulators*: This primitive offers a mechanism of producing a short value, called *accumulator*, that commits to a large storage. Further, accumulators should also be updatable – if a small portion of storage changes, then only a correspondingly small change is required to update the accumulator value. In the security proof, accumulators allow for programming the parameters with respect to a particular location in such a way that the accumulator uniquely determines the value at that location. However, such programming requires to know ahead of time all the changes the storage undergoes since its initialization. Henceforth, we refer to the hybrids to be in ENFORCE-MODE when the accumulator parameters are programmed and the setting when it is not programmed to be REAL-MODE.
– *"Puncturable" Oblivious RAM*: Oblivious RAM (ORAM) is a randomized compiler that compiles any RAM program into one with a fixed distribution of random access pattern to hide its actual (logic) access pattern. CCC+ relies on stronger "puncturable" property of specific ORAM construction of [21], which roughly says the compiled access pattern of a particular logic memory access can be simulated if certain local ORAM randomness is information theoretically "punctured out," and this local randomness is determined at the time

the logic memory location is last accessed. Henceforth, we refer to the hybrids to be in PUNCTURING-MODE when the ORAM randomness is punctured out.

We show that the security proof of CCC+ has a nice proof. We denote the set of hybrids in CCC+ to be $H_1, \ldots, H_\ell$. Correspondingly, we denote the reductions that argue indistinguishability of H_i and H_{i+1} to be R_i. We consider the following three cases depending on the type of neighboring hybrids H_i and H_{i+1}:

1. ORAM IS IN PUNCTURING-MODE IN ONE OR BOTH OF THE NEIGHBORING HYBRIDS: In this case, the hybrid challenger needs to know which ORAM local randomness to puncture out to hide the logic memory access to location q at a particular time point t. As mentioned, this local randomness appears for the first time at the last time point t' that location q is accessed, possibly by a previous machine. As a result, in the proof, some machine components need to be programmed depending on the memory access of later machines. In this case, G_i or G_{i+1} need to contain information about q, t and t', which can be described in poly(λ) bits.
2. POSITIONAL ACCUMULATOR IS IN ENFORCE-MODE IN ONE OR BOTH OF THE NEIGHBORING HYBRIDS: Here, the adversary is supposed to declare all its inputs in the beginning of experiment. The reason being that in the enforce-mode, the accumulator parameters need to be programmed. As remarked earlier, programming the parameters is possible only with the knowledge of the entire computation.
3. REMAINING CASES: In remaining cases, the indistinguishability of neighboring hybrids reduces to the security of other cryptographic primitives, such as, iterators, splittable signatures, indistinguishability obfuscation and others. We note that in these cases, we simply have $G_i = G_{i+1} = \mathsf{null}$, which outputs an empty string.

As seen from the above description, only the second case is problematic for us since the information to be declared by the adversary in the beginning of the experiment is too long. Hence, we need to think of alternate variants to positional accumulators where the enforce-mode can be implemented without the knowledge of the computation history.

History-less Accumulators. To this end, we introduce a primitive called *history-less accumulators*. As the name is suggestive, in this primitive, programming the parameters requires only the location being information-theoretically bound to be known ahead of time. And note that the location can be represented using only logarithmic bits and satisfies the size requirements. That is, the output length of G_i is now short. By plugging this into the CCC+ scheme, we obtain a "nice" security proof.

All that remains is to construct history-less accumulators. The construction of this primitive can be found in the full version.

3   Abstract Proof

In this section, we present our abstract proof that turns "nice" selective security proofs, to adaptive security proofs. As discussed in the introduction, we use generalized security experiments and games to describe our transformation. We present small-loss complexity leveraging in Sect. 3.3 and how to locally apply it in Sect. 3.4. In the latter, we focus our attention on proofs of indistinguishability against selective adversaries, as opposed to proofs of arbitrary security properties.

3.1   Cryptographic Experiments and Games

We recall standard cryptographic experiments and games between two parties, a challenger CH and an adversary A. The challenger defines the procedure and output of the experiment (or game), whereas the adversary can be any probabilistic interactive machine.

Definition 1 (Canonical Experiments). *A canonical experiment between two probabilistic interactive machines, the* challenger CH *and the* adversary A, *with security parameter* $\lambda \in \mathbb{N}$, *denoted as* $\mathsf{Exp}(\lambda, CH, A)$, *has the following form:*

- *CH and A receive common input 1^λ, and interact with each other.*
- *After the interaction, A writes an output γ on its output tape. In case A aborts before writing to its output tape, its output is set to $\perp$.*
- *CH additionally receives the output of A (receiving $\perp$ if A aborts), and outputs a bit b indicating accept or reject. (CH never aborts.)*

We say A wins whenever CH outputs 1 in the above experiment.
    A canonical game (CH, τ) has additionally a threshold $\tau \in [0, 1)$. We say A has advantage γ if A wins with probability $\tau + \gamma$ in $\mathsf{Exp}(\lambda, CH, A)$.

For machine $\star \in \{CH, A\}$, we denote by $\mathsf{Out}_\star(\lambda, CH, A)$ and $\mathsf{View}_\star(\lambda, CH, A)$ the random variables describing the output and view of machine $\star$ in $\mathsf{Exp}(\lambda, CH, A)$.

Definition 2 (Cryptographic Experiments and Games). *A cryptographic experiment is defined by an ensemble of PPT challengers $\mathcal{CH} = \{CH_\lambda\}$. And a cryptographic game $(\mathcal{CH}, \tau)$ has additionally a threshold $\tau \in [0, 1)$. We say that a non-uniform adversary $\mathcal{A} = \{A_\lambda\}$ wins the cryptographic game with advantage $\mathsf{Advt}(\star)$, if for every $\lambda \in \mathbb{N}$, its advantage in $\mathsf{Exp}(\lambda, CH_\lambda, A_\lambda)$ is $\tau + \mathsf{Advt}(\lambda)$.*

Definition 3 (Intractability Assumptions). *An intractability assumption $(\mathcal{CH}, \tau)$ is the same as a cryptographic game, but with* potentially *unbounded challengers. It states that the advantage of every non-uniform PPT adversary $\mathcal{A}$ is negligible.*

3.2   Generalized Cryptographic Games

In the literature, experiments (or games) for selective security and adaptive security are often defined separately: In the former, the challenger requires the adversary to choose certain information at the beginning of the interaction, whereas in the latter, the challenger does not require such information.

We generalize standard cryptographic experiments so that the same experiment can work with both selective and adaptive adversaries. This is achieved by separating information necessary for the execution of the challenger and information an adversary chooses statically, which can be viewed as a property of the adversary. More specifically, we consider adversaries that have a *special output tape*, and write information α it chooses statically at the beginning of the execution on it; and only the necessary information specified by a function, $G(\alpha)$, is sent to the challenger. (See Fig. 3.)

Definition 4 (Generalized Experiments). *A generalized experiment between a challenger CH and an adversary A with respect to a function G, with security parameter $\lambda \in \mathbb{N}$, denoted as* $\mathsf{Exp}(\lambda, CH, G, A)$, *has the following form:*

1. *The adversary A on input 1^λ writes on its special output tape string α at the beginning of its execution, called the initial choice of A, and then proceeds as a normal probabilistic interactive machine. (α is set to the empty string ε if A does not write on the special output tape at the beginning.)*
2. *Let A[G] denote the adversary that on input 1^λ runs A with the same security parameter internally; upon A writing α on its special output tape, it sends out message $m_1 = G(\alpha)$, and later forwards messages A sends, $m_2, m_3, \cdots$*
3. *The generalized experiment proceeds as a standard experiment between CH and A[G], $\mathsf{Exp}(\lambda, CH, A[G])$.*

We say that A wins whenever CH outputs 1.

   Furthermore, for any function $F : \{0,1\}^* \to \{0,1\}^*$, we say that A is F-selective in $\mathsf{Exp}(\lambda, CH, G, A)$, *if it holds with probability 1 that either A aborts or its initial choice α and messages it sends satisfy that $F(\alpha) = F(m_2, m_3, \cdots)$. We say that A is adaptive, in the case that F is a constant function.*

Similar to before, we denote by $\mathsf{Out}_\star(\lambda, CH, G, A)$ and $\mathsf{View}_\star(\lambda, CH, G, A)$ the random variables describing the output and view of machine $\star \in \{CH, A\}$ in $\mathsf{Exp}(\lambda, CH, G, A)$. In this work, we restrict our attention to all the functions G that are efficiently computable, as well as, *reversely computable*, meaning that given a value y in the domain of G, there is an efficient procedure that can output an input x such that $G(x) = y$.

Definition 5 (Generalized Cryptographic Experiments and $\mathcal{F}$-Selective Adversaries). *A generalized cryptographic experiment is a tuple $(\mathcal{CH}, \mathcal{G})$, where $\mathcal{CH}$ is an ensemble of PPT challengers $\{CH_\lambda\}$ and $\mathcal{G}$ is an ensemble of efficiently computable functions $\{G_\lambda\}$. Furthermore, for any ensemble of functions $\mathcal{F} = \{F_\lambda\}$ mapping $\{0,1\}^*$ to $\{0,1\}^*$, we say that a non-uniform adversary $\mathcal{A}$ is $\mathcal{F}$-selective in cryptographic experiments $(\mathcal{CH}, \mathcal{G})$ if for every $\lambda \in \mathbb{N}$, A_λ is F_λ-selective in experiment $\mathsf{Exp}(\lambda, CH_\lambda, G_\lambda, A_\lambda)$.*

Similar to Definition 2, a generalized cryptographic experiment can be extended to a *generalized cryptographic game* $(\mathcal{CH}, \mathcal{G}, \tau)$ by adding an additional threshold $\tau \in [0, 1)$, where the advantage of any non-uniform probabilistic adversary $\mathcal{A}$ is defined identically as before.

We can now quantify the level of selective/adaptive security of a generalized cryptographic game.

Definition 6 ($\mathcal{F}$-Selective Security). *A generalized cryptographic game* $(\mathcal{CH}, \mathcal{G}, \tau)$ *is* $\mathcal{F}$-*selective secure if the advantage of every non-uniform* PPT $\mathcal{F}$-*selective adversary* $\mathcal{A}$ *is negligible.*

3.3   Small-loss Complexity Leveraging

In this section, we present our small-loss complexity leveraging technique to lift fully selective security to fully adaptive security for a generalized cryptographic game $\Pi = (\mathcal{CH}, \mathcal{G}, \tau)$, provided that the game and its (selective) security proof satisfies certain niceness properties. We will focus on the following class of *guessing* games, which captures indistinguishability security. We remark that our technique also applies to generalized cryptographic games with arbitrary threshold (See Remark 1).

Definition 7 (Guessing Games). *A generalized game* (CH, G, τ) *(for a security parameter λ) is a* guessing game *if it has the following structure.*

- *At beginning of the game, CH samples a uniform bit $b \leftarrow \{0, 1\}$.*
- *At the end of the game, the adversary guesses a bit $b' \in \{0, 1\}$, and he wins if $b = b'$.*
- *When the adversary aborts, his guess is a uniform bit $b' \leftarrow \{0, 1\}$.*
- *The threshold $\tau = 1/2$.*

The definition extends naturally to a sequence of games $\Pi = (\mathcal{CH}, \mathcal{G}, 1/2)$. Our technique consists of two modular steps: First reach $\mathcal{G}$-selective security, and then adaptive security, where the first step applies to any generalized cryptographic game.

Step 1: $\mathcal{G}$-Selective Security. In general, a fully selectively secure Π may not be $\mathcal{F}$-selective secure for $\mathcal{F} \neq \mathcal{F}_{\mathrm{id}}$, where $\mathcal{F}_{\mathrm{id}}$ denotes the identity function. We restrict our attention to the following case: The security is proved by a straight-line black-box security reduction from Π to an intractability assumption $(\mathcal{CH}', \tau')$, where the reduction is an ensemble of PPT machines $\mathcal{R} = \{R_\lambda\}$ that interacts simultaneously with an adversary for Π and $\mathcal{CH}'$, the reduction is syntactically well-defined with respect to any class of $\mathcal{F}$-selective adversary. This, however, does not imply that R is a correct reduction to prove $\mathcal{F}$-selective security of Π. Here, we identify a sufficient condition on the "niceness" of reduction that implies $\mathcal{G}$-selective security of Π. We start by defining the syntax of a straight-line black-box security reduction.

Standard straight-line black-box security reduction from a cryptographic game to an intractability assumption is a PPT machine R that interacts simultaneously with an adversary and the challenger of the assumption. Since our generalized cryptographic games can be viewed as standard cryptographic games with adversaries of the form $\mathcal{A}[\mathcal{G}] = \{A_\lambda[G_\lambda]\}$, the standard notion of reductions extends naturally, by letting the reductions interact with adversaries of the form $\mathcal{A}[\mathcal{G}]$.

Definition 8 (Reductions). *A probabilistic interactive machine R is a (straight-line black-box) reduction from a generalized game (CH, G, τ) to a (canonical) game (CH', τ') for security parameter λ, if it has the following syntax:*

- *Syntax: On common input 1^λ, R interacts with CH' and an adversary $A[G]$ simultaneously in a straight-line—referred to as "left" and "right" interactions respectively. The left interaction proceeds identically to the experiment $\mathsf{Exp}(\lambda, CH', R \leftrightarrow A[G])$, and the right to experiment $\mathsf{Exp}(\lambda, CH' \leftrightarrow R, A[G])$.*

A (straight-line black-box) reduction from an ensemble of generalized cryptographic game $(\mathcal{CH}, \mathcal{G}, \tau)$ to an intractability assumption $(\mathcal{CH}', \tau')$ is an ensemble of PPT reductions $\mathcal{R} = \{R_\lambda\}$ from game $(CH_\lambda, G_\lambda, \tau)$ to (CH'_λ, τ') (for security parameter λ).

At a high-level, we say that a reduction is μ-nice, where μ is a function, if it satisfies the following syntactical property: R (together with the challenger CH' of the assumption) generates messages and output that are statistically close to the messages and output of the challenger CH of the game, *at every step*.

More precisely, let $\rho = (m_1, a_1, m_2, a_2, \cdots, m_t, a_t)$ denote a transcript of messages and outputs in the interaction between CH and an adversary (or in the interaction between $CH' \leftrightarrow R$ and an adversary) where $\boldsymbol{m} = m_1, m_2, \cdots, m_{t-1}$ and m_t correspond to the messages and output of the adversary ($m_t = \bot$ if the adversary aborts) and $\boldsymbol{a} = a_1, a_2, \cdots, a_{t-1}$ and a_t corresponds to the messages and output of CH (or $CH' \leftrightarrow R$). A transcript ρ possibly appears in an interaction with CH (or $CH' \leftrightarrow R$) if when receiving $\boldsymbol{m}$, CH (or $CH' \leftrightarrow R$) generates $\boldsymbol{a}$ with non-zero probability. The syntactical property requires that for every prefix of a transcript that possibly appear in both interaction with CH and interaction with $CH' \leftrightarrow R$, the distributions of the next message or output generated by CH and $CH' \leftrightarrow R$ are statistically close. In fact, for our purpose later, it suffices to consider the prefixes of transcripts that are *G-consistent*: A transcript ρ is G-consistent if $\boldsymbol{m}$ satisfies that either $m_t = \bot$ or $m_1 = G(m_2, m_3, \cdots, m_{t-1})$; in other words, ρ could be generated by a G-selective adversary.

Definition 9 (Nice Reductions). *We say that a reduction R from a generalized game (CH, G, τ) to a (canonical) game (CH', τ) (with the same threshold) for security parameter λ is μ-nice, if it satisfies the following property:*

- *$\mu(\lambda)$-statistical emulation for G-consistent transcripts:*
  For every prefix $\rho = (m_1, a_1, m_2, a_2, \cdots, m_{\ell-1}, a_{\ell-1}, m_\ell)$ of a G-consistent

transcript of messages that possibly appears in interaction with both CH and $CH' \leftrightarrow R$, *the following two distributions are* $\mu(\lambda)$*-close:*

$$\Delta(\mathsf{D}_{CH' \leftrightarrow R}(\lambda, \rho), \ \mathsf{D}_{CH}(\lambda, \rho)) \le \mu(\lambda)$$

where $\mathsf{D}_M(\lambda, \rho)$ *for* $M = CH' \leftrightarrow R$ *or* CH *is the distribution of the next message or output* a_ℓ *generated by* $M(1^\lambda)$ *after receiving messages* $\boldsymbol{m}$ *in* ρ, *and conditioned on* $M(1^\lambda)$ *having generated* $\boldsymbol{a}$ *in* ρ.

Moreover, we say that a reduction $\mathcal{R} = \{R_\lambda\}$ *from a generalized cryptographic game* $(\mathcal{CH}, \mathcal{G}, \tau)$ *to a intractability assumption* $(\mathcal{CH}', \tau)$ *is nice if there is a negligible function* μ, *such that,* R_λ *is* $\mu(\lambda)$*-nice for every* λ.

When a reduction is μ-nice with negligible μ, it is sufficient to imply $\mathcal{G}$-selective security of the corresponding generalized cryptographic game. We defer the proofs to the full version.

Lemma 2. *Suppose* R *is a* μ*-nice reduction from* (CH, G, τ) *to* (CH', τ) *for security parameter* λ, *and* A *is a deterministic* G*-semi-selective adversary that wins* (CH, G, τ) *with advantage* $\gamma(\lambda)$, *then* $R \leftrightarrow A[G]$ *is an adversary for* (CH', τ) *with advantage* $\gamma(\lambda) - t(\lambda) \cdot \mu(\lambda)$, *where* $t(\lambda)$ *is an upper bound on the run-time of* R.

By a standard argument, Lemma 2 implies the following asymptotic version theorem.

Theorem 3. *If there exists a nice reduction* $\mathcal{R}$ *from a generalized cryptographic game* $(\mathcal{CH}, \mathcal{G}, \tau)$ *to an intractability assumption* $(\mathcal{CH}', \tau)$, *then* $(\mathcal{CH}, \mathcal{G}, \tau)$ *is* $\mathcal{G}$*-selectively secure.*

Step 2: Fully Adaptive Security. We now show how to move from $\mathcal{G}$-selective security to fully adaptive security for the class of guessing games with security loss $2^{L_G(\lambda)}$, where $L_G(\lambda)$ is the output length of $\mathcal{G}$, provided that the challenger's messages hide the information of $G(\alpha)$ computationally. We start with formalizing this hiding property.

Roughly speaking, the challenger CH of a generalized experiment (CH, G) is G-hiding, if for any α and α', interactions with CH receiving $G(\alpha)$ or $G(\alpha')$ at the beginning are indistinguishable. Denote by $CH(x)$ the challenger with x hardcoded as the first message.

Definition 10 (G-hiding). *We say that a generalized guessing game* (CH, G, τ) *is* $\mu(\lambda)$*-G-hiding for security parameter* λ, *if its challenger* CH *satisfies that for every* α *and* α', *and every non-uniform* PPT *adversary* A,

$$|\Pr[\mathsf{Out}_A(\lambda, CH(G(\alpha)), A) = 1] - \Pr[\mathsf{Out}_A(\lambda, CH(G(\alpha')), A) = 1]| \le \mu(\lambda)$$

Moreover, we say that a generalized cryptographic guessing game $(\mathcal{CH}, \mathcal{G}, \tau)$ *is* $\mathcal{G}$*-hiding, if there is a negligible function* μ, *such that,* $(CH_\lambda, G_\lambda, \tau(\lambda))$ *is* $\mu(\lambda)$*-* G_λ*-hiding for every* λ.

The following lemma says that if a generalized guessing game $(CH, G, 1/2)$ is G-selectively secure and G-hiding, then it is fully adaptively secure with 2^{L_G} security loss. Its formal proof is deferred to the full version.

Lemma 3. *Let* $(CH, G, 1/2)$ *be a generalized cryptographic guessing game for security parameter* λ. *If there exists a fully adaptive adversary A for* $(CH, G, 1/2)$ *with advantage* $\gamma(\lambda)$ *and* $(CH, G, 1/2)$ *is* $\mu(\lambda)$-G-hiding with $\mu(\lambda) \leq \gamma/2^{L_G(\lambda)+1}$, *then there exists a* G-selective adversary A' *for* $(CH, G, 1/2)$ *with advantage* $\gamma(\lambda)/2^{L_G(\lambda)+1}$, *where* L_G *is the output length of* G.

Therefore, for a generalized cryptographic guessing game $(\mathcal{CH}, \mathcal{G}, \tau)$, if $\mathcal{G}$ has logarithmic output length $L_G(\lambda) = O(\log \lambda)$ and the game is $\mathcal{G}$-hiding, then its $\mathcal{G}$-selective security implies fully adaptive security.

Theorem 4. *Let* $(\mathcal{CH}, \mathcal{G}, \tau)$ *be a* $\mathcal{G}$-selectively secure generalized cryptographic guessing game. If $(\mathcal{CH}, \mathcal{G}, \tau)$ *is* $\mathcal{G}$-hiding and $L_G(\lambda) = O(\log \lambda)$, *then* $(\mathcal{CH}, \mathcal{G}, \tau)$ *is fully adaptively secure.*

Remark 1. The above proof of small-loss complexity leveraging can be extended to a more general class of security games, beyond the guessing games. The challenger with an arbitrary threshold τ has the form that if the adversary aborts, the challenger toss a biased coin and outputs 1 with probability τ. The same argument above goes through for games with this class of challengers.

3.4 Nice Indistinguishability Proof

In this section, we characterize an abstract framework of proofs—called "nice" proofs—for showing the indistinguishability of two ensembles of (standard) cryptographic experiments. We focus on a common type of indistinguishability proof, which consists of a sequence of hybrid experiments and shows that neighboring hybrids are indistinguishable via a reduction to a intractability assumption. We formalize required nice properties of the hybrids and reductions such that a fully selective security proof can be lifted to prove fully adaptive security by local application of small-loss complexity leveraging technique to neighboring hybrids. We start by describing common indistinguishability proofs using the language of generalized experiments and games.

Consider two ensembles of standard cryptographic experiments $\mathcal{RL}_0$ and $\mathcal{RL}_1$. They are special cases of generalized cryptographic experiments with a function $G = \text{null} : \{0, 1\}^* \rightarrow \{\varepsilon\}$ that always outputs the empty string, that is, $(\mathcal{RL}_0, \text{null})$ and $(\mathcal{RL}_1, \text{null})$; we refer to them as the "real" experiments.

Consider a proof of indistinguishability of $(\mathcal{RL}_0, \text{null})$ and $(\mathcal{RL}_1, \text{null})$ against fully selective adversaries via a sequence of hybrid experiments. As discussed in the overview, the challenger of the hybrids often depends non-trivially on partial information of the adversary's initial choice. Namely, the hybrids are generalized cryptographic experiments with non-trivial $\mathcal{G}$ function. Since small-loss complexity leveraging has exponential security loss in the output length of $\mathcal{G}$, we require all hybrid experiments have logarithmic-length $\mathcal{G}$ function. Below, for convenience, we use the notation $\mathcal{X}_i$ to denote an ensemble of the form $\{X_{i,\lambda}\}$, and the notation $\mathcal{X}_I$ with a function I, as the ensemble $\{X_{I(\lambda),\lambda}\}$.

1. **Security via hybrids with logarithmic-length $\mathcal{G}$ function:** The proof involves a sequence of *polynomial* number $\ell(\star)$ of hybrid experiments. More precisely, for every $\lambda \in \mathbb{N}$, there is a sequence of $\ell(\lambda)+1$ hybrid (generalized) experiments $(H_{0,\lambda}, G_{0,\lambda}), \cdots (H_{\ell(\lambda),\lambda}, G_{\ell(\lambda),\lambda})$, such that, the "end" experiments matches the real experiments,

$$(\mathcal{H}_0, \mathcal{G}_0) = (\{H_{0,\lambda}\}, \{G_{0,\lambda}\}) = (\mathcal{RL}_0, \mathsf{null})$$
$$(\mathcal{H}_\ell, \mathcal{G}_\ell) = (\{H_{\ell(\lambda),\lambda}\}, \{G_{\ell(\lambda),\lambda}\}) = (\mathcal{RL}_1, \mathsf{null}),$$

Furthermore, there exists a function $L_G(\lambda) = O(\log \lambda)$ such that for every λ and i, the output length of $G_{i,\lambda}$ is at most $L_G(\lambda)$.

We next formalize required properties to lift security proof of neighboring hybrids. Towards this, we formulate indistinguishability of two generalized cryptographic experiments as a generalized cryptographic guessing game. The following is a known fact.

Fact. Let $(\mathcal{CH}_0, \mathcal{G}_0)$ and $(\mathcal{CH}_1, \mathcal{G}_1)$ be two ensembles of generalized cryptographic experiments, $\mathcal{F}$ be an ensemble of efficiently computable functions, and $\mathcal{C}_\mathcal{F}$ denote the class of non-uniform PPT adversaries $\mathcal{A}$ that are $\mathcal{F}$-selective in $(\mathcal{CH}_b, \mathcal{G}_b)$ for both $b = 0, 1$. Indistinguishability of $(\mathcal{CH}_0, \mathcal{G}_0)$ and $(\mathcal{CH}_1, \mathcal{G}_1)$ against (efficient) $\mathcal{F}$-selective adversaries is equivalent to $\mathcal{F}$-selective security of a generalized cryptographic guessing game $(\mathcal{D}, \mathcal{G}_0 \| \mathcal{G}_1, 1/2)$, where $\mathcal{G}_0 \| \mathcal{G}_1 = \{G_{0,\lambda} \| G_{1,\lambda}\}$ are the concatenations of functions $G_{0,\lambda}$ and $G_{1,\lambda}$, and the challenger $\mathcal{D} = \{D_\lambda[CH_{0,\lambda}, CH_{1,\lambda}]\}$ proceeds as follows: For every security parameter $\lambda \in \mathbb{N}$, $D = D_\lambda[CH_{0,\lambda}, CH_{1,\lambda}]$, $G_b = G_{b,\lambda}$, $CH_b = CH_{b,\lambda}$, in experiment $\mathsf{Exp}(\lambda, D, G_0 \| G_1, \star)$,

- D tosses a random bit $b \xleftarrow{\$} \{0,1\}$.
- Upon receiving $g_0 \| g_1$ (corresponding to $g_d = G_d(\alpha)$ for $d = 0, 1$ where α is the initial choice of the adversary), D internally runs challenger CH_b by feeding it g_b and forwarding messages to and from CH_b.
- If the adversary aborts, D output 0. Otherwise, upon receiving the adversary's output bit b', it output 1 if and only if $b = b'$.

By the above fact, indistinguishability of neighboring hybrids $(\mathcal{H}_i, \mathcal{G}_i)$ and $(\mathcal{H}_{i+1}, \mathcal{G}_{i+1})$ against $\mathcal{F}$-selective adversary is equivalent to $\mathcal{F}$-selective security of the generalized cryptographic guessing game $(\mathcal{D}_i, \mathcal{G}_i \| \mathcal{G}_{i+1}, 1/2)$, where $\mathcal{D}_i = \{D_{i,\lambda}[H_{i,\lambda}, H_{i+1,\lambda}]\}$. We can now state the required properties for every pair of neighboring hybrids:

2. **Indistinguishability of neighboring hybrids via nice reduction:** For every neighboring hybrids $(\mathcal{H}_i, \mathcal{G}_i)$ and $(\mathcal{H}_{i+1}, \mathcal{G}_{i+1})$, their indistinguishability proof against fully selective adversary is established by a nice reduction $\mathcal{R}_i$ from the corresponding guessing game $(\mathcal{D}_i, \mathcal{G}_i \| \mathcal{G}_{i+1}, 1/2)$ to some intractability assumption.

3. $\mathcal{G}_i\|\mathcal{G}_{i+1}$-hiding: For every neighboring hybrids $(\mathcal{H}_i,\mathcal{G}_i)$ and $(\mathcal{H}_{i+1},\mathcal{G}_{i+1})$, their corresponding guessing game $(\mathcal{D}_i,\mathcal{G}_i\|\mathcal{G}_{i+1},1/2)$ is $\mathcal{G}_i\|\mathcal{G}_{i+1}$-hiding.

In summary,

Definition 11 (Nice Indistinguishability Proof). *A "nice" proof for the indistinguishability of two real experiments $(\mathcal{RL}_0,\mathsf{null})$ and $(\mathcal{RL}_1,\mathsf{null})$ is one that satisfy properties 1, 2, and 3 described above.*

It is now straightforward to lift security of nice indistinguishability proof by local application of small-loss complexity leveraging for neighboring hybrids. Please refer to the full version for its proof.

Theorem 5. *A "nice" proof for the indistinguishability of two real experiments $(\mathcal{RL}_0,\mathsf{null})$ and $(\mathcal{RL}_1,\mathsf{null})$ implies that these experiments are indistinguishable against fully adaptive adversaries.*

4   Adaptive Delegation for RAM Computation

In this section, we introduce the notion of adaptive delegation for RAM computation $(\mathcal{DEL})$ and state our formal theorem. In a $\mathcal{DEL}$ scheme, a client outsources the database encoding and then generates a sequence of program encodings. The server will evaluate those program encodings with intended order on the database encoding left over by the previous one. For security, we focus on *full privacy* where the server learns nothing about the database, delegated programs, and its outputs. Simultaneously, $\mathcal{DEL}$ is required to provide *soundness* where the client has to receive the correct output encoding from each program and current database.

We first give a brief overview of the structure of the delegation scheme. First, the setup algorithm DBDel, which takes as input the database, is executed. The result is the database encoding and the secret key. PDel is the program encoding procedure. It takes as input the secret key, session ID and the program to be encoded. Eval takes as input the program encoding of session ID sid along with a memory encoding associated with sid. The result is an encoding which is output along with a proof. Along with this the updated memory state is also output. We employ a verification algorithm Ver to verify the correctness of computation using the proof output by Eval. Finally, Dec is used to decode the output encoding.

We present the formal definition below.

4.1   Definition

Definition 12 ($\mathcal{DEL}$ with Persistent Database). *A $\mathcal{DEL}$ scheme with persistent database, consists of PPT algorithms $\mathcal{DEL} = \mathcal{DEL}.\{$DBDel, PDel, Eval, Ver, Dec$\}$, is described below. Let* sid *be the program session identity where $1 \leq$ sid $\leq l$. We associate $\mathcal{DEL}$ with a class of programs $\mathcal{P}$.*

- $\mathcal{DEL}.\mathsf{DBDel}(1^\lambda, \mathsf{mem}^0, S) \rightarrow (\widetilde{\mathsf{mem}}^1, \mathsf{sk})$: *The database delegation algorithm* DBDel *is a randomized algorithm which takes as input the security parameter* 1^λ, *database* mem^0, *and a space bound* S. *It outputs a garbled database* $\widetilde{\mathsf{mem}}^1$ *and a secret key* sk.
- $\mathcal{DEL}.\mathsf{PDel}(1^\lambda, \mathsf{sk}, \mathsf{sid}, P_\mathsf{sid}) \rightarrow \widetilde{P}_\mathsf{sid}$: *The algorithm* PDel *is a randomized algorithm which takes as input the security parameter* 1^λ, *the secret key* sk, *the session ID* sid *and a description of a* RAM *program* $P_\mathsf{sid} \in \mathcal{P}$. *It outputs a program encoding* $\widetilde{P}_\mathsf{sid}$.
- $\mathcal{DEL}.\mathsf{Eval}\left(1^\lambda, T, S, \widetilde{P}_\mathsf{sid}, \widetilde{\mathsf{mem}}^\mathsf{sid}\right) \rightarrow \left(c_\mathsf{sid}, \sigma_\mathsf{sid}, \widetilde{\mathsf{mem}}^{\mathsf{sid}+1}\right)$: *The evaluating algorithm* Eval *is a deterministic algorithm which takes as input the security parameter* 1^λ, *time bound* T, *space bound* S, *a garbled program* $\widetilde{P}_\mathsf{sid}$, *and the database* $\widetilde{\mathsf{mem}}^\mathsf{sid}$. *It outputs* $(c_\mathsf{sid}, \sigma_\mathsf{sid}, \widetilde{\mathsf{mem}}^{\mathsf{sid}+1})$ *or* $\bot$, *where* c_sid *is the encoding of the output* y_sid, σ_sid *is a proof of* c_sid, *and* $(y_\mathsf{sid}, \mathsf{mem}^{\mathsf{sid}+1}) = P_\mathsf{sid}(\mathsf{mem}^\mathsf{sid})$.
- $\mathcal{DEL}.\mathsf{Ver}(1^\lambda, \mathsf{sk}, c_\mathsf{sid}, \sigma_\mathsf{sid}) \rightarrow b_\mathsf{sid} \in \{0,1\}$: *The verification algorithm takes as input the security parameter* 1^λ, *secret key* sk, *encoding* c_sid, *proof* σ_sid *and returns* $b_\mathsf{sid} = 1$ *if* σ_sid *is a valid proof for* c_sid, *or returns* $b_\mathsf{sid} = 0$ *if not*.
- $\mathcal{DEL}.\mathsf{Dec}(1^\lambda, \mathsf{sk}, c_\mathsf{sid}) \rightarrow y_\mathsf{sid}$: *The decoding algorithm* Dec *is a deterministic algorithm which takes as input the security parameter* 1^λ, *secret key* sk, *output encoding* c_sid. *It outputs* y_sid *by decoding* c_sid *with* sk.

Associated to the above scheme are correctness, (adaptive) security, (adaptive) soundness and efficiency properties.

Correctness. A delegation scheme $\mathcal{DEL}$ is said to be *correct* if both verification and decryption are correct: for all $\mathsf{mem}^0 \in \{0,1\}^{\mathrm{poly}(\lambda)}, 1 \leq \mathsf{sid} \leq \ell, P_\mathsf{sid} \in \mathcal{P}$, consider the following process:

- $(\widetilde{\mathsf{mem}}^1, \mathsf{sk}) \leftarrow \mathcal{DEL}.\mathsf{DBDel}(1^\lambda, \mathsf{mem}^0, S)$;
- $\widetilde{P}_\mathsf{sid} \leftarrow \mathcal{DEL}.\mathsf{PDel}(1^\lambda, \mathsf{sk}, \mathsf{sid}, P_\mathsf{sid})$;
- $(c_\mathsf{sid}, \sigma_\mathsf{sid}, \widetilde{\mathsf{mem}}^{\mathsf{sid}+1}) \leftarrow \mathcal{DEL}.\mathsf{Eval}(1^\lambda, T, S, \widetilde{P}_\mathsf{sid}, \widetilde{\mathsf{mem}}^\mathsf{sid})$;
- $b_\mathsf{sid} = \mathcal{DEL}.\mathsf{Ver}(1^\lambda, \mathsf{sk}, c_\mathsf{sid}, \sigma_\mathsf{sid})$;
- $y_\mathsf{sid} = \mathcal{DEL}.\mathsf{Dec}(1^\lambda, \mathsf{sk}, c_\mathsf{sid})$;
- $(y'_\mathsf{sid}, \mathsf{mem}^{\mathsf{sid}+1}) \leftarrow P_\mathsf{sid}(\mathsf{mem}^\mathsf{sid})$;

The following holds:

$$\Pr\left[(y_\mathsf{sid} = y'_\mathsf{sid} \wedge b_\mathsf{sid} = 1) \; \forall \mathsf{sid}, 1 \leq \mathsf{sid} \leq l\right] = 1.$$

Adaptive Security (full privacy). This property is designed to protect the privacy of the database and the programs from the adversarial server. We formalize this using a simulation based definition. In the real world, the adversary is supposed to declare the database at the beginning of the game. The challenger computes the database encoding and sends it across to the adversary. After this, the adversary can submit programs to the challenger and in return it receives the corresponding program encodings. We emphasize the program queries can be made adaptively. On the other hand, in the simulated world, the simulator does not get to see either the database or the programs submitted by the adversary. But instead it receives as input the length of the database, the lengths of the individual programs and runtimes of all the corresponding computations.[6] It then generates the simulated database and program encodings. The job of the adversary in the end is to guess whether he is interacting with the challenger (real world) or whether he is interacting with the simulator (ideal world).

Definition 13. *A delegation scheme* $\mathcal{DEL} = \mathcal{DEL}.\{\mathsf{DBDel}, \mathsf{PDel}, \mathsf{Eval}, \mathsf{Ver}, \mathsf{Dec}\}$ *with persistent database is said to be* adaptively secure *if for all sufficiently large* $\lambda \in \mathbb{N}$, *for all total round* $l \in \mathrm{poly}(\lambda)$, *time bound* T, *space bound* S, *for every interactive PPT adversary* $\mathcal{A}$, *there exists an interactive PPT simulator* $\mathcal{S}$ *such that* $\mathcal{A}$*'s advantage in the following security game* Exp-Del-Privacy$(1^\lambda, \mathcal{DEL}, \mathcal{A}, \mathcal{S})$ *is at most negligible in* λ.

Exp-Del-Privacy$(1^\lambda, \mathcal{DEL}, \mathcal{A}, \mathcal{S})$

1. *The challenger* $\mathcal{C}$ *chooses a bit* $b \in \{0, 1\}$.
2. $\mathcal{A}$ *chooses and sends database* mem^0 *to challenger* $\mathcal{C}$.
3. *If* $b = 0$, *challenger* $\mathcal{C}$ *computes* $(\widetilde{\mathsf{mem}}^1, \mathsf{sk}) \leftarrow \mathcal{DEL}.\mathsf{DBDel}(1^\lambda, \mathsf{mem}^0, S)$. *Otherwise,* $\mathcal{C}$ *simulates* $(\widetilde{\mathsf{mem}}^1, \mathsf{sk}) \leftarrow \mathcal{S}(1^\lambda, |\mathsf{mem}^0|)$, *where* $|\mathsf{mem}^0|$ *is the length of* mem^0. $\mathcal{C}$ *sends* $\widetilde{\mathsf{mem}}^1$ *back to* $\mathcal{A}$.
4. *For each round* sid *from* 1 *to* l,
   (a) $\mathcal{A}$ *chooses and sends program* P_{sid} *to* $\mathcal{C}$.
   (b) *If* $b = 0$, *challenger* $\mathcal{C}$ *sends* $\widetilde{P}_{\mathsf{sid}} \leftarrow \mathcal{DEL}.\mathsf{PDel}(1^\lambda, \mathsf{sk}, \mathsf{sid}, P_{\mathsf{sid}})$ *to* $\mathcal{A}$. *Otherwise,* $\mathcal{C}$ *simulates and sends* $\widetilde{P}_{\mathsf{sid}} \leftarrow \mathcal{S}(1^\lambda, \mathsf{sk}, \mathsf{sid}, 1^{|P_{\mathsf{sid}}|}, 1^{|c_{\mathsf{sid}}|}, T, S)$ *to* $\mathcal{A}$.
5. $\mathcal{A}$ *outputs a bit* b'. $\mathcal{A}$ *wins the security game if* $b = b'$.

We notice that an unrestricted *adaptive adversary* can adaptively choose RAM programs P_i depending on the program encodings it receives, whereas a restricted *selective adversary* can only make the choice of programs statically at the beginning of the execution.

Adaptive Soundness. This property is designed to protect the clients against adversarial servers producing invalid output encodings. This is formalized in the form of a security experiment: the adversary submits the database to the challenger. The challenger responds with the database encoding. The adversary then chooses programs to be encoded adaptively. In response, the challenger sends the corresponding program encodings. In the end, the adversary is required to submit the output encoding and the corresponding proof. The soundness property requires that the adversary can only submit a convincing "false" proof only with negligible probability.

---

[6] Note that unlike the standard simulation based setting, the simulator does not receive the output of the programs. This is because the output of the computation is never revealed to the adversary.

Definition 14. *A delegation scheme $\mathcal{DEL}$ is said to be adaptively sound if for all sufficiently large $\lambda \in \mathbb{N}$, for all total round $l \in \text{poly}(\lambda)$, time bound T, space bound S, there exists an interactive PPT adversary $\mathcal{A}$, such that the probability of $\mathcal{A}$ win in the following security game Exp-Del-Soundness$(1^\lambda, \mathcal{DEL}, \mathcal{A})$ is at most negligible in λ.*

Exp-Del-Soundness$(1^\lambda, \mathcal{DEL}, \mathcal{A})$

1. *$\mathcal{A}$ chooses and sends database mem^0 to challenger $\mathcal{C}$.*
2. *The challenger $\mathcal{C}$ computes $(\widetilde{\text{mem}}^1, \text{sk}) \leftarrow \mathcal{DEL}.\text{DBDel}(1^\lambda, \text{mem}^0, S)$. $\mathcal{C}$ sends $\widetilde{\text{mem}}^1$ back to $\mathcal{A}$.*
3. *For each round sid from 1 to l,*
   (a) $\mathcal{A}$ chooses and sends program P_{sid} to $\mathcal{C}$.
   (b) $\mathcal{C}$ sends $\widetilde{P}_{\text{sid}} \leftarrow \mathcal{DEL}.\text{PDel}(1^\lambda, \text{sk}, \text{sid}, P_{\text{sid}})$ to $\mathcal{A}$.
4. *$\mathcal{A}$ outputs a triplet (k, c_k^*, σ_k^*). $\mathcal{A}$ wins the security game if $1 \leftarrow \mathcal{DEL}.\text{Ver}(1^\lambda, \text{sk}, c_k^*, \sigma_k^*)$ and $c_k^* \neq c_k$ for the k-th round, where c_k is generated as follows: for $\text{sid} = 1, \ldots, k$, $(c_{\text{sid}}, \sigma_{\text{sid}}, \widetilde{\text{mem}}^{\text{sid}+1}) \leftarrow \mathcal{DEL}.\text{Eval}(1^\lambda, T, S, \widetilde{P}_{\text{sid}}, \widetilde{\text{mem}}^{\text{sid}})$.*

Efficiency. For every session with session ID sid, we require that DBDel and PDel execute in time $\text{poly}(\lambda, |\text{mem}^0|)$ and $\text{poly}(\lambda, |P_{\text{sid}}|)$ respectively. Furthermore we require that Eval run in time $\text{poly}(\lambda, t^*_{\text{sid}})$, where t^*_{sid} denotes the running time of P_{sid} on mem^{sid}. We require that both Ver and Dec run in time $\text{poly}(\lambda, |y_{\text{sid}}|)$. Finally, the length of $c_{\text{sid}}, \sigma_{\text{sid}}$ should depend only on $|y_{\text{sid}}|$.

A construction of adaptive delegation is provided in the full version [2] with its security proof.

Theorem 6. *Assuming the existence of $i\mathcal{O}$ for circuits and DDH, there exists an efficient RAM delegation scheme $\mathcal{DEL}$ with persistent database with adaptive security and soundness.*

Acknowledgements. We thank Yael Kalai for insightful discussions in the early stages of this project.

This work was done in part while the authors were visiting the Simons Institute for the Theory of Computing, supported by the Simons Foundation and by the DIMACS/Simons Collaboration in Cryptography through NSF grant CNS-1523467.

Prabhanjan Ananth is supported in part by grant #360584 from the Simons Foundation and supported in part from a DARPA/ARL SAFEWARE award, NSF Frontier Award 1413955, NSF grants 1228984, 1136174, 1118096, and 1065276. This material is based upon work supported by the Defense Advanced Research Projects Agency through the ARL under Contract W911NF-15-C-0205. The views expressed are those of the author and do not reflect the official policy or position of the Department of Defense, the National Science Foundation, or the U.S. Government.

Kai-Min Chung was partially supported by Ministry of Science and Technology, Taiwan, under Grant no. MOST 103-2221-E-001-022-MY3.

Huijia Lin was partially supported by NSF grants CNS-1528178 and CNS-1514526.

References

1. Ananth, P., Brakerski, Z., Segev, G., Vaikuntanathan, V.: From selective to adaptive security in functional encryption. In: Gennaro, R., Robshaw, M. (eds.) CRYPTO 2015. LNCS, vol. 9216, pp. 657–677. Springer, Heidelberg (2015). doi:10.1007/978-3-662-48000-7_32

2. Ananth, P., Chen, Y.-C., Chung, K.-M., Lin, H., Lin, W.-K.: Delegating RAM computations with adaptive soundness and privacy. Cryptology ePrint Archive, Report 2015/1082 (2015). http://eprint.iacr.org/2015/1082

3. Ananth, P., Jain, A., Naor, M., Sahai, A., Eylon Y.: Universal obfuscation and witness encryption: boosting correctness and combining security. In: CRYPTO (2016)

4. Ananth, P., Jain, A.: Indistinguishability obfuscation from compact functional encryption. In: Gennaro, R., Robshaw, M. (eds.) CRYPTO 2015. LNCS, vol. 9215, pp. 308–326. Springer, Heidelberg (2015). doi:10.1007/978-3-662-47989-6_15

5. Ananth, P., Jain, A., Sahai, A.: Achieving compactness generically: indistinguishability obfuscation from non-compact functional encryption. IACR Cryptology ePrint Archive 2015:730 (2015)

6. Ananth, P., Jain, A., Sahai, A.: Patchable obfuscation. IACR Cryptology ePrint Archive 2015:1084 (2015)

7. Ananth, P., Sahai, A.: Functional encryption for turing machines. In: Kushilevitz, E., Malkin, T. (eds.) TCC 2016. LNCS, vol. 9562, pp. 125–153. Springer, Heidelberg (2016). doi:10.1007/978-3-662-49096-9_6

8. Applebaum, B.: Bootstrapping obfuscators via fast pseudorandom functions. In: Sarkar, P., Iwata, T. (eds.) ASIACRYPT 2014. LNCS, vol. 8874, pp. 162–172. Springer, Heidelberg (2014). doi:10.1007/978-3-662-45608-8_9

9. Barak, B., Garg, S., Kalai, Y.T., Paneth, O., Sahai, A.: Protecting obfuscation against algebraic attacks. In: Nguyen, P.Q., Oswald, E. (eds.) EUROCRYPT 2014. LNCS, vol. 8441, pp. 221–238. Springer, Heidelberg (2014). doi:10.1007/978-3-642-55220-5_13

10. Bitansky, N., Garg, S., Lin, H., Pass, R., Telang, S.: Succinct randomized encodings and their applications. In: STOC (2015)

11. Bitansky, N., Vaikuntanathan, V.: Indistinguishability obfuscation from functional encryption. In: IEEE 56th Annual Symposium on Foundations of Computer Science, FOCS 2015, Berkeley, CA, USA, 17–20 October 2015, pp. 171–190 (2015)

12. Boyle, E., Chung, K.-M., Pass, R.: Oblivious parallel RAM and applications. In: Kushilevitz, E., Malkin, T. (eds.) TCC 2016. LNCS, vol. 9563, pp. 175–204. Springer, Heidelberg (2016). doi:10.1007/978-3-662-49099-0_7

13. Boyle, E., Chung, K.-M., Pass, R.: On extractability obfuscation. In: Lindell, Y. (ed.) TCC 2014. LNCS, vol. 8349, pp. 52–73. Springer, Heidelberg (2014). doi:10.1007/978-3-642-54242-8_3

14. Brakerski, Z., Rothblum, G.N.: Virtual black-box obfuscation for all circuits via generic graded encoding. In: Lindell, Y. (ed.) TCC 2014. LNCS, vol. 8349, pp. 1–25. Springer, Heidelberg (2014). doi:10.1007/978-3-642-54242-8_1

15. Canetti, R., Chen, Y., Holmgren, J., Raykova, M.: Succinct adaptive garbled RAM. In: TCC 2016-B

16. Canetti, R., Holmgren, J.: Fully succinct garbled RAM. In: ITCS (2016)

17. Canetti, R., Holmgren, J., Jain, A., Vaikuntanathan, V.: Indistinguishability obfuscation of iterated circuits and RAM programs. In: STOC (2015)

18. Canetti, R., Lin, H., Tessaro, S., Vaikuntanathan, V.: Obfuscation of probabilistic circuits and applications. In: Dodis, Y., Nielsen, J.B. (eds.) TCC 2015. LNCS, vol. 9015, pp. 468–497. Springer, Heidelberg (2015). doi:10.1007/978-3-662-46497-7_19

19. Chen, B., Lin, H., Tessaro, S.: Oblivious parallel RAM: improved efficiency and generic constructions. In: Kushilevitz, E., Malkin, T. (eds.) TCC 2016. LNCS, vol. 9563, pp. 205–234. Springer, Heidelberg (2016). doi:10.1007/978-3-662-49099-0_8

20. Chen, Y.-C., Chow, S.S.M., Chung, K.-M., Lai, R.W.F., Lin, W.-K., Zhou, H.-S.: Cryptography for parallel RAM from indistinguishability obfuscation. In: ITCS (2016)

21. Chung, K.-M., Pass, R.: A simple ORAM. IACR Cryptology ePrint Archive 2013:243 (2013)

22. Coron, J.-S., Gentry, C., Halevi, S., Lepoint, T., Maji, H.K., Miles, E., Raykova, M., Sahai, A., Tibouchi, M.: Zeroizing without low-level zeroes: new MMAP attacks and their limitations. In: Gennaro, R., Robshaw, M. (eds.) CRYPTO 2015. LNCS, vol. 9215, pp. 247–266. Springer, Heidelberg (2015). doi:10.1007/978-3-662-47989-6_12

23. Coron, J.-S., Lepoint, T., Tibouchi, M.: Practical multilinear maps over the integers. In: Canetti, R., Garay, J.A. (eds.) CRYPTO 2013. LNCS, vol. 8042, pp. 476–493. Springer, Heidelberg (2013). doi:10.1007/978-3-642-40041-4_26

24. Coron, J.-S., Lepoint, T., Tibouchi, M.: New multilinear maps over the integers. In: Gennaro, R., Robshaw, M. (eds.) CRYPTO 2015. LNCS, vol. 9215, pp. 267–286. Springer, Heidelberg (2015). doi:10.1007/978-3-662-47989-6_13

25. Garg, S., Gentry, C., Halevi, S.: Candidate multilinear maps from ideal lattices. In: Johansson, T., Nguyen, P.Q. (eds.) EUROCRYPT 2013. LNCS, vol. 7881, pp. 1–17. Springer, Heidelberg (2013). doi:10.1007/978-3-642-38348-9_1

26. Garg, S., Gentry, C., Halevi, S., Raykova, M., Sahai, A., Waters, B.: Candidate indistinguishability obfuscation and functional encryption for all circuits. In: FOCS (2013)

27. Garg, S., Lu, S., Ostrovsky, R.: Black-box garbled RAM. In: FOCS (2015)

28. Garg, S., Lu, S., Ostrovsky, R., Scafuro, A.: Garbled RAM from one-way functions. In: STOC (2015)

29. Gentry, C., Gorbunov, S., Halevi, S.: Graph-induced multilinear maps from lattices. In: Dodis, Y., Nielsen, J.B. (eds.) TCC 2015. LNCS, vol. 9015, pp. 498–527. Springer, Heidelberg (2015). doi:10.1007/978-3-662-46497-7_20

30. Gentry, C., Halevi, S., Lu, S., Ostrovsky, R., Raykova, M., Wichs, D.: Garbled RAM revisited. In: Nguyen, P.Q., Oswald, E. (eds.) EUROCRYPT 2014. LNCS, vol. 8441, pp. 405–422. Springer, Heidelberg (2014). doi:10.1007/978-3-642-55220-5_23

31. Gentry, C., Halevi, S., Raykova, M., Wichs, D.: Outsourcing private RAM computation. In: FOCS (2014)

32. Gorbunov, S., Vaikuntanathan, V., Wee, H.: Functional encryption with bounded collusions via multi-party computation. In: Safavi-Naini, R., Canetti, R. (eds.) CRYPTO 2012. LNCS, vol. 7417, pp. 162–179. Springer, Heidelberg (2012). doi:10.1007/978-3-642-32009-5_11

33. Goyal, V., Ishai, Y., Sahai, A., Venkatesan, R., Wadia, A.: Founding cryptography on tamper-proof hardware tokens. In: Micciancio, D. (ed.) TCC 2010. LNCS, vol. 5978, pp. 308–326. Springer, Heidelberg (2010). doi:10.1007/978-3-642-11799-2_19

34. Hemenway, B., Jafargholi, Z., Ostrovsky, R., Scafuro, A., Wichs, D.: Adaptively secure garbled circuits from one-way functions. In: Robshaw, M., Katz, J. (eds.) CRYPTO 2016. LNCS, vol. 9816, pp. 149–178. Springer, Heidelberg (2016). doi:10.1007/978-3-662-53015-3_6

35. Kalai, Y.T., Paneth, O.: Delegating RAM computations. IACR Cryptology ePrint Archive 2015: 957 (2015)
36. Kalai, Y.T., Raz, R., Rothblum, R.D.: How to delegate computations: the power of no-signaling proofs. In: STOC (2014)
37. Koppula, V., Bishop Lewko, A., Waters, B.: Indistinguishability obfuscation for turing machines with unbounded memory. In: STOC (2015)
38. Lin, H.: Indistinguishability obfuscation from constant-degree graded encoding schemes. In: Fischlin, M., Coron, J.-S. (eds.) EUROCRYPT 2016. LNCS, vol. 9665, pp. 28–57. Springer, Heidelberg (2016). doi:10.1007/978-3-662-49890-3_2
39. Lin, H., Vaikuntanathan, V.: Indistinguishability obfuscation from ddh-like assumptions on constant-degree graded encodings. Cryptology ePrint Archive, Report 2016/795 (2016). http://eprint.iacr.org/2016/795
40. Lu, S., Ostrovsky, R.: How to garble RAM programs? In: Johansson, T., Nguyen, P.Q. (eds.) EUROCRYPT 2013. LNCS, vol. 7881, pp. 719–734. Springer, Heidelberg (2013). doi:10.1007/978-3-642-38348-9_42
41. Lu, S., Ostrovsky, R.: Black-box parallel garbled RAM. Cryptology ePrint Archive, Report 2015/1068 (2015). http://eprint.iacr.org/2015/1068
42. Miles, E., Sahai, A., Zhandry, M.: Annihilation attacks for multilinear maps: cryptanalysis of indistinguishability obfuscation over GGH13. In: CRYPTO (2016)
43. Naor, M.: On cryptographic assumptions and challenges. In: Boneh, D. (ed.) CRYPTO 2003. LNCS, vol. 2729, pp. 96–109. Springer, Heidelberg (2003). doi:10.1007/978-3-540-45146-4_6
44. Waters, B.: A punctured programming approach to adaptively secure functional encryption. In: Gennaro, R., Robshaw, M. (eds.) CRYPTO 2015. LNCS, vol. 9216, pp. 678–697. Springer, Heidelberg (2015). doi:10.1007/978-3-662-48000-7_33

Interactive Oracle Proofs

Eli Ben-Sasson[1], Alessandro Chiesa[2(✉)], and Nicholas Spooner[3]

[1] Technion, Haifa, Israel
eli@cs.technion.ac.il
[2] UC Berkeley, Berkeley, USA
alexch@berkeley.edu
[3] University of Toronto, Toronto, Canada
spooner@cs.toronto.edu

Abstract. We initiate the study of a proof system model that naturally combines interactive proofs (IPs) and probabilistically-checkable proofs (PCPs), and generalizes interactive PCPs (which consist of a PCP followed by an IP). We define an *interactive oracle proof* (IOP) to be an interactive proof in which the verifier is not required to read the prover's messages in their entirety; rather, the verifier has oracle access to the prover's messages, and may probabilistically query them. IOPs retain the expressiveness of PCPs, capturing NEXP rather than only PSPACE, and also the flexibility of IPs, allowing multiple rounds of communication with the prover. IOPs have already found several applications, including unconditional zero knowledge [BCGV16], constant-rate constant-query probabilistic checking [BCG+16], and doubly-efficient constant-round IPs for polynomial-time bounded-space computations [RRR16].

We offer two main technical contributions. First, we give a compiler that maps any public-coin IOP into a non-interactive proof in the random oracle model. We prove that the soundness of the resulting proof is tightly characterized by the soundness of the IOP against *state restoration attacks*, a class of rewinding attacks on the IOP verifier that is reminiscent of, but incomparable to, resetting attacks.

Second, we study the notion of state-restoration soundness of an IOP: we prove tight upper and lower bounds in terms of the IOP's (standard) soundness and round complexity; and describe a simple adversarial strategy that is optimal, in expectation, across all state restoration attacks.

Our compiler can be viewed as a generalization of the Fiat–Shamir paradigm for public-coin IPs (CRYPTO '86), and of the "CS proof" constructions of Micali (FOCS '94) and Valiant (TCC '08) for PCPs. Our analysis of the compiler gives, in particular, a unified understanding of these constructions, and also motivates the study of state restoration attacks, not only for IOPs, but also for IPs and PCPs.

Parts of this paper appear in the third author's master's thesis (April 2015) in the Department of Computer Science at ETH Zurich, supervised by Alessandro Chiesa and Thomas Holenstein. Independent of our work, [RRR16] introduce the notion of *Probabilistically Checkable Interactive Proofs*, which is the same as our notion of Interactive Oracle Proofs.

M. Hirt and A. Smith (Eds.): TCC 2016-B, Part II, LNCS 9986, pp. 31–60, 2016.
DOI: 10.1007/978-3-662-53644-5_2

When applied to known IOP constructions, our compiler implies, e.g., blackbox unconditional ZK proofs in the random oracle model with quasilinear prover and polylogarithmic verifier, improving on a result of [IMSX15].

1   Introduction

The notion of *proof* is central to modern cryptography and complexity theory. The class NP, for example, is the set of languages whose membership can be decided by a deterministic polynomial-time verifier by reading proof strings of polynomial length; this class captures the traditional notion of a mathematical proof. Over the last three decades, researchers have introduced and studied proof systems that generalize the above traditional notion, and investigations from these points of view have led to breakthroughs in cryptography, hardness of approximation, and other areas. In this work we introduce and study a new model of proof system.

1.1   Models of Proof Systems

We give some context by recalling three of the most well-known among alternative models of proof systems.

Interactive Proofs (IPs). Interactive proofs were introduced by Goldwasser, Micali, and Rackoff [GMR89]: in a k-round interactive proof, a probabilistic polynomial-time verifier exchanges k messages with an all-powerful prover, and then accepts or rejects; IP[k] is the class of languages with a k-round interactive proof. Independently, Babai [Bab85] introduced Arthur–Merlin games: a k-round Arthur–Merlin game is a k-round *public-coin* interactive proof (i.e., the verifier messages are uniformly and independently random); AM[k] is the class of languages with a k-round Arthur–Merlin game. Goldwasser and Sipser [GS86] showed that the two models are equally powerful: for polynomial k, IP[k] $\subseteq$ AM[$k+2$]. Shamir [Sha92], building on the "sum-check" interactive proof of Lund, Fortnow, Karloff, and Nisan [LFKN92], proved that interactive proofs correspond to languages decidable in polynomial space: IP[poly(n)] = PSPACE. (Also see [Bab90].)

Multi-prover Interactive Proofs (MIPs). Multi-prover interactive proofs were introduced by Ben-Or, Goldwasser, Kilian, and Wigderson [BGKW88]: in a k-round p-prover interactive proof, a probabilistic polynomial-time verifier interacts k times with p non-communicating all-powerful provers, and then accepts or rejects; MIP[p, k] is the class of languages that have a k-round p-prover interactive proof. In [BGKW88], the authors show that two provers always suffice (i.e., MIP[p, k] = MIP[2, k]), and that all languages in NP have perfect zero knowledge proofs in this model. Fortnow, Rompel, and Sipser [FRS88] show that interaction with two provers is equivalent to interaction with one prover plus oracle access to a proof string, and from there obtain that MIP[poly(n), poly(n)] $\subseteq$ NEXP; Babai,

Fortnow and Lund [BFL90] show that NEXP has 1-round 2-prover interactive proofs, thus showing that MIP[2, 1] = NEXP.

Probabilistically Checkable Proofs (PCPs). Probabilistically checkable proofs were introduced by [FRS88, BFLS91, AS98, ALM+98]: in a probabilistically-checkable proof, a probabilistic polynomial-time verifier has oracle access to a proof string; PCP[r, q] is the class of languages for which the verifier uses at most r bits of randomness, and queries at most q locations of the proof (note that the proof length is at most 2^r). The above results on MIPs imply that PCP[poly(n), poly(n)] = NEXP. Later works "scaled down" this result to NP: Babai, Fortnow, Levin and Szegedy [BFLS91] show that NP = PCP[$O(\log n)$, poly($\log n$)]; Arora and Safra [AS98] show that NP = PCP[$O(\log n)$, $O(\sqrt{\log n})$]; and Arora, Lund, Motwani, Sudan, and Szegedy [ALM+92] show that NP = PCP[$O(\log n)$, $O(1)$]. This last is known as the *PCP Theorem*.

Researchers have studied other models of proof systems, and here we name only a few: *linear IPs* [BCI+13], *no-signaling MIPs* [IKM09, Ito10, KRR13, KRR14], *linear PCPs* [IKO07, Gro10, Lip12, BCI+13, GGPR13, PGHR13, BCI+13, SBW11, SMBW12, SVP+12, SBV+13], *interactive PCPs* [KR08, KR09, GIMS10].

We introduce **interactive oracle proofs** (IOPs), a model of proof system that combines aspects of IPs and PCPs, and also generalizes interactive PCPs (which consist of a PCP followed by an IP). Our work focuses on cryptographic applications of this proof system, as we discuss next.

1.2   Compiling Proof Systems into Argument Systems

The proof systems mentioned so far share a common feature: they make no assumptions on the computational resources of a (malicious) prover trying to convince the verifier. Instead, many proof systems make "structural" assumptions on the prover: MIPs assume that the prover is a collection of non-communicating strategies (each representing a "sub-prover"); PCPs assume that the prover is non-adaptive (the answer to a message does not depend on previous messages); linear IPs assume that the prover is a linear function; and so on.

In contrast, in cryptography, one often considers *argument systems* [BC86, BCC88, Kil92, Mic00]: these are proof systems where soundness holds only against provers that have a bound on computational resources (e.g., provers that run in probabilistic polynomial time). The relaxation from statistical soundness to computational soundness allows circumventing various limitations of IPs [BHZ87, GH98, GVW02, PSSV07], while also avoiding "structural" assumptions on the prover, which can be hard to enforce in applications.

Constructing Argument Systems. A common methodology to construct argument systems with desirable properties (e.g., sublinear communication complexity) follows these two steps: (1) give a proof system that achieves these properties in a model with structural restrictions on (all-powerful) provers; (2) use cryptographic tools to compile that proof system into an argument system, i.e., one where the only restriction on the prover is that it is an efficient algorithm.

Thus, the compilation trades any structural assumptions for computational ones. This methodology has been highly productive.

Proofs in the Random Oracle Model. An idealized model for studying computationally-bounded provers is the random oracle model [FS86, BR93], where every party has access to the same random function. A protocol proved secure in this model can potentially be instantiated in practice by replacing the random function with a concrete "random-looking" efficient function. While this intuition fails in the general case [CGH04, BBP04, GK03, BDG+13], the random oracle model is nonetheless a useful testbed for cryptographic primitives. In this paper we focus on proof systems in this model for which the proof consists of a single message from the prover to the verifier. A **non-interactive random-oracle argument** (NIROA) for a relation $\mathscr{R}$ is a pair of probabilistic polynomial-time algorithms, the prover $\mathbb{P}$ and verifier $\mathbb{V}$, that satisfy the following. (1) *Completeness:* for every instance-witness pair (x, w) in the relation $\mathscr{R}$, $\Pr[\mathbb{V}^\rho(x, \mathbb{P}^\rho(x, w)) = 1] = 1$, where the probability is taken over the random oracle ρ as well as any randomness of $\mathbb{P}$ and $\mathbb{V}$. (2) *Soundness:* for every instance x not in the language of $\mathscr{R}$ and every malicious prover $\tilde{\mathbb{P}}$ that asks at most a polynomial number of queries to the random oracle, it holds that $\Pr[\mathbb{V}^\rho(x, \tilde{\mathbb{P}}^\rho) = 1]$ is negligible in the security parameter.

Prior NIROAs and Our Focus. Prior work uses the above 2-step methodology to obtain NIROAs with desirable properties. For example, the Fiat–Shamir paradigm maps 3-message public-coin IPs to corresponding NIROAs [FS86, PS96]; when invoked on suitable IP constructions, this yields efficient zero knowledge non-interactive proofs. As another example, Micali's "CS proof" construction, building on [Kil92], transforms PCPs to corresponding NIROAs; Valiant [Val08] revisits Micali's construction and proves that it is a proof of knowledge; when invoked on suitable PCPs, these yield non-interactive arguments of knowledge that are short and easy to verify. In this work we study the question of how to compile IOPs (which generalize IPs and PCPs) into NIROAs;[1] our work ultimately leads to formulating and studying a game-theoretic property of IOPs, which in turn motivates similar questions for IPs and PCPs. We now discuss our results.

1.3 Results

We present three main contributions: one is definitional and the other two are technical in nature.

---

[1] We do not study the question of avoiding assuming random oracles: this is not our focus. Reducing assumptions when compiling constant-round IPs is the subject of much research, obtaining arguments with non-programmable random oracles and a common random string [Lin15, CPSV16], obfuscation [KRR16, MV16], and others. Extending such ideas to IOPs is an interesting direction.

Interactive Oracle Proofs A New Proof System Model. We introduce a new proof system model: *interactive oracle proofs* (IOPs).[2] This model naturally combines aspects of IPs and PCPs, and also generalizes IPCPs (see comparison in Remark 1 below); namely, an IOP is a "multi-round PCP" that generalizes an interactive proof as follows: the verifier has oracle access to the prover's messages, and may probabilistically query them (rather than having to read them in full). In more detail, a k-round IOP comprises k rounds of interaction. In the i-th round of interaction: the verifier sends a message m_i to the prover, which he reads in full; then the prover replies with a message f_i to the verifier, which he can query, as an oracle proof string, in this and all later rounds. After the k rounds of interaction, the verifier either accepts or rejects.

Like the PCP model, two fundamental measures of efficiency in the IOP model are the *proof length* p, which is the total number of bits in all of the prover's messages, and the *query complexity* q, which is the total number of locations queried by the verifier across all of the prover's messages. Unlike the PCP model, another fundamental measure of efficiency is the round complexity k; the PCP model can then be viewed as a special case where $k = 1$ (and the first verifier message is empty).

We show that IOPs characterize NEXP (like PCPs); both sequential and parallel repetition of IOPs yield (perfect) exponential soundness error reduction (like IPs); and any IOP can be converted into a public-coin one (like IPs). These basic complexity-theoretic properties confirm that our definition of IOP is a natural way to combine aspects of PCPs and IPs, and to generalize IPCPs.

Motivation: Efficiency. IOPs extend IPs, by treating the prover's messages as oracle strings, and PCPs, by allowing for more than 1 round. These additional degrees of freedom enable IOPs to retain the expressive power of PCP while also allowing for additional efficiency, as already demonstrated in several works.

For example, [BCGV16] obtain unconditional zero knowledge via a 2-round IOP with quasilinear proof length; such a result is not known for PCPs (or even IPCPs [KR08]). Moreover, when combined with our compiler (see next contribution) we obtain blackbox unconditional zero-knowledge with quasilinear prover and polylogarithmic verifier in the random-oracle model, improving prover runtime of [IMSX15, Sect. 2.3];

As another example, [BCG+16] obtain 3-round IOPs for circuit satisfiability *with linear proof length and constant query complexity*, while for PCPs prior work only achieves sublinear query complexity [BKK+13]. To do so, [BCG+16] show that *sumcheck* [LFKN92, Sha92] and *proof composition* [AS98] (used in many PCP constructions such as [ALM+98, HS00, BGH+04]) have more efficient "IOP analogues", which in turn imply a number of probabilistic checking results that are more efficient than corresponding ones that only rely on PCPs. We briefly sketch the intuition for why interactive proof composition, via IOPs, is more efficient. In a composed proof, the prover first writes a part π_0 of the proof (e.g., in [ALM+98] π_0 is an evaluation of a low-degree multivariate polynomial, and

---

[2] Independent of our work, [RRR16] introduce *Probabilistically Checkable Interactive Proofs*, which are equivalent to our IOPs.

in [BS08] it is an evaluation of a low-degree univariate polynomial). Then, to demonstrate that π_0 has certain good properties (e.g., it is low degree), the prover also appends a (long) sequence of sub-proofs, where each sub-proof allegedly demonstrates to the verifier that a subset of entries of π_0 is "good". Afterwards, in another invocation of the recursion, the prover appends to each sub-proof a sequence of sub-sub-proofs, and so on. A crucial observation is that the verifier typically queries locations of only a small number of such sub-proofs; moreover, once the initial proof π_0 is fixed, soundness is not harmed if the verifier randomly selects the set of sub-proofs he wants to see and tells this to the prover. In sum, in many PCP constructions (including the aforementioned ones), *the proof length can be greatly reduced via interaction between the prover and verifier*, via an IOP.

As yet another example, [RRR16] use IOPs to obtain doubly-efficient constant-round IPs for polynomial-time bounded-space computations. The result relies on an "amortization theorem" for IOPs that states that, for a so-called *unambiguous* IOPs, batch verification of multiple statements can be more efficient than simply running an independent IOP for each statement.

Remark 1 (comparison with IPCP). Kalai and Raz [KR08] introduce and study *interactive PCPs* (IPCPs), a model of proof system that also combines aspects of IPs and PCPs, but in a different way: an IPCP is a PCP followed by an IP, i.e., the prover sends to the verifier a PCP and then the prover and verifier engage in an interactive proof. An IPCP can be viewed as a special case of an IOP, i.e., it is an IOP in which the verifier has oracle access to the first prover message, but must read in full subsequent prover messages. The works of [KR08, GKR08] show that boolean formulas with n variables, size m, and depth d have IPCPs where the PCP's size is polynomial in d and n and the communication complexity of the subsequent IP is polynomial in d and $\log m$. This shows that even IPCPs give efficiency advantages over both IPs and PCPs given separately.

From Interactive Oracle Proofs to Non-interactive Random-Oracle Arguments. We give a polynomial-time transformation that maps any public-coin interactive oracle proof (IOP) to a corresponding non-interactive random-oracle argument (NIROA). We prove that the soundness of the output proof is tightly characterized by the soundness of the IOP verifier against *state restoration attacks*, a class of rewinding attacks on the verifier that we now describe.

At a high level, a state restoration attack against an IOP verifier works as follows: the malicious prover and the verifier start interacting, as they normally would in an IOP; at any moment, however, the prover can choose to set the verifier to any state at which the verifier has previously been, and the verifier then continues onwards from that point *with fresh randomness*. Of course, if the prover could restore the verifier's state an unbounded number of times, the prover would eventually succeed in making the verifier accept. We thus only consider malicious provers that interact with the verifier for at most a certain number of rounds: for $b \in \mathbb{N}$, we say a prover is *b-round* if it plays at most b rounds during any interaction with any verifier. Then, we say that an IOP has state restoration soundness $s_{\mathrm{sr}}(\mathbf{x}, b)$ if every b-round state-restoring prover cannot make the IOP verifier accept an instance x (not in the language) with probability greater than $s_{sr}(x, b)$. This notion is reminiscent of, *but incomparable to*, the notion of resettable soundness [BGGL01]; see Remark 2 below.

Informally, our result about transforming IOPs into NIROAs can be stated as follows.

Theorem 1 (IOP → NIROA). *There exists a polynomial-time transformation T such that, for every relation $\mathscr{R}$, if (P, V) is a public-coin interactive oracle proof system for $\mathscr{R}$ with state restoration soundness $s_{sr}(x, b)$, then $(\mathbb{P}, \mathbb{V}) := T(P, V)$ is a non-interactive random-oracle argument system for $\mathscr{R}$ with soundness*

$$s_{sr}(x, m) + O(m^2 2^{-\lambda}) \ ,$$

where m is an upper bound on the number of queries to the random oracle that a malicious prover can make, and λ is a security parameter. The aforementioned soundness is tight up to small factors. (Good state restoration soundness can be obtained, e.g., via parallel repetition as in Remark 4.)

Moreover, we prove that the transformation T is benign in the sense that it preserves natural properties of the IOP. Namely, (1) the runtimes of the NIROA prover and verifier are linear in those of the IOP prover and verifier (up to a polynomial factor in λ); (2) the NIROA is a proof of knowledge if the IOP is a proof of knowledge (and the extractor strategy straight-line, which has desirable properties [BW15]); and (3) the NIROA is (malicious-verifier) statistical zero knowledge if the IOP is honest-verifier statistical zero knowledge.[3] See Theorem 3 for the formal statement; the statement employs the notion of *restricted state restoration soundness* as it allows for a tighter lower bound on soundness.

An immediate application is obtained by plugging the work of [BCGV16] into our compiler, thereby achieving a variant of the black-box ZK results of [IMSX15, Sect. 2.3] where the prover runs in quasilinear (rather than merely polynomial) time.

Corollary 1 (informal). *There is a blackbox non-interactive argument system for NP, in the random-oracle model, with unconditional zero knowledge, quasilinear-time prover, and polylogarithmic-time verifier.*

Our compiler can be viewed as a generalization of the Fiat–Shamir paradigm for public-coin IPs [FS86, PS96], and of the "CS proof" constructions of Micali [Mic00] and Valiant [Val08] for PCPs. Our analysis of the compiler gives, in particular, a *unified understanding of these constructions*, and motivates the study of state restoration attacks, not only for IOPs, but also for IPs and PCPs. (Indeed, we are not aware of works that study the security of the Fiat–Shamir

---

[3] Security in the random oracle model sometimes does *not* imply security when the oracle is substituted with a hash function, e.g., when applying the Fiat–Shamir paradigm to zero-knowledge proofs/arguments [HT98, DNRS03, GOSV14]. However, our transformation T only assumes that the IOP is zero knowledge against the honest verifier, seemingly avoiding the above limitations.

paradigm, in the random oracle model, applied to a public-coin IP with arbitrary number of rounds; the analyses that we are aware of focus on the case of 2 rounds.)

Our next contribution is a first set of results about such kinds of attacks, as described in the next section.

Remark 2 (resetting, backtracking). We compare state restoration soundness with other soundness notions:

- State restoration attacks are reminiscent of, *but incomparable to*, resetting attacks [BGGL01]. In the latter, the prover invokes multiple verifier incarnations with independent randomness, and may interact multiple times with each incarnation; also, this notion does not assume that the verifier is public-coin. Instead, in a state restoration attack, the verifier must be public-coin and its randomness is not fixed at the start but, instead, a new fresh random message is sampled each time the prover restores to a previously-seen state.
- State restoration is closely related to backtracking [BD16] (independent work). The two notions differ in that: (1) backtracking "charges" more for restoring verifier states that are further in the past, and (2) backtracking also allows the verifier to restore states of the prover (as part of the completeness property of the protocol); backtracking soundness is thus polynomially related to state restoration soundness.

  Bishop and Dodis [BD16] give a compiler from a public-coin IP to an error-resilient IP, whose soundness is related to the backtracking soundness of the original IP; essentially, they use hashing techniques to limit a malicious prover impersonating an adversarial channel to choosing when to backtrack the protocol. Their setting is a completely different example in which backtracking, and thus state restoration, plays a role.

Remark 3 (programmability). As in most prior works, soundness and proof of knowledge do *not* rely on programming the random oracle. As for zero knowledge, the situation is more complicated: there are several notions of zero knowledge in the random oracle model, depending on "how programmable" the random oracle is (see [Wee09]). The notion that we use is zero knowledge in the explicitly-programmable random oracle (EPRO) model; the stronger notion in the non-programmable random oracle model is not achievable for NIROAs. Such a limitation can sometimes be avoided by also using a common random string [Lin15, CPSV16], and extending such techniques to the setting of IOPs is an interesting problem.

State Restoration Attacks on Interactive Oracle Proofs. The analysis of our transformation from public-coin IOPs to NIROAs highlights state restoration soundness as a notion that merits further study. We provide two results in this direction. First, we prove tight upper and lower bounds on state restoration soundness in terms of the IOP's (standard) soundness and round complexity.

Theorem 2. *For any relation $\mathscr{R}$, public-coin k-round IOP for $\mathscr{R}$, and instance $\mathbf{x}$ not in the language of $\mathscr{R}$,*

$$\forall b \geq k(\mathbf{x}) + 1, \quad \left\lfloor \frac{b}{k(\mathbf{x}) + 1} \right\rfloor s(\mathbf{x})(1 - o(1)) \leq s_{\mathrm{sr}}(\mathbf{x}, b) \leq \binom{b}{k(\mathbf{x}) + 1} s(\mathbf{x}) ,$$

[4]*where $s_{\mathrm{sr}}(\mathbf{x}, b)$ is the state restoration soundness of IOP and $s(\mathbf{x})$ its (standard) soundness for the instance $\mathbf{x}$. Also, the bounds are tight: there are IOPs that meet the lower bound and IOPs that meet the upper bound.*

Remark 4 (good state restoration soundness). A trivial way to obtain state restoration soundness $2^{-\lambda}$ in the general case is to apply r-fold parallel repetition to the IOP with $r = \Omega(\frac{k \log b + \lambda}{\log s(\mathbf{x})})$; note that r is polynomially bounded for natural choices of k, b, λ. This choice of r is pessimistic, because for IOPs that do not meet the upper bound (i.e., are "robust" against such attacks) a smaller r suffices. This use of parallel repetition is analogous to its use in achieving the incomparable notion of resettable soundness [PTW09, COPV13].

Second, we study the structure of optimal state restoration attacks: we prove that, for any public-coin IOP, there is a simple state restoration attack that has optimal expected cost, where cost is the number of rounds until the prover wins. This result relies on a correspondence that we establish between IOP verifiers and certain games, which we call *tree exploration games*, pitting one player against Nature. We go in more detail about this result in later sections (see Sect. 1.4 and full version [BCS16].). A better understanding of state restoration soundness may enable us to avoid trivial soundness amplification (see Remark 4) for IOPs of interest.

1.4   Techniques

We summarize the techniques that we use to prove our technical contributions.

The Transformation. Our transformation maps any public-coin IOP to a corresponding NIROA, and it generalizes two transformations that we now recall.

The first transformation is the Fiat–Shamir paradigm [FS86, PS96], which maps any public-coin IP to a corresponding NIROA, and it works as follows. The NIROA prover runs the interaction between the IP prover and the IP verifier "in his head", by setting the IP verifier's next message to be the output of the random oracle on the query that equals the transcript of previously exchanged messages. The NIROA prover sends a non-interactive proof that contains the final transcript of interaction; the NIROA verifier checks the proof's validity by checking that all the IP verifier's messages are computed correctly via the random oracle.

---

[4] We note that [BGGL01] prove an analogous upper bound for the *incomparable* notion of resettable soundness (see Remark 2). Also, [BD16] prove an analogous, weaker upper bound on the related notion of backtracking soundness (see Remark 2). Neither of the two studies lower bounds, or tightness of bounds.

The second transformation is the "CS proof" construction of Micali [Mic00] and Valiant [Val08], which maps any PCP to a corresponding NIROA, and it works as follows. The NIROA prover first commits to the PCP via a Merkle tree [Mer89a] based on the random oracle, then queries the random oracle with the root of this tree to obtain randomness for the PCP verifier, and finally sends a non-interactive proof that contains the root as well as authentication paths for each query by the PCP verifier to the PCP; the NIROA verifier checks the proof's validity by checking that the PCP verifier's randomness is computed correctly through the random oracle, and that all authentication paths are valid. (The transformation can be viewed as a non-interactive variant of Kilian's protocol [Kil92, BG08] that uses ideas from the aforementioned Fiat–Shamir paradigm.)

Our transformation takes as input IOPs, for which both IPs and PCPs are special cases, and hence must support both (i) multiple rounds of interaction between the IOP prover and IOP verifier, as well as (ii) oracle access by the IOP verifier to the IOP prover messages. Given an instance x, the NIROA prover thus uses the random oracle ρ to run the interaction between the IOP prover and the IOP verifier "in his head" in a way that combines the aforementioned two approaches, as follows. First, the NIROA prover computes an initial value $\sigma_0 := \rho(x)$. Then, for $i = 1, 2, \ldots$, it simulates the i-th round by deriving the IOP verifier's i-th message m_i as $\rho(x\|\sigma_{i-1})$, compressing the IOP prover's i-th message f_i via a Merkle tree to obtain the root rt_i, and computing the new value $\sigma_i := \rho(rt_i\|\sigma_{i-1})$. The values $\sigma_0, \sigma_1, \ldots$ are related by the Merkle–Damgård transform [Dam89, Mer89b] that, intuitively, enforces ordering between rounds. If there are $k(x)$ rounds of interaction, then $\rho(x\|\sigma_{k(x)})$ is used as randomness for the queries to $f_1, \ldots, f_{k(x)}$. The NIROA prover provides in the non-interactive proof all the roots rt_i, the final value $\sigma_{k(x)}$, the answers to the queries, and an authentication path for each query. This sketch omits several details; see Sect. 5.

Soundness Analysis of the Transformation. We prove that the soundness of the NIROA produced by the above transformation is tightly characterized by the state restoration soundness of the underlying IOP. This characterization comprises two arguments: an upper bound and a lower bound on the NIROA's soundness. We only discuss the upper bound here: proving that the soundness (error) of the NIROA is at most the soundness (error) of the IOP against state restoration attacks, up to small additive factors.

The upper bound essentially implies that all that a malicious prover $\tilde{\mathbb{P}}$ can do to attack the NIROA verifier is to conduct a state restoration attack against the underlying IOP verifier "in his own head": roughly, $\tilde{\mathbb{P}}$ can provide multiple inputs to the random oracle in order to induce multiple fresh samples of verifier messages for a given round so to find a lucky one, or instead go back to previous rounds and do the same there.

In more detail, the proof itself relies on a reduction: given a malicious prover $\tilde{\mathbb{P}}$ against the NIROA verifier, we show how to construct a corresponding malicious prover $\tilde{P}$ that conducts a state restoration attack against the underlying IOP verifier. We prove that the winning probability of $\tilde{P}$ is essentially the same as that of $\tilde{\mathbb{P}}$; moreover, we also prove that the reduction preserves the resources needed for the attack in the sense that if $\tilde{\mathbb{P}}$ asks at most m queries to the random oracle, then $\tilde{P}$ plays at most m rounds during the attack.

Intuitively, the construction of $\tilde{P}$ in terms of $\tilde{\mathbb{P}}$ must use some form of extraction: $\tilde{\mathbb{P}}$ outputs a non-interactive proof that contains only (i) the roots that (allegedly) are commitments to underlying IOP prover's messages, and (ii) answers to the IOP verifier's queries and corresponding authentication paths; in contrast, $\tilde{P}$ needs to actually output these IOP prover's messages. In principle, the malicious prover $\tilde{\mathbb{P}}$ may not have "in mind" any underlying IOP prover, and we must prove that, nevertheless, there is a way for $\tilde{P}$ to extract some IOP prover message for each round that convince the verifier with the claimed probability.

Our starting point is the extractor algorithm of Valiant [Val08] for the "CS proof" construction of Micali [Mic00]: Valiant proves that Micali's NIROA construction is a proof of knowledge by exhibiting an algorithm, let us call it *Valiant's extractor*, that recovers the underlying PCP whenever the NIROA prover convinces the NIROA verifier with sufficient probability. (In particular, our proof is not based on a "forking lemma" [PS96].) Our setting differs from Valiant's in that the IOP prover $\tilde{P}$ obtained from the NIROA prover $\tilde{\mathbb{P}}$ needs to be able to extract multiple times, "on the fly", while interacting with the IOP verifier; this more complex setting can potentially cause difficulties in terms of extractor size (e.g., if relying on rewinding the NIROA prover) or correlations (e.g., when extracting multiple times from the same NIROA prover). We tackle the more complex setting in two steps.

First, we prove an extractability property of Valiant's extractor and state it as a property of Merkle trees in the random oracle model (see Sect. A.1). Informally, we prove that, except with negligible probability, whenever an algorithm with access to a random oracle outputs multiple Merkle tree roots each accompanied with some number of (valid) authentication paths, it holds that Valiant's extractor run separately on each of these roots outputs a decommitment that is consistent with each of the values revealed in authentication paths relative to that root. We believe that distilling and proving this extractability property of Valiant's extractor is of independent interest.

Second, we show how the IOP prover $\tilde{P}$ can interact with an IOP verifier, by successively extracting messages to send, throughout the interaction, by invoking Valiant's extractor multiple times on $\tilde{\mathbb{P}}$ relative to different roots. The IOP prover $\tilde{P}$ does not rely on rewinding $\tilde{\mathbb{P}}$, and its complexity is essentially that of a single run of $\tilde{\mathbb{P}}$ plus a small amount of work.

Preserving Proof of Knowledge. We prove that the above soundness analysis can be adapted so that, if the underlying IOP is a proof of knowledge, then we can construct an extractor to show that the resulting NIROA is also a proof of knowledge. Moreover, the extractor algorithm only needs to inspect the queries and answers of one execution of $\tilde{\mathbb{P}}$ if the underlying IOP extractor does not use rewinding (known IOP constructions are of this type [BCGV16, BCG+16]); such extractors are known as *straight line* [Pas03] or *online* [Fis05], and have very desirable properties [BW15].

Preserving Zero Knowledge. We prove that, if the underlying IOP is *honest-verifier* statistical zero knowledge, then the resulting NIROA is statistical zero knowledge (i.e., is a non-interactive statistical zero knowledge proof in the explicitly-programmable random oracle model). This is because the transformation uses a Merkle tree with suitable privacy guarantees (see Sect. A.2) to construct the NIROA. Indeed, the authentication path for a leaf in the Merkle tree reveals the sibling leaf, so one must ensure that the sibling leaf does not leak information about other values; this follows by letting leaves be commitments to the underlying values. A Merkle tree with privacy is similarly used by [IMS12, IMSX15], along with honest-verifier PCPs, to achieve zero knowledge in modifications of Kilian's [Kil92, BG08] and Micali's [Mic00] constructions. (Note that the considerations [HT98, DNRS03, GOSV14] seem to only apply to compilation of malicious-verifier IOPs, which neither [IMS12, IMSX15] nor we require.)

Understanding State Restoration Attacks. We prove tight upper and lower bounds to state restoration soundness in terms of the IOP's (standard) soundness and round complexity k. The upper bound takes the form of a reduction: given a b-round state-restoring malicious prover $\tilde{P}_{sr}$ that makes the IOP verifier accept with probability s_{sr}, we construct a (non state-restoring) malicious prover $\tilde{P}$ that makes the IOP verifier accept with probability at least $\binom{b}{k+1}^{-1} s_{sr}$. Informally, $\tilde{P}$ internally simulates $\tilde{P}_{sr}$, while interacting with the "real" IOP verifier, as follows: $\tilde{P}$ first selects a random subset S of $\{1, \ldots, b\}$ with cardinality $k + 1$, and lets $S[i]$ be the i-th smallest value in S; then, $\tilde{P}$ runs $\tilde{P}_{sr}$ and simulates its state restoration attack on a "virtual" IOP verifier, executing round j (a) by interacting with the real verifier if $j = S[i]$ for some i; (b) by sampling fresh randomness otherwise. While this reduction appears wasteful (since it relies on S being a good guess), we show that there are IOPs for which the upper bound is tight. In other words, the sharp degradation as a function of round complexity (for large b, $\binom{b}{k+1} \approx b^{k+1}/(k+1)!$) is inherent for some choices of IOPs; this also gives a concrete answer to the intuition that compiling IOPs with large round complexity to NIROAs is "harder" (i.e., incurs in a greater soundness loss) than for IOPs with small round complexity. As for the lower bound on state restoration soundness, it takes the form of a universal state restoration attack that always achieves the lower bound; this bound is also tight.

While state restoration soundness may be far, in the worst case, from (standard) soundness for IOPs with large round complexity, it need not always be far. We thus investigate state restoration soundness for any particular IOP, and derive a simple attack strategy (which depends on the IOP) that we prove has optimal expected cost, where cost is the number of rounds until the prover wins. To do so, we "abstract away" various details of the proof system to obtain a simple game-theoretic notion, which we call *tree exploration games*, that pits a single player against Nature in reaching a node of a tree with label 1. Informally, such a game is specified by a rooted tree T and a predicate function ϕ that maps T's vertices to $\{0, 1\}$. The game proceeds in rounds: in the i-th round, a subtree $S_{i-1} \subseteq T$ is *accessible* to the player; the player picks a node $v \in S_{i-1}$, and Nature randomly samples a child u of v; the next accessible subtree is $S_i := S_{i-1} \cup \{u\}$.

The initial S_0 is the set consisting of T's root vertex. The player wins in round r if there is $v \in S_r$ with $\phi(v) = 1$.

We establish a correspondence between state restoration attacks and strategies for tree exploration games, and then show a simple greedy strategy for such games with optimal expected cost. Via the correspondence, a strategy's cost determines whether the underlying IOP is strong or weak against sate restoration attacks.

2 Preliminaries

2.1 Basic Notations

We denote the security parameter by λ. For $f : \{0,1\}^* \to \mathbb{R}$, we define $\hat{f} : \mathbb{N} \to \mathbb{R}$ as $\hat{f}(n) := \max_{x \in \{0,1\}^n} f(x)$.

Languages and Relations. We denote by $\mathscr{R}$ a relation consisting of pairs $(\mathbf{x}, \mathbf{w})$, where $\mathbf{x}$ is the *instance* and $\mathbf{w}$ is the *witness*, and by $\mathscr{R}_n$ the restriction of $\mathscr{R}$ to instances of size n. We denote by $\mathscr{L}(\mathscr{R})$ the language corresponding to $\mathscr{R}$. For notational convenience, we define $\bar{\mathscr{L}}(\mathscr{R}_n) := \{\mathbf{x} \in \{0,1\}^n \mid \mathbf{x} \notin \mathscr{L}(\mathscr{R})\}$.

Random Oracles. We denote by $\mathcal{U}(\lambda)$ the uniform distribution over all functions $\rho : \{0,1\}^* \to \{0,1\}^\lambda$ (implicitly defined by the probabilistic algorithm that assigns, uniformly and independently at random, a λ-bit string to each new input). If ρ is sampled from $\mathcal{U}(\lambda)$, then we write $\rho \leftarrow \mathcal{U}(\lambda)$ and say that ρ is a *random oracle*. Given an oracle algorithm A, NumQueries(A, ρ) is the number of oracle queries that A^ρ makes. We say that A is *m-query* if NumQueries$(A, \rho) \le m$ for any $\rho \in \mathcal{U}(\lambda)$ (i.e., for any ρ in $\mathcal{U}(\lambda)$'s support).

Statistical Distance. The statistical distance between two discrete random variables X and Y with support V is $\Delta(X; Y) := \frac{1}{2} \sum_{v \in V} |\Pr[X = v] - \Pr[Y = v]|$. We say that X and Y are δ-*close* if $\Delta(X; Y) \le \delta$.

Remark 5. An oracle $\rho \in \mathcal{U}(\lambda)$ outputs λ bits. Occasionally we need ρ to output more than λ bits; in such cases (we point out where), we implicitly extend ρ's output via a simple strategy, e.g., we set $y := y_1 \| y_2 \| \cdots$ where $y_i := \rho(i\|x)$ and prefix 0 to all inputs that do not require an output extension.

2.2 Merkle Trees

We use Merkle trees [Mer89a] based on random oracles as succinct commitments to long lists of values for which one can cheaply decommit to particular values in the list. Concretely, a *Merkle-tree scheme* is a tuple MERKLE = (MERKLE.GetRoot, MERKLE.GetPath, MERKLE.CheckPath) that uses a random oracle ρ sampled from $\mathcal{U}(\lambda)$ and works as follows.

- MERKLE.GetRoot$^\rho(\mathbf{v}) \to$ rt. Given input list $\mathbf{v} = (v_i)_{i=1}^n$, the *root generator* MERKLE.GetRoot computes, in time $O_\lambda(n)$, a root rt of the Merkle tree over $\mathbf{v}$.

- MERKLE.GetPath$^\rho(\mathbf{v}, i) \to$ ap. Given input list $\mathbf{v}$ and index i, the *authentication path generator* MERKLE.GetPath computes the authentication path ap for the i-th value in $\mathbf{v}$.
- MERKLE.CheckPath$^\rho(\mathsf{rt}, i, v, \mathsf{ap}) \to b$. Given root rt, index i, input value v, and authentication path ap, the *path checker* MERKLE.CheckPath outputs $b = 1$ if ap is a valid path for v as the i-th value in a Merkle tree with root rt; the check can be carried out in time $O_\lambda(\log_2 n)$.

We assume that an authentication path ap contains the root rt, position i, and value v; accordingly, we define $\mathsf{Root}(\mathsf{ap}) := \mathsf{rt}$, $\mathsf{Position}(\mathsf{ap}) := i$, and $\mathsf{Value}(\mathsf{ap}) := v$.

Merkle trees are well known, so we do not review their construction. Less known, however, are the hiding and extractability properties of Merkle trees that we rely on in this work; we describe these in Appendix A.

2.3   Non-interactive Random-Oracle Arguments

A *non-interactive random-oracle argument system* for a relation $\mathscr{R}$ with soundness $s\colon \{0,1\}^* \to [0,1]$ is a tuple $(\mathbb{P}, \mathbb{V})$, where $\mathbb{P}, \mathbb{V}$ are (oracle) probabilistic algorithms, that satisfies the following properties.

1. COMPLETENESS. For every $(\mathbf{x}, \mathbf{w}) \in \mathscr{R}$ and $\lambda \in \mathbb{N}$,

$$\Pr\left[\mathbb{V}^\rho(\mathbf{x}, \pi) = 1 \ \middle| \ \begin{matrix} \rho \leftarrow \mathcal{U}(\lambda) \\ \pi \leftarrow \mathbb{P}^\rho(\mathbf{x}, \mathbf{w}) \end{matrix}\right] = 1 \ .$$

2. SOUNDNESS. For every $\mathbf{x} \notin \mathscr{L}(\mathscr{R})$, m-query $\tilde{\mathbb{P}}$, and $\lambda \in \mathbb{N}$,

$$\Pr\left[\mathbb{V}^\rho(\mathbf{x}, \pi) = 1 \ \middle| \ \begin{matrix} \rho \leftarrow \mathcal{U}(\lambda) \\ \pi \leftarrow \tilde{\mathbb{P}}^\rho \end{matrix}\right] \leq s(\mathbf{x}, m, \lambda) \ .$$

Complexity Measures. Beyond soundness, we consider other complexity measures. Given $p\colon \{0,1\}^* \to \mathbb{N}$, we say that $(\mathbb{P}, \mathbb{V})$ has proof length p if π has length $p(\mathbf{x}, \lambda)$. Given $t_{\mathrm{prv}}, t_{\mathrm{ver}}\colon \{0,1\}^* \to \mathbb{N}$, we say that $(\mathbb{P}, \mathbb{V})$ has prover time complexity t_{prv} and verifier time complexity t_{ver} if $\mathbb{P}^\rho(\mathbf{x}, \mathbf{w})$ runs in time $t_{\mathrm{prv}}(\mathbf{x}, \lambda)$ and $\mathbb{V}^\rho(\mathbf{x}, \pi)$ runs in time $t_{\mathrm{ver}}(\mathbf{x}, \lambda)$. In sum, we say that $(\mathbb{P}, \mathbb{V})$ has complexity $(s, p, t_{\mathrm{prv}}, t_{\mathrm{ver}})$ if $(\mathbb{P}, \mathbb{V})$ has soundness s, proof length p, prover time complexity t_{prv}, and verifier time complexity t_{ver}.

Proof of Knowledge. Given $e\colon \{0,1\}^* \to [0,1]$, we say that $(\mathbb{P}, \mathbb{V})$ has proof of knowledge e if there exists a probabilistic polynomial-time algorithm $\mathbb{E}$ (the *extractor*) such that, for every $\mathbf{x}$, m-query $\tilde{\mathbb{P}}$, and $\lambda \in \mathbb{N}$,

$$\Pr\left[(\mathbf{x}, \mathbf{w}) \in \mathscr{R} \ \middle| \ \mathbf{w} \leftarrow \mathbb{E}^{\tilde{\mathbb{P}}}(\mathbf{x}, 1^m, 1^\lambda)\right] \geq \Pr\left[\mathbb{V}^\rho(\mathbf{x}, \pi) = 1 \ \middle| \ \begin{matrix} \rho \leftarrow \mathcal{U}(\lambda) \\ \pi \leftarrow \tilde{\mathbb{P}}^\rho \end{matrix}\right] - e(\mathbf{x}, m, \lambda) \ .$$

The notation $\mathbb{E}^{\tilde{\mathbb{P}}}(\mathbf{x}, 1^m, 1^\lambda)$ means that $\mathbb{E}$ receives as input $(\mathbf{x}, 1^m, 1^\lambda)$ and may obtain an output of $\tilde{\mathbb{P}}^\rho$ for choices of oracles ρ, as we now describe. At any time, $\mathbb{E}$ may send a λ-bit string z to $\tilde{\mathbb{P}}$; then $\tilde{\mathbb{P}}$ interprets z as the answer to its last query to ρ (if any) and then continues computing until it reaches either its next query θ or its output π; then this query or output is sent to $\mathbb{E}$ (distinguishing the two cases in some way); in the latter case, $\tilde{\mathbb{P}}$ goes back to the start of its computation (with the same randomness and any auxiliary inputs). Throughout, the code, randomness, and any auxiliary inputs of $\tilde{\mathbb{P}}$ are not available to $\mathbb{E}$.

Zero Knowledge. Given $z \colon \{0,1\}^* \to [0,1]$, we say that $(\mathbb{P}, \mathbb{V})$ has z-statistical zero knowledge (in the explicitly-programmable random oracle model) if there exists a probabilistic polynomial-time algorithm $\mathbb{S}$ (the *simulator*) such that, for every $(\mathbf{x}, \mathbf{w}) \in \mathscr{R}$ and unbounded distinguisher D, the following two probabilities are $z(\mathbf{x}, \lambda)$-close:

$$\Pr\left[D^{\rho[\mu]}(\pi) = 1 \ \middle| \ \begin{matrix} \rho \leftarrow \mathcal{U}(\lambda) \\ (\pi, \mu) \leftarrow \mathbb{S}^\rho(\mathbf{x}) \end{matrix} \right] \text{ and } \Pr\left[D^\rho(\pi) = 1 \ \middle| \ \begin{matrix} \rho \leftarrow \mathcal{U}(\lambda) \\ \pi \leftarrow \mathbb{P}^\rho(\mathbf{x}, \mathbf{w}) \end{matrix} \right] .$$

Above, $\rho[\mu]$ is the function such that, given an input x, equals $\mu(x)$ if μ is defined on x, or $\rho(x)$ otherwise.

3   Interactive Oracle Proofs

We first define *interactive oracle protocols* and then *interactive oracle proof systems*.

3.1   Interactive Oracle Protocols

A k-round interactive oracle protocol between two parties, call them Alice and Bob, comprises k rounds of interaction. In the i-th round of interaction: Alice sends a message m_i to Bob, which he reads in full; then Bob replies with a message f_i to Alice, which she can query (via random access) in this and all later rounds. After the k rounds of interaction, Alice either accepts or rejects.

More precisely, let k be in $\mathbb{N}$ and A, B be two interactive probabilistic algorithms. A k-round interactive oracle protocol between A and B, denoted $\langle B, A \rangle$, works as follows. Let r_A, r_B denote the randomness for A, B and, for notational convenience, set $f_0 := \bot$ and $\mathsf{state}_0 := \bot$. For $i = 1, \ldots, k$, in the i-th round: (i) Alice sends a message $m_i \in \{0,1\}^{u_i}$, where $(m_i, \mathsf{state}_i) := A^{f_0, \ldots, f_{i-1}}(\mathsf{state}_{i-1}; r_A)$ and $u_i \in \mathbb{N}$; (ii) Bob sends a message $f_i \in \{0,1\}^{\ell_i}$, where $f_i := B(m_1, \ldots, m_i; r_B)$ and $\ell_i \in \mathbb{N}$. The output of the protocol is $m_{\mathsf{fin}} := A^{f_0, \ldots, f_k}(\mathsf{state}_k; r_A)$, and belongs to $\{0,1\}$.

The accepting probability of $\langle B, A \rangle$ is the probability that $m_{\mathsf{fin}} = 1$ for a random choice of r_A, r_B; this probability is denoted $\Pr[\langle B, A \rangle = 1]$ (leaving r_A, r_B implicit). The query complexity of $\langle B, A \rangle$ is the number of queries asked by A to any of the oracles during the k rounds. The proof complexity of $\langle B, A \rangle$ is the number of bits communicated by Bob to Alice (i.e., $\sum_{i=1}^k \ell_i$). The view of A

in $\langle B, A \rangle$, denoted $\text{View}_{\langle B,A \rangle}(A)$, is the random variable $(a_1, \ldots, a_q, r_A)$ where a_j denotes the answer to the j-th query.

Public Coins. An interactive oracle protocol is *public-coin* if Alice's messages are uniformly and independently random and Alice postpones any query to after the k-th round (i.e., all queries are asked when running $A^{f_0, \ldots, f_k}(\text{state}_k; r_A)$). We can thus take the randomness r_A to be of the form $(m_1, \ldots, m_k, r)$, where r is additional randomness that A may use of to compute m_{fin} after the last round.

3.2 Interactive Oracle Proof Systems

An *interactive oracle proof system* for a relation $\mathscr{R}$ with round complexity $k \colon \{0,1\}^* \to \mathbb{N}$ and soundness $s \colon \{0,1\}^* \to [0,1]$ is a tuple (P, V), where P, V are probabilistic algorithms, that satisfies the following properties.

1. COMPLETENESS. For every $(\mathbf{x}, \mathbf{w}) \in \mathscr{R}$, $\langle P(\mathbf{x}, \mathbf{w}), V(\mathbf{x}) \rangle$ is a $k(\mathbf{x})$-round interactive oracle protocol with accepting probability 1.
2. SOUNDNESS. For every $\mathbf{x} \notin \mathscr{L}(\mathscr{R})$ and $\tilde{P}$, $\langle \tilde{P}, V(\mathbf{x}) \rangle$ is a $k(\mathbf{x})$-round interactive oracle protocol with accepting probability at most $s(\mathbf{x})$.

Message Lengths. We assume the existence of polynomial-time functions that determine the message lengths. Namely, for any instance $\mathbf{x}$ and malicious prover $\tilde{P}$, when considering the interactive oracle protocol $\langle \tilde{P}, V(\mathbf{x}) \rangle$, the i-th messages m_i (from $V(\mathbf{x})$) and f_i (to $V(\mathbf{x})$) lie in $\{0,1\}^{u_i(\mathbf{x})}$ and $\{0,1\}^{\ell_i(\mathbf{x})}$ respectively.

Complexity Measures. Beyond round complexity and soundness, we consider other complexity measures. Given $p, q \colon \{0,1\}^* \to \mathbb{N}$, we say that (P, V) has proof length p and query complexity q if the proof length and query complexity of $\langle \tilde{P}, V(\mathbf{x}) \rangle$ are $p(\mathbf{x})$ and $q(\mathbf{x})$ respectively. (Note that $q(\mathbf{x}) \leq p(\mathbf{x})$ and $p(\mathbf{x}) = \sum_{i=1}^{k(\mathbf{x})} \ell_i(\mathbf{x})$.) Given $t_{\text{prv}}, t_{\text{ver}} \colon \{0,1\}^* \to \mathbb{N}$, we say that (P, V) has prover time complexity t_{prv} and verifier time complexity t_{ver} if $P(\mathbf{x}, \mathbf{w})$ runs in time $t_{\text{prv}}(\mathbf{x})$ and $V(\mathbf{x})$ runs in time $t_{\text{ver}}(\mathbf{x})$. In sum, we say that (P, V) has complexity $(k, s, p, q, t_{\text{prv}}, t_{\text{ver}})$ if (P, V) has round complexity k, soundness s, proof length p, query complexity q, prover time complexity t_{prv}, and verifier time complexity t_{ver}.

Proof of Knowledge. Given $e \colon \{0,1\}^* \to [0,1]$, we say that (P, V) has proof of knowledge e if there exists a probabilistic polynomial-time oracle algorithm E (the *extractor*) such that, for every $\mathbf{x}$ and $\tilde{P}$, $\Pr[(\mathbf{x}, E^{\tilde{P}}(\mathbf{x})) \in \mathscr{R}] \geq \Pr[\langle \tilde{P}, V(\mathbf{x}) \rangle = 1] - e(\mathbf{x})$. [5] The notation $E^{\tilde{P}}(\mathbf{x})$ means that E receives as input $\mathbf{x}$ and may interact with $\tilde{P}$ via rewinding, as we now describe. At any time, E may send a partial prover-verifier transcript to $\tilde{P}$ and then receive $\tilde{P}$'s next

---

[5] Proof of knowledge e implies soundness $s := e$. The definition that we use is equivalent to the one in [BG93, Section 6] except that: (a) we use extractors that run in strict, rather than expected, probabilistic polynomial time; and (b) we extend the condition to hold for all $\mathbf{x}$, rather than for only those in $\mathscr{L}(\mathscr{R})$, so that proof of knowledge implies soundness.

message (which is empty for invalid transcripts) in the subsequent computation step; the code, randomness, and any auxiliary inputs of $\tilde{P}$ are not available to E.

Honest-Verifier Zero Knowledge. Given $z\colon \{0,1\}^* \to [0,1]$, we say that (P, V) has z-statistical honest-verifier zero knowledge if there exists a probabilistic polynomial-time algorithm S (the *simulator*) such that, for every $(\mathbf{x}, \mathbf{w}) \in \mathscr{R}$, $S(\mathbf{x})$ is $z(\mathbf{x})$-close to $\mathrm{View}_{\langle P(\mathbf{x},\mathbf{w}), V(\mathbf{x}) \rangle}(V(\mathbf{x}))$.

Public Coins. We say that (P, V) is *public-coin* if the underlying interactive oracle protocol is public-coin.

4   State Restoration Attacks on Interactive Oracle Proofs

We introduce state restoration attacks on interactive oracle proofs.

In an interactive oracle proof, a malicious prover $\tilde{P}$ works as follows: for each round i, $\tilde{P}$ receives the i-th verifier message m_i and then sends to the verifier a message f_i computed as a function of his own randomness and all the verifier messages received so far, i.e., $m_1, \ldots, m_i$.

For the case of public-coin interactive oracle proof systems, we also consider a larger class of malicious provers, called *state-restoring provers*. Informally, a state-restoring prover receives in each round a verifier message as well as a *complete verifier state*, and then sends to the verifier a message and a previously-seen complete verifier state, which sets the verifier to that state; this forms a state restoration attack on the verifier.

More precisely, let (P, V) be a k-round public-coin interactive proof system (see Sect. 3.2) and $\mathbf{x}$ an instance. A complete verifier state cvs of $V(\mathbf{x})$ takes one of three forms: (1) the symbol null, which denotes the "empty" complete verifier state; (2) a tuple of the form $(m_1, f_1, \ldots, m_i)$, with $i \in \{1, \ldots, k(\mathbf{x})\}$, where each m_j is in $\{0,1\}^{u_j(\mathbf{x})}$ and each f_j is in $\{0,1\}^{\ell_j(\mathbf{x})}$; (3) a tuple of the form $(m_1, f_1, \ldots, m_{k(\mathbf{x})}, f_{k(\mathbf{x})}, r)$ where each m_j and f_j is as in the previous case and r is the additional randomness of the verifier $V(\mathbf{x})$.

The interaction between a state-restoring prover $\tilde{P}$ and the verifier $V(\mathbf{x})$ is mediated through a game:

---

1. The game initializes the list SeenStates to be (null).
2. Repeat the following until the game halts and outputs:
   (a) The prover chooses a complete verifier state cvs in the list SeenStates.
   (b) The game sets the verifier to cvs.
   (c) If cvs = null: the verifier samples a message m_1 in $\{0,1\}^{u_1(\mathbf{x})}$ and sends it to the prover; the game appends $\mathsf{cvs}' := (m_1)$ to the list SeenStates.
   (d) If cvs $= (m_1, f_1, \ldots, m_{i-1})$ with $i \in \{2, \ldots, k(\mathbf{x})\}$: the prover outputs a message f_{i-1} in $\{0,1\}^{\ell_{i-1}(\mathbf{x})}$; the verifier samples a message m_i in $\{0,1\}^{u_i(\mathbf{x})}$ and sends it to the prover; the game appends $\mathsf{cvs}' := \mathsf{cvs}\|f_{i-1}\|m_i$ to the list SeenStates.
   (e) If cvs $= (m_1, f_1, \ldots, m_{k(\mathbf{x})})$: the prover outputs a message $f_{k(\mathbf{x})}$ in $\{0,1\}^{\ell_{k(\mathbf{x})}(\mathbf{x})}$; the verifier samples additional randomness r; the game appends $\mathsf{cvs}' := \mathsf{cvs}\|f_{k(\mathbf{x})}\|r$ to the list SeenStates.

(f) If $\mathsf{cvs} = (m_1, f_1, \ldots, m_{k(\mathbf{x})}, f_{k(\mathbf{x})}, r)$: the verifier computes his decision $b := V^{f_0, \ldots, f_{k(\mathbf{x})}}(\mathbf{x}, \mathsf{state}_{k(\mathbf{x})}; r_V)$ where $\mathsf{state}_{k(\mathbf{x})} := \emptyset$ and $r_V := (m_1, \ldots, m_k, r)$; then the game halts and outputs b.

Note that there are two distinct notions of a round. *Verifier rounds* are the rounds played by the verifier within a single execution, as tracked by a complete verifier state cvs; the number of such rounds lies in the set $\{0, \ldots, k(\mathbf{x}) + 1\}$ (the extra $(k(\mathbf{x}) + 1)$-th round represents the verifier V sampling r after receiving the last prover message). *Prover rounds* are all verifier rounds played by the prover across different verifier executions; the number of such rounds is the number of states in SeenStates above. Accordingly, for $b \in \mathbb{N}$, we say a prover is b-*round* if it plays at most b prover rounds during any interaction with any verifier.

Also note that the prover is not able to set the verifier to arbitrary states but only to previously-seen ones (starting with the empty state null); naturally, setting the verifier multiple times to the same state may yield distinct new states, because the verifier samples his message afresh each time. After being set to a state cvs, the verifier does one of three things: (i) if the number of verifier rounds in cvs is less than $k(\mathbf{x})$ (see Step 2c and Step 2d), the verifier samples a fresh next message; (ii) if the number of verifier rounds in cvs is $k(\mathbf{x})$ (see Step 2e), the verifier samples his additional randomness r; (iii) if cvs contains a full protocol execution (see Step 2f), the verifier outputs the decision corresponding to this execution. The second case means that the prover can set the verifier even *after* the conclusion of the execution (after r is sampled and known to the prover). The game halts only in the third case.

The above game between a state-restoring prover and a verifier yields corresponding notions of soundness and proof of knowledge. Below, we denote by $\Pr[\langle \tilde{P}, V(\mathbf{x}) \rangle_{\mathrm{sr}} = 1]$ the probability that the state-restoring prover $\tilde{P}$ makes V accept $\mathbf{x}$ in this game.

Definition 1. *Given $s_{\mathrm{sr}}, e_{\mathrm{sr}} \colon \{0,1\}^* \to [0,1]$, a public-coin interactive oracle proof system (P, V) has*

- STATE RESTORATION SOUNDNESS s_{sr} *if, for every $\mathbf{x} \notin \mathscr{L}(\mathscr{R})$ and b-round state-restoring prover $\tilde{P}$, $\Pr[\langle \tilde{P}, V(\mathbf{x}) \rangle_{\mathrm{sr}} = 1] \leq s_{\mathrm{sr}}(\mathbf{x}, b)$.*
- STATE RESTORATION PROOF OF KNOWLEDGE e_{sr} *if there exists a probabilistic polynomial-time algorithm E_{sr} (the extractor) such that, for every $\mathbf{x}$ and b-round state-restoring prover $\tilde{P}$, $\Pr[(\mathbf{x}, E_{\mathrm{sr}}^{\tilde{P}}(\mathbf{x})) \in \mathscr{R}] \geq \Pr[\langle \tilde{P}, V(\mathbf{x}) \rangle_{\mathrm{sr}} = 1] - e_{\mathrm{sr}}(\mathbf{x}, b)$.*

Due to space limitations, our bounds on state restoration and our results on the corresponding tree exploration games are in the full version [BCS16].

5   From IOPs to Non-interactive Random-Oracle Arguments

We describe a transformation T such that if (P, V) is a public-coin interactive oracle proof system for a relation $\mathscr{R}$ then $(\mathbb{P}, \mathbb{V}) := T(P, V)$ is a non-interactive random-oracle argument system for $\mathscr{R}$. The transformation T runs in polynomial time: given as input code for P and V, it runs in time polynomial in the size of this code and then outputs code for $\mathbb{P}$ and $\mathbb{V}$.

Notation. For convenience, we split the random oracle ρ into two random oracles, denoted ρ_1 and ρ_2, as follows: $\rho_1(x) := \rho(1\|x)$ and $\rho_2(x) := \rho(2\|x)$. At a high level, we use ρ_1 for the verifier's randomness, and ρ_2 for Merkle trees and other hashing purposes. When counting queries, we count queries to both ρ_1 and ρ_2.

Construction of $\mathbb{P}$. The algorithm $\mathbb{P}$, given input $(\mathbf{x}, \mathbf{w})$ and oracle access to ρ:

1. Set $k := k(\mathbf{x})$, $q := q(\mathbf{x})$, $f_0 := \bot$, and $\sigma_0 := \rho_2(\mathbf{x})$.
2. Start running $P(\mathbf{x}, \mathbf{w})$ and, for $i = 1, \dots, k$:
   (a) Compute the verifier message $m_i := \rho_1(\mathbf{x}\|\sigma_{i-1})$.
   (b) Give m_i to $P(\mathbf{x}, \mathbf{w})$ to obtain f_i.
   (c) Compute the Merkle-tree root $\mathsf{rt}_i := \mathsf{MERKLE.GetRoot}^{\rho_2}(f_i)$.
   (d) Compute the "root hash" $\sigma_i := \rho_2(\mathsf{rt}_i\|\sigma_{i-1})$.
3. Set $\mathsf{state}_k := \emptyset$ and $r_V := (m_1, \dots, m_k, r)$, where $r := \rho_1(\mathbf{x}\|\sigma_k)$.
4. Run $V^{f_0, \dots, f_k}(\mathbf{x}, \mathsf{state}_k; r_V)$ and compute an authentication path for each query. Namely, for $j = 1, \dots, q$: if the j-th query is to the x_j-th bit of the y_j-th oracle, then compute $\mathsf{ap}_j := \mathsf{MERKLE.GetPath}^{\rho_2}(f_{y_j}, x_j)$. (If $\mathsf{MERKLE.GetRoot}$ is probabilistic, then give the same randomness to $\mathsf{MERKLE.GetPath}$ as well.)
5. Set $\pi := ((\mathsf{rt}_1, \dots, \mathsf{rt}_k), (\mathsf{ap}_1, \dots, \mathsf{ap}_q), \sigma_k)$. That is, π comprises the Merkle-tree roots, an authentication path for each query, and the final root hash.
6. Output π.

Construction of $\mathbb{V}$. The algorithm $\mathbb{V}$, given input $(\mathbf{x}, \tilde{\pi})$ and oracle access to ρ:

1. Set $k := k(\mathbf{x})$, $q := q(\mathbf{x})$, $f_0 := \bot$, and $\sigma_0 := \rho_2(\mathbf{x})$.
2. Parse $\tilde{\pi}$ as a tuple $((\tilde{\mathsf{rt}}_1, \dots, \tilde{\mathsf{rt}}_k), (\tilde{\mathsf{ap}}_1, \dots, \tilde{\mathsf{ap}}_q), \tilde{\sigma}_k)$.
3. For $i = 1, \dots, k$:
   (a) Compute $m_i := \rho_1(\mathbf{x}\|\sigma_{i-1})$.
   (b) Compute $\sigma_i := \rho_2(\tilde{\mathsf{rt}}_i\|\sigma_{i-1})$.
4. Set $\mathsf{state}_k := \emptyset$ and $r_V := (m_1, \dots, m_k, r)$, where $r := \rho_1(\mathbf{x}\|\sigma_k)$.
5. Compute $m_{\mathsf{fin}} := V^{f_0, \dots, f_k}(\mathbf{x}, \mathsf{state}_k; r_V)$, answering the j-th query with the answer a_j in the path $\tilde{\mathsf{ap}}_j$.
6. If $\sigma_k \neq \tilde{\sigma}_k$, halt and output 0.
7. For $j = 1, \dots, q$: if the j-th query is to the x_j-th bit of the y_j-th oracle and $\mathsf{MERKLE.CheckPath}^{\rho_2}(\mathsf{rt}_{y_j}, x_j, a_j, \tilde{\mathsf{ap}}_j) \neq 1$, halt and output 0.
8. Output m_{fin}.

---

6 Analysis of the Transformation T

The theorem below specifies guarantees of the transformation T, described in Sect. 5.

Theorem 3. (IOP → NIROA). *For every relation $\mathscr{R}$, if (P, V) is a public-coin interactive oracle proof system for $\mathscr{R}$ with*

$$
\begin{aligned}
\text{round complexity} \quad & k(\mathsf{x}) \\
\text{restricted state restoration soundness} \quad & \bar{s}_{\mathrm{sr}}(\mathsf{x}, b) \\
\text{proof length} \quad & p(\mathsf{x}) \\
\text{prover time} \quad & t_{\mathrm{ver}}(\mathsf{x}) \\
\text{verifier time} \quad & t_{\mathrm{prv}}(\mathsf{x})
\end{aligned}
$$

then $(\mathbb{P}, \mathbb{V}) := T(P, V)$ *is a non-interactive random-oracle argument system for* $\mathscr{R}$ *with*

$$
\begin{aligned}
\text{soundness} \quad & s'(\mathsf{x}, m, \lambda) := \bar{s}_{\mathrm{sr}}(\mathsf{x}, m) + 3(m^2 + 1)2^{-\lambda} \\
\text{proof length} \quad & p'(\mathsf{x}, \lambda) := \big(k(\mathsf{x}) + q(\mathsf{x}) \cdot (\lceil \log_2 p(\mathsf{x}) \rceil + 2) + 1\big) \cdot \lambda \quad {}^9 \\
\text{prover time} \quad & t'_{\mathrm{prv}}(\mathsf{x}, \lambda) := O_\lambda(k(\mathsf{x}) + p(\mathsf{x})) + t_{\mathrm{prv}}(\mathsf{x}) + t_{\mathrm{ver}}(\mathsf{x}) \\
\text{verifier time} \quad & t'_{\mathrm{ver}}(\mathsf{x}, \lambda) := O_\lambda(k(\mathsf{x}) + q(\mathsf{x})) + t_{\mathrm{ver}}(\mathsf{x})
\end{aligned}
$$

By construction, if $\langle P(\mathsf{x}, \mathsf{w}), V(\mathsf{x}) \rangle$ has accepting probability δ, then the probability that $\mathbb{V}^\rho(\mathsf{x}, \mathbb{P}^\rho(\mathsf{x}, \mathsf{w}))$ accepts is δ. The complexities $p', t'_{\mathrm{prv}}, t'_{\mathrm{ver}}$ above also directly follow from the construction. Therefore, we are left to discuss soundness. Due to space limitations, the discussion of the soundness lower bound, as well as proof of knowledge and zero knowledge, are left to the full version [BCS16].

Let $\mathsf{x} \notin \mathscr{L}(\mathscr{R})$ and let $\tilde{\mathbb{P}}$ be an m-query prover for the non-interactive random-oracle argument system $(\mathbb{P}, \mathbb{V})$. We construct a prover $\tilde{P}$ (depending on x and $\tilde{\mathbb{P}}$) for the interactive oracle proof system (P, V), and show that $\tilde{P}$'s ability to cheat in a (restricted) state restoration attack is closely related to $\tilde{\mathbb{P}}$'s ability to cheat.

Construction of $\tilde{P}$. Given no inputs or oracles, the prover $\tilde{P}$ works as follows.

1. Let ρ_1, ρ_2 be tables mapping $\{0,1\}^*$ to $\{0,1\}^\lambda$, and let α be a table mapping λ-bit strings to verifier states. The tables are initially empty and are later populated with suitable values, during the simulation of $\tilde{\mathbb{P}}$. Intuitively, ρ_1, ρ_2 are used to simulate $\tilde{\mathbb{P}}$'s access to a random oracle, while α is used to keep track of which verifier states $\tilde{\mathbb{P}}$ has "seen in his mind".
2. Draw $\sigma_0 \in \{0,1\}^\lambda$ at random, and define $\rho_2(\mathsf{x}) := \sigma_0$ (i.e., the oracle ρ_2 replies the query x with the answer σ_0). After receiving V's first message m_1, also define $\rho_1(\mathsf{x} \| \sigma_0) := m_1$ and $\alpha(\sigma_0) := (m_1)$.
3. Begin simulating $\tilde{\mathbb{P}}^\rho$ and, for $i = 1, \ldots, m$:
   (a) Let θ_i be the i-th query made by $\tilde{\mathbb{P}}^\rho$.
   (b) If θ_i is a query to a location of ρ_1 that is defined, respond with $\rho_1(\theta_i)$. Otherwise (if θ_i to an undefined location of ρ_1), draw a string in $\{0,1\}^\lambda$ at random and respond with it. Then go to the next iteration of Step 3.
   (c) If θ_i is a query to a location of ρ_2 that is defined, respond with $\rho_2(\theta_i)$; then go to the next iteration of Step 3. Otherwise (if θ_i is to an undefined location of ρ_2), draw a string $\sigma' \in \{0,1\}^\lambda$ at random and respond with it; then continue as follows.

(d) Let rt be the first λ bits of θ_i, and σ be the second λ bits. (If the length of θ_i is not 2λ bits, go to the next iteration of Step 3.) If $\alpha(\sigma)$ is defined, let cvs := $\alpha(\sigma)$ and let j be the number of verifier rounds in the state cvs. If $\alpha(\sigma)$ is not defined, go to the next iteration of Step 3 .
(e) Find the query θ_{i*} whose result is rt. If this query is not unique, or there is no such query, then answer the verifier V with some dummy message (e.g., an all zero message of the correct length) and skip to Step 3g. Otherwise, note the index i^* and continue.
(f) Compute $f := \mathsf{VE}^{\rho_2}(\tilde{\mathbb{P}}, \ell_j(\mathbf{x}), i^*, i)$; if VE aborts, set $f := 0^{\ell_j(\mathbf{x})}$. Recall that $\ell_j(\mathbf{x})$ is the length of the prover message in the j-th verifier round, and VE is Valiant's extractor (see Sect. A.1). Also note that VE does not query ρ_2 on any value outside the table, because we have already simulated the first i queries of $\tilde{\mathbb{P}}$ (see Remark 6).
(g) Send the message f to the verifier and tell the game to set the verifier to the state cvs. (Whether cvs lies in the set SeenStates is a matter of analysis further below.) If the game is not over, the verifier replies with a new message m'. (If $j = k(\mathbf{x}) + 1$, for the purposes of the proof, we interpret m' as the additional randomness r.) The game adds cvs' := cvs$\|f\|m'$ to SeenStates. The prover defines $\rho_1(\mathbf{x}\|\sigma') := m'$ and $\alpha(\sigma') := $ cvs'.

Analysis of $\tilde{P}$. We now analyze $\tilde{P}$. We first prove a simple lemma, and then discuss $\tilde{P}$'s ability to cheat.

Lemma 1. *Let A be an m-query algorithm. Define:*

1. *E_1 to be the event that A^{ρ_2} outputs $\mathbf{x} \in \{0,1\}^n$, $\mathsf{rt}_1, \ldots, \mathsf{rt}_{k(\mathbf{x})} \in \{0,1\}^\lambda$, and $\sigma_{k(\mathbf{x})} \in \{0,1\}^\lambda$ that satisfy the recurrence $\sigma_0 = \rho_2(\mathbf{x})$ and $\sigma_i = \rho_2(\mathsf{rt}_i\|\sigma_{i-1})$ for all $i \in \{1, \ldots, k(\mathbf{x})\}$;*
2. *E_2 to be the event that A^{ρ_2} queries ρ_2 at $\mathbf{x}, \mathsf{rt}_1\|\sigma_0, \ldots, \mathsf{rt}_{k(\mathbf{x})}\|\sigma_{k(\mathbf{x})-1}$ (in order) and, if any rt_i is the result of a query, this query first occurs before $\mathsf{rt}_i\|\sigma_{i-1}$.*

Then
$$\Pr\left[(\neg E_1) \vee E_2 \mid \rho_2 \leftarrow \mathcal{U}(\lambda)\right] \geq 1 - (m^2 + 1)2^{-\lambda} .$$

Proof. Let rt_0 be $\mathbf{x}$ and σ_{-1} be the empty string. Suppose, by contradiction, that E_1 occurs and E_2 does not. Then there exists $i \in \{0, \ldots, k(\mathbf{x})\}$ for which at least one of the following holds: (i) A^{ρ_2} does not query $\mathsf{rt}_i\|\sigma_{i-1}$; (ii) A^{ρ_2} queries $\mathsf{rt}_{i+1}\|\sigma_i$ before it queries $\mathsf{rt}_i\|\sigma_{i-1}$; (iii) rt_i is the result of a query but this query first occurs after $\mathsf{rt}_i\|\sigma_{i-1}$. Consider the largest index i for which one of the above holds.

In case (i), the behavior of A^{ρ_2} is independent of $\rho_2(\mathsf{rt}_i\|\tilde{\sigma}_{i-1})$. If $i = k(\mathbf{x})$, then the output $\sigma_{k(\mathbf{x})}$ of A^{ρ_2} equals $\rho_2(\mathsf{rt}_{k(\mathbf{x})}\|\sigma_{k(\mathbf{x})-1})$ with probability $2^{-\lambda}$. If $i < k(\mathbf{x})$, then there is a sequence of queries $\mathsf{rt}_{i+1}\|\tilde{\sigma}_i, \ldots, \mathsf{rt}_{k(\mathbf{x})}\|\tilde{\sigma}_{k(\mathbf{x})-1}$ for which $\tilde{\sigma}_i = \rho_2(\mathsf{rt}_i\|\tilde{\sigma}_{i-1})$ for $i = 1, \ldots, k(\mathbf{x}) - 1$ and $\rho_2(\mathsf{rt}_{k(\mathbf{x})}\|\tilde{\sigma}_{k(\mathbf{x})-1}) = \sigma_{k(\mathbf{x})}$. If this sequence is not unique, then A^{ρ_2} has found a collision. Otherwise, the unique sequence has $\tilde{\sigma}_i = \sigma_i$ for each i, which occurs with probability at most $2^{-\lambda}$.

In cases (ii) and (iii), A^{ρ_2} has found a collision, since $\sigma_i = \rho_2(\mathsf{rt}_i\|\sigma_{i-1})$. The fraction of oracles ρ_2 for which A^{ρ_2} finds a collision is at most $m^2 2^{-\lambda}$.

Overall, the probability that E_2 does not occur and E_1 does is, by the union bound, at most $(m^2 + 1)2^{-\lambda}$.

We now state and prove the lemma about the soundness s' as stated in Theorem 3.

Lemma 2. *Define* $\epsilon := \Pr\left[\mathbb{V}^\rho(\mathbf{x}, \pi) = 1 \,\middle|\, \begin{array}{c} \rho \leftarrow \mathcal{U}(\lambda) \\ \pi \leftarrow \tilde{\mathbb{P}}^\rho \end{array}\right]$. *Then there exists* $b \in \mathbb{N}$ *with* $b \leq m$ *such that* $\tilde{P}$ *is a* b-*round state-restoring prover that makes* V *accept with probability at least* $\epsilon - 3(m^2 + 1)2^{-\lambda}$.

Proof. We first note that $\tilde{P}$ described plays no more than m rounds, because $\tilde{P}$ sends a message to the verifier V only in response to $\tilde{\mathbb{P}}$ making a query. Next, we define some useful notions, and use them to prove three claims which together imply the lemma.

Definition 4. We say $\rho \in \mathcal{U}(\lambda)$ is *good* if

1. The verifier accepts relative to ρ, i.e., $\mathbb{V}^\rho(\mathbf{x}, \pi) = 1$ where $\pi \leftarrow \tilde{\mathbb{P}}^\rho$.
2. Parsing π as $\left((\tilde{\mathsf{rt}}_1, \ldots, \tilde{\mathsf{rt}}_{k(\mathbf{x})}), (\tilde{\mathsf{ap}}_1, \ldots, \tilde{\mathsf{ap}}_q), \tilde{\sigma}_{k(\mathbf{x})}\right)$ and setting $\sigma_0 := \mathbf{x}$, for each $i \in \{1, \ldots, k(\mathbf{x})\}$, where $\sigma_i := \rho_2(\tilde{\mathsf{rt}}_i \| \sigma_{i-1})$, there exist indices $1 \leq j_1 < \cdots < j_k \leq m$ such that:
   (a) $\tilde{\mathbb{P}}^\rho$'s j_i-th query is to ρ_2 at $\tilde{\mathsf{rt}}_i \| \sigma_{i-1}$;
   (b) if rt_i is the result of a query, this query first occurs before j_i;
   (c) if $\tilde{\mathbb{P}}^\rho$ queries ρ_1 at $\mathbf{x} \| \sigma_i$, then this query occurs *after* query j_i;
   (d) if there exists l such that $\mathsf{Root}(\tilde{\mathsf{ap}}_l) = \tilde{\mathsf{rt}}_i$, there is a unique (up to duplicate queries) $a_i \in \{0, \ldots, j_i\}$ such that $\rho_2(\theta_{a_i}) = \tilde{\mathsf{rt}}_i$ and, for every $i_{\max} \in \{a_i, \ldots, j_i\}$, $\mathbf{v} := \mathsf{VE}^{\rho_2}(A, \ell_i, a_i, i_{\max})$ is such that, for all l with $\mathsf{Root}(\tilde{\mathsf{ap}}_l) = \tilde{\mathsf{rt}}_i$, $\mathsf{Value}(\tilde{\mathsf{ap}}_l)$ equals the $\mathsf{Position}(\tilde{\mathsf{ap}}_l)$-th value in $\mathbf{v}$; we say $\mathbf{v}$ is *extracted at* i if this holds.
3. $\tilde{\sigma}_{k(\mathbf{x})} = \sigma_{k(\mathbf{x})}$.

Definition 5. We say that $\tilde{\mathbb{P}}$ *chooses* $\rho \in \mathcal{U}(\lambda)$ if for every query θ made by $\tilde{\mathbb{P}}^\rho$ to its oracle, $\tilde{P}$ supplies it with $\rho(\theta)$ (ignoring whether this response comes from $\tilde{P}$ itself or the messages sent by V; this choice is fixed for a given ρ).

Claim 6. $(\tilde{P}, V)$ *chooses* $\rho \in \mathcal{U}(\lambda)$ *uniformly at random.*

Whenever the simulation of $\tilde{\mathbb{P}}$ makes a query, $\tilde{P}$ responds consistently, either with a uniformly randomly drawn string of its own, or the uniform randomness provided by V. This is equivalent in distribution to drawing ρ uniformly at random at the beginning of the protocol. ∎

Claim 7. *For any choice of randomness such that* $\tilde{P}$ *chooses a good* ρ, $\tilde{P}$ *makes* $V(\mathbf{x})$ *accept with a state restoration attack.*

We begin by defining a property of the map α.

Definition 8. For $i = 0, \ldots, k$, we say that α is *correct at i* if, immediately before $\tilde{\mathbb{P}}$'s j_{i+1}-th query is simulated (for $i = k$, at the end of the simulation), it holds that $\alpha(\sigma_i) = (\rho_1(\mathbf{x}\|\sigma_0), f_1, \ldots, \rho_1(\mathbf{x}\|\sigma_i))$, where for each $l \in \{1, \ldots, i\}$, f_l is extracted at l (see Condition 2d above), and $\alpha(\sigma_i) \in$ SeenStates.

We show by induction that α is correct at i for every $i \in \{0, \ldots, k\}$. First, α is correct at 0 since $\alpha(\sigma_0) = (\rho_1(\mathbf{x}\|\sigma_0))$ by construction. Suppose that α is correct at $i - 1$. When $\tilde{\mathbb{P}}^\rho$ queries $\tilde{\mathsf{rt}}_i\|\sigma_{i-1}$ (i.e., query θ_{j_i}), $\tilde{P}$ restores $\alpha(\sigma_{i-1}) \in$ SeenStates. By Condition 2d, f_i is extracted at i. In Step 3g, $\rho_1(\mathbf{x}\|\sigma_i)$ is set to the message (or, similarly, internal randomness) sent by V in this round, which is possible by Condition 2c. The newly stored state is then $\alpha(\sigma_i) = (\rho_1(\mathbf{x}\|\sigma_0), f_1, \ldots, \rho_1(\mathbf{x}\|\sigma_{i-1}), f_i, \rho_1(\mathbf{x}\|\sigma_i)) \in$ SeenStates. This state is stored before query j_{i+1} by Condition 2a, and so α is correct at i.

Hence $\tilde{\mathbb{P}}$ sends a state $\alpha(\sigma_k) = (\rho_1(\mathbf{x}\|\sigma_1), f_1, \ldots, \rho_1(\mathbf{x}\|\sigma_k)) \in$ SeenStates. Since $\mathbb{V}$'s simulation of V accepts with this state, so does the real V when interacting with $\tilde{\mathbb{P}}$. ∎

Claim 9. The probability that $\rho \in \mathcal{U}(\lambda)$ is good is at least $\epsilon - 3(m^2 + 1)2^{-\lambda}$.

By assumption, the density of oracles satisfying Condition 1 is ϵ. Lemma 1 implies that the density of oracles satisfying Condition 1 but not satisfying Condition 2a, Condition 2b, and Condition 3 is at most $(m^2 + 1)2^{-\lambda}$.[6] The density of oracles failing to satisfy Condition 2c is at most $m^2 2^{-\lambda}$, since this implies a 'collision' (in the sense of Lemma 3) between ρ_1 and ρ_2. Finally, the density of oracles satisfying Condition 1, Condition 2a, and Condition 2b, but not Condition 2d is at most $(m^2 + 1)2^{-\lambda}$, by Lemma 3 and Condition 2b (where Condition 2b allows us to restrict the possible values for a_i to $0 \leq a_i < j_i$).

By the union bound, the density of good oracles ρ is at least $\epsilon - 3(m^2 + 1)2^{-\lambda}$. ∎ Combining the claims, we deduce that $\tilde{P}$ makes V accept with probability at least $\epsilon - 3(m^2 + 1)2^{-\lambda}$ with a state restoration attack. Finally, note that this state restoration attack is restricted because $\tilde{P}$ never requests to set V to the empty verifier state null.

A   Extractability and Privacy of Merkle Trees

We describe the specific extractability and privacy properties of Merkle trees that we rely on in this work.

A.1   Extractability

We rely on a certain extractability property of Merkle trees: there is an efficient procedure for extracting the committed list in a Merkle-tree scheme. We call the

---

[6] More precisely, we apply Lemma 1 to an algorithm $\tilde{\mathbb{P}}$ that does not itself output $\mathbf{x}$ but this does not affect the lemma's validity because we can substitute into the definition of the event E_1 the fixed instance $\mathbf{x}$.

procedure *Valiant's extractor*, and denote it by VE, because it is described in [Val08]. Our presentation of the extractor and its guarantee differs from [Val08] because our use of it in this work requires "distilling" a more general property; see Lemma 3 below.

The Extractor. For any oracle algorithm A, integers $\ell, i^\star, i_{max} > 0$ with $i^\star \in \{1, \ldots, i_{max}\}$, and ρ sampled from $\mathcal{U}(\lambda)$, the procedure VE, given input $(A, \ell, i^\star, i_{max})$ and with oracle access to ρ, works as follows.

1. Run A^ρ until it has asked i_{max} unique queries to ρ (and abort if A^ρ asks fewer than i_{max}). Along the way, record the queries $\theta_1, \ldots, \theta_{i_{max}}$ and answers $\rho(\theta_1), \ldots, \rho(\theta_{i_{max}})$, in order and omitting duplicates.
2. Parse each query θ_i as $\theta_i^0 \| \theta_i^1$ where θ_i^0 are the first λ bits of θ_i and θ_i^1 the second λ bits. For brevity, we write $z \in \theta_i$ if $z = \theta_i^0$ or $z = \theta_i^1$. (If a query has length not equal to 2λ, then $z \notin \theta_i$ for all z.)
3. If there exist indices i, j such that $i \neq j$ and $\rho(\theta_i) = \rho(\theta_j)$, abort.
4. If there exist indices i, j such that $i \leq j$ and $\rho(\theta_j) \in \theta_i$, abort.
5. Construct a directed graph G with nodes $V = \{\theta_1, \ldots, \theta_{i_{max}}\}$ and edges $E = \{(\theta_i, \theta_j) : \rho(\theta_j) \in \theta_i\}$. Note that G is acyclic, every node has out-degree ≤ 2, and $\theta_1, \ldots, \theta_{i_{max}}$ is a (reverse) topological ordering.
6. Output $\mathbf{v}$, the string obtained by traversing in order the first ℓ leaf nodes of the depth-$\lceil \log_2 \ell \rceil$ binary tree rooted at $\theta_{i^\star}$ and recording the first bit of each node. If any such node does not exist, set this entry to 0.

A sample execution of the extractor is depicted in Fig. 1.

Remark 6. The queries to ρ asked by $VE^\rho(A, \ell, i^\star, i_{max})$ equals the first i_{max} queries to ρ asked by A^ρ (provided that A does not ask fewer than i_{max} queries). Later on we use this fact.

The Extractor's Guarantee. We interpret A's output as containing a (possibly empty) list of tuples of the form $(\mathsf{rt}, i, v, \mathsf{ap})$, where rt is a root, i an index, v a value, and ap an authentication path.[7] We define the following events:

(i) E_1 is the event that, for each tuple $(\mathsf{rt}, i, v, \mathsf{ap})$ output by A^ρ, MERKLE.CheckPath$(\mathsf{rt}, i, v, \mathsf{ap}) = 1$;
(ii) E_2 is the event that, for each $\mathsf{rt} \in \{0,1\}^\lambda$, there exists $\ell_{\mathsf{rt}} \in \mathbb{N}$ such that if A^ρ outputs a tuple of the form $(\mathsf{rt}, \cdot, \cdot, \mathsf{ap})$ then ap is an authentication path having the correct length for a ℓ_{rt}-leaf Merkle tree;
(iii) E_3 is the event that, for every $\mathsf{rt} \in \{0,1\}^\lambda$ such that A^ρ outputs some tuple of the form $(\mathsf{rt}, \cdot, \cdot, \cdot)$, there is a unique $j_{\mathsf{rt}} \in \{0, \ldots, \mathsf{NumQueries}(A, \rho)\}$ such that $\rho(\theta_{j_{\mathsf{rt}}}) = \mathsf{rt}$ and, for every $i_{max} \in \{j_{\mathsf{rt}}, \ldots, \mathsf{NumQueries}(A, \rho)\}$, $\mathbf{v} := VE^\rho(A, \ell_{\mathsf{rt}}, j_{\mathsf{rt}}, i_{max})$ is such that $\mathbf{v}$'s i-th entry equals v_i for any tuple of the form $(\mathsf{rt}, i, v, \mathsf{ap})$ output by A^ρ.

The extractability property that we rely on is the following.

---

[7] Note that A's output may contain additional information not of the above form; if so, we simply ignore it for now.

Lemma 3. *Let A^ρ be a m-query algorithm. Then*

$$\Pr\left[(\neg(E_1 \wedge E_2)) \vee E_3 \mid \rho \leftarrow \mathcal{U}(\lambda)\right] \geq 1 - (m^2 + 1)2^{-\lambda} \ .$$

Proof. Observe the following.

- By the union bound, the probability that there exist indices i, j such that $(i \neq j) \wedge (\rho(\theta_i) = \rho(\theta_j))$ or $(i \leq j) \wedge (\rho(\theta_j) \in \theta_i)$ is at most $m^2 2^{-\lambda}$. If this occurs, we say that A^ρ has found a *collision*.
- The probability that, for a tuple $(\mathsf{rt}, i, v, \mathsf{ap})$ output by A^ρ such that $\mathsf{MERKLE.CheckPath}(\mathsf{rt}, i, v, \mathsf{ap}) = 1$, the authentication path ap contains a node with no corresponding query is at most $2^{-\lambda}$, since this would mean that A^ρ has 'guessed' the answer to the query. In other words, no matter what strategy A uses to generate the result, if it does not query the oracle on this input then it can perform no better than chance.

Now suppose that $E_1 \wedge E_2$ occurs with probability δ. Then, with probability at least $\delta - (m^2 + 1)2^{-\lambda}$: (a) for each root rt output by A^ρ there is a unique query θ_{i^*} such that $\rho(\theta_{i^*}) = \mathsf{rt}$; (b) for each root rt output by A^ρ, if an authentication path ap claims to have root rt then ap appears in the tree rooted at θ_{i^*} in G; and (c) the condition in the VE's Step 3 or Step 4 does not hold. In such a case we may take $j_{\mathsf{rt}} := i^*$, and then $\mathsf{VE}^\rho(A, \ell_{\mathsf{rt}}, j_{\mathsf{rt}}, i_{\mathsf{max}})$ outputs a list v with the desired property. Hence, $\Pr[E_1 \wedge E_2 \wedge E_3] \geq \delta - (m^2 + 1)2^{-\lambda}$. The predicate is also satisfied if $\neg(E_1 \wedge E_2)$ occurs, which is the case with probability $1 - \delta$ and is disjoint from $E_1 \wedge E_2 \wedge E_3$. The lemma follows.

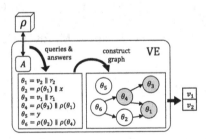

Fig. 1. A diagram of an execution of Valiant's extractor VE, with input parameters $\ell = 2$, $i^* = 4$, and $i_{\mathsf{max}} = 6$.

A.2 Privacy

We rely not only on the fact that the root rt of a Merkle tree is hiding, but also on the fact that an authentication path ap reveals no information about values other than the decommitted one. The latter property can be ensured via a slight tweak of the standard construction of Merkle trees: when committing to a list $v = (v_i)_{i=1}^n$, the i-th leaf is not v_i but, instead, is a hiding commitment

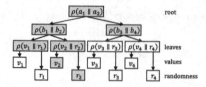

Fig. 2. A diagram of the data structure of a Merkle tree with privacy. An authentication path for v_2 is shaded; the corresponding truncated authentication path is the same minus r_2 and v_2.

to v_i. In our case, we will store the value $\rho(v_i\|r_i)$ in the i-th leaf, where $r_i \in \{0,1\}^{2\lambda}$ is drawn uniformly at random; see Fig. 2. (An authentication path for v_i then additionally includes r_i, and path verification is modified accordingly.) In what follows, we regard $\rho(v_i\|r_i)$ as a *leaf*, rather than v_i; moreover, a *truncated authentication path* ap'_i is identical to ap_i except that it does not contain r_i or v_i, and the *truncated Merkle tree* for $\mathbf{v}$ is $T'_{\mathbf{v}} := (\mathsf{ap}'_i)_{1\leq i\leq n}$. Note that the *same* randomness $\mathbf{r} \in \{0,1\}^{2\lambda n}$ is used by MERKLE.GetRoot and MERKLE.GetPath (to be "in sync").

We summarize the privacy property of Merkle trees as above via the following definition and lemma.

Definition 2. *A Merkle-tree scheme has $z(n,\lambda)$-statistical privacy if there exists a probabilistic polynomial-time simulator S such that, for every list $\mathbf{v} = (v_i)_{i=1}^{n}$ and unbounded distinguisher D, the following two probabilities are $z(n,\lambda)$-close:*

$$\Pr_{\mathbf{r}}\left[\begin{array}{c} I \subseteq \{1,\ldots,n\} \\ D^\rho(\mathsf{rt},(\mathsf{ap}_i)_{i\in I}) = 1 \end{array} \middle| \begin{array}{r} \rho \leftarrow \mathcal{U}(\lambda) \\ \mathsf{rt} \leftarrow \text{MERKLE.GetRoot}^\rho(\mathbf{v};\mathbf{r}) \\ I \leftarrow D^\rho \\ \forall i \in I,\ \mathsf{ap}_i \leftarrow \text{MERKLE.GetPath}^\rho(\mathbf{v},i;\mathbf{r}) \end{array}\right]$$

and

$$\Pr\left[\begin{array}{c} I \subseteq \{1,\ldots,n\} \\ D^\rho(\mathsf{rt},(\mathsf{ap}_i)_{i\in I}) = 1 \end{array} \middle| \begin{array}{r} \rho \leftarrow \mathcal{U}(\lambda) \\ I \leftarrow D^\rho \\ (\mathsf{rt},(\mathsf{ap}_i)_{i\in I}) \leftarrow S^\rho(n,(i,v_i)_{i\in I}) \end{array}\right] \,.$$

We make no assumption on the power of the distinguisher D in the definition above. In particular, D may query the random oracle ρ at every input, and use the information to attempt to learn v_i for some $i \notin I$. For example, for some ρ, it is the case that $\Pr_r[v = 1 \mid \rho(v\|r) = x] \gg \Pr_r[v = 0 \mid \rho(v\|r) = x]$ for $x = \rho(v_2\|r_2)$, in which case D can determine v_2 from ap_1 with good accuracy. The next (easy to prove) lemma shows that the probability that D gains a significant statistical advantage in this way (or otherwise) is negligible in λ.

Lemma 4. *There exists a Merkle-tree scheme having $z(n,\lambda)$-statistical privacy with $z(n,\lambda) := n2^{-\lambda/4+2}$.*

References

[ALM+92] Arora, S., Lund, C., Motwani, R., Sudan, M., Szegedy, M.: Proof verification and hardness of approximation problems (1992)

[ALM+98] Arora, S., Lund, C., Motwani, R., Sudan, M., Szegedy, M.: Proof verification and the hardness of approximation problems. JACM **45**(3), 501–555 (1998)

[AS98] Arora, S., Safra, S.: Probabilistic checking of proofs: a new characterization of NP. JACM **45**(1), 70–122 (1998)

[Bab85] Babai, L.: Trading group theory for randomness. In: STOC 1985 (1985)

[Bab90] Babai, L.: E-mail and the unexpected power of interaction. Technical report, University of Chicago, Chicago, IL, USA (1990)

[BBP04] Bellare, M., Boldyreva, A., Palacio, A.: An uninstantiable random-oracle-model scheme for a hybrid-encryption problem. In: Cachin, C., Camenisch, J.L. (eds.) EUROCRYPT 2004. LNCS, vol. 3027, pp. 171–188. Springer, Heidelberg (2004). doi:10.1007/978-3-540-24676-3_11

[BC86] Brassard, G., Crépeau, C.: Non-transitive transfer of confidence: a perfect zero-knowledge interactive protocol for SAT and beyond. In: FOCS 1986 (1986)

[BCC88] Brassard, G., Chaum, D., Crépeau, C.: Minimum disclosure proofs of knowledge. J. Comput. Syst. Sci. **37**(2), 156–189 (1988)

[BCG+16] Ben-Sasson, E., Chiesa, A., Gabizon, A., Riabzev, M., Spooner, N.: Short interactive oracle proofs with constant query complexity, via composition and sumcheck (2016). Crypto ePrint 2016/324

[BCGV16] Ben-Sasson, E., Chiesa, A., Gabizon, A., Virza, M.: Quasilinear-size zero knowledge from linear-algebraic PCPs. In: TCC 2016-A (2016)

[BCI+13] Bitansky, N., Chiesa, A., Ishai, Y., Ostrovsky, R., Paneth, O.: Succinct non-interactive arguments via linear interactive proofs. In: TCC 2013 (2013)

[BCS16] Ben-Sasson, E., Chiesa, A., Spooner, N.: Interactive oracle proofs, 2016. Crypto ePrint 2016/116

[BD16] Bishop, A., Dodis, Y.: Interactive coding for interactive proofs. In: Kushilevitz, E., Malkin, T. (eds.) TCC 2016. LNCS, vol. 9563, pp. 352–366. Springer, Heidelberg (2016). doi:10.1007/978-3-662-49099-0_13

[BDG+13] Bitansky, N., Dachman-Soled, D., Garg, S., Jain, A., Kalai, Y.T., López-Alt, A., Wichs, D.: Why "Fiat-Shamir for proofs" lacks a proof. In: TCC 2013 (2013)

[BFL90] Babai, L., Fortnow, L., Lund, C.: Nondeterministic exponential time has two-prover interactive protocols. In: SFCS 1990 (1990)

[BFLS91] Babai, L., Fortnow, L., Levin, L.A., Szegedy, M.: Checking computations in polylogarithmic time. In: STOC 1991 (1991)

[BG93] Bellare, M., Goldreich, O.: On defining proofs of knowledge. In: Brickell, E.F. (ed.) CRYPTO 1992. LNCS, vol. 740, pp. 390–420. Springer, Heidelberg (1993). doi:10.1007/3-540-48071-4_28

[BG08] Barak, B., Goldreich, O.: Universal arguments and their applications. SIAM J. Comput. **38**(5), 1661–1694 (2008)

[BGGL01] Barak, B., Goldreich, O., Goldwasser, S., Lindell, Y.: Resettably-sound zero-knowledge and its applications. In: FOCS 2001 (2001)

[BGH+04] Ben-Sasson, E., Goldreich, O., Harsha, P., Sudan, M., Vadhan, S.: Robust PCPs of proximity, shorter PCPs and applications to coding. In: STOC 2004 (2004)

[BGKW88] Ben-Or, M., Goldwasser, S., Kilian, J., Wigderson, A.: Multi-prover interactive proofs: how to remove intractability assumptions. In: STOC 1988 (1988)

[BHZ87] Boppana, R.B., Håstad, J., Zachos, S.: Does co-NP have short interactive proofs? Inf. Process. Lett. **25**(2), 127–132 (1987)

[BKK+13] Ben-Sasson, E., Kaplan, Y., Kopparty, S., Meir, O., Stichtenoth, H.: Constant rate PCPs for circuit-SAT with sublinear query complexity. In: FOCS 2013 (2013)

[BR93] Bellare, M., Rogaway, P.: Random oracles are practical: a paradigm for designing efficient protocols. In: CCS 1993 (1993)

[BS08] Ben-Sasson, E., Sudan, M.: Short PCPs with polylog query complexity. SIAM J. Comput. **38**(2), 551–607 (2008)

[BW15] Bernhard, D., Warinschi, B.: On limitations of the Fiat-Shamir transformation. ePrint 2015/712 (2015)

[CGH04] Canetti, R., Goldreich, O., Halevi, S.: The random oracle methodology, revisited. JACM **51**(4), 557–594 (2004)

[COPV13] Chung, K.-M., Ostrovsky, R., Pass, R., Visconti, I.: Simultaneous resettability from one-way functions. In: FOCS 2013 (2013)

[CPSV16] Ciampi, M., Persiano, G., Siniscalchi, L., Visconti, I.: A transform for NIZK almost as efficient and general as the Fiat-Shamir transform without programmable random oracles. In: TCC 2016-A (2016)

[Dam89] Damgård, I.B.: A design principle for hash functions. In: Brassard, G. (ed.) CRYPTO 1989. LNCS, vol. 435, pp. 416–427. Springer, Heidelberg (1990). doi:10.1007/0-387-34805-0_39

[DNRS03] Dwork, C., Naor, M., Reingold, O., Stockmeyer, L.J.: Magic functions. JACM **50**(6), 852–921 (2003)

[Fis05] Fischlin, M.: Communication-efficient non-interactive proofs of knowledge with online extractors. In: Shoup, V. (ed.) CRYPTO 2005. LNCS, vol. 3621, pp. 152–168. Springer, Heidelberg (2005). doi:10.1007/11535218_10

[FRS88] Fortnow, L., Rompel, J., Sipser, M.: On the power of multi-prover interactive protocols (1988)

[FS86] Fiat, A., Shamir, A.: How to prove yourself: practical solutions to identification and signature problems. In: Odlyzko, A.M. (ed.) CRYPTO 1986. LNCS, vol. 263, pp. 186–194. Springer, Heidelberg (1987). doi:10.1007/3-540-47721-7_12

[GGPR13] Gennaro, R., Gentry, C., Parno, B., Raykova, M.: Quadratic span programs and succinct NIZKs without PCPs. In: Johansson, T., Nguyen, P.Q. (eds.) EUROCRYPT 2013. LNCS, vol. 7881, pp. 626–645. Springer, Heidelberg (2013). doi:10.1007/978-3-642-38348-9_37

[GH98] Goldreich, O., Håstad, J.: On the complexity of interactive proofs with bounded communication. Inf. Process. Lett. **67**(4), 205–214 (1998)

[GIMS10] Goyal, V., Ishai, Y., Mahmoody, M., Sahai, A.: Interactive locking, zero-knowledge PCPs, and unconditional cryptography. In: Rabin, T. (ed.) CRYPTO 2010. LNCS, vol. 6223, pp. 173–190. Springer, Heidelberg (2010). doi:10.1007/978-3-642-14623-7_10

[GK03] Goldwasser, S., Kalai, Y.T.: On the (in)security of the Fiat-Shamir paradigm. In: FOCS 2003 (2003)

[GKR08] Goldwasser, S., Kalai, Y.T., Rothblum, G.N.: Delegating computation: interactive proofs for Muggles. In: STOC 2008 (2008)

[GMR89] Goldwasser, S., Micali, S., Rackoff, C.: The knowledge complexity of interactive proof systems. SIAM J. Comput. **18**(1), 186–208 (1989)

[GOSV14] Goyal, V., Ostrovsky, R., Scafuro, A., Visconti, I.: Black-box non-black-box zero knowledge. In: STOC 2014 (2014)

[Gro10] Groth, J.: Short pairing-based non-interactive zero-knowledge arguments. In: Abe, M. (ed.) ASIACRYPT 2010. LNCS, vol. 6477, pp. 321–340. Springer, Heidelberg (2010). doi:10.1007/978-3-642-17373-8_19

[GS86] Goldwasser, S., Sipser, M.: Private coins versus public coins in interactive proof systems. In: STOC 1986 (1986)

[GVW02] Goldreich, O., Vadhan, S., Wigderson, A.: On interactive proofs with a laconic prover. Comput. Complex. **11**(1–2), 1–53 (2002)

[HS00] Harsha, P., Sudan, M.: Small PCPs with low query complexity. Comput. Complex. **9**(3–4), 157–201 (2000)

[HT98] Hada, S., Tanaka, T.: On the existence of 3-round zero-knowledge protocols. In: Krawczyk, H. (ed.) CRYPTO 1998. LNCS, vol. 1462, pp. 408–423. Springer, Heidelberg (1998). doi:10.1007/BFb0055744

[IKM09] Ito, T., Kobayashi, H., Matsumoto, K.: Oracularization and two-prover one-round interactive proofs against nonlocal strategies. In: CCC 2009 (2009)

[IKO07] Ishai, Y., Kushilevitz, E., Ostrovsky, R.: Efficient arguments without short PCPs. In: CCC 2007 (2007)

[IMS12] Ishai, Y., Mahmoody, M., Sahai, A.: On efficient zero-knowledge PCPs. In: Cramer, R. (ed.) TCC 2012. LNCS, vol. 7194, pp. 151–168. Springer, Heidelberg (2012). doi:10.1007/978-3-642-28914-9_9

[IMSX15] Ishai, Y., Mahmoody, M., Sahai, A., Xiao, D.: On zero-knowledge PCPs: limitations, simplifications, and applications (2015). http://www.cs.virginia.edu/mohammad/files/papers/ZKPCPs-Full.pdf

[Ito10] Ito, T.: Polynomial-space approximation of no-signaling provers. In: ICALP 2010 (2010)

[Kil92] Kilian, J.: A note on efficient zero-knowledge proofs and arguments. In: STOC 1992 (1992)

[KR08] Kalai, Y., Raz, R.: Interactive PCP. In: ICALP 2008 (2008)

[KR09] Kalai, Y.T., Raz, R.: Probabilistically checkable arguments. In: Halevi, S. (ed.) CRYPTO 2009. LNCS, vol. 5677, pp. 143–159. Springer, Heidelberg (2009). doi:10.1007/978-3-642-03356-8_9

[KRR13] Kalai, Y., Raz, R., Rothblum, R.: Delegation for bounded space. In: STOC 2013 (2013)

[KRR14] Kalai, Y.T., Raz, R., Rothblum, R.D.: How to delegate computations: the power of no-signaling proofs. In: STOC 2014 (2014)

[KRR16] Kalai, Y.T., Rothblum, G.N., Rothblum, R.D.: From obfuscation to the security of Fiat-Shamir for proofs. ePrint 2016/303 (2016)

[LFKN92] Lund, C., Fortnow, L., Karloff, H.J., Nisan, N.: Algebraic methods for interactive proof systems. JACM **39**(4), 859–868 (1992)

[Lin15] Lindell, Y.: An efficient transform from sigma protocols to NIZK with a CRS and non-programmable random oracle. In: Dodis, Y., Nielsen, J.B. (eds.) TCC 2015. LNCS, vol. 9014, pp. 93–109. Springer, Heidelberg (2015). doi:10.1007/978-3-662-46494-6_5

[Lip12] Lipmaa, H.: Progression-free sets and sublinear pairing-based non-interactive zero-knowledge arguments. In: Cramer, R. (ed.) TCC 2012. LNCS, vol. 7194, pp. 169–189. Springer, Heidelberg (2012). doi:10.1007/978-3-642-28914-9_10

[Mer89a] Merkle, R.C.: A certified digital signature. In: Brassard, G. (ed.) CRYPTO 1989. LNCS, vol. 435, pp. 218–238. Springer, Heidelberg (1990). doi:10. 1007/0-387-34805-0_21

[Mer89b] Merkle, R.C.: One way hash functions and DES. In: Brassard, G. (ed.) CRYPTO 1989. LNCS, vol. 435, pp. 428–446. Springer, Heidelberg (1990). doi:10.1007/0-387-34805-0_40

[Mic00] Micali, S.: Computationally sound proofs. SIAM J. Comput. **30**(4), 1253–1298 (2000)

[MV16] Mittelbach, A., Venturi, D.: Fiat-shamir for highly sound protocols is instantiable. ePrint 2016/313 (2016)

[Pas03] Pass, R.: On deniability in the common reference string and random oracle model. In: Boneh, D. (ed.) CRYPTO 2003. LNCS, vol. 2729, pp. 316–337. Springer, Heidelberg (2003). doi:10.1007/978-3-540-45146-4_19

[PGHR13] Parno, B., Gentry, C., Howell, J., Raykova, M.: Pinocchio: nearly practical verifiable computation. In: Oakland 2013 (2013)

[PS96] Pointcheval, D., Stern, J.: Security proofs for signature schemes. In EURO-CRYPT '96, 1996

[PSSV07] Pavan, A., Selman, A.L., Sengupta, S., Vinodchandranm, N.V.: Polylogarithmic-round interactive proofs for coNP collapse the exponential hierarchy. Theoret. Comput. Sci. **385**(1), 167–178 (2007)

[PTW09] Pass, R., Tseng, W.-L.D., Wikström, D.: On the composition of public-coin zero-knowledge protocols. In: Halevi, S. (ed.) CRYPTO 2009. LNCS, vol. 5677, pp. 160–176. Springer, Heidelberg (2009). doi:10.1007/978-3-642-03356-8_10

[RRR16] Reingold, O., Rothblum, R., Rothblum, G.: Constant-round interactive proofs for delegating computation. In: STOC 2016 (2016)

[SBV+13] Setty, S., Braun, B., Victor, V., Blumberg, A.J., Parno, B., Walfish, M.: Resolving the conflict between generality and plausibility in verified computation. In: EuroSys 2013 (2013)

[SBW11] Setty, S., Blumberg, A.J., Walfish, M.: Toward practical and unconditional verification of remote computations. In: HotOS 2011 (2011)

[Sha92] Shamir, A.: IP = PSPACE. JACM **39**(4), 869–877 (1992)

[SMBW12] Setty, S., McPherson, M., Blumberg, A.J., Walfish, M.: Making argument systems for outsourced computation practical (sometimes). In: NDSS 2012 (2012)

[SVP+12] Setty, S., Victor, V., Panpalia, N., Braun, B., Blumberg, A.J., Walfish, M.: Taking proof-based verified computation a few steps closer to practicality. In: Security 2012 (2012)

[Val08] Valiant, P.: Incrementally verifiable computation or proofs of knowledge imply time/space efficiency. In: Canetti, R. (ed.) TCC 2008. LNCS, vol. 4948, pp. 1–18. Springer, Heidelberg (2008). doi:10.1007/978-3-540-78524-8_1

[Wee09] Wee, H.: Zero knowledge in the random oracle model, revisited. In: Matsui, M. (ed.) ASIACRYPT 2009. LNCS, vol. 5912, pp. 417–434. Springer, Heidelberg (2009). doi:10.1007/978-3-642-10366-7_25

Adaptive Succinct Garbled RAM
or: How to Delegate Your Database

Ran Canetti[1,2(✉)], Yilei Chen[1], Justin Holmgren[3], and Mariana Raykova[4]

[1] Boston University, Boston, USA
{`canetti,chenyl`}`@bu.edu`
[2] Tel Aviv University and CPIIS, Tel Aviv, Israel
[3] MIT, Cambridge, USA
`holmgren@mit.edu`
[4] Yale University and SRI, New Haven, USA
`mariana.raykova@yale.edu`

Abstract. We show how to garble a large persistent database and then garble, one by one, a sequence of adaptively and adversarially chosen RAM programs that query and modify the database in arbitrary ways. The garbled database and programs reveal only the outputs of the programs when run in sequence on the database. Still, the runtime, space requirements and description size of the garbled programs are proportional only to those of the plaintext programs and the security parameter. We assume indistinguishability obfuscation for circuits and somewhat-regular collision-resistant hash functions. In contrast, all previous garbling schemes with persistent data were shown secure only in the static setting where all the programs are known in advance.

As an immediate application, we give the first scheme for efficiently outsourcing a large database and computations on the database to an untrusted server, then delegating computations on this database, where these computations may update the database.

Our scheme extends the non-adaptive RAM garbling scheme of Canetti and Holmgren [ITCS 2016]. We also define and use a new primitive of independent interest, called adaptive accumulators. The primitive extends the positional accumulators of Koppula et al. [STOC 2015] and somewhere statistical binding hashing of Hubáček and Wichs [ITCS 2015] to an adaptive setting.

1 Introduction

Database delegation. Alice is embarking on a groundbreaking experiment that involves collecting huge amounts of data over several months and then querying and running analytics on the data in ways to be determined as the data accumulates. Alas she does not have sufficient storage and processing power.

This paper was presented jointly with "Delegating RAM Computations with Adaptive Soundness and Privacy" by Prabhanjan Ananth, Yu-Chi Chen, Kai-Min Chung, Huijia Lin and Wei-Kai Lin.

© International Association for Cryptologic Research 2016
M. Hirt and A. Smith (Eds.): TCC 2016-B, Part II, LNCS 9986, pp. 61–90, 2016.
DOI: 10.1007/978-3-662-53644-5_3

Eve, who runs a large competing lab, offers servers for rent, but charges proportionally to storage and computing time. Can Alice make use of Eve's offer while being guaranteed that Eve does not learn or modify Alice's data and algorithms? Can she do it at a cost that's reasonably proportional to the size of the actual data and resource requirements?

The rich literature on verifiable delegation of computation, e.g. [4,12,20,24, 25,31,34], provides Alice with ways to guarantee the correctness of the results on her weak machines, while paying Eve only a relatively moderate cost. In particular, with [24] the cost is proportional only to the unprotected database size, the complexity of her unprotected queries and the security parameter. However, these schemes do not provide secrecy for Alice's data and computations. Searchable encryption schemes such as [6,26,32] provide varying levels of secrecy at a reasonable cost, but no verifiability.

So Alice turns to delegation schemes based on garbling. Such schemes, starting with [16], can indeed provide both verifiability and privacy. Here the client garbles its input and program (along with some authentication information) and hands them to the server, who evaluates the garbled program on the garbled input and returns the result to the client. In Alice's case the garbling scheme should be *persistent,* namely it should be possible to garble multiple programs that operate on the same garbled data, possibly updating the data over time. Alice would also like the scheme to be *succinct,* in the sense that the overhead of garbling each new query should be proportional to the description size of that query as a RAM program, independently of on the size of the database. Furthermore, the evaluation process should be *efficiency preserving,* namely it should preserve the RAM efficiency of the underlying computation.

Gennaro et al. [16] use the original Yao circuit-garbling scheme [33,35], which is neither succinct, efficiency-preserving, nor persistent. [15,17,18,29] describe garbling schemes that operate on persistent memory, improve on efficiency, but haven't yet achieved succinctness. Succinct, efficiency-preserving and persistent garbling schemes, based on indistinguishability obfuscation for circuits and one way functions are constructed in [9,11], building on techniques from [5,10,28].

However, the security of these schemes is only analyzed in a static setting, where all the queries and data updates are fixed beforehand. Given the dynamic and on-going character of Alice's research, a static guarantee is hardly adequate. Instead, Alice needs to consider a setting where new queries and updates may depend on the public information released so far. The dependence may be arbitrary and potentially adversarially influenced. Adaptive security is considered in [3,20,22] in the context of *one-time, non-succinct* garbling. An adaptive garbling scheme for Turing machines is constructed in [2]. Still, adaptive security has not been achieved in the pertinent setting of succinct and persistent RAM garbling.

1.1   This Work

We construct an adaptively secure, efficiency-preserving, succinct and persistent garbling scheme for RAM programs. That is, the scheme allows its user to garble an initial memory, and then garble RAM programs that arrive one by one in sequence. The machines can read from and update the memory, and also have local output. It is guaranteed that:

(1) Running the garbled programs one after another in sequence on the garbled memory results in the same sequence of outputs as running the plaintext machines one by one in sequence on the plaintext memory.
(2) The view of any adversary that generates a database and programs and obtains their garbled versions is simulatable by a machine that sees only the initial database size and sequence of outputs of the plaintext programs when run in sequence on the plaintext database. This holds even when the adversary chooses new plaintext programs adaptively, based on the garbled memory and garbled programs seen so far.
(3) The time to garble the memory is proportional to the plaintext memory. Up to polynomial factors in the security parameter, the garbling time and size of the garbled program are proportional only to the size of the plaintext RAM program. The runtime and space usage of each garbled machine are comparable to those of the plaintext machine.

Given such a scheme, constructing a database delegation scheme as specified above is straightforward: Alice sends Eve a garbled version of her database. To delegate a query, she garbles the program that executes the query. To guarantee (public) verifiability Alice can use the following technique from [18]: Each program signs its outputs using an embedded signing key, and Alice publishes the corresponding public key. To hide the query results from the server, the program encrypts its output under a secret key known to Alice. We provide herein a more complete definition (within the UC framework), as well as an explicit construction and analysis.

1.2   Overview of the Construction

Our starting point is the statically-secure garbling scheme of Canetti and Holmgren [9]. We briefly sketch their construction, and then explain where the issues with adaptivity come up and how we solve them.

Statically-secure garbling scheme for RAMs - an overview. The Canetti-Holmgren construction consists of four main steps. They first build a *fixed-transcript garbling scheme*, i.e. a garbling scheme which guarantees indistinguishability of the garbled machines and inputs as long as the entire transcripts of the communication with the external memory, as well as the local states kept between the RAM computation steps are the same in the two computations. In other words, if the computation of machine M_1 on input x_1 has the same transcript as that of M_2 on input x_2, then the garbled machines $\tilde{M}_1$, $\tilde{M}_2$ and the garbled inputs $\tilde{x}_1$, $\tilde{x}_2$ are computationally indistinguishable: $(\tilde{M}_1, \tilde{x}_1) \approx (\tilde{M}_2, \tilde{x}_2)$. This step closely follows the scheme of Koppula, Lewko and Waters [28] for garbling of Turing machines. The garbled program is essentially an obfuscated RAM/CPU-step circuit, which takes as input a local state and a memory symbol, and outputs an updated local state, as well as a memory operation. The main challenge here is to guarantee the authenticity and freshness of the values read from the memory. This is done using a number of mechanisms, namely splittable signatures, iterators and positional accumulators.

The second step extends the construction to *fixed-access garbling scheme*, which allows the intermediate local states of the two transcripts to differ while everything else stays the same. This is achieved by encrypting the state in an obfuscation-friendly way. The third step is to obtain a *fixed-address garbling scheme*, namely a scheme that guarantees indistinguishability of the garbled machines as long as only the sequence of *addresses* of memory accesses is the same in the two computations. Here they apply the same type of encryption used for the local state also to the memory content. The final step is to use an obfuscation-friendly ORAM in order to hide the program's memory access pattern. (Specifically, they use the ORAM of Chung and Pass [13].)

The challenge of adaptive security. The first (and biggest) challenge has to do with the *positional accumulator*, which is an iO-friendly variant of a Merkle-hash-tree built on top of the memory. That is, the contents of the memory is hashed down until a short root (called the accumulator value ac) is obtained. Then this value is signed together with the current local state by the CPU and is kept (in memory) for subsequent verification of database accesses. Using the accumulator, the evaluator is later able to efficiently convince the CPU that the contents of a certain memory location L is v. We call this operation "opening" the accumulator value ac to contents v at location L. Intuitively, the main security property is that it should be computationally infeasible to open an accumulator to an incorrect value.

However, to be useful with indistinguishability obfuscation, the accumulator needs an additional property, called *enforceability*. In [28], this property allows to generate, given memory location L^* and symbol v^*, a "rigged" public key for the accumulator along with a "rigged" accumulator value ac*. The rigged public key and accumulator look indistinguishable from honestly generated public key and accumulator value, and also have the property that *there does not exist* a way to open ac* at location L^* to any value other than v^*.[1]

The fact that the special values v^*, L^*, and ac* are encoded in the rigged public key forces these values to be known before the adversary sees the public key.

---

[1] To get an idea of why enforceability is needed, consider two programs C_0 and C_1, such that $C_0(L^*, v^*) = C_1(L^*, v^*)$, but whose functionality may differ elsewhere, and let $C_i'(L, v)$ be the program "if L, v are consistent with ac* then run C_i, else output $\perp$". Let iO be an indistinguishability obfuscator, i.e. it is guaranteed that $iO(A) \approx iO(B)$ whenever equal sized programs A, B have the same functionality everywhere. We would like to argue that $iO(C_0') \approx iO(C_1')$; however, we cannot do it directly using a plain Merkle hash tree, since collisions exist and so C_0' and C_1' have very different functionalities. Positional accumulators get around this difficulty: Using enforceability it is possible to argue that, when C_0' and C_1' use the rigged public key for the accumulator, the two programs have exactly the same functionality, and so indistinguishability holds. Due to the indistinguishability of rigged public accumulator keys from honest ones, indistinguishability holds even for the case of non-rigged accumulator keys.

This suffices for the case of static garbling, since the special values depend only on the underlying computation, and this computation is fixed in advance and does not depend on adversary's view. However, in the adaptive setting, this is not the case. This is so since the adversary can choose new computations — and thus new special values v^*, L^* — depending on its view so far, which includes the public key of the accumulator.

Adaptive Accumulators. We get around this problem by defining and constructing a new primitive, called *adaptive accumulators*, which are an adaptive alternative to positional accumulators. In our adaptive accumulators there are no "rigged" public keys. Instead, correctness of an opening of a hash value at some location is verified using a *verification key* which can be generated later. In addition to the usual computational binding guarantees, it should be possible to generate, given a special accumulator value ac^*, value v^* and location L^*, a "rigged" verification key vk^* that looks indistinguishable from an honestly generated one, and such that vk^* does not verify an opening of ac^* at location L^* to any value other than v^*. Furthermore, it is possible to generate multiple verification keys, that are all rigged to enforce the same accumulator value ac^* to different values v^* at different locations L^*, where all are indistinguishable from honest verification keys.

We then use adaptive accumulators as follows: There is a single set of public parameters that is posted together with the garbled database and is used throughout the lifetime of the system. Now, each new garbled machine is given a different, independently generated verification key. This allows us, at the proof of security, to use a different rigged verification key for each machine. Since the key is determined only when a machine is being garbled (and its computation and output values are already fixed), we can use a rigged verification key that enforces the correct values, and obtain the same tight security reduction as in the static setting.

Adaptively accumulators from adaptive puncturable hash functions. We build adaptive accumulators from a new primitive called *adaptively puncturable (AP)* hash function ensembles. In this primitive a standard collision resistant hash function $h(x)$ is augmented with three algorithms Verify, GenVK, GenBindingVK. GenVK generates a verification key vk, which can be later used in $\mathsf{Verify}(\mathsf{vk}, x, y)$ to check that $h(x) = y$. $\mathsf{GenBindingVK}(x^*)$ produces a binding key vk^* such that $\mathsf{Verify}(\mathsf{vk}^*, x, y = h(x^*))$ accepts only if $x = x^*$. Finally, we require that real and binding verification keys should be indistinguishable even for the adversary which chooses x^* adaptively after seeing h.

The construction of adaptive accumulators from AP hash functions proceeds as follows. The public key is an AP hash function h, and the initial accumulator value ac_0 is the root of a Merkle tree on the initial data store (which can be thought of as empty, or the all-0 string) using h. We maintain the invariant that at every moment the root value ac is the result of hashing down the memory store. In order to write a new symbol v to a position L the evaluator recomputes all hashes on the path from the root to L. The "opening information" for v at L is all hashes of siblings on the path from the root to L.

The verification key is a sequence of $d = \log |S|$ (honest) verification keys for h - one for each level of the tree. The "rigged" verification key for accumulator value ac* and value v^* at location L^* consists of a sequence of d rigged verification keys for the AP hash, where each key forces the opening of a single value along the path from the root to leaf L^*. Security of the adaptive accumulator follows from the security of the AP hash via standard reduction.

Constructing AP hash. We construct adaptively puncturable hash function ensembles from indistinguishability obfuscation for circuits, plus collision-resistant hash functions with the property that any image has at most polynomially many preimages. (This implies that the CRHF shrinks at most logarithmically many bits). We say that a hash function is *c-bounded* if the number of preimages for any image is no more than c. To be usable in the Merkle-Damgård construction, we will also need that the hash functions have domain $\{0,1\}^\lambda$ and range $\{0,1\}^{\lambda'}$ for some $\lambda' < \lambda$. For simplicity we focus on the setting where $\lambda = \lambda' + 1$. We construct 4-bounded CRHFs assuming hardness of discrete log and 64-bounded CRHFs assuming hardness of factoring.

Our construction of an AP hash ensemble can be understood in two steps.

1. First we construct a c-bounded AP hash ensemble from any c-bounded hash ensemble $\{h_k\}$. This is done as follows: The public key is a hash function h_k. A verification key vk is $i\mathcal{O}(V)$, where V is the program that on input x, y outputs 1 if $h_k(x) = y$. A "rigged" verification key vk* that is binding for input x^* is $i\mathcal{O}(V_{x^*})$ where V_{x^*} is the program that on input (x, y) does the following:
   - if $y = f_{h_k}(x^*)$, it accepts if and only if $x = x^*$;
   - otherwise it accepts if and only if $y = h_k(x)$.
   Since h_k is c-bounded, the functionality of V and V_{x^*} differ only on polynomially many inputs. Therefore, the real and "rigged" verification keys are indistinguishable following the di$\mathcal{O}$-i$\mathcal{O}$ equivalence for circuits with polynomially many differing inputs [7].
2. Next we construct AP hash functions which are length halving (and are thus not polynomially bounded) from bounded AP hashing. This is done in the natural way by extending the hash function's domain using Merkle-Damgård. Suppose we start with a function $h' : \{0,1\}^{\lambda+1} \to \lambda$, and build $h : \{0,1\}^{2\lambda} \to \{0,1\}^\lambda$. A verification key vk for h is an obfuscated circuit C which takes x and y, and directly checks that $h(x) = y$.

The proof of security involves a sequence of hybrids, in which C is modified to contain a verification key for h'. This implies that in the real world, C must also be padded to this same size. In other words, the verification key vk must be as large as twice-obfuscated circuit computing h'. We note that it is possible to avoid this overhead by instead distributing λ different verification keys for h', but we avoid this approach for conceptual simplicity.

From adaptive accumulators to adaptively secure fixed-transcript garbling. We return to the challenges encountered when trying to use the [9] construction in our adaptive setting. With adaptive accumulators in hand, the additional modifications made on the use of iterator and splittable signatures are relatively local. Since these primitives do not access the long-lived shared memory, it suffices to generate a fresh instance of each primitive for each new query.

Adaptively secure fixed-access and fixed-address garbling. Next we upgrade the next two layers in the [9] construction, namely the fixed-access and fixed-address garbling schemes, to adaptively secure ones. This is done with relatively local changes from the original construction. Specifically we include the index and time step in the domain of puncturable PRF that is used to derive the randomness of the one-time-pad-like encryption on the state and memory. The technical details can be found in the main construction.

Adaptive full garbling. Recall that in [9] full garbling is achieved by applying an Oblivious RAM scheme on top of the fixed-access garbling. The randomness for the ORAM accesses is sampled using a PRF. This leads to a situation where a PRF key is first used inside a program M_i for some execution i. Later, the key needs to be punctured at a point that may depend on the PRF values. This leads to another adaptivity problem.

We get around this problem by noticing that the Chung-Pass ORAM has a special property which allows us to guess which points to puncture with only polynomial security loss. This property, which we call *strong localized randomness*, is sketched as follows. Let R be the randomness used by the ORAM. Let $A_i = a_{i1}, \ldots, a_{im}$ be a set of locations accessed by the ORAM during emulation of access i. The strong localized randomness property guarantees that there exists a set of intervals $I_{11}, \ldots, I_{Tm}$, $I_{ij} \subset [1, |R|]$, such that:

1. Each a_{ij} depends only on $R_{I_{ij}}$, i.e., the part of the randomness R indexed with I_{ij}; furthermore, a_{ij} is efficiently computable from I_{ij};
2. All I_{ij} are mutually disjoint;
3. All I_{ij} are efficiently computable given the sequence of memory operations.

To see that the Chung-Pass ORAM has strong localized randomness, observe that in its non-recursive form, each virtual access of addr touches two paths: one is the path used for the eviction, which is purely random, and the other is determined by the randomness chosen in the previous virtual access of addr. Therefore, the set of accessed locations is determined by two randomness intervals. When the ORAM is applied recursively, each virtual access consists of $O(\log S)$ phases, each of whose physical addresses are determined by two randomness intervals. Since the number of intervals in the range $[1, \ldots, |R|]$ is only polynomial in the security parameter, the reduction can guess the intervals for a phase (and therefore the points to puncture at) with only polynomial security loss.

In contrast, the localized randomness property used in [9] differs in property 1 above, requiring only that each A_i depends on polylogarithmically many bits of R. This does not suffice for us, because there are superpolynomially many possible dependencies, and so the reduction cannot guess correctly with any non-negligible probability.

Concurrent and independent work. A potential alternative to our adaptive positional accumulators is to build on the *somewhere statistically binding (SSB) hash* of Hubáček and Wichs [23] or Okamoto et al. [30]. SSB hashes have a similar flavor to positional accumulators, but they allow rigging to be statistically binding at a hidden location L^*. However it turns out that SSB hashes alone do not suffice for positional accumulators, even in the non-adaptive case! In concurrent and independent work, Ananth et al. [1] give a stronger definition of SSBs which *does* suffice, and then show that a known construction [30] satisfies this stronger property. Their reduction can then be made adaptive by guessing L^*, at the price of reducing the reduction's winning probability by a factor proportional to the database size. In all, their construction uses a somewhat stronger assumption than ours (DDH vs. discrete log) and their security reduction is somewhat less efficient than ours.

Organization. The rest of the paper is organized as follows. Section 2 provides definitions of RAM and adaptively secure garbled RAM. Section 3, 4 and 5 define and construct bounded hashing, adaptively puncturable hashing and adaptively secure positional accumulator. Sections 6, 7, 8 and 9 provide the definitions and constructions of fixed-transcript, fixed-access, fixed-address, and fully secure garbling. Section 10 includes the definition of secure database delegation within the UC framework and our construction and proof.

Due to the page limitation, some missing details are only available in the full version of this paper [8]. Those missing details include (1) The other primitives used in our work; (2) The proofs of fixed-transcript, fixed-access, fixed-address and fully secure garbling; (3) A construction of a (stateful) reusable GRAM with persistent data.

2   Definitions

RAM Programs. A RAM M is defined as a tuple (Σ, Q, Y, C), where Σ is the set of memory symbols, Q is the set of possible local states, Y is the output space, and C is the transition function.

Memory Configurations. A memory configuration on alphabet Σ is a function $s : \mathbb{N} \to \Sigma \cup \{\epsilon\}$, where ϵ denotes the contents of an empty memory cell. Let $\|s\|_0$ denote $|\{a : s(a) \neq \epsilon\}|$ and, in an abuse of notation, let $\|s\|_\infty$ denote $\max(\{a : s(a) \neq \epsilon\})$, which we will call the *length* of the memory configuration. A memory configuration s can be implemented (say with a balanced binary tree) by a data structure of size $O(\|s\|_0)$, supporting updates to any index in $O(\log \|s\|_\infty)$ time.

We can naturally identify a string $x = x_1 \dots x_n \in \Sigma^*$ with the memory configuration s_x, where $s_x(i) = x_i$ if $i \leq |x|$ and $s_x(i) = \epsilon$, otherwise. Looking ahead, efficient representations of sparse memory configurations (in which $\|s\|_0 < \|s\|_\infty$) are convenient for succinctly garbling computations where the space usage is larger than the input length.

Execution. A RAM $M = (\Sigma, Q, Y, C)$ is executed on an initial memory $s_0 \in \Sigma^{\mathbb{N}}$ to obtain $M(s_0)$ by iteratively computing $(q_i, a_i, v_i) = C(q_{i-1}, s_{i-1}(a_{i-1}))$, where $a_0 = 0$, and defining $s_i(a) = v$ if $a = a_i$ and $s_i(a) = s_{i-1}$ otherwise.

When $M(s_0) \neq \bot$, it is convenient to define the following functions:

Time(M, s_0): *runtime of M on s_0*, i.e., the number of iterations of C.

Space(M, s_0): *space usage* of M on s_0, i.e., $\max_{i=0}^{t-1}(\|s_i\|_\infty)$.

$\mathcal{T}(M, s_0)$: *execution transcript* of M on s_0 defined as $((q_0, a_0, v_0), \ldots, (q_{t-1}, a_{t-1}, v_{t-1}), y)$.

Addr(M, s_0): *addresses accessed* by M on s_0, i.e., $(a_0, \ldots, a_{t-1})$.

NextMem(M, s_0): *resultant memory configuration* s_t after executing M on s_0.

Garbled RAM
Syntax. A garbling scheme for RAM programs is a tuple of p.p.t. algorithms (Setup, GbPrg, GbMem, Eval).

Setup$(1^\lambda, S)$ takes the security parameter λ in unary and a space bound S, and outputs a secret key SK.

GbMem(SK, s) takes a secret key SK and a memory configuration s, and then outputs a memory configuration $\tilde{s}$.

GbPrg(SK, M_i, T_i, i) takes a secret key SK, a RAM machine M_i, a running time bound T_i, and a sequence number i, and outputs a garbled RAM machine $\tilde{M}_i$.

Eval$(\tilde{M}, \tilde{x})$: takes a garbled RAM $\tilde{M}$ and gabled input $\tilde{x}$ and evaluates the machine on the input, which we denote $\tilde{M}(\tilde{x})$.

Remark 1. The index number i given as input to GbPrg enforces defines a fixed order, so that $M_1, \ldots, M_\ell$ cannot be executed in any other order.

We are interested in garbling schemes which are *correct*, *efficient*, and *secure*.

Correctness. A garbling scheme is said to be correct if for all p.p.t. adversaries $\mathcal{A}$ and every $t = \text{poly}(\lambda)$

$$\Pr\left[\tilde{M}_t(\tilde{s}_{t-1}) = M_t(s_{t-1}) \,\middle|\, \begin{array}{l} (s_0, S) \leftarrow \mathcal{A}(1^\lambda) \\ SK \leftarrow \text{Setup}(1^\lambda, S) \\ \tilde{s}_0 \leftarrow \text{GbMem}(SK, s_0) \\ \text{for } i = 1, \ldots, t \\ \quad M_i, T_i \leftarrow \mathcal{A}(\tilde{s}_0, \tilde{M}_1, \ldots \tilde{M}_{i-1}) \\ \quad \tilde{M}_i \leftarrow \text{GbPrg}(SK, M_i, T_i, i) \\ \quad s_i = \text{NextMem}(M_i, s_{i-1}) \\ \quad \tilde{s}_i = \text{NextMem}(\tilde{M}_i, \tilde{s}_{i-1}) \end{array}\right] \geq 1 - \text{negl}(\lambda),$$

where

- $\sum T_i \leq \text{poly}(\lambda)$, $|s_0| \leq S \leq \text{poly}(\lambda)$;
- Space$(M_i, s_{i-1}) \leq S$ and Time$(M_i, s_{i-1}) \leq T_i$ for each i.

Efficiency. A garbling scheme is said to be efficient if:

1. Setup, GbPrg, and GbMem are probabilistic polynomial-time algorithms. Furthermore, GbMem runs in time linear in $\|s_0\|$. We require *succinctness* for the garbled programs, which means that the size of a garbled program $\tilde{M}$ is linear in the description length of the plaintext program M. The bounds T_i and S are encoded in binary, so the time to garble does not significantly depend on either of these quantities.
2. With $\tilde{M}_i$ and $\tilde{s}_i$ defined as above, it holds that $\mathsf{Space}(\tilde{M}_i, \tilde{s}_{i-1}) = \tilde{O}(S)$ and $\mathsf{Time}(\tilde{M}_i, \tilde{s}_{i-1}) = \tilde{O}(\mathsf{Time}(M_i, s_{i-1}))$ (hiding polylogarithmic factors in S).

Security. We define the security property of GRAM as follows.

Definition 1. *Let* $\mathcal{GRAM} = (\mathsf{Setup}, \mathsf{GbMem}, \mathsf{GbPrg})$ *be a garbling scheme. We define the following two experiments, where each M_i is a program with time and space complexity bounded by T_i and S. We denote $y_i = M_i(s_{i-1})$, $s_i = \mathsf{NextMem}(M_i, s_{i-1})$, and $t_i = \mathsf{Time}(M_i, s_{i-1})$.*

Experiment $REAL_{\mathcal{A}}(1^\lambda)$

$(s_0, S) \leftarrow \mathcal{A}(1^\lambda)$

$SK \leftarrow \mathsf{Setup}(1^\lambda, S), \tilde{s}_0 \leftarrow \mathsf{GbMem}(SK, s_0)$

$(M_1, 1^{T_1}) \leftarrow \mathcal{A}(\tilde{s}_0)$

$\tilde{M}_1 \leftarrow \mathsf{GbPrg}(SK, M_1, T_1, 1)$

for $i = 1$ *to* $\ell = \mathsf{poly}(\lambda)$

$\quad (M_{i+1}, 1^{T_{i+1}}) \leftarrow \mathcal{A}(\tilde{M}_{[i,\dots,1]}, \tilde{s}_0)$

$\quad \tilde{M}_{i+1} \leftarrow \mathsf{GbPrg}(SK, M_{i+1}, T_{i+1}, i+1)$

Output : $b \leftarrow \mathcal{A}(\tilde{M}, \tilde{s}_0)$

Experiment $IDEAL_{\mathcal{A}}(1^\lambda)$

$(s_0, S) \leftarrow \mathcal{A}(1^\lambda)$

$\tilde{s}_0 \leftarrow \mathsf{Sim}(1^\lambda, \|s_0\|_0)$

$(M_1, 1^{T_1}) \leftarrow \mathcal{A}(\tilde{s}_0)$

$\tilde{M}_1 \leftarrow \mathsf{Sim}(y_1, |M_1|, t_1)$

for $i = 1$ *to* $\ell = \mathsf{poly}(\lambda)$

$\quad (M_{i+1}, 1^{T_{i+1}}) \leftarrow \mathcal{A}(\tilde{M}_{[i,\dots,1]}, \tilde{s}_0)$

$\quad \tilde{M}_{i+1} \leftarrow \mathsf{Sim}(y_{i+1}, |M_{i+1}|, t_{i+1})$

Output : $b' \leftarrow \mathcal{A}(\tilde{M}, \tilde{s}_0)$

The garbling scheme $\mathcal{GRAM}$ is $\epsilon(\cdot)$-adaptively secure if

$$\left| \Pr[1 \leftarrow REAL_{\mathcal{A}}(1^\lambda)] - \Pr[1 \leftarrow IDEAL_{\mathcal{A}}(1^\lambda)] \right| < \epsilon(\lambda).$$

3 c-Bounded Collision-Resistant Hash Functions

We say that a hash function ensemble $\mathcal{H} = \{\mathcal{H}_\lambda\}_{\lambda \in \mathbb{N}}$ with $\mathcal{H}_\lambda = \{h_k : D_\lambda \to R_\lambda\}_{k \in \mathcal{K}_\lambda}$ is $c(\cdot)$-bounded if

$$\Pr{}_{h \leftarrow \mathcal{H}_\lambda} [\forall y \in R_\lambda, \#\{x : h(x) = y\} \leq c(\lambda)] \geq 1 - \mathsf{negl}(\lambda)$$

That is, with high probability, every element in the codomain of h has at most $c(\lambda)$ pre-images. In our adaptively secure garbling scheme, we need $c(\cdot)$ to be any polynomial (smaller is better for the security reduction), and we need $D_\lambda = \{0,1\}^{\lambda'}$ and $R_\lambda = \{0,1\}^{\lambda'-1}$ for some $\lambda' = \text{poly}(\lambda)$. For both of the constructions in this section, we obtain constant $c(\cdot)$.

The starting point for our constructions is the construction of [14], using a claw-free pair of permutations (π_0, π_1) on a domain D_λ, where for some fixed y_0, the hash $h(x)$ is defined as $(\pi_{x_0} \circ \cdots \circ \pi_{x_n})(y_0)$. Unfortunately, while this construction allows an arbitrarily-compressing hash function, it in general may not be $\text{poly}(n)$-bounded even if $n = \log |D_\lambda| + O(1)$.

However, a slight modification of this construction allows us to take any injective functions $\iota_{in} : \{0,1\}^n \hookrightarrow D_\lambda$ and $\iota_{out} : D_\lambda \hookrightarrow \{0,1\}^m$, and produce a 2^k-bounded collision-resistant function mapping $\{0,1\}^{n+k} \to \{0,1\}^m$. As long there is such injections exist with $m - n = O(\log \lambda)$, this yields a $\text{poly}(\lambda)$-bounded collision-resistant hash family.

Theorem 1. *If for a random λ-bit prime p, it is hard to solve the discrete log problem in $\mathbb{Z}_p^*$, then there exists a 4-bounded CRHF ensemble $\mathcal{H} = \{\mathcal{H}_\lambda\}_{\lambda \in \mathbb{N}}$ where $\mathcal{H}_\lambda$ consists of functions mapping $\{0,1\}^{\lambda+1} \to \{0,1\}^\lambda$.*

Proof. Let p be a random λ-bit prime, and let g and h be randomly chosen generators of $\mathbb{Z}_p^*$. Our hash function is keyed by p, g, h. It is well-known that the permutations $\pi_0(x) = g^x$ and $\pi_1(x) = g^x h$ are claw-free. It is easy to see there is an injection $\iota_{in} : \{0,1\}^{\lambda-1} \to \mathbb{Z}_p^*$ and an injection $\iota_{out} : \mathbb{Z}_p^* \to \{0,1\}^\lambda$. Define a hash function

$$f : \{0,1\}^{\lambda-1} \times \{0,1\} \times \{0,1\} \to \{0,1\}^\lambda$$

$$a, b, c \mapsto \iota_{out}(\pi_c(\pi_b(\iota_{in}(a))))$$

Clearly given $x \neq x'$ such that $f(x) = f(x')$, one can find a claw (and therefore find $\log_g h$), so f is collision-resistant. Also for any given image, there is at most one corresponding pre-image per choice of b, c, so f is 4-bounded.

Theorem 2. *If for random λ-bit primes p and q, with $p \equiv 3 \pmod 8$ and $q \equiv 7 \pmod 8$, it is hard to factor $N = pq$, then there exists a 64-bounded CRHF ensemble $\mathcal{H} = \{\mathcal{H}_\lambda\}_{\lambda \in \mathbb{N}}$ where $\mathcal{H}_\lambda$ consists of functions mapping $\{0,1\}^{2\lambda+1} \to \{0,1\}^{2\lambda}$.*

Proof. First, we construct injections $\iota_0 : \{0,1\}^{2\lambda-4} \to [N/6]$ and $\iota_1 : [N/6] \to \mathbb{Z}_N^* \cap [N/2]$, using the fact that for sufficiently large p and q, for any integer $x \in [N/6]$, at least one of $3x, 3x+1$, and $3x+2$ is relatively prime to N. $\iota_1(x)$ is therefore well-defined as the smallest of $\{3x, 3x + 1, 3x + 2\} \cap \mathbb{Z}_n^*$. Let $\iota_{in} : \{0,1\}^{2\lambda-4} \to \mathbb{Z}_N^* \cap [N/2]$ denote $\iota_1 \circ \iota_0$. Let ι_{out} denote an injection from $\mathbb{Z}_N^* \to \{0,1\}^{2\lambda}$.

Next, following [21], we define the claw-free pair of permutations $\pi_0(x) = x^2 \pmod N$ and $\pi_1(x) = 4x^2 \pmod N$, where the domain of π_0 and π_1 is the set of quadratic residues mod N.

Now we define the hash function

$$f : \{0,1\}^{2\lambda-4} \times \{0,1\}^5 \to \{0,1\}^{2\lambda}$$

$$f(x, y) = (\iota_{out} \circ \pi_{y_5} \circ \cdots \circ \pi_{y_1})(\iota_{in}(x)^2 \mod N)$$

This is 64-bounded because for any given image, there is at most one preimage under $\iota_{out} \circ \pi_{y_5} \circ \cdots \circ \pi_{y_1}$ per possible y value. This accounts for a factor of 32. The remaining factor of 2 comes from the fact that every quadratic residue has four square roots, two of which are in $[N/2]$ (the image of ι_{in}). The collision resistance of $x \mapsto \iota_{in}(x)^2 \pmod{N}$ follows from the fact that the two square roots are nontrivially related, i.e., neither is the negative of the other, so given both it would be possible to factor N.

Notation. For a function $h : \{0,1\}^{\lambda+1} \to \{0,1\}^\lambda$, we let h^0 denote the identity function and for $k > 0$ inductively define

$$h^k : \{0,1\}^{\lambda+k} \to \{0,1\}^\lambda$$
$$h^k(x) = h(x_1\|h^{k-1}(x_2\|\cdots\|x_{\lambda+k}))$$

4   Adaptively Puncturable Hash Functions

We say that an ensemble $\mathcal{H}$ is *adaptively puncturable* if there are algorithms Verify, GenVK, and ForceGenVK such that:

Correctness
    For all x, y, and $h \in \mathcal{H}$, $\mathsf{Verify}(vk, x, y) = 1$ iff $y = h(x)$, where $vk \leftarrow \mathsf{GenVK}(1^\lambda, h)$.
Forced Verification
    For all x^* and $h \in \mathcal{H}$, let $y^* = h(x^*)$. $\mathsf{Verify}(vk, x, y^*) = 1$ iff $x = x^*$, where $vk \leftarrow \mathsf{ForceGenVK}(1^\lambda, h, x^*)$.
Indistinguishability
    For all p.p.t. $\mathcal{A}_1, \mathcal{A}_2$

$$\Pr\left[\mathcal{A}_2(s, vk_b) = b \;\middle|\; \begin{array}{l} h \leftarrow \mathcal{H}_\lambda \\ x^*, s \leftarrow \mathcal{A}_1(1^\lambda, h) \\ vk_0 \leftarrow \mathsf{GenVK}(1^\lambda, h) \\ vk_1 \leftarrow \mathsf{ForceGenVK}(1^\lambda, h, x^*) \\ b \leftarrow \{0,1\} \end{array}\right] \le \frac{1}{2} + \mathrm{negl}(\lambda)$$

Theorem 3. *If iO exists and there is a* $\mathrm{poly}(\lambda)$*-bounded CRHF ensemble mapping* $\{0,1\}^{\lambda'+1} \to \{0,1\}^{\lambda'}$*, then there is an adaptively puncturable hash function ensemble mapping* $\{0,1\}^{2\lambda'}$ *to* $\{0,1\}^{\lambda'}$*.*

Let $\mathcal{H} = \{\mathcal{H}_\lambda\}$ be a $\mathrm{poly}(\lambda)$-bounded CRHF ensemble, where $\mathcal{H}_\lambda$ is a family of functions mapping $\{0,1\}^{\lambda+1} \to \{0,1\}^\lambda$. We define an adaptively puncturable hash function ensemble $\mathcal{F} = \{\mathcal{F}_\lambda\}$, where $\mathcal{F}_\lambda$ is a family of functions mapping $\{0,1\}^{2\lambda} \to \{0,1\}^\lambda$.

Setup
    The key space for $\mathcal{F}_\lambda$ is the same as the key space for $\mathcal{H}_\lambda$.
Evaluation
    For a key $h \in \mathcal{H}_\lambda$ and a string $x \in \{0,1\}^{2\lambda}$, we define

$$f_h(x) = h^\lambda(x)$$

Verification

GenVK$(1^\lambda, f_h)$ outputs an $i\mathcal{O}$-obfuscation of a circuit which directly computes

$$x, y \mapsto \begin{cases} 1 & \text{if } f_h(x) = y \\ 0 & \text{otherwise} \end{cases}$$

ForceGenVK$(1^\lambda, f_h, x^*)$ outputs an $i\mathcal{O}$-obfuscation of a circuit which directly computes

$$x, y \mapsto \begin{cases} 1 & \text{if } y \neq f_h(x^*) \wedge y = f_h(x) \\ 1 & \text{if } (x, y) = (x^*, f_h(x^*)) \\ 0 & \text{otherwise} \end{cases}$$

Verify(vk, x, y) simply evaluates and outputs $vk(x, y)$.

Claim. No p.p.t. adversary which adaptively chooses x^* after seeing h can distinguish between GenVK$(1^\lambda, h)$ and ForceGenVK$(1^\lambda, h, x^*)$.

Proof. We present $\lambda + 1$ hybrid games $H_0, \ldots, H_\lambda$. In each game h is sampled from $\mathcal{H}_\lambda$, but the circuit given by the challenger to the adversary depends on the game and on x^*. In hybrid H_i, the challenger computes $y^* = h^\lambda(x^*)$ and $y_{\lambda-i} = h^{\lambda-i}(x^*_{i+1} \| \cdots \| x^*_{2\lambda})$. The challenger then sends $i\mathcal{O}(C_i)$ to the adversary, where C_i has y^*, $y_{\lambda-i}$, and $x^*_1, \ldots, x^*_i$ hard-coded and is defined as

$$C_i(x, y) = \begin{cases} 1 & \text{if } y \neq y^* \wedge y = h^\lambda(x) \\ 1 & \text{if } y = y^* \wedge x_1 = x^*_1 \wedge \cdots \wedge x_i = x^*_i \wedge h^{\lambda-i}(x_{i+1} \| \cdots \| x_{2\lambda}) = y_{\lambda-i} \\ 0 & \text{otherwise} \end{cases}$$

The challenger sends $i\mathcal{O}(C_i)$ to the adversary.

It is easy to see that C_0 is functionally equivalent to the circuit produced by GenVK, and C_λ is functionally equivalent to the circuit produced by ForceGenVK. So we only need to show that $H_i \approx H_{i+1}$ for $0 \leq i < \lambda$. We give a sequence of indistinguishable changes to the challenger, by which we transform the circuit C given to the adversary from C_i to C_{i+1}.

1. We first change C so that when $y = y^*$, it computes the intermediate value $y' = h^{\lambda-i-1}(x_{i+2} \| \cdots \| x_{2\lambda})$ and outputs 1 if:
   - $h(x_{i+1} \| y') = y_{\lambda-i}$
   - For all $1 \leq j \leq i$, $x_i = x^*_i$.
   When $y \neq y^*$, the behavior of C is unchanged.
   This change preserves functionality (we only introduced a name y' for an intermediate value in the computation) and hence is indistinguishable by $i\mathcal{O}$.
2. Now we change C so that instead of directly checking whether $h(x_{i+1} \| y') = y_{\lambda-i}$, it uses a hard-coded helper circuit $\tilde{V} = i\mathcal{O}(V)$, where

$$V : \{0, 1\} \times \{0, 1\}^\lambda \times \{0, 1\}^\lambda \to \{0, 1\}$$
$$V(a, b, c) = \begin{cases} 1 & \text{if } c = h(a \| b) \\ 0 & \text{otherwise} \end{cases}$$

This is functionally equivalent and hence indistinguishable by $i\mathcal{O}$.

3. Now we change V. The challenger computes $y_{\lambda-i-1} = h^{\lambda-i-1}(x^*_{i+2}\| \cdots \|x^*_{2\lambda})$ and $y_{\lambda-i} = h(x^*_{i+1}\|y_{\lambda-i-1})$, and define

$$V(a,b,c) = \begin{cases} 1 & \text{if } c \neq y_{\lambda-i} \wedge c = h(a\|b) \\ 1 & \text{if } (a,b,c) = (x^*_{i+1}, y_{\lambda-i-1}, y_{\lambda-i}), \\ 0 & \text{otherwise} \end{cases}$$

with $y_{\lambda-i}$, $y_{\lambda-i-1}$, and x^*_{i+1} hard-coded. The old and new $\tilde{V}$'s are indistinguishable because:

- By the collision-resistance of h, it is difficult to find an input on which they differ.
- Because $\mathcal{H}_\lambda$ is poly(λ)-bounded, they differ on only polynomially many points.
- i$\mathcal{O}$ is equivalent to di$\mathcal{O}$ for circuits which differ on polynomially many points.

4. C is now functionally equivalent to C_{i+1} and hence is indistinguishable by i$\mathcal{O}$.

5   Adaptively Secure Positional Accumulators

In this section we define and construct adaptive positional accumulators (APA). We use this primitive for memory authentication in our garbling construction. A garbled program will be an obfuscated functionality where one input is a succinct commitment ac to some memory contents, another is a piece of data v allegedly resulting from a memory operation op, and another is a commitment ac', allegedly to the resulting memory configuration. Informally, APAs provide a way for the garbled program to check the consistency of v and ac' with ac (given a short proof),

As described so far, Merkle trees satisfy our needs, and indeed our construction is built around a Merkle tree. However, we require more. As in the positional accumulators of [28], we need a way to indistinguishably "rig" the public parameters so that for some ac and op, there is exactly one (ac', v) with any accepting proof. We differ from [27] by separating the parameters used for proof verification from those used for updating the accumulator, and allowing the rigged (ac, op) to be chosen adaptively as an adversarial function of the update parameters.

We now formally define the algorithms of the APA primitive.

SetupAcc($1^\lambda, S$) $\to$ PP, ac$_0$, store$_0$

> The setup algorithm takes as input the security parameter λ in unary and a bound S (in binary) on the memory addresses accessed. SetupAcc produces as output public parameters PP, an initial accumulator value ac$_0$, and an initial data store store$_0$.

Update(PP, store, op) $\to$ store', ac', v, π

> The update algorithm takes as input the public parameters PP, a data store store, and a memory operation op. Update then outputs a new store store', a memory value v, a succinct accumulator ac', and a succinct proof π.

Verify(vk, ac, op, ac′, v, π) → {0, 1}

The verification algorithm takes as inputs a verification key vk, an initial accumulator value ac, a memory operation op, a resulting accumulator ac′, a memory value v, and a proof π. Verify then outputs 0 or 1. Intuitively, Verify checks the following statement:

π *is a proof that the operation* op, *when applied to the memory config-uration corresponding to* ac, *yields a value* v *and results in a memory configuration corresponding to* ac′.

Verify is run by a garbled program to authenticate the memory values that the evaluator gives it.

SetupVerify(PP) → vk

SetupVerify generates a regular verification key for checking Update's proofs. This is the verification key that is used in the "real world" garbled programs.

SetupEnforceVerify(PP, (op$_1$, ..., op$_k$)) → vk

SetupEnforceVerify takes a sequence of memory operations, and generates a verification key which is perfectly sound when verifying the action of op$_k$ in the sequence (op$_1$, ..., op$_k$). This type of verification key is used in the hybrid garbled programs in our security proof.

An adaptive positional accumulator must satisfy the following properties.

Correctness

Let op$_0$, ..., op$_k$ be any arbitrary sequence of memory operations, and let v_i^* denote the result of the ith memory operation when (op$_0$, ..., op$_{k-1}$) are sequentially executed on an initially empty memory.

Correctness requires that, when sampling

$$\text{PP}, \text{ac}_0, \text{store}_0 \leftarrow \text{SetupAcc}(1^\lambda, S)$$
$$\text{vk} \leftarrow \text{SetupVerify}(\text{PP})$$
$$\text{For } i = 0, \ldots, k:$$
$$\quad \text{store}_{i+1}, \text{ac}_{i+1}, v_i, \pi_i \leftarrow \text{Update}(\text{PP}, \text{store}_i, \text{op}_i)$$
$$\quad b_i \leftarrow \text{Verify}(\text{vk}, \text{ac}_i, \text{op}_i, \text{ac}_{i+1}, v_i, \pi_i)$$

it holds (with probability 1) that for all $j \in \{0, \ldots, k\}$, $v_j = v_j^*$ and $b_j = 1$

Enforcing

Enforcing requires that for all space bounds S, all sequences of operations op$_0$, ..., op$_{k-1}$, when sampling

$$\text{PP}, \text{ac}_0, \text{store}_0 \leftarrow \text{SetupAcc}(1^\lambda, S)$$
$$\text{vk} \leftarrow \text{SetupEnforceVerify}(\text{PP}, (\text{op}_0, \ldots, \text{op}_{k-1}))$$
$$\text{For } i = 0, \ldots, k - 1$$
$$\quad \text{store}_{i+1}, \text{ac}_{i+1}, v_i, \pi_i \leftarrow \text{Update}(\text{PP}, \text{store}_i, \text{op}_i)$$

it holds (with probability 1) that for all accumulators $\hat{\text{ac}}$, all values $\hat{v}$, and all proofs $\hat{\pi}$, if Verify(vk, ac$_{k-1}$, op$_{k-1}$, $\hat{\text{ac}}$, $\hat{v}$, $\hat{\pi}$) = 1, then $(\hat{v}, \hat{\text{ac}}) = (v_{k-1}, \text{ac}_k)$

Indistinguishability of Enforcing Verify

Now we require that the output of SetupVerify(PP) is indistinguish-able from the output of SetupEnforceVerify(PP, (op$_1$, ..., op$_k$)), even when (op$_1$, ..., op$_k$) are chosen adaptively as a function of PP.

More formally, for all p.p.t. $\mathcal{A}_1$ and $\mathcal{A}_2$,

$$
\Pr\left[\mathcal{A}_2(s,\mathsf{vk}_b)=b \;\middle|\;
\begin{array}{l}
\mathsf{PP}, \mathsf{ac}_0, \mathsf{store}_0 \leftarrow \mathsf{SetupAcc}(1^\lambda, S) \\
(\mathsf{op}_0, \ldots, \mathsf{op}_{k-1}), s \leftarrow \mathcal{A}_1(1^\lambda, \mathsf{PP}) \\
\mathsf{vk}_0 \leftarrow \mathsf{SetupVerify}(\mathsf{PP}) \\
\mathsf{vk}_1 \leftarrow \mathsf{SetupEnforceVerify}(\mathsf{PP}, \\
\qquad\qquad (\mathsf{op}_0, \ldots, \mathsf{op}_{k-1})) \\
b \leftarrow \{0,1\}
\end{array}
\right] \leq \frac{1}{2} + \mathrm{negl}(\lambda)
$$

Efficiency

In addition to all the algorithms being polynomial-time, we require that:

–The size of an accumulator is $\mathrm{poly}(\lambda)$.

–The size of proofs is $\mathrm{poly}(\lambda, \log S)$.

–The size of a store is $O(S)$

Theorem 4. *If there is an adaptively puncturable hash function ensemble $\mathcal{H} = \{\mathcal{H}_\lambda\}_{\lambda \in \mathbb{N}}$ with $\mathcal{H}_\lambda = \{H_k : \{0,1\}^{2\lambda} \to \{0,1\}^\lambda\}_{k \in \mathcal{K}_\lambda}$, then there exists an adaptive positional accumulator.*

Proof. We construct an adaptive positional accumulator in which stores are low-depth binary trees, each node of which contains a λ-bit value. The accumulator corresponding to a given store is the value held by the root node. The public parameters for the accumulator consist of an adaptively puncturable hash $h : \{0,1\}^{2\lambda} \to \{0,1\}^\lambda$, and we preserve the invariant that the value in any internal node is equal to the hash h applied to its children's values. It will be convenient for us to assume the existence of a $\bot$, which is represented as a λ-bit string not in the image of h. Without loss of generality, h can be chosen to have such a value.

$\mathsf{Setup}(1^\lambda, S) \to \mathsf{PP}, \mathsf{ac}_0, \mathsf{store}_0$

    Setup samples $h \leftarrow \mathcal{H}_\lambda$, and sets $\mathsf{PP} = h$, $\mathsf{ac}_0 = h(\bot\|\bot)$, and store_0 to be a root node with value $h(\bot\|\bot)$.

$\mathsf{Update}(h, \mathsf{store}, \mathsf{op}) \to \mathsf{store}', \mathsf{ac}', v, \pi$

    Suppose op is $\mathsf{ReadWrite}(\mathsf{addr} \mapsto v')$. There is a unique leaf node in store which is indexed by a prefix of addr. Let v be the value of that leaf, and let π be the values of all siblings on the path from the root to that leaf.

    Update adds a leaf node indexed by the entirety of addr to store if no such node already exists, and sets the value of the leaf to v'. Then Update updates the value of ancestor of that leaf to preserve the invariant.

$\mathsf{SetupVerify}(h) \to \mathsf{vk}$

    For $i = 1, \ldots, \log S$, SetupVerify samples

$$
\mathsf{vk}_i \leftarrow \mathsf{GenVK}(1^\lambda, h)
$$

and sets $\mathsf{vk} = (\mathsf{vk}_1, \ldots, \mathsf{vk}_{\log S})$.

$\mathsf{Verify}((\mathsf{vk}_1, \ldots, \mathsf{vk}_{\log S}), \mathsf{ac}, \mathsf{op}, \mathsf{ac}', v, (w_1, \ldots, w_d)) \to \{0, 1\}$

Define $z_d := v$. Let $b_1 \cdots b_{d'}$ denote the bit representation of the address on which op acts. For $0 \le i < d$, Verify computes

$$z_i = \begin{cases} h(w_{i+1} \| z_{i+1}) & \text{if } b_{i+1} = 1 \\ h(z_{i+1} \| w_{i+1}) & \text{otherwise} \end{cases}$$

For all i such that $b_i = 1$, Verify checks that $\mathsf{vk}_i(w_{i+1} \| z_{i+1}, z_i) = 1$. For all i such that $b_i = 0$, Verify checks that $\mathsf{vk}_i(z_{i+1} \| w_{i+1}, z_i) = 1$. If all these checks pass, then Verify outputs 1; otherwise, Verify outputs 0.

$\mathsf{SetupEnforceVerify}(h, (\mathsf{op}_1, \ldots, \mathsf{op}_k)) \to \mathsf{vk}$

Computes the store_{k-1} which would result from processing $\mathsf{op}_1, \ldots, \mathsf{op}_{k-1}$. Suppose op_k accesses address $\mathsf{addr}_k \in \{0, 1\}^{\log S}$. Then there is a unique leaf node in store_{k-1} which is indexed by a prefix of addr_k; write this prefix as $b_1 \cdots b_d$.

For each $i \in \{1, \ldots, d\}$, define z_i as the value of the node indexed by $b_1 \cdots b_i$, and let w_i denote the value of that node's sibling. If $b_i = 0$, sample

$$\mathsf{vk}_i \leftarrow \mathsf{ForceGenVK}(1^\lambda, h, z_i \| w_i).$$

Otherwise, sample

$$\mathsf{vk}_i \leftarrow \mathsf{ForceGenVK}(1^\lambda, h, w_i \| z_i).$$

For $i \in \{d + 1, \ldots, \log S\}$, just sample $\mathsf{vk}_i \leftarrow \mathsf{GenVK}(1^\lambda, h)$.
Finally we define the total verification key to be $(\mathsf{vk}_1, \ldots, \mathsf{vk}_{\log S})$.

All the requisite properties of this construction are easy to check.

6 Fixed-Transcript Garbling

Next we present the first step in our construction, a garbling scheme that provides adaptive security for RAM programs that have the same transcript. The notion extends the first stage of Canetti-Holmgren scheme into the adaptive setting, and the construction employs the adaptive positional accumulators plus local changes in the other primitives.

We define fixed-transcript security via the following game.

1. The challenger samples $\mathsf{SK} \leftarrow \mathsf{Setup}(1^\lambda, S)$ and $b \leftarrow \{0, 1\}$.
2. The adversary sends a memory configuration s to the challenger. The challenger sends back $\mathsf{GbMem}(\mathsf{SK}, s)$.
3. The adversary repeatedly sends pairs of RAM programs (M_i^0, M_i^1) along with a time bound T_i, and the challenger sends back $\tilde{M}_i^b \leftarrow \mathsf{GbPrg}(\mathsf{SK}, M_i^b, T_i, i)$. Each pair (M_i^0, M_i^1) is chosen adaptively after seeing $\tilde{M}_{i-1}^b$.
4. The adversary outputs a guess b'.

Let $((M_1^0, M_1^1), \ldots, (M_\ell^0, M_\ell^1))$ denote the sequence of pairs of machines output by the adversary. The adversary is said to win if $b' = b$ and:

- Sequentially executing $M_1^0, \ldots, M_\ell^0$ on initial memory configuration s yields the same transcript as executing $M_1^1, \ldots, M_\ell^1$.
- Each M_i^b runs in time at most T_i and space at most S.
- For each i, $|M_i^0| = |M_i^1|$.

Definition 2. *A garbling scheme is* fixed-transcript secure *if for all p.p.t. algorithms $\mathcal{A}$, there is a negligible function* negl *so that $\mathcal{A}$'s probability of winning the game is at most $\frac{1}{2} + \mathrm{negl}(\lambda)$.*

Theorem 5. *Assuming the existence of indistinguishability obfuscation and an adaptive positional accumulator, there is a fixed-transcript secure garbling scheme.*

Proof. Our construction follows closely the fixed-transcript garbling scheme of [9], using our *adaptive* positional accumulator in place of [28]'s positional accumulator. We also rely on puncturable PRFs (PPRFs), splittable signatures and cryptographic iterators defined in the full version.

Setup$(1^\lambda, S) \to$ SK: sample (Acc.PP, ac$_{\mathsf{init}}$, store$_{\mathsf{init}}$) $\leftarrow$ Acc.Setup$(1^\lambda, S)$, a PPRF
  F. Set SK $=$ (Acc.PP, ac$_{\mathsf{init}}$, store$_{\mathsf{init}}$, Itr.PP, itr$_{\mathsf{init}}$, F), and (Itr.PP, itr$_{\mathsf{init}}$) $\leftarrow$ Itr.Setup(1^λ).

GbMem$(\mathsf{SK}, s) \to \tilde{s}$: GbMem updates the APA (ac$_{\mathsf{init}}$, store$_{\mathsf{init}}$) to set the underlying memory to s (via a sequence of calls to Update) and let ac$_0$, store$_0$ denote the result. It then generates (sk, vk) $\leftarrow$ Spl.Setup$(1^\lambda; F(1,0))$, where $(1,0)$ represents the initial index number i and initial time-step number 0^2. Finally, GbMem computes $\sigma_0 \leftarrow$ Spl.Sign(sk, $(\bot, \bot, \mathsf{ac}_0, \mathsf{ReadWrite}(0 \mapsto 0)))$. Here the first $\bot$ represents an initial local state q_0 for M_1, and the second $\bot$ represents an initial iterator value itr$_0$. GbMem outputs $\tilde{s} = (\sigma_0, \mathsf{ac}_0, \mathsf{store}_0)$.

GbPrg$(\mathsf{SK}, M_i, T_i, i) \to \tilde{M}_i$: GbPrg first transforms M_i so that its initial state is $\bot$. Note this can be done without loss of generality by hard-coding the "real" initial state in the transition function. GbPrg then computes $\tilde{C}_i \leftarrow i\mathcal{O}(C_i)$, where C_i is described in Algorithm 1.
  Finally, GbPrg defines and outputs a RAM machine $\tilde{M}_i$, which has $\tilde{C}_i$ hardcoded as part of its transition function, such that $\tilde{M}_i$ does the following:
  1. Reads (ac$_0, \sigma_0$) from memory. Define op$_0 = \mathsf{ReadWrite}(0 \mapsto 0)$, $q_0 = \bot$, and itr$_0 = \bot$.
  2. For $t = 0, 1, 2, \ldots$:
     (a) Compute store$_{t+1}$, ac$_{t+1}, v_t, \pi_t \leftarrow$ Acc.Update(Acc.PP, store$_t$, op$_t$).
     (b) Compute out$_t \leftarrow \tilde{C}_i(t, q_t, \mathsf{itr}_t, \mathsf{ac}_t, \mathsf{op}_t, \sigma_t, v_t, \mathsf{ac}_{t+1}, \pi_t)$.
     (c) If out$_t$ parses as (y, σ), then write (ac$_{t+1}, \sigma$) to memory, output y, and terminate.
     (d) Otherwise, parse out$_t$ as $(q_{t+1}, \mathsf{itr}_{t+1}, \mathsf{ac}_{t+1}, \mathsf{op}_{t+1}), \sigma_{t+1}$ or terminate if out$_t$ is not of this form.

---

^{2} Looking ahead, all the intermediate (sk, vk) key pairs are generated by applying F to the (index, time-step) tuple.

Input: Time t, state q, iterator itr, accumulator ac, operation op, signature σ,
  memory value v, new accumulator ac$'$, proof π
Data: Puncturable PRF F, RAM machine M_i with transition function δ_i,
  Accumulator verification key vk$_{\text{Acc}}$, index i, iterator public parameters
  Itr.PP, time bound T_i

1 $(\text{sk}, \text{vk}) \leftarrow \text{Spl.Setup}(1^\lambda; \text{F}(i, t))$;
2 **if** $t > T_i$ *or* $\text{Spl.Verify}(\text{vk}, (q, \text{itr}, \text{ac}, \text{op}), \sigma) = 0$ *or*
   $\text{Acc.Verify}(\text{vk}_{\text{Acc}}, \text{ac}, \text{op}, \text{ac}', v, \pi) = 0$ **then return** $\perp$;
3 $\text{out} \leftarrow \delta_i(q, v)$;
4 **if** out $\in Y$ **then**
5 $\quad$ $(\text{sk}', \text{vk}') \leftarrow \text{Spl.Setup}(1^\lambda; \text{F}(i+1, 0))$;
6 $\quad$ **return** out, $\text{Sign}(\text{sk}', (\perp, \perp, \text{ac}', \text{ReadWrite}(0 \mapsto 0))$
7 **else**
8 $\quad$ Parse out as (q', op');
9 $\quad$ $\text{itr}' \leftarrow \text{Itr.Iterate}(\text{Itr.PP}, (q, \text{itr}, \text{ac}, \text{op}))$;
10 $\quad$ $(\text{sk}', \text{vk}') \leftarrow \text{Spl.Setup}(1^\lambda; \text{F}(i, t+1))$;
11 $\quad$ **return** $(q', \text{itr}', \text{ac}', \text{op}'), \text{Sign}(\text{sk}', (q', \text{itr}', \text{ac}', \text{op}'))$

Algorithm 1: Transition function for M_i, with memory verified by a signed accumulator.

We note that GbPrg can efficiently produce $\tilde{M}_i$ from $\tilde{C}_i$ and Acc.PP. This means that later, when we prove security, it will suffice to analyze a game in which the adversary receives $\tilde{C}_i$ instead of $\tilde{M}_i$.

Eval$(\tilde{M}, \tilde{s})$ The evaluation algorithm runs $\tilde{M}$ on the garbled memory $\tilde{s}$, and outputs $\tilde{M}(\tilde{s})$.

Correctness and efficiency are easy to verify. For the proof of security we refer the readers to the full version.

7   Fixed-Access Garbling

Fixed-access security is defined in the same way as fixed-transcript security, but the left and right machines produced by $\mathcal{A}$ do not need to have the same transcripts for $\mathcal{A}$ to win - they may not have the same intermediate states, but only need to perform the same memory operations.

Definition 3 (Fixed-access security)
We define fixed-access security via the following game.

1. *The challenger samples* $SK \leftarrow \text{Setup}(1^\lambda, S)$ *and* $b \leftarrow \{0, 1\}$.
3. *The adversary sends a memory configuration* s *to the challenger. The challenger sends back* $\text{GbMem}(SK, s)$.
4. *The adversary repeatedly sends pairs of RAM programs* (M_i^0, M_i^1) *to the challenger, together with a time bound* 1^{T_i}, *and the challenger sends back* $\tilde{M}_i^b \leftarrow \text{GbPrg}(SK, M_i^b, T_i, i)$. *Each pair* (M_i^0, M_i^1) *is chosen adaptively after seeing* $\tilde{M}_{i-1}^b$.
5. *The adversary outputs a guess* b'.

Let $((M_1^0, M_1^1), \ldots, (M_\ell^0, M_\ell^1))$ denote the sequence of pairs of machines output by the adversary. The adversary is said to win if $b' = b$ and:

- Sequentially executing $M_1^0, \ldots, M_\ell^0$ on initial memory configuration s yields the same transcript as executing $M_1^1, \ldots, M_\ell^1$, except that the local states can be different.
- Each M_i^b runs in time at most T_i and space at most S.

A garbling scheme is said to have fixed-access security if all p.p.t. adversaries $\mathcal{A}$ win in the game above with probability less than $1/2 + \text{negl}(\lambda)$.

To achieve fixed-access security, we adapt the exact same technique from [9]: xoring the state with a pseudorandom function applied on the local time t. The PRF keys used in different machines are sampled independently.

Theorem 6. If there is a fixed-transcript garbling scheme, then there is a fixed-access garbling scheme.

Proof. From a fixed-transcript garbling scheme (Setup', GbMem', GbPrg', Eval'), we construct a fixed-access garbling scheme (Setup, GbMem, GbPrg, Eval).

Setup($1^\lambda, S$) samples $SK' \leftarrow$ Setup'($1^\lambda, S$), sets it as SK.
GbMem(SK, s) outputs $\tilde{s}' \leftarrow$ GbMem'(SK', s).
GbPrg(SK, M_i, T_i, i) samples a PPRF F_i, outputs $\tilde{M}_i' \leftarrow$ GbPrg'(SK', M_i', T_i, i), where M_i' is defined as in Algorithm 2. If M_i's initial state is q_0, the initial state of M_i' is $(0, q_0 \oplus F_i(0))$.
Eval($\tilde{M}, \tilde{s}$) outputs Eval'($\tilde{M}', \tilde{s}'$).

Input: State (t, c_q), memory symbol σ
Data: RAM machine M_i, puncturable PRF F_i
1 $q \leftarrow c_q \oplus F_i(t)$;
2 out $\leftarrow M_i(q, \sigma)$;
3 if out $\in Y$ then return out;
4 Parse out as (q', op);
5 return $((t+1, q' \oplus F_i(t+1)), \text{op})$;

Algorithm 2: M_i', the modified version of M_i which encrypts its state.

For the proof that this construction satisfies the requisite security, we refer the readers to the full version.

8 Fixed-Address Garbling

Fixed-address security is defined in the same way as fixed-access security, but the left and right machines produced by $\mathcal{A}$ do not need to make the same memory operations for $\mathcal{A}$ to win - their memory operations only need to access the same addresses. Additionally, the adversary $\mathcal{A}$ now provides not only a single memory configuration s_0, but two memory configurations s_0^0 and s_0^1. The challenger returns GbMem(SK, s_0^b). In keeping with the spirit of fixed-address garbling, we require s_0^0 and s_0^1 to have the same set of addresses storing non-ϵ values.

Definition 4 (Fixed-address security). *We define fixed-address security via the following game.*

1. *The challenger samples $SK \leftarrow \mathsf{Setup}(1^\lambda, S)$ and $b \leftarrow \{0,1\}$.*
2. *The adversary sends the initial memory configurations s_0^0, s_0^1 to the challenger. The challenger sends back $\tilde{s}_0^b \leftarrow \mathsf{GbMem}(SK, s_0^b)$.*
3. *The adversary repeatedly sends pairs of RAM programs (M_i^0, M_i^1) to the challenger, together with a time bound 1^{T_i}, and the challenger sends back $\tilde{M}_i^b \leftarrow \mathsf{GbPrg}(SK, M_i^b, T_i, i)$. Each pair (M_i^0, M_i^1) is chosen adaptively after seeing $\tilde{M}_{i-1}^b$.*
4. *The adversary outputs a guess b'.*

Let $((s_0^0, s_0^1), (M_1^0, M_1^1), \ldots, (M_\ell^0, M_\ell^1))$ denote the sequence of pairs of memory configurations and machines output by the adversary. The adversary is said to win if $b' = b$ and:

- *$\{a : s_0^0(a) \neq \epsilon\} = \{a : s_0^1(a) \neq \epsilon\}$.*
- *The sequence of addresses accessed and the outputs during the sequential execution of $M_1^0, \ldots, M_\ell^0$ on initial memory configuration s_0^0 are the same as when executing $M_1^1, \ldots, M_\ell^1$ on s_0^1.*
- *Each M_i^b runs in time at most T_i and space at most S.*
- *For each i, $|M_i^0| = |M_i^1|$.*

A garbling scheme is said to have fixed-address security if all p.p.t. adversaries $\mathcal{A}$ win in the game above with probability less than $1/2 + \mathrm{negl}(\lambda)$.

Our construction of fixed-address garbling is almost the same with the two-track solution in [9], with a slight modification at the way to "encrypt" the memory configuration. In [9], the memory configurations are xored with different puncturable PRF values in the two tracks, where the PRFs are applied on the time t and address a. In this work, the PRFs are applied on the execution index i and time t, not on the address a. This is enough for our purpose, because in each execution index i and step t, the machine only writes on a single address (for the initial memory configuration, the index is assigned as 0, and different timestamps will be assigned on different addresses). By this modification, we are able to prove adaptive security based on selective secure puncturable PRF, and adaptively secure fixed-access garbling.

We note that, even if the address a is included in the domain of PRF, as in [9], the construction is still adaptively secure if the underlying PRF is based on GGM's tree construction. Here we choose to present the simplified version which suffices for our purpose.

Construction 7. Suppose $(\mathsf{Setup'}, \mathsf{GbMem'}, \mathsf{GbPrg'}, \mathsf{Eval'})$ is a fixed-access garbling scheme, we construct a fixed-address garbling scheme $(\mathsf{Setup}, \mathsf{GbMem}, \mathsf{GbPrg}, \mathsf{Eval})$:

$\mathsf{Setup}(1^\lambda, S)$ samples $SK' \leftarrow \mathsf{Setup'}(1^\lambda, S)$ and puncturable PRFs F_A and F_B.

GbMem(SK, s) outputs $\tilde{s}_0' \leftarrow$ GbMem$'(SK', s_0')$, where

$$s_0'(a) = \begin{cases} (0, -a, F_A(0, -a) \oplus s_0(a), F_B(0, -a) \oplus s_0(a)) & \text{if } s_0(a) \neq \epsilon \\ \epsilon & \text{otherwise} \end{cases}$$

GbPrg(SK, M_i, T_i, i) outputs $\tilde{M}_i' \leftarrow$ GbPrg$'(SK', M_i', T_i, i)$, where M_i' is defined as in Algorithm 3. If the initial state of M_i was q_0, the initial state of M_i' is $(0, q_0, q_0)$.
Eval$(\tilde{M}, \tilde{s})$ outputs Eval$'(\tilde{M}', \tilde{s}_0')$.

Input: State (t_q, q_A, q_B), memory symbol $(i_{in}, t_{in}, c_A, c_B)$
Data: RAM machine M_i, puncturable PRFs F_A, F_B
1 out $\leftarrow M_i(q_A, F_A(i_{in}, t_{in}) \oplus c_A)$;
2 **if** out $\in Y$ **then return** out;
3 Parse out as $(q', \text{ReadWrite}(\text{addr}' \mapsto v'))$;
4 op$' := \text{ReadWrite}(\text{addr}' \mapsto (i, t_q, F_A(i, t_q) \oplus v', F_B(i, t_q) \oplus v'))$;
5 **return** $(t_q + 1, q', q'), \text{op}'$;

Algorithm 3: M_i': Modified version of M_i which encrypts its memory twice in parallel.

Theorem 8. *If* (Setup$'$, GbMem$'$, GbPrg$'$) *is a fixed-access garbling scheme, then Construction 7 is a fixed-address garbling scheme.*

Proof. For the proof of security we refer the readers to the full version.

9   Full Garbling

In order to construct a fully secure garbling scheme, we will need to make use of an oblivious RAM (ORAM) [19] to hide the addresses accessed by the machine.

9.1   Oblivious RAMs with Strong Localized Randomness

We require that the ORAM has a strong localized randomness property[3], which is satisfied by the ORAM construction of [13]. Below we give a brief definition of ORAM and the property we need.

An ORAM is a probabilistic scheme for memory storage and access that provides obliviousness for access patterns with sublinear access complexity. It is convenient for us to model an ORAM scheme as follows. We define a deterministic algorithm OProg so that for a security parameter 1^λ, a memory operation op, and a space bound S, OProg$(1^\lambda, \text{op}, S)$ outputs a probabilistic RAM machine M_{op}. More generally, for a RAM machine M, we can define OProg$(1^\lambda, M, S)$ as the (probabilistic) machine which executes OProg$(1^\lambda, \text{op}, S)$ for every operation op output by M.

---

[3] This notion is similar but stronger to the "localized randomness" defined in [9].

We also define OMem, a procedure for making a memory configuration oblivious, in terms of OProg, as follows: Given a memory configuration s with n non-empty addresses $a_1, \ldots, a_n$, all less than or equal to a space bound S, OMem$(1^\lambda, s, S)$ iteratively samples

$$s_0' \leftarrow \epsilon^{\mathbb{N}}$$

and

$$s_i' = \mathsf{NextMem}(\mathsf{OProg}(1^\lambda, \mathsf{ReadWrite}(a_i \mapsto s(a_i)), S), s_{i-1}')$$

and outputs s_n'.

Security (Strong Localized Randomness). Informally, we consider obliviously executing operations $\mathsf{op}_1, \ldots, \mathsf{op}_t$ on a memory of size S, i.e. executing machines $M_{\mathsf{op}_1}; \ldots; M_{\mathsf{op}_t}$ using a random tape $R \in \{0,1\}^{\mathbb{N}}$. This yields a sequence of addresses $\boldsymbol{A} = \boldsymbol{a}_1 \| \cdots \| \boldsymbol{a}_t$. There should be a natural way to decompose each $\boldsymbol{a}_i$ (in the Chung-Pass ORAM, we consider each recursive level of the construction) such that we can write $\boldsymbol{a}_i = \boldsymbol{a}_{i,1} \| \cdots \| \boldsymbol{a}_{i,m}$. Our notion of strong localized randomness requires that (after having fixed $\mathsf{op}_1, \ldots, \mathsf{op}_t$), each $\boldsymbol{a}_{i,j}$ depends on some small substring of R, which does not influence any other $\boldsymbol{a}_{i',j'}$. In other words:

- There is some $\alpha_{i,j}, \beta_{i,j} \in \mathbb{N}$ such that $0 < \beta_{i,j} - \alpha_{i,j} \leq \mathrm{poly}(\log S)$ and such that $\boldsymbol{a}_{i,j}$ is a function of $R_{\alpha_{i,j}}, \ldots, R_{\beta_{i,j}}$.
- The collection of intervals $[\alpha_{i,j}, \beta_{i,j}]$ for $i \in \{1, \ldots, t\}$, $j \in \{1, \ldots, m\}$ is pairwise disjoint.

Formally, we say that an ORAM with multiplicative time overhead η has strong localized randomness if:

- For all λ and S, there exists m and $\tau_1 < \tau_2 < \cdots < \tau_m$ with $\tau_1 = 1$ and $\tau_m = \eta(S, \lambda) + 1$, and there exist circuits $C_1, \ldots, C_m$, such that for all memory operations $\mathsf{op}_1, \ldots, \mathsf{op}_t$, there exist pairwise disjoint intervals $I_1, \ldots, I_m \subset \mathbb{N}$ such that:
  - If we write
  $$\boldsymbol{A}_1 \| \cdots \| \boldsymbol{A}_t \leftarrow \mathsf{addr}(M_{\mathsf{op}_1}^{R_1}; \ldots; M_{\mathsf{op}_t}^{R_t}, \epsilon^{\mathbb{N}})$$

    where $R = R_1 \| \cdots \| R_t$ denotes the randomness used by the oblivious accesses and each $\boldsymbol{A}_i$ denotes the addresses accessed by $M_{\mathsf{op}_i}^{R_i}$, then $(\boldsymbol{A}_t)_{[\tau_j, \tau_{j+1})} = C_j(R_{I_j})$ with high probability over R. Here R_{I_j} denotes the contiguous substring of R indexed by the interval $I_j \subset [\|R\|]$.
  - With high probability over the choice of $R_{\mathbb{N} \setminus I_j}$, $\boldsymbol{A}_1, \ldots, \boldsymbol{A}_{t-1}$ does not depend on R_{I_j} as a function.
- τ_j and the circuits C_j are computable in polynomial time given 1^λ, S, and j.
- I_j is computable in polynomial time given 1^λ, S, $\mathsf{op}_1, \ldots, \mathsf{op}_t$, and j.

A full exposure, including the full definition and proof that Chung-Pass ORAM satisfies the strong localized randomness property can be found in the full version.

9.2 Full Garbling Construction

Theorem 9. *If there is an efficient fixed-address garbling scheme, then there is an efficient full garbling scheme.*

Proof. Given a fixed-address garbling scheme $(\mathsf{Setup}', \mathsf{GbMem}', \mathsf{GbPrg}', \mathsf{Eval}')$ and an oblivious RAM OProg with space overhead ζ and time overhead η. We construct a full garbling scheme $(\mathsf{Setup}, \mathsf{GbMem}, \mathsf{GbPrg}, \mathsf{Eval})$.

$\mathsf{Setup}(1^\lambda, T, S)$ samples $SK' \leftarrow \mathsf{Setup}'(1^\lambda, \eta(S, \lambda) \cdot T, \zeta(S, \lambda) \cdot S)$ and samples a PPRF $F : \{0,1\}^\lambda \times \{0,1\}^\lambda \to \{0,1\}^{\ell_R}$, where ℓ_R is the length of randomness needed to obliviously execute one memory operation. We will sometimes think of the domain of F as $[2^{2\lambda}]$.

$\mathsf{GbMem}(SK, s_0)$ outputs $\tilde{s}'_0 \leftarrow \mathsf{GbMem}'(SK', \mathsf{OMem}(1^\lambda, s_0, S))$.

$\mathsf{GbPrg}(SK, M_i, i)$ outputs $\tilde{M}'_i \leftarrow \mathsf{GbPrg}'(SK', \mathsf{OProg}(1^\lambda, M_i, S)^{\mathsf{F}(i, \cdot)}, i)$.

$\mathsf{Eval}(\tilde{M}, \tilde{s})$ outputs $\mathsf{Eval}'(\tilde{M}', \tilde{s}'_0)$.

Simulator To show security of this construction, we define the following simulator.

1. The adversary provides S, and an initial memory configuration s_0. Say that s_0 has n non-ϵ addresses. The simulator is given S and n, and samples $SK' \leftarrow \mathsf{Setup}'(1^\lambda, \zeta(S, \lambda) \cdot S)$ and sends $\mathsf{GbMem}'(SK', \mathsf{OMem}(1^\lambda, 0^n, S))$ to the adversary.
2. When the adversary makes a query $M_i, 1^{T_i}$, the simulator is given $y_i = M_i(s_{i-1})$ and $t_i = \mathsf{Time}(M_i, s_{i-1})$, where $s_i = \mathsf{NextMem}(M_i, s_{i-1})$, and outputs $\mathsf{GbPrg}'(SK', D_i, \eta(S, \lambda) \cdot T_i, i)$, where D_i is a "dummy program". As described in Algorithm 4, D_i independently samples addresses to access for t_i steps, and then outputs y_i.

Data: Underlying running time t_i, output value y_i, PPRF G_i, circuits
      $C_1, \ldots, C_m$ guaranteed by localized randomness

1 **for** $t = 1, \ldots, t_i$ **do**
2   **for** $k = 1, \ldots, m$ **do**
3         $r_k \leftarrow G_i(t, k)$;
4   Access addresses given by $C_k(r_k)$

5 **return** y_i.

Algorithm 4: Pseudocode for a dummy RAM machine which simulates pseudorandom addresses to access using the circuits $C_1, \ldots, C_m$ given in the definition of localized randomness, and then outputs y_i.

We refer the readers to the full version for the proof.

10   Database Delegation

We define security for the task of delegating a database to an untrusted server. Here we have a database owner that wishes to keep the database on a remote server. Over time, the owner wishes to update the database and query it. Furthermore, the owner wishes to enable other parties to do so as well, perhaps under some restrictions. Informally, the security requirements from the scheme are:

Verifiability: The data owner should be able to verify the correctness of the answers to its queries, relative to the up-to-date state of the database following all the updates made so far.

Secrecy of database and queries: For queries made by the database owner and honest third parties, the adversary does not learn anything other than the size of the database, the sizes and runtimes of the queries, and the sizes of the answers. This holds even if the answers to the queries become partially or fully known by other means.

For queries made by adversarially controlled third parties, the adversary learns in addition only the answers to the queries.

(We stress that the secrecy requirement for the case of a corrupted third party is incomparable to the secrecy requirement in the case of an honest third party. In particular, the case of corrupted third parties guarantees secrecy even when the entire evaluation and verification processes are completely exposed.)

More precisely, a database delegation scheme (or, protocol) consists of the following algorithms:

DBDelegate: Initial delegation of the database. Takes as input a plain database, and outputs an encrypted database (to be sent to the server), public verification key vk and private master key msk to be kept secret.

Query: Delegation of a query or database update. Takes a RAM program and the master secret key msk, and outputs a delegated program to be sent to the server and a secret key sk_{enc} that allows recovering the result of the evaluation from the returned response.

Eval: Evaluation of a query or update. Takes a delegated database $\tilde{D}$ and a delegated program $\tilde{M}$, runs $\tilde{M}$ on $\tilde{D}$. Returns a response value a and an updated database $\tilde{D}'$.

AnsDecVer: Local processing of the server's answer. Takes the public verification key vk, the private decryption key sk_{enc} and outputs either an answer value or $\perp$.

Security. The security requirement from a database delegation scheme $\mathcal{S} = $ (DBDelegate, Query, Eval, AnsDecVer) is that it UC-realize the *database delegation ideal functionality* $\mathcal{F}_{dd}$ defined as follows. (For simplicity we assume that the database owner is uncorrupted, and that the communication channels are authenticated.)

1. When activated for the first time, $\mathcal{F}_{dd}$ expects to obtain from the activating party (the database owner) a database D. It then records D and discloses $\|D\|_0$ to the adversary.
2. In each subsequent activation by the owner, that specifies a program M and party P, run M on D, obtain an answer a and a modified database D', store D' and disclose $|M|$, the running time of M, and the length of a to the adversary. If the adversary returns ok then output (M, a) to P.

To make the requirements imposed by $\mathcal{F}_{dd}$ more explicit, we also provide an alternative (and equivalent) formulation of the definition in terms of a distinguishability game. Specifically, we require that there exists a simulator Sim such that no adversary (environment) $\mathcal{A}$ will be able to distinguish whether it is interacting with the real or the ideal games as described here:

Real game $REAL_{\mathcal{A}}(1^\lambda)$:

1. $\mathcal{A}$ provides a database D, receives the public outputs of DBDelegate(D).
2. $\mathcal{A}$ repeatedly provides a program M_i and a bit that indicates either *honest* or *dishonest*. In response, Query is run to obtain sk^i_{enc} and $\tilde{M}_i$. $\mathcal{A}$ obtains $\tilde{M}_i$, and in the dishonest case also the decryption key sk^i_{enc}.
3. In the honest case $\mathcal{A}$ provides the server's output out_i for the execution of M_i, and obtains in response the result of AnsDecVer($\mathsf{vk}, \mathsf{sk}_{enc}, \mathsf{out}_i$).

Ideal game $IDEAL_{\mathcal{A}}(1^\lambda)$:

1. $\mathcal{A}$ provides a database D, receives the output of Sim($\|D\|_0$).
2. $\mathcal{A}$ repeatedly provides a program M_i and either *honest* or *dishonest*. In response, M_i runs on the current state of the database D to obtain output a and modified database D'. D' is stored instead of D. In the case of dishonest, $\mathcal{A}$ obtains Sim(a, s, t), where s is the description size of M and t is the runtime of M. In the case of honest, $\mathcal{A}$ obtains Sim(s, t).
3. In the honest case $\mathcal{A}$ provides the server's output out_i for the execution of M_i, and obtains in response Sim(out_i), where here Sim(out_i) can take one out of only two values: either a or $\perp$.

Definition 5. *A delegation scheme $\mathcal{S} = (\mathsf{DBDelegate}, \mathsf{Query}, \mathsf{Eval}, \mathsf{AnsDecVer})$ is secure if it UC-realizes $\mathcal{F}_{dd}$. Equivalently, it is secure if there exists a simulator Sim such that no $\mathcal{A}$ can guess with non-negligible advantage whether it is interacting in the real interaction with $\mathcal{S}$ or in the ideal interaction with Sim.*

Theorem 10. *If there exist adaptive succinct garbled RAMs with persistent memory, unforgeable signature schemes and symmetric encryption schemes with pseudorandom ciphertexts, then there exist secure database delegation schemes with succinct queries and efficient delegation, query preparation, query evaluation, and response verification.*

Proof. Let (Setup, GbMem, GbPrg, Eval) be an adaptively secure garbling scheme for RAM with persistent memory. We construct a database delegation scheme as follows:

DBDelegate(1^λ): Run $SK \leftarrow$ Setup($1^\lambda, D$) and $\tilde{D} \leftarrow$ GbMem($SK, D, |D|$). Generate signing and verification keys ($\mathsf{vk_{sign}}, \mathsf{sk_{sign}}$) for the signature scheme. Set msk $\leftarrow (SK, \mathsf{sk_{sign}})$ and vk $\leftarrow \mathsf{vk_{sign}}$.

Query(M_i, msk, pk): Generate a symmetric encryption key $\mathsf{sk_{enc}}$. Generate the extended version of M_i' of M_i as in Algorithm 5.

Output $\tilde{M} \leftarrow$ GbPrg($SK, M_i'[\mathsf{sk_{sign}}, \mathsf{sk_{enc}}], i$)

Input: State q, memory value v
Data: RAM program M_i with transition function δ_i and output space Y, and signing and encryption keys $\mathsf{sk_{sign}}, \mathsf{sk_{enc}}$

1 out $\leftarrow \delta_i(q, v)$;
2 if out $\in Y$ then
3      $\mathsf{ct_{out}} \leftarrow$ Enc($\mathsf{sk_{enc}}$, out)
4      $\sigma_{out} \leftarrow$ Sign($\mathsf{sk_{sign}}, \mathsf{ct_{out}} \| i$)
5   **return** ($\mathsf{ct_{out}}, \sigma_{out}$);
6 **return** out

Algorithm 5: M_i': modified version of M_i which encrypts and signs its final output

Eval: Run $\tilde{M}$ on $\tilde{D}$ and return the output value a and an updated database $\tilde{D}'$.

AnsDecVer(i, out, vk, sk): Parse out $= (\mathsf{ct}, \sigma)$. If Verify(vk, $\mathsf{ct} \| i, \sigma) \neq 1$, output $\perp$. Else output Dec(sk, ct).

We construct a simulator Sim for the delegation scheme as follows:

- DBDelegate: Sim generates signing and verifications keys $\mathsf{sk_{sign}}, \mathsf{vk_{sign}}$. Sim runs the simulator $\mathsf{Sim_{GRAM}}$ for a GRAM scheme to obtain a simulated garbled database $\tilde{D}$. It provides $\tilde{D}$ and $\mathsf{vk_{sign}}$ as output to the adversary $\mathcal{A}$.
- Query: If Sim is executed with inputs (a, s, t) on the i-th iteration, it generates symmetric encryption key $\mathsf{sk_{enc}}$. It computes $\mathsf{ct} =$ Enc($\mathsf{sk_{enc}}, a$), $\sigma \leftarrow$ Sign($\mathsf{sk_{sign}}, \mathsf{ct} \| i$) and runs the simulator $\mathsf{Sim_{GRAM}}$ with inputs ($\mathsf{ct} \| i, \sigma$) to obtain simulated garbled RAM $\tilde{M}_i$. It returns $\tilde{M}_i$ and $\mathsf{sk_{enc}}$ to $\mathcal{A}$.

  If Sim is executed with inputs (s, t) on the i-th iteration, it generates a random value ct, computes $\sigma \leftarrow$ Sign($\mathsf{sk_{sign}}, \mathsf{ct} \| i$) and runs the simulator $\mathsf{Sim_{GRAM}}$ with inputs ($\mathsf{ct} \| i, \sigma$) to obtain simulated garbled RAM $\tilde{M}_i$. It returns $\tilde{M}_i$ to $\mathcal{A}$.
- AnsDecVer: If Sim executes on input out_i then it outputs AnsDecVer(vk, $\mathsf{sk_{enc}}$, out_i).

To show validity of Sim, we construct the following hybrids.

$\mathbf{H_0}$: This is the real world execution.

$\mathbf{H_1}$: In this hybrid we start using the simulator for the GRAM $\mathsf{Sim_{GRAM}}$ to generate simulated database $\tilde{D}'$. We generate the signature scheme keys ($\mathsf{vk_{sign}}, \mathsf{sk_{sign}}$) honestly. We also use $\mathsf{Sim_{GRAM}}$ to generate the garbling for the programs M_i' given inputs $\mathsf{ct}_i \leftarrow$ Enc($\mathsf{pk_{enc}}$, out)$\| i$, $\sigma_i \leftarrow$ Sign($\mathsf{sk_{sign}}, \mathsf{ct}_i$) and out is the result of the evaluation of M_i with the memory state after the previous $i - 1$ evaluations.

The indistinguishability of $\mathbf{H_0}$ and $\mathbf{H_1}$ follows from the simulation security of the GRAM scheme.

$\mathbf{H_2}$: In this hybrid for all honest executions for machines M_i where the adversary $\mathcal{A}$ does not get $\mathsf{sk_{enc}}$, we run $\mathsf{Sim_{GRAM}}$ to generate the garbling for the programs M_i' with inputs $\mathsf{ct}_i \leftarrow r$, where r is a random value, and $\sigma_i \leftarrow \mathsf{Sign}(\mathsf{sk_{sign}}, \mathsf{ct}_i \| i)$.

The indistinguishability of $\mathbf{H_1}$ and $\mathbf{H_2}$ follows from the pseudorandom property of symmetric encryption ciphertexts.

Now, consider the event where, in execution $\mathbf{H_2}$, the adversary provides a value out_i such that $\mathsf{AnsDecVer}(\mathsf{vk}, \mathsf{sk_{enc}}, \mathsf{out}_i) = a'$ and $a \neq a' \neq \bot$, where a is the correct answer for the i-th query in this execution. We argue that:

- Conditioned on this event not happening, $\mathcal{A}$'s view of $\mathbf{H_2}$ is identical to its view in the ideal interaction.
- The event happens with at most negligible probability. Otherwise $\mathcal{A}$ can be used to break the unforgeability of the signature scheme. To see this consider an interaction between $\mathcal{A}$ and Sim that is the same as $\mathbf{H_2}$ except that Sim queries the signature scheme challenger $\mathcal{C}$ to obtain verification key $\mathsf{vk_{sign}}$ and signatures σ_i for the values ct_i. Then out_i, which $\mathcal{A}$ returns, contains a signature of a message that Sim has not queried. Hence, Sim breaks the unforgeability property of the signature scheme.

Acknowledgments. We would like to thank Oxana Poburinnaya. Although she preferred not to co-author this paper, her constructive suggestions and criticisms played an essential role throughout the creation of this work.

This work is supported by US NSF grants 1413920, 1218461, 1012798, 1012910, 1421102, 1562888, 1565208, 1633282, ISF grant 1523/14, and DARPA SafeWare W911NF-15-C-0236. Part of the research by Y.C. was conducted while at SRI funded by the NSF grant 1421102.

References

1. Ananth, P., Chen, Y.-C., Chung, K.-M., Lin, H., Lin, W.-K.: Delegating ram computations with adaptive soundness and privacy. Cryptology ePrint Archive, Report 2015/1082 (2015)
2. Ananth, P., Sahai, A.: Functional encryption for turing machines. IACR Cryptology ePrint Archive 2015, p. 776 (2015)
3. Bellare, M., Hoang, V.T., Rogaway, P.: Adaptively secure garbling with applications to one-time programs and secure outsourcing. In: Wang, X., Sako, K. (eds.) ASIACRYPT 2012. LNCS, vol. 7658, pp. 134–153. Springer, Heidelberg (2012). doi:10.1007/978-3-642-34961-4_10
4. Bitansky, N., Canetti, R., Chiesa, A., Goldwasser, S., Lin, H., Rubinstein, A., Tromer, E.: The hunting of the SNARK. IACR Cryptology ePrint Archive 2014, p. 580 (2014)
5. Bitansky, N., Garg, S., Lin, H., Pass, R., Telang, S.: Succinct randomized encodings and their applications. In: Rubinfeld, R. (ed.) Symposium on the Theory of Computing (STOC) (2015)
6. Bösch, C.T., Hartel, P.H., Jonker, W., Peter, A.: A survey of provably secure searchable encryption. ACM Comput. Surv. **47**(2), 18:1–18:51 (2014)

7. Boyle, E., Chung, K.-M., Pass, R.: On extractability obfuscation. In: Lindell, Y. (ed.) TCC 2014. LNCS, vol. 8349, pp. 52–73. Springer, Heidelberg (2014). doi:10. 1007/978-3-642-54242-8_3

8. Canetti, R., Chen, Y., Holmgren, J., Raykova, M.: Succinct adaptive garbled ram. Cryptology ePrint Archive, Report 2015/1074 (2015)

9. Canetti, R., Holmgren, J.: Fully succinct garbled ram. In: ITCS (2016)

10. Canetti, R., Holmgren, J., Jain, A., Vaikuntanathan, V.: Indistinguishability obfuscation of iterated circuits and ram programs. Cryptology ePrint Archive, Report 2014/769 (2014)

11. Chen, Y.-C., Chow, S.S., Chung, K.-M., Lai, R.W., Lin, W.-K., Zhou, H.-S.: Computation-trace indistinguishability obfuscation and its applications. IACR Cryptology ePrint Archive (2015)

12. Chung, K.-M., Kalai, Y., Vadhan, S.: Improved delegation of computation using fully homomorphic encryption. In: Rabin, T. (ed.) CRYPTO 2010. LNCS, vol. 6223, pp. 483–501. Springer, Heidelberg (2010). doi:10.1007/978-3-642-14623-7_26

13. Chung, K.-M., Pass, R.: A simple ORAM. IACR Cryptology ePrint Archive 2013, p. 243 (2013)

14. Damgård, I.B.: Collision free hash functions and public key signature schemes. In: Chaum, D., Price, W.L. (eds.) EUROCRYPT 1987. LNCS, vol. 304, pp. 203–216. Springer, Heidelberg (1988). doi:10.1007/3-540-39118-5_19

15. Garg, S., Steve, L., Ostrovsky, R., Scafuro, A.: Garbled ram from one-way functions. In: STOC (2015)

16. Gennaro, R., Gentry, C., Parno, B.: Non-interactive verifiable computing: outsourcing computation to untrusted workers. In: Rabin, T. (ed.) CRYPTO 2010. LNCS, vol. 6223, pp. 465–482. Springer, Heidelberg (2010). doi:10.1007/ 978-3-642-14623-7_25

17. Gentry, C., Halevi, S., Lu, S., Ostrovsky, R., Raykova, M., Wichs, D.: Garbled RAM revisited. In: Nguyen, P.Q., Oswald, E. (eds.) EUROCRYPT 2014. LNCS, vol. 8441, pp. 405–422. Springer, Heidelberg (2014). doi:10.1007/978-3-642-55220-5_23

18. Gentry, C., Halevi, S., Raykova, M., Wichs, D.: Outsourcing private ram computation. In: FOCS (2014)

19. Goldreich, O., Ostrovsky, R.: Software protection and simulation on oblivious rams. J. ACM 43(3), 431–473 (1996)

20. Goldwasser, S., Kalai, Y.T., Rothblum, G.N.: One-time programs. In: Wagner, D. (ed.) CRYPTO 2008. LNCS, vol. 5157, pp. 39–56. Springer, Heidelberg (2008). doi:10.1007/978-3-540-85174-5_3

21. Goldwasser, S., Micali, S., Rivest, R.L.: A digital signature scheme secure against adaptive chosen-message attacks. SIAM J. Comput. 17(2), 281–308 (1988)

22. Hemenway, B., Jafargholi, Z., Ostrovsky, R., Scafuro, A., Wichs, D.: Adaptively secure garbled circuits from one-way functions

23. Hubáček, P., Wichs, D.: On the communication complexity of secure function evaluation with long output. In: ITCS (2015)

24. Kalai, Y.T., Paneth, O.: Delegating ram computations. Cryptology ePrint Archive, Report 2015/957 (2015)

25. Kalai, Y.T., Raz, R., Rothblum, R.D.: How to delegate computations: the power of no-signaling proofs. In: Symposium on Theory of Computing, STOC 2014, New York, NY, USA, May 31 - June 03, 2014, pp. 485–494 (2014)

26. Kamara, S.: Encrypted search. ACM Crossroads 21(3), 30–34 (2015)

27. Koppula, V., Lewko, A.B., Waters, B.: Indistinguishability obfuscation for turing machines with unbounded memory. Cryptology ePrint Archive, Report 2014/925 (2014)

28. Koppula, V., Lewko, A.B., Waters, B.: Indistinguishability obfuscation for turing machines with unbounded memory. In: STOC (2015)
29. Lu, S., Ostrovsky, R.: How to garble RAM programs? In: Johansson, T., Nguyen, P.Q. (eds.) EUROCRYPT 2013. LNCS, vol. 7881, pp. 719–734. Springer, Heidelberg (2013). doi:10.1007/978-3-642-38348-9_42
30. Okamoto, T., Pietrzak, K., Waters, B., Wichs, D.: New realizations of somewhere statistically binding hashing and positional accumulators. IACR Cryptology ePrint Archive 2015, p. 869 (2015)
31. Papamanthou, C., Tamassia, R., Triandopoulos, N.: Optimal verification of operations on dynamic sets. In: Rogaway, P. (ed.) CRYPTO 2011. LNCS, vol. 6841, pp. 91–110. Springer, Heidelberg (2011). doi:10.1007/978-3-642-22792-9_6
32. Popa, R.A., Redfield, C., Zeldovich, N., Balakrishnan, H.: Cryptdb: protecting confidentiality with encrypted query processing. In: SOSP 2011, Cascais, Portugal, October 23–26, 2011, pp. 85–100 (2011)
33. Rogaway, P.: The round complexity of secure protocols. Ph.D. thesis, Massachusetts Institute of Technology (1991)
34. Walfish, M., Blumberg, A.J.: Verifying computations without reexecuting them. Commun. ACM 58(2), 74–84 (2015)
35. Yao, A.C.-C.: How to generate and exchange secrets. In: FOCS, pp. 162–167 (1986)

Delegating RAM Computations

Yael Kalai[1(✉)] and Omer Paneth[2]

[1] Microsoft Research, Redmond, USA
yael@microsoft.com
[2] Boston University, Boston, USA

Abstract. In the setting of cloud computing a user wishes to delegate its data, as well as computations over this data, to a cloud provider. Each computation may read and modify the data, and these modifications should persist between computations. Minding the computational resources of the cloud, delegated computations are modeled as RAM programs. In particular, the delegated computations' running time may be sub-linear, or even exponentially smaller than the memory size.

We construct a two-message protocol for delegating RAM computations to an untrusted cloud. In our protocol, the user saves a short digest of the delegated data. For every delegated computation, the cloud returns, in addition to the computation's output, the digest of the modified data, and a proof that the output and digest were computed correctly. When delegating a T-time RAM computation M with security parameter k, the cloud runs in time $\mathrm{poly}(\mathsf{T}, k)$ and the user in time $\mathrm{poly}(|M|, \log \mathsf{T}, k)$.

Our protocol is secure assuming super-polynomial hardness of the Learning with Error (LWE) assumption. Security holds even when the delegated computations are chosen adaptively as a function of the data and output of previous computations.

We note that RAM delegation schemes are an improved variant of memory delegation schemes [Chung et al. CRYPTO 2011]. In memory delegation, computations are modeled as Turing machines, and therefore, the cloud's work always grows with the size of the delegated data.

1 Introduction

In recent years, with the growing popularity of cloud computing platforms, more and more users store data and run computations on the cloud. This raises many concerns. As cryptographers, our first concern is that of secrecy: users may wish to hide their confidential data and computations from the cloud. But perhaps a more fundamental concern is that of *integrity*: ensuring that the cloud is doing what it is supposed to do. In this paper we focus on the latter.

We ask the following question: how can a cloud provider convince a user that a delegated computation was performed correctly? We believe that the adoption of

O. Paneth—Supported by the Simons award for graduate students in theoretical computer science and an NSF Algorithmic foundations grant 1218461.

M. Hirt and A. Smith (Eds.): TCC 2016-B, Part II, LNCS 9986, pp. 91–118, 2016.
DOI: 10.1007/978-3-662-53644-5_4

cloud computing services depends on the existence of such mechanisms. Indeed, even if not every computation is explicitly checked, the mere ability to check computations may be desirable.

RAM Delegation. We model the above problem as follows. Initially the user owns some memory D containing the data it wishes to delegate. In order to verify the correctness of future computations over this memory, the user must save some short digest of the memory D. We therefore allow the user to pre-process the memory once, before delegating it, and compute a digest d. We also allow the cloud to pre-process the memory before storing it. During this pre-processing the cloud can compute auxiliary information that will be stored together with the memory and used to construct proofs efficiently.

To compute on the memory, the user specifies a program M and sends its description to the cloud. We model the program M as a RAM program. We believe that this is the most realistic choice when the outsourced memory is very large and the computation may not access it all.[1] The cloud sends back to the user the output y of the program M when executed on the memory D. The user can delegate multiple computations sequentially where each computation may modify the memory. We require that the state of the memory persists between computations. Therefore, after every computation, the cloud sends back to the user, together with the output y, the new digest d_{new} corresponding to the new digest of the memory.

The cloud also provides a proof that the output y and the new digest d_{new} are correct with respect to the program M and the digest d of the original memory. We require that this proof proceeds in two messages, namely, together with the program M, the user sends a challange ch, and together with y and d_{new}, the cloud sends a proof pf. Thus, the proof of correctness does not require additional rounds of interaction. We refer to such a protocol as a *two-message delegation scheme for RAM computations*.

1.1   Our Results

We construct a two-message delegation scheme for RAM computations based on the Learning with Errors (LWE) assumption.

Efficiency. For security parameter k and for initial memory of size n such that $n < 2^k$, the user's and the cloud's pre-processing time is $n \cdot \text{poly}(k)$, and the digest is of size $\text{poly}(k)$. If the running time of the delegated RAM program is T (we assume that $T < 2^k$), then the running time of the cloud is $T^3 \cdot \text{poly}(k)$. The communication complexity of the proof, and the time it takes the user to generate a challenge and verify a proof are $\text{poly}(k)$, and are independent of the computation time.

---

[1] For example, consider the setting where the user wishes to simply retrieve an element from the outsourced database D. We would like the runtime of the user in this case to be proportional to $\log |D|$, as opposed to proportional to D.

Adaptive Soundness. The soundness of our scheme holds even if the adversary (acting as the cloud) can choose the program to be delegated adaptively depending on the memory and on the outcome of previously delegated computations. This feature is especially important in applications where the pre-processing step is performed once and then used and reused to delegate many computations over time. We emphasize that our protocol may not be sound if the adversary chooses the program adaptively depending on the user's challenge ch.

Public Pre-processing. In a two-message delegation scheme for RAM computations the user must pre-process the memory before delegating it. In our scheme the pre-processing step is public – it does not require any secret randomness. In particular, the user is not required to keep any secret state between computations. This feature also allows a single execution of the pre-processing step to serve multiple users, as long as they all trust the generated memory digest.[2]

Security with Adversarial Digest. We prove that our scheme is sound even in the setting where the pre-processing step is executed by an untrusted party. In this setting honest users cannot be sure that the digest they hold corresponds to some "correct" memory, or even that it is indeed the digest of any memory string. The soundness we require is that an adversary cannot prove that the same computation with the same digest leads to two different outcomes. We note that soundness for digests that are honestly computed follows from this stronger formulation.

Efficient Pre-processing. Another feature of our scheme is that the efficiency of the pre-processing step only depends on the *initial* memory size and does not depend on the amount of memory required to execute future computations. In particular, if there is no initial memory to delegate, the pre-processing step can be skipped.[3]

Informal Theorem 1.1. *There exists a two-message delegation scheme for RAM computations, with efficiency, adaptive soundness and public pre-processing, as described above, assuming the existence of a collision resistant hash family that is sub-exponentially secure and assuming that the LWE problem (with security parameter k) is hard to break in time quasi-polynomial in T, where T is an upper bound on the running time of the delegated computations.*

We note that the existence of a sub-exponentially secure collision resistant hash family follows from the sub-exponential hardness of the LWE problem.

---

[2] When different users delegate different computations that may *change* the memory, there should be an external mechanism to synchronize these computations, and make sure that every computation is verified with respect to the most recent digest of the memory.

[3] In fact, we can always replace the pre-processing step with an initial delegation round where the user delegates a program that initializes the memory.

On the Necessity of Cryptographic Assumptions. Since the user does not store its memory locally, and only stores a short digest, we cannot hope to get information-theoretic soundness. An all powerful malicious cloud can always cheat by finding a fake memory D' with the same digest as the original memory, and perform computations using the fake memory. Therefore, the soundness of our scheme must rely on some hardness assumption (such as the hardness of finding digest collisions).

On Delegation with Secrecy. Our delegation protocol does not achieve secrecy. That is, it does not hide the user's data and computations from the cloud. One method for achieving secrecy is to execute the entire delegation protocol under fully-homomorphic encryption. However, this method is not applicable when delegating RAM computations, since it increases the cloud's running time proportionally to the size of the entire memory.

1.2   Previous Work

We compare our result with previous results on delegating computation in various models based on various computational assumptions.

Delegating Non-deterministic Computations. Previous works constructed delegation schemes for non-deterministic computations in the random oracle model or based on strong "knowledge" assumptions. As we observe in this work (see Sect. 1.3), any delegation scheme for non-deterministic computations, combined with a collision-resistant hash function, can be used to delegate RAM computations.

The Random Oracle Model. Based on the interactive arguments of Kilian [Kil92], Micali [Mic94] gave the first construction of a non-interactive delegation scheme in the random oracle model. Micali's scheme supports non-deterministic computations and can therefore be used to delegate RAM computations assuming also the existence of a collision-resistant hash family.[4] The main advantage of Micali's scheme over the scheme presented in this work is that it is completely non-interactive (it requires one message rather than two). In particular, Micali's scheme is also *publicly verifiable*. However, our scheme can be proven secure in the standard model based on standard cryptographic assumptions.

Knowledge Assumptions. In a sequence of recent works, non-interactive (one message) delegation schemes in the *common reference string* (CRS) model, were constructs based on strong and non-standard "knowledge" assumptions such as variants of the Knowledge of Exponent assumption [Gro10, Lip12, DFH12, GGPR13,

---

[4] The solution described in Sect. 1.3 makes non-black-box use of the collision-resistant hash function, and therefore we cannot replace the hash function with the random oracle.

BCI+13, BCCT13, BCC+14]. These schemes support non-deterministic computations and can therefore be used to delegate RAM computations. Some of the above schemes are also publicly verifiable (the user does not need any secret trapdoor on the CRS). The main advantage of our scheme is that it can be based on standard cryptographic assumptions.

Indistinguishability Obfuscation. Several recent results construct non-interactive (one message) delegation schemes for RAM computations in the CRS model based on indistinguishability obfuscation [GHRW14, BGL+15, CHJV15, CH15, CCC+15]. Next we compare our scheme to the obfuscation based schemes.

The advantage of their schemes is that they achieve secrecy. In fact, they construct stronger objects such as garbling and obfuscation schemes for RAM computations. In addition, their schemes are publicly verifiable. The advantages of our scheme, compared to the obfuscation based schemes, are the following:

Assumptions. Our scheme is based on the hardness of the LWE problem – a standard and well studied cryptographic assumption. In particular, the LWE problem is known to be as hard as certain worst-case lattice problems.

Adaptivity. In our scheme security holds even against an adaptive adversary that chooses the delegated computations as a function of the delegated memory. In contrast, the obfuscation based schemes only have static security. That is, in the analysis all future delegated computations must be fixed before the memory is delegated. We note that using complexity leveraging and sub-exponential hardness assumptions it is possible to prove that obfuscation based schemes are secure against a *bounded* number of adaptively chosen computations, where the bound on the number of computations depends on the size of the CRS.

Security with adversarial digest. In our scheme the pre-processing step is public and soundness holds even in the setting where the pre-processing step is executed by an untrusted party. In the obfuscation based schemes however, the pre-processing step requires private randomness and if it is not carried out honestly the cloud may be able to prove arbitrary statements.

Following our work, Canetti et al. [CCHR15] and Ananth et al. [ACC+15] gave a delegation scheme for RAM computations from indistinguishability obfuscation that satisfies the same notion of adaptivity as our scheme. These constructions do not have public digest and they are not secure with adversarial digest.

Learning with Errors. We review existing delegation protocols based on the hardness of the LWE problem. These protocols are less efficient than our delegation protocols for RAM computations.

Deterministic Turing Machine delegation. The work of [KRR14] gives a two-message delegation scheme for deterministic Turing machine computations based on the quasi-polynomial hardness of the LWE problem. The main differences between delegation of RAM computations and delegation of deterministic Turing machine computations are as follows:

1. In deterministic Turing machine delegation, the user needs to save the entire memory (thought of as the input to the computation), while in RAM delegation, the user only needs to save a short digest of the memory.
2. In deterministic Turing machine delegation, the cloud's running time depends on the running time of the computation when described as a Turing machine, rather than a RAM program. In particular, the cloud's running time always grows with the memory size, even if the delegated computation does not access the entire memory.

We mention that our scheme has better asymptotic efficiency than the scheme of [KRR14] even for Turing machine computations. For delegated computations running in time T and space S the cloud's running time in our scheme is $T^3 \cdot \text{poly}(k)$ instead of $(T \cdot S)^3 \cdot \text{poly}(k)$ as in [KRR14].

Memory Delegation. As mentioned in [KRR14], the techniques of Chung et al. [CKLR11] can be used to convert the [KRR14] scheme into a memory delegation scheme that overcomes the first difference above, but not the second one.

Fully-Homomorphic Signatures. The work of Gorbunov et al. [GVW15] on fully-homomorphic signatures gives a non-interactive, publicly verifiable protocol in the CRS model, overcoming both differences above. However, while their protocol has small communication, the user's work is still proportional to computation's running time. Additionally, their protocol does not support computations that write to the memory.

Proofs of Proximity. Finally, we mention a recent line of works on proofs of proximity [RVW13, GR15, KR15, GGR15]. These proofs can be verified much faster than the size of the memory, however, unlike in RAM delegation, in their model the user does not get to pre-process the memory. Instead the user has oracle access to the memory during proof verification. In proofs of proximity the user is only convinced that the computation output is consistent with some memory that is close to the real memory. Additionally, in proofs of proximity the verification takes time at least $\Omega(\sqrt{n})$ where n is the memory size [KR15].

1.3   Technical Overview

We start with a high level description of our scheme.

Pre-processing. In the pre-processing step, the user computes a hash-tree [Mer87] over the memory D and saves the root of this tree as the digest d. The cloud also pre-processes the delegated memory D by computing the same hash-tree and stores the entire tree. The hash-tree allows the cloud to efficiently access the memory in an "authenticated" way. Specifically, the cloud performs the following operations:

1. Read a bit from memory.
2. Write a bit to memory, update the hash tree, and obtain a new digest.

The cloud can then compute a short certificate (in the form of an authenticated path), authenticating the value of the bit read or the value of the updated digest. The time required to access the memory and compute the certificate depends only logarithmically on the memory size.

Emulated Computations and their Transcript. When the user delegates a computation given by a RAM program M, the cloud starts by emulating the execution of M on the memory D as described in [BEG+91]: whenever M accesses the memory, the cloud performs an authenticated memory access via the hash tree. When the emulation of M terminates, the cloud obtains the program output y and the updated memory digest d_{new}. The cloud also compiles a *transcript* of the memory accessed during the computation. This transcript contains an ordered list of M's memory accesses. For every memory access, the transcript contains the memory location, the bit that was read or written, the new memory digest (in case the memory changed), and the certificate of authenticity. This transcript allows to "re-execute" the computation of the program M and obtain y and d_{new}, without accessing the memory D directly. Moreover, it is computationally hard to find a valid transcript (containing only valid certificates) that yields the wrong output or digest $(y', d'_{new}) \neq (y, d_{new})$. For security parameter k and a RAM program M executing in time $T \leq 2^k$, the time to generate the transcript and to re-execute the program based on the transcript is $T \cdot \text{poly}(k)$.

Proof of Correctness. After emulating the execution of M, the cloud sends the output y and the new digest d_{new} to the user. The cloud also proves to the user that it knows a valid computation transcript which is consistent with y and d_{new}. More formally, we consider a non-deterministic Turing machine TVer that accepts an input tuple (M, d, y, d_{new}) if and only if there exists a valid transcript Trans with respect to d such that the emulation of the program M with Trans produces the output y and the digest d_{new}.

Proving knowledge of a witness Trans that makes TVer accept (M, d, y, d_{new}) requires a delegation scheme supporting *non-deterministic computations*. The problem with this approach is that currently, two-message delegation schemes for non-deterministic computations are only known in the random oracle model or based on strong knowledge assumptions (see Sect. 1.2). However, it turns out that for the specific computation TVer, we can construct a two-message delegation scheme based on standard cryptographic assumptions.

Re-purposing the KRR Proof System. Our solution is based on the delegation scheme of Kalai et al. [KRR14]. While in general, their proof system only supports deterministic computations, we extend their security proof so it also applies to non-deterministic computations of a certain form.

We start with a brief overview of the [KRR14] proof system and explain why it does not support general non-deterministic computations. Then we describe the extended security proof and the type of non-deterministic computations it does support.

The [KRR14] proof system can be used to prove that a deterministic Turing machine M is accepting. The soundness proof of [KRR14] has two steps. In the first step M is translated into a 3-SAT formula ϕ that is satisfiable if and only if M is accepting. The analysis of [KRR14] shows that if the cloud convinces the user to accept, then the formula ϕ satisfies a relaxed notion of satisfiability called *local satisfiability* (See [KRR14, Lemma 7.29]). In the second step, the specific structure of the formula ϕ is exploited to prove that if ϕ is locally satisfiable it must also be satisfiable.

The work of Paneth and Rothblum [PR14] further abstracts the notion of local satisfiability, redefining it in a way that is independent of the protocol of [KRR14]. Based on this abstraction, they separate the *construction* of [KRR14] into two steps. In the first step, the main part of the [KRR14] proof system is converted into a protocol for proving local satisfiability of formulas. In the second step, the cloud uses this protocol to convince the user that the formula ϕ is locally satisfiable. As before, the structure of the formula ϕ is exploited to prove that ϕ is satisfiable.

Local Satisfiability. Unlike full-fledged satisfiability, the notion of local satisfiability only considers assignments to ℓ variables at a time, where ℓ is a locality parameter that may be much smaller than the total number of variables in the formula. Formally, we say that a 3-SAT formula ϕ is ℓ-locally satisfiable if for every set Q of ℓ variables there exists a distribution D_Q over assignments to the variables in Q such that the following conditions are satisfied:

Everywhere local consistency. For every set Q of ℓ variables, a random assignment in D_Q satisfies all local constraints in ϕ over the variables in Q with high probability.

No-signaling. For every set Q of ℓ variables and for every subset $Q' \subseteq Q$, the distribution of an assignment sampled from D_Q restricted to the variables in Q' is independent of the other variables in $Q \setminus Q'$.

From Local Satisfiability to Full-Fledged Satisfiability. In the [KRR14] proof system, ℓ is a fixed polynomial in the security parameter, independent of the size of the formula ϕ (the communication complexity of the proof grows with ℓ). In this setting, local satisfiability does not generally imply full-fledged satisfiability. However, the analysis of [KRR14] exploits the specific structure of ϕ to go from local satisfiability to full-fledged satisfiability. The proof of this step crucially relies on the fact that the formula ϕ describes a deterministic computation. We show how to extend this proof for non-deterministic computations of a specific form.

Roughly, we require that (computationally) there exists a unique "correct" witness that can be verified *locally*. Namely, for any proposed witness (that can be found efficiently) and any bit of this proposed witness, it is possible to verify that the value of this bit agrees with the correct witness in time that is independent of the running time of the entire computation.

More on the Analysis of KRR. We describe the argument of [KRR14] and explain why it fails for non-deterministic computations. To go from local satisfiability to full-fledged satisfiability, the proof of [KRR14] relies on the fact that the formula ϕ describing an accepting deterministic computation has a unique satisfying assignment. We call this the *correct* assignment to ϕ. The rest of the proof uses the fact that the variables of ϕ can be partitioned into "layers" such that variables in the i-th layer correspond to the computation's state immediately before the i-th computation step. The proof proceeds by induction over the layers. In the inductive step we assume that local assignments to any ℓ variables in the i-th layer are correct (agrees with the correct assignment) with high probability and prove that the same holds for the $(i + 1)$-st layer. Indeed, if the local assignment to some set of ℓ variables in the $(i + 1)$-st layer is correct with a significantly lower probability, the special structure of ϕ and the no-signaling property of the assignments can be used to argue that there must exist a set of ℓ variables whose assignment violates ϕ's local constraints with some significant probability.

Non-deterministic Computations. The above argument does not extend to non-deterministic computations, since the notion of a "correct" assignment is not well defined in this setting. Moreover, even if there is a unique witness that makes the computation accept, and we consider the correct assignment defined by this witness, the above argument still fails. The issue is that even if every local assignment to any set of variables in the i-th layer is correct, there could still be more than one assignment to variables in the $(i + 1)$-st layer satisfying all of ϕ's local constraints.

We show how to overcome this problem for non-deterministic computations where (computationally) there exists a unique "correct" witness that can be verified *locally*, as described above. Consider for example the computation of the Turing machine TVer on input $(M, \mathsf{d}, y, \mathsf{d_{new}})$ where d is the digest of the initial memory D. The (computationally) unique witness for this computation is a transcript of the program execution that can be verified locally – one step at a time.

In more details, let Trans be the correct transcript defined by the execution of M on memory D. Let ϕ be the formula describing the computation of $\mathsf{TVer}(M, \mathsf{d}, y, \mathsf{d_{new}})$. We prove that any accepting local assignment to variables of ϕ must agree with the global correct assignment to ϕ defined by the execution of TVer with the (well defined) transcript Trans. As in the case of deterministic computations, we partition ϕ's variables into layers. In the i-th inductive step we assume that local assignments to any ℓ variables in the i-th layer are correct with high probability. If the local assignment to some set of ℓ variables in the $(i + 1)$-th layer is correct with a significantly lower probability then we prove that the assignment must describe an incorrect transcript. Since both the correct transcript and the incorrect one contain valid certificates, we can use these certificates to break the security of the hash tree.

Multi-prover Arguments. The presentation of the construction in [KRR14], as well as the presentation in the body of this work, goes through the intermediate step of constructing a no-signaling multi-prover proof-system. In more details, [KRR14] first construct a no-signaling multi-prover interactive proof for local-satisfiability. They then leverage local-satisfiability to prove full-fledged satisfiability, resulting in a no-signaling multi-prover interactive proof (with unconditional soundness) for deterministic computations. Finally, they transform any no-signaling multi-prover interactive proof into a delegation scheme assuming fully-homomorphic encryption.

Our construction follows the same blueprint. We first construct a no-signaling multi-prover interactive argument for RAM computations, and then transform it into a delegation scheme. (Due to space limitations, we do not describe the transformation from a multi-prover interactive argument into a delegation scheme which can be found in the full version of this work [KP15].) Unlike in [KRR14], the soundness of our multi-prover arguments is conditional on the existence of collision-resistant hashing. We note that for RAM delegation, computational assumptions are necessary even in the multi-prover model.

2   Tools and Definitions

2.1   Notation

For sets B, S, we denote by B^S the set of vectors of elements in B indexed by the elements of S. That is, every vector $\mathbf{a} \in B^S$ is of the form $\mathbf{a} = (\mathsf{a}_i \in B)_{i \in S}$. For a vector $\mathbf{a} \in B^S$ and a subset $Q \subseteq S$, we denote by $\mathbf{a}[Q] \in B^Q$ the vector that contains only the elements in $\mathbf{a}$ with indices in Q, that is, $\mathbf{a}[Q] = (\mathsf{a}_i)_{i \in Q}$.

2.2   RAM Computation

We consider the standard model of RAM computation where a program M can access an initial memory string $D \in \{0,1\}^n$. For an input x, we denoted by $M^D(x)$ an execution of the program M with input x and initial memory D. For a bit $y \in \{0,1\}$ and for a string $D_{\mathsf{new}} \in \{0,1\}^n$ we also use the notation $y \leftarrow M^{(D \to D_{\mathsf{new}})}(x)$ to denote that y is the output of the program M on input x and initial memory D, and D_{new} is the final memory string after the execution. For simplicity we think only of RAM programs that output a single bit.[5] The computation of M is carried out one step at a time by a CPU algorithm STEP. STEP is a polynomial-time algorithm that takes as input a description of a program M, an input x, a state of size $O(\log n)$, and a bit that was supposedly read from memory, and it outputs a quadruple

$$(\mathsf{state}_{\mathsf{new}}, i^{\mathsf{r}}, i^{\mathsf{w}}, b^{\mathsf{w}}) \leftarrow \mathsf{STEP}(M, x, \mathsf{state}, b^{\mathsf{r}}),$$

where $\mathsf{state}_{\mathsf{new}}$ is the updated state, i^{r} denotes the location in memory to be read next, the location i^{w} denotes the location in memory to write to next,

---

[5] A program that outputs multiple bits can be simulated by executing several programs in parallel, or by writing the output directly to the memory.

and the bit b^w denotes the bit to be written in location i^w. The execution $M^D(x)$ proceeds as follows. The program starts with some empty initial state state_1. By convention we set the first memory location read by the program to be $i_1^r = 1$. Starting from $j = 1$, the j-th execution step proceeds as follows:

1. Read from memory the bit $b_j^r \leftarrow D[i_j^r]$.
2. Compute $(\mathsf{state}_{j+1}, i_{j+1}^r, i_{j+1}^w, b_{j+1}^w) \leftarrow \mathsf{STEP}(M, x, \mathsf{state}_j, b_j^r)$.
3. Write a bit to memory $D[i_{j+1}^w] \leftarrow b_{j+1}^w$. (If $i_{j+1}^w = \bot$ no writing is performed in this step.)

The execution terminates when the program STEP outputs a special terminating state. We assume that the terminating state includes the value of the output bit y. Note that after the last step was executed and an output has been produced, the memory is written to one last time. We say that a machine M is *read only*, if for every (x, state, b^r), $\mathsf{STEP}(M, x, \mathsf{state}, b^r)$ outputs $(\mathsf{state}_{\mathsf{new}}, i^r, i^w, b^w)$ where $i^w = \bot$.

Remark 2.1 (Space complexity of STEP). We assume without loss of generality that the RAM program M reads the input x once and copies it to memory. Therefore the space complexity of the algorithm STEP is polylog(n).

2.3   Hash Tree

Let $D \in \{0,1\}^n$ be a string. Let k be a security parameter such that $n < 2^k$.
    A hash-tree scheme consists of algorithms:

$$(\mathsf{HT.Gen}, \mathsf{HT.Hash}, \mathsf{HT.Read}, \mathsf{HT.Write}, \mathsf{HT.VerRead}, \mathsf{HT.VerWrite}),$$

with the following syntax and efficiency:

- $\mathsf{HT.Gen}(1^k) \rightarrow \mathsf{key}$:
  A randomized polynomial-time algorithm that outputs a hash key, denoted by key.
- $\mathsf{HT.Hash}(\mathsf{key}, D) \rightarrow (\mathsf{tree}, \mathsf{rt})$:
  A deterministic polynomial-time algorithm that outputs a hash tree denoted by tree, and a hash root rt of size poly(k) (we assume that both strings tree and rt include key).
- $\mathsf{HT.Read}^{\mathsf{tree}}(i^r) \rightarrow (b^r, \mathsf{pf})$:
  A deterministic read-only RAM program that accesses the initial memory string tree, runs in time poly(k), and outputs a bit, denoted by b^r, and a proof, denoted by pf.
- $\mathsf{HT.Write}^{\mathsf{tree}}(i^w, b^w) \rightarrow (\mathsf{rt}_{\mathsf{new}}, \mathsf{pf})$:
  A deterministic RAM program that accesses the initial memory string tree, runs in time poly(k), and outputs a new hash root, denoted by $\mathsf{rt}_{\mathsf{new}}$, and a proof, denoted by pf.
- $\mathsf{HT.VerRead}(\mathsf{rt}, i^r, b^r, \mathsf{pf}) \rightarrow b$:
  A deterministic polynomial-time algorithm that outputs an acceptance bit b.

– HT.VerWrite$(\mathsf{rt}, i^{\mathsf{w}}, b^{\mathsf{w}}, \mathsf{rt}_{\mathsf{new}}, \mathsf{pf}) \to b$:
   A deterministic polynomial-time algorithm that outputs an acceptance bit b.

Definition 2.1 (Hash-Tree). *A hash-tree scheme*

$$(\mathsf{HT.Gen}, \mathsf{HT.Hash}, \mathsf{HT.Read}, \mathsf{HT.Write}, \mathsf{HT.VerRead}, \mathsf{HT.VerWrite}),$$

satisfies the following properties.

– *Completeness of Read. For every $k \in \mathbb{N}$ and for every $D \in \{0,1\}^n$ such that $n \le 2^k$, and for every $i^{\mathsf{r}} \in [n]$*

$$\Pr\left[\begin{array}{c} 1 = \mathsf{HT.VerRead}(\mathsf{rt}, i^{\mathsf{r}}, b^{\mathsf{r}}, \mathsf{pf}) \\ D[i^{\mathsf{r}}] = b^{\mathsf{r}} \end{array} \middle| \begin{array}{l} \mathsf{key} \leftarrow \mathsf{HT.Gen}(1^k) \\ (\mathsf{tree}, \mathsf{rt}) \leftarrow \mathsf{HT.Hash}(\mathsf{key}, D) \\ (b^{\mathsf{r}}, \mathsf{pf}) \leftarrow \mathsf{HT.Read}^{\mathsf{tree}}(i^{\mathsf{r}}) \end{array} \right] = 1.$$

– *Completeness of Write. For every $k \in \mathbb{N}$ and for every $D \in \{0,1\}^n$ such that $n \le 2^k$, for every $i^{\mathsf{w}} \in [n], b^{\mathsf{w}} \in \{0,1\}$, and for $D_{\mathsf{new}} \in \{0,1\}^n$ that is equal to the string D except that $D_{\mathsf{new}}[i^{\mathsf{w}}] = b^{\mathsf{w}}$*

$$\Pr\left[\begin{array}{c} 1 = \mathsf{HT.VerWrite}(\mathsf{rt}, i^{\mathsf{w}}, b^{\mathsf{w}}, \mathsf{rt}'_{\mathsf{new}}, \mathsf{pf}) \\ \mathsf{rt}'_{\mathsf{new}} = \mathsf{rt}_{\mathsf{new}} \end{array} \middle| \begin{array}{l} \mathsf{key} \leftarrow \mathsf{HT.Gen}(1^k) \\ (\mathsf{tree}, \mathsf{rt}) \leftarrow \mathsf{HT.Hash}(\mathsf{key}, D) \\ (\mathsf{tree}_{\mathsf{new}}, \mathsf{rt}_{\mathsf{new}}) \leftarrow \mathsf{HT.Hash}(\mathsf{key}, D_{\mathsf{new}}) \\ (\mathsf{rt}'_{\mathsf{new}}, \mathsf{pf}) \leftarrow \mathsf{HT.Write}^{\mathsf{tree}}(i^{\mathsf{w}}, b^{\mathsf{w}}) \end{array} \right] = 1.$$

– *Soundness of Read. For every polynomial size adversary Adv there exists a negligible function μ such that for every $k \in \mathbb{N}$*

$$\Pr\left[\begin{array}{c} (b', \mathsf{pf}') \ne (b, \mathsf{pf}) \\ 1 = \mathsf{HT.VerRead}(\mathsf{rt}, i, b, \mathsf{pf}) \\ 1 = \mathsf{HT.VerRead}(\mathsf{rt}, i, b', \mathsf{pf}') \end{array} \middle| \begin{array}{l} \mathsf{key} \leftarrow \mathsf{HT.Gen}(1^k) \\ (\mathsf{rt}, i, b, \mathsf{pf}, b', \mathsf{pf}') \leftarrow \mathsf{Adv}(\mathsf{key}) \end{array} \right] \le \mu(k).$$

– *Soundness of Write. For every poly-size adversary Adv there exists a negligible function μ such that for every $k \in \mathbb{N}$*

$$\Pr\left[\begin{array}{c} (\mathsf{rt}'_{\mathsf{new}}, \mathsf{pf}') \ne (\mathsf{rt}_{\mathsf{new}}, \mathsf{pf}) \\ 1 = \mathsf{HT.VerWrite}(\mathsf{rt}, i, b, \mathsf{rt}_{\mathsf{new}}, \mathsf{pf}) \\ 1 = \mathsf{HT.VerWrite}(\mathsf{rt}, i, b, \mathsf{rt}'_{\mathsf{new}}, \mathsf{pf}') \\[4pt] \begin{array}{l} \mathsf{key} \leftarrow \mathsf{HT.Gen}(1^k) \\ (\mathsf{rt}, i, b, \mathsf{rt}_{\mathsf{new}}, \mathsf{pf}, \mathsf{rt}'_{\mathsf{new}}, \mathsf{pf}') \leftarrow \mathsf{Adv}(\mathsf{key}) \end{array} \end{array} \right] \le \mu(k).$$

We say that the hash-tree scheme is (S, ϵ)-secure, for a function $S(k)$ and a negligible function $\epsilon(k)$, if for every constant $c > 0$, the soundness of read and soundness of write properties hold for every adversary of size $S(k)^c$ with probability at most $\epsilon(k)^c$. We say that the hash-tree scheme has sub-exponential security if it is $(2^{k^\delta}, 2^{-k^\delta})$-secure for some constant $\delta > 0$.

Remark 2.2 (Unique proofs in Definition 2.1). In the soundness properties of Definition 2.1 we make the strong requirement that it is hard to find two different proofs for any statement (even a correct one). This strong requirement simplifies the proof of Theorem 4.1, however the proof can be modified to rely on a weaker soundness requirement.

Theorem 2.1 (*[Mer87]*). *A hash-tree scheme satisfying Definition 2.1 can be constructed from any family of collision-resistant hash functions. Moreover, the hash-tree scheme is sub-exponentially secure if the underlying collision-resistant hash family is sub-exponentially secure.*

2.4   Delegation for RAM Computations

Let M be a T-time RAM program, let $x \in \{0,1\}^m$ be an input to the program, and let $D \in \{0,1\}^n$ be some initial memory string. Let k be a security parameter such that $|M|, \mathsf{T}(m), n < 2^k$. A two-message delegation scheme for RAM computations consists of algorithms:

$$(\mathsf{ParamGen}, \mathsf{MemGen}, \mathsf{QueryGen}, \mathsf{Output}, \mathsf{Prover}, \mathsf{Verifier}),$$

with the following syntax and efficiency:

- $\mathsf{ParamGen}(1^k) \to \mathsf{pp}$:
  A randomized polynomial-time algorithm that outputs public parameters pp.
- $\mathsf{MemGen}(\mathsf{pp}, D) \to (\mathsf{dt}, \mathsf{d})$:
  A deterministic polynomial-time algorithm that outputs the processed memory dt, and a digest of the memory d of size $\mathrm{poly}(k)$.
- $\mathsf{QueryGen}(1^k) \to (\mathsf{q}, \mathsf{st})$:
  A randomized polynomial-time algorithm that outputs a query q and a secret state st.
- $\mathsf{Output}^{\mathsf{dt}}(1^{\mathsf{T}(m)}, n, M, x) \to (y, \mathsf{d}_{\mathsf{new}}, \mathsf{Trans})$:
  A deterministic RAM program running in time $\mathsf{T}(m) \cdot \mathrm{poly}(k)$ that accesses the processed memory dt, and outputs the output bit y, and a new digest $\mathsf{d}_{\mathsf{new}}$ of size $\mathrm{poly}(k)$ and a computation transcript Trans.
- $\mathsf{Prover}((M, x, \mathsf{T}(m), \mathsf{d}, y, \mathsf{d}_{\mathsf{new}}), \mathsf{Trans}, \mathsf{q}) \to \mathsf{pf}$:
  A deterministic algorithm running in time $\mathrm{poly}(\mathsf{T}(m), k)$ that outputs a proof pf of size $\mathrm{poly}(k)$.
- $\mathsf{Verifier}((M, x, \mathsf{T}(m), \mathsf{d}, y, \mathsf{d}_{\mathsf{new}}), \mathsf{st}, \mathsf{pf}) \to b$:
  A deterministic algorithm running in time $m \cdot \mathrm{poly}(k)$ that outputs an acceptance bit b.

Remark 2.3 (Statement-independent queries). In the above, the queries generated by the algorithm QueryGen are independent of the program, the input and the memory digest. We could consider a more liberal definition that allows such a dependency, however, in our construction this is not needed.

Remark 2.4 (Verifier efficiency). We note that the dependence of the verification time on the input length m can be improved. In particular, in our construction, given oracle access to a *low-degree extension* encoding of the input x, the verifier's running time is poly(k).

Remark 2.5 (The Output algorithm). In the above interface we separated the prover computation into two algorithms. The first algorithm, Output, accesses the memory, carries out the computation, and produces the output as well as a *transcript* of the computation. This transcript may include all the memory accessed during the RAM computation or any other information. We only restrict the size of the transcript to be related to the running time of the RAM computation. The second algorithm, Prover, is given the transcript and the challenge query and outputs the proof. This separation ensures that the memory locations accessed by the prover are independent of the challenge query. This property is used in the transformation from no-signaling multi-prover arguments to delegationin [KP15].

Definition 2.2 (Two-Message Argument for RAM computations).

A two-message delegation scheme (ParamGen, MemGen, QueryGen, Prover, Verifier) *for RAM computations satisfies the following properties.*

- *Completeness. For every security parameter $k \in \mathbb{N}$, every T-time RAM program M, every input $x \in \{0,1\}^m$, every $D \in \{0,1\}^n$, and every (y, D_{new}) such that $\mathsf{T}(m), n \leq 2^k$ and $y \leftarrow M^{(D \to D_{new})}(x)$*

$$\Pr\left[\begin{array}{l} 1 = \mathsf{Verifier}((M, x, \mathsf{T}(m), \mathsf{d}, y, \mathsf{d}_{new}), \mathsf{st}, \mathsf{pf}) \\ (\mathsf{dt}_{new}, \mathsf{d}_{new}) = \mathsf{MemGen}(\mathsf{pp}, D_{new}) \end{array} \right|$$

$$\left. \begin{array}{l} \mathsf{pp} \leftarrow \mathsf{ParamGen}(1^k) \\ (\mathsf{dt}, \mathsf{d}) \leftarrow \mathsf{MemGen}(\mathsf{pp}, D) \\ (\mathsf{q}, \mathsf{st}) \leftarrow \mathsf{QueryGen}(1^k) \\ (y, \mathsf{d}_{new}, \mathsf{Trans}) \leftarrow \mathsf{Output}^{\mathsf{dt} \to \mathsf{dt}_{new}}(1^{\mathsf{T}(m)}, n, M, x) \\ \mathsf{pf} \leftarrow \mathsf{Prover}((M, x, \mathsf{T}(m), \mathsf{d}, y, \mathsf{d}_{new}), \mathsf{Trans}, \mathsf{q}) \end{array} \right] = 1.$$

- *Soundness. For every pair of polynomial-size adversaries $(\mathsf{Adv}_1, \mathsf{Adv}_2)$ there exists a negligible function μ such that for every $k \in \mathbb{N}$*

$$\Pr\left[\begin{array}{l} (y, \mathsf{d}_{new}) \neq (y', \mathsf{d}'_{new}) \\ 1 = \mathsf{Verifier}((M, x, \mathsf{T}, \mathsf{d}, y, \mathsf{d}_{new}), \mathsf{st}, \mathsf{pf}) \\ 1 = \mathsf{Verifier}((M, x, \mathsf{T}, \mathsf{d}, y', \mathsf{d}'_{new}), \mathsf{st}, \mathsf{pf}') \end{array} \right|$$

$$\left. \begin{array}{l} \mathsf{pp} \leftarrow \mathsf{ParamGen}(1^k) \\ (M, x, 1^{\mathsf{T}}, \mathsf{d}, y, \mathsf{d}_{new}, y', \mathsf{d}'_{new}) \leftarrow \mathsf{Adv}_1(1^k, \mathsf{pp}) \\ (\mathsf{q}, \mathsf{st}) \leftarrow \mathsf{QueryGen}(1^k) \\ (\mathsf{pf}, \mathsf{pf}') \leftarrow \mathsf{Adv}_2(1^k, \mathsf{pp}, \mathsf{q}) \end{array} \right] \leq \mu(k).$$

We say that the delegation scheme is (S, ϵ)-secure, for a function $S(k)$ and a negligible function $\epsilon(k)$, if for every constant $c > 0$, the soundness property holds for every pair of adversaries of size $S(k)^c$ with probability at most $\epsilon(k)^c$.

2.5   Multi-prover Arguments for RAM Computations

Let ℓ be a polynomial, M be a T-time RAM program, let $x \in \{0,1\}^m$ be an input to the program, and let $D \in \{0,1\}^n$ be some initial memory string. Let k be a security parameter such that $|M|, \mathsf{T}(m), n < 2^k$. An ℓ-*prover argument* for RAM computations consists of algorithms:

$$(\mathsf{ParamGen}, \mathsf{MemGen}, \mathsf{QueryGen}, \mathsf{Output}, \mathsf{Prover}, \mathsf{Verifier}),$$

with the following syntax and efficiency:

- $\mathsf{ParamGen}(1^k) \to \mathsf{pp}$:
  A randomized polynomial-time algorithm that outputs public parameters pp.
- $\mathsf{MemGen}(\mathsf{pp}, D) \to (\mathsf{dt}, \mathsf{d})$:
  A deterministic polynomial-time algorithm that outputs the processed memory dt and a digest of the memory d of size $\mathrm{poly}(k)$.
- $\mathsf{QueryGen}(1^k) \to ((\mathsf{q}_1, \ldots, \mathsf{q}_\ell), \mathsf{st})$:
  A randomized polynomial-time algorithm that outputs a set of $\ell = \ell(k)$ queries $(\mathsf{q}_1, \ldots, \mathsf{q}_\ell)$, and a secret state st.
- $\mathsf{Output}^{\mathsf{dt}}(1^{\mathsf{T}(m)}, n, M, x) \to (y, \mathsf{d}_{\mathsf{new}}, \mathsf{Trans})$:
  A deterministic RAM program running in time $\mathsf{T}(m) \cdot \mathrm{poly}(k)$ that accesses the processed memory dt, and outputs the output bit y, a new digest $\mathsf{d}_{\mathsf{new}}$ of size $\mathrm{poly}(k)$, and a computation transcript Trans.
- $\mathsf{Prover}((M, x, \mathsf{T}(m), \mathsf{d}, y, \mathsf{d}_{\mathsf{new}}), \mathsf{Trans}, \mathsf{q}) \to \mathsf{a}$:
  A deterministic algorithm running in time $\mathrm{poly}(\mathsf{T}(m), k)$ that outputs an answer a of size $\mathrm{poly}(k)$ to a single query q.
- $\mathsf{Verifier}((M, x, \mathsf{T}(m), \mathsf{d}, y, \mathsf{d}_{\mathsf{new}}), \mathsf{st}, (\mathsf{a}_1, \ldots, \mathsf{a}_\ell)) \to b$:
  A deterministic algorithm running in time $m \cdot \mathrm{poly}(k)$ that outputs an acceptance bit b.

Remark 2.6 (Statement-independent queries). In the above, the queries generated by the algorithm $\mathsf{QueryGen}$ are independent of the program, the input and the memory digest. We could consider a more liberal definition that allows such a dependency, however, in our construction this is not needed.

Remark 2.7 (Verification efficiency). We note that the dependence of the verification time on the input length m can be improved. In particular, in our construction, given oracle access to a *low-degree extension* encoding of the input x, the verifier's running time is $\mathrm{poly}(k)$.

Remark 2.8 (The Output algorithm). In the above interface we separated the prover computation into two algorithms. The first algorithm, Output, accesses the memory, carries out the computation, and produces the output as well as a *transcript* of the computation. This transcript may include all the memory accessed during the RAM computation or any other information. We only restrict the size of the transcript to be related to the running time of the RAM computation. The second algorithm, Prover, is given the transcript and a challenge query and outputs an answer. This separation ensures that the memory locations accessed by the prover are independent of the challenge queries. This property is used in the transformation from no-signaling multi-prover arguments to delegationin [KP15].

Definition 2.3 (Multi-Prover Argument for RAM computations).

  Let $\ell = \ell(k)$ be a polynomial in the security parameter. An ℓ-prover argument system (ParamGen, MemGen, QueryGen, Output, Prover, Verifier) *for RAM computations satisfies the following properties.*

- *Completeness. For every security parameter $k \in \mathbb{N}$, every T-time RAM program M, every input $x \in \{0,1\}^m$, every $D \in \{0,1\}^n$, and every (y, D_{new}), such that $T(m), n \leq 2^k$ and $y \leftarrow M^{(D \to D_{\mathsf{new}})}(x)$*

$$\Pr \left[\begin{array}{l} 1 = \mathsf{Verifier}((M, x, \mathsf{T}(m), \mathsf{d}, y, \mathsf{d}_{\mathsf{new}}), \mathsf{st}, (\mathsf{a}_1, \ldots, \mathsf{a}_\ell)) \\ (\mathsf{dt}_{\mathsf{new}}, \mathsf{d}_{\mathsf{new}}) = \mathsf{MemGen}(\mathsf{pp}, D_{\mathsf{new}}) \end{array} \middle| \begin{array}{l} \mathsf{pp} \leftarrow \mathsf{ParamGen}(1^k) \\ (\mathsf{dt}, \mathsf{d}) \leftarrow \mathsf{MemGen}(\mathsf{pp}, D) \\ ((\mathsf{q}_1, \ldots, \mathsf{q}_\ell), \mathsf{st}) \leftarrow \mathsf{QueryGen}(1^k) \\ (y, \mathsf{d}_{\mathsf{new}}, \mathsf{Trans}) \leftarrow \mathsf{Output}^{\mathsf{dt} \to \mathsf{dt}_{\mathsf{new}}}(1^{\mathsf{T}(m)}, n, M, x) \\ \forall i \in [\ell] : \mathsf{a}_i \leftarrow \mathsf{Prover}((M, x, \mathsf{T}(m), \mathsf{d}, y, \mathsf{d}_{\mathsf{new}}), \mathsf{Trans}, \mathsf{q}_i) \end{array} \right] = 1.$$

- *Soundness. For every pair of polynomial-size adversaries* $(\mathsf{Adv}_1, \mathsf{Adv}_2)$ *there exists a negligible function μ such that for every $k \in \mathbb{N}$ and for $\ell = \ell(k)$*

$$\Pr \left[\begin{array}{l} (y, \mathsf{d}_{\mathsf{new}}) \neq (y', \mathsf{d}'_{\mathsf{new}}) \\ 1 = \mathsf{Verifier}((M, x, \mathsf{T}, \mathsf{d}, y, \mathsf{d}_{\mathsf{new}}), \mathsf{st}, (\mathsf{a}_1, \ldots, \mathsf{a}_\ell)) \\ 1 = \mathsf{Verifier}((M, x, \mathsf{T}, \mathsf{d}, y', \mathsf{d}'_{\mathsf{new}}), \mathsf{st}, (\mathsf{a}'_1, \ldots, \mathsf{a}'_\ell)) \end{array} \middle| \begin{array}{l} \mathsf{pp} \leftarrow \mathsf{ParamGen}(1^k) \\ (M, x, 1^{\mathsf{T}}, \mathsf{d}, y, \mathsf{d}_{\mathsf{new}}, y', \mathsf{d}'_{\mathsf{new}}) \leftarrow \mathsf{Adv}_1(1^k, \mathsf{pp}) \\ ((\mathsf{q}_1, \ldots, \mathsf{q}_\ell), \mathsf{st}) \leftarrow \mathsf{QueryGen}(1^k) \\ \forall i \in [\ell] : (\mathsf{a}_i, \mathsf{a}'_i) \leftarrow \mathsf{Adv}_2(1^k, \mathsf{pp}, \mathsf{q}_i) \end{array} \right] \leq \mu(k).$$

  We say that the argument system is (S, ϵ)-secure, for a function $S(k)$ and a negligible function $\epsilon(k)$, if for every constant $c > 0$, the soundness property holds for every pair of adversaries of size $S(k)^c$ with probability at most $\epsilon(k)^c$.

2.6 No-Signaling Multi-prover Arguments for RAM Computations

No signaling multi-prover arguments are multi-prover arguments, where the cheating provers are given extra power. In multi-prover arguments (or proofs), each prover answers its own query *locally*, without knowing anything about the queries that were sent to the other provers.

  In the no-signaling model we allow the *malicious* provers' answers to depend on all the queries, as long as for any subset $Q \subset [\ell]$ and for every two query vectors $\mathbf{q}^1 = (\mathsf{q}_1^1, \ldots, \mathsf{q}_\ell^1)$ and $\mathbf{q}^2 = (\mathsf{q}_1^2, \ldots, \mathsf{q}_\ell^2)$, such that $\mathbf{q}^1[Q] = \mathbf{q}^2[Q]$, the corresponding vectors of answers $\mathbf{a}^1, \mathbf{a}^2$ (as random variables) satisfy that $\mathbf{a}^1[Q]$

and $a^2[Q]$ are identically distributed. Intuitively, this means that the answers of the provers in the set Q do not contain information about the queries to the provers outside Q, except for the information that is already found in the queries to the provers in Q.

Definition 2.4. *For a set B and for $\ell \in \mathbb{N}$, we say that a pair of vectors of correlated random variables*

$$q = (q_1, \ldots, q_\ell), a = (a_1, \ldots, a_\ell) \in B^{[\ell]}.$$

is no-signaling *if for every subset $Q \subset [\ell]$ and every two vectors q^1, q^2 in the support of q such that $q^1[Q] = q^2[Q]$, the random variables $a[Q]$ conditioned on $q = q^1$ and $a[Q]$ conditioned on $q = q^2$ are identically distributed.*

If these random are not identical, but rather, the statistical distance between them is at most δ, we say that the pair (q, a) is δ-no-signaling.

Definition 2.5. *An ℓ-prover argument system* (ParamGen, MemGen, QueryGen, Output, Prover, Verifier) *for RAM computations is said to be sound against δ-no-signaling strategies (or provers) if the following (more general) soundness property is satisfied:*

For every pair of polynomial-size adversaries $(\mathsf{Adv}_1, \mathsf{Adv}_2)$ *satisfying a δ-no-signaling condition (specified below), there exists a negligible function μ such that for every $k \in \mathbb{N}$ and for $\ell = \ell(k)$:*

$$\Pr \left[\begin{array}{l} (y, \mathsf{d_{new}}) \neq (y', \mathsf{d'_{new}}) \\ 1 = \mathsf{Verifier}((M, x, \mathsf{T}, \mathsf{d}, y, \mathsf{d_{new}}), \mathsf{st}, (a_1, \ldots, a_\ell)) \\ 1 = \mathsf{Verifier}((M, x, \mathsf{T}, \mathsf{d}, y', \mathsf{d'_{new}}), \mathsf{st}, (a'_1, \ldots, a'_\ell)) \\[4pt] \mathsf{pp} \leftarrow \mathsf{ParamGen}(1^k) \\ (M, x, 1^\mathsf{T}, \mathsf{d}, y, \mathsf{d_{new}}, y', \mathsf{d'_{new}}) \leftarrow \mathsf{Adv}_1(1^k, \mathsf{pp}) \\ ((q_1, \ldots, q_\ell), \mathsf{st}) \leftarrow \mathsf{QueryGen}(1^k) \\ ((a_1, a'_1), \ldots, (a_\ell, a'_\ell)) \leftarrow \mathsf{Adv}_2(1^k, \mathsf{pp}, (q_1, \ldots, q_\ell)) \end{array} \right] \leq \mu(k).$$

where $(\mathsf{Adv}_1, \mathsf{Adv}_2)$ *satisfy the δ-no-signaling condition if the random variables $(q_1, \ldots, q_\ell)$ and*
$((a_1, a'_1), \ldots, (a_\ell, a'_\ell))$ *are δ-no-signaling.*

We say that the argument system is (S, ϵ)-secure against δ-no-signaling strategies, for a function $S(k)$ and a negligible function $\epsilon(k)$, if for every constant $c > 0$, the soundness property holds with probability at most $\epsilon(k)^c$ for every pair of adversaries of size $S(k)^c$ satisfying the δ-no-signaling condition.

3 Local Satisfiability

In this section we introduce the notion of *local satisfiability* for formulas, and state a result of [KRR14] providing a no-signaling multi-prover argument for the *local* satisfiability of any *non-deterministic* Turing machine computation. This presentation is based on an abstraction of [PR14].

We start by describing, for every non-deterministic Turing machine M and input x, a formula $\varphi_{M,x}$ of a specific structure that is satisfiable if and only if M accepts x. Then we define the notion of local satisfiability for formulas. Finally we state a result of [KRR14] providing a no-signaling multi-prover argument for the local satisfiability of formulas of the form $\varphi_{M,x}$.

3.1 A Formula Describing Non-Deterministic Computations

The machine M. Let M be a T-time S-space non-deterministic Turing machine. We can think of M as a two-input machine, such that M accepts the input x if and only if there exists a witness w such that $M(x, w)$ accepts. In what follows, we consider a machine M and an input x such that $|x|$ is smaller than the machine's space S. Therefore, we can assume that M copies the entire input x to its work tape. However, the witness w we consider may be such that $|w|$ is much larger than S and therefore w must be given on a separate read-only read-once witness tape.

The Machine's State. For $i \in [\mathsf{T}]$ let $\mathsf{st}_i \in \{0, 1\}^{O(\mathsf{S})}$ denote the state of the computation $M(x, w)$ immediately before the i-th step. The state st_i includes:

- the machine's state.
- the entire content of the work tape, including the reading head's location.
- the reading head's location j on the witness tape, and the witness bit w_j.

Note that st_i does not include the entire content of the witness tape which may be much longer than S.

The following theorem states that the decision of whether a non-deterministic Turing machine M accepts an inputs x can be converted into a 3-CNF formula $\varphi_{M,x}$ of a specific structure. Loosely speaking, the variables of $\varphi_{M,x}$ correspond to the entire tableau of the computation of $M(x, w)$, and the formula verifies the consistency of all the states of this computation. Thus, $\varphi_{M,x}$ can be separated into sub-formulas, where each sub-formula verifies the consistency of two adjacent states of the computation. This intuition is formalized in the following theorem.

Theorem 3.1. *For any* T*-time* S*-space non-deterministic Turing machine M and any input x there exists a 3-CNF Boolean formula $\varphi_{M,x}$ of size $O(\mathsf{T} \cdot \mathsf{S})$ such that the following holds:*

1. *$\varphi_{M,x}$ is satisfiable if and only if M accepts x. Moreover, given a witness for the fact that M accepts x there is an efficient way to find a satisfying assignment to $\varphi_{M,x}$.*
2. *The formula $\varphi_{M,x}$ can be written as*

$$\varphi_{M,x} = \bigwedge_{i \in [\mathsf{T}-1]} \varphi_{M,x}^i,$$

and the set of the input variables of $\varphi_{M,x}$, denoted by V, can be divided into subsets

$$V = \bigcup_{i \in [T]} V_i,$$

such that each formula $\varphi_{M,x}^i$ is over the variables $V_i \cup V_{i+1}$, and each $V_i \subseteq V$ is of size $S' = O(S)$.

3. There exists an efficient algorithm State such that given an assignment to the variables V_i, outputs a state st_i of the computation of $M(x)$ immediately before the i-th step,

$$\mathsf{st}_i = \mathsf{State}(\mathsf{a}[V_i]).$$

The algorithm State satisfies the following properties:

- For every $i \in [T-1]$ and for every assignment $\mathsf{a} \in \{0,1\}^{V_i \cup V_{i+1}}$, if $\varphi_{M,x}^i(\mathsf{a}) = 1$ then the states

$$\mathsf{st}_i = \mathsf{State}(\mathsf{a}[V_i]), \quad \mathsf{st}_{i+1} = \mathsf{State}(\mathsf{a}[V_{i+1}])$$

are consistent with the program M.
- For every assignment $\mathsf{a} \in \{0,1\}^{V_1 \cup V_2}$, if $\varphi_{M,x}^1(\mathsf{a}) = 1$ then the state

$$\mathsf{st}_1 = \mathsf{State}(\mathsf{a}[V_1])$$

is the initial state of the machine M with the input x.
- For every assignment $\mathsf{a} \in \{0,1\}^{V_{T-1} \cup V_T}$, if $\varphi_{M,x}^{T-1}(\mathsf{a}) = 1$ then the state

$$\mathsf{st}_T = \mathsf{State}(\mathsf{a}[V_T])$$

is an accepting state.

Remark 3.1 (On the formula size). It is well known that there exists a formula of size only $\tilde{O}(T)$ (independent of S) that is satisfiable if and only if M accepts x. Such a formula can be obtained by first making the machine M oblivious [PF79]. However such a formula will not have the desired structure described in Theorem 3.1.

3.2   Definition of Local Satisfiability

In this section we define the notion of local satisfiability for formulas.

Definition 3.1 (Local Assignment Generator [PR14]). *Let φ be a 3-CNF formula over a set of variables V. An (ℓ, ϵ, δ)-local assignment generator Assign for φ is a probabilistic algorithm running in time $\mathrm{poly}(|V|)$ that takes as input a set of at most ℓ queries $Q \subseteq V, |Q| \leq \ell$, and outputs an assignment $\mathbf{a} \in \{0,1\}^Q$, such that the following two properties hold.*

- **Everywhere Local Consistency.** *For every set $Q \subseteq V, |Q| \leq \ell$, with probability $1 - \epsilon$ over a draw*

$$\mathbf{a} \leftarrow \mathsf{Assign}(Q),$$

the assignment is locally consistent with the formula φ. That is, for every variables $q_1, q_2, q_3 \in Q$, every clause in φ over the variables q_1, q_2, q_3 is satisfied by the assignment $\mathbf{a}[\{q_1, q_2, q_3\}]$.

– **No-signaling.** *For every (all powerful) distinguisher D and every pair of sets Q, Q' such that $Q' \subseteq Q \subseteq V, |Q| \leq \ell$:*

$$\left| \Pr_{\mathsf{a} \leftarrow \mathsf{Assign}(Q)} [D(\mathsf{a}[Q']) = 1] - \Pr_{\mathsf{a}' \leftarrow \mathsf{Assign}(Q')} [D(\mathsf{a}') = 1] \right| \leq \delta.$$

Remark 3.2 (On ordered queries). In [PR14], the notion of local satisfiability is formalized using an ordered vector of queries. In Definition 3.1 however, the queries are given as an unordered set. We note that these formulations are equivalent.

3.3   No-Signaling Multi-prover Arguments for Local Satisfiability

To obtain our results we use a multi-prover proof system satisfying a no-signaling local soundness property (see Theorem 3.2 below). Such a proof system was constructed in [KRR14].

Let k be the security parameter and let $\ell = \ell(k)$ be a polynomial. Let M be a non-deterministic Turing machine running in time T and space S, let $x \in \{0,1\}^m$ be an input to M such that $\mathsf{T}(m) < 2^k$ and let w be a witness. We consider an ℓ-prover proof system $(\mathsf{LS.QueryGen}, \mathsf{LS.Prover}, \mathsf{LS.Verifier})$ with the following syntax and efficiency:

– $\mathsf{LS.QueryGen}(1^k) \rightarrow ((\mathsf{q}_1, \ldots, \mathsf{q}_\ell), \mathsf{st})$:
  A randomized polynomial-time algorithm that outputs a set of $\ell = \ell(k)$ queries $(\mathsf{q}_1, \ldots, \mathsf{q}_\ell)$, and a secret state st.
– $\mathsf{LS.Prover}(1^{\mathsf{T}(m)}, M, x, w, \mathsf{q}) \rightarrow \mathsf{a}$:
  A deterministic algorithm running in time $(\mathsf{T}(m) \cdot \mathsf{S}(m))^3 \cdot \mathsf{poly}(k)$ that outputs an answer a to a single query q where $|\mathsf{a}| = O(\log(k))$.
– $\mathsf{LS.Verifier}(M, x, \mathsf{st}, (\mathsf{a}_1, \ldots, \mathsf{a}_\ell)) \rightarrow b$:
  A deterministic algorithm running in time $m \cdot \mathsf{poly}(k)$, that outputs an acceptance bit b.

The completeness and no-signaling local soundness properties of the above proof system are given by Theorem 3.2 proved in [KRR14].[6]

Theorem 3.2 (*[KRR14]*). *There exists a polynomial ℓ_0, such that for every polynomial ℓ' and for $\ell = \ell_0 \cdot \ell'$ there exists an ℓ-prover proof system $(\mathsf{LS.QueryGen}, \mathsf{LS.Prover}, \mathsf{LS.Verifier})$ that satisfies the following properties.*

– *Completeness. For every T-time (two input) Turing machine M, every input $x \in \{0,1\}^m$ and witness w such that $M(x, w) = 1$, every $k \in \mathbb{N}$ such that $\mathsf{T}(m) < 2^k$, and for $\ell = \ell(k)$,*

$$\Pr\left[1 = \mathsf{LS.Verifier}(M, x, \mathsf{st}, (\mathsf{a}_1, \ldots, \mathsf{a}_\ell)) \mid \begin{array}{l} ((\mathsf{q}_1, \ldots, \mathsf{q}_\ell), \mathsf{st}) \leftarrow \mathsf{LS.QueryGen}(1^k) \\ \forall i \in [\ell] : \mathsf{a}_i \leftarrow \mathsf{LS.Prover}(1^{\mathsf{T}(m)}, M, x, w, \mathsf{q}_i) \end{array}\right] = 1.$$

---

[6] The proof of Theorem 3.2 follows by combining Lemmas 14.1, 6.1 and 7.29 in [KRR14] together with the fact that all the claims and lemmas in Sects. 7.1–7.5 hold for arbitrary setting of parameters, and in particular for any ϵ and δ.

– *No-Signaling Local Soundness. There exists a probabilistic polynomial-time oracle machine* Assign *such that the following holds. For every* T*-time (two input) Turing machine* M, *every input* $x \in \{0,1\}^m$, *every security parameter* $k \in \mathbb{N}$ *such that* $\mathsf{T}(m) < 2^k$ *and* $\ell = \ell(k)$, *every* $\epsilon = \epsilon(k)$, *every* $\delta = \delta(k)$, *and every* δ*-no-signaling cheating prover* Prover* *such that*

$$\Pr\left[1 = \mathsf{LS.Verifier}(M, x, \mathsf{st}, (\mathsf{a}_1, \ldots, \mathsf{a}_\ell)) \,\middle|\, \begin{array}{l} ((\mathsf{q}_1, \ldots, \mathsf{q}_\ell), \mathsf{st}) \leftarrow \mathsf{LS.QueryGen}(1^k) \\ (\mathsf{a}_1, \ldots, \mathsf{a}_\ell) \leftarrow \mathsf{Prover}^*(\mathsf{q}_1, \ldots, \mathsf{q}_\ell) \end{array}\right] \geq \epsilon,$$

Assign$^{\mathsf{Prover}^*}$ *is an* $(\ell', \delta', \epsilon')$-*local assignment generator for the 3-CNF formula* $\varphi_{M,x}$ *given by Theorem 3.1, with*

$$\delta' = \frac{\delta \cdot 2^{k \cdot \mathrm{polylog}(\mathsf{T}(m))}}{\epsilon}, \quad \epsilon' = \frac{\delta \cdot \mathrm{polylog}(\mathsf{T}(m))}{\epsilon}.$$

As before, we say that Prover* *is* δ-*no-signaling if the random variables* $(\mathsf{q}_1, \ldots, \mathsf{q}_\ell)$ *and* $(\mathsf{a}_1, \ldots, \mathsf{a}_\ell)$ *are* δ-*no-signaling.*

Remark 3.3. The oracle machine Assign constructed in [KRR14] has a super-polynomial runtime.[7] However, by carefully observing the proof, it is easy to see that this super-polynomial blowup is unnecessary. This was formally proved in a followup work of [BHK16].

4 No-Signaling Multi-prover Arguments for RAM Computations

4.1 Verifying RAM Computations via Local Satisfiability

In this section we translate any RAM computation into a non-deterministic Turing machine such that the RAM computation is correct if and only if the Turing machine's computation is locally satisfiable. Consider an execution of a RAM program M that on input x and initial memory string D outputs y and results in memory D_{new} within time T. Consider also a hash-tree of the initial memory D rooted at rt and a hash-tree of the final memory D_{new} rooted at $\mathsf{rt}_{\mathsf{new}}$.

We describe a Turing machine TVer that takes as input tuples of the form $(M, x, \mathsf{T}, \mathsf{rt}, y, \mathsf{rt}_{\mathsf{new}})$, together with a corresponding witness, which is a *transcript* of the RAM computation. We start by describing the algorithm TGen which generates the transcript. Roughly, the transcript contains a hash-tree proof of consistency for every memory access made by M (the precise structure of the transcript is described below). We then describe the algorithm TVer. The running time of TVer and TGen is proportional to the running time of the RAM computation (up to polynomial factors in the security parameter) and is independent of the size of the memory. In terms of soundness we argue that for any $(M, x, \mathsf{T}, \mathsf{rt})$ (even if rt is not computed honestly as the hash-tree root of some memory) and for every $(y', \mathsf{rt}'_{\mathsf{new}}) \neq (y, \mathsf{rt}_{\mathsf{new}})$, any cheating prover that passes

---

[7] This blowup is due to the soundness amplification lemma of [KRR14].

the no-signaling local soundness criterion for the computation of TVer with both the input $(M, x, \mathsf{T}, \mathsf{rt}, y, \mathsf{rt}_{\mathsf{new}})$ and the input $(M, x, \mathsf{T}, \mathsf{rt}, y', \mathsf{rt}'_{\mathsf{new}})$ can be used to break the soundness of the hash tree.

Let M be a RAM program, $x \in \{0, 1\}^m$ be an input, and $D \in \{0, 1\}^n$ be an initial memory string. Let

$$(\mathsf{HT.Gen}, \mathsf{HT.Hash}, \mathsf{HT.Read}, \mathsf{HT.Write}, \mathsf{HT.VerRead}, \mathsf{HT.VerWrite})$$

be a hash-tree scheme and let

$$\mathsf{key} \leftarrow \mathsf{HT.Gen}(1^k),$$
$$(\mathsf{tree}, \mathsf{rt}) \leftarrow \mathsf{HT.Hash}(\mathsf{key}, D).$$

The Transcript Generation Program TGen. We start by describing a program TGen that creates the transcript of the computation $M^D(x)$. Let

$$\mathsf{TGen}^{(\mathsf{tree} \rightarrow \mathsf{tree}_{\mathsf{new}})}(1^k, 1^\mathsf{T}, n, M, x) \rightarrow (y, \mathsf{rt}_{\mathsf{new}}, \mathsf{Trans})$$

be the following RAM program. TGen emulates the execution of $M^D(x)$ step by step as described in Sect. 2.2. The emulation begins with the initial memory containing the hash tree $\mathsf{tree}_1 = \mathsf{tree}$ with the initial root $\mathsf{rt}_1 = \mathsf{rt}$, the empty initial state state_1 and the read location $i_1^r = 1$. Starting from $j = 1$, the j-th emulation step proceeds as follows:

1. Read from the hash tree the bit:

$$(b_j^r, \mathsf{pf}_j^r) \leftarrow \mathsf{HT.Read}^{\mathsf{tree}_j}(i_j^r).$$

2. Compute $(\mathsf{state}_{j+1}, i_{j+1}^r, i_{j+1}^w, b_{j+1}^w) \leftarrow \mathsf{STEP}(M, x, \mathsf{state}_j, b_j^r)$.
3. If $i_{j+1}^w \neq \bot$, write a bit to the hash tree:

$$(\mathsf{rt}_{j+1}, \mathsf{pf}_{j+1}^w) \leftarrow \mathsf{HT.Write}^{(\mathsf{tree}_j \rightarrow \mathsf{tree}_{j+1})}(i_{j+1}^w, b_{j+1}^w).$$

The program M terminates after T emulation steps were completed with the terminating state $\mathsf{state}_{\mathsf{T}+1}$, which contains the output bit y. TGen then outputs y, $\mathsf{rt}_{\mathsf{new}} = \mathsf{rt}_{\mathsf{T}+1}$ and the transcript:

$$\mathsf{Trans} = \left((i_j^r, b_j^r, \mathsf{pf}_j^r), (i_{j+1}^w, b_{j+1}^w, \mathsf{rt}_{j+1}, \mathsf{pf}_{j+1}^w) \right)_{j \in [\mathsf{T}]}.$$

The running time of the program TGen is $\mathsf{T} \cdot \mathsf{poly}(k)$.

The transcript verification program TVer. Let

$$\mathsf{TVer}((M, x, \mathsf{T}, \mathsf{rt}, y, \mathsf{rt}_{\mathsf{new}}), \mathsf{Trans}) \rightarrow b$$

be the following Turing machine. TVer verifies the emulation of $M^D(x)$ based on the transcript:

$$\mathsf{Trans} = \left((i_j^r, b_j^r, \mathsf{pf}_j^r), (i_{j+1}^w, b_{j+1}^w, \mathsf{rt}_{j+1}, \mathsf{pf}_{j+1}^w) \right)_{j \in [\mathsf{T}']},$$

produced by TGen. The program first verifies that $\mathsf{T}' = \mathsf{T}$ Then, starting from the initial root $\widetilde{\mathsf{rt}}_1 = \mathsf{rt}$, the empty initial state state_1, the read location $i_1^r = 1$, and from $j = 1$, the j-th verification step proceeds as follows:

1. Verify that $\widetilde{i_j^r} = i_j^r$ and that

$$1 = \mathsf{HT.VerRead}(\widetilde{\mathsf{rt}_j}, i_j^r, b_j^r, \mathsf{pf}_j^r).$$

2. Compute $(\mathsf{state}_{j+1}, \widetilde{i_{j+1}^r}, \widetilde{i_{j+1}^w}, \widetilde{b_{j+1}^w}) \leftarrow \mathsf{STEP}(M, x, \mathsf{state}_j, b_j^r).$
3. Verify that $(\widetilde{i_{j+1}^w}, \widetilde{b_{j+1}^w}) = (i_{j+1}^w, b_{j+1}^w).$
4. If $i_{j+1}^w = \bot$ then verify that $\widetilde{\mathsf{rt}_j} = \mathsf{rt}_{j+1}$. Else, verify that

$$1 = \mathsf{HT.VerWrite}(\widetilde{\mathsf{rt}_j}, i_{j+1}^w, b_{j+1}^w, \mathsf{rt}_{j+1}, \mathsf{pf}_{j+1}^w).$$

5. If $j = \mathsf{T}$ verify that $\mathsf{rt}_{\mathsf{T}+1} = \mathsf{rt}_{\mathsf{new}}$ and that $\mathsf{state}_{\mathsf{T}+1}$ is terminating and includes the output y.
6. $\widetilde{\mathsf{rt}_{j+1}} \leftarrow \mathsf{rt}_{j+1}.$

The program outputs 1 if and only if all the verifications were successful. The running time of the program TVer is $\mathsf{T} \cdot \mathsf{poly}(k)$ and its space complexity is $\mathsf{poly}(k) \cdot \mathsf{polylog}(n) = \mathsf{poly}(k)$ (see Remark 2.1).

Additional structure of TVer. In order to prove Theorem 4.1 below, we make additional assumptions on the structure of the Turing machine TVer. We start by introducing some notation.

Verification Blocks. We assume that the execution of the machine can be divided into *blocks* where the computation in the j-th block is executing the j-th verification step. This assumption is satisfied by some "natural" implementation of TVer.

Formally, let $b = b(k) \leq \mathsf{poly}(k)$ be the block size. For every input $\tilde{x} = (M, x, \mathsf{T}, \mathsf{rt}, y, \mathsf{rt}_{\mathsf{new}})$ and for every transcript

$$\mathsf{Trans} = \left((i_j^r, b_j^r, \mathsf{pf}_j^r), (i_{j+1}^w, b_{j+1}^w, \mathsf{rt}_{j+1}, \mathsf{pf}_{j+1}^w) \right)_{j \in [\mathsf{T}]},$$

(not neccessarily such that $\mathsf{TVer}(\tilde{x}, \mathsf{Trans})$ accepts) let $\mathsf{T}' = \mathsf{T} \cdot b$ be the running time of $\mathsf{TVer}(\tilde{x}, \mathsf{Trans})$. For $i \in [\mathsf{T}']$ let st_i be the state of the computation $\mathsf{TVer}(\tilde{x}, \mathsf{Trans})$ immediately before the i-th step, and let $\mathsf{st}_{\mathsf{T}'+1}$ be the final state of the computation. The variables st_i describe the states of the computation of the program TVer, as defined by Theorem 3.1. (Note that these states are different from the local variables state_j used by the program TVer to emulate the RAM computation M.) For $j \in [\mathsf{T}]$, let B_j be the set of states in the j-th computation block.

$$B_j = \{ \mathsf{st}_i : (j-1) \cdot b < i \leq j \cdot b \}.$$

For notational convenience, we also define the block $B_{\mathsf{T}+1} = \{\mathsf{st}_{\mathsf{T}'+1}\}$ which describes the state of the computation after the final verification stap.

Additional requirements on the structure of TVer. Using the notion of blocks we formulate some additional requirements on the structure of TVer.

1. For every $j \in [\mathsf{T}]$, the bits of the transcript read in the j-th computation block contain the j-th entry of the transcript. Formally, there exists an efficient algorithm TVer.Transcript such that given the set of states B_j, outputs the j-th entry of the transcript

$$\left(i_j^{\mathsf{r}}, b_j^{\mathsf{r}}, \mathsf{pf}_j^{\mathsf{r}}\right), \left(i_{j+1}^{\mathsf{w}}, b_{j+1}^{\mathsf{w}}, \mathsf{rt}_{j+1}, \mathsf{pf}_{j+1}^{\mathsf{w}}\right) = \mathsf{TVer.Transcript}(B_j).$$

We also require that $\perp = \mathsf{TVer.Transcript}(B_{\mathsf{T}+1})$.

2. For every $j \in [\mathsf{T}]$, the j-th computation block contains the j-th state in the emulation of M. Formally, there exists an efficient algorithm TVer.State such that given the set of states B_j, outputs the state of M, the location of the next read and the root of the hash-tree before the j-th step of the emulation

$$\left(\mathsf{state}_j, \widetilde{i_j^{\mathsf{r}}}, \widetilde{\mathsf{rt}_j}\right) = \mathsf{TVer.State}(B_j).$$

On the final block $B_{\mathsf{T}+1}$, TVer.State outputs the terminating state of M, the last read location (TVer never reads the bit in this location), and the root of the final memory state.

$$\left(\mathsf{state}_{\mathsf{T}+1}, \widetilde{i_{\mathsf{T}+1}^{\mathsf{r}}}, \widetilde{\mathsf{rt}_{\mathsf{T}+1}}\right) = \mathsf{TVer.State}(B_{\mathsf{T}+1}).$$

3. When one of the tests performed by TVer fails, the machine transitions into a "rejecting state". Once TVer is in a rejecting state, we require that all its future states are rejecting and TVer rejects. Formally, there exists an efficiently computable predicate Reject such that
   (a) If in the j-th verification step test 1, 3 or 4 fails, or if $j = \mathsf{T}$ and test 5 fails, then $\mathsf{Reject}(B_j) = 1$.
   (b) For every $j \in [\mathsf{T}]$ if $\mathsf{Reject}(B_j) = 1$ then $\mathsf{Reject}(B_{j+1}) = 1$.
   (c) The computation $\mathsf{TVer}(\tilde{x}, \mathsf{Trans})$ rejects if and only if $\mathsf{Reject}(B_{\mathsf{T}+1}) = 1$.

Theorem 4.1. *The machines* TGen *and* TVer *satisfy the following properties:*

- *Completeness. For every $k \in \mathbb{N}$, every T-time RAM program M, every input $x \in \{0,1\}^m$, every initial memory $D \in \{0,1\}^n$ and every (y, D_{new}) such that $\mathsf{T}(m), n \leq 2^k$ and $y \leftarrow M^{(D \to D_{\mathsf{new}})}(x)$*

$$\Pr\left[\begin{matrix} 1 = \mathsf{TVer}((M, x, \mathsf{T}(m), \mathsf{rt}, y', \mathsf{rt}'_{\mathsf{new}}), \mathsf{Trans}) \\ (y', \mathsf{rt}'_{\mathsf{new}}) = (y, \mathsf{rt}_{\mathsf{new}}) \end{matrix} \middle| \begin{matrix} \mathsf{key} \leftarrow \mathsf{HT.Gen}(1^k) \\ (\mathsf{tree}, \mathsf{rt}) \leftarrow \mathsf{HT.Hash}(\mathsf{key}, D) \\ (\mathsf{tree}_{\mathsf{new}}, \mathsf{rt}_{\mathsf{new}}) \leftarrow \mathsf{HT.Hash}(\mathsf{key}, D_{\mathsf{new}}) \\ (y', \mathsf{rt}'_{\mathsf{new}}, \mathsf{Trans}) \leftarrow \mathsf{TGen}^{(\mathsf{tree} \to \mathsf{tree}_{\mathsf{new}})}(1^k, 1^{\mathsf{T}(m)}, n, M, x) \end{matrix}\right] = 1.$$

- *Soundness*

  Assume HT *is an* (S, ϵ)-*secure hash-tree scheme for a function* $S(k)$ *and a negligible function* $\epsilon(k)$. *There exists a polynomial* ℓ' *such that for every constant* $c > 0$ *and every triplet of adversaries* $(\mathsf{Adv}_1, \mathsf{Adv}_2, \mathsf{Adv}_3)$ *of size* $S(k)^c$, *there exist constants* $c_1, c_2 > 0$ *such that for every large enough* $k \in \mathbb{N}$

$$\Pr\left[\begin{array}{c} (y, \mathsf{rt}_{\mathsf{new}}) \neq (y', \mathsf{rt}'_{\mathsf{new}}) \\ \mathsf{CHEAT} \end{array} \,\middle|\, \begin{array}{l} \mathsf{key} \leftarrow \mathsf{HT.Gen}(1^k) \\ (M, x, 1^\mathsf{T}, \mathsf{rt}, y, \mathsf{rt}_{\mathsf{new}}, y', \mathsf{rt}'_{\mathsf{new}}) \leftarrow \mathsf{Adv}_1(1^k, \mathsf{key}) \end{array} \right] \leq \epsilon(k)^{c_2},$$

  where CHEAT *is the event that:*

  - $\mathsf{Adv}_2(\mathsf{key}, \cdot)$ *is an* $(\ell'(k), S(k)^{-c_1}, S(k)^{-c_1})$-*local assignment generator for the 3-CNF formula* $\varphi_{\mathsf{TVer}, \tilde{x}_2}$ *where* $\tilde{x}_2 = (M, x, \mathsf{T}, \mathsf{rt}, y, \mathsf{rt}_{\mathsf{new}})$ *and* $\varphi_{\mathsf{TVer}, \tilde{x}}$ *is as defined in Theorem 3.1.*
  - $\mathsf{Adv}_3(\mathsf{key}, \cdot)$ *is an* $(\ell'(k), S(k)^{-c_1}, S(k)^{-c_1})$-*local assignment generator for the 3-CNF formula* $\varphi_{\mathsf{TVer}, \tilde{x}_3}$ *where* $\tilde{x}_3 = (M, x, \mathsf{T}, \mathsf{rt}, y', \mathsf{rt}'_{\mathsf{new}})$ *and* $\varphi_{\mathsf{TVer}, \tilde{x}'}$ *is as defined in Theorem 3.1.*

The proof of Theorem 4.1 can be found in the full version of this work [KP15].

4.2   The Protocol

In this section we describe our no-signaling multi-prover argument for RAM computations. The construction uses the following components.

- A hash-tree scheme (HT.Gen, HT.Hash, HT.Read, HT.Write, HT.VerRead, HT.VerWrite), given by Theorem 2.1.
- The ℓ-prover proof system (LS.QueryGen, LS.Prover, LS.Verifier) for local satisfiability given by Theorem 3.2 in Sect. 3.3, where $\ell = \ell' \cdot \ell_0$, and ℓ' is the polynomial given by Theorem 4.1 and ℓ_0 is the polynomial given by Theorem 3.2.
- The transcript generation and verification programs TGen, TVer described in Sect. 4.1. We only rely on the following facts
  - The programs TGen, TVer satisfy Theorem 4.1.
  - For security parameter k and for a T-time computation, the running time of the transcript generation program TGen is $\mathsf{T} \cdot \mathsf{poly}(k)$. The running time of the transcript verification program TVer (on the transcript generated by TGen) is $\mathsf{T} \cdot \mathsf{poly}(k)$ and its space complexity is $\mathsf{poly}(k)$.

The multi-prover argument is given by the following procedures:

- ParamGen(1^k) generates a key for the hash-tree:

$$\mathsf{key} \leftarrow \mathsf{HT.Gen}(1^k),$$

and outputs pp = key.
- MemGen(pp, D), given pp = key, computes a hash-tree for the memory D:

$$(\mathsf{tree}, \mathsf{rt}) \leftarrow \mathsf{HT.Hash}(\mathsf{key}, D),$$

and outputs (dt, d) = (tree, rt).

- QueryGen(1^k) executes the query generation algorithm of the local-satisfiability proof system:

$$((q_1, \ldots, q_\ell), st) \leftarrow LS.QueryGen(1^k),$$

and outputs $((q_1, \ldots, q_\ell), st)$.
- Output$^{dt}(1^T, n, M, x)$, given access to the memory $dt = tree$, executes the transcript generation program:

$$(y, rt_{new}, Trans) \leftarrow TGen^{(tree \rightarrow tree_{new})}(1^k, 1^T, n, M, x),$$

and outputs $(y, d_{new}, Trans) = (y, rt_{new}, Trans)$.
- Prover$((M, x, T, d, y, d_{new}), Trans, q)$, where $(d, d_{new}) = (rt, rt_{new})$, does the following:
  1. Let $T' = T \cdot poly(k)$ and $S' = poly(k)$ be the time and space complexity of the computation

$$TVer((M, x, T, rt, y, rt_{new}), Trans).$$

  2. Execute the local-satisfiability prover for the above computation:

$$a \leftarrow LS.Prover(1^{T'}, TVer, (M, x, T, rt, y, rt_{new}), Trans, q).$$

  3. Output a.
- Verifier$((M, x, T, d, y, d_{new}), st, (a_1, \ldots, a_\ell))$, where $(d, d_{new}) = (rt, rt_{new})$, executes the local-satisfiability verifier:

$$b \leftarrow LS.Verifier(TVer, (M, x, T, rt, y, rt_{new}), st, (a_1, \ldots, a_\ell)),$$

and outputs b.

Theorem 4.2. *Assume* HT *is an* (S, ϵ)*-secure hash-tree scheme for a function* $S(k)$ *and a negligible function* $\epsilon(k)$*. Then* (ParamGen, MemGen, QueryGen, Output, Prover, Verifier) *is an* ℓ*-prover argument system for RAM computations that is* (S, ϵ)*-secure against* δ*-no-signaling provers for* $\delta(k) = 2^{-k \cdot polylog(S(k))}$*.*

The proof of Theorem 4.2 follows by combining Theorems 3.2 and 4.1 and can be found in the full version of this work [KP15].

References

[ACC+15] Ananth, P., Chen, Y.-C., Chung, K.-M., Lin, H., Lin, W.-K.: Delegating RAM computations with adaptive soundness and privacy. IACR Cryptology ePrint Archive, 2015:1082 (2015)
[BCC+14] Bitansky, N., Canetti, R., Chiesa, A., Goldwasser, S., Lin, H., Rubinstein, A., Tromer, E.: The hunting of the SNARK. IACR Cryptology ePrint Archive, 2014:580 (2014)
[BCCT13] Bitansky, N., Canetti, R., Chiesa, A., Tromer, E.: Recursive composition and bootstrapping for snarks and proof-carrying data. In: STOC, pp. 111–120 (2013)

[BCI+13] Bitansky, N., Chiesa, A., Ishai, Y., Paneth, O., Ostrovsky, R.: Succinct non-interactive arguments via linear interactive proofs. In: Sahai, A. (ed.) TCC 2013. LNCS, vol. 7785, pp. 315–333. Springer, Heidelberg (2013). doi:10.1007/978-3-642-36594-2_18

[BEG+91] Blum, M., Evans, W.S., Gemmell, P., Kannan, S., Naor, M.: Checking the correctness of memories. In: 32nd Annual Symposium on Foundations of Computer Science, San Juan, Puerto Rico, 1–4 October 1991, pp. 90–99 (1991)

[BGL+15] Bitansky, N., Garg, S., Lin, H., Pass, R., Telang, S.: Succinct random-ized encodings and their applications. In: Proceedings of the Forty-Seventh Annual ACM on Symposium on Theory of Computing, STOC 2015, Portland, OR, USA, June 14–17, 2015, pp. 439–448 (2015)

[BHK16] Brakerski, Z., Holmgren, J., Kalai, Y.T.: Non-interactive RAM and batch NP delegation from any PIR. Electron. Colloquium Comput. Complex. (ECCC) **23**, 77 (2016)

[CCC+15] Chen, Y.-C., Chow, S.S.M., Chung, K.-M., Lai, R.W.F., Lin, W.-K., Zhou, H.-S.: Computation-trace indistinguishability obfuscation and its applica-tions. IACR Cryptology ePrint Archive,2015:406 (2015)

[CCHR15] Canetti, R., Chen, Y., Holmgren, J., Raykova, M.: Succinct adaptive gar-bled ram. Cryptology ePrint Archive, Report 2015/1074 (2015). http://eprint.iacr.org/

[CH15] Canetti, R., Holmgren, J.: Fully succinct garbled RAM. IACR Cryptology ePrint Archive, 2015:388 (2015)

[CHJV15] Canetti, R., Holmgren, J., Jain, A., Vaikuntanathan, V.: Succinct garbling and indistinguishability obfuscation for RAM programs. In: Proceedings of the Forty-Seventh Annual ACM on Symposium on Theory of Computing, STOC 2015, Portland, OR, USA, June 14–17, 2015, pp. 429–437 (2015)

[CKLR11] Chung, K.-M., Kalai, Y.T., Liu, F.-H., Raz, R.: Memory delegation. In: Rogaway, P. (ed.) CRYPTO 2011. LNCS, vol. 6841, pp. 151–168. Springer, Heidelberg (2011). doi:10.1007/978-3-642-22792-9_9

[DFH12] Damgård, I., Faust, S., Hazay, C.: Secure two-party computation with low communication. In: Cramer, R. (ed.) TCC 2012. LNCS, vol. 7194, pp. 54–74. Springer, Heidelberg (2012). doi:10.1007/978-3-642-28914-9_4

[GGPR13] Gennaro, R., Gentry, C., Parno, B., Raykova, M.: Quadratic span programs and succinct NIZKs without PCPs. In: Johansson, T., Nguyen, P.Q. (eds.) EUROCRYPT 2013. LNCS, vol. 7881, pp. 626–645. Springer, Heidelberg (2013). doi:10.1007/978-3-642-38348-9_37

[GGR15] Goldreich, O., Gur, T., Rothblum, R.: Proofs of proximity for context-free languages and read-once branching programs. Electron. Colloquium Comput. Complex. (ECCC) **22**, 24 (2015)

[GHRW14] Gentry, C., Halevi, S., Raykova, M., Wichs, D.: Outsourcing private RAM computation. In: 55th IEEE Annual Symposium on Foundations of Com-puter Science, FOCS 2014, Philadelphia, PA, USA, October 18–21, 2014, pp. 404–413 (2014)

[GR15] Gur, T., Rothblum, R.D.: Non-interactive proofs of proximity. In: Pro-ceedings of the 2015 Conference on Innovations in Theoretical Computer Science, ITCS 2015, Rehovot, Israel, January 11–13, 2015, pp. 133–142 (2015)

[Gro10] Groth, J.: Short pairing-based non-interactive zero-knowledge arguments. In: Abe, M. (ed.) ASIACRYPT 2010. LNCS, vol. 6477, pp. 321–340. Springer, Heidelberg (2010). doi:10.1007/978-3-642-17373-8_19

[GVW15] Gorbunov, S., Vaikuntanathan, V., Wichs, D.: Leveled fully homomorphic signatures from standard lattices. In: Proceedings of the Forty-Seventh Annual ACM on Symposium on Theory of Computing, STOC 2015, Portland, OR, USA, June 14–17, 2015, pp. 469–477 (2015)

[Kil92] Kilian, J.: A note on efficient zero-knowledge proofs and arguments. In: Proceedings of the 24th Annual ACM Symposium on Theory of Computing, pp. 723–732 (1992)

[KP15] Kalai, Y.T., Paneth, O.: Delegating RAM computations. IACR Cryptology ePrint Archive, 2015:957 (2015)

[KR15] Kalai, Y.T., Rothblum, R.D.: Arguments of proximity (extended abstract). In: Gennaro, R., Robshaw, M. (eds.) CRYPTO 2015. LNCS, vol. 9216, pp. 422–442. Springer, Heidelberg (2015). doi:10.1007/978-3-662-48000-7_21

[KRR14] Kalai, Y.T., Raz, R., Rothblum, R.D.: How to delegate computations: the power of no-signaling proofs. In: Symposium on Theory of Computing, STOC 2014, New York, NY, USA, May 31 - June 03, 2014, pp. 485–494 (2014)

[Lip12] Lipmaa, H.: Progression-free sets and sublinear pairing-based non-interactive zero-knowledge arguments. In: Cramer, R. (ed.) TCC 2012. LNCS, vol. 7194, pp. 169–189. Springer, Heidelberg (2012). doi:10.1007/978-3-642-28914-9_10

[Mer87] Merkle, R.C.: A digital signature based on a conventional encryption function. In: Pomerance, C. (ed.) CRYPTO 1987. LNCS, vol. 293, pp. 369–378. Springer, Heidelberg (1988). doi:10.1007/3-540-48184-2_32

[Mic94] Micali, S.: CS proofs (extended abstracts). In': 35th Annual Symposium on Foundations of Computer Science, Santa Fe, New Mexico, USA, 20–22 November 1994, pp. 436–453 (1994)

[PF79] Pippenger, N., Fischer, M.J.: Relations among complexity measures. J. ACM **26**(2), 361–381 (1979)

[PR14] Paneth, O., Rothblum, G.N.: Publicly verifiable non-interactive arguments for delegating computation. Cryptology ePrint Archive, Report 2014/981 (2014). http://eprint.iacr.org/

[RVW13] Rothblum, G.N., Vadhan, S.P., Wigderson, A.: Interactive proofs of proximity: delegating computation in sublinear time. In: Symposium on Theory of Computing Conference, STOC 2013, Palo Alto, CA, USA, June 1–4, 2013, pp. 793–802 (2013)

Public-Key Encryption

Standard Security Does Not Imply Indistinguishability Under Selective Opening

Dennis Hofheinz[1]([⊠]), Vanishree Rao[2], and Daniel Wichs[3]

[1] Karlsruhe Institute of Technology, Karlsruhe, Germany
dennis.hofheinz@kit.edu
[2] PARC, a Xerox Company, Palo Alto, USA
[3] Northeastern University, Massachusetts, USA

Abstract. In a *selective opening attack* (SOA) on an encryption scheme, the adversary is given a collection of ciphertexts and she selectively chooses to see some subset of them "opened", meaning that the messages and the encryption randomness are revealed to her. A scheme is SOA secure if the data contained in the unopened ciphertexts remains hidden. A fundamental question is whether every CPA secure scheme is necessarily also SOA secure. The work of Bellare et al. (EUROCRYPT'12) gives a partial negative answer by showing that some CPA secure schemes do not satisfy a simulation-based definition of SOA security called SIM-SOA. However, until now, it remained possible that every CPA-secure scheme satisfies an indistinguishability-based definition of SOA security called IND-SOA.

In this work, we resolve the above question in the negative and construct a highly contrived encryption scheme which is CPA (and even CCA) secure but is not IND-SOA secure. In fact, it is broken in a very obvious sense by a selective opening attack as follows. A random value is secret-shared via Shamir's scheme so that any t out of n shares reveal no information about the shared value. The n shares are individually encrypted under a common public key and the n resulting ciphertexts are given to the adversary who selectively chooses to see t of the ciphertexts opened. Counter-intuitively, by the specific properties of our encryption scheme, this suffices for the adversary to completely recover the shared value. Our contrived scheme relies on strong assumptions: public-coin differing inputs obfuscation and a certain type of correlation intractable hash functions.

We also extend our negative result to the setting of SOA attacks with *key opening* (IND-SOA-K) where the adversary is given a collection of ciphertexts under different public keys and selectively chooses to see some subset of the secret keys.

1 Introduction

When it comes to defining the security of encryption schemes, the standard definitions of chosen-plaintext attack (CPA) and chosen-ciphertext attack (CCA) security are generally thought of as the gold standard. Nevertheless, there are

© International Association for Cryptologic Research 2016
M. Hirt and A. Smith (Eds.): TCC 2016-B, Part II, LNCS 9986, pp. 121–145, 2016.
DOI: 10.1007/978-3-662-53644-5_5

scenarios in which these notions do not appear to provide sufficient guarantees. One such scenario is that of *selective opening attacks* (SOA) [4,11].

Selective Opening Attacks. In a selective opening attack, the adversary gets a collection of n ciphertexts $(c_i = \mathsf{Enc}_{pk}(m_i; r_i))_{i \in [n]}$ encrypting messages m_i with randomness r_i under a common public key pk. The adversary can adaptively choose to see some subset $\mathcal{I} \subseteq [n]$ of the ciphertexts "opened", meaning that she gets $(m_i, r_i)_{i \in \mathcal{I}}$. For example, this could model a scenario where these ciphertexts are created by different senders and the adversary adaptively corrupts some subset of them. Intuitively, a scheme is SOA secure if the data contained in the unopened ciphertexts remains hidden. Formalizing this notion requires great care, and several definitions have been proposed.

Simulation-Based SOA Security. Perhaps the strongest notion of SOA security is a *simulation-based* definition, which we denote SIM-SOA. It was originally proposed for commitments by Dwork et al. [11] and later adapted to encryption by Bellare et al. [4]. This definition requires that for any n-tuple of messages $\mathbf{m} = (m_1, \ldots, m_n)$ the view of the adversary in the above SOA scenario is indistinguishable from a simulated view created as follows: the simulator selects a message subset I, obtains $(m_i)_{i \in I}$, and is then supposed to output a view of a selective opening attack with ciphertexts, random coins, and an adversary as above. At when constructing the simulator in a black-box fashion out of a given adversary, this means that the simulator must initially creates a collection of simulated ciphertexts $\mathbf{c} = (c_1, \ldots, c_n)$ without knowing anything about the messages. The adversary then gets $\mathbf{c}$ and specifies a subset $\mathcal{I} \subseteq [n]$ of the ciphertexts to be opened. At this point, the simulator learns the messages $(m_i)_{i \in \mathcal{I}}$ and has to produce simulated openings $(m_i, r_i)_{i \in \mathcal{I}}$ to give to the adversary.

On the positive side, this definition is easy to use in applications and clearly captures the intuitive goal of SOA security, since the adversary's view can be simulated without using any knowledge of the unopened messages. Moreover, we have constructions that achieve SIM-SOA security from a wide variety of number theoretic assumptions [4,12,14,18,19,21].

On the negative side, this definition might be overkill in many applications and therefore also unnecessarily hard to achieve. The work of Bellare et al. [3] shows that many natural encryption schemes are *not* SIM-SOA secure, in the sense that there is no efficient simulator that would satisfy the given definition. The lack of a simulator already constitutes an attack on SIM-SOA security in the formal sense. However, these schemes are also not "obviously broken" by a selective opening attack in the intuitive sense. In particular, it is not clear how to, e.g., extract an unopened plaintext in a selective opening attack. At the very least, it remains unclear what exactly can go wrong when using such schemes in the context of the SOA scenario described above.

Indistinguishability-Based SOA Security. The work of Bellare et al. [4] also proposes an indistinguishability-based security definition, which we denote IND-SOA. The definition requires that we have an "efficiently re-samplable" distribution on n-tuples of messages $\mathbf{m} = (m_1, \ldots, m_n)$ such that for any set

$\mathcal{I} \subseteq [n]$ we can efficiently sample from the correct conditional distribution with a fixed choice of $(m_i)_{i \in \mathcal{I}}$. For any such distribution we consider the SOA scenario where the adversary initially gets encryptions of the messages $\mathbf{m} = (m_1, \ldots, m_n)$ chosen from the distribution, and selectively gets to see an opening of a subset $\mathcal{I}$ of the ciphertext. At the end of the game the adversary either gets the initially encrypted message vector $\mathbf{m}$ or a freshly re-sampled message vector $\mathbf{m}' = (m_1', \ldots, m_n')$ conditioned on $m_i' = m_i$ matching in the opened positions $i \in \mathcal{I}$. The adversary should not be able to distinguish these two cases.

On the negative side, the definition of IND-SOA security is more complex and its implications are harder to interpret. However, it can already provide sufficient security guarantees in many interesting applications and might be significantly easier to achieve than SIM-SOA security. Prior to this work, we did not know whether it is the case that every CPA secure encryption scheme is also IND-SOA secure. The work of Hofheinz and Rupp [20] shows that, if one considers a definition that combines IND-SOA security with CCA security, denoted by IND-SO-CCA, then there are schemes that are CCA secure but are not IND-SO-CCA secure. However, this result crucially relies on the embedding of an attack in the decryption oracle, and does not appear to extend to the standard IND-SOA. In fact, the same work of [20] gave a partial positive result showing that CPA security implies IND-SOA security for a large class of encryption schemes in a generic group model, but it was unclear what the situation is in the standard model.

More Related Work. The relations between different definitions of SOA security have also been investigated by Böhl et al. [6]. It turns out that the notion of IND-SOA security we consider is the weakest known notion of SOA security among the ones studied (and that the "efficient resamplability" condition is essential for this property). Hazay et al. [17] recently studied SOA for keys (where the adversary receives secret keys of corresponding chosen subset of ciphertexts) and showed that the indistinguishability-based security is strictly weaker than the simulation-based counterpart. Furthermore, there exist several efficient constructions of IND-SOA secure encryption schemes that are *not* known to be SIM-SOA secure. Most prominently, every lossy encryption scheme is IND-SOA secure [4], which opens the door for efficient IND-SOA secure schemes from various computational assumptions [26–28]. In that sense, the notion of IND-SOA we consider is very attractive from a practical point of view. In an orthogonal direction, Fuchsbauer et al. [13], recently showed that standard security implies IND-SOA for certain specific graph-induced distributions; it is interesting to note that, while we used dependencies of messages to show our negative result, [13] used the lack of dependencies to show a positive result.

Secret Sharing: A Concrete SOA Scenario. At this point, an intuitive definition of SOA security might appear elusive, with strong definitions like SIM-SOA that could be overkill and weaker definitions like IND-SOA that are hard to interpret. Instead of trying to pin down a general notion of SOA security, we will focus on defining a concrete and easy to understand security goal, which any reasonable definition of SOA security should satisfy. We call this goal *secret-sharing selective-opening attack* (SecShare-SOA) security, and define it via the following game.

The challenger chooses a random polynomial F of degree $\leq t$ and sets $m_i = F(i)$ for $i \in [n]$. We can think of this as a Shamir secret sharing of a random value $F(0)$ where any t of the n shares preserve privacy. The adversary is given encryptions of the shares $(c_i = \mathsf{Enc}_{pk}(m_i; r_i))_{i \in [n]}$ and can selectively choose to get openings $(m_i, r_i)_{i \in \mathcal{I}}$ for a subset $\mathcal{I}$ of the ciphertexts where $|\mathcal{I}| = t$. The adversary should not be able to predict $F(0)$.

It is easy to show that SecShare-SOA security is implied by IND-SOA (and therefore also SIM-SOA) security. At first thought, it may seem that SecShare-SOA security should also follow from standard CPA security. However, upon some reflection, it becomes clear that natural reductions fail. In particular, there is no easy way to embed the challenge ciphertext c^* into a correctly distributed vector $(c_i)_{i \in [n]}$ while maintaining the ability to provide openings for a large subset of the ciphertexts.

Our Results. In this work, we construct a contrived encryption scheme which is CPA (and even CCA) secure, but is not SecShare-SOA secure (and therefore also not IND-SOA secure). In particular, we have an attack against the SecShare-SOA security of the scheme where the attacker always recovers the shared secret with probability 1. This is the first example of a CPA secure scheme which is obviously broken in the SOA setting. As a corollary, this shows that not every CPA secure scheme is IND-SOA secure.

We also extend our results to selective opening attacks on receiver keys (IND-SOA-K), also known as selective opening under receiver corruption. In this setting, the adversary is given a collection of ciphertexts under different public keys and he can selectively chose to see some subset of the secret keys. We give an analogous example of a scheme which is CCA secure but is not IND-SOA-K secure.

Our results rely on strong assumptions: public-coin differing inputs obfuscation [23] and a certain type of correlation-intractable hash functions [8].

1.1 Our Techniques

We construct a scheme which is CCA secure but for which there is an attack on the SecShare-SOA security. For concreteness, we will show an attack on the SecShare-SOA game using a secret sharing scheme with parameters $t = k$ (degree of polynomial) and $n = 3k$ (number of shares) where k is the security parameter.

An SOA Helper Oracle. As our starting point, we consider the construction of Hofheinz and Rupp [20] which gives a CCA secure scheme that is not IND-SO-CCA secure. Their construction starts with any CCA secure scheme and, as an implicit first step, defines a (stateful and interactive) "SOA helper oracle" that has knowledge of the secret key sk of the scheme. The way that the oracle is defined ensures that the scheme remains CCA secure but is not SecShare-SOA secure relative to this oracle. They then show how to embed this oracle into the decryption procedure of the scheme to get a scheme which is not IND-SO-CCA secure.

The SOA helper oracle gets as input ciphertexts $(c_i)_{i \in [3k]}$ and it randomly chooses a subset $\mathcal{I} \subset [3k]$ of size $|\mathcal{I}| = k$ of them to open. It then receives the openings $(m_i, r_i)_{i \in \mathcal{I}}$ and decrypts the remaining ciphertexts using knowledge of sk. It checks that there is a (unique) degree $\leq k$ polynomial F such that $F(i) = m_i$ for $> 2k$ of the indices $i \in [3k]$ and that this polynomial also satisfies $F(i) = m_i$ for all of the indices $i \in \mathcal{I}$. If so, it outputs $F(0)$ and else $\bot$.

It is easy to see that this oracle breaks SecShare-SOA security. The harder part is showing that the scheme remains CPA/CCA secure relative to the oracle. In particular, we want to show that this oracle will not help the adversary decrypt some challenge ciphertext c^*. We do so by defining an "innocuous SOA helper oracle" that functions the same way as the real SOA helper oracle but it never decrypts c^*. Instead, it just pretends that the decryption of c^* is $\bot$. The only time that innocuous SOA helper and the real SOA helper give a different answer is when the ciphertexts $(c_i)_{i \in [3k]}$ encrypt messages (m_i) such that there is a unique degree $\leq k$ polynomial F with $F(i) = m_i$ for exactly $2k+1$ of the indices $i \in [3k]$, and this polynomial satisfies $F(i) = m_i$ for all $i \in \mathcal{I}$. Only in this case, there is a possibility that the SOA helper correctly outputs F while the innocuous SOA helper outputs $\bot$ when the decryption of c^* is replaced by $\bot$. However, since the set $\mathcal{I} \subseteq [3k]$ of size $|\mathcal{I}| = k$ is chosen randomly and independently of (c_i), the probability that it is fully contained in the set of $2k + 1$ indices for which $F(i) = m_i$ is negligible. Therefore, the SOA helper and the innocuous SOA helper give the same answer with all but negligible probability, meaning that the former cannot break CCA security.

Obfuscating the SOA Helper. Our main idea is that, instead of embedding the SOA helper in the secret-key decryption procedure, we obfuscate the SOA helper and include the obfuscated code in the public key of the scheme. (We note that a similar technique of "obfuscating a helper oracle that aids an attacker" has been used in the key-dependent message setting [24, 25].) There are two main difficulties that we must take care of.

The first difficulty is that the SOA helper is stateful/interactive whereas we can only obfuscate a stateless program. We squash the interactive helper into a non-interactive one by choosing the set of indices $\mathcal{I} \subseteq [3k], |\mathcal{I}| = k$ via a hash function h applied to the ciphertexts $(c_i)_{i \in [3k]}$. One can think of this as an analogue of the Fiat-Shamir heuristic which is used to squash a 3 move Σ-protocol into a non-interactive argument. (We stress, however, that we do not rely on random oracles, as in the Fiat-Shamir heuristic. Instead, we use a suitable standard-model hash function.) The squashed SOA helper now expects to get the ciphertexts $(c_i)_{i \in [3k]}$ and the opening $(m_i, r_i)_{i \in \mathcal{I}}$ where $\mathcal{I} = h((c_i)_{i \in [3k]})$ in one shot. Previously, we used the fact that the set $\mathcal{I}$ is random to argue that the SOA helper and the innocuous SOA helper are indistinguishable. We now instead rely on correlation intractability [8] of the hash function h to argue that it is hard to find an input on which the two oracles would give a different answer (even given the entire code and secrets of the oracles).

The second difficulty is how to use reasonable notions of obfuscation to argue that the obfuscated SOA helper, which contains the decryption key inside it, does not break CPA/CCA security. We rely on public-coin differing inputs obfuscation (PdiO) [23]. This security notion says that, given two programs represented as circuits C, C', together with all the random coins used to sample them, if it is hard to find an input x such that $C(x) \neq C'(x)$ then the obfuscations of C and C' are indistinguishable. We can rely on public-coin differing-inputs obfuscation and the correlation intractability of h to replace the obfuscated SOA oracle with an obfuscated "innocuous SOA oracle" that never decrypts the challenge ciphertext c^*. However, even the latter oracle still has the secret key sk hard-coded and therefore it is not clear if an obfuscated version of the innocuous oracle remains innocuous. To solve this problem, we will need the underlying CCA encryption scheme to be "puncturable" meaning that we can create a punctured secret key $sk[c^*]$ which correctly decrypts all ciphertexts other than c^* but preserves the semantic security of c^*. Such encryption schemes were constructed in the work of [9] from indistinguishability obfuscation. With this approach we can argue that security of the challenge ciphertext c^* is preserved.

Discussion on our Assumptions. We recall that two of the main assumptions behind our results are public-coin differing inputs obfuscation and correlation-intractable hash functions.

The notion of public-coin differing inputs obfuscation (PdiO) is stronger than indistinguishability obfuscation (iO) but weaker than differing-inputs obfuscation (diO) [1]. There is some evidence that diO is unachievable in its full generality [5,16], but no such evidence exists for PdiO. Indeed, at present we do not have much more evidence for the existence of iO than we do for PdiO. We note that if PdiO exists, then by the "best-possible" nature of iO, any iO obfuscator (with sufficient padding) is already also a PdiO obfuscator as well. All that said, we view it as an intriguing open problem to base our results on iO rather than PdiO.

The correlation intractability assumption that we need is in a parameter regime with no known counter-examples and has been conjectured to be achievable. As evidence, a recent work [7] constructs such correlation-intractable hash functions under obfuscation-based assumptions. However, the description of the hash functions is not public-coin samplable, whereas we need a hash function that is. We simply conjecture that standard hash function constructions such as SHA-3 achieve this property. We note that the notion of correlation intractability that we need is also a special case of *entropy-preserving* hashing [2,10] which is sufficient to guarantee the soundness of the Fiat-Shamir heuristic for all proof (but not argument) systems and has been conjectured to exist.

On this note, an interesting direction for future work is to re-establish the results based on weaker assumptions.

2   Preliminaries

General Notation. For $n \in \mathbb{N}$ we define $[n] := \{1, \ldots, n\}$. Throughout the paper, $k \in \mathbb{N}$ denotes the security parameter. For any function $g(\cdot)$, we let $g(k) = \mathsf{negl}(k)$ denote that $g(\cdot)$ is a negligible function. For any two distributions $\mathcal{D}_0, \mathcal{D}_1$ parameterized by k, we denote that they are computationally (resp., statistically) indistinguishable by $\mathcal{D}_0 \approx_c \mathcal{D}_1$ (resp., $\mathcal{D}_0 \approx_s \mathcal{D}_1$); we denote that they are identical by $\mathcal{D}_0 \equiv \mathcal{D}_1$.

Interpolation, Error Decoding. Let $\mathbb{F}$ be the finite field. For pairwise different $X_i \in \mathbb{F}$ we let $\mathsf{ipol}((X_i, Y_i)_{i \in [k+1]})$ denote the unique degree $\leq k$ polynomial $F \in \mathbb{F}[X]$ with $F(X_i) = Y_i$ for all $i \in [k+1]$. We note that ipol can be efficiently computed, e.g., via Lagrange interpolation. Also, let $\mathsf{decc}_k((X_i, Y_i)_{i \in [n]})$ denote the the unique degree $\leq k$ polynomial $F \in \mathbb{F}[X]$ such that $F(X_i) = Y_i$ for $> n - (n-k)/2$ of the indices $i \in [n]$, or $\perp$ if no such polynomial exists. Evaluating decc amounts to performing error correction for the Reed-Solomon code with distance $d = (n-k)$ when there are $< d/2$ errors, which can be done efficiently. Let $\mathcal{S}_\ell^S$ denote the set of all ℓ-sized subsets of S.

PKE Schemes. A public-key encryption (PKE) scheme PKE with message space $\mathcal{M}$ (parameterized by the security parameter k) consists of three PPT algorithms $\mathsf{Gen}, \mathsf{Enc}, \mathsf{Dec}$. Key generation $\mathsf{Gen}(1^k)$ outputs a public key pk and a secret key sk. Encryption $\mathsf{Enc}(pk, m)$ takes pk and a message $m \in \mathcal{M}$, and outputs a ciphertext c. Decryption $\mathsf{Dec}(sk, c)$ takes sk and a ciphertext c, and outputs a message m. For correctness, we want $\mathsf{Dec}(sk, c) = m$ for all $m \in \mathcal{M}$, all $(pk, sk) \leftarrow \mathsf{Gen}(1^k)$, and all $c \leftarrow \mathsf{Enc}(pk, m)$.

CCA Security. We recall the standard definition of IND-CCA security from the literature.

Definition 1 (IND-CCA security.). *We say that a scheme* PKE *is* IND-CCA *secure if for all PPT attackers A the advantage*

$$\mathsf{Adv}_{\mathsf{PKE},A}^{\mathsf{ind\text{-}cca}}(k) := \left| \Pr\left[\mathsf{Exp}_{\mathsf{PKE},A}^{\mathsf{ind\text{-}cca}}(k) = 1 \right] - \frac{1}{2} \right|$$

is negligible in the security parameter k, where the experiment $\mathsf{Exp}_{\mathsf{PKE},A}^{\mathsf{ind\text{-}cca}}$ *is defined in Fig. 1, and* $\mathsf{Dec}_{c^*}(sk, \cdot)$ *is an oracle that outputs* $\mathsf{Dec}(sk, c)$ *for every input $c \neq c^*$ and $\perp$ for input c^*.*

IND-SOA Security. We now recall the definition of indistinguishability-based SOA security from [4]. By default, we will consider the weakest variant where the adversary specifies an efficiently re-samplable distribution.

Definition 2 (Efficiently re-samplable). *Let $n = n(k) > 0$, and let $\mathcal{D}$ be a joint distribution over $\mathcal{M}^n$. We say that $\mathcal{D}$ is efficiently re-samplable if there is a PPT algorithm $\mathsf{msamp}_\mathcal{D}$ such that for any $\mathcal{I} \subseteq [n]$ and any partial vector $\mathbf{m}'_\mathcal{I} := (m'_i)_{i \in \mathcal{I}} \in \mathcal{M}^{|\mathcal{I}|}$, $\mathsf{msamp}_\mathcal{D}(\mathbf{m}'_\mathcal{I})$ samples from $\mathcal{D} \mid \mathbf{m}'_\mathcal{I}$, i.e., from the distribution $\mathbf{m} \leftarrow \mathcal{D}$, conditioned on $m_i = m'_i$ for all $i \in \mathcal{I}$. Note that in particular, $\mathsf{msamp}_\mathcal{D}()$ samples from $\mathcal{D}$.*

Definition 3 (IND-SOA Security). *For a PKE scheme* PKE = (Gen, Enc, Dec), *a polynomially bounded function $n = n(k) > 0$, and a stateful PPT adversary A, consider the experiment in Fig. 1. We only allow A that always output*

Experiment $\mathsf{Exp}_{\mathsf{PKE},A}^{\mathsf{ind\text{-}soa}}$

> $b \leftarrow \{0,1\}$
> $(pk, sk) \leftarrow \mathsf{Gen}(1^k)$
> $\mathsf{msamp}_\mathcal{D}(\cdot) \leftarrow A(pk)$
> $\mathbf{m}_0 := (m_i)_{i \in [n]} \leftarrow \mathsf{msamp}_\mathcal{D}()$
> $\mathbf{R} := (R_i)_{i \in [n]} \leftarrow (\mathcal{R}_{\mathsf{Enc}})^n$
> $\mathbf{c} := (c_i)_{i \in [n]} := (\mathsf{Enc}(pk, m_i; R_i))_{i \in [n]}$
> $\mathcal{I} \leftarrow A(pk, \mathbf{c})$
> $\mathbf{m}_1 \leftarrow \mathsf{msamp}_\mathcal{D}(\mathbf{m}_\mathcal{I})$
> $out_A \leftarrow A((R_i)_{i \in \mathcal{I}}, \mathbf{m}_b)$
> return 1 if $out_A = b$, and 0 otherwise

Experiment $\mathsf{Exp}_{\mathsf{PKE},A}^{\mathsf{ind\text{-}cca}}$

> $b \leftarrow \{0,1\}$
> $(pk, sk) \leftarrow \mathsf{Gen}(1^k)$
> $(m_0, m_1) \leftarrow A^{\mathsf{Dec}(sk, \cdot)}(pk)$
> $c^* \leftarrow \mathsf{Enc}(pk, m_b)$
> $out_A \leftarrow A^{\mathsf{Dec}_{c^*}(sk, \cdot)}(c^*)$
> return 1 if $out_A = b$, and 0 otherwise

Fig. 1. IND-CCA and IND-SOA experiments.

re-sampling algorithms as in Definition 2. We call PKE *IND-SOA secure if for all polynomials n and all PPT A, we have*

$$\mathsf{Adv}_{\mathsf{PKE},A}^{\mathsf{ind\text{-}soa}}(k) := \left| \Pr\left[\mathsf{Exp}_{\mathsf{PKE},A}^{\mathsf{ind\text{-}soa}}(k) = 1 \right] - \frac{1}{2} \right| = \mathsf{negl}(k).$$

Public-Coin Differing-Inputs Obfuscation. In this paper, we require a strengthening [22,23] of the notion of indistinguishability obfuscation [1,15].

We shall first define the notion of a public-coin differing-inputs sampler.

Definition 4 (Public-Coin Differing-Inputs Sampler). *A (circuit) sampling algorithm* csamp *is an algorithm that takes as input random coins* $r \in \{0,1\}^{\ell(k)}$ *for a suitable polynomial* $\ell = \ell(k)$, *and outputs the description of two circuits* C_0 *and* C_1. *We call* csamp *a public-coin differing-inputs sampler for the parameterized collection of circuits* $\mathcal{C} = \{\mathcal{C}_k\}$ *if the output of* csamp *is distributed over* $\mathcal{C}_k \times \mathcal{C}_k$, *and for every PPT adversary A, we have*

$$\Pr_r[C_0(x) \neq C_1(x) : (C_0, C_1) \leftarrow \mathsf{csamp}(1^k; r), x \leftarrow A(1^k, r)] = \mathsf{negl}(k).$$

Observe that the sampler and the attacker both receive the same random coins as input. Therefore, csamp cannot keep any "secret" from A. We now define the notion of a public-coin differing-inputs obfuscator.

Definition 5 (Public-Coin Differing-Inputs Obfuscator). *A uniform PPT algorithm* PdiO *is a public-coin differing-inputs obfuscator for the parameterized collection of circuits* $\mathcal{C} = \{\mathcal{C}_k\}$ *if the following requirements hold:*

Correctness: $\forall k, \forall C \in \mathcal{C}_k, \forall x$, it is $\Pr[C'(x) = C(x) : C' \leftarrow \mathsf{PdiO}(1^k, C)] = 1$.

Security: *for every public-coin differing-inputs sampler* csamp *for the collection* $\mathcal{C}$, *every PPT (distinguishing) algorithm D, we have*

$$\mathsf{Adv}^{\mathsf{pdio}}_{\mathsf{PdiO},D} :=$$
$$\left| \Pr[D(1^k, r, C') = 1 : (C_0, C_1) \leftarrow \mathsf{csamp}(1^k; r), C' \leftarrow \mathsf{PdiO}(1^k, C_0)] - \right.$$
$$\left. \Pr[D(1^k, r, C') = 1 : (C_0, C_1) \leftarrow \mathsf{csamp}(1^k; r), C' \leftarrow \mathsf{PdiO}(1^k, C_1)] \right| = \mathsf{negl}(k)$$

Correlation-intractable Hash Functions. We begin by reviewing the definition of correlation-intractable hash function from [8].

Definition 6 (Hash Function Ensembles). *A family of functions* $\mathcal{H} = \{h_s : D_k \to R_k\}_{k \in \mathbb{N}, s \in \{0,1\}^{\ell(k)}}$ *with domain* D_k, *range* R_k, *and seed length* $\ell(k)$ *is said to be an* efficient hash function ensemble, *if there exists a PPT algorithm that given* $x \in D_k$ *and* s, *outputs* $h_s(x)$.

In the sequel, we shall simply denote this computation by $h_s(x)$. Furthermore, we shall often call s the *description* or the *seed* of the function h_s.

Definition 7 (Binary Relations). *A class of* efficient binary relations *consists of* $\mathcal{REL} = \{\mathrm{Rel}_r \subseteq (D_k, R_k)\}_{k \in \mathbb{N}, r \in \{0,1\}^{\ell'(k)}}$, *where membership in* Rel_r *is testable in polynomial time given* r.
  The relation $\mathcal{REL}$ *is said to be* evasive *if for any* $r \in \{0,1\}^{\ell'(k)}, x \in D_k$ *we have:*

$$\Pr_{y \leftarrow R_k} [(x, y) \in \mathrm{Rel}_r] = \mathsf{negl}(k).$$

Definition 8 (Correlation Intractability). *Assume an efficient hash function ensemble* $\mathcal{H} = \{h_s : D_k \to R_k\}_{k \in \mathbb{N}, s \in \{0,1\}^{\ell(k)}}$. *Furthermore, let* $\mathcal{REL} = \{\mathrm{Rel}_r \subseteq (D_k, R_k)\}_{k \in \mathbb{N}, r \in \{0,1\}^{\ell'(k)}}$ *be a class of efficient binary relations. We say that* $\mathcal{H}$ *is* correlation intractable *with respect to* $\mathcal{REL}$ *if for every PPT A,*

$$\Pr_{s \leftarrow \{0,1\}^{\ell(k)}, r \leftarrow \{0,1\}^{\ell'(k)}} [(x, h_s(x)) \in \mathrm{Rel}_r : x \leftarrow A(s, r)] = \mathsf{negl}(k)$$

The work of [8] showed that no hash function ensemble is correlation-intractable with respect to *all* evasive binary relations $\mathcal{REL}$. However, for any fixed domains/ranges D_k, R_k it is plausible that there is a correlation-intractable hash function $\mathcal{H}$ for all evasive relations over D_k, R_k as long as the seed length $\ell(k)$ of the hash function is made sufficiently large relative to D_k, R_k. This would be sufficient for our needs. For concreteness, we define a specific class of relations $\mathcal{REL}$ for which we need correlation intractability.

Definition 9 (Special Class of Evasive Binary Relations). *Let* PKE = (Gen, Enc, Dec) *be a PKE scheme with plaintext space* $\mathbb{F}$ *(a field), ciphertext space* $\mathcal{C}$ *(parametrized by the security parameter k), and which uses* $\ell'(k)$ *bits of randomness in key-generation. We define a special class of binary relations* $\mathcal{REL}^{\mathsf{PKE}} = \{\mathrm{Rel}_r \subseteq (\mathcal{C}^{3k}, \mathcal{S}_k^{[3k]})\}_{k \in \mathbb{N}, r \in \{0,1\}^{\ell'(k)}}$, *as follows.*

- To determine if $((c_i)_{i\in[3k]}, \mathcal{I}) \in \text{Rel}_r$: Let $(pk, sk) = \text{Gen}(1^k; r)$, $m_i = \text{Dec}(sk, c_i)$, $F = \text{decc}_k((i, m_i)_{i\in[3k]})$ and $Q = \{i \in [3k] : F(i) = m_i\}$. The tuple is in the relation if $F \neq \bot$, $|Q| = 2k + 1$ and $\mathcal{I} \subseteq Q$.

Intuitively, the above says that a tuple $((c_i)_{i\in[3k]}, \mathcal{I}) \in \text{Rel}_r$ if the decrypted messages $(m_i)_{i\in[3k]}$ agree with the evaluations of a degree $\leq k$ polynomial F in *exactly* $2k+1$ positions and the set $\mathcal{I}$ only contains these positions. It is easy to see that this is an evasive relation as shown below (following [20, Lemma 3.3]).

Lemma 1. *The relation* $\mathcal{REL}^{\text{PKE}}$ *is evasive. In particular, for any* $r \in \{0, 1\}^{\ell'(k)}$ *any* $(c_i)_{i\in[3k]} \in \mathcal{C}^{3k}$ *we have* $\Pr_{\mathcal{I}\leftarrow\mathcal{S}_k^{[3k]}}\left[((c_i)_{i\in[3k]}, \mathcal{I}) \in \text{Rel}_r\right] = \text{negl}(k)$.

Proof. Let $(pk, sk) = \text{Gen}(1^k; r)$, $m_i = \text{Dec}(sk, c_i)$, $F = \text{decc}_k((i, m_i)_{i\in[3k]})$ and $Q = \{i \in [3k] : F(i) = m_i\}$. If $F = \bot$ or $|Q| \neq 2k + 1$ then the probability in the lemma is 0. Otherwise

$$\Pr_{\mathcal{I}\leftarrow\mathcal{S}_k^{[3k]}}\left[((c_i)_{i\in[3k]}, \mathcal{I}) \in \text{Rel}_r\right] = \Pr_{\mathcal{I}\leftarrow\mathcal{S}_k^{[3k]}}\left[\mathcal{I} \subseteq Q\right] = \frac{\binom{2k+1}{k}}{\binom{3k}{k}} \leq \left(\frac{5}{6}\right)^k$$

for all $k \geq 2$, which proves the lemma.

Special Correlation-intractable Hash Functions. Let $\mathcal{REL}^{\text{PKE}}$ be a special class of binary relations for PKE scheme PKE, like in Definition 9. We define special correlation-intractable hash functions $\mathcal{H} = \{h_s : \mathcal{C}^{3k} \to \mathcal{S}_k^{[3k]}\}_{k\in\mathbb{N}, s\in\{0,1\}^{\ell(k)}}$ as a function ensemble that is correlation intractable with respect to the relation $\mathcal{REL}^{\text{PKE}}$. We reiterate that this is a special case of correlation intractability with respect to all evasive relations, which is conjectured to be possible as long as the seed length $\ell(k)$ of the hash function is made sufficiently large relative to the domain/range. In our case, we allow $\ell(k)$ to be an arbitrarily large polynomial.

2.1 Puncturable Encryption Schemes

We will rely on the notion of puncturable encryption from [9]. Let PKE = (Gen, Enc, Dec, Puncture) be a tuple of PPT algorithms. PKE is said to be a puncturable encryption scheme, if the following holds.

Syntax. (Gen, Enc, Dec) is a PKE scheme with message space $\mathcal{M} = \{\mathcal{M}_k\}_{k\in\mathbb{N}}$ and ciphertext space $\mathcal{C} = \{\mathcal{C}_k\}$ which are efficiently sampleable.
Correctness. For all $m \in \mathcal{M}$ it holds that $\Pr[\text{Dec}(sk, c) = m : (pk, sk) \leftarrow \text{Gen}(1^k), c \leftarrow \text{Enc}(pk, m)] = 1$.
Puncturability. $\forall (pk, sk)$ in the support of $\text{Gen}(1^k)$, $\forall c_0, c_1 \in \mathcal{C}$, $\forall sk[\{c_0, c_1\}]$ in the support of $\text{Puncture}(sk, \{c_0, c_1\})$, and $\forall c \notin \{c_0, c_1\}$, it holds that: $\text{Dec}(sk[\{c_0, c_1\}], c) = \text{Dec}(sk, c)$.
Security. For every PPT adversary A,

$$\text{Adv}_{\text{PKE}, A}^{\text{punc-ind-cca}}(k) := \left|\Pr\left[\text{Exp}_{\text{PKE}, A}^{\text{punc-ind-cca}}(k) = 1\right] - \frac{1}{2}\right| = \text{negl}(k)$$

where the experiment $\text{Exp}_{\text{PKE}, A}^{\text{punc-ind-cca}}(k)$ is defined in Fig. 2.

Ciphertext sparseness. $\forall (pk, sk)$ we have

$$\Pr\left[\mathsf{Dec}(sk, c) \neq \perp\right] = \mathsf{negl}(k),$$

where the probability is over $c \leftarrow \mathcal{C}_{pk}$.

Note that the puncturing algorithm Puncture takes as input a *set* of two ciphertexts, so that the distribution of $\mathsf{Puncture}(sk, \{c_0, c_1\})$ is identical to that of $\mathsf{Puncture}(sk, \{c_1, c_0\})$.

3   Secret-Sharing Selective Opening Attack (SecShare-SOA)

We now define a special case of IND-SOA security that we call SecShare-SOA. It corresponds to the case where the encrypted values are shares in a t-out-of-n secret sharing scheme.

Secret-sharing Message Distribution. Let $\mathbb{F}$ be a field of cardinality p. We consider a distribution $\mathcal{D}$ which chooses a polynomial in $F \in \mathbb{F}[X]$ of degree at most t and sets the messages to be $m_i = F(i)$ for $i \in [n]$. We let t and n be two polynomials in the security parameter, such that $t < n \leq p$. More formally,

$$\mathcal{D}_{\mathbb{F}, t, n} = \left\{(F(1), \dots, F(n)) \,\middle|\, F \in \mathbb{F}[X] \text{ uniformly chosen degree-}\leq t \text{ polynomial}\right\}$$

Note that there exists an efficient re-sampling algorithm msamp for the above distribution. In particular, for any $\mathcal{I}$, msamp can randomly extend its input $(F(i))_{i \in \mathcal{I}}$ to $t + 1$ evaluation points as necessary and then use polynomial interpolation to retrieve F and thus all $F(i)$.

Note that, conditioned on any choice of $F(i)$ for $i \in \mathcal{I}$ where $|\mathcal{I}| \leq t$, the value $F(0)$ is uniformly random.

Definition 10 (SecShare-SOA Security). *Let $\mathbb{F}$ be a field of size determined by the security parameter and let* $\mathsf{PKE} = (\mathsf{Gen}, \mathsf{Enc}, \mathsf{Dec})$ *be a PKE scheme, with message space $\mathcal{M} = \mathbb{F}$. For any polynomials parameters $t = t(k), n = n(k)$ such that $t < n \leq |\mathbb{F}|$, consider the experiment in Fig. 2 with a stateful PPT adversary A. We say that* PKE *secret-sharing selective opening attack secure if*

$$\mathsf{Adv}_{\mathsf{PKE}, A}^{\mathsf{secsh\text{-}soa}}(k) := \left| \Pr\left[\mathsf{Exp}_{\mathsf{PKE}, A}^{\mathsf{secsh\text{-}soa}}(k) = 1\right] - \frac{1}{|\mathbb{F}|} \right|$$

is negligible for all PPT A.

3.1   IND-SOA Implies SecShare-SOA

Theorem 1. *If a PKE scheme* PKE *is IND-SOA secure, then it is SecShare-SOA secure.*

Experiment $\mathsf{Exp}_{\mathsf{PKE},A}^{\mathsf{secsh\text{-}soa}}$
$(pk, sk) \leftarrow \mathsf{Gen}(1^k)$
$\mathbf{m} := (m_i)_{i \in [n]} \leftarrow \mathcal{D}_{\mathbb{F},t,n}$
$\mathbf{R} := (R_i)_{i \in [n]} \leftarrow (\mathcal{R}_{\mathsf{Enc}})^n$
$\mathbf{c} := (c_i)_{i \in [n]} := (\mathsf{Enc}(pk, m_i; R_i))_{i \in [n]}$
$\mathcal{I} \leftarrow A(pk, \mathbf{c})$, where, $\mathcal{I} \in [n]$ and $|\mathcal{I}| = t$
$out_A \leftarrow A((m_i, R_i)_{i \in \mathcal{I}})$
return 1 if $out_A = F(0)$, and 0 otherwise

Experiment $\mathsf{Exp}_{\mathsf{PKE},A}^{\mathsf{punc\text{-}ind\text{-}cca}}$
$b \leftarrow \{0, 1\}$
$m^* \leftarrow A(1^k)$
$(pk, sk) \leftarrow \mathsf{Gen}(1^k)$
$c_0 \leftarrow \mathsf{Enc}(pk, m^*)$
$c_1 \leftarrow \mathcal{C}_{pk}$
$sk[\{c_0, c_1\}] \leftarrow \mathsf{Puncture}(sk, \{c_0, c_1\})$
$out_A \leftarrow A(pk, c_b, c_{1-b}, sk[\{c_0, c_1\}])$
return 1 if $out_A = b$, and 0 otherwise

Fig. 2. SecShare-SOA and Punc-IND-CCA experiments.

Proof. Let PKE be a PKE scheme with message space $\mathcal{M} = \mathbb{F}$ which is a field. Suppose there exists an adversary A that breaks the SecShare-SOA security of PKE with probability $\varepsilon = \mathsf{Adv}_{\mathsf{PKE},A}^{\mathsf{secsh\text{-}soa}}(k)$. Then we construct an adversary B that, given access to A, breaks the IND-SOA security of PKE. We describe the adversary below.

Adversary B. By using A, B interacts with its challenger in the IND-SOA game as follows. Upon receiving a public key pk, B presents the secret-sharing message distribution $\mathcal{D}$ to its challenger. To recall,

$$\mathcal{D}_{\mathbb{F},t,n} = \{(F(1), \ldots, F(n)) \mid F \in \mathbb{F}[X] \text{ uniformly chosen degree-}\leq t \text{ polynomial}\}$$

for some $t < n$. Upon receiving a tuple of ciphertexts $\mathbf{c} := (c_i)_{i \in [n]}$, B forwards $(pk, \mathbf{c})$ to A. Upon receiving $\mathcal{I}$ from A, B forwards it to the challenger. Recall that $\mathcal{I} \in [n]$ and $|\mathcal{I}| = t$. Upon receiving a message vector $\mathbf{m}$ and the openings $(R_i)_{i \in \mathcal{I}}$ of $(c_i)_{i \in \mathcal{I}}$ to $(m_i)_{i \in \mathcal{I}}$, B proceeds as follows. It computes $F = \mathsf{ipol}((i, m_i)_{i \in [n]})$. Thereafter, it forwards the messages and openings just for $i \in \mathcal{I}$; namely, $(m_i, R_i)_{i \in \mathcal{I}}$. Let out_A be the value output by A. B compares whether $out_A = F(0)$. If so, then it outputs 0, else it outputs 1.

Analysis. We shall now analyze the success probability of B in the IND-SOA game. Intuitively, B succeeds in the IND-SOA game whenever the A succeeds in the SSSOA game except when the resampling results in the same message vector as the original plaintext message. More formally, we have:

$$\Pr\left[\mathsf{Exp}_{\mathsf{PKE},B}^{\mathsf{ind\text{-}soa}}(k) = 1 | b = 0\right] = \Pr\left[\mathsf{Exp}_{\mathsf{PKE},A}^{\mathsf{secsh\text{-}soa}}(k) = 1\right]$$

$$\Pr\left[\mathsf{Exp}_{\mathsf{PKE},B}^{\mathsf{ind\text{-}soa}}(k) = 1 | b = 1\right] = \left(1 - \frac{1}{|\mathbb{F}|}\right)$$

Thus,

$$\left| \Pr\left[\mathsf{Exp}^{\mathsf{ind\text{-}soa}}_{\mathsf{PKE},B}(k) = 1\right] - \frac{1}{2} \right| = \left| \frac{1}{2}\left(\Pr\left[\mathsf{Exp}^{\mathsf{secsh\text{-}soa}}_{\mathsf{PKE},A}(k) = 1\right] + \left(1 - \frac{1}{|\mathbb{F}|}\right)\right) - \frac{1}{2}\right|$$

$$= \frac{1}{2}\left| \Pr\left[\mathsf{Exp}^{\mathsf{secsh\text{-}soa}}_{\mathsf{PKE},A}(k) = 1\right] - \frac{1}{|\mathbb{F}|}\right| = \frac{\varepsilon}{2}$$

which is non-negligible by assumption.

4   CCA Secure, SOA Insecure Encryption

In this section, we describe a PKE scheme that is IND-CCA secure, but not IND-SOA secure.

4.1   The Scheme

Let $\mathsf{PKE}' = (\mathsf{Gen}', \mathsf{Enc}', \mathsf{Dec}', \mathsf{Puncture}')$ be a puncturable encryption scheme with message space $\mathbb{F}$ for some field of size $|\mathbb{F}| \geq 3k$ and $|\mathbb{F}| = O(k)$ and with ciphertext space $\mathcal{C}$. Let $\mathcal{H} = \{h_s : \mathcal{C}^{3k} \to \mathcal{S}_k^{[3k]}\}_{k \in \mathbb{N}, s \in \{0,1\}^{\ell(k)}}$ be a special correlation-intractable hash function ensemble with respect to $\mathcal{REL}^{\mathsf{PKE}'}$ and with seed length $\ell(k)$. Let PdiO be a public-coin differing-inputs obfuscator.

We construct a scheme $\mathsf{PKE} = (\mathsf{Gen}, \mathsf{Enc}, \mathsf{Dec})$ as follows.

- $\mathsf{Gen}(1^k)$: Run $(pk', sk') \leftarrow \mathsf{Gen}'(1^k)$. Sample $s \leftarrow \{0,1\}^{\ell(k)}$ as a seed of the hash function $h_s \in \mathcal{H}$. Then construct the program SOA-Helper in Fig. 3. Set secret key $sk = sk'$ and public key $pk = (pk', s, \mathsf{PdiO}(\text{SOA-Helper}))$.
- $\mathsf{Enc}(pk, m)$: Parse $pk = (pk', s, \mathsf{PdiO}(\text{SOA-Helper}))$. Output $\mathsf{Enc}'(pk', m)$.
- $\mathsf{Dec}(sk, c)$: Output $\mathsf{Dec}'(sk', c)$.

SOA-Helper

Constants: sk', seed s.
Input: $Z = ((c'_i)_{i \in [3k]}, (m_i, R_i)_{i \in \mathcal{I}})$.

1. If there are indices $i \neq j$ with $c'_i = c'_j$, then return $\perp$.
2. Set $\mathcal{I} := h_s((c'_i)_{i \in [3k]})$. If there is an $i \in \mathcal{I}$ with $\mathsf{Enc}'(pk', m_i; R_i) \neq c'_i$, then return $\perp$.
3. Decrypt $m_i = \mathsf{Dec}'(sk', c'_i)$ for $i \in [3k] \setminus \mathcal{I}$.
4. Let $F = \mathsf{decc}_k((i, m_i)_{i \in [3k]})$. If $F = \perp$ or $F(i) \neq m_i$ for some $i \in \mathcal{I}$ then return $\perp$ else return F.

Fig. 3. Program SOA-Helper

4.2 PKE is Not SecShare-SOA Secure

We now formally show that PKE allows for a simple SecShare-SOA attack.

Theorem 2. *The PKE scheme* PKE *from Sect. 4.1 is not SecShare-SOA secure.*

Proof. We construct a PPT algorithm A that breaks the SecShare-SOA security of PKE with non-negligible probability.

Adversary A: Upon receiving a public key $pk = (pk', s, \mathsf{PdiO}(\mathsf{SOA\text{-}Helper}))$ and a tuple of ciphertexts $\mathbf{c} := (c_i)_{i \in [3k]}$, A computes $\mathcal{I} = h_s((c_i)_{i \in [3k]})$. This $\mathcal{I}$ is the subset that A submits to its SecShare-SOA experiment. By the security of PKE', we may assume that $c'_i \neq c'_j$ for all $i \neq j$, except with some negligible probability $\nu(k)$ (otherwise, an adversary on PKE' could simply encrypt a challenge message and hope for a collision).

Upon receiving openings $(m_i, R_i)_{i \in \mathcal{I}}$, A runs the program $\mathsf{PdiO}(\mathsf{SOA\text{-}Helper})$ on input $((c_i)_{i \in [3k]}, (m_i, R_i)_{i \in \mathcal{I}})$. By construction of the program, the output is a polynomial F with $m_i = F(i)$ for all $i \in [3k]$. Thus, the adversary can finally compute $F(0)$ and give it to the challenger.

Analysis: The analysis is pretty straight-forward, as, by design, the output of the program is exactly what the adversary needs to break the SecShare-SOA security. Thus, we have that,

$$\mathsf{Adv}_{\mathsf{PKE},A}^{\mathsf{secsh\text{-}soa}}(k) = 1 - \frac{1}{|\mathbb{F}|} - \nu(k),$$

which is non-negligible by assumption about $\nu(k)$.

4.3 PKE is Still IND-CCA Secure

We show that PKE inherits PKE''s IND-CCA security.

Theorem 3. *Suppose that* PKE' *is puncturably IND-CCA secure,* $\mathcal{H}$ *is a special correlation-intractable hash function ensemble with respect to* $\mathcal{REL}^{\mathsf{PKE'}}$, *and* PdiO *be a secure public-coin differing-inputs obfuscator. Then, the PKE scheme* PKE *from Sect. 4.1 is IND-CCA secure.*

Proof. Recall that our scheme has polynomial-size message space. We consider a variation to the IND-CCA game where the challenger himself chooses the pair of challenge messages. We call this modified game the $-IND-CCA game (defined formally in Definition 13). The IND-CCA game and the $-IND-CCA game are polynomially equivalent (as proved in Theorem 7) for polynomially-sized message spaces. Thus, it suffices to show here that our PKE scheme is $-IND-CCA secure.

Assume for contradiction that there exists an adversary A that breaks the $-IND-CCA security of PKE with some non-negligible advantage ε. We shall arrive at a contradiction through a sequence of hybrid arguments defined below. Let us denote the event that a hybrid Hyb_i outputs 1 by $\mathsf{Hyb}_i \to 1$. The first hybrid corresponds to the original $-IND-CCA security game.

– Hyb_0: In the first hybrid the following game is played.
  - Sample $b \leftarrow \{0, 1\}$.
  - Sample $m_0, m_1 \leftarrow \mathcal{M}$
  - Sample $s \leftarrow \{0, 1\}^{\ell(k)}$.
  - Run $(pk', sk') \leftarrow \mathsf{Gen}'(1^k)$. Let $sk = sk'$.
  - $c^* \leftarrow \mathsf{Enc}(pk, m_b)$
  - Construct the program SOA-Helper in Fig. 3 and obfuscate it to get PdiO(SOA-Helper).
  - Let $pk = (pk', s, \mathsf{PdiO}(\mathsf{SOA\text{-}Helper}))$.
  - Give pk, m_0, m_1, c^* to the adversary, and offer the adversary access to a decryption oracle $\mathsf{Dec}_{c^*}(sk, \cdot)$.
  - Finally, let b' be the output of A. Output 1 if $b' = b$ and 0 otherwise.

We note here that $\mathsf{Exp}_{\mathsf{PKE},A}^{\$\text{-ind-cca}}(k) \equiv \mathsf{Hyb}_0$.

– Hyb_1: This hybrid is the same as Hyb_0 with the exception of the following modifications. Just before generating the program SOA-Helper, sample $c_r \leftarrow \mathcal{C}$. Compute $sk[\{c^*, c_r\}] \leftarrow \mathsf{Puncture}(sk, \{c^*, c_r\})$. We then make the following modifications in Step 3 of the program obfuscated, and denote the resulting program by SOA-Helper$_2$:

   3. Decrypt $m_i = \mathsf{Dec}'(sk[\{c^*, c_r\}], c_i')$ for $i \in [3k] \setminus \mathcal{I}$ and for $c_i' \notin \{c^*, c_r\}$.
   Use m_b and $\perp$ as the plaintext values when $c_i' = c^*$ and $c_i' = c_r$, resp.
   Furthermore, we now use the punctured key $sk[\{c^*, c_r\}]$ to answer A's decryption queries, and we output $\perp$ on inputs c^*, c_r.

Claim. $\mathsf{Hyb}_1 \approx_c \mathsf{Hyb}_0$.

Proof. We observe that the input/output functionality of the program has not changed with overwhelming probability (over the choice of c_r), thanks to the puncturability and ciphertext sparseness properties of PKE$'$. Furthermore, there is also no change in the functionality of the decryption oracle. Thus, by relying on the indistinguishability obfuscation security (which follows from public-coin differing inputs security) of PdiO, we have $\mathsf{Hyb}_1 \approx_c \mathsf{Hyb}_0$.

– Hyb_2: This hybrid is the same as Hyb_1 with the exception of the following modifications in Step 3 of the program obfuscated. Denote the resulting program by Program SOA-Helper$_2$.
   3. Decrypt $m_i = \mathsf{Dec}'(sk[\{c^*, c_r\}], c_i')$ for $i \in [3k] \setminus \mathcal{I}$ and for $c_i' \notin \{c^*, c_r\}$.
   Use $\perp$ and $\perp$ as the plaintext values when $c_i' = c^*$ and $c_i' = c_r$, respectively.

Claim. $\mathsf{Hyb}_2 \approx_c \mathsf{Hyb}_1$.

Proof. We employ the public-coins differing-inputs obfuscation security of PdiO and the special correlation intractability of $\mathcal{H}$ to prove this claim. More specifically, consider an algorithm $\mathsf{csamp}(1^k)$ that generates two circuits SOA-Helper$_1$ and SOA-Helper$_2$.

We first show that csamp is a public-coin differing-inputs sampler, by employing the special correlation intractability of $\mathcal{H}$. Recall the special class of binary relations $\mathcal{REL}^{\mathsf{PKE}'} = \{\mathrm{Rel}_r\}_{k \in \mathbb{N}, r \in \{0,1\}^{\ell'(k)}}$ from Definition 9.

Let $Z = ((c'_i)_{i \in [3k]}, (m_i, R_i)_{i \in \mathcal{I}})$ be an input to the two programs. We will now argue that the only time that $\mathsf{SOA\text{-}Helper}_1(Z) \neq \mathsf{SOA\text{-}Helper}_2(Z)$ is if $((c'_i)_{i \in [3k]}, \mathcal{I}) \in \mathrm{Rel}_r$ where r is the randomness of the key-generation procedure used to create (pk', sk') and $\mathcal{I} = h_s((c'_i)_{i \in [3k]})$.

In order to do so, let $m'_i = \mathsf{Dec}_{sk'}(c'_i)$ for $i \in [3k] \setminus \mathcal{I}$. Now if $m'_i \neq m_i$ for some $i \in \mathcal{I}$ then both programs output $\perp$ since the check in line 2 will fail (by correctness of decryption). Let $F_1 = \mathsf{decc}_k((i, m'_i)_{i \in [3k]})$ and let $F_2 = \mathsf{decc}_k((i, m''_i)_{i \in [3k]})$ where $m''_i = m'_i$ unless $c'_i = c^*$ and $i \notin \mathcal{I}$ in which case $m''_i = \perp$. These are the two polynomials that are used in line 3 of the execution of $\mathsf{SOA\text{-}Helper}_1(Z), \mathsf{SOA\text{-}Helper}_2(Z)$ respectively. Let $Q = \{i \ : \ F_1(i) = m'_i\}$. If $F_1 = F_2$ then both programs have the same output. The only case where this does not happen is if $F_1 \neq \perp$, $|Q| = 2k + 1$ and $F_2 = \perp$. Moreover, even in this case both programs output $\perp$ unless it is the case that $\mathcal{I} \subseteq Q$. Therefore, the only case where the two programs might produce differing outputs is if $F_1 \neq \perp$, $|Q| = 2k + 1$ and $\mathcal{I} \subseteq Q$. This means that $((c_i)_{i \in [3k]}, \mathcal{I}) \in \mathrm{Rel}_r$ where $\mathcal{I} = h_s((c_i)_{i \in [3k]})$.

By the special correlation intractability property of $\mathcal{H}$ such inputs Z such that $\mathsf{SOA\text{-}Helper}_1(Z) \neq \mathsf{SOA\text{-}Helper}_2(Z)$ are computationally hard to find, even given all of the random coins used to generate the two programs, including the hash-seed s and the randomness r used to generate (pk', sk'). In other words this shows that algorithm csamp that generates two circuits $\mathsf{SOA\text{-}Helper}_1$ and $\mathsf{SOA\text{-}Helper}_2$ defines a public-coin differing-inputs family. We can therefore rely on the public-coin differing-input security of PdiO to see that Hyb_2 and Hyb_1 are indistinguishable.

- Hyb_3: This hybrid is the same as Hyb_2 with the following exception. Instead of giving c^* to the adversary, give c_r to the adversary as a challenge.

Claim. $\mathsf{Hyb}_3 \approx_c \mathsf{Hyb}_2$.

Proof. Assume that for an adversary A, $\Pr[\mathsf{Hyb}_2 \to 1]$ and $\Pr[\mathsf{Hyb}_3 \to 1]$ differ by a non-negligible amount ε. Then we shall construct an adversary B that breaks the puncturability of PKE' with probability ε. B behaves the same as the challenger in Hyb_2 and interacts with A except the following modifications. It first samples b, m_0, m_1, and gives m_b to its challenger. Upon receiving two ciphertexts $c_{\hat{b}}, c_{1-\hat{b}}$, proceed by giving $c_{\hat{b}}$ to A as the challenge ciphertext. Finally, output the output of the experiment.

Observe that if $\hat{b} = 0$, then we are in Hyb_2. Else, we are in Hyb_3. Thus,

$$\mathsf{Adv}^{\mathsf{punc\text{-}ind\text{-}cca}}_{\mathsf{PKE}', B}(k) \geq \varepsilon$$

- Hyb_4: This hybrid is the same as Hyb_3 with the exception of the following modifications in Step 3 of the program obfuscated. Denote the resulting program by Program $\mathsf{SOA\text{-}Helper}_4$.

3. Decrypt $m_i = \mathsf{Dec}'(sk[\{c^*, c_r\}], c_i')$ for $i \in [3k] \setminus \mathcal{I}$ and for $c_i' \notin \{c^*, c_r\}$. Use $\underline{m_b}$ and $\bot$ as the plaintext values when $c_i' = c^*$ and $c_i' = c_r$, respectively.

Claim. $\mathsf{Hyb}_4 \approx_c \mathsf{Hyb}_3$.

Proof. The modification introduced in Hyb_4 is similar to the modification introduced in Hyb_2 earlier. Hence, the proof here follows on the same lines as the proof of Claim 4.3.

- Hyb_5: This hybrid is the same as Hyb_4 with the following exception. In the obfuscated program, instead of hardcoding the punctured secret key, we shall hardcode again the original secret key.

Claim. $\mathsf{Hyb}_5 \approx_c \mathsf{Hyb}_4$.

Proof. Note that the input/output functionalities of the programs have not changed as we moved from Hyb_4 to Hyb_5. By applying the indistinguishability obfuscation security of PdiO, we have that $\mathsf{Hyb}_5 \approx_c \mathsf{Hyb}_4$.

Finally, note that in Hyb_5, the adversary's view does not depend on the challenge bit b anymore: neither the obfuscated circuit SOA-Helper nor the challenge ciphertext c_r depend on b. We get that Hyb_5 outputs 1 with probability exactly $1/2$. The theorem follows.

5   Extension to Selective Opening of Keys (SOA-K)

We now show how to extend our main result to selective opening *of keys* (SOA-K), where the adversary gets ciphertexts under many different public keys and can selectively request to see some of the secret keys. This corresponds to a setting where there are multiple receivers and the adversary can corrupt some subset of them and get their keys (rather than the previous setting where there were multiple senders and the adversary could corrupt some subset of them and get their encryption randomness).

For this notion, we will consider PKE schemes where the public/secret key pairs are generated dependent on some common public parameters. More specifically, in addition to the triple of algorithms (Gen, Enc, Dec), we introduce an algorithm PGen that takes the security parameter and outputs some public parameters params $\leftarrow$ PGen(1^k). All the other algorithms take params as an additional input. We show how to construct such PKE schemes which are CCA secure but are IND-SOA-K insecure. We leave it as an open problem to construct such examples in the setting without public parameters.

SOA-K has been considered before [3,17]; while [3] only treated the the simulation-based definition, [17] treated the indistinguishability-based definition that we will also consider in this work.

Definition 11 (IND-SOA-K Security). *For a PKE scheme* PKE = (PGen, Gen, Enc, Dec), *a polynomially bounded function* $n = n(k) > 0$, *and a stateful PPT adversary A, consider the experiment in Fig. 4. We only allow A that always output re-sampling algorithms as in Definition 2. We call* PKE IND-SOA-K *(for "indistinguishable under selective-opening key attacks") secure if for all PPT A, we have*

$$\mathsf{Adv}^{\mathsf{ind\text{-}soa\text{-}k}}_{\mathsf{PKE},A}(k) := \left| \Pr\left[\mathsf{Exp}^{\mathsf{ind\text{-}soa\text{-}k}}_{\mathsf{PKE},A}(k) = 1 \right] - \frac{1}{2} \right| = \mathsf{negl}(k).$$

Secret Sharing SOA-K Security. We now define the dual of SecShare-SOA-K security for key corruption. The only difference from the SecShare-SOA-K security security is that each secret share is encrypted with an independently sampled public key (instead of one public key being used to encrypt all shares), and corruption would reveal the corresponding secret keys (instead of the random coins used to generate the ciphertexts). Details follow.

Definition 12 (SecShare-SOA-K Security). *Let* $\mathbb{F}$ *be a field of size determined by the security parameter and let* PKE = (PGen, Gen, Enc, Dec) *be a PKE scheme, with message space* $\mathcal{M} = ([n] \times \mathbb{F})$. *For any parameters* t, n *that are polynomial in the security parameter, consider the experiment in Fig. 4 with a stateful PPT adversary A. We say that* PKE *receiver-corruption secret-sharing selective opening attack secure if*

$$\mathsf{Adv}^{\mathsf{secsh\text{-}soa\text{-}k}}_{\mathsf{PKE},A}(k) := \left| \Pr\left[\mathsf{Exp}^{\mathsf{secsh\text{-}soa\text{-}k}}_{\mathsf{PKE},A}(k) = 1 \right] - \frac{1}{|\mathbb{F}|} \right|$$

is negligible for all PPT A.

Theorem 4. *If a PKE scheme* PKE *is IND-SOA-K secure, then it is SecShare-SOA-K secure.*

The proof of Theorem 4 follows with the same argument as Theorem 1.

In this section, we describe a PKE scheme PKE* that is IND-CCA secure, but not IND-SOA-K secure.

5.1   A CCA Secure, SecShare-SOA-K Insecure Encryption

Let PKE' = (Gen', Enc', Dec') be any encryption scheme that is IND-CPA secure but is not SecShare-SOA secure (with sender-randomness corruption), and whose message space is a field $\mathbb{F}$ and randomness space is $\mathcal{R}_{\mathsf{Enc'}}$. In particular, this can be the scheme that we constructed in Sect. 4. Let PKE = (Gen, Enc, Dec) be an IND-CCA secure encryption scheme with message space $\mathbb{F} \times \mathcal{R}_{\mathsf{Enc'}}$. We construct a PKE scheme PKE* = (PGen*, Gen*, Enc*, Dec*) as follows.

Experiment $\mathsf{Exp}_{\mathsf{PKE},A}^{\mathsf{ind\text{-}soa\text{-}k}}$

$b \leftarrow \{0,1\}$

$\mathsf{params} \leftarrow \mathsf{PGen}(1^k)$

$((pk_i, sk_i))_{i=1}^n \leftarrow \mathsf{Gen}^n(1^k, \mathsf{params})$

$\mathsf{msamp}_{\mathcal{D}}(\cdot) \leftarrow A(\mathsf{params}, \mathbf{pk})$ for $\mathbf{pk} := (pk_1, \ldots, pk_n)$

$\mathbf{m}_0 := (m_i)_{i \in [n]} \leftarrow \mathsf{msamp}_{\mathcal{D}}()$

$\mathbf{c} := (c_i)_{i \in [n]}$ for $c_i := \mathsf{Enc}(\mathsf{params}, pk_i, m_i)$

$\mathcal{I} \leftarrow A(\mathbf{c})$

$\mathbf{m}_1 \leftarrow \mathsf{msamp}_{\mathcal{D}}(\mathbf{m}_{\mathcal{I}})$

$out_A \leftarrow A((sk_i)_{i \in \mathcal{I}}, \mathbf{m}_b)$

return 1 if $out_A = b$, and 0 otherwise

Experiment $\mathsf{Exp}_{\mathsf{PKE},A}^{\mathsf{secsh\text{-}soa\text{-}k}}$

$\mathsf{params} \leftarrow \mathsf{PGen}(1^k)$

$((pk_i, sk_i))_{i=1}^n \leftarrow \mathsf{Gen}^n(1^k, \mathsf{params})$

$\mathbf{m} := (m_i)_{i \in [n]} \leftarrow \mathcal{D}_{\mathbb{F}, t, n}$

$\mathbf{c} := (c_i)_{i \in [n]}$

    for $c_i := \mathsf{Enc}(\mathsf{params}, pk_i, m_i)$

$\mathcal{I} \leftarrow A(\mathsf{params}, \mathbf{pk}, \mathbf{c})$

    for $\mathbf{pk} := (pk_1, \ldots, pk_n)$

$out_A \leftarrow A((m_i, sk_i)_{i \in \mathcal{I}})$

return 1 if $out_A = F(0)$, and 0 else

Fig. 4. The IND-SOA-K and SecShare-SOA-K experiments.

- $\mathsf{PGen}^*(1^k)$: Sample $(pk', sk') \leftarrow \mathsf{Gen}'(1^k)$ and set public parameters $\mathsf{params} = pk'$.
- $\mathsf{Gen}^*(1^k, \mathsf{params})$: Run $(pk, sk) \leftarrow \mathsf{Gen}(1^k)$. Output (pk, sk) as the public-key secret-key pair.
- $\mathsf{Enc}^*(\mathsf{params}, pk, m)$: Parse $\mathsf{params} = pk'$. Sample $r \leftarrow \mathcal{R}_{\mathsf{Enc}'}$ and compute $c' \leftarrow \mathsf{Enc}'(pk', m; r)$. Compute $c \leftarrow \mathsf{Enc}(pk, (m, r))$. Output the ciphertext $c^* = (c', c)$.
- $\mathsf{Dec}^*(\mathsf{params}, sk, c^*)$: Parse $\mathsf{params} = pk'$ and $c^* = (c', c)$. Compute $(m, r) = \mathsf{Dec}(sk, c)$. Verify if $c' \leftarrow \mathsf{Enc}'(pk', m; r)$. If so, then output m, else output $\perp$.

PKE* is not SecShare-SOA-K secure. We now formally show that PKE* allows for a simple SecShare-SOA-K attack. The idea is straight-forward. An SecShare-SOA-K adversary, upon learning secret keys can decrypt the ciphertexts and learn the random coins used to compute the PKE' part of ciphertexts. This amounts to an SecShare-SOA attack on PKE', which is SecShare-SOA insecure.

Theorem 5. *If* PKE' *is not SecShare-SOA secure then* PKE* *is not SecShare-SOA-K secure. In particular, if there is a polynomial-time attack with advantage*

ε against the SecShare-SOA security of PKE$'$ then there is also a polynomial time attack with the same advantage ε against the SecShare-SOA-K security of PKE*.

Proof. Assume that there exists a PPT adversary A with non-negligible advantage ε against SecShare-SOA security of PKE$'$. We construct a PPT algorithm B that breaks the SecShare-SOA-K security of PKE* also with probability ε.

Adversary B: Upon receiving params $= pk'$, B gives pk' to A. Upon receiving a tuple of public keys $\mathbf{pk} := (pk_1, \ldots, pk_n)$ and a tuple of ciphertexts $\mathbf{c}^* := (c_i^*)_{i \in [n]}$, where $c_i^* = (c_i', c_i)$, give $(c_1', \ldots, c_n')$ to A. When A responds with a subset $\mathcal{I} \in [n]$, give $\mathcal{I}$ to the challenger. Then, upon receiving $(m_i, sk_i)_{i \in \mathcal{I}}$, compute $(m_i, r_i) = \mathsf{Dec}(sk_i, c_i)$ for every $i \in \mathcal{I}$. Then, give $(m_i, r_i)_{i \in \mathcal{I}}$ to A. Since this emulates the SecShare-SOA attack on PKE$'$ to A, A's output is such that it breaks SecShare-SOA security of PKE$'$ with probability ε. In turn, by outputting A's output, B also breaks SecShare-SOA-K security of PKE* with probability ε.

PKE* is IND-CCA secure. We now show that PKE* inherits PKE's IND-CCA security.

Theorem 6. *Suppose that PKE is an IND-CCA-secure encryption scheme, PKE$'$ is an IND-CPA secure encryption scheme. Then, PKE* is IND-CCA secure.*

Proof. Assume for contradiction that there exists an adversary A that breaks the IND-CCA security of PKE* with some advantage ε. We shall show that ε is negligible through a hybrid argument as follows. Let us denote the event that a hybrid Hyb_i outputs 1 by $\mathsf{Hyb}_i = 1$. The first hybrid corresponds to the original IND-CCA security game.

- Hyb_0: In the first hybrid the following game is played.
  1. Sample $b \leftarrow \{0, 1\}$.
  2. Sample $(pk', sk') \leftarrow \mathsf{Gen}'(1^k)$ and set params $= pk'$.
  3. Sample $(pk, sk) \leftarrow \mathsf{Gen}(1^k)$.
  4. Give (params, pk) to A and answer its decryption queries using sk.
  5. Upon receiving m_0, m_1, proceed as follows.
  6. Sample $r \leftarrow \mathcal{R}_{\mathsf{Enc}'}$ and compute $\tilde{c}' \leftarrow \mathsf{Enc}'(pk', m_b; r)$.
  7. Compute $\tilde{c} \leftarrow \mathsf{Enc}(pk, (m_b, r))$.
  8. Give $(\tilde{c}', \tilde{c})$ to A and continue to answer A's decryption queries using sk.
  9. Finally, let b' be the output of A. Output 1 if $b' = b$ and 0 otherwise.
- Hyb_1: This hybrid is the same as Hyb_0, except that all of A's decryption queries $c^* = (c', c)$ with $c = \tilde{c}$ are automatically rejected.
  We note that this change is purely conceptual: any such query would have been rejected already in Hyb_0 (by the decryption check for $c' = \mathsf{Enc}(pk', m; r)$, where $(m, r) \leftarrow \mathsf{Dec}(sk, c)$).

- Hyb_2: This hybrid is the same as Hyb_1 except for the following modification. In constructing $\tilde{c}$, instead of using (m_b, r) as the plaintext, we use $(m_{\tilde{b}}, \tilde{r})$ for an indendently sampled random bit $\tilde{b}$, and a freshly uniformly sampled random string $\tilde{r}$.
   7. Sample $\tilde{b} \leftarrow \{0,1\}$ and $\tilde{r} \leftarrow \mathcal{R}_{\mathsf{Enc'}}$. Compute $\tilde{c} \leftarrow \mathsf{Enc}(pk, \underline{(m_{\tilde{b}}, \tilde{r})})$.

Claim. $\mathsf{Hyb}_2 \approx_c \mathsf{Hyb}_1$.

Proof. We shall establish this claim by relying on the CCA security of PKE. Assume for contradiction that $|\Pr[\mathsf{Hyb}_2 = 1] - \Pr[\mathsf{Hyb}_1 = 1]| = \varepsilon$ is non-negligible. Then we construct an adversary B that breaks CCA security of PKE with advantage ε.

Adversary B. B simultaneously interacts with its CCA challenger for PKE and A as follows. Upon receiving pk from the challenger, sample $(pk', sk') \leftarrow \mathsf{Gen'}(1^k)$, set $\mathsf{params} = pk'$, and give params, pk to A. Answer A's decryption queries by using its own decryption oracle as follows. Upon given a query $c^* = (c', c)$ by A, decrypt c using its own oracle to get (m, r); verify whether $c' \leftarrow \mathsf{Enc'}(pk', m; r)$. If so, then give m to A and give $\bot$ otherwise. Next, upon receiving (m_0, m_1) from A, sample $r \leftarrow \mathcal{R}_{\mathsf{Enc'}}$ and compute $\tilde{c}' \leftarrow \mathsf{Enc'}(pk', m_b; r)$. Also sample $\tilde{b} \leftarrow \{0,1\}$ and $\tilde{r} \leftarrow \mathcal{R}_{\mathsf{Enc'}}$. Give $(m_b, r), (m_{\tilde{b}}, \tilde{r})$ to the challenger. Upon receiving $\tilde{c}$, give $\tilde{c}^* = (\tilde{c}', \tilde{c})$ to the A. Continue to answer decryption queries in the same manner, except that all of A's decryption queries with $c = \tilde{c}$ are automatically rejected. (Note that B cannot decrypt the corresponding c on its own; however, by our change from Hyb_1, this is not necessary.)

Analysis. Observe that A perfectly simulates Hyb_2 or Hyb_1, depending on $\tilde{c}$.

- Hyb_3: This hybrid is the same as Hyb_2 except for the following modification. In constructing $\tilde{c}'$, instead of encrypting m_b as the plaintext, we encrypt $m_{\tilde{b}}$.
   6. Sample $r \leftarrow \mathcal{R}_{\mathsf{Enc'}}$ and compute $\tilde{c}' \leftarrow \mathsf{Enc'}(pk', \underline{m_{\tilde{b}}}; r)$.

Claim. $\mathsf{Hyb}_3 \approx_c \mathsf{Hyb}_2$.

Proof. We shall establish this claim by relying on the CPA security of PKE'. Assume for contradiction that $|\Pr[\mathsf{Hyb}_3 = 1] - \Pr[\mathsf{Hyb}_2 = 1]| = \varepsilon$ is non-negligible. Then we construct an adversary B that breaks CPA security of PKE' with advantage ε.

Adversary B. B simultaneously interacts with its CPA challenger for PKE' and A as follows. B behaves the same way as the Hyb_2 challenger, except for the following modification. Instead of generating pk' and $\tilde{c}'$ by himself, he uses pk' from the challenger and generates $\tilde{c}'$ as follows. Upon A giving (m_0, m_1), send $(m_b, m_{\tilde{b}})$ to the challenger and use the response as $\tilde{c}'$ in the interaction with A. Finally, output the output of A.

Analysis. Firstly, we argue that the above description of B is sound: note that the random coins used in computing $\tilde{c}'$ is not used in any part of B's interaction with A. Next, we observe that when $\tilde{c}'$ encrypts 0 then the view of A is identical to that in Hyb_2; when $\tilde{c}'$ encrypts 1 then the view of A is identical to that in Hyb_3. Thus, $\mathsf{Adv}^{\text{ind-cpa}}_{\mathsf{PKE}',B}(k) = \varepsilon$, a contradiction.

We note here that in Hyb_3, A's view is independent of the challenge bit b, and thus the theorem follows.

6   Conclusions

In this paper, we show that there are schemes which are CPA and even CCA secure, but which are clearly insecure in the selective opening scenario. Several open questions remain. Most importantly, it would be interesting to get such examples under weaker assumptions. As a first step, one could hope for an example that only relies on indistinguihsability obfuscation rather than public-coin differing-inputs obfuscation and correlation-intractable hash functions. Ideally, one would get rid of obfuscation altogether. Alternatively, it would be interesting if such examples can lead to surprising positive results or perhaps can imply (some variant of) obfuscation. Another open question is to construct a counterexample for SOA-K security without relying on a scheme with common public parameters.

A   IND-CCA Game with Random Challenge Messages

We shall now define a variation of the IND-CCA game. The modification at a high level is that, in the new game, the challenger himself samples the challenge message pair uniformly at random.

Definition 13 ($-IND-CCA Secure PKE). *Let* $\mathsf{PKE} = (\mathsf{Gen}, \mathsf{Enc}, \mathsf{Dec})$ *be a tuple of PPT algorithms.* PKE *is said to be a $-IND-CCA-secure encryption, if it for every PPT adversary A,*

$$\mathsf{Adv}^{\$\text{-ind-cca}}_{\mathsf{PKE},A}(k) := \left| \Pr\left[\mathsf{Exp}^{\$\text{-ind-cca}}_{\mathsf{PKE},A}(k) = 1 \right] - \frac{1}{2} \right|$$

is negligible.

Theorem 7. *Let* $\mathsf{PKE} = (\mathsf{Gen}, \mathsf{Enc}, \mathsf{Dec})$ *is a PKE scheme with polynomial-size message ciphertext. If* PKE *is $-IND-CCA secure as per Definition 13, then* PKE *is IND-CCA secure.*

Proof. Let the message space of PKE be $\mathcal{M}$, with $|\mathcal{M}| = \ell(k)$, for a polynomial ℓ. Assume for contradiction that there exists an adversary A that breaks the IND-CCA security of PKE with advantage ε. Then we shall show an adversary B that breaks the $-IND-CCA security of PKE with advantage $\varepsilon/\ell(k)^2$. With the help of A, B interacts with its $-IND-CCA challenger as follows (Fig. 5).

Adversary B: Upon receiving $pk, (m_0, m_1), c^*$, give pk to A. Let (m'_0, m'_1) be the pair of message given in response by A. Check if $(m'_0, m'_1) = (m_0, m_1)$. If not,

---

Experiment $\mathsf{Exp}^{\text{\$-ind-cca}}_{\text{PKE},A}$

$b \leftarrow \{0,1\}$

$(pk, sk) \leftarrow \mathsf{Gen}(1^k)$

$(m_0, m_1) \leftarrow A^{\mathsf{Dec}(sk,\cdot)}(pk)$ ~~(struck through)~~

$m_0, m_1 \leftarrow \mathcal{M}$

$c^* \leftarrow \mathsf{Enc}(pk, m_b)$

$out_A \leftarrow A^{\mathsf{Dec}_{c^*}(sk,\cdot)}(pk, (m_0, m_1), c^*)$

return 1 if $out_A = b$, and 0 otherwise

Fig. 5. $-IND-CCA experiment.

sample $b' \leftarrow \{0,1\}$ and respond to the challenger with b'. Otherwise, give c^* to A. Output whatever A outputs.

Analysis: Let $\mathsf{E_{SameChal}}$ denote the event that $(m_0', m_1') = (m_0, m_1)$. Note that, since m_0, m_1 are chosen uniformly at random, we have that $\Pr[\mathsf{E_{SameChal}}] = \frac{1}{\ell(k)^2}$. Furthermore,

$$\Pr\left[\mathsf{Exp}^{\text{\$-ind-cca}}_{\text{PKE},B}(k) = 1 | \neg\mathsf{E_{SameChal}}\right] = \frac{1}{2}$$

On the other hand,

$$\Pr\left[\mathsf{Exp}^{\text{\$-ind-cca}}_{\text{PKE},B}(k) = 1 | \mathsf{E_{SameChal}}\right] = \frac{1}{2} + \varepsilon$$

Putting them together, we have that, Thus,

$$\mathsf{Adv}^{\text{\$-ind-cca}}_{\text{PKE},A}(k) = \left| \left(\frac{1}{2} + \varepsilon\right)\frac{1}{\ell(k)^2} + \frac{1}{2}\left(1 - \frac{1}{\ell(k)^2}\right) - \frac{1}{2} \right|$$

$$= \frac{\varepsilon}{\ell(k)^2}$$

leading to a contradiction.

References

1. Barak, B., Goldreich, O., Impagliazzo, R., Rudich, S., Sahai, A., Vadhan, S.P., Yang, K.: On the (im)possibility of obfuscating programs. J. ACM **59**(2), 6 (2012)
2. Barak, B., Lindell, Y., Vadhan, S.P.: Lower bounds for non-black-box zero knowledge. In: Proceedings of 44th Symposium on Foundations of Computer Science, FOCS 2003, 11–14 October 2003, Cambridge, MA, USA, pp. 384–393. IEEE Computer Society (2003). http://dx.doi.org/10.1109/SFCS.2003.1238212
3. Bellare, M., Dowsley, R., Waters, B., Yilek, S.: Standard security does not imply security against selective-opening. In: Pointcheval, D., Johansson, T. (eds.) EUROCRYPT 2012. LNCS, vol. 7237, pp. 645–662. Springer, Heidelberg (2012). doi:10.1007/978-3-642-29011-4_38

4. Bellare, M., Hofheinz, D., Yilek, S.: Possibility and impossibility results for encryption and commitment secure under selective opening. In: Joux, A. (ed.) EURO-CRYPT 2009. LNCS, vol. 5479, pp. 1–35. Springer, Heidelberg (2009). doi:10.1007/978-3-642-01001-9_1

5. Bellare, M., Stepanovs, I., Waters, B.: New negative results on differing-inputs obfuscation. In: Fischlin, M., Coron, J.-S. (eds.) EUROCRYPT 2016. LNCS, vol. 9666, pp. 792–821. Springer, Heidelberg (2016). doi:10.1007/978-3-662-49896-5_28

6. Böhl, F., Hofheinz, D., Kraschewski, D.: On definitions of selective opening security. In: Fischlin, M., Buchmann, J., Manulis, M. (eds.) PKC 2012. LNCS, vol. 7293, pp. 522–539. Springer, Heidelberg (2012). doi:10.1007/978-3-642-30057-8_31

7. Canetti, R., Chen, Y., Reyzin, L.: On the correlation intractability of obfuscated pseudorandom functions. IACR ePrint Archive, report 2015/334 (2015). http://eprint.iacr.org/2015/334

8. Canetti, R., Goldreich, O., Halevi, S.: The random oracle methodology, revisited (preliminary version). In: Proceedings of STOC 1998, pp. 209–218. ACM (1998)

9. Cohen, A., Holmgren, J., Vaikuntanathan, V.: Publicly verifiable software watermarking. IACR ePrint Archive, report 2015/373 (2015). http://eprint.iacr.org/2015/373

10. Dodis, Y., Ristenpart, T., Vadhan, S.: Randomness condensers for efficiently samplable, seed-dependent sources. In: Cramer, R. (ed.) TCC 2012. LNCS, vol. 7194, pp. 618–635. Springer, Heidelberg (2012). doi:10.1007/978-3-642-28914-9_35

11. Dwork, C., Naor, M., Reingold, O., Stockmeyer, L.J.: Magic functions. In: Proceedings of FOCS 1999, pp. 523–534. IEEE Computer Society (1999)

12. Fehr, S., Hofheinz, D., Kiltz, E., Wee, H.: Encryption schemes secure against chosen-ciphertext selective opening attacks. In: Gilbert, H. (ed.) EUROCRYPT 2010. LNCS, vol. 6110, pp. 381–402. Springer, Heidelberg (2010). doi:10.1007/978-3-642-13190-5_20

13. Fuchsbauer, G., Heuer, F., Kiltz, E., Pietrzak, K.: Standard security does imply security against selective opening for markov distributions. In: Kushilevitz, E., Malkin, T. (eds.) TCC 2016. LNCS, vol. 9562, pp. 282–305. Springer, Heidelberg (2016). doi:10.1007/978-3-662-49096-9_12

14. Fujisaki, E.: All-but-many encryption - a new framework for fully-equipped uc commitments. In: Sarkar, P., Iwata, T. (eds.) ASIACRYPT 2014. LNCS, vol. 8874, pp. 426–447. Springer, Heidelberg (2014). doi:10.1007/978-3-662-45608-8_23

15. Garg, S., Gentry, C., Halevi, S., Raykova, M., Sahai, A., Waters, B.: Candidate indistinguishability obfuscation and functional encryption for all circuits. In: 54th Annual IEEE Symposium on Foundations of Computer Science, FOCS 2013, 26–29 October 2013, Berkeley, CA, USA, pp. 40–49. IEEE Computer Society (2013). http://dx.doi.org/10.1109/FOCS.2013.13

16. Garg, S., Gentry, C., Halevi, S., Wichs, D.: On the implausibility of differing-inputs obfuscation and extractable witness encryption with auxiliary input. In: Garay, J.A., Gennaro, R. (eds.) CRYPTO 2014. LNCS, vol. 8616, pp. 518–535. Springer, Heidelberg (2014). doi:10.1007/978-3-662-44371-2_29

17. Hazay, C., Patra, A., Warinschi, B.: Selective opening security for receivers. In: Iwata, T., Cheon, J.H. (eds.) ASIACRYPT 2015. LNCS, vol. 9452, pp. 443–469. Springer, Heidelberg (2015). doi:10.1007/978-3-662-48797-6_19

18. Hemenway, B., Libert, B., Ostrovsky, R., Vergnaud, D.: Lossy encryption: constructions from general assumptions and efficient selective opening chosen ciphertext security. In: Lee, D.H., Wang, X. (eds.) ASIACRYPT 2011. LNCS, vol. 7073, pp. 70–88. Springer, Heidelberg (2011). doi:10.1007/978-3-642-25385-0_4

19. Hofheinz, D.: All-but-many lossy trapdoor functions. In: Pointcheval, D., Johansson, T. (eds.) EUROCRYPT 2012. LNCS, vol. 7237, pp. 209–227. Springer, Heidelberg (2012). doi:10.1007/978-3-642-29011-4_14

20. Hofheinz, D., Rupp, A.: Standard versus selective opening security: separation and equivalence results. In: Lindell, Y. (ed.) TCC 2014. LNCS, vol. 8349, pp. 591–615. Springer, Heidelberg (2014). doi:10.1007/978-3-642-54242-8_25

21. Huang, Z., Liu, S., Qin, B.: Sender-equivocable encryption schemes secure against chosen-ciphertext attacks revisited. In: Kurosawa, K., Hanaoka, G. (eds.) PKC 2013. LNCS, vol. 7778, pp. 369–385. Springer, Heidelberg (2013). doi:10.1007/978-3-642-36362-7_23

22. Ishai, Y., Pandey, O., Sahai, A.: Public-coin differing-inputs obfuscation and its applications. IACR ePrint Archive, report 2014/942 (2014). http://eprint.iacr.org/2014/942

23. Ishai, Y., Pandey, O., Sahai, A.: Public-coin differing-inputs obfuscation and its applications. In: Dodis, Y., Nielsen, J.B. (eds.) TCC 2015. LNCS, vol. 9015, pp. 668–697. Springer, Heidelberg (2015). doi:10.1007/978-3-662-46497-7_26

24. Koppula, V., Ramchen, K., Waters, B.: Separations in circular security for arbitrary length key cycles. In: Dodis, Y., Nielsen, J.B. (eds.) TCC 2015, Part II. LNCS, vol. 9015, pp. 378–400. Springer, Heidelberg (2015). doi:10.1007/978-3-662-46497-7_15

25. Marcedone, A., Orlandi, C.: Obfuscation $\Rightarrow$ (IND-CPA security $\nRightarrow$ circular security). In: Abdalla, M., Prisco, R. (eds.) SCN 2014. LNCS, vol. 8642, pp. 77–90. Springer, Heidelberg (2014). doi:10.1007/978-3-319-10879-7_5

26. Naor, M., Pinkas, B.: Efficient oblivious transfer protocols. In: Proceedings of SODA 2001, pp. 448–457. ACM/SIAM (2001)

27. Peikert, C., Vaikuntanathan, V., Waters, B.: A framework for efficient and composable oblivious transfer. In: Wagner, D. (ed.) CRYPTO 2008. LNCS, vol. 5157, pp. 554–571. Springer, Heidelberg (2008). doi:10.1007/978-3-540-85174-5_31

28. Peikert, C., Waters, B.: Lossy trapdoor functions and their applications. In: Proceedings of STOC 2008, pp. 187–196. ACM (2008)

Public-Key Encryption with Simulation-Based Selective-Opening Security and Compact Ciphertexts

Dennis Hofheinz[1]([⊠]), Tibor Jager[2], and Andy Rupp[1]

[1] Karlsruhe Institute of Technology, Karlsruhe, Germany
{dennis.hofheinz,andy.rupp}@kit.edu
[2] Ruhr-University Bochum, Bochum, Germany
tibor.jager@rub.de

Abstract. In a selective-opening (SO) attack on an encryption scheme, an adversary A gets a number of ciphertexts (with possibly related plaintexts), and can then adaptively select a subset of those ciphertexts. The selected ciphertexts are then opened for A (which means that A gets to see the plaintexts and the corresponding encryption random coins), and A tries to break the security of the unopened ciphertexts.

Two main flavors of SO security notions exist: indistinguishability-based (IND-SO) and simulation-based (SIM-SO) ones. Whereas IND-SO security allows for simple and efficient instantiations, its usefulness in larger constructions is somewhat limited, since it is restricted to special types of plaintext distributions. On the other hand, SIM-SO security does not suffer from this restriction, but turns out to be significantly harder to achieve. In fact, all known SIM-SO secure encryption schemes either require $O(|m|)$ group elements in the ciphertext to encrypt $|m|$-bit plaintexts, or use specific algebraic properties available in the DCR setting.

In this work, we present the first SIM-SO secure PKE schemes in the discrete-log setting with compact ciphertexts (whose size is $O(1)$ group elements plus plaintext size). The SIM-SO security of our constructions can be based on, e.g., the k-linear assumption for any k.

Technically, our schemes extend previous IND-SO secure schemes by the property that simulated ciphertexts can be *efficiently* opened to arbitrary plaintexts. We do so by encrypting the plaintext in a bitwise fashion, but such that each encrypted bit leads only to a single ciphertext bit (plus $O(1)$ group elements that can be shared across many bit encryptions). Our approach leads to rather large public keys (of $O(|m|^2)$ group elements), but we also show how this public key size can be reduced (to $O(|m|)$ group elements) in pairing-friendly groups.

Keywords: Public-key encryption · Selective-opening security · Lossy encryption · Matrix assumptions

© International Association for Cryptologic Research 2016
M. Hirt and A. Smith (Eds.): TCC 2016-B, Part II, LNCS 9986, pp. 146–168, 2016.
DOI: 10.1007/978-3-662-53644-5_6

1   Introduction

Selective-opening (SO) attacks. A selective-opening (SO) attack on an encryption scheme models the adaptive corruption of multiple senders. More formally, an SO adversary A first receives many ciphertexts $c_1, \ldots, c_n$ for respective plaintexts $m_1, \ldots, m_n$ that are jointly sampled (and may thus be related). A may then ask for the opening of an arbitrary subset of the c_i.[1] Finally, A is asked to break the security of the unopened ciphertexts.

Different flavors of SO security notions. Note that it is not entirely clear what "breaking the security of the unopened ciphertexts" should mean. For instance, since the plaintexts are related, it is possible that *all* plaintexts (including those from unopened ciphertexts) can be efficiently computed from the opened plaintexts. Furthermore, to achieve greater generality, usually the joint *distribution* from which the $m_1, \ldots, m_n$ are sampled is adversarially chosen, so A may already have some a-priori (partial or even full) knowledge about the unopened m_i.

Hence, two different flavors of SO security have developed: simulation-based (SIM-SO [2,10]) and indistinguishability-based (IND-SO [2,5]) security. Intuitively, SIM-SO security requires that the output of A above can be simulated by a simulator that sees only the opened m_i (and no ciphertexts at all). In particular, all information A can extract about the unopened m_i can also be generated by a simulator from the opened m_i alone.

On the other hand, IND-SO security requires that the unopened plaintexts look indistinguishable from independently sampled plaintexts. Because the plaintexts may be related, this independent sampling must be conditioned on the already opened plaintexts to avoid trivial attacks. Hence, if, e.g., the opened plaintexts already fully determine all plaintexts, conditional sampling will lead to the originally encrypted plaintexts, and IND-SO security is trivially achieved.

As a consequence, the IND-SO experiment itself is only efficient for plaintext distributions that are "efficiently (conditionally) re-samplable" in the above sense. In fact, usually IND-SO security is only considered for such plaintext distributions [2,18,19], which limits its applicability to scenarios with such distributions; there is no known encryption scheme that is IND-SO secure against arbitrary (i.e., only efficiently *samplable*) plaintext distributions.

The difficulty of achieving simulation-based SO security. Hence, from an application point of view, SIM-SO security is the preferable notion of SO security. Unfortunately, while IND-SO security (restricted to efficiently re-samplable plaintext distributions and in the chosen-plaintext case) is already achieved by any lossy encryption scheme [2,29], SIM-SO security seems much harder to obtain. For instance, [1] show (under mild computational assumptions) that there are encryption schemes that are IND-CPA but not SIM-SO secure. Furthermore, known constructions of SIM-SO secure encryption schemes follow dedicated

---

[1] In this paper, we consider sender corruptions, in which case the opening of a c_i consists of the plaintext m_i and the encryption random coins used to construct c_i.

(and somewhat nonstandard) design strategies [2,3,12,15,18,19,22]. As a result, all known SIM-SO secure schemes fall into one of the following two categories:

Large ciphertexts. The SIM-SO secure schemes from [2,3,12,22] have ciphertexts of $O(|m|)$ group elements, where $|m|$ is the bitsize of the plaintext.

DCR-based. The schemes from [15,18,19][2] have more compact ciphertexts, but are limited to the decisional composite residuosity (DCR) setting [9,28] (and rely on its specific algebraic features).

Below, when explaining our technical approach, we will also comment on the technical obstacles that need to be overcome for SIM-SO security.

Our results. In this work, we offer the first SIM-SO secure encryption schemes with compact ciphertexts in the discrete-log setting. Specifically, ciphertexts in our scheme carry $O(1)$ group elements (plus $|m|$ bits, where $|m|$ is the plaintext bitsize), and SIM-SO security can be proved under any matrix assumption [11] (thus, in particular under, e.g., the k-linear assumption for any $k \geq 1$). Our construction is simple, works in the standard model, and does not require pairings.

The price we pay for these features is a rather large public key size (of $O(|m|^2)$ group elements, and computationally expensive encryption and decryption procedures. Specifically, our encryption proceeds bitwise, and requires $O(|m|)$ exponentiations for each message bit. (Alternatively, the operation needed to encrypt one bit could also be viewed as one multi-exponentiation with respect to $O(|m|)$ fixed bases. So there is room for some small improvements in runtime by a constant factor, e.g., using interleaving multi-exponentiation [23].) Concerning the key size, we show how a technique of [7] can be used to at least compress the public key to $O(|m|)$ group elements by using a pairing. Still, in particular in light of the relatively inefficient encryption and decryption in our scheme, we view our result mainly as a feasibility result (Table 1).

In the following, we give a brief overview over our approach.

Our starting point. Our starting point is the lossy (and thus IND-SO secure) PKE scheme of [25] (see also [2,17,29]). In this scheme, public keys and ciphertexts are of the form

$$pk = (g, g^x, g^y, g^z) \qquad c = (u, v) = (g^{r+sx}, g^{ry+sz} \cdot m) \tag{1}$$

for random exponents x, y, r, s, for $z = xy$, and a plaintext m. Note that if we switch z to an independently random value (however with $z \neq xy$), then encryption becomes lossy: ciphertexts are tuples of random group elements, independently of m. Furthermore, such a switch can be justified with the decisional Diffie-Hellman (DDH) assumption.

Efficient openability. In order to achieve SIM-SO security, we additionally require a property called "efficient openability" of ciphertexts [2,12]. In a nutshell, efficient openability requires that ciphertexts generated under lossy public keys can be opened to arbitrary messages with a special trapdoor. (Note that such an arbitrary opening is always possible inefficiently in the lossy case.)

---

[2] In fact, [18] also offers a scheme with large ciphertexts in the discrete-log setting.

Table 1. Comparison of our construction with other SO-secure PKE schemes. (We omit schemes that do not achieve SIM-SO-CPA security in any more efficient way than the ones mentioned, e.g., because they focus on CCA security [15,18,19] or on the IBE setting [3].) $|G|$ denotes the description (bit-)size of elements of a group in the discrete-log setting, and $|G|$ and $|G_T|$ denote the corresponding sizes in a pairing-friendly setting with source group G and target group G_T. λ denotes the security parameter. The entry poly(λ) in the $|m|$ column means that the message size is not restricted and might be set arbitrarily (and especially independent of the group size). QR denotes the quadratic residuosity assumption, DCR denotes Paillier's decisional composite residuosity assumption, and $|N|$ denotes the length of a suitable composite number (determining the modulus) for such schemes. TDOWP denotes an arbitrary trapdoor one-way permutation, and $|img|$ denotes the (bit-)size of elements in the corresponding image. $|c| - |m|$ denotes the ciphertext overhead (i.e., the bitlength of the ciphertext minus the plaintext bitlength).

Scheme	Security	Assumption	$|pk|$	$|m|$	$|c| - |m|$
BHY09 [2]	IND-SO-CPA	DDH	$2 \times |G|$	$|G|$	$|G|$
BHY09 [2]	SIM-SO-CPA	QR	$1 \times |N|$	n	$n(|N| - 1)$
BY12 [4]^a	SIM-SO-CPA	DDH	$2 \times |G|$	1	$2|G| - 1$
FHKW10 [12]	SIM-SO-CPA	TDOWP	TDOWP-pk	1	$|img| - 1$
FHKW10 [12]	SIM-SO-CCA	DDH	$2 \times |G|$	poly(λ)	$2|m||G| + |m|\lambda$
HLOV11 [18]	SIM-SO-CPA	DCR	$2 \times |N|$	$|N|$	$|N|$
Ours	SIM-SO-CPA	DDH	$(|m| + 1)^2$	poly(λ)	$1 \times |G|$
Ours	SIM-SO-CPA	DLIN	$(|m| + 2)^2$	poly(λ)	$2 \times |G|$
Ours	SIM-SO-CPA	k-linear	$(|m| + k)^2$	poly(λ)	$k \times |G|$
Ours	SIM-SO-CPA	BDDH	$|m| \cdot (4|G| + |G_T|)$	poly(λ)	$1 \times |G_T|$

^{a}This denotes a scheme present in a September 23, 2012 update of the eprint report [4]. This scheme operates bitwise, and the message length n can be chosen arbitrarily.

We note that efficient openability implies SIM-SO security [2]. In fact, all mentioned SIM-SO secure schemes achieve (a suitable variant of) efficient open-ability.[3] Unfortunately, this strong property is not achieved easily. For instance, consider the PKE scheme from (1) (with lossy public keys, i.e., with $z \neq xy$). In order to open a given ciphertext $c = (u, v)$ as an encryption of an externally given plaintext m, a simulator would have to supply random coins (r, s) satisfying $r + sx = \text{dlog}_g(u)$ and $ry + sz = \text{dlog}_g(v) - \text{dlog}_g(m)$. Hence, the ability to open to arbitrary m implies the ability to compute discrete logarithms (which would seem to require special trapdoors in standard discrete-log groups).[4]

---

[3] However, it should also be noted that neither efficient openability nor lossiness (in the sense of [2,29]) may be necessary for SIM-SO security (see [27] for the lossiness case). Still, our construction is easiest to explain by following this path.

[4] One reason why the DCR settings seems much more suitable for SO security is that certain DCR subgroups allow to easily compute discrete logarithms. Put differently: in DCR-based encryption schemes [9,18,28], both plaintexts and encryption random coins are exponents. Hence, encryption random coins can be computed from plaintexts (as required for a SIM-SO simulation) much more easily.

A bitwise scheme. Our first observation is that the situation changes if only bits (or messages from a small domain) are encrypted. Concretely, consider the following slightly modified scheme that encrypts only bits:

$$pk = (g, g^x, g^y, g^z) \qquad c = (u, v) = (g^{r+sx}, H(g^{ry+sz}) \oplus m) \qquad (2)$$

where x, y, z, r, s are as before, H is a universal hash function that maps group elements to bits, and $m \in \{0, 1\}$. This scheme allows for an efficient opening operation (if $z \neq xy$). Namely, to open a ciphertext $c = (u, v)$ (as in (2)) to a message m, using as trapdoor x, y, z, r, s, simply sample r', s' randomly subject to $r' + s'x = r + sx$ until $H(g^{r'y+s'z}) \oplus m = v$. (On average, it takes 2 such samplings until suitable r', s' are found.) This scheme can be generalized to messages $m \in \{0, 1\}^{O(\log(\lambda))}$ (where λ denotes the security parameter), using hash functions with output length $|m|$, at the cost of a less efficient opening algorithm. In the following, however, we will focus on the bitwise processing of messages for simplicity.

The scheme from (2) hence achieves efficient openability (and thus SIM-SO security), but suffers from a small message space. Of course, its message space can be expanded by concatenating several ciphertexts, which would however increase the ciphertext size to $O(|m|)$ group elements.

Compressing ciphertexts. Hence, instead of concatenating ciphertexts, we reuse the value of u across several bit encryptions. Doing so naively (e.g., by setting $u = g^{r+sx}$ and $v_i = H(g_i^{ry+sz}) \oplus m_i$ for different generators g_i) would however interfere with our efficient opening strategy. Specifically, it is not obvious how to efficiently sample r', s' as above that would lead to $H(g_i^{r'y+s'z}) \oplus m_i = v_i$ for all i simultaneously.

We resolve this issue by adding more encryption random coins (and thus more "degrees of freedom" for our efficient opening procedure). That is, we set

$$pk = (g, (g^{x_i}, g^{y_j})_{j=1}^{\mu}, (g^{z_{i,j}})_{i,j=1}^{\mu})$$
$$c = (u, (v_i)_{i=1}^{\mu}) = (g^{r+\sum_{j=1}^{\mu} s_j x_j}, (H(g^{ry_j + \sum_{j=1}^{\mu} s_j z_{i,j}}) \oplus m_i)_{i=1}^{\mu}) \qquad (3)$$

for random exponents $x_i, y_j, z_{i,j}, r, s_j$ and $z_{i,j} = x_i y_j$, and an μ-bit plaintext $m = (m_i)_{i=1}^{\mu}$. Since $z_{i,j} = x_i y_j$, knowledge of all x_i, y_j allows to decrypt. However, switching to random $z_{i,j} \neq x_i y_j$ (which can be justified with the DDH assumption) implies that encryption becomes lossy (as with (1) and (2)).

Moreover, in case $z_{i,j} \neq x_i y_j$, a ciphertext $c = (u, (v_i)_{i=1}^{\mu})$ can be efficiently opened as follows. First, select "target exponents" $t_1, \ldots, t_\mu$ randomly subject to $H(g^{t_i}) \oplus m_i = v_i$ for all i. (The t_i can be sampled individually, one after the other, and so this step requires 2μ samplings on average.) Next, solve the system that consists of the linear equations $r'y_j + \sum_{j=1}^{\mu} s'_j z_{i,j} = t_i$ (with $1 \leq i \leq \mu$) and $r' + \sum_{j=1}^{\mu} s'_j x_j = r + \sum_{j=1}^{\mu} s_j x_j$ for the variables r', s'_j. (Since the $z_{i,j} \neq x_i y_j$ are random, this system is solvable using linear algebra with high probability.) Finally, output $(r', (s'_j)_{j=1}^{\mu})$ as the desired random coins that open c to m.

Extensions and open problems. Inside, we also show how to generalize this idea to weaker assumptions than DDH (in the same spirit in which [14] generalize the DDH-based lossy trapdoor function of [30]). In particular, we obtain constructions based on any Matrix Diffie-Hellman (MDDH) assumption [11] (at the price of somewhat larger ciphertexts, but whose overhead is still independent of $|m|$, and somewhat larger public keys), including the k-linear assumption [20,31]. Furthermore, we show how to compress the public key of our scheme from $\mathbf{O}(|m|^2)$ to $\mathbf{O}(|m|)$ group elements using a pairing-based technique used to compress the public key of lossy trapdoor functions [7].

In this work, we focus on chosen-plaintext (CPA) security. One interesting open problem is to extend our techniques to the chosen-ciphertext (CCA) setting to obtain a SIM-SO-CCA secure scheme with compact ciphertexts in the discrete-log regime. Besides, of course a further compression of the public key in our schemes or an improvement in computational efficiency would be desirable.

Relation to a scheme of Bellare and Yilek. Our "bitwise" scheme from (2) above is very similar to a scheme of Bellare and Yilek (from Sect. 5.4 of the September 23, 2012 update of [4]). (We thank one TCC reviewer for pointing us to that scheme, which we were not aware of previously.) The main difference is that we use the use the term $H(g^{ry+sz}) \oplus m$ to hide the message, whereas Bellare and Yilek use $g^{ry+sz}) \cdot g^m$. This entails (conceptually not very significant) differences in the respective opening algorithms. However, the more important difference in these schemes is that our scheme from (2) only has one group element (plus one hidden message bit) in the ciphertext, while Bellare and Yilek use a whole group element to hide a one-bit message. Hence, our main trick above (namely, to modify and then reuse the first ciphertext element g^{ry+sz} for many bit encryptions) would not lead to compact ciphertexts when applied to the scheme of Bellare and Yilek.

SO security against corrupted receivers, and relation to non-committing encryption. Traditionally, SO security models a setting in which only senders are corrupted (and thus, an opening only reveals the corresponding encryption random coins). However, some works (e.g., [1,21]) *additionally* consider SO security against corrupted receivers (in which case there are many public keys, and an opening consists of the respective secret key). In this setting, strong impossibility results hold [1], which provide a fixed upper limit the number of secure encryptions under any given public key. The arising technical problems are *commitment problems*, and are very related to the inherent problems of non-committing encryption (NCE, [8]). Indeed, NCE schemes can be seen as encryption schemes that are SO secure both against corrupted senders and corrupted receivers.

In contrast, the more commonly considered notion of SO security against corrupted senders (which we also consider here) allows for more efficient schemes, that in particular tolerate an arbitrary number of encryptions and corruptions. The price to pay here is of course that only corruptions of senders (but not of receivers) are considered.

Roadmap. After fixing some notation and basic definitions in Sect. 2, we introduce our construction of lossy encryption with efficient weak opening in Sect. 3.

The construction is generic and relies on what we call a matrix rank assumption. In Sect. 4, we then instantiate those assumptions with the family of MDDH assumptions from [11] (and thus in particular with the k-linear assumption). Finally, Sect. 5 presents a matrix rank assumption with a linear-size representation which is implied by the BDDH assumption in pairing groups. This results in a scheme with a public key size that is linear in $|m|$.

2   Preliminaries

Notation. Throughout the paper, $\lambda \in \mathbb{N}$ denotes the security parameter. For a finite set $\mathcal{S}$, we denote by $s \leftarrow \mathcal{S}$ the process of sampling s uniformly from $\mathcal{S}$. For a probabilistic algorithm A, we denote with $\mathcal{R}_A$ the space of A's random coins. $y \leftarrow A(x; R)$ denotes the process of running A on input x and with randomness $R \leftarrow \mathcal{R}_A$, and assigning y the result. We write $y \leftarrow A(x)$ for $y \leftarrow A(x; R)$ with uniform R. If A's running time is polynomial in λ, then A is called probabilistic polynomial-time (PPT). We call a positive function η negligible if for every polynomial p there exists λ_0 such that for all $\lambda \geq \lambda_0$ holds $\eta(\lambda) \leq \frac{1}{p(\lambda)}$. We call η overwhelming if $\eta(\lambda) \geq 1 - \nu(\lambda)$, where ν is a negligible function. The statistical distance between two random variables X and Y over a finite common domain D is defined by $\Delta(X, Y) = \frac{1}{2} \sum_{z \in D} |\Pr[X = z] - \Pr[Y = z]|$. We say that two families $X = (X_\lambda)_{\lambda \in \mathbb{N}}$ and $Y = (Y_\lambda)_{\lambda \in \mathbb{N}}$ of random variables are statistically close or statistically indistinguishable, denoted by $X \approx_s Y$, if $\Delta(X_\lambda, Y_\lambda)$ is negligible in λ.

2.1   Groups and Matrix Assumptions

Prime-order k-linear group generators. We use the following formal definition of a k-linear prime-order group generator for our constructions.

Remark 1. We stress that *our constructions do not require multilinear maps* in the sense of [16]. We rather want to capture both single-group settings and bilinear group settings in one unified definition, because this will be helpful in the sequel for the exposition of results that apply to both settings. Hence, one should have $k = 1$ or $k = 2$ in mind in the following definition.

Definition 1. *A* prime-order k-linear group generator *is a PPT algorithm* $\mathcal{G}_k$ *that on input of a security parameter* 1^λ *outputs a tuple of the form*

$$\mathcal{MG}_k := (k, G_1, \dots, G_k, G_{k+1}, g_1, \dots, g_k, e, p) \leftarrow \mathcal{G}_k(1^\lambda)$$

where $G_1, \dots, G_{k+1}$ *are descriptions of cyclic groups of prime order* p, $\log p = \Theta(\lambda)$, g_i *is a generator of* G_i *for* $1 \leq i \leq k$, *and* $e \colon G_1 \times \dots \times G_k \to G_{k+1}$ *is a map which satisfies the following properties:*

- *k-linearity: For all* $a_1 \in G_1, \dots, a_k \in G_k$, $\alpha \in \mathbb{Z}_p$, *and* $i \in \{1, \dots, k\}$ *we have*
  $e(a_1, \dots, a_{i-1}, \alpha a_i, a_{i+1}, \dots, a_k) = \alpha e(a_1, \dots, a_k).$

– Non-degeneracy: $g_{k+1} := e(g_1, \ldots, g_k)$ *generates* G_{k+1}.

If $G_1 = \ldots = G_k$, *we call* $\mathcal{G}_k$ a symmetric k-*linear group generator.*

Note that Definition 1 captures both ordinary single group generators and symmetric bilinear group generators:

– In the single-group setting, $\mathcal{G}_1(1^\lambda)$ would output $\mathcal{MG}_1 := (1, G_1, G_2, g_1, e, p)$, where $G_1 = G_2$ and $e : G_1 \rightarrow G_2$ is the identity mapping.
– In the symmetric bilinear group setting, $\mathcal{G}_2(1^\lambda)$ would output $\mathcal{MG}_2 := (1, G_1, G_2, G_3, g_1, g_2, e, p)$, where $G_1 = G_2$ and $g_1 = g_2$ and $e : G_1 \times G_2 \rightarrow G_3$ is a pairing.

Implicit Representation. Following [11], we introduce the notion of implicit representations. Let G_i be a cyclic group of order p generated by g_i. Then by $[a]_i := g_i^a$ we denote the *implicit representation* of $a \in \mathbb{Z}_p$ in G_i. More generally, we also define such representations for vectors $\vec{b} \in \mathbb{Z}_p^n$ by $[\vec{b}]_i := ([b_j]_i)_j \in G_i^n$ and for matrices $\mathbf{A} = (a_{j,k})_{j,k} \in \mathbb{Z}_p^{n \times \ell}$ by $[\mathbf{A}]_i := ([a_{j,k}]_i)_{j,k} \in G_i^{n \times \ell}$.

Matrix-vector operations in implicit representation. If a matrix $[\mathbf{A}] = [(a_{i,j})_{i,j}] \in G^{n \times \ell}$ is known "in the exponent", and a vector $\vec{u} = (u_i)_i \in \mathbb{Z}_p^\ell$ is known "in clear", then the product $[\mathbf{A} \cdot \vec{u}] \in G^n$ can be efficiently computed as $[(v_i)_i]$ for $[v_i] = \prod_{j=1}^\ell [a_{i,j}]^{u_j}$. Similarly, $[\mathbf{A} \cdot \mathbf{B}] \in G^{n \times k}$ can be computed given $[\mathbf{A}] = [(a_{i,j})_{i,j}] \in G^{n \times \ell}$ and $\mathbf{B} \in \mathbb{Z}_p^{\ell \times k}$. If only $[\mathbf{A}]_1$ and $[\mathbf{B}]_2$ are known (i.e., only "in the exponent") and a bilinear map $e : G_1 \times G_2 \rightarrow G_3$ is given, we can still compute the matrix product $[\mathbf{A} \cdot \mathbf{B}]_3$ in the target group G_3, as $[(c_{i,j})_{i,j}]_3$ for $[c_{i,j}]_3 = \prod_{t=1}^\ell e([a_{i,t}]_1, [b_{t,j}]_2)$.

Matrix distributions and MDDH assumptions. For instantiating our construction we will make use of matrix distributions and the Matrix Diffie-Hellman assumption family as introduced in [11].

Let $n, \ell \in \mathbb{N}$, $n > \ell$. We call $\mathcal{D}_{n,\ell}$ a *matrix distribution* if it outputs (in probabilistic polynomial time and with overwhelming probability in $\log(p)$) matrices $\mathbf{A} \in \mathbb{Z}_p^{n \times \ell}$ of full rank ℓ. We define $\mathcal{D}_\ell := \mathcal{D}_{\ell+1,\ell}$.

Definition 2. *We say that the* $\mathcal{D}_{n,\ell}$-*Matrix Diffie-Hellman assumption, or just* $\mathcal{D}_{n,\ell}$-*MDDH assumption for short, holds in* G_i *and relative to the* k-*linear group generator* $\mathcal{G}_k$, *if for all PPT adversaries* D, *we have that*

$$\mathbf{Adv}_{\mathcal{D}_{n,\ell}, \mathcal{G}_k}(D) = |\Pr[D(\mathcal{MG}_k, [\mathbf{A}]_i, [\mathbf{A}\vec{w}]_i) = 1] - \Pr[D(\mathcal{MG}_k, [\mathbf{A}]_i, [\vec{u}]_i) = 1]|$$

is negligible, where the probability is taken over the output

$$\mathcal{MG}_k = (k, G_1, \ldots, G_k, G_{k+1}, g_1, \ldots, g_k, e, p) \leftarrow \mathcal{G}_k(1^\lambda),$$

$\mathbf{A} \leftarrow \mathcal{D}_{n,\ell}$, $\vec{w} \leftarrow \mathbb{Z}_p^\ell$, $\vec{u} \leftarrow \mathbb{Z}_p^n$ *and the coin tosses of the adversary* D.

In particular, we will refer to the following examples of matrix distributions, all for $n = \ell + 1$:

$$
\mathcal{SC}_\ell : \mathbf{A} = \begin{pmatrix} s & 0 & \cdots & 0 & 0 \\ 1 & s & \cdots & 0 & 0 \\ 0 & 1 & & 0 & 0 \\ & & \ddots & & \\ 0 & 0 & \cdots & 1 & s \\ 0 & 0 & \cdots & 0 & 1 \end{pmatrix}, \quad \mathcal{L}_\ell : \mathbf{A} = \begin{pmatrix} s_1 & 0 & 0 & \cdots & 0 \\ 0 & s_2 & 0 & \cdots & 0 \\ \vdots & \vdots & \vdots & & \vdots \\ 0 & 0 & 0 & \cdots & s_\ell \\ 1 & 1 & 1 & \cdots & 1 \end{pmatrix}, \quad \mathcal{U}_\ell : \mathbf{A} \leftarrow \mathbb{Z}_p^{(\ell+1) \times \ell},
$$

where $s, s_i \leftarrow \mathbb{Z}_p$. The $\mathcal{SC}_\ell$ assumption, introduced in [11], is the ℓ-*symmetric cascade assumption* (ℓ-SCasc). The $\mathcal{L}_\ell$ assumption is actually the well-known ℓ-*linear assumption* (ℓ-Lin, [6]) in matrix language (DDH equals 1-Lin), and the $\mathcal{U}_\ell$ assumption is the ℓ-*uniform assumption*. Moreover, ℓ-SCasc, ℓ-Lin, and the ℓ-uniform assumption hold in the generic group model [32] relative to a k-linear group generator if $k \leq \ell$ [11].

The circulant matrix assumption

$$
\mathcal{C}_{\ell+d,\ell} : \mathbf{A} = \begin{pmatrix} s_1 & & & & 0 \\ \vdots & s_1 & & & \\ s_d & & \ddots & & \\ 1 & s_d & & s_1 & \\ & 1 & & \ddots & \vdots \\ & & \ddots & & s_d \\ 0 & & & 1 \end{pmatrix},
$$

has very recently been proposed in [24] as a $\mathcal{D}_{n,\ell}$-MDDH assumption with optimal representation size among all assumptions with $n > \ell + 1$. This assumption has been shown to hold in the ℓ-linear generic group model [24]. More generally, we can also define the $\mathcal{U}_{n,\ell}$ assumption for arbitrary $n > \ell$. Note that the $\mathcal{U}_{n,\ell}$ assumption is the weakest MDDH assumption (with the worst representation size) and implied by any other $\mathcal{D}_{n,\ell}$ assumption [11]. In particular ℓ-Lin implies the ℓ-uniform assumption as shown by Freeman [13].

Bilinear Decisional Diffie-Hellman. We will make use of the bilinear decisional Diffie-Hellman (BDDH) assumption for our construction with linear-size public keys.

Definition 3. *Let* $\mathcal{MG}_2 := (2, G_1, G_2, G_3, g_1, g_2, e, p) \leftarrow \mathcal{G}_2(1^\lambda)$, *where* $\mathcal{G}_2$ *is a symmetric bilinear group generator (i.e.,* $G_1 = G_2$ *and* $g_1 = g_2$*), and let* $a, b, c \leftarrow \mathbb{Z}_p$, $b \leftarrow \{0, 1\}$, $T_0 := abc$ *and* $T_1 \leftarrow \mathbb{Z}_p$. *We say that the bilinear decisional Diffie-Hellman (BDDH) assumption holds relative to* $\mathcal{G}_2$, *if*

$$
\mathsf{Adv}_{B,\mathcal{G}_2}^{\mathsf{bddh}}(1^\lambda) := \left| \begin{array}{l} \Pr\left[1 \leftarrow B(1^\lambda, \mathcal{MG}_2, [(a,b,c)]_1, [T_0]_3)\right] \\ - \Pr\left[1 \leftarrow B(1^\lambda, \mathcal{MG}_2, [(a,b,c)]_1, [T_1]_3)\right] \end{array} \right|
$$

is a negligible function for all PPT adversaries B.

2.2   Selective-Opening Secure Encryption

Public-Key Encryption. A public-key encryption (PKE) scheme PKE with message space $\mathcal{M}$ consists of three PPT algorithms Gen, Enc, Dec. The key

Experiment $\mathsf{Exp}_{\mathsf{PKE},A,\mathcal{T},n}^{\mathsf{sim\text{-}so\text{-}cpa\text{-}real}}(\lambda)$	Experiment $\mathsf{Exp}_{\mathsf{PKE},S,\mathcal{T},n}^{\mathsf{sim\text{-}so\text{-}cpa\text{-}ideal}}(\lambda)$
$(pk, sk) \leftarrow \mathsf{Gen}(1^\lambda)$	$\mathfrak{D}_{\mathsf{so}} \leftarrow S(\mathtt{dist})$
$\mathfrak{D}_{\mathsf{so}} \leftarrow A(\mathtt{dist}, pk)$	$(m_i)_{i\in[n]} \leftarrow \mathfrak{D}_{\mathsf{so}}$
$(m_i)_{i\in[n]} \leftarrow \mathfrak{D}_{\mathsf{so}}$	$\mathcal{I} \leftarrow S(\mathtt{sel}, 1^{\lvert m_i \rvert})$
$(R_i)_{i\in[n]} \leftarrow (\mathcal{R}_{\mathsf{Enc}})^n$	$out_A \leftarrow S(\mathtt{out}, (m_i)_{i\in\mathcal{I}})$
$(c_i)_{i\in[n]} := (\mathsf{Enc}(pk, m_i; R_i))_{i\in[n]}$	return $\mathcal{T}(\mathfrak{D}_{\mathsf{so}}, (m_i)_{i\in[n]}, out_A)$
$\mathcal{I} \leftarrow A(\mathtt{sel}, (c_i)_{i\in[n]})$	
$out_A \leftarrow A(\mathtt{out}, (m_i)_{i\in\mathcal{I}}, (R_i)_{i\in\mathcal{I}})$	
return $\mathcal{T}(\mathfrak{D}_{\mathsf{so}}, (m_i)_{i\in[n]}, out_A)$	

Fig. 1. SIM-SO-CPA security experiments.

generation algorithm $\mathsf{Gen}(1^\lambda)$ outputs a public key pk and a secret key sk. Encryption algorithm $\mathsf{Enc}(pk, m)$ takes pk and a message $m \in \mathcal{M}$, and outputs a ciphertext c. Decryption algorithm $\mathsf{Dec}(sk, c)$ takes sk and a ciphertext c, and outputs a message m. For correctness, we want $\mathsf{Dec}(sk, \mathsf{Enc}(pk, m)) = m$ for all $m \in \mathcal{M}$ and all $(pk, sk) \leftarrow \mathsf{Gen}(1^\lambda)$.

Simulation-Based Selective Opening Security. We use the definition of SO-security against chosen-plaintext attacks of Fehr *et al.* [12], which refines the definition of [2,4] (by letting the adversary choose the message distribution).

Definition 4 (Simulation-based security against selective opening attacks). *For a PKE scheme* PKE = (Gen, Enc, Dec), *a polynomially bounded function* $n = n(\lambda) > 0$, *a function* $\mathcal{T}$ *and a stateful PPT adversary A, consider the experiments in Fig. 1. We call* PKE *SIM-SO-CPA secure if for any PPT adversary A and PPT function* $\mathcal{T}$ *there is a stateful PPT simulator S such that*

$$\mathsf{Adv}_{\mathsf{PKE},A}^{\mathsf{sim\text{-}so\text{-}cpa}}(\lambda) := \left| \Pr\left[\mathsf{Exp}_{\mathsf{PKE},A,\mathcal{T},n}^{\mathsf{sim\text{-}so\text{-}cpa\text{-}real}}(\lambda) = 1\right] - \Pr\left[\mathsf{Exp}_{\mathsf{PKE},S,\mathcal{T},n}^{\mathsf{sim\text{-}so\text{-}cpa\text{-}ideal}}(\lambda) = 1\right] \right|$$

is negligible. As usual, we require that the distribution $\mathfrak{D}_{\mathsf{so}}$ *that A outputs is encoded as a circuit. Since A is PPT, this enforces efficient samplability of* $\mathfrak{D}_{\mathsf{so}}$.

2.3 Selective Opening Security from Lossy Encryption

In [2,4], Bellare et al. show that any lossy encryption scheme where ciphertexts can be *efficiently* opened to arbitrary messages is indeed SIM-SO-CPA secure. The following definition essentially repeats the definition of lossy encryption with efficient opening from [4] with one small change: the Opener algorithm may receive an additional input, the random coins used to generate the ciphertext (that should now be opened to a different message). We call a scheme satisfying this definition, a lossy encryption scheme with efficient weak opening.

Definition 5 (Lossy encryption with efficient weak opening). *A lossy encryption scheme with efficient weak opening and message space $\mathcal{M}$ is a tuple of PPT algorithms* $\mathsf{LPKE} = (\mathsf{Gen}, \mathsf{LGen}, \mathsf{Enc}, \mathsf{Dec})$ *such that*

- $\mathsf{Gen}(1^\lambda)$ *takes as input the security parameter* 1^λ *and outputs a keypair* (pk, sk). *We call* pk *a real or injective public key.*
- $\mathsf{LGen}(1^\lambda)$ *takes as input the security parameter* 1^λ *and outputs a keypair* (pk, sk). *We call* pk *a lossy public key.*
- $\mathsf{Enc}(pk, m)$ *takes as input a (real or lossy) public key* pk *and a message* $m \in \mathcal{M}$ *and outputs a ciphertext* c
- $\mathsf{Dec}(sk, c)$ *takes as input a secret key* sk *and a ciphertext* c *and outputs either a message* $m \in \mathcal{M}$ *or* $\bot$ *in case of a failure.*

Additionally, LPKE *needs to satisfy the following properties:*

1. *Correctness for real keys: For all* $\lambda \in \mathbb{N}$, $(pk, sk) \leftarrow \mathsf{Gen}(1^\lambda)$, *messages* $m \in \mathcal{M}$, *and ciphertexts* $c \leftarrow \mathsf{Enc}(pk, m)$, *it always holds that* $m \leftarrow \mathsf{Dec}(sk, c)$.
2. *Indistinguishability of real keys from lossy keys: For any PPT algorithm* D *it holds that the advantage*

$$\mathsf{Adv}^{\mathsf{ind\text{-}lossy\text{-}key}}_{\mathsf{LPKE}, D}(\lambda) := \left| \begin{array}{l} \Pr[1 \leftarrow D(1^\lambda, pk) \mid (pk, sk) \leftarrow \mathsf{Gen}(1^\lambda)] \\ - \Pr[1 \leftarrow D(1^\lambda, pk) \mid (pk, sk) \leftarrow \mathsf{LGen}(1^\lambda)] \end{array} \right|$$

is negligible in λ.

3. *Lossiness of encryption with lossy keys: Let* $\lambda \in \mathbb{N}$. *For any* $(pk, sk) \leftarrow \mathsf{LGen}(1^\lambda)$ *and distinct messages* $m_0 \neq m_1 \in \mathcal{M}$, *holds that*

$$(sk, \mathsf{Enc}(pk, m_0)) \approx_s (sk, \mathsf{Enc}(pk, m_1))$$

4. *Efficient weak openability: Let* $\mathcal{R}_{\mathsf{Enc}}$ *denote the space of random coins for encryption. There exists a PPT algorithm* Opener *such that for any two messages* $m_0, m_1 \in \mathcal{M}$, *the probability that* Opener *on input of a lossy public and secret key* $(pk, sk) \leftarrow \mathsf{LGen}(1^\lambda)$, *a ciphertext* $c \leftarrow \mathsf{Enc}(pk, m_0; r')$, *where* $r' \leftarrow \mathcal{R}_{\mathsf{Enc}}$, *the corresponding random coins* r', *and a message* m_1, *outputs uniform random coins* r *from* $\{r \in \mathcal{R}_{\mathsf{Enc}} \mid \mathsf{Enc}(pk, m_1; r) = c\}$ *is overwhelming.*

Despite our small changes with respect to the definition of lossy encryption and SIM-SO-CPA compared to the definitions in [4], the following theorem still follows from the corresponding proof in [4]: It does not matter for the proof if the message distribution is some arbitrary but fixed distribution (where we quantify over all efficiently samplable distributions) or if it is the output of the adversary after seeing the (lossy) public key. Moreover, the simulator which uses the Opener algorithm knows the encryption randomness of the (dummy) ciphertexts (that should be opened differently) as it has generated these ciphertexts itself.[5]

Theorem 1. *([2,4]). If* LPKE *is a lossy encryption scheme with efficient weak opening then* LPKE *is SIM-SO-CPA secure.*

---

[5] Note that the los-ind2 adversary C in the proof of Theorem 5.2 in [4] is unbounded and thus may find the appropriate encryption randomness required for our Opener algorithm itself.

3   Lossy Encryption from Matrix Rank Assumptions

First, we would like to stress that although we use k-linear group generators $\mathcal{G}_k$ in the following definitions and constructions for generality, the existence of k-linear maps for $k > 2$ is not required to instantiate our constructions. For the instantiations based on MDDH assumptions (Sect. 4), an ordinary group generator $\mathcal{G}_1$ or bilinear group generator $\mathcal{G}_2$ can be assumed (where the pairing is not used for encryption). For the instantiation based on the BDDH assumption (Sect. 5), a bilinear group generator $\mathcal{G}_2$ is required where the pairing is needed in the encryption routine. Hence, for the remainder of this paper, it might be best to have $k = 1$ or $k = 2$ in mind.

In the following, we show how to build efficient lossy encryption with efficient weak opening for multiple bits from rank problems. Roughly speaking, this problem asks to distinguish a $n \times n$ matrix of rank $\ell < n$ chosen according to some (not necessarily uniform) distribution from a matrix of full rank n chosen according to some (not necessarily uniform) distribution, where both matrices are given in implicit representation. The following definition captures rank assumptions and additionally allows the considered matrices to be given in some "compressed form" (which, e.g., can be decompressed efficiently using a pairing).

Definition 6. Let $\mathcal{MG}_k := (k, G_1, \ldots, G_{k+1}, g_1, \ldots, g_k, e, p) \leftarrow \mathcal{G}_k(1^\lambda)$ be a k-linear group generator. A (n, ℓ)-indistinguishable matrix constructor MCon for G_i, where $1 \leq i \leq k+1$, is a tuple MCon = (SetupNFR, SetupFR, Constr) of PPT algorithms with the following properties.

Setup of non-full rank matrix description. SetupNFR($\mathcal{MG}_k$) returns a matrix $\mathbf{A} \in \mathbb{Z}_p^{n \times n}$ of rank ℓ, where we assume that $\mathbf{A}$'s first ℓ rows are linearly independent, as well as a (compact) description $mat \in \{0,1\}^*$ of the implicit representation $[\mathbf{A}]_i$ of $\mathbf{A}$.

Setup of full rank matrix description. SetupFR($\mathcal{MG}_k$) returns a matrix $\mathbf{A} \in \mathbb{Z}_p^{n \times n}$ of rank n as well as a (compact) description $mat \in \{0,1\}^*$ of the implicit representation $[\mathbf{A}]_i$ of $\mathbf{A}$.[6]

Reconstruction of matrix from matrix description.
Constr($\mathcal{MG}_k, mat$) returns $[\mathbf{A}]_i \in G_i^{n \times n}$ on input of a matrix description mat.

Correctness. MCon is called correct relative to $\mathcal{G}_k$ if for all $\lambda \in \mathbb{N}$, $\mathcal{MG}_k := (k, G_1, \ldots, G_{k+1}, g_1, \ldots, g_k, e, p) \leftarrow \mathcal{G}_k(1^\lambda)$, and $(\mathbf{A}, mat_\mathbf{A}) \leftarrow$ SetupNFR $(\mathcal{MG}_k)$, $(\mathbf{B}, mat_\mathbf{B}) \leftarrow$ SetupFR($\mathcal{MG}_k$), the matrices $\mathbf{A}$ and $\mathbf{B}$ are of rank ℓ and of rank n with probability 1, respectively, and $[\mathbf{A}]_i \leftarrow$ Constr($\mathcal{MG}_k, mat_\mathbf{A}$) and $[\mathbf{B}]_i \leftarrow$ Constr($\mathcal{MG}_k, mat_\mathbf{B}$).

---

[6] This description mat can always be set to $[\mathbf{A}]_i$. In some cases (e.g., in case of the ℓ-linear distribution), however, $[\mathbf{A}]_i$ has more structure can be represented with fewer group elements, see also [11].

Security. MCon *is called* secure *relative to* $\mathcal{G}_k$, *if for all PPT algorithms* A *and for* $\mathcal{MG}_k \leftarrow \mathcal{G}_k(1^\lambda)$, $(\mathbf{A}, mat) \leftarrow$ SetupNFR$(\mathcal{MG}_k)$, *and* $(\mathbf{A}', mat') \leftarrow$ SetupFR$(\mathcal{MG}_k)$ *holds that the advantage*

$$\mathsf{Adv}^{\text{ind-matrix-rank}}_{\mathsf{MCon},A}(1^\lambda) := \left| \Pr[1 \leftarrow A(1^\lambda, \mathcal{MG}_k, mat)] - \Pr[1 \leftarrow A(1^\lambda, \mathcal{MG}_k, mat')] \right|$$

is negligible in λ.

Construction of the LPKE scheme with efficient weak opening. Apart from an (n, ℓ)-indistinguishable matrix constructor for G_i, we additionally need a hash function $H : G_i \rightarrow \{0, 1\}$ such that $H(a)$, for uniformly random $a \leftarrow G_i$, is statistically indistinguishable from the uniform distribution on $\{0, 1\}$. By writing $H(\vec{b})$, where $\vec{b}$ is a vector of group elements from G_i, we refer to the component-wise application of the hash function, which results in a (bit-)vector of hash values of the same length as $\vec{b}$.

Based on these ingredients, we can define a lossy encryption scheme with efficient weak opening LPKE = (Gen, LGen, Enc, Dec) with message space $\{0, 1\}^{n-\ell}$ and ciphertexts consisting of ℓ group elements and $n - \ell$ bits. Note that the parameter ℓ reflects the strength of the assumption we are willing to make, the smaller ℓ, the stronger the underlying assumption. For instance, the assumption that random rank ℓ matrices are indistinguishable from random full rank matrices is implied by the assumption that random rank $\ell - 1$ matrices are indistinguishable from random full rank matrices. (Furthermore, rank ℓ vs. n indistinguishability is implied by the ℓ-linear assumption.) Hence, to make ciphertexts as compact as possible, one would choose $\ell = 1$ and could, e.g., base security on the 1-linear assumption which equals DDH.

The idea underlying encryption (with a real key) in our construction is as follows: a message bit is encrypted using the hash of a randomized linear dependent row vector of $\mathbf{A}$ given in implicit representation. Additionally, the linear independent row vectors of $\mathbf{A}$ are randomized the same way and given in implicit representation as part of the ciphertext. Decryption then boils down to recomputing the (implicit representation of the) linear dependent vector from the (implicit representations of the) linear independent vectors. As all row vectors are randomized the same way (which is a linear operation), the dependencies are not changed by the randomization. The details of LPKE are given below.

- Gen(1^λ) runs the group generator $\mathcal{MG}_k := (k, G_1, \ldots, G_{k+1}, g_1, \ldots, g_k, e, p) \leftarrow \mathcal{G}_k(1^\lambda)$ as well as $(\mathbf{A}, mat) \leftarrow$ SetupNFR$(\mathcal{MG}_k)$ to choose a matrix of rank ℓ. Let $\mathbf{A}_0$ denote the first ℓ rows of $\mathbf{A}$ and $\mathbf{A}_1$ the remaining $n - \ell$ rows. Then it computes a matrix $\mathbf{T} \in \mathbb{Z}_p^{(n-\ell) \times \ell}$ satisfying

$$\mathbf{T}\mathbf{A}_0 = \mathbf{A}_1 \tag{4}$$

As the rows of $\mathbf{A}_1$ linearly depend on the rows of $\mathbf{A}_0$, $\mathbf{T}$ always exists and can be computed efficiently (e.g., using Gaussian Elimination). The algorithm returns $pk := (\mathcal{MG}_k, mat)$ and $sk := (\mathcal{MG}_k, \mathbf{T})$.

- $\mathsf{LGen}(1^\lambda)$ runs the group generator $\mathcal{MG}_k := (k, G_1, \ldots, G_{k+1}, g_1, \ldots, g_k, e, p) \leftarrow \mathcal{G}_k(1^\lambda)$ as well as $(\mathbf{A}, mat) \leftarrow \mathsf{SetupFR}(\mathcal{MG}_k)$ to choose a matrix of rank n. The algorithm returns $pk := (\mathcal{MG}_k, mat)$ and $sk := (\mathcal{MG}_k, \mathbf{A})$.
- $\mathsf{Enc}(pk, \vec{m})$ reconstructs the matrix $[\mathbf{A}]_i \leftarrow \mathsf{Constr}(\mathcal{MG}_k, mat)$. Let $[\mathbf{A}_0]_i$ denote the first ℓ rows of $[\mathbf{A}]_i$ and $[\mathbf{A}_1]_i$ the remaining $n - \ell$ rows. Then it chooses $\vec{w} \leftarrow \mathbb{Z}_p^n$, computes

$$[\vec{c}_0]_i := [\mathbf{A}_0 \vec{w}]_i$$
$$\vec{c}_1 := H([\mathbf{A}_1 \vec{w}]_i) \oplus \vec{m} \tag{5}$$

(using exponentiations with the entries of $\vec{w}$), and returns ciphertext $c := ([\vec{c}_0]_i, \vec{c}_1) \in G_i^\ell \times \{0,1\}^{n-\ell}$.
- $\mathsf{Dec}(sk, c)$ recomputes $\vec{m}$ as $\vec{m} := H([\mathbf{T}\vec{c}_0]_i) \oplus \vec{c}_1$.

We show that LPKE indeed satisfies the four properties of a lossy encryption scheme with efficient weak opening.

Theorem 2. *If* MCon *is secure and the output of* H *statistically indistinguishable from uniform for random input then* LPKE *is a lossy encryption scheme with efficient weak opening.*

Proof.

Correctness for real keys. Given a real public key $pk := (\mathcal{MG}_k, mat)$ and secret key $sk := (\mathcal{MG}_k, \mathbf{T})$ returned by $\mathsf{Gen}(1^\lambda)$ as well as a ciphertext $c := ([\vec{c}_0]_i, \vec{c}_1)$, correctness of decryption follows from the equation

$$\begin{aligned} H([\mathbf{T}\vec{c}_0]_i) \oplus \vec{c}_1 &= H([\mathbf{T}\vec{c}_0]_i) \oplus H([\mathbf{A}_1 \vec{w}]_i) \oplus \vec{m} \\ &= H([\mathbf{T}\mathbf{A}_0 \vec{w}]_i) \oplus H([\mathbf{A}_1 \vec{w}]_i) \oplus \vec{m} \\ &= H([\mathbf{A}_1 \vec{w}]_i) \oplus H([\mathbf{A}_1 \vec{w}]_i) \oplus \vec{m} \end{aligned} \tag{6}$$

Indistinguishability of real keys from lossy keys. It follows from the security of MCon that a real public key $(\mathcal{MG}_k, mat)$ generated by $\mathsf{Gen}(1^\lambda)$ is indistinguishable from a lossy one $(\mathcal{MG}_k, mat')$ generated by $\mathsf{LGen}(1^\lambda)$.

Lossiness of encryption with lossy keys. Consider the matrix $[\mathbf{A}]_i \leftarrow \mathsf{Constr}(\mathcal{MG}_k, mat)$, where mat is computed by $\mathsf{LGen}(1^\lambda)$. This matrix has full rank, so the linear map defined by $\mathbf{A}$ as $\vec{w} \mapsto \mathbf{A}\vec{w}$ is bijective. Thus, for uniformly random $\vec{w}$, $[\vec{c}_0]_i = [\mathbf{A}_0 \vec{w}]_i$ is uniformly random over G_i^ℓ and $[\mathbf{A}_1 \vec{w}]_i$ is uniformly random over $G_i^{n-\ell}$ (even when $\mathbf{A}$ is given).

Now, since by assumption the output of H is statistically close to uniform for uniformly random input, $H([\mathbf{A}_1 \vec{w}]_i) \oplus \vec{m}$ will also be statistically close to uniform over $\{0,1\}^{n-\ell}$ for any string $\vec{m}$.

Hence, for uniformly random $\vec{w} \leftarrow \mathbb{Z}_p^n$, the distributions of

$$(\mathbf{A}, ([\mathbf{A}_0 \vec{w}]_i, H([\mathbf{A}_1 \vec{w}]_i) \oplus \vec{m})) \quad \text{and} \quad (\mathbf{A}, ([\mathbf{A}_0 \vec{w}]_i, H([\mathbf{A}_1 \vec{w}]_i) \oplus \vec{m}'))$$

are statistically close for any two distinct message vectors $\vec{m} \neq \vec{m}' \in \{0,1\}^{n-\ell}$.

Efficient weak openability. Let a lossy keypair $(pk = (\mathcal{MG}_k, mat), sk = (\mathcal{MG}_k, \mathbf{A})) \leftarrow \mathsf{LGen}(1^\lambda)$, message vector $\vec{m}$, a ciphertext $c := ([\vec{c}_0]_i, \vec{c}_1) \leftarrow \mathsf{Enc}(pk, \vec{m}'; \vec{w}')$, as well as the corresponding encryption randomness $\vec{w}'$ be given. Then Opener should efficiently determine some encryption randomness $\vec{w}$ such that $\mathsf{Enc}(pk, \vec{m}; \vec{w}) = ([\vec{c}_0]_i, \vec{c}_1)$. This can be done by setting up a linear system of equations in the exponent

$$\mathbf{A}\vec{w} = \vec{b}, \tag{7}$$

where the right-hand side vector

$$\vec{b} = \begin{pmatrix} \vec{b}_0 \\ \vec{b}_1 \end{pmatrix} \tag{8}$$

satisfies $\vec{b}_0 = \vec{c}_0$ and $H([\vec{b}_1]_i) \oplus \vec{c}_1 = \vec{m}$.

First, Opener can easily determine $\vec{b}_0 := \vec{c}_0 \in \mathbb{Z}_p^\ell$, i.e., the discrete logarithms of $[\vec{c}_0]_i$ to the base g_i, by computing $\mathbf{A}\vec{w}'$. Second, it can efficiently find a vector $\vec{b}_1 \in \mathbb{Z}_p^{n-\ell}$ satisfying $H([\vec{b}_1]_i) \oplus \vec{c}_1 = \vec{m}$ by randomly guessing one component of $\vec{b}_1$ after another and verifying the equation for this component. As the output of H is close to uniform for random input, this will require about $2(n-\ell)$ steps. After that, Opener can solve the system of linear equations from Eq. 7 by multiplying with the inverse of $\mathbf{A}$ as this matrix is of full rank.

It is not hard to see that the determined randomness $\vec{w}$ has the correct distribution, i.e., $\vec{w}$ is uniformly chosen from

$$\mathsf{Coins}(\vec{m}, c) := \{\vec{w} \in \mathbb{Z}_p^n \mid \mathsf{Enc}(pk, \vec{m}; \vec{w}) = c\} \tag{9}$$

Note that each $\vec{w} \in \mathsf{Coins}(\vec{m}, c)$ uniquely determines a right-hand side $\vec{b}$ in (7), i.e., a vector from

$$\mathsf{KENCs}(\vec{m}, c) := \left\{\vec{b} = \begin{pmatrix} \vec{b}_0 \\ \vec{b}_1 \end{pmatrix} \;\middle|\; \vec{b}_0 = \vec{c}_0 \wedge H([\vec{b}_1]_i) \oplus \vec{c}_1 = \vec{m}\right\} \tag{10}$$

Hence, to uniformly sample $\vec{w}$ from $\mathsf{Coins}(\vec{m}, c)$ it suffices to uniformly sample $\vec{b}$ from $\mathsf{KENCs}(\vec{m}, c)$ and invert the bijective mapping by computing $\mathbf{A}^{-1}\vec{b}$. This is exactly what Opener does.

4   From MDDH Assumptions to Matrix Rank Assumptions

We have seen in Sect. 3 that in order to build an $(n - \ell)$-bit LPKE scheme with efficient weak opening, it suffices to define a secure (n, ℓ)-indistinguishable matrix constructor. In the following, we first show that such a constructor is generically given by any $\mathcal{D}_{n,\ell}$-MDDH assumption (including DDH, ℓ-Lin, ℓ-SCasc, (n, ℓ)-circulant matrix assumption, etc.). Then, we consider the size of the public key when using different members of MDDH assumption family.

Generic construction from MDDH assumptions. Let $\mathcal{G}_k$ be a k-linear group generator and $\mathcal{MG}_k := (k, G_1, \ldots, G_{k+1}, g_1, \ldots, g_k, e, p) \leftarrow \mathcal{G}_k(1^\lambda)$. Furthermore, let $\mathcal{D}_{n,\ell}$ be a matrix distribution over $\mathbb{Z}_p^{n\times\ell}$, where $n > \ell$. We assume that the first ℓ rows of an output of $\mathcal{D}_{n,\ell}$ forms a regular matrix with overwhelming probability. A (n, ℓ)-indistinguishable matrix constructor $\mathsf{MCon}_{\mathcal{D}_{n,\ell}\text{-MDDH}}$ for G_i can then be defined based on $\mathcal{D}_{n,\ell}$-MDDH as follows:

- $\mathsf{SetupNFR}(\mathcal{MG}_k)$ samples a matrix $\mathbf{A}' \leftarrow \mathcal{D}_{n,\ell}$ of rank ℓ according to the given matrix distribution. If $\mathbf{A}'$ is not of rank ℓ the sampling is repeated. (Note that since $\mathcal{D}_{n,\ell}$ outputs full rank matrices with overwhelming probability this case should virtually never happen.) Furthermore, a random matrix $\mathbf{R} \leftarrow \mathbb{Z}_p^{\ell \times (n-\ell)}$ is sampled. Then it computes $\mathbf{A} := \mathbf{A}'(\mathbf{I}_\ell \| \mathbf{R}) = (\mathbf{A}' \| \mathbf{A}'\mathbf{R})$, where $\mathbf{I}_\ell$ is the $\ell \times \ell$ identity matrix, and returns $(\mathbf{A}, [\mathbf{A}]_i)$.
- $\mathsf{SetupFR}(\mathcal{MG}_k)$ samples a matrix $\mathbf{A}' \leftarrow \mathcal{D}_{n,\ell}$ of rank ℓ (if the rank of $\mathbf{A}'$ is smaller sampling is repeated). After that, random matrices $\mathbf{U} \leftarrow \mathbb{Z}_p^{n \times (n-\ell)}$ are sampled until $\mathbf{A} := (\mathbf{A}' \| \mathbf{U})$ is of full rank n. (Note that $\mathbf{A}$ will be of rank n with overwhelming probability of at least $1 - \frac{n-\ell}{p^{n-\ell}}$ for uniform $\mathbf{U}$.) It then returns $(\mathbf{A}, [\mathbf{A}]_i)$.
- $\mathsf{Constr}(\mathcal{MG}_k, mat)$ returns mat (as the matrix is not compressed).

Remark 2. Consider the matrix $\mathbf{A}' \leftarrow \mathcal{D}_{n,\ell}$ generated during $\mathsf{SetupNFR}(\mathcal{MG}_k)$. Let $\mathbf{A}_0'$ denote the first ℓ rows of $\mathbf{A}'$ and $\mathbf{A}_1'$ the last $n - \ell$ rows of $\mathbf{A}'$. Then the transformation matrix $\mathbf{T}$ from Eq. 4, which is used as the secret key, can be set to $\mathbf{T} := \mathbf{A}_1'(\mathbf{A}_0')^{-1}$. Correctness follows from

$$\begin{aligned}
\mathbf{TA}_0 &= \mathbf{A}_1'(\mathbf{A}_0')^{-1}\mathbf{A}_0 \\
&= \mathbf{A}_1'(\mathbf{A}_0')^{-1}\mathbf{A}_0'(\mathbf{I}_\ell \| \mathbf{R}) \\
&= \mathbf{A}_1'(\mathbf{I}_\ell \| \mathbf{R}) \\
&= \mathbf{A}_1
\end{aligned} \tag{11}$$

Correctness. Consider $(\mathbf{A}, mat_\mathbf{A}) \leftarrow \mathsf{SetupNFR}(\mathcal{MG}_k)$ and $(\mathbf{B}, mat_\mathbf{B}) \leftarrow \mathsf{SetupFR}(\mathcal{MG}_k)$. Obviously, $\mathbf{A} = (\mathbf{A}' \| \mathbf{A}'\mathbf{R})$ will be of rank ℓ as this is the case for $\mathbf{A}'$. Similarly, $\mathbf{B} := (\mathbf{B}' \| \mathbf{U})$ will be of rank n by construction. Furthermore, clearly, it holds that $[\mathbf{A}]_i \leftarrow \mathsf{Constr}(\mathcal{MG}_k, mat_\mathbf{A})$ and $[\mathbf{B}]_i \leftarrow \mathsf{Constr}(\mathcal{MG}_k, mat_\mathbf{B})$.

Security. As for security we show

Lemma 1. *If the $\mathcal{D}_{n,\ell}$-MDDH assumption holds relative to $\mathcal{G}_k$, then the scheme $\mathsf{MCon}_{\mathcal{D}_{n,\ell}\text{-MDDH}}$ is secure.*

Proof. First note that the distribution of $\mathbf{A}$ returned by $\mathsf{SetupNFR}$ and the distribution of $\mathbf{B}$ returned by $\mathsf{SetupFR}$ are statistically indistinguishable from the distribution of $(\mathbf{A}' \| \mathbf{A}'\mathbf{R})$ and $(\mathbf{A}' \| \mathbf{U})$, respectively, where $\mathbf{A}' \leftarrow \mathcal{D}_{n,\ell}$, $\mathbf{R} \leftarrow \mathbb{Z}_p^{\ell \times (n-\ell)}$, and $\mathbf{U} \leftarrow \mathbb{Z}_p^{n \times (n-\ell)}$.

Then considering the latter distributions, the lemma immediately follows from the $\mathcal{D}_{n,\ell}$-Matrix Diffie-Hellman assumption and its random self-reducibility. More concretely, the $\mathcal{D}_{n,\ell}$-MDDH assumption demands that for all PPT adversaries D holds that

$$|\Pr[D(\mathcal{MG}_k, [\mathbf{A}']_i, [\mathbf{A}'\vec{r}]_i) = 1] - \Pr[D(\mathcal{MG}_k, [\mathbf{A}']_i, [\vec{u}]_i) = 1]|$$

is negligible, where $\mathcal{MG}_k \leftarrow \mathcal{G}_k(1^\lambda)$, $\mathbf{A}' \leftarrow \mathcal{D}_{n,\ell}$, $\vec{r} \leftarrow \mathbb{Z}_p^\ell$ and $\vec{u} \leftarrow \mathbb{Z}_p^n$. Hence, $[\mathbf{A}'\|\mathbf{A}'\vec{r}]_i$ is computationally indistinguishable from $[\mathbf{A}'\|\vec{u}]_i$. As any matrix assumption is random self-reducible (Lemma 1 in [11]), it follows that

$$|\Pr[D(\mathcal{MG}_k, [\mathbf{A}']_i, [\mathbf{A}'\mathbf{R}]_i) = 1] - \Pr[D(\mathcal{MG}_k, [\mathbf{A}']_i, [\mathbf{U}]_i) = 1]|$$

is negligible, where $\mathbf{R} \leftarrow \mathbb{Z}_p^{\ell \times (n-\ell)}$ and $\mathbf{U} \leftarrow \mathbb{Z}_p^{n \times (n-\ell)}$. Thus, $[\mathbf{A}'\|\mathbf{A}'\mathbf{R}]_i$ is computationally indistinguishable from $[\mathbf{A}'\|\mathbf{U}]_i$.

Concrete instantiations. Let us now consider what we get from different members of the MDDH assumption family.

1-bit LPKE from standard assumptions. From standard assumptions like DDH and ℓ-Lin, we can immediately obtain a one bit lossy encryption scheme by means of the corresponding indistinguishable matrix constructor. More precisely, for ℓ-Lin we would consider the $\mathcal{L}_{\ell+1,\ell}$ matrix distribution which samples $(\ell + 1) \times \ell$ matrices of the form

$$\mathbf{A}' = \begin{pmatrix} s_1 & 0 & 0 & \dots & 0 \\ 0 & s_2 & 0 & \dots & 0 \\ \vdots & \vdots & & \ddots & \vdots \\ 0 & 0 & 0 & \dots & s_\ell \\ 1 & 1 & 1 & \dots & 1 \end{pmatrix} \tag{12}$$

Hence, this results in a public key of the form

$$[\mathbf{A}]_i = \left[\begin{pmatrix} s_1 & 0 & 0 & \dots & 0 & s_1 r_1 \\ 0 & s_2 & 0 & \dots & 0 & \\ \vdots & \vdots & & \ddots & \vdots & \vdots \\ 0 & 0 & 0 & \dots & s_\ell & s_\ell r_\ell \\ 1 & 1 & 1 & \dots & 1 & r_1 + \dots + r_\ell \end{pmatrix} \right]_i , \tag{13}$$

where $r_i \leftarrow \mathbb{Z}_p$, which can be represented using $2(\ell + 1)$ group elements.

Multi-bit LPKE from standard assumptions. Note that the number of bits we can encrypt equals the number of linearly dependent row vectors of $\mathbf{A} \in \mathbb{Z}_p^{n \times n}$, i.e., $n - \ell$. Thus, if we had a distribution $\mathcal{D}_{n,\ell}$ that yields matrices with more than one linearly dependent vector, i.e., $n > \ell + 1$, our construction would be able to encrypt more than one bit. Hence, we could obtain a scheme for multiple bits from a standard assumption by finding a $\mathcal{D}_{n,\ell}$-MDDH assumption with $n > \ell+1$ which is implied by this standard assumption. For instance, the ℓ-Lin assumption implies $\mathcal{U}_{n,\ell}$-MDDH for arbitrary n, where $\mathcal{U}_{n,\ell}$ samples uniform $n \times \ell$ matrices of rank ℓ (this follows from Lemma A.1 in [26]). Hence, from DDH, for example, we can get a scheme for $(n - 1)$-bit messages with arbitrary $n \in \mathbb{N}$ by means of the uniform distribution $\mathcal{U}_{n,1}$ which samples a matrix of the form

$$\mathbf{A}' = \begin{pmatrix} s_1 \\ \vdots \\ s_n \end{pmatrix} \tag{14}$$

and, thus, yields a public key of the form

$$[\mathbf{A}]_i = \left[\begin{pmatrix} s_1 & s_1 r_1 & \dots & s_1 r_{n-1} \\ \vdots & & & \vdots \\ s_n & s_n r_1 & \dots & s_n r_{n-1} \end{pmatrix} \right]_i , \tag{15}$$

where $r_i \leftarrow \mathbb{Z}_p$. Note that the resulting scheme is essentially the DDH-based scheme sketched in the introduction (with the minor difference that s_n is set to 1 instead of being uniformly chosen).

It is interesting to observe that ℓ-Lin is a family of assumptions which (at least in the generic group model) become strictly weaker as ℓ grows and that we can get an LPKE scheme for messages of arbitrary size for each member of this family (by means of $\mathcal{U}_{n,\ell}$).

On the downside, if make the detour to the $\mathcal{U}_{n,\ell}$ distribution (instead of directly building on ℓ-Lin), the public key will consist of n^2 group elements to represent $[\mathbf{A}]_i$. Alternatively, we can take a more direct approach and extend a (standard) $\mathcal{D}_{\ell+1,\ell}$-MDDH assumption (like ℓ-Lin) to the $\mathcal{D}_{n,\ell}$-MDDH assumption, where the first $\ell+1$ rows of $\mathbf{A}' \leftarrow \mathcal{D}_{n,\ell}$ are sampled as by $\mathcal{D}_{\ell+1,\ell}$ and the remaining $n-\ell-1$ are sampled uniformly. In this case, $\mathcal{D}_{n,\ell}$-MDDH is implied by $\mathcal{D}_{\ell+1,\ell}$-MDDH [24]. The representation of $[\mathbf{A}]_i$ will consist of $E + (n-\ell-1)\ell + n(n-\ell)$ group elements to encrypt $n-\ell$ bits, where E is the number of elements required to represent a matrix sampled by the $\mathcal{D}_{\ell+1,\ell}$ distribution (e.g., 1 for ℓ-SCasc).

Multi-bit LPKE from a new $\mathcal{D}_{n,\ell}$-MDDH assumption. A $\mathcal{D}_{n,\ell}$-MDDH for $n > \ell+1$ with an optimal representation size has recently been proposed in [24]. The circulant matrix distribution $\mathcal{C}_{\ell+d,\ell}$ outputs matrices $\mathbf{A}' \in \mathbb{Z}_p^{(\ell+d)\times\ell}$ which can be represented using d group elements. The assumption has been shown to hold in the ℓ-linear generic group model [24]. Plugging this distribution into our scheme, we obtain a public key consisting of $d + (\ell + d)d$ group elements (representing $[\mathbf{A}]_i$) to encrypt d bits.

5   From the BDDH Assumption to a Compact Matrix Rank Assumption

In this section, we show how to leverage the lossy trapdoor function construction of Boyen and Waters [7] to obtain a $(n, 1)$-indistinguishable matrix constructor MCon$_\mathsf{BDDH}$ with a linear-size matrix description *mat*. This translates to an $(n-1)$-bit lossy encryption scheme featuring a linear public key size. (Note that the size of the secret key is also linear.)

Essentially, the idea is to generate the quadratic number of group elements in the matrix from a linear number of group elements, by applying a bilinear map. A technical hurdle is to do this in a way such that matrices computed in this way have either rank 1 or full rank, in a computationally indistinguishable way. Here we apply the "linear equations" technique of Boyen and Waters, which enables an algorithm to re-compute the full matrix by applying the bilinear map, *except for the diagonal*. The diagonal entries of the matrix are given additionally in the matrix description *mat*, and set-up such that the resulting matrix has either rank 1 or full rank. Interestingly, the *lossy* trapdoor function of Boyen and Waters corresponds to our *injective* encryption scheme, and vice versa.

Let $\mathcal{MG}_2 := (2, G_1, G_2, G_3, g_1, g_2, g_3, e, p) \leftarrow \mathcal{G}_2(1^\lambda)$, where $G_1 = G_2$ and $g_1 = g_2$, be a symmetric bilinear group generator. Then a $(n, 1)$-indistinguishable matrix constructor MCon$_\mathsf{BDDH}$ for G_1 can be defined as follows:

- SetupNFR($\mathcal{MG}_2$) samples two uniformly random elements $h, k \leftarrow \mathbb{Z}_p^*$, and two exponent vectors $\vec{r} = (r_1, \ldots, r_n)^\top \leftarrow (\mathbb{Z}_p^*)^n$ and $\vec{u} = (u_1, \ldots, u_n)^\top \leftarrow (\mathbb{Z}_p^*)^n$. Then it sets $\mathbf{A} := (a_{i,j}) \in (\mathbb{Z}_p^*)^{n \times n}$ with $a_{i,j} := hr_i u_j$. Furthermore, it computes
  - $[\vec{s}]_1 := [(s_1, \ldots, s_n)^\top]_1 \in G_1^n$ where $s_i := (hi + k)r_i$
  - $[\vec{v}]_1 := [(v_1, \ldots, v_n)^\top]_1 \in G_1^n$ where $v_j := (hj + k)u_j$
  - $[\vec{d}]_3 := [(d_1, \ldots, d_n)^\top]_3 \in G_3^n$ where $d_i := hr_i u_i$
  and sets $mat := ([\vec{r}]_1, [\vec{s}]_1, [\vec{u}]_1, [\vec{v}]_1, [\vec{d}]_3)$. It returns $(\mathbf{A}, mat)$.
- SetupFR($\mathcal{MG}_2$) samples elements $h, k \leftarrow \mathbb{Z}_p^*$ and vectors $\vec{r}, \vec{u} \leftarrow (\mathbb{Z}_p^*)^n$ the same way as SetupNFR. It sets $\mathbf{A} := (a_{i,j}) \in \mathbb{Z}_p^{n \times n}$ with $a_{i,j} := hr_i u_j$ for $i \neq j$ and $a_{i,i} := hr_i u_i + 1$. Accordingly, $[\vec{s}]_1$ and $[\vec{v}]_1$ are defined as in SetupNFR but d_i is set to $d_i := hr_i u_i + 1$. It sets $mat := ([\vec{r}]_1, [\vec{s}]_1, [\vec{u}]_1, [\vec{v}]_1, [\vec{d}]_3)$ and returns $(\mathbf{A}, mat)$.
- Constr($\mathcal{MG}_2, mat$) computes $[\mathbf{A}]_3 := ([a_{i,j}]_3)_{i,j}$ for $1 \leq i, j \leq n$ as follows:
  - For $i \neq j$, it uses the pairing to compute

$$[a_{i,j}]_3 := e([r_i]_1, [v_j]_1)^{1/(j-i)} e([u_j]_1, [s_i]_1)^{-1/(j-i)} = [(r_i \cdot v_j - u_j \cdot s_i)/(j - i)]_3$$

  - For $i = j$ it sets $[a_{i,i}]_3 := [d_i]_3$

Remark 3. The transformation matrix $\mathbf{T}$ from Eq. 4 can be set to $\mathbf{T} := (r_2/r_1, \ldots, r_n/r_1)^\top$.

Correctness. Consider $(\mathbf{A}, mat_\mathbf{A}) \leftarrow$ SetupNFR($\mathcal{MG}_2$) and $(\mathbf{B}, mat_\mathbf{B}) \leftarrow$ SetupFR($\mathcal{MG}_2$). Let $\mathbf{A}_0$ be the first row of $\mathbf{A}$ and $\mathbf{A}_1$ be the remaining $n - 1$ rows. It is easy to see that $\mathbf{T}\mathbf{A}_0 = \mathbf{A}_1$, where $\mathbf{T}$ is defined as described above. Moreover, $\mathbf{A}$ cannot be the zero-matrix, because h and all r_i and u_j are non-zero. So $\mathbf{A}$ is of rank 1.

Note also that by construction of SetupFR we have $\mathbf{B} = \mathbf{A} + \mathbf{I}_n$, where $\mathbf{A}$ has rank 1 (as above) and $\mathbf{I}_n$ is the $(n \times n)$-identity matrix. Thus, since $\mathbf{A}$ has rank 1, $\mathbf{B}$ is row-equivalent to $\mathbf{I}_n$, which is equivalent to $\mathbf{B}$ having full rank.

To see that for $[\mathbf{A}']_3 :=$ Constr($\mathcal{MG}_2, mat_\mathbf{A}$) and $[\mathbf{B}']_3 :=$ Constr($\mathcal{MG}_2, mat_\mathbf{B}$) we have $[\mathbf{A}']_3 = [\mathbf{A}]_3$ and $[\mathbf{B}']_3 = [\mathbf{B}]_3$, first observe that the diagonal entries are correct, since $[a'_{i,i}]_3 = hr_i u_i$ and $[b'_{i,i}]_3 = hr_i u_i + 1$. Moreover, in either case we have for $i \neq j$ that

$$\begin{aligned}[a'_{i,j}]_3 = [b'_{i,j}]_3 &= [(r_i v_j - u_j s_i)/(j - i)]_3 \\ &= [(r_i(hj + k)u_j - u_j(hi + k)r_i)/(j - i)]_3 \\ &= [(hr_i u_j j + kr_i u_j - hr_i u_j i - kr_i u_j)/(j - i)]_3 \qquad (16) \\ &= [hr_i u_j(j - i)/(j - i)]_3 \\ &= [hr_i u_j]_3 \end{aligned}$$

Security. Following [7], we prove security under the bilinear decisional Diffie-Hellman assumption (cf. Definition 3). However, to simplify the security proof of MCon$_{\mathsf{BDDH}}$, we first define the following slightly modified BDDH* assumption, which is implied by the standard BDDH assumption from Definition 3 by a straightforward reduction.

Definition 7. *Let* $\mathcal{MG}_2 := (2, G_1, G_2, G_3, g_1, g_2, e, p) \leftarrow \mathcal{G}_2(1^\lambda)$, $a, b, c \leftarrow \mathbb{Z}_p^*$, $b \leftarrow \{0, 1\}$, $T_0 := abc$ *and* $T_1 := abc + 1$. *We say that the BDDH* assumption holds relative to* $\mathcal{G}_2$, *if*

$$\mathsf{Adv}_{B,\mathcal{G}_2}^{\mathsf{bddh*}}(1^\lambda) := \left| \begin{array}{c} \Pr\left[1 \leftarrow B(1^\lambda, \mathcal{MG}_2, [(a, b, c)]_1, [T_0]_3)\right] \\ - \Pr\left[1 \leftarrow B(1^\lambda, \mathcal{MG}_2, [(a, b, c)]_1, [T_1]_3)\right] \end{array} \right|$$

is a negligible function for all PPT adversaries B.

Remark 4. A straightforward reduction allows to show that $\mathsf{Adv}_{B,\mathcal{G}_2}^{\mathsf{bddh*}}(1^\lambda) \leq 2 \cdot \mathsf{Adv}_{B,\mathcal{G}_2}^{\mathsf{bddh}}(1^\lambda)$ for all PPT algorithms B.

Theorem 3. *If the BDDH* assumption holds relative to* $\mathcal{G}_2$, *then* $\mathsf{MCon}_{\mathsf{BDDH}}$ *is secure.*

Proof. We will show that one can construct an adversary B against the BDDH* assumption from each adversary A against MCon such that

$$\mathsf{Adv}_{\mathsf{MCon},A}^{\mathsf{ind\text{-}matrix\text{-}rank}}(1^\lambda) \leq n \cdot \mathsf{Adv}_{B,\mathcal{G}_2}^{\mathsf{bddh*}}(1^\lambda) \tag{17}$$

To this end, we describe a hybrid argument which consists of $n + 1$ hybrid games $H_0, \ldots, H_n$. In Hybrid H_δ, $\delta \in \{0, \ldots, n\}$, we run A on input $mat := ([\vec{r}]_1, [\vec{s}]_1, [\vec{u}]_1, [\vec{v}]_1, [\vec{d}]_3)$, where all values are computed exactly as in SetupNFR, except that

$$d_i := \begin{cases} hr_i u_i + 1 & \text{for } i < \delta \\ hr_i u_i & \text{for } i \geq \delta \end{cases}$$

Note that the input mat of A in H_0 is identically distributed to the matrix descriptions computed by $(\mathbf{A}, mat) \leftarrow \mathsf{SetupNFR}(\mathcal{MG}_2)$. In H_n, A receives a matrix description mat which is distributed exactly as a matrix description computed by $(\mathbf{A}, mat) \leftarrow \mathsf{SetupFR}(\mathcal{MG}_2)$.

Let X_δ denote the event that A outputs "1" in Hybrid H_δ. We show that for each $\delta \in \{1, \ldots, n\}$ we can construct an adversary B such that

$$\mathsf{Adv}_{B,\mathcal{G}_2}^{\mathsf{bddh*}} \geq |\Pr[X_{\delta-1}] - \Pr[X_\delta]| \tag{18}$$

which proves (17). B receives as input a BDDH*-instance $(\mathcal{MG}_2, [(a, b, c)]_1, [T])$. It creates $mat = ([\vec{r}]_1, [\vec{s}]_1, [\vec{u}]_1, [\vec{v}]_1, [\vec{d}]_3)$ as follows.

- $[\vec{r}]_1 := [(r_1, \ldots, r_n)^\top]_1$, where $[r_\delta]_1 := [a]_1$ and $r_i \leftarrow \mathbb{Z}_p^*$ for all $i \in \{1, \ldots, n\}$ with $i \neq \delta$
- $[\vec{u}]_1 := [(u_1, \ldots, u_n)^\top]_1$, where $[u_\delta]_1 := [b]_1$ and $u_i \leftarrow \mathbb{Z}_p^*$ for all $i \in \{1, \ldots, n\}$ with $i \neq \delta$
- $[h]_1 := [c]_1$ and $[k]_1 := [-h\delta + y]$ for $y \leftarrow \mathbb{Z}_p \setminus \{h\delta\}$
- $[\vec{s}]_1 := [(s_1, \ldots, s_n)^\top]_1$, where $[s_i]_1 := [(hi + k)r_i]_1 = [(h(i - \delta) + y)r_i]_1$. Note that all the $[s_i]_1$ can efficiently be computed by B, due to the above setup of $[h]_1, [k]_1, [\vec{r}]_1$.

– $[\vec{v}]_1 := [(v_1, \dots, v_n)^\top]_1$, where $[v_j]_1 = [(hj + k)u_j]_1 = [(h(j - \delta) + y)u_j]_1$. As above, all the $[v_i]_1$ can efficiently be computed by B, due to the setup of $[h]_1$, $[k]_1$, $[\vec{u}]_1$.

Finally, B sets $[\vec{d}]_3 := [(d_1, \dots, d_n)^\top]_3$, where

$$[d_i]_3 = \begin{cases} [hr_iu_i + 1]_3 & \text{for } i < \delta \\ [T]_3 & \text{for } i = \delta \\ [hr_iu_i]_3 & \text{for } i > \delta \end{cases}$$

Then it runs A on input $(\mathcal{MG}_2, mat)$ and outputs whatever A outputs.

Note that if $[T]_3 = [abc]_3 = [hr_\delta u_\delta]_3$, then the view of A when interacting with B is identical to its view in hybrid $H_{\delta-1}$. Thus, the probability that A outputs "1" in this case is equal to $\Pr[X_{\delta-1}]$. If $[T]_3 = [abc + 1]_3 = [hr_\delta u_\delta + 1]_3$, then it is identical to H_δ, so that the the probability that A outputs "1" in this case is equal to $\Pr[X_\delta]$. This yields (18) and thus concludes the proof.

Shortcut evaluation. We remark that it is possible to reduce the number of pairing computations required to compute $[\mathbf{A}\vec{w}]_3$ for $\vec{w} \in \mathbb{Z}_p^n$, given mat. In the naïve approach sketched above, one first has to recompute $[\mathbf{A}]_3$ from mat, which requires $\mathbf{O}(n^2)$ pairing evaluations, and then $[\mathbf{A}]_3\vec{w}$.

Following the "shortcut evaluation" approach described in [7], we note that the number of pairing evaluations can be reduced to $2n = \mathbf{O}(n)$, by computing $([z_1]_3, \dots, [z_n]_3)$ from $mat = ([\vec{r}]_1, [\vec{s}]_1, [\vec{u}]_1, [\vec{v}]_1, [\vec{d}]_3)$ and $\vec{w} \in \mathbb{Z}_p^n$ as

$$[z_j]_3 := \frac{\left[\sum_{i \neq j} \frac{w_i r_i}{j - i}\right]_1 \cdot [v_j]_1}{\left[\sum_{i \neq j} \frac{w_i u_i}{j - i}\right]_1 \cdot [s_j]_1} + [w_j d_j]_3$$

Indeed, as shown by Boyen and Waters [7], it is easy to verify that

$$[z_j]_3 = \left[\sum_{i=1}^{n} r_i u_i h w_i\right]_3$$

for all $j \in \{1, \dots, n\}$, and thus it holds that $([z_1]_3, \dots, [z_n]_3)^\top = [\mathbf{A}\vec{w}]_3$. Note that this "shortcut evaluation" takes only two pairing evaluations for each $j \in \{1, \dots, n\}$, which amounts to only $2n$ pairing evaluations in total.

References

1. Bellare, M., Dowsley, R., Waters, B., Yilek, S.: Standard security does not imply security against selective-opening. In: Pointcheval, D., Johansson, T. (eds.) EURO-CRYPT 2012. LNCS, vol. 7237, pp. 645–662. Springer, Heidelberg (2012). doi:10.1007/978-3-642-29011-4_38

2. Bellare, M., Hofheinz, D., Yilek, S.: Possibility and impossibility results for encryption and commitment secure under selective opening. In: Joux, A. (ed.) EUROCRYPT 2009. LNCS, vol. 5479, pp. 1–35. Springer, Heidelberg (2009). doi:10.1007/978-3-642-01001-9_1

3. Bellare, M., Waters, B., Yilek, S.: Identity-based encryption secure against selective opening attack. In: Ishai, Y. (ed.) TCC 2011. LNCS, vol. 6597, pp. 235–252. Springer, Heidelberg (2011). doi:10.1007/978-3-642-19571-6_15

4. Bellare, M., Yilek, S.: Encryption schemes secure under selective opening attack. Cryptology ePrint Archive, Report 2009/101 (2009). http://eprint.iacr.org/2009/101

5. Böhl, F., Hofheinz, D., Kraschewski, D.: On definitions of selective opening security. In: Fischlin, M., Buchmann, J., Manulis, M. (eds.) PKC 2012. LNCS, vol. 7293, pp. 522–539. Springer, Heidelberg (2012). doi:10.1007/978-3-642-30057-8_31

6. Boneh, D., Boyen, X., Shacham, H.: Short group signatures. In: Franklin, M. (ed.) CRYPTO 2004. LNCS, vol. 3152, pp. 41–55. Springer, Heidelberg (2004). doi:10.1007/978-3-540-28628-8_3

7. Boyen, X., Waters, B.: Shrinking the keys of discrete-log-type lossy trapdoor functions. In: Zhou, J., Yung, M. (eds.) ACNS 2010. LNCS, vol. 6123, pp. 35–52. Springer, Heidelberg (2010). doi:10.1007/978-3-642-13708-2_3

8. Canetti, R., Feige, U., Goldreich, O., Naor, M.: Adaptively secure multi-party computation. In: 28th ACM STOC, pp. 639–648. ACM Press, May 1996

9. Damgård, I., Jurik, M.: A generalisation, a simplification and some applications of paillier's probabilistic public-key system. In: Kim, K. (ed.) PKC 2001. LNCS, vol. 1992, pp. 119–136. Springer, Heidelberg (2001). doi:10.1007/3-540-44586-2_9

10. Dwork, C., Naor, M., Reingold, O., Stockmeyer, L.J.: Magic functions. In: 40th FOCS. pp. 523–534. IEEE Computer Society Press, October 1999

11. Escala, A., Herold, G., Kiltz, E., Ràfols, C., Villar, J.: An algebraic framework for Diffie-Hellman assumptions. In: Canetti, R., Garay, J.A. (eds.) CRYPTO 2013. LNCS, vol. 8043, pp. 129–147. Springer, Heidelberg (2013). doi:10.1007/978-3-642-40084-1_8

12. Fehr, S., Hofheinz, D., Kiltz, E., Wee, H.: Encryption schemes secure against chosen-ciphertext selective opening attacks. In: Gilbert, H. (ed.) EUROCRYPT 2010. LNCS, vol. 6110, pp. 381–402. Springer, Heidelberg (2010). doi:10.1007/978-3-642-13190-5_20

13. Freeman, D.M.: Converting pairing-based cryptosystems from composite-order groups to prime-order groups. In: Gilbert, H. (ed.) EUROCRYPT 2010. LNCS, vol. 6110, pp. 44–61. Springer, Heidelberg (2010). doi:10.1007/978-3-642-13190-5_3

14. Freeman, D.M., Goldreich, O., Kiltz, E., Rosen, A., Segev, G.: More constructions of lossy and correlation-secure trapdoor functions. J. Cryptology **26**(1), 39–74 (2013)

15. Fujisaki, E.: All-but-many encryption. In: Sarkar, P., Iwata, T. (eds.) ASIACRYPT 2014. LNCS, vol. 8874, pp. 426–447. Springer, Heidelberg (2014). doi:10.1007/978-3-662-45608-8_23

16. Garg, S., Gentry, C., Halevi, S.: Candidate multilinear maps from ideal lattices. In: Johansson, T., Nguyen, P.Q. (eds.) EUROCRYPT 2013. LNCS, vol. 7881, pp. 1–17. Springer, Heidelberg (2013). doi:10.1007/978-3-642-38348-9_1

17. Groth, J., Sahai, A.: Efficient non-interactive proof systems for bilinear groups. In: Smart, N. (ed.) EUROCRYPT 2008. LNCS, vol. 4965, pp. 415–432. Springer, Heidelberg (2008). doi:10.1007/978-3-540-78967-3_24

18. Hemenway, B., Libert, B., Ostrovsky, R., Vergnaud, D.: Lossy encryption: constructions from general assumptions and efficient selective opening chosen ciphertext security. In: Lee, D.H., Wang, X. (eds.) ASIACRYPT 2011. LNCS, vol. 7073, pp. 70–88. Springer, Heidelberg (2011). doi:10.1007/978-3-642-25385-0_4

19. Hofheinz, D.: All-but-many lossy trapdoor functions. In: Pointcheval, D., Johansson, T. (eds.) EUROCRYPT 2012. LNCS, vol. 7237, pp. 209–227. Springer, Heidelberg (2012). doi:10.1007/978-3-642-29011-4_14

20. Hofheinz, D., Kiltz, E.: Secure hybrid encryption from weakened key encapsulation. In: Menezes, A. (ed.) CRYPTO 2007. LNCS, vol. 4622, pp. 553–571. Springer, Heidelberg (2007). doi:10.1007/978-3-540-74143-5_31

21. Hofheinz, D., Rao, V., Wichs, D.: Standard security does not imply indistinguishability under selective opening. IACR Cryptology ePrint Archive 2015, 792 (2015). http://eprint.iacr.org/2015/792

22. Huang, Z., Liu, S., Qin, B.: Sender-equivocable encryption schemes secure against chosen-ciphertext attacks revisited. In: Kurosawa, K., Hanaoka, G. (eds.) PKC 2013. LNCS, vol. 7778, pp. 369–385. Springer, Heidelberg (2013). doi:10.1007/978-3-642-36362-7_23

23. Möller, B.: Algorithms for multi-exponentiation. In: Vaudenay, S., Youssef, A.M. (eds.) SAC 2001. LNCS, vol. 2259, pp. 165–180. Springer, Heidelberg (2001). doi:10.1007/3-540-45537-X_13

24. Morillo, P., Ràfols, C., Villar, J.L.: Matrix computational assumptions in multi-linear groups. Cryptology ePrint Archive, Report 2015/353 (2015). http://eprint.iacr.org/

25. Naor, M., Pinkas, B.: Efficient oblivious transfer protocols. In: Kosaraju, S.R. (ed.) 12th SODA, pp. 448–457. ACM-SIAM, January 2001

26. Naor, M., Segev, G.: Public-key cryptosystems resilient to key leakage. In: Halevi, S. (ed.) CRYPTO 2009. LNCS, vol. 5677, pp. 18–35. Springer, Heidelberg (2009). doi:10.1007/978-3-642-03356-8_2

27. Ostrovsky, R., Rao, V., Scafuro, A., Visconti, I.: Revisiting lower and upper bounds for selective decommitments. In: Sahai, A. (ed.) TCC 2013. LNCS, vol. 7785, pp. 559–578. Springer, Heidelberg (2013). doi:10.1007/978-3-642-36594-2_31

28. Paillier, P.: Public-key cryptosystems based on composite degree residuosity classes. In: Stern, J. (ed.) EUROCRYPT 1999. LNCS, vol. 1592, pp. 223–238. Springer, Heidelberg (1999). doi:10.1007/3-540-48910-X_16

29. Peikert, C., Vaikuntanathan, V., Waters, B.: A framework for efficient and composable oblivious transfer. In: Wagner, D. (ed.) CRYPTO 2008. LNCS, vol. 5157, pp. 554–571. Springer, Heidelberg (2008). doi:10.1007/978-3-540-85174-5_31

30. Peikert, C., Waters, B.: Lossy trapdoor functions and their applications. In: Ladner, R.E., Dwork, C. (eds.) 40th ACM STOC, pp. 187–196. ACM Press, May 2008

31. Shacham, H.: The BBG HIBE has limited delegation. Cryptology ePrint Archive, Report 2007/201 (2007). http://eprint.iacr.org/2007/201

32. Shoup, V.: Lower bounds for discrete logarithms and related problems. In: Fumy, W. (ed.) EUROCRYPT 1997. LNCS, vol. 1233, pp. 256–266. Springer, Heidelberg (1997). doi:10.1007/3-540-69053-0_18

Towards Non-Black-Box Separations of Public Key Encryption and One Way Function

Dana Dachman-Soled$^{(\boxtimes)}$

University of Maryland, College Park, USA
danadach@ece.umd.edu

Abstract. Separating public key encryption from one way functions is one of the fundamental goals of complexity-based cryptography. Beginning with the seminal work of Impagliazzo and Rudich (STOC, 1989), a sequence of works have ruled out certain classes of reductions from public key encryption (PKE)—or even key agreement—to one way function. Unfortunately, known results—so called *black-box separations*—do not apply to settings where the construction and/or reduction are allowed to directly access the code, or circuit, of the one way function. In this work, we present a meaningful, *non-black-box* separation between public key encryption (PKE) and one way function.

Specifically, we introduce the notion of BBN$^-$ reductions (similar to the BBNp reductions of Baecher et al. (ASIACRYPT, 2013)), in which the construction E accesses the underlying primitive in a black-box way, but wherein the universal reduction $\mathbb{R}$ receives the efficient code/circuit of the underlying primitive as input and is allowed oracle access to the adversary Adv. We additionally require that the functions describing the number of oracle queries made to Adv, and the success probability of $\mathbb{R}$ are independent of the run-time/circuit size of the underlying primitive. We prove that there is no *non-adaptive*, BBN$^-$ *reduction* from PKE to one way function, under the assumption that certain types of strong one way functions exist. Specifically, we assume that there exists a regular one way function f such that there is no Arthur-Merlin protocol proving that $z \notin \mathsf{Range}(f)$, where soundness holds with high probability over "no instances," $y \sim f(U_n)$, and Arthur may receive polynomial-sized, non-uniform advice. This assumption is related to the average-case analogue of the widely believed assumption coNP $\not\subseteq$ **NP**/poly.

1 Introduction

Complexity-based cryptography seeks to formalize generic assumptions, such as the existence of one way functions or trapdoor functions, and then determine

This work is supported in part by an NSF CAREER Award #CNS-1453045 and by a Ralph E. Powe Junior Faculty Enhancement Award. This work was done in part while the author was visiting the Simons Institute for the Theory of Computing, supported by the Simons Foundation and by the DIMACS/Simons Collaboration in Cryptography through NSF grant #CNS-1523467.

© International Association for Cryptologic Research 2016
M. Hirt and A. Smith (Eds.): TCC 2016-B, Part II, LNCS 9986, pp. 169–191, 2016.
DOI: 10.1007/978-3-662-53644-5_7

which cryptographic primitives can be constructed from these assumptions. For example, it has been shown that the existence of one way functions implies the existence of pseudorandom generators [24], pseudorandom functions [20], digital signatures [26,31] and symmetric key encryption. For other primitives, such as public key encryption, it is believed that stronger assumptions are necessary. Indeed, a gap between symmetric key and public key encryption schemes also emerges in practice: Practical symmetric key encryption schemes, such as AES, are far more efficient and have proven to be less susceptible to attack, than practical public key encryption schemes, such as RSA. Understanding whether this gap in security and efficiency is inherent seems tied to determining whether public key encryption requires stronger complexity assumptions than one way functions. Unfortunately, even formalizing this question is difficult: We cannot hope to prove that one way function does not imply public key encryption in the *logical* sense, i.e. OWF $\not\to$ PKE, since if public key encryption exists then the logical statement OWF $\to$ PKE is always true. Therefore, one approach has been to ask whether there exists a *black-box* reduction of public key encryption to one way function, wherein the construction and security proof (reduction) only access the one way function in an input/output manner, but cannot make use of its code. The answer turns out to be negative as shown by the seminal work of Impagliazzo and Rudich [25] (who proved that even key agreement cannot be black-box reduced to one way function) and, in fact, their oracle separation technique was subsequently used to rule out black-box reductions between various primitives such as collision resistant hash functions to one way functions [34], oblivious transfer to public key encryption [18] and many more. But what about *non-black-box* reductions between these primitives, where the construction/reduction may use the *code* of the underlying primitive?

Pass et al. [29] initiated a systematic study of this question, ruling out a type of non-black-box reduction called a *Turing-reductions*—where the code of the underlying primitive is used in an arbitrary manner, but the adversary is used in a black-box manner only—under the assumption that one way functions with very strong properties exist. Briefly, languages coupled with an efficiently samplable distribution over the no instances are considered to be in $\mathsf{Heur}_{1/\mathrm{poly}}\mathsf{AM}$ if there exists an AM (constant-round) protocol that accepts the language, with the relaxation that soundness only needs to hold with high probability over the no instances. For efficiently computable f, Pass et al. [29] consider the distributional language $\overline{\mathsf{Range}(f)} = \{z : \forall x \in \{0,1\}^*, f(x) \neq z\}$ along with the distribution $f(U_n)$ over the "No" instances. Their assumption is that there exists an efficiently computable function f such that $\overline{\mathsf{Range}(f)} \notin \mathsf{Heur}_{1/\mathrm{poly}}\mathsf{AM}$. Pass et al. [29] justify their assumption by arguing that it is a natural average-case analogue of the widely believed assumption $\mathsf{coNP} \not\subseteq \mathsf{AM}$. Based on this assumption, [29] rule out various Turing reductions including, reductions from one-way permutations to one-way functions. Additionally, based on other newly introduced complexity assumptions, Pass et al. [29] prove separations among various other primitives. However, none of their results address the case of constructing key agreement (or even public key encryption) from one way function. Separating key agreement from one way function is especially significant, since it implies a separation of public key cryptography from private key cryptography. Indeed, resolving this question is one of the fundamental goals of complexity-based cryptography.

In order to make progress towards this goal, we seek to formalize a meaningful, *non-black-box* separation between one way function and public key encryption (PKE). To the best of our knowledge, the only known separations to date between one way function and public key encryption (PKE) are *oracle* separations. Such separations instantiate the one way function with a random oracle and so do not apply to settings where the construction and/or reduction are allowed to access the code of the one way function. We first define Turing reductions and discuss why it seems hard to rule out all Turing reductions from PKE to one-way function based on the assumption that there exist (classes of) one-way functions f for which $\overline{\mathrm{Range}(f)} \notin \mathrm{Heur}_{1/\mathrm{poly}}\mathrm{AM}$. We then introduce and motivate a new, more restricted type of non-black-box reduction, BBN$^-$ reductions, which are related to the BBNp reductions considered in the taxonomy of Baecher et al. [3]. Looking ahead, our main theorem will rule out non-adaptive, BBN$^-$ reductions from public key encryption to one-way functions based on the assumption that there exists a regular one-way-function f such that $\overline{\mathrm{Range}(f)} \notin \mathrm{Heur}_{1/\mathrm{poly}}\mathrm{AM}^{\mathrm{poly}}$, where $\mathrm{AM}^{\mathrm{poly}}$ is the non-uniform analogue of AM (i.e. A is allowed to receive polynomial-sized, non-uniform advice).

1.1 Turing Reductions and the Difficulty of Ruling Them Out

We begin by recalling the definition of a type of non-black-box reduction known in the literature as a Turing reduction. The formal definition below will be useful when we define our new class of non-black-box reductions (BBN$^-$ reductions) and compare to the notion of a Turing reduction.

Turing reductions. A Turing reduction from a primitive Q to a primitive P is a pair of *oracle* PPT Turing machines $(E, \mathbb{R})$ such that the following two properties hold:

Construction. For every efficient implementation f of primitive P, $E(f)$ implements Q.

Reduction. For every efficient implementation f of P, and every (inefficient) adversary Adv who breaks $E(f)$ with probability $\varepsilon = \varepsilon(n)$, on security parameter n, we have that $\mathbb{R}^{\mathsf{Adv}}(1^n, 1^\varepsilon, f)$ breaks f with probability $1/t(\max(n, 1/\varepsilon(n)))$ and $\mathbb{R}^{\mathsf{Adv}}(1^n, 1^{1/\varepsilon}, f)$ makes at most $v(\max(n, 1/\varepsilon(n)))$ oracle queries to Adv, for polynomials t, v.

Difficulty of ruling out Turing reductions. To rule out Turing reductions from PKE to one-way function based on the assumption that there exist efficiently computable f for which $\overline{\mathrm{Range}(f)} \notin \mathrm{Heur}_{1/\mathrm{poly}}\mathrm{AM}$, one must construct an AM protocol proving $z \notin \mathrm{Range}(f)$ (i.e. that z is "invalid") for any efficiently computable f, assuming there exists a Turing reduction from PKE to one-way function. The following is the natural way to do this: Let $\mathbb{R}$ be the assumed

Turing reduction from PKE to one-way function. The protocol does the following: A emulates the reduction $\mathbb{R}^{\mathsf{Adv}}(f, z)$, using the all-powerful M to respond to queries made to the adversary Adv. Queries made to Adv will be of the form (pk, e) where pk is a public key and e is a ciphertext and in return, M should return the message m encrypted in e, along with a proof (e.g. the coins for Gen and Enc showing that this is a correct decryption). If the emulation of $\mathbb{R}^{\mathsf{Adv}}$ outputs a value x such that $f(x) = z$, then A rejects; otherwise, A accepts. Intuitively, the reason this should work, is that if $\mathbb{R}$ is a "good" reduction, then $\mathbb{R}$ should invert w.h.p. for most $z \sim f(U_n)$, whereas if $z \notin \mathsf{Range}(f)$, no matter what $\mathbb{R}$ does, it cannot invert.

Of course, there is a huge hole in the above argument: The reduction $\mathbb{R}$ may send queries to Adv, that "look like" valid transcripts (pk, e), but actually do not correspond to the output of Gen, Enc on any valid input and randomness. So, we must allow M to claim to A that (pk, e) is invalid, but to prevent M* from cheating, we must also demand a proof of invalidity. But note that whatever protocol we use to prove that (pk, e) is invalid should not work for proving $z \notin \mathsf{Range}(f)$ for general one way functions f, since this would contradict our assumption. On the other hand, since the AM protocol must work for *every* construction of PKE from one-way function, it is not clear how to restrict the class of functions.

Nevertheless, there is a difference between the two settings: When proving $z \notin \mathsf{Range}(f)$, M* knows the "statement," i.e., the value of z. On the other hand, *during* the proof of the statement $z \notin \mathsf{Range}(f)$, A samples (pk, e) by running the randomized reduction $\mathbb{R}(f, z)$ and outputting its queries to Adv. Moreover, if M* cannot distinguish transcripts (pk, e) sampled using the reduction $\mathbb{R}$, from (pk, e) sampled honestly using Gen and Enc, then M* cannot "cheat." At first glance, it seems that, indeed, the two distributions must be close, since if $\mathbb{R}$'s output is far from the output of Gen and Enc, then Adv can always reject (and thus is useless for inverting f). However, there is a subtle issue here: For Adv to be useful for breaking f, we only need that the output queries of $\mathbb{R}(f, f(U_n))$ (over random variable $f(U_n)$) is close to the output of Gen and Enc; whereas in order to force M* to behave honestly, we need that the output queries of $\mathbb{R}(f, z)$, *with fixed input z*, are close to the output of Gen and Enc. Thus, in order to force honest behavior from M*, we would need to show that with high probability over choice of $z \sim f(U_n)$, the queries output by $\mathbb{R}(f, f(U_n))$ are distributed closely to the queries outputted by $\mathbb{R}(f, z)$. In other words, the queries made by $\mathbb{R}(f, z)$ should be (close to) independent of z. But this seems highly implausible since in order for $\mathbb{R}$ to invert z, given oracle access to Adv, a successful $\mathbb{R}$ should embed z in the transcripts (pk, e) it submits to Adv, and so the queries to Adv will clearly depend on z!

Unfortunately, we do not know how to get around this problem for the case of general Turing reductions. However, for the restricted class of non-adaptive, BBN⁻ reductions, which we introduce next, we will show how to overcome this apparent contradiction.

BBN⁻ *reductions.* A BBN⁻ reduction from a primitive $\mathcal{Q}$ to a primitive $\mathcal{P}$ is a pair of *oracle* PPT Turing machines $(E, \mathbb{R})$ such that the following two properties hold[1]:

Construction. For every implementation f of primitive $\mathcal{P}$, E^f implements $\mathcal{Q}$.

Reduction. There exist polynomials $t(\cdot), v(\cdot)$ such that: For every efficient implementation f of $\mathcal{P}$, and every (inefficient) adversary Adv who breaks E^f with probability $\varepsilon = \varepsilon(n)$, on security parameter n, we have that $\mathbb{R}^{\mathsf{Adv}}(1^n, 1^\varepsilon, f)$ breaks f with probability $1/t(\max(n, \varepsilon(n)))$ and $\mathbb{R}^{\mathsf{Adv}}(1^n, 1^{1/\varepsilon}, f)$ makes at most $v(\max(n, \varepsilon(n)))$ oracle queries to Adv.

We remark that an implementation of a primitive is any specific scheme that meets the requirements of that primitive (e.g., an implementation of a public-key encryption scheme provides samplability of key pairs, encryption with the public-key, and decryption with the private key).

In the above definition, the construction E makes only black-box calls to f, but the reduction $\mathbb{R}^{\mathsf{Adv}}(f)$ receives the description of f as input and so is *non-black-box*. Allowing only $\mathbb{R}$ access to the code of f already thwarts known techniques (e.g., oracle separations) for proving impossibility results. We also require that the functions describing the number of oracle queries made to Adv, and the success probability of $\mathbb{R}$ are independent of the run-time/circuit size of f.

1.2   Necessity of the Restrictions

The notion of BBN⁻ reductions is supposed to capture the setting where the construction is "black-box" in the underlying primitive, but the proof is "non-black-box" in the underlying primitive but "black-box" in the adversary. This is a natural subclass of Turing reductions, in which the construction/reduction may both be "non-black-box" in the underlying primitive, but the reduction is "black-box" in the adversary.

However, a careful reader will notice that we placed additional restrictions when defining BBN⁻ reductions (this was why we called our notion "BBN minus" in that the polynomials $t(\cdot), v(\cdot)$ are independent of the particular function f and so specifically, the polynomials $t(\cdot), v(\cdot)$ must be *independent* of the run-time (i.e. circuit size) of f. Specifically, consider the following alternative definition, which we call BBN':

An Alternative Definition BBN':

Construction. For every implementation f of primitive $\mathcal{P}$, E^f implements $\mathcal{Q}$.

Reduction. For every efficient implementation f of $\mathcal{P}$, and every (inefficient) adversary Adv who breaks E^f with probability $\varepsilon = \varepsilon(n)$, on security parameter n, we have that $\mathbb{R}^{\mathsf{Adv}}(1^n, 1^\varepsilon, f)$ breaks f with probability $1/t(\max(n, \varepsilon(n)))$

---

[1] We may also consider families of primitives—e.g. families of one-way functions $\mathcal{F}$ with uniform generation algorithms. Here, the generation algorithm is represented as a Turing Machine and each function $f \in \mathcal{F}$ is represented as a circuit.

and $\mathbb{R}^{\mathsf{Adv}}(1^n, 1^{1/\varepsilon}, f)$ makes at most $v(\max(n, 1/\varepsilon(n)))$ oracle queries to Adv, for polynomials t, v.

In the following, we argue that the more restrictive notion of BBN^- is necessary in the following sense: If there exists a Turing reduction from PKE to OWF, then there also exists a BBN' reduction from PKE to OWF. Therefore, ruling out BBN' reductions from PKE to OWF also implies ruling out Turing reductions from PKE to OWF. Since our goal is to relax the notion of Turing reduction in a meaningful way, in order to make progress on this fundamental question, it is necessary to restrict $t(\cdot), v(\cdot)$ as in the definition of BBN^-.

Theorem 1 (Informal). *If there exists a Turing reduction from PKE to (uniform) OWF, then there also exists a* BBN' *reduction from PKE to OWF.*

We sketch the proof of the above theorem.

Proof of Theorem 1 (Sketch): Assume there exists a Turing reduction $(E, \mathbb{R})$ from PKE to one way function, then (using the reduction from one way function to weak one-way function), there also exists a Turing reduction $(E, \mathbb{R})$ from PKE to weak-one-way-function, where an efficient adversary can invert the one way function with probability at most $1 - 1/\text{poly}(n)$, where n is security parameter (i.e. input/output length). We will use this to build a BBN' reduction $(E', \mathbb{R}')$ from PKE to one way function.[2] We first define E': We completely ignore oracle f and set $E' := E(f_{univ})$, where f_{univ} is the "weak" universal one-way function described in [19]. Namely, on input Turing machine f' and string x, $f_{univ}(f', x)$ outputs $f'||f'^{|x|^2}(x)$, where $f'^{|x|^2}(x)$ denotes the output of f' after running on input x for $|x|^2$ number of steps. Now, we define the reduction $\mathbb{R}'$: On input (f, y), where f has (polynomial) running time n^c on inputs of length n, and oracle access to adversary Adv breaking E'^f, $\mathbb{R}'^{\mathsf{Adv}}(f, y)$ does the following: Define the new one-way-function f' that runs in time $\tilde{n}^2$ on inputs of length $\tilde{n}$ in the following way: f' parses its input as $x||a$, where x has length $\tilde{n}^{1/c}$ and outputs $f(x) = y$ in time $\tilde{n}^2$. $\mathbb{R}'$ then runs $\mathbb{R}^{\mathsf{Adv}}$ on inputs $(f_{univ}, (f', y||a))$, where a is a dummy string of length $n^c - n$. Note that the input/output length $\mathbb{R}^{\mathsf{Adv}}$ gets run on is now n^c. Since $\mathbb{R}$ is a good Turing reduction, $\mathbb{R}$ inverts f_{univ} with $1 - 1/\text{poly}(n)$ probability, which means that $\mathbb{R}$ will return an element in $f_{univ}^{-1}((f', y))$ with non-negligible probability. Using this information $\mathbb{R}'$ can then recover an element in $f^{-1}(y)$ with non-negligible probability. However, note that the functions describing the number of times $\mathbb{R}$ runs the adversary Adv and the success probability of $\mathbb{R}$ depend on the input/output length of $(f_{univ}, (f', y||a))$, which is n^c and thus depends on the run time of f. This means that the functions describing the number of times $\mathbb{R}'$ runs Adv and the success probability of $\mathbb{R}'$ depends on the runtime of f.

---

[2] Note that the argument also holds in the case that the Turing reduction works for a family $\mathcal{F}$ of one way functions f with a uniform generation algorithm. Specifically, if Gen is a uniform, public-coin, generation algorithm that samples a circuit $f \in \mathcal{F}$, then we can construct a single one way function $\tilde{f}$ that on input randomness r and input x, first runs Gen(r) to select f, then evaluates $y = f(x)$ and then outputs (r, y).

1.3 Our Main Result

We are now ready to state our main theorem:

Theorem 2 (Informal). *Under the assumption that there exists a regular one-way function f such that the distributional language $\overline{\mathsf{Range}(f)} \notin \mathsf{Heur}_{1/\mathrm{poly}}\mathsf{AM}^{\mathrm{poly}}$, there is no non-adaptive, BBN^{-} reduction from PKE to one way function.*

In the above, $\mathsf{Heur}_{1/\mathrm{poly}}\mathsf{AM}^{\mathrm{poly}}$ is the same as the class $\mathsf{Heur}_{1/\mathrm{poly}}\mathsf{AM}$, except that A is allowed to receive polynomial-sized, non-uniform advice. Note that our result is restricted to non-adaptive reductions $\mathbb{R}$ which make $v(\max(n, 1/\varepsilon(n)))$ *parallel* oracle queries to the adversary Adv.

We conjecture that using techniques of Akavia et al. [2], Theorem 2 can be proven under the assumption that there exists a regular one-way function f such that the distributional language $\overline{\mathsf{Range}(f)} \notin \mathsf{Heur}_{1/\mathrm{poly}}\mathsf{AM}$ (i.e. without requiring the non-uniform advice). The requirement for regularity of f in the assumption comes from our use of the *randomized iterate* (see [21]) whose hardness amplification properties only hold for (nearly) regular functions f. Recently, the analysis of the randomized iterate was extended to a more general class of functions called "weakly-regular" functions [37]. We conjecture that our results hold for this broader class of functions as well. Extending our results to general one-way functions seems tied to the development of security-preserving hardness amplification techniques for general one-way functions. We leave these as opens problem for future work.

1.4 Our Techniques

A key insight of our work is the relationship between our newly introduced notion of BBN^{-} reductions and the problem of *instance compression*. Instance compression [7,11,14,23] is the fundamental complexity-theoretic problem of taking an instance of a hard problem and compressing it into a smaller, equivalent instance, of the same or different problem.[3] The relationship between BBN^{-} reductions and instance compression is the following: The reduction $\mathbb{R}$ takes as input an instance (y, c), where y is a random image of c, and submits queries to Adv, which take the form of transcripts (pk, e) where pk is the public key and e is a ciphertext. Since the public key encryption scheme uses the underlying one-way function in a black-box manner, the size of the transcript (pk, e) must be a fixed polynomial in the security parameter n (i.e. the input-output size of the one-way function). Thus, as long as $\mathbb{R}$ (on input security parameter n) does not query Adv with security parameter $\tilde{n}$ that is too large and depends on the circuit size (i.e.

---

[3] [23] showed that strong instance compression algorithms imply a *non-black-box construction* of public key encryption from one-way function. It was later shown that, under standard complexity assumptions, instance compression for certain NP-hard problems is impossible [11,14], indicating that the approach of [23] is unlikely to succeed.

runtime) of c, then it must be the case that the total length of the messages sent from $\mathbb{R}$ to Adv is *independent* of the size of the circuit c. In order to force $\mathbb{R}$ to have this behavior, we instantiate Adv in such a way that queries submitted by $\mathbb{R}$ with security parameter which is too large are "useless" due to the restrictions of the BBN$^-$ reduction. Now, in the AM protocol proving statement $z \notin \text{Range}(f)$, instead of using (z, f) itself as the input to $\mathbb{R}$, we construct a new one-way function instance (c, y) (with the same input-output length) using $k = k(n)$ instances $(x_1, y_1), \ldots, (x_k, y_k)$. I.e., $(c, y) \leftarrow \Phi(x_1, y_1), \ldots, (x_k, y_k)$, where Φ is some randomized function, each (x_i, y_i) is an input-output pair of f, and one of the y_i's is set to the common input z. The requirement on $(c, y) \leftarrow \Phi(x_1, y_1), \ldots, (x_k, y_k)$ is that inverting c (i.e. finding x such that $c(x) = y$) implies inverting y_i (i.e. finding x_i such that $f(x_i) = y_i$) with probability $1/\text{poly}(k)$. By choosing $k = k(n)$ to be a sufficiently large polynomial, it is possible to ensure that there is not enough room for all individual instances $y_1, \ldots, y_k$ to be embedded in the interaction with Adv. Thus, the reduction $\mathbb{R}$ itself which takes as input (c, y) and produces queries to Adv can be viewed as an instance compression algorithm. Using techniques of Drucker [11] (similar to techniques that appeared previously in [30,33]), we will now be able to circumvent the problem with the naive attempt to rule out Turing reductions discussed above, which was that with high probability over $z \sim f(U_n)$, the distribution over $\mathbb{R}(f, f(U_n))$ will be far from the distribution over $\mathbb{R}(f, z)$. We elaborate further below on the necessary steps of our proof and in the discussion below, we point out where each restriction we place on the class of reductions is being used:

Eliminating security parameter blow-up. We construct an adversary Adv that has the following property: When the one-way function has input/output length $\tilde{n}$, Adv flips a coin and returns $\bot$ with probability $1 - 1/\tilde{n}$. Note that this means that we can replace any reduction $\mathbb{R}$ that on security parameter n makes queries to Adv with extremely large security parameter (i.e. input/output length) greater than $\tilde{n} := 2 \cdot t(\max(n, 1/\varepsilon(n))) \cdot v(\max(n, 1/\varepsilon(n)))$, with another reduction $\mathbb{R}'$ that simulates all answers of Adv to queries with security parameter greater than $\tilde{n}$ with $\bot$ without actually making the query. The probability that the view of $\mathbb{R}$ and $\mathbb{R}'$ differs is at most $v(\max(n, 1/\varepsilon(n))) \cdot 1/(2 \cdot t(\max(n, 1/\varepsilon(n))) \cdot v(\max(n, 1/\varepsilon(n))))$ and thus $\mathbb{R}'$ should still succeed with probability at least $1/2t(\varepsilon(n))$. This means that the length of the total output of $\mathbb{R}'$ to Adv depends only on n, but not on the size (runtime) of c and so $\mathbb{R}'$ is indeed a compression function, when we choose appropriate circuit c. Here we use the restriction that $t(), v(), \varepsilon()$ are all independent of the runtime of c.

Designing a circuit-oblivious adversary. The adversary $\text{Adv} = (\text{Adv}_1, \text{Adv}_2, \text{Adv}_3)$ will have the property that $\text{Adv}_1, \text{Adv}_3$ are *efficient* algorithms, whereas Adv_2 is inefficient but *does not require access* to the one-way function c. Looking ahead, M will be used to implement Adv_2 *only*. The fact that Adv_2 does not require access to c is crucial, since otherwise, the size of the interaction would be at least $|c|$ and there would be no compression. The techniques of [5,25,35] are crucial for constructing such Adv. Allowing the construction only black-box access to the underlying one-way function is necessary for this step in the proof, since Adv_2 will essentially emulate the adversary from the black-box separation of OWF and PKE of [5,25,35]. See Sect. 3.

Applying instance compression techniques. For a fixed f, denote by $\Phi((x^1, y^1), \ldots, (x^k, y^k))$ the randomized mapping that derives (c, y) from $(x^1, y^1), \ldots, (x^k, y^k)$, where $y^i = f(x^i)$ and view $\mathbb{R} \circ \Phi$ as a compression algorithm. For $z \sim f(U_n)$, we would like to embed $(x^i, y^i) = (x, z)$, where $f(x) = z$, for a random position $i \in [k]$. Call this randomized mapping Φ_z. Using techniques of Drucker [11], we will choose Φ so that with high probability over $z \sim f(U_n)$, the distribution over the output of $\mathbb{R} \circ \Phi_z$, denoted $\mathcal{T}(z)$, where a fixed z is embedded in a random position and the remaining inputs are random, is statistically close to the distribution over the output of $\mathbb{R} \circ \Phi$, denoted $\mathcal{T}$ when all (x^i, y^i) are sampled at random. Here also it is crucial to allow the construction only black-box access to the underlying one-way function since otherwise the length of the transcript could depend on the size of c, instead of just the input-output length. We also use here the fact that $\mathbb{R}$'s success probability is independent of the size/run-time of c. This is because the closeness in distributions that we are able to show using techniques of [11], will be significantly larger than $1/|c|$. If $\mathbb{R}$ only achieved success probability smaller than $1/|c|$ to begin with, then switching the distributions as discussed above would lead to a "useless" $\mathbb{R}$, which might never succeed in inverting the one-way function.

Designing an AM *verifier—first stage.* Unfortunately, even in the "no case," when $z \sim f(U_n)$, A will not be able to sample directly from $\mathcal{T}(z)$ since it will not know a preimage x such that $z = f(x)$. Instead, A will sample from a simulated distribution, denoted by $\widetilde{\mathcal{T}}(z)$. We use techniques of Haitner et al. [21] to show that $\widetilde{\mathcal{T}}(z)$ and $\mathcal{T}(z)$ are somewhat close.

Designing an AM *prover.* On input an instance z, where z is not in the image of f, we must provide an AM prover who uses $\mathbb{R}$ to prove that z is not in the image. This will yield a contradiction to the existence of $\mathbb{R}$. To construct the AM proof, we use the fact that $\widetilde{\mathcal{T}}(z)$ and $\mathcal{T}$ are somewhat close to allow A to run a rejection sampling protocol with the help of M. This allows A to essentially output transcripts to M that are sampled as in the "honest" distribution $\mathcal{T}$. Using techniques of Bogdanov and Trevisan [8] and Akavia et al. [2], we can then provide A with non-uniform advice in the form of statistics on the distribution $\mathcal{T}$, which allows him to force M^* to respond to queries honestly.

Designing an AM *verifier—second stage.* The above steps guarantee that on input (c, y), the reduction $\mathbb{R}^{\mathsf{Adv}}$ (with M assisting A in the simulation of Adv) succeeds in recovering x such that $c(x) = y$ with noticeable probability. However, we must now show that given x, A can also recover x^* such that $f(x^*) = z$ with noticeable probability. Since the circuit c output by Φ is a slight modification of the k-th randomized iterate, defined by Haitner et al. [21], we can now leverage hardness amplification properties of the k-th randomized iterate to show that A recovers x^* with $1/\text{poly}$ probability for most $z \sim f(U_n)$ We must also be careful since the argument above guarantees that x can be recovered when the adversary is stateless. It is possible that a stateful M^* can respond in such a way that A recovers x such that $c(x) = y$, but cannot recover x^* such that $f(x^*) = z$. The key to ruling out such a case is that, because of the nature of public key encryption wherein ciphertexts encrypt either a 0 or a 1, for almost all transcripts output to M*, there is actually a single "correct" response and we force M* to respond with this "correct" response with very high probability over the transcripts outputted by the reduction.

1.5   Related Work

In their seminal work, Impagliazzo and Rudich [25] ruled out black-box reductions from key agreement to one-way function. Their oracle separation technique was subsequently used to rule out black-box reductions between various primitives such as collision resistant hash functions to one way functions [34], oblivious transfer to public key encryption [18] and many more. The oracle separation technique cannot be used to rule out non-black-box reductions, since the underlying primitive is modeled as an oracle with an exponentially large description size.

The meta-reduction technique (cf. [1,6,10,13,15–17,27,28,32]) has been useful for ruling out Turing reductions—reductions where the construction is arbitrary, but the reduction must use the adversary in a black-box manner. Often these techniques are used to give evidence that a construction of primitive P along with a security proof of the above form is impossible under "standard assumptions" (e.g. falsifiable assumptions or non-interactive assumptions). This differs from our setting of separating one-way function from public key encryption, since in this case we can construct public key encryption from most well-studied, concrete assumptions for which we can construct one-way functions (such as factoring, Diffie-Hellman assumptions, and lattice assumptions).

The power of non-black-box usage of the adversary in security reductions has been well-studied since the seminal work of Barak [4]. In this case it is well-known that non-black-box techniques are more powerful than black-box techniques. However, in our work, we are interested in non-black-box use of the underlying *primitive*, as opposed to non-black-box use of the *adversary*. Several recent works have dealt with the systematic study of the power of non-black-box reductions in such settings. These include the aforementioned work of Pass et al. [29] as well as a work of Brakerski et al. [9], which, among other results, addresses the question of whether zero knowledge proofs can help to construct key agreement from one-way function. However, the results of Brakerski et al. hold only in an oracle setting, where an oracle is added to simulate the power of a zero-knowledge proof. Baecher et al. [3] gave a taxonomy of black-box and non-black-box reductions. Indeed, the term BBN that we use is borrowed from Baecher et al. [3], who used BBN to indicate reductions wherein the construction uses the primitive in a **B**lack-box manner, the reduction uses the adversary in a **B**lack-box manner, but the reduction uses the primitive in a **N**on-black-box manner. Our notion of BBN⁻ differs from the notion of Baecher et al. [3] in that we require the reduction $\mathbb{R}$ to be universal, but allow $\mathbb{R}$ to receive the description of the code/circuit of f as input. Moreover, we allow the query complexity and success probability of $\mathbb{R}$ to depend on the success probability of the adversary Adv, but require it to be independent of the run-time/circuit size of f.

2 Preliminaries and Background

Notation. We use capital letters for random variables, standard letters for variables and calligraphic letters for sets. We adopt the convention that when the same random variable appears multiple times in an expression, all occurrences refer to the same instantiation. Given a distribution X and an event E, we denote by $X \mid E$ the conditional distribution over X, conditioned on the event E occurring. Let X be a random variable taking values in a finite set $\mathcal{U}$. If $\mathcal{S}$ is a subset of $\mathcal{U}$, then $x \sim \mathcal{S}$ means that x is selected according to the uniform distribution on $\mathcal{S}$. We write U_n to denote the random variable distributed uniformly over $\{0,1\}^n$ and $U_{[0,1]}$ to denote the continuous random variable distributed uniformly over $[0,1]$. In general, for a finite set S, we denote by U_S the uniform distribution over S.

Two distributions X and Y over $\mathcal{U}$ are ε close, denoted $\Delta(X,Y) \leq \varepsilon$, if $\frac{1}{2}\sum_{x\in\mathcal{U}} |\mathrm{Pr}_X[x] - \mathrm{Pr}_Y[x]| \leq \varepsilon$. For a set $\mathcal{S} \subseteq \mathcal{U}$, we denote by $\mathrm{Pr}_X[\mathcal{S}] := \sum_{x\in\mathcal{S}} \mathrm{Pr}_X[x]$, i.e. the weight placed on $\mathcal{S}$ by the distribution X.

For functions $f : \{0,1\}^n \to \{0,1\}^n$ and $y \in \{0,1\}^n$, we denote by $f(U_n)$ the distribution induced by f operating on U_n and we denote by $f^{-1}(y)$ the set $f^{-1}(y) := \{x \in \{0,1\}^n : f(x) = y\}$. For a distribution X with (implicit) sampling algorithm Samp, that takes n coins, we denote by $X(r)$ for $r \in \{0,1\}^n$, the output x of Samp(r). For an element x in the support of X, we denote by $X^{-1}(x)$ the set of random coins $r \in \{0,1\}^n$ such that $X(r) = x$.

Let $\mathcal{C} = \{\mathcal{C}_{k,n}\}$ be a parametrized collection of uniformly generated polynomially-sized circuits, indexed by $n \in \mathbb{N}$ and $k = k(n) = \mathrm{poly}(n)$. For a fixed (n,k) pair, let $C_{k,n}$ denote the random variable representing the choice of circuit $c_{k,n} \sim \mathcal{C}_{k,n}$, where $\mathcal{C}_{k,n}$ is a family of one-way functions. We require that with probability 1, $C_{k,n}$ implements a one-way function.

Definition 1 (BBN⁻ reduction from PKE to OWF). *A BBN⁻ reduction from public key encryption (PKE) to one-way function (OWF) is a pair of oracle PPT Turing machines $(E, \mathbb{R})$ with the following properties:*

Construction. *With all but negligible probability over $C_{k,n}$, $E^{C_{k,n}}(1^n)$ implements a PKE scheme.*

Reduction. *There exist polynomials $t(\cdot), v(\cdot)$ such that: For every (inefficient) adversary Adv who, with probability $\varepsilon_1 = \varepsilon_1(n) = 1/\mathrm{poly}(n)$ over $c_{k,n} \sim C_{k,n}$, breaks $E^{c_{k,n}}(1^n)$ with probability $\varepsilon_2 = \varepsilon_2(n) = 1/\mathrm{poly}(n)$, we have:*

$$\Pr_{c_{k,n}\sim C_{k,n}}\left[\Pr\left[\mathbb{R}^{\mathsf{Adv}}(1^n, 1^{\frac{1}{\varepsilon_2}}, c_{k,n}, c_{k,n}(U_n)) \in c_{k,n}^{-1}(c_{k,n}(U_n)) \right] \geq \frac{1}{t(\max(n, \frac{1}{\varepsilon_2(n)}))} \right] \geq \varepsilon_1,$$

and $\mathbb{R}^{\mathsf{Adv}}(1^n, 1^{1/\varepsilon}, c_{k,n}, y)$ makes at most $v(\max(n, 1/\varepsilon_2(n)))$ oracle queries to the adversary Adv.

Definition 2 (BBN⁻ reduction from PKE to $(1 - \delta/2)$-weak one way function). *A* BBN⁻ *reduction from public key encryption (PKE) to $(1-\delta)$-weak one-way function (for $q = \text{poly}(n)$) is a pair of* oracle *PPT Turing machines $(E, \mathbb{R})$ with the following properties:*

Construction. *With all but negligible probability over $C_{k,n}$, $E^{C_{k,n}}(1^n)$ implements a PKE scheme.*

Reduction. *There exists a polynomial $v(\cdot)$ such that: For every (inefficient) adversary* Adv *who, with probability $\varepsilon_1 = \varepsilon_1(n) = 1/\text{poly}(n)$ over $c_{k,n} \sim C_{k,n}$, breaks $E^{c_{k,n}}(1^n)$ with probability $\varepsilon_2 = \varepsilon_2(n) = 1/\text{poly}(n)$, we have:*

$$\Pr_{c_{k,n} \sim C_{k,n}} \left[\Pr\left[\mathbb{R}^{\text{Adv}}(1^n, 1^{1/\varepsilon_2}, c_{k,n}, c_{k,n}(U_n)) \in c_{k,n}^{-1}(c_{k,n}(U_n))\right] \geq 1 - \delta/2 \right] \geq \varepsilon_1,$$

and $\mathbb{R}^{\text{Adv}}(1^n, 1^{1/\varepsilon_2}, c_{k,n}, y)$ makes at most $v(\max(n, 1/\varepsilon_2(n)))$ oracle queries to the adversary Adv.

Definition 3 (Non-adaptive Reductions $\mathbb{R}$). *The reduction $\mathbb{R} = (\mathbb{R}_1, \mathbb{R}_2)$ is non-adaptive if it interacts with the adversary* Adv *in the following way:*

– *On input $(1^n, 1^{1/\varepsilon_2}, c_{k,n}, y)$ and random coins, $\mathbb{R}_1$ produces a transcript* tr *consisting of $v(\max(n, 1/\varepsilon_2(n)))$ parallel queries to* Adv, *as well as the intermediate state* st.
– *On input* tr, Adv *returns responses $d_1, \ldots, d_{v(\max(n,1/\varepsilon_2(n)))}$. $\mathbb{R}_2(\text{st}, d_1, \ldots, d_{v(\max(n,1/\varepsilon_2(n)))})$ returns either x such that $c_{k,n}(x) = y$ or returns $\perp$.*

For fixed (tr, st) pair, we also denote the output of $\mathbb{R}_2$ with respect to an oracle Adv and a fixed (tr, st) output by $\mathbb{R}_1$, by $\mathbb{R}^{\text{Adv}}(c, y, \text{tr}, \text{st}; r)$ or $\mathbb{R}^{\text{Adv}}(c, y, \text{tr}, \text{st})$ (depending on whether the coins of $\mathbb{R}_2$ are explicit or implicit). Note that in the above, r denotes the coins used by $\mathbb{R}_2$ only (and not the coins of $\mathbb{R}_1$ or Adv).

Constant-round interactive protocols with advice. An interactive protocol with advice consists of a pair of interactive machines $\langle P, V \rangle$, where P is a computationally unbounded prover and V is a PPT verifier which receive a common input x and advice string a. Feigenbaum and Fortnow [12] define the class AM^{poly} as the class of languages L for which there exists a constant c, a polynomial p and an interactive protocol $\langle P, V \rangle$ with advice such that for every n, there exists an advice string a of length $p(n)$ such that for every x of length n, on input x and advice a, $\langle P, V \rangle$ produces an output after c rounds of interaction and, for small constant ε':

– If $x \in L$, then $\Pr[\langle P, V \rangle$ accepts x with advice $a] \geq 1 - \varepsilon'$.
– If $x \notin L$, then for every prover P^*, $\Pr[\langle P^*, V \rangle$ accepts x with advice $a] \leq \varepsilon'$.

It was shown by [12] that AM^{poly} is equal to **NP**/poly. Thus, coNP $\subseteq$ AM^{poly} implies coNP $\subseteq$ **NP**/poly, which gives $\Sigma_3 = \Pi_3$ [36]. We use the terms M, "prover" and P (resp. A, "verifier" and V) interchangeably.

Definition 4. *A distributional language (L, D) is in $\mathsf{Heur}_{1/\mathrm{poly}}\mathsf{AM}^{\mathrm{poly}}$ if for every inverse polynomial q, there exists an AM (i.e., constant-round public-coin) protocol (P, V) where A receives advice of length polynomial in the input length such that, for small constant ε':*

- **Completeness:** *If $x \in L$, $\Pr[\langle P, V \rangle(x) = 1] \geq 1 - \varepsilon'$.*
- **Soundness:** *For every $n \in \mathbb{N}$ and every machine P^*, with probability $1 - q(n)$, and $x \in \{0,1\}^n$ sampled from D_n conditioned on $x \notin L$ satisfies $\Pr[\langle P^*, V \rangle(x) = 1] \leq \varepsilon'$.*

Our protocols will use the AM protocol $\mathsf{RandSamp}^m$ (the multi-query variant of $\mathsf{RandSamp}$) with the following properties as a subroutine. $\mathsf{RandSamp}$ has been used extensively in the literature; the formalization below is due to [22].

Lemma 3. *Let $w = g(r)$ for a $\mathrm{poly}(n)$-time computable, randomized function g and random coins r. Assume w has bit length $\widehat{n}$. Then there exists an AM protocol $\mathsf{RandSamp}^m$ with an efficient verifier V that gets as input a security parameter 1^n, $\delta' = 1/\mathrm{poly}(n)$ (as the approximation and soundness parameter), $s_1, \ldots, s_m$ (as size of $f^{-1}(w_1), \ldots, f^{-1}(w_m)$) such that for all $i \in [m]$, $s_i \in (1 \pm \lambda)|f^{-1}(w_i)|$ (for $\lambda = \mathrm{poly}(1/m, 1/(\widehat{n} \cdot m), \delta')$) and returns $(r_1, \ldots, r_m)$ such that:*

- **Completeness:** *There is a prover strategy (the honest prover) s.t. V aborts with probability at most δ.*
- **Soundness:** *For any prover P^* either*
  - $\langle P^*, V \rangle$ *aborts with probability $1 - \delta'$ OR*
  - $\Delta((U_{f^{-1}(w_1)}, \ldots, U_{f^{-1}(w_m)}), (r_1, \ldots, r_m))) \leq \delta' + \Pr[\langle P^*, V \rangle \text{ aborts}].$

The following fact will be useful when protocol $\mathsf{RandSamp}^m$ is employed:

Fact 4. *Let X, Y be random variables distributed over the set $\mathcal{S} \cup \{\bot\}$ such that $\Pr[Y = \bot] = 0$ and $\Delta(X, Y) \leq \Pr[X = \bot] + \delta'$. Then for any event $T \subset \mathcal{S}$ it holds that:*

$$\Pr[X \in T] = \Pr_{x \sim X}[x \neq \bot \wedge x \in T] \leq \Pr[Y \in T] + \delta'.$$

and so

$$\Pr_{x \sim X | (x \neq \bot)}[x \in T] \leq (\Pr[Y \in T] + \delta') \cdot \frac{1}{\Pr_{x \sim X}[x \neq \bot]}.$$

Definition 5 (Enhanced Randomized Iterate). *Let $f : \{0,1\}^n \to \{0,1\}^n$, let $\mathcal{H}$ be a family of pairwise-independent length-preserving hash functions over strings of length n and let $\widehat{\mathcal{H}}$ be a family of $p' \cdot p_q(\widetilde{n}) + p(\widetilde{n})$-wise independent length-preserving hash functions (where $p', p_q(\widetilde{n}), p(\widetilde{n})$ are polynomials in n that will be defined later) over strings of length n. Define the k-th enhanced randomized iterate $F : \{0,1\}^n \times \mathcal{H}^{k-1} \times \widehat{\mathcal{H}}^2 \to \{0,1\}^n$ as*

$$F(x, \overline{h}, \widehat{h}_1, \widehat{h}_2) = \widehat{h}_2(f(h_{k-1}(f(h_{k-2}(\cdots(f(\widehat{h}_1(x)))\cdots))))).$$

We denote by H_j (resp. $\hat{H}_b$, $b \in \{1,2\}$) random variables uniformly distributed over $\mathcal{H}$ (resp. $\widehat{\mathcal{H}}$).

Let $c_{k,n} = c_{k,n}(\cdot, \overline{h} = h_1, \ldots, h_{k-1}, \widehat{h}_1, \widehat{h}_2)$ denote the circuit which has $\overline{h}, \widehat{h}_1, \widehat{h}_2$ hardwired and on input x computes $y = F(x, \overline{h}, \widehat{h}_1, \widehat{h}_2)$. Let $\mathcal{C}_{k,n}$ denote the set of circuits $c_{k,n}$ obtained when taking $h_1, \ldots, h_{k-1} \in \mathcal{H}, \widehat{h}_1, \widehat{h}_2 \in \widehat{\mathcal{H}}$. Let $C_{k,n}$ be the random variable defined as $C_{k,n} = c_{k,n}(\cdot, H_1, \ldots, H_{k-1}, \hat{H}_b, \hat{H}_b)$.

Lemma 5 [21]. *For $i \in [k]$, let $c_{k,n}^i = c_{k,n}^i(\cdot, \overline{h} = h_1, \ldots, h_{i-1}, \widehat{h}_1)$ denote the circuit which has $\overline{h}, \widehat{h}_1$ hardwired and on input x computes $y^i = \mathrm{Y}^i(c_{k,n}, x) = F^i(x, \overline{h}, \widehat{h}_1)$. Let the random variable $C_{k,n}^i$ denote the distribution over circuits $c^i := c_{k,n}^i$ as above.*
Then for any set $\mathcal{L} \subseteq \{0,1\}^n \times \mathcal{C}_{k,n}^i$ with

$$\Pr[(C_{k,n}^i(U_n), C_{k,n}^i) \in \mathcal{L}] \geq \delta,$$

it holds that

$$\Pr[(f(U_n), C_{k,n}^i) \in \mathcal{L}] \geq \frac{\delta^2}{i}.$$

We now describe a transformation (folklore, formalized by Haitner et al. [21]), of an arbitrary one-way function into a length-preserving one-way function.

Lemma 6. *Let $f : \{0,1\}^n \to \{0,1\}^{\ell(n)}$ be a $(T = T(n), \varepsilon = \varepsilon(n))$-OWF and let $\mathcal{H}$ be an efficient family of 2^{-2n}-almost pairwise-independent hash functions from $\{0,1\}^{\ell(n)}$ to $\{0,1\}^{2n}$. We define $\overline{f}$ as*

$$\overline{f}(x_a, x_b, h) = (h(f(x_a)), h),$$

where $x_a, x_b \in \{0,1\}^n$ and $h \in \mathcal{H}$. Then $\overline{f}$ is a length-preserving $(T - n^{O(1)}, \varepsilon + 2^{-n+1})$-one-way function.

If the original function f is regular, then the output function $\overline{f}$ is *nearly regular*: There is some fixed s such that with all but negligible probability over $y \sim \overline{f}(U_{2n}, H)$, the number of pre-images of y is exactly s. It turns out that nearly regular functions are sufficient for all of our results.

3   The Circuit-Oblivious Adversary Adv

Let $E^f = (\mathsf{Gen}^f, \mathsf{Enc}^f, \mathsf{Dec}^f)$ be a public key encryption scheme making oracle calls to one-way function f. Assume polynomial $p_q(n)$ is an upperbound on the total number of queries made by $\mathsf{Gen}^f, \mathsf{Enc}^f, \mathsf{Dec}^f$ on input security parameter n and message of length n. We consider the following two distributions corresponding to sampling the function f from two different distributions.

In the following, $\mathcal{F}_n$ denotes the set of all functions from $\{0,1\}^n \to \{0,1\}^n$. Note that when $c \sim C_{k,n}$ is fixed, we write $\mathcal{E}^c$ to denote the distribution $\mathcal{E}^C$, with a fixed oracle c (whereas C denotes a random variable).

We next describe a modification (folklore and formally proved in [35]) of the well-known Eve algorithm, which is tailored for breaking public key encryption

Distribution $\mathcal{E}^C$	**Distribution $\mathcal{E}^O$**
To sample from this distribution:	To sample from this distribution:
– $c \sim \mathcal{C}_{k,n}$; $m \sim \{0,1\}$.	– $\mathcal{O} \sim \mathcal{F}_n$; $m \sim \{0,1\}$.
– $(pk, sk) \sim \mathsf{Gen}^{C(\cdot)}(1^n)$; $e \sim \mathsf{Enc}^c_{pk}(m)$.	– $(pk, sk) \sim \mathsf{Gen}^{\mathcal{O}(\cdot)}(1^n)$; $e \sim \mathsf{Enc}^\mathcal{O}_{pk}(m)$.
– Output (c, m, sk, pk, e).	– Output $(\mathcal{O}, m, sk, pk, e)$.

in the random oracle model. The advantage of this $\mathsf{Eve} = (\mathsf{Eve}_1, \mathsf{Eve}_2, \mathsf{Eve}_3)$ algorithm is that $\mathsf{Eve}_1, \mathsf{Eve}_3$ are polynomial-time and Eve_2 is inefficient but *does not* require oracle access to $\mathcal{O}$.

Eve runs on transcripts of the form (pk, e), where pk is the public key and e is the ciphertext. Eve's goal is to correctly decrypt e. Sotakova [35] proves the existence of an Eve with the following properties:

Eve_1 is an *efficient* oracle algorithm which takes input pk and outputs $\mathcal{Q}_{\mathsf{Eve}}$:

– Initialize $\mathcal{Q}_{\mathsf{Eve}} := \emptyset$. Choose $\hat{p}$ random strings $r^1, \ldots, r^{\hat{p}}$ and messages $m^1, \ldots, m^{\hat{p}}$.
– For $1 \leq i \leq \hat{p}$, run $\mathsf{Enc}^\mathcal{O}_{pk}(m^i; r^i)$. Add all queries and responses to $\mathcal{Q}_{\mathsf{Eve}}$. Let $p(n) = \mathrm{poly}(n)$ be the total number of queries made. $p(\cdot)$ depends only on $p_q(n)$ and the desired success probability $1 - \delta/8$.

Eve_2 takes $(pk, \mathcal{Q}_{\mathsf{Eve}})$ as input and outputs $[(\mathcal{I}_i, r_i)]_{i \in [p'(n)]}$ (note that Eve_2 *does not* have oracle access):

– Return p' number of elements $\{(\mathcal{I}_1, r_1), \ldots, (\mathcal{I}_{p'}, r_{p'})\}$ chosen uniformly at random from the set $\mathcal{S}(pk, \mathcal{Q}_{\mathsf{Eve}}) := \{(\mathcal{I}, r) \mid \mathsf{Gen}^\mathcal{I}(r) = (*, pk) \wedge \mathsf{Eve}_1^\mathcal{I}(pk) = \mathcal{Q}_{\mathsf{Eve}} \wedge |\mathcal{I}| = p_q(n) + p(n)\}$.[4]

Eve_3 is an *efficient* oracle algorithm which takes $[(\mathcal{I}_i, r_i, e)]_{i \in [p']}$ as input and outputs a bit d.

– For $i \in [p']$, run $\mathsf{Gen}^{\mathcal{I}_i}(r_i)$ to generate a (sk_i, pk_i)-pair and compute $\tilde{d}_i := \mathsf{Dec}^{\mathcal{I}_i, \mathcal{O}}_{sk}(e)$. By this notation we mean that whenever Dec queries the oracle, if the query is in $\mathcal{I}$, respond according to $\mathcal{I}$. Otherwise, respond according to $\mathcal{O}$.
– Given the resulting set of decryptions $\{\tilde{d}_1, \ldots, \tilde{d}_{p'}\}$, let num_0 denote the number of decryptions equaling 0 and num_1 denote the number of decryptions equaling 1. Let $b = 0$ if $\mathrm{num}_0 > \mathrm{num}_1$ and $b = 1$ otherwise.
– If $V := \mathrm{num}_0/p' \in [3/8 + (\ell - 1)/4p'', 3/8 + (\ell + 1)/4p'']$, return $d := 0$. Otherwise, return $d := b$.

We define parameters p', p'', ℓ in the full version. The exact setting will depend on properties of the given BBN^- reduction $\mathbb{R}$.

We next turn to proving success of the adversary.

---

[4] $\mathcal{I}$ is an ordered set of $p_q(n) + p(n)$ length n strings. Whenever Gen or Eve_1 make a query, if the query has not been made before, the next string is used to respond to the query. If the query has previously been made, the same string is returned.

Lemma 7 ([35], restated). *For $(\mathcal{O}, m, sk, pk, e) \sim \mathcal{E}^{\mathcal{O}}$, $\mathsf{Eve}^{\mathcal{O}}(pk, e)$ outputs m with probability at least $1 - \delta/8$.*

The basic intuition is the following: Given the first message pk sent from receiver to the sender, w.h.p, the set $\mathcal{Q}_{\mathsf{Eve}}$ will contain all queries made by the sender when computing the second message (the ciphertext e) with probability greater than some threshold $1/p_{th}(n)$. Now, we sample a view for the receiver consistent with $(pk, \mathcal{Q}_{\mathsf{Eve}})$, which will contain a secret key sk and use this secret key sk to decrypt the *real* ciphertext e sent by the real sender. Loosely speaking, sk should only decrypt e "incorrectly" if there is a query q to the random oracle that is answered *inconsistently* in the sampled receiver's view and the real sender's view. However, note that any individual query q that is made in the sampled receiver's view but is not contained in $\mathcal{Q}_{\mathsf{Eve}}$, is made by the real sender with probability less than $1/p_{th}(n)$. Now, since we choose p_{th} far larger than the number of queries contained in the receiver's view, it is unlikely that there are *any* queries in the sampled receiver's view that were also made by the sender, but do not appear in $\mathcal{Q}_{\mathsf{Eve}}$. Thus, w.h.p, there are no queries q answered inconsistently in the sampled receiver's view and real sender's view and thus with high probability, the sampled sk decrypts e correctly.

We now describe the actual adversary $\mathsf{Adv} = (\mathsf{Adv}_1, \mathsf{Adv}_2, \mathsf{Adv}_3)$:

- Adv_1: On input pk, and oracle access to c, Adv_1 computes $\mathcal{Q}_{\mathsf{Eve}} \leftarrow \mathsf{Eve}_1^c(pk)$, where Eve's queries are answered according to c (instead of the random oracle $\mathcal{O}$). Adv_1 outputs $\mathcal{Q}_{\mathsf{Eve}}$.
- Adv_2: On input $(pk, \mathcal{Q}_{\mathsf{Eve}})$, Adv_2 runs $\mathsf{Eve}_2(pk, \mathcal{Q}_{\mathsf{Eve}})$ and outputs $[(\mathcal{I}_i, r_i)]_{i \in [p'(n)]}$.
- Adv_3: On input $[(\mathcal{I}_i, r_i, e)]_{i \in [p'(n)]}$, Adv_3 runs $\mathsf{Eve}_3^c([(\mathcal{I}_i, r_i, e)]_{i \in [p'(n)]})$ where Eve's queries are answered according to c (instead of the random oracle $\mathcal{O}$). Adv_3 flips a coin and outputs $\perp$ with probability $1 - 1/n$. With probability $1/n$, Adv outputs the same bit d that is outputted by Eve_3^c.

We purposely "weaken" the adversary, by defining Adv such that it outputs $\perp$ with probability $1 - 1/n$—where n is the input/output length of the one-way function—in order to argue that queries made by the reduction, $\mathbb{R}$, to Adv with security parameter n set too large are "useless." See Sect. 1.4 for further discussion. We next turn to proving success of the adversary:

Lemma 8. *For $(c, m, sk, pk, e) \sim \mathcal{E}^C$, d computed by $\mathsf{Adv}^c(pk, e)$ is equal to m with probability at least $1 - \delta/4$.*

Intuitively, Lemma 8 holds since Adv^c makes at most $p(n) + p' \cdot p_q(n)$ queries and so since $\hat{h}_1$, $\hat{h}_2$ are $p(n) + p' \cdot p_q(n)$-wise independent, the view of the adversary is nearly the same when interacting with a random oracle $\mathcal{O}$ or with the randomly sampled circuit C. For the full proof, see the full version.

Now, using Markov's inequality and the fact that Adv_3 tosses a coin independently of all its other coins to decide whether to output $\perp$ at the final stage with probability $1 - 1/n$, we have that:

Corollary 9. *With probability $\varepsilon_1 := 1 - \delta/2$ over choice of $c \sim C$, we have that for $(m, sk, pk, e) \sim \mathcal{E}^c$, the output of $\mathsf{Adv}^c(pk, e)$ is equal to m with probability is at least $\varepsilon_2 := \delta/4n$.*

4 The Mapping $\varPhi$

Instead of sampling $c \sim C_{k,n}$, $x \sim U_n$, and outputting $(c, x, y := c(x))$, we can alternatively sample $(x^1, y^1), \ldots, (x^k, y^k) \sim (U_n, f(U_n))$ and $r \sim \{0,1\}^*$, set $(x, c, y) := \varPhi((x^1, y^1), \ldots, (x^k, y^k); r)$, for $\varPhi$ defined below:

The randomized mapping $\varPhi((x^1, y^1), \ldots, (x^k, y^k))$

- Sample $x \sim U_n$, $\widehat{h}_2 \sim \widehat{\mathcal{H}}$.
- Sample $h_1, \ldots, h_{k-1} \sim \mathcal{H}$, such that $h_i(y^i) = x^{i+1}$, for $i \in [k-1]$; $\widehat{h}_1 \sim \widehat{\mathcal{H}}$ such that $\widehat{h}_1(x) = x^1$.
- Return the tuple (x, c, y) where $c = c_{k,n}(\cdot, h_1, \ldots, h_{k-1}, \widehat{h}_1, \widehat{h}_2)$ and $y = c(x)$.

It is straightforward to see that the two sampling methods described above induce the same distribution. We additionally introduce the notation $\varPhi_2$ to denote the second and third coordinates of the output of $\varPhi$ (i.e. (c, y)).

5 Useful Distributions

For public key encryption scheme $E^{\mathcal{O}} = (\mathsf{Gen}^{\mathcal{O}}, \mathsf{Enc}^{\mathcal{O}}, \mathsf{Dec}^{\mathcal{O}})$, relative to random oracle $\mathcal{O}$, the following distribution (Fig. 1) corresponds to sampling a partial random oracle and running Gen.

To sample from distribution $\mathcal{PK}_n$:

- Sample a set $\mathcal{I}$ of $p' \cdot q_p(n) + p(n)$ random strings of length n. Sample randomness $r \sim \{0,1\}^*$.
- Compute $(pk, sk) := \mathsf{Gen}^{\mathcal{I}}(1^n; r)$, where queries are answered using $\mathcal{I}$.
- Compute $\mathcal{Q}_{\mathsf{Eve}} := \mathsf{Adv}_1^{\mathcal{I}}(pk)$, where queries are answered using $\mathcal{I}$.
- Output $(pk, \mathcal{Q}_{\mathsf{Eve}})$.

Fig. 1. The distribution $\mathcal{PK}_n$.

We assume that security parameter n can be determined given the generated pk. We slightly abuse notation and for a fixed $(pk, \mathcal{Q}_{\mathsf{Eve}})$, we denote by $\mathcal{PK}^{-1}(pk, \mathcal{Q}_{\mathsf{Eve}})$ the set of pairs $(\mathcal{I}, r)$ that yield output $(pk, \mathcal{Q}_{\mathsf{Eve}})$ when sampling from $\mathcal{PK}_n$, for appropriate n.

For each of the following distributions χ, we refer by χ_2 to the marginal distribution over the final coordinate, the transcript $\widetilde{tr}$. For marginal distributions (e.g. the marginal distribution over the second, third and sixth coordinates) we use full-length tuples with $*$ symbols in the "don't care" positions (e.g.

To sample from distribution $\mathcal{T}$:
- $i \sim [k]$; $x^1, \ldots, x^k \sim U_n$. Set $y^1 := f(x^1), \ldots, y^k := f(x^k)$ Choose $r, r' \sim \{0,1\}^*$.
- Set $(x, c, y) := \Phi((x^1, y^1), \ldots, (x^k, y^k); r)$, where $c := c(\cdot, h_1, \ldots, h_{k-1}, \widehat{h}_1, \widehat{h}_2)$, $c^i := c^i(\cdot, h_1, \ldots, h_{i-1}, \widehat{h}_1)$.
- Compute $(\mathsf{st}, \mathsf{tr}) = \mathbb{R}_1(c, y; r')$, where $\mathsf{tr} = ((\mathsf{pk}_1, \mathsf{e}_1), \ldots, (\mathsf{pk}_{v'(n)}, \mathsf{e}_{v'(n)}))$.
- For each $j \in [v'(n)]$, set $\mathcal{Q}^j_{\mathsf{Eve}} = \mathsf{Adv}^c_1(\mathsf{pk}_j)$.
- Ouput $(x, c, y, c^i, y^i, i, \mathsf{st}, \widetilde{\mathsf{tr}})$, where $\widetilde{\mathsf{tr}} = ((\mathsf{pk}_1, \mathcal{Q}^1_{\mathsf{Eve}}, \mathsf{e}_1), \ldots, (\mathsf{pk}_{v'(n)}, \mathcal{Q}^{v'(n)}_{\mathsf{Eve}}, \mathsf{e}_{v'(n)}))$.

Fig. 2. The distribution $\mathcal{T}$.

To sample from distribution $\widetilde{\mathcal{T}}$:
- $i \sim [k]$, $h_1, \ldots, h_{k-1} \sim \mathcal{H}$, $\widehat{h}_1, \widehat{h}_2 \sim \widehat{\mathcal{H}}$, $r' \sim \{0,1\}^*$. Choose $x^* \sim U_n$, $y^i := f(x^*)$.
- Set $y := F_i^{k-i}(y^i, h_{i+1}, \ldots, h_{k-1}, \widehat{h}_2)$. Set the circuit $c := c(\cdot, h_1, \ldots, h_{k-1}, \widehat{h}_1, \widehat{h}_2)$ and $c^i = c(\cdot, h_1, \ldots, h_{i-1}, \widehat{h}_1)$.
- Compute $(\mathsf{st}, \mathsf{tr}) = \mathbb{R}_1(c, y; r')$, where $\mathsf{tr} = ((\mathsf{pk}_1, \mathsf{e}_1), \ldots, (\mathsf{pk}_{v'(n)}, \mathsf{e}_{v'(n)}))$.
- For each $j \in [v'(n)]$, set $\mathcal{Q}^j_{\mathsf{Eve}} = \mathsf{Adv}^c_1(\mathsf{pk}_j)$.
- Ouput $(c, y, c^i, y^i, i, \mathsf{st}, \widetilde{\mathsf{tr}})$, where $\widetilde{\mathsf{tr}} = ((\mathsf{pk}_1, \mathcal{Q}^1_{\mathsf{Eve}}, \mathsf{e}_1), \ldots, (\mathsf{pk}_{v'(n)}, \mathcal{Q}^{v'(n)}_{\mathsf{Eve}}, \mathsf{e}_{v'(n)}))$.

Fig. 3. The distribution $\widetilde{\mathcal{T}}$.

$(*, c, y, *, *, i, *, *) \sim \mathcal{T}$ or $\Pr_{\mathcal{T}}[(*, c, y, *, *, i, *, *)])$. To denote the distribution χ, conditioned on one of the tuple coordinates fixed to some value v, we write $\chi \mid v$, where it is understood from context which coordinate is fixed (e.g. $\mathcal{T} \mid c$ means that the second coordinate is fixed to constant c). Note that if χ is a distribution over tuples with t number of coordinates, then $\chi \mid v$ is a distribution over tuples with $t - 1$ number of coordinates.

Henceforth, we fix a particular BBN^- reduction $\mathbb{R}$ with parameters $(v(\cdot), t(\cdot))$ and use the particular adversary Adv with success probability $(\varepsilon_1 = 1 - \delta/4, \varepsilon_2 = \delta)$ defined in Sect. 3. We denote by $v'(\cdot), t'(\cdot)$ the following polynomials: $v'(n) := v(\max(n, 1/\varepsilon_2(n)))$ and $t'(n) := t(\max(n, 1/\varepsilon_2(n)))$.

We next define the distributions $\mathcal{T}$ and $\widetilde{\mathcal{T}}$ in Figs. 2 and 3. Let $\mathsf{num}_{\mathcal{T}}$ be the number of random coins to sample from $\mathcal{T}$. Let $N_{\mathcal{T}} := 2^{\mathsf{num}_{\mathcal{T}}}$.

We additionally define the distribution $\mathcal{T}^{i^*}$ (resp. $\widetilde{\mathcal{T}}^{i^*}$) for $i^* \in [k]$ as the distribution $\mathcal{T}$ (resp. $\widetilde{\mathcal{T}}$), conditioned on $i := i^*$, and the distribution $\mathcal{T}(z)$ (resp. $\widetilde{\mathcal{T}}(z)$) for $z \in \mathsf{Range}(f)$ as the distribution $\mathcal{T}$ (resp. $\widetilde{\mathcal{T}}$), conditioned on $y^i := z$. Even when $z \notin \mathsf{Range}(f)$, we still use the notation $\widetilde{\mathcal{T}}(z)$. This refers to a distribution which is sampled with the same sampling algorithm as the one used for $\widetilde{\mathcal{T}}$, except $y^i := z$ is always fixed to a constant value (not necessarily in the range of f). Let $\mathsf{num}_{\widetilde{\mathcal{T}}}$ be the number of random coins to sample from $\mathcal{T}(z)$. Let $N_{\widetilde{\mathcal{T}}} := 2^{\mathsf{num}_{\widetilde{\mathcal{T}}}}$.

6   The AM Protocol

We begin with a high-level overview of the protocol: Recall that we fix a particular BBN^- reduction $\mathbb{R}$ with parameters $(v(\cdot), t(\cdot))$ and use the particular adversary Adv with success probability $(\varepsilon_1 = 1 - \delta/4, \varepsilon_2 = \delta)$ defined in Sect. 3. Additionally, recall that we denote by $v'(\cdot), t'(\cdot)$ the following polynomials: $v'(n) := v(\max(n, 1/\varepsilon_2(n))$ and $t'(n) := t(\max(n, 1/\varepsilon_2(n)))$ and that we assume WLOG (see discussion in Sect. 1.4) that $\mathbb{R}$ never makes calls to Adv with security parameter $\widetilde{n} > 2 \cdot t'(n) \cdot v'(n)$. On input z, A constructs many (c, y) pairs and runs many copies of the BBN^- reduction $\mathbb{R}^{\mathsf{Adv}}(c, y)$, using Merlin to help simulate the adversary Adv.

Our AM protocol uses the $\mathsf{HidProt}$ and CBC protocols of Akavia et al. [2] (see also the full version for more details.) and the $\mathsf{RandSamp}$ protocol (See Lemma 3) as subroutines. Parameters $\widetilde{\delta} := (\varepsilon')^2/2, \lambda := 1/k^{1/11}$ are both of order $1/\mathrm{poly}(n)$. For $\widetilde{\mathsf{tr}}$ sampled from $\widetilde{T}(z)_2$, $\mathsf{HidProt}$ will be used to determine the size of the sets $\widetilde{T}(z)_2^{-1}(\widetilde{\mathsf{tr}})$ and $T_2^{-1}(\widetilde{\mathsf{tr}})$. For $(pk^w, \mathcal{Q}_{\mathsf{Eve}}^w) \in \widetilde{\mathsf{tr}}$, CBC (along with the non-uniform advice provided to A) will be used to determine the size α of the set $\mathcal{PK}^{-1}(pk^w, \mathcal{Q}_{\mathsf{Eve}}^w)$. Given α, $\mathsf{RandSamp}$ will be used to sample preimages from the set $\mathcal{PK}^{-1}(pk^w, \mathcal{Q}_{\mathsf{Eve}}^w)$, thus simulating the adversary's (Adv_2's) response. Note that soundness of $\mathsf{HidProt}$ and CBC only hold under specific conditions (see the full version for more details.). Indeed, a key technical part of the proof is showing that the necessary conditions hold. The purpose of the *testing for goodness* subroutine is the following: We show in the analysis that w.h.p when $z \sim f(U_n)$, the distribution $\widetilde{T}(z)$ is "good," i.e. somewhat close to the distribution T, so the rejection sampling procedure can be employed. On the other hand, if $\widetilde{T}(z)$ is not "good" (i.e. very far from T), then A can safely output ACCEPT. Our AM protocol is presented in Fig. 4. We next state our main technical result.

Theorem 10. *Assume that there exists a non-adaptive,* BBN^- *reduction* $(E, \mathbb{R})$ *from PKE to* $(1 - \delta/2)$*-weak one way function. Then for any efficiently computable, length-preserving, (nearly) regular function* f*, the above non-uniform AM protocol* Π_f *has completeness* $1 - \varepsilon'$ *and soundness* $1 - \varepsilon'$ *(for small constant* ε'*), for the distributional language* $\overline{\mathrm{Range}(f)}$*, where soundness holds with probability* $1 - 7\delta$ *over* $z \sim f(U_n)$*.*

We note that if f is not length-preserving, it can be made length-preserving, while (nearly) preserving regularity, via the transformation described in Lemma 6.

To rule out non-adaptive, BBN^- reductions from PKE to one way function, recall that there is a non-adaptive, black-box reduction from OWF to $(1-\delta/2)$-weak OWF, where the parameters of the reduction depend only on the input-output size and on δ. but not on the description size of the function. Therefore, if there exists a non-adaptive BBN^- reduction $(E, \mathbb{R})$ from PKE to OWF, then for every polynomial q there also exists a non-adaptive, BBN^- reduction $(E, \mathbb{R})$ from PKE to $(1 - \delta/2)$-weak OWF, where $\delta = 1/7q$. By Theorem 10 (and the

The AM Protocol Π_f

Common Input: $z \in \{0,1\}^n$. M is proving to A that $z \notin \mathsf{Range}(f)$.
Non-Uniform Advice: The values $[\mu_\ell^w]_{w \in [v'(n)], \ell \in [p_2(n)/\widetilde{\delta}]}$, where $\mu_\ell^w = \mathsf{E}_{\mathcal{T}_2}[\lfloor(\log_{1+\lambda} \frac{|\mathcal{PK}^{-1}(PK^w, \mathcal{Q}_{\mathsf{Eve}}^w)|}{1 + \ell \frac{\delta\lambda}{p_2(n)}})\rfloor]]$, and $(PK^w, \mathcal{Q}_{\mathsf{Eve}}^w)$ is the random variable corresponding to the w-th pair in $\widetilde{\mathsf{TR}}$.

1. **A $\leftrightarrow$ M:**
    (a) Let $M := 2k^8$ and let $p_1(n) = p_2(n) \cdot (p^*(n) + 1)$. Set $\mathsf{ctr}, \mathsf{ctr}_1 = 0$.
    (b) **Testing for Goodness:**
        i. A samples $\widetilde{\mathsf{tr}}_1, \ldots, \widetilde{\mathsf{tr}}_{p_1(n)} \sim \widetilde{\mathcal{T}}(z)_2$.
        ii. A, M run $\mathsf{HidProt}^{p_1(n)}(\widetilde{\mathsf{tr}}_1, \ldots, \widetilde{\mathsf{tr}}_{p_1(n)})$, returning sizes $[(\rho_i, \tau_i)]_{i \in p_1(n)}$.
        iii. For $j \in [p_1(n)]$, A does the following local computation:
            A. Set $\zeta_i' := \rho_i/\mathsf{N}_{\mathcal{T}}, \gamma_i' := \tau_i/\mathsf{N}_{\widetilde{\mathcal{T}}}$ (approximating $\mathsf{Pr}_{\mathcal{T}_2}[\widetilde{\mathsf{tr}}]$, $\mathsf{Pr}_{\widetilde{\mathcal{T}}(z)_2}[\widetilde{\mathsf{tr}}]$).
            B. If the number of $i \in [p_1(n)]$ such that $\zeta_i' > 2M/3 \cdot \gamma_i'$ or $\gamma_i' > 2M/3 \cdot \zeta_i'$ is at least $\frac{3p_1(n)}{2k^{1/6}}$, A outputs ACCEPT.
    (c) **Rejection Sampling:**
        i. A samples $[(c_j, y_j, c_j^i, y_j^i, i_j, \mathsf{st}_j, \widetilde{\mathsf{tr}}_j)]_{j \in p_1(n)} \sim \widetilde{\mathcal{T}}(z)$.
        ii. A, M run $\mathsf{HidProt}^{p_1(n)}(\widetilde{\mathsf{tr}}_1, \ldots, \widetilde{\mathsf{tr}}_{p_1(n)})$, returning set sizes $[(\rho_j, \tau_j)]_{j \in [p_1(n)]}$. For $j \in [p_1(n)]$, set $\zeta_j' := \rho_j/\mathsf{N}_{\mathcal{T}}, \gamma_j' := \tau_j/\mathsf{N}_{\widetilde{\mathcal{T}}}$.
        iii. A sets $j = 0$ and repeats the following local computation until $j = p_1(n) + 1$:
            A. Set $j = j + 1, \mathsf{ctr}_1 = \mathsf{ctr}_1 + 1$.
            B. If $\zeta_j'/M\gamma_j' \le 1$, sample $u \sim U_{[0,1]}$ and check whether $u < \zeta_j'/M\gamma_j'$.
            C. If yes, set $\mathsf{ctr}_1 = 0, \mathsf{ctr} := \mathsf{ctr} + 1, c_{\mathsf{ctr}} := c, \widetilde{\mathsf{tr}}_{\mathsf{ctr}} := \widetilde{\mathsf{tr}}_j$ and $j = \mathsf{ctr} \cdot (p^*(n) + 1)$, where $\widetilde{\mathsf{tr}}_{\mathsf{ctr}} := [(\mathsf{pk}_{\mathsf{ctr}}^w, \mathcal{Q}_{\mathsf{Eve},\mathsf{ctr}}^w, \mathsf{e}_{\mathsf{ctr}}^w)]_{w \in [v'(n)]}$.
            D. If $\mathsf{ctr}_1 = p^*(n) + 1$, set $\mathsf{ctr}_1 = 0, \mathsf{ctr} := \mathsf{ctr} + 1, c_{\mathsf{ctr}} := \bot, \widetilde{\mathsf{tr}}_{\mathsf{ctr}} := \bot$.
    (d) **Simulating Adv_2:**
        i. A, M run $\mathsf{CBC}([\mathsf{pk}_j^w, \mathcal{Q}_{\mathsf{Eve},j}^w]_{j \in [p_2(n)], w \in [v'(n)]})$, using non-uniform advice $[\mu_\ell^w]_{w \in [v'(n)], \ell \in [p_2(n)/\widetilde{\delta}]}$, returning set sizes $[\alpha_j^w]_{j \in [p_2(n)], w \in [v'(n)]}$.
        ii. A, M simulate $\mathsf{Adv}_2(\mathsf{pk}_j^w, \mathcal{Q}_{\mathsf{Eve},j}^w)$ by running $p_2(n)$ parallel copies of $\mathsf{RandSamp}^{p'(n) \cdot v'(n)}([(\mathsf{pk}_j^w, \mathcal{Q}_{\mathsf{Eve},j}^w, \alpha_j^w)^{p'(n)}]_{j \in [p_2(n)], w \in [v'(n)]})$, returning values $[\mathcal{I}_{j,m}^w, r_{j,m}^w]_{m \in p'(n), j \in [p_2(n)], w \in [v'(n)]}$.

2. **Generating A's Output:** A does the following local computation:
    (a) For $1 \le j \le p_2(n)$, $1 \le w \le v'(n)$, compute $d_j^w := \mathsf{Adv}_3^{c_j}([\mathcal{I}_{j,m}^w, r_{j,m}^w, \mathsf{e}_{j,m}^w]_{m \in p'(n)})$.
    (b) For each $1 \le j \le p_2(n)$, run $\mathbb{R}_2(\mathsf{st}_j, d_j^1, \ldots, d_j^{v'(n)})$, returning outputs $x_1, \ldots, x_{p_2(n)}$.
    (c) Evaluate each $c_j(x_j)$. If during the evaluation some x^* such that $f(x^*) = z$ is queried to f, output REJECT. Otherwise, output ACCEPT.

Fig. 4. AM protocol for proving that z is not in the image of f.

extension to non-length-preserving f discussed above) this means that that for every efficiently computable, regular function f and every polynomial q, there exists a non-uniform AM protocol for proving $z \notin \mathsf{Range}(f)$, where soundness holds with probability $1 - 7\delta = 1 - 1/q$ over $z \in f(U_n)$. This contradicts our assumption that there exists an efficiently computable, (nearly) regular function f such that $\overline{\mathsf{Range}(f)} \notin \mathsf{Heur}_{1/\mathrm{poly}}\mathsf{AM}^{\mathrm{poly}}$.

Theorem 11. *Under the assumption that there exists an efficiently computable, regular function f such that $\overline{\mathsf{Range}(f)} \notin \mathsf{Heur}_{1/\mathrm{poly}}\mathsf{AM}^{\mathrm{poly}}$, there is no non-adaptive, BBN^- reduction from PKE to one way function.*

It remains to prove Theorem 10, which we defer to the full version.

Acknowledgements. We thank Tal Malkin for insightful discussions on the notion of BBN^- reductions and the anonymous reviewers for TCC B-2016 for their many helpful comments.

References

1. Abe, M., Groth, J., Ohkubo, M.: Separating short structure-preserving signatures from non-interactive assumptions. In: Lee, D.H., Wang, X. (eds.) ASIACRYPT 2011. LNCS, vol. 7073, pp. 628–646. Springer, Heidelberg (2011). doi:10.1007/978-3-642-25385-0_34

2. Akavia, A., Goldreich, O., Goldwasser, S., Moshkovitz, D.: On basing one-way functions on NP-hardness. In: Kleinberg, J.M. (ed.) 38th Annual ACM Symposium on Theory of Computing, pp. 701–710. ACM Press, May 2006

3. Baecher, P., Brzuska, C., Fischlin, M.: Notions of black-box reductions, revisited. In: Sako, K., Sarkar, P. (eds.) ASIACRYPT 2013. LNCS, vol. 8269, pp. 296–315. Springer, Heidelberg (2013). doi:10.1007/978-3-642-42033-7_16

4. Barak, B.: How to go beyond the black-box simulation barrier. In: 42nd Annual Symposium on Foundations of Computer Science, pp. 106–115. IEEE Computer Society Press, October 2001

5. Barak, B., Mahmoody-Ghidary, M.: Merkle puzzles are optimal — an $O(n^2)$-query attack on any key exchange from a random Oracle. In: Halevi, S. (ed.) CRYPTO 2009. LNCS, vol. 5677, pp. 374–390. Springer, Heidelberg (2009). doi:10.1007/978-3-642-03356-8_22

6. Bitansky, N., Dachman-Soled, D., Garg, S., Jain, A., Kalai, Y.T., López-Alt, A., Wichs, D.: Why "Fiat-Shamir for Proofs" lacks a proof. In: Sahai, A. (ed.) TCC 2013. LNCS, vol. 7785, pp. 182–201. Springer, Heidelberg (2013). doi:10.1007/978-3-642-36594-2_11

7. Bodlaender, H.L., Downey, R.G., Fellows, M.R., Hermelin, D.: On problems without polynomial kernels. J. Comput. Syst. Sci. **75**(8), 423–434 (2009)

8. Bogdanov, A., Trevisan, L.: On worst-case to average-case reductions for NP problems. In: 44th Annual Symposium on Foundations of Computer Science, pp. 308–317. IEEE Computer Society Press, October 2003

9. Brakerski, Z., Katz, J., Segev, G., Yerukhimovich, A.: Limits on the power of zero-knowledge proofs in cryptographic constructions. In: Ishai, Y. (ed.) TCC 2011. LNCS, vol. 6597, pp. 559–578. Springer, Heidelberg (2011). doi:10.1007/978-3-642-19571-6_34

10. Coron, J.-S.: Optimal security proofs for PSS and other signature schemes. In: Knudsen, L.R. (ed.) EUROCRYPT 2002. LNCS, vol. 2332, pp. 272–287. Springer, Heidelberg (2002). doi:10.1007/3-540-46035-7_18

11. Drucker, A.: New limits to classical and quantum instance compression. In: 53rd Annual Symposium on Foundations of Computer Science, pp. 609–618. IEEE Computer Society Press, October 2012

12. Feigenbaum, J., Fortnow, L.: Random-self-reducibility of complete sets. SIAM J. Comput. 22(5), 994–1005 (1993)

13. Fischlin, M., Schröder, D.: On the impossibility of three-move blind signature schemes. In: Gilbert, H. (ed.) EUROCRYPT 2010. LNCS, vol. 6110, pp. 197–215. Springer, Heidelberg (2010). doi:10.1007/978-3-642-13190-5_10

14. Fortnow, L., Santhanam, R.: Infeasibility of instance compression and succinct PCPs for NP. In: Ladner, R.E., Dwork, C. (eds.) 40th Annual ACM Symposium on Theory of Computing, pp. 133–142. ACM Press, May 2008

15. Fuchsbauer, G., Konstantinov, M., Pietrzak, K., Rao, V.: Adaptive security of constrained PRFs. In: Sarkar, P., Iwata, T. (eds.) ASIACRYPT 2014. LNCS, vol. 8874, pp. 82–101. Springer, Heidelberg (2014). doi:10.1007/978-3-662-45608-8_5

16. Garg, S., Bhaskar, R., Lokam, S.V.: Improved bounds on security reductions for discrete log based signatures. In: Wagner, D. (ed.) CRYPTO 2008. LNCS, vol. 5157, pp. 93–107. Springer, Heidelberg (2008). doi:10.1007/978-3-540-85174-5_6

17. Gentry, C., Wichs, C.: Separating succinct non-interactive arguments from all falsifiable assumptions. In: Fortnow, L., Vadhan, S.P. (eds.) 43rd Annual ACM Symposium on Theory of Computing, pp. 99–108. ACM Press, June 2011

18. Gertner, Y., Kannan, S., Malkin, T., Reingold, O., Viswanathan, M.: The relationship between public key encryption and oblivious transfer. In: 41st Annual Symposium on Foundations of Computer Science, pp. 325–335. IEEE Computer Society Press, November 2000

19. Goldreich, O.: Foundations of Cryptography: Basic Tools, vol. 1. Cambridge University Press, Cambridge (2001)

20. Goldreich, O., Goldwasser, S., Micali, S.: On the cryptographic applications of random functions. In: Blakley, G.R., Chaum, D. (eds.) CRYPTO 1984. LNCS, vol. 196, pp. 276–288. Springer, Heidelberg (1985). doi:10.1007/3-540-39568-7_22

21. Haitner, I., Harnik, D., Reingold, O.: On the power of the randomized iterate. In: Dwork, C. (ed.) CRYPTO 2006. LNCS, vol. 4117, pp. 22–40. Springer, Heidelberg (2006). doi:10.1007/11818175_2

22. Haitner, I., Mahmoody, M., Xiao, D.: A new sampling protocol and applications to basing cryptographic primitives on the hardness of NP. In: Proceedings of the 25th Annual IEEE Conference on Computational Complexity, CCC 2010, 9–12 June 2010, Cambridge, Massachusetts, pp. 76–87 (2010)

23. Harnik, D., Naor, M.: On the compressibility of NP instances and cryptographic applications. In: 47th Annual Symposium on Foundations of Computer Science, pp. 719–728. IEEE Computer Society Press, October 2006

24. Håstad, J., Impagliazzo, R., Levin, L.A., Luby, M.: A pseudorandom generator from any one-way function. SIAM J. Comput. 28(4), 1364–1396 (1999)

25. Impagliazzo, R., Rudich, S.: Limits on the provable consequences of one-way permutations. In: 21st Annual ACM Symposium on Theory of Computing, pp. 44–61. ACM Press, May 1989

26. Lamport, L.: Constructing digital signatures from a one-way function. Technical report SRI-CSL-98, SRI International Computer Science Laboratory, October 1979

27. Paillier, P., Vergnaud, D.: Discrete-log-based signatures may not be equivalent to discrete log. In: Roy, B. (ed.) ASIACRYPT 2005. LNCS, vol. 3788, pp. 1–20. Springer, Heidelberg (2005). doi:10.1007/11593447_1

28. Pass, R.: Limits of provable security from standard assumptions. In: Fortnow, L., Vadhan, S.P. (eds.) 43rd Annual ACM Symposium on Theory of Computing, pp. 109–118. ACM Press, June 2011

29. Pass, R., Tseng, W.-L.D., Venkitasubramaniam, M.: Towards non-black-box lower bounds in cryptography. In: Ishai, Y. (ed.) TCC 2011. LNCS, vol. 6597, pp. 579–596. Springer, Heidelberg (2011). doi:10.1007/978-3-642-19571-6_35

30. Raz, R.: A parallel repetition theorem. SIAM J. Comput. **27**(3), 763–803 (1998)

31. Rompel, J.: One-way functions are necessary and sufficient for secure signatures. In: 22nd Annual ACM Symposium on Theory of Computing, pp. 387–394. ACM Press, May 1990

32. Seurin, Y.: On the exact security of Schnorr-Type signatures in the random Oracle model. In: Pointcheval, D., Johansson, T. (eds.) EUROCRYPT 2012. LNCS, vol. 7237, pp. 554–571. Springer, Heidelberg (2012). doi:10.1007/978-3-642-29011-4_33

33. Shaltiel, R.: Derandomized parallel repetition theorems for free games. In: Proceedings of the 25th Annual IEEE Conference on Computational Complexity, CCC 2010, 9–12 June 2010, Cambridge, Massachusetts, pp. 28–37 (2010)

34. Simon, D.R.: Finding collisions on a one-way street: can secure hash functions be based on general assumptions? In: Nyberg, K. (ed.) EUROCRYPT 1998. LNCS, vol. 1403, pp. 334–345. Springer, Heidelberg (1998). doi:10.1007/BFb0054137

35. Sotakova, M.: Breaking one-round key-agreement protocols in the random Oracle model. Cryptology ePrint Archive, Report 2008/053 (2008). http://eprint.iacr.org/2008/053

36. Yap, C.-K.: Some consequences of non-uniform conditions on uniform classes. Theoret. Comput. Sci. **26**, 287–300 (1983)

37. Yu, Y., Gu, D., Li, X., Weng, J.: The randomized iterate, revisited - almost linear seed length PRGs from a broader class of one-way functions. In: Dodis, Y., Nielsen, J.B. (eds.) TCC 2015. LNCS, vol. 9014, pp. 7–35. Springer, Heidelberg (2015). doi:10.1007/978-3-662-46494-6_2

Post-Quantum Security of the Fujisaki-Okamoto and OAEP Transforms

Ehsan Ebrahimi Targhi$^{(\boxtimes)}$ and Dominique Unruh

University of Tartu, Tartu, Estonia
{ehsan.ebrahimi.targhi,unruh}@ut.ee

Abstract. In this paper, we present a hybrid encryption scheme that is chosen ciphertext secure in the quantum random oracle model. Our scheme is a combination of an asymmetric and a symmetric encryption scheme that are secure in a weak sense. It is a slight modification of the Fujisaki-Okamoto transform that is secure against classical adversaries. In addition, we modify the OAEP-cryptosystem and prove its security in the quantum random oracle model based on the existence of a partial-domain one-way injective function secure against quantum adversaries.

Keywords: Quantum · Random oracle · Indistinguishability against chosen ciphertext attacks

1 Introduction

The interest in verifying the security of cryptosystems in the presence of a quantum adversary increased after the celebrated paper of Shor [10]. Shor showed that any cryptosystem based on the factoring problem and the discrete logarithm problem is breakable in the presence of a quantum adversary. Also, many efficient classical cryptosystems are proved to be secure in the random oracle model [3] and many of them still lack an equivalent proof in the quantum setting. Therefore, even if we find a cryptographic primitive immune to quantum attacks, to construct an efficient cryptosystem secure against quantum adversaries, we may have to consider its security in the quantum random oracle model in which the adversary has quantum access to the random oracle.

Fujisaki and Okamoto [8] constructed a hybrid encryption scheme that is secure against chosen ciphertext attacks (IND-CCA) in the random oracle model. Their scheme is a combination of a symmetric and an asymmetric encryption scheme using two hash functions where the symmetric and asymmetric encryption schemes are secure in a very weak sense. However, their proof of security works against only classical adversaries and it is not clear how one can fix their proof in the quantum setting. In the following, we mention the parts of the classical proof that may not work in the quantum setting.

(a) The classical proof uses the list of all queries made to the random oracles to simulate the decryption algorithm without possessing the secret key of the

© International Association for Cryptologic Research 2016
M. Hirt and A. Smith (Eds.): TCC 2016-B, Part II, LNCS 9986, pp. 192–216, 2016.
DOI: 10.1007/978-3-662-53644-5_8

asymmetric encryption scheme. In the quantum case, where the adversary has quantum access to the random oracles and submits queries in superpositions, such a list is not a well-defined concept.

(b) Also, the classical proof uses the fact that using a random value h^* instead of a given random oracle output $H(x)$ cannot be noticed by the adversary, provided that the adversary never queries x from the random oracle. In the quantum setting, the adversary may in a certain sense always query all values x by querying the random oracle on the superposition $\sum_x |x\rangle$ of all values. The situation gets especially difficult since the value x depends in turn on messages produced by the adversary.

(c) Finally, the classical proof uses the fact that for a randomized encryption scheme, it is hard to find values $x \neq x'$ such that encrypting a message m with randomness $H(x)$ or $H(x')$ leads to the same ciphertext. (Note: this does not follow directly from the collision resistance of the random oracle H.)

Consequently, the quantum security of the scheme is left as an open problem by Boneh et al. [6] and Zhandry [17].

We show how to circumvent those problems. Problem (c) is solved by using a recent result showing the collision resistance of random functions with outputs sampled from a non-uniform distribution [12]. Problem (b) is solved by the "one-way to hiding" lemmas from [13,14] which gives us a tool for handling the reprogramming of the random oracle. Problem (a) remains. In fact, we do not have a proof for the unmodified Fujisaki-Okamoto scheme. However, we show how to solve the problem by adding one more hash value $H'(\delta)$ to the ciphertext. Although in general, it may not be well-defined in the quantum setting what the list of queries to the random oracle is, we can show it to be well-defined in this case, using the fact that range and domain of H' have the same size. (A similar idea was used by [15] for the construction of quantum-secure non-interactive zero-knowledge proofs.)

Bellare and Rogaway [4] proposed another method, named OAEP, for converting a trapdoor permutation into an encryption scheme. It was believed that the OAEP-cryptosystem is provable secure in the random oracle model based on one-wayness of trapdoor permutation, but Shoup [11] showed it is an unjustified belief. Later, Fujisaki et al. [9] proved IND-CCA security of the OAEP-cryptosystem based on a stronger assumption, namely, partial-domain one-wayness of the underlying permutation. As pointed out by [6], the proof of OAEP security uses preimage awareness (i.e., that the preimage of a random oracle query is well-defined and known to the algorithm making it), a technique that does not seem to work in the quantum setting. This problem is the same as problem (a) above, we show that a similar approach works also in the case of OAEP.

Our Contribution. We modify the hybrid encryption scheme presented by Fujisaki and Okamoto using an extra hash function H'. We prove that our scheme is indistinguishable secure against chosen ciphertext attacks in the quantum random oracle model. For a message m, the encryption algorithm of our scheme, Enc_{pk}^{hy}, works as follows:

$$Enc_{pk}^{hy}(m; \delta) = \left(Enc_{pk}^{asy}\left(\delta; H\left(\delta \| Enc_{G(\delta)}^{sy}(m) \right) \right), \ Enc_{G(\delta)}^{sy}(m), \ H'(\delta) \right)$$

where pk and sk are the public key and the secret key of the asymmetric encryption scheme. Enc_{pk}^{asy} and Enc_{sk}^{sy} are the asymmetric and symmetric encryption algorithms respectively and δ is a random element from the message space of the asymmetric encryption scheme. H, G and H' are random oracles. The asymmetric encryption scheme is one-way secure, that is, the adversary can not decrypt the encryption of a random message. The symmetric encryption scheme is one-time secure, that is, the adversary can not distinguish between the encryptions of two messages when a fresh key is used for every encryption. In addition, the asymmetric encryption scheme is well-spread, i.e. any message can lead to at least $2^{\omega(\log n)}$ potential ciphertexts.

Note that our modification increases the ciphertext size by only a single hash value $H'(\delta)$ and is computationally inexpensive.

As already mentioned above, the added hash value $H'(\delta)$ solves problem (a) because given $H'(\delta)$, it is well-defined what δ is. This is because H' is chosen to have the same domain and range size, and hence is indistinguishable from a permutation [16]. However, in the formal proof, we do not directly use that fact, instead our proof goes along the following lines: We replace H' with a random polynomial to force the adversary to submit the input that has been used to obtain the ciphertext. This can be done due to a result by Zhandry [17] that shows a random oracle is indistinguishable from a $2q$-wise independent function where q is the number of queries that the adversary makes to the oracle function. In addition, we use the "one way to hiding" lemmas presented in [13,14]. As soon as H' is implemented as a polynomial, we can use the fact that roots of a polynomial can be found in polynomial-time; this allows us to efficiently get all candidates for δ given $H'(\delta)$.

Also, we modify OAEP-cryptosystem and prove its security in the quantum random oracle model based on the existence of a partial-domain one-way trapdoor injective function secure against quantum adversaries. This will remain theoretical until a candidate for a quantum secure partial-domain one-way trapdoor injective function is discovered. The proof follows similar lines as that of the Fujisaki-Okamoto transform.

A note on superposition queries. Following [6], we use the quantum random oracle model in which the adversary can make queries to the random oracle in superposition (that is, given a superposition of inputs, he can get a superposition of output values). This is necessary since a quantum adversary attacking a scheme based on a real hash function is necessarily able to evaluate that function in superposition. Hence the random oracle model must reflect that ability.

However, we do not model superposition queries to the encryption and decryption oracles. (As was done, for example, in [7].) We do strive to achieve security for the case where the encryption is used within a classical protocol (this is modeled by the fact that plaintexts and ciphertexts are classical, while the adversary is quantum), which is probably the most important use case for post-quantum secure encryption schemes.

In contrast, [7] considers security where an encryption scheme intended for classical plaintexts is used with a quantum superposition of plaintexts. And [1] considers the case where an encryption scheme intended for encrypting quantum data is used.

On the necessity of our modifications. We have slightly modified both the Fujisaki-Okamoto and the OAEP-cryptosystem by adding one additional hash to the ciphertexts. Although these additions are not very costly, it is a natural question whether they are necessary, especially in light of the question whether existing implementations are post-quantum secure. Although it is clear that our proof technique strongly relies on these additional hashes, this does not mean that the original schemes are insecure. However, we urge the reader not to assume that they are post-quantum secure just because they are classically secure. For example, in [2] it was shown that (at least relative to a specific oracle) the Fiat-Shamir transform is insecure in the quantum setting (using quantum random oracles). Their setting is similar to ours, so while there are no known attacks on Fujisaki-Okamoto or OAEP, we should not rely on their security until a security proof is found. We leave finding either an attack or a proof as a (highly non-trivial) open problem.

Organization. In Sect. 2, we present the required security definitions and other definitions, as well as various theorems related to random oracles that we import from the prior works. In Sect. 3, we define our variant of the Fujisaki-Okamoto transform and prove its security. In Sect. 5, we define our variant of OAEP and present its security proof.

2 Preliminaries

Let KSP and MSP stand for the key space and the message space respectively. The notation $x \xleftarrow{\$} X$ means that x is chosen uniformly at random from the set X. A symmetric encryption scheme and an asymmetric encryption scheme are defined as follows:

A symmetric encryption scheme Π consists of two polynomial time (in the security parameter n) algorithms, $\Pi = (Enc, Dec)$, such that:

1. Enc, the encryption algorithm, is a probabilistic algorithm which takes as input a key $k \in$ KSP and a message $m \in$ MSP and outputs a ciphertext $c \leftarrow Enc_k(m)$. The message space can be infinite and may depend on the security parameter.
2. Dec, the decryption algorithm, is a deterministic algorithm that takes as input a key k and a ciphertext c and returns message the $m := Dec_k(c)$. It is required that decryption algorithm returns the original message, i.e., $Dec_k(Enc_k(m)) = m$, for every $k \in$ KSP and every $m \in$ MSP.

An asymmetric encryption scheme Π consists of three polynomial time (in the security parameter n) algorithms, $\Pi = (Gen, Enc, Dec)$, such that:

1. Gen, the key generation algorithm, is a probabilistic algorithm which on input 1^n outputs a pair of keys, $(pk, sk) \leftarrow Gen(1^n)$, called the public key and the secret key for the encryption scheme, respectively.
2. Enc, the encryption algorithm, is a probabilistic algorithm which takes as input a public key pk and a message $m \in$ MSP and outputs a ciphertext $c \leftarrow Enc_{pk}(m)$. The message space, MSP, may depend on pk.
3. Dec, the decryption algorithm, is a deterministic algorithm that takes as input a secret key sk and a ciphertext c and returns message the $m := Dec_{sk}(c)$. It is required that the decryption algorithm returns the original message, i.e., $Dec_{sk}(Enc_{pk}(m)) = m$, for every $(pk, sk) \leftarrow Gen(1^n)$ and every $m \in$ MSP. The algorithm Dec returns $\bot$ if ciphertext c is not decryptable.

Let $y := Enc_{pk}(x; h)$ be the encryption of message x using the public key pk and the randomness $h \in$ COIN where COIN stands for the coin space of the encryption scheme. $Pr[P : G]$ is the probability that the predicate P holds true where free variables in P are assigned according to the program in G.

Definition 1 (γ-spread, Definition 5.2 [8]). *An asymmetric encryption scheme $\Pi = (Gen, Enc, Dec)$ is γ-spread if for every pk generated by $Gen(1^n)$ and every $x \in$ MSP,*

$$\max_{y \in \{0,1\}^*} Pr[y = Enc_{pk}(x; h) : h \xleftarrow{\$} COIN] \leq \frac{1}{2^\gamma}.$$

Particularly, we say that the encryption scheme Π is well-spread if $\gamma = \omega(\log(n))$.

Definition 2. *We say that a function $f : \{0,1\}^{n_1} \rightarrow \{0,1\}^{n_2}$ has min-entropy k if*

$$-\log \max_{y \in \{0,1\}^{n_2}} Pr[y = f(x) : x \xleftarrow{\$} \{0,1\}^{n_1}] = k.$$

2.1   Security Definitions

Let $\mathsf{negl}(n)$ be any non-negative function that is smaller than the inverse of any non-negative polynomial $p(n)$ for sufficiently large n. That is, $\lim_{n \to \infty} \mathsf{negl}(n)p(n) = 0$ for any polynomial $p(n)$. In the following, we present the security definitions that are needed in this paper. Note that the definitions are the same as the security definitions in [8], except they have been represented in the presence of a **quantum** adversary in this paper. As the following two security definitions will both be used in the security proof of our scheme, we differentiate between them by using $\mathsf{negl}(n)^{sy}$ and $\mathsf{negl}(n)^{asy}$ in the definitions.

Definition 3 (One-time secure). *A symmetric encryption scheme $\Pi = (Enc, Dec)$ is one-time secure if no **quantum** polynomial time adversary $\mathcal{A}$ can win in the $PrivK^{OT}_{\mathcal{A},\Pi}(n)$ game, except with probability at most $1/2 + negl(n)^{sy}$:*

$PrivK^{OT}_{\mathcal{A},\Pi}(n) game$:

Key Gen: *The challenger picks up a key k from KSP uniformly at random, i.e., $k \xleftarrow{\$} KSP$.*

Query: *The adversary $\mathcal{A}$ on input (1^n) chooses two messages m_0, m_1 of the same length and sends them to the challenger. The challenger chooses $b \xleftarrow{\$} \{0, 1\}$ and responds with $c^* \leftarrow Enc_k(m_b)$.*

Guess: *The adversary $\mathcal{A}$ produces a bit b', and wins if $b = b'$.*

Definition 4 (One-way secure). *An asymmetric encryption scheme $\Pi = (Gen, Enc, Dec)$ is one-way secure if no **quantum** polynomial time adversary $\mathcal{A}$ can win in the $PubK^{OW}_{\mathcal{A},\Pi}(n)$ game, except with probability at most $negl(n)^{asy}$:*

$PubK^{OW}_{\mathcal{A},\Pi}(n) game$:

Key Gen: *The challenger runs $Gen(1^n)$ to obtain a pair of keys (pk, sk).*

Challenge Query: *The challenger picks a uniformly random x from the message space, i.e., $x \xleftarrow{\$} MSP$, and encrypts it using the encryption algorithm Enc_{pk} to obtain the ciphertext $y \leftarrow Enc_{pk}(x)$, and sends y to the adversary $\mathcal{A}$.*

Guess: *The adversary $\mathcal{A}$ on input (pk, y) produces a bit string x', and wins if $x' = x$.*

In the next definition, we say that the quantum algorithm $\mathcal{A}$ has quantum access to the random oracle H if $\mathcal{A}$ can submit queries in superposition and the oracle H answers to these queries by applying a unitary transformation that maps $|x, y\rangle$ to $|x, y \oplus H(x)\rangle$.

Definition 5 (IND-CCA in the quantum random oracle model). *An asymmetric encryption scheme $\Pi^{asy} = (Gen, Enc, Dec)$ is IND-CCA secure if no **quantum** polynomial time adversary $\mathcal{A}$ can win in the $PubK^{CCA-QRO}_{\mathcal{A},\Pi}(n)$ game, except with probability at most $1/2 + negl(n)$:*

$PubK^{CCA-QRO}_{\mathcal{A},\Pi}(n)$ game:

Key Gen: *The challenger runs $Gen(1^n)$ to obtain a pair of keys (pk, sk) and chooses random oracles.*

Query: *The adversary $\mathcal{A}$ is given the public key pk and with **classical** oracle access to the decryption oracle and **quantum** access to the random oracles chooses two messages m_0, m_1 of the same length and sends them to the challenger. The challenger chooses $b \xleftarrow{\$} \{0, 1\}$ and responds with $c^* \leftarrow Enc_{pk}(m_b)$.*

Guess: *The adversary $\mathcal{A}$ continues to query the decryption oracle and the random oracles, but may not query the ciphertext c^* in a decryption query. Finally, the adversary $\mathcal{A}$ produces a bit b', and wins if $b = b'$.*

2.2 Quantum Accessible Random Oracles

In this section, we present some existing results about random oracles that we need to prove the security of our scheme.

Lemma 1 (One way to hiding (O2H)) [14]). *Let $H : \{0,1\}^n \rightarrow \{0,1\}^m$ be a random oracle. Consider an oracle algorithm A_1 that makes at most q_1 queries to H. Let C be an oracle algorithm that on input x does the following: pick $i \xleftarrow{\$} \{1,\dots,q_1\}$ and $y \xleftarrow{\$} \{0,1\}^m$, run $A_1^H(x,y)$ until (just before) the i-th query, measure the argument of the query in the computational basis, and output the measurement outcome. (When A_1 makes less than i queries, C outputs $\perp \notin \{0,1\}^n$.)*
Let

$$P_A^1 := \Pr[b' = 1 : H \xleftarrow{\$} (\{0,1\}^n \rightarrow \{0,1\}^m), x \xleftarrow{\$} \{0,1\}^n, b' \leftarrow A_1^H(x, H(x))]$$

$$P_A^2 := \Pr[b' = 1 : H \xleftarrow{\$} (\{0,1\}^n \rightarrow \{0,1\}^m), x \xleftarrow{\$} \{0,1\}^n, y \xleftarrow{\$} \{0,1\}^m,$$
$$b' \leftarrow A_1^H(x, y)]$$

$$P_C := \Pr[x' = x : H \xleftarrow{\$} (\{0,1\}^n \rightarrow \{0,1\}^m), x \xleftarrow{\$} \{0,1\}^n, x' \leftarrow C^H(x, i)]$$

Then

$$\left| P_A^1 - P_A^2 \right| \leq 2q_1 \sqrt{P_C}.$$

Lemma 2 (One way to hiding, adaptive (O2HA)) [13]). *Let $H : \{0,1\}^* \rightarrow \{0,1\}^n$ be a random oracle. Consider an oracle algorithm A_0 that makes at most q_0 queries to H. Consider an oracle algorithm A_1 that uses the final state of A_0 and makes at most q_1 queries to H. Let C be an oracle algorithm that on input (j, B, x) does the following: run $A_1^H(x, B)$ until (just before) the j-th query, measure the argument of the query in the computational basis, and output the measurement outcome. (When A_1 makes less than j queries, C outputs $\perp \notin \{0,1\}^\ell$.)*
Let

$$P_A^1 := \Pr[b' = 1 : H \xleftarrow{\$} (\{0,1\}^* \rightarrow \{0,1\}^n), m \leftarrow A_0^H(), x \xleftarrow{\$} \{0,1\}^\ell,$$
$$b' \leftarrow A_1^H(x, H(x\|m))]$$

$$P_A^2 := \Pr[b' = 1 : H \xleftarrow{\$} (\{0,1\}^* \rightarrow \{0,1\}^n), m \leftarrow A_0^H(), x \xleftarrow{\$} \{0,1\}^\ell,$$
$$B \xleftarrow{\$} \{0,1\}^n, b' \leftarrow A_1^H(x, B)]$$

$$P_C := \Pr[x = x' \wedge m = m' : H \xleftarrow{\$} (\{0,1\}^* \rightarrow \{0,1\}^n), m \leftarrow A_0^H(), x \xleftarrow{\$} \{0,1\}^\ell,$$
$$B \xleftarrow{\$} \{0,1\}^n, j \xleftarrow{\$} \{1, \cdots, q_1\}, x'\|m' \leftarrow C^H(j, B, x)]$$

Then

$$\left| P_A^1 - P_A^2 \right| \leq 2q_1 \sqrt{P_C} + q_0 2^{-\ell/2+2}.$$

Lemma 3 (Corollary 6 of [12]). *Let $f : \{0,1\}^{n_1} \to \{0,1\}^{n_2}$ be a function with min-entropy k. Let $H : \{0,1\}^* \to \{0,1\}^{n_1}$ be a random oracle. Then any quantum algorithm A making q queries to H returns a collision for $f \circ H$ with probability at most $O\left(\frac{q^{9/5}}{2^{k/5}}\right)$.*

3 The Hybrid Scheme and Its Security

In this section, we combine an asymmetric encryption scheme with a symmetric encryption scheme by using three hash functions in order to gain an IND-CCA secure public encryption scheme $\Pi^{hy} = (Gen^{hy}, Enc^{hy}, Dec^{hy})$ in the quantum random oracle model.

Let $\Pi^{asy} = (Gen^{asy}, Enc^{asy}, Dec^{asy})$ be an asymmetric encryption scheme with the message space $\mathrm{MSP}^{asy} = \{0,1\}^{n_1}$ and the coin space $\mathrm{COIN}^{asy} = \{0,1\}^{n_2}$. Let $\Pi^{sy} = (Enc^{sy}, Dec^{sy})$ be a symmetric encryption scheme where MSP^{sy} and $\mathrm{KSP}^{sy} = \{0,1\}^m$ are its message space and key space, respectively. The parameters n_1, n_2 and m depend on the security parameter n. We define three hash functions:

$$G : \mathrm{MSP}^{asy} \to \mathrm{KSP}^{sy}, \quad H : \{0,1\}^* \to \mathrm{COIN}^{asy} \quad \text{and} \quad H' : \mathrm{MSP}^{asy} \to \mathrm{MSP}^{asy}.$$

These hash functions will be modeled as random oracles in the following.

The hybrid scheme $\Pi^{hy} = (Gen^{hy}, Enc^{hy}, Dec^{hy})$ is constructed as follows, with MSP^{hy} as its message space:

1. Gen^{hy}, the key generation algorithm, on input 1^n runs Gen^{asy} to obtain a pair of keys (pk, sk).
2. Enc^{hy}, the encryption algorithm, on input pk and message $m \in \mathrm{MSP}^{hy} := \mathrm{MSP}^{sy}$ does the following:
   - Select $\delta \xleftarrow{\$} \mathrm{MSP}^{asy}$.
   - Compute $c \leftarrow Enc_a^{sy}(m)$, where $a := G(\delta)$.
   - Compute $e := Enc_{pk}^{asy}(\delta; h)$, where $h := H(\delta \| c)$.
   - Finally, output (e, c, d) as $Enc_{pk}^{hy}(m; \delta)$, where $d := H'(\delta)$.
3. Dec^{hy}, the decryption algorithm, on input sk and ciphertext (e, c, d) does the following:
   - Compute $\hat{\delta} := Dec_{sk}^{asy}(e)$.
   - If $\hat{\delta} = \perp$: abort and output $\perp$.
   - Otherwise set $\hat{h} := H(\hat{\delta} \| c)$.
   - If $e \neq Enc_{pk}^{asy}(\hat{\delta}; \hat{h})$: abort and output $\perp$.
   - Else if $d = H'(\hat{\delta})$:
     - Compute $\hat{a} := G(\hat{\delta})$ and output $Dec_{\hat{a}}^{sy}(c)$.
   - Else output $\perp$.

Note that our construction is the same as the Fujisaki-Okamoto construction, except that we use an extra random oracle H'. Consequently, the ciphertext has one more component, the encryption algorithm has an additional instruction to compute $H'(\delta)$ and the decryption algorithm has an additional check corresponding to H'.

Theorem 1. *The hybrid scheme Π^{hy} constructed above is IND-CCA secure in the quantum random oracle model if Π^{sy} is an one-time secure symmetric encryption scheme and Π^{asy} is a well-spread one-way secure asymmetric encryption scheme.*

Proof. Let A_{hy} be a quantum polynomial time adversary that attacks Π^{hy} in the sense of IND-CCA in the quantum random oracle model. Suppose that A_{hy} makes at most q_H, q_G and $q_{H'}$ quantum queries to the random oracles H, G and H', respectively, and q_{dec} classical decryption queries. Set $q_{hy} := q_H + q_G + q_{H'} + q_{dec} + 1$, i.e., the total number of queries that the adversary A_{hy} may make, including the challenge query. Let Ω_H, Ω_G, $\Omega_{H'}$ be the set of all function $H : \{0,1\}^* \to \{0,1\}^{n_2}$, $G : \{0,1\}^{n_1} \to \{0,1\}^m$ and $H' : \{0,1\}^{n_1} \to \{0,1\}^{n_1}$, respectively. The following game shows the chosen ciphertext attack by the adversary A_{hy} in the quantum setting where the adversary A_{hy} has quantum access to the random oracles H, G and H' and classical access to the decryption algorithm Dec^{hy}.

Game 0:

> **let** $H \xleftarrow{\$} \Omega_H, G \xleftarrow{\$} \Omega_G, H' \xleftarrow{\$} \Omega_{H'}, \delta^* \xleftarrow{\$} \mathsf{MSP}^{asy}, (pk, sk) \leftarrow Gen^{asy}(1^n)$
> **let** $m_0, m_1 \leftarrow A_{hy}^{H,G,H',Dec^{hy}}(pk)$
> **let** $b \xleftarrow{\$} \{0,1\}, c^* \leftarrow Enc_{G(\delta^*)}^{sy}(m_b), e^* \leftarrow Enc_{pk}^{asy}(\delta^*; H(\delta^*\|c^*)),$
> $d^* := H'(\delta^*)$
> **let** $b' \leftarrow A_{hy}^{H,G,H',Dec^{hy}}(e^*, c^*, d^*)$
> **return** $[b = b']$

In order to show that the success probability of Game 0 is at most $1/2 + \mathsf{negl}(n)$, we shall introduce a sequence of games and compute the difference between their success probabilities. For simplicity, we omit the definitions of random variables that appear with the same distribution and without any changes in all of the following games. These random variables are: $H \xleftarrow{\$} \Omega_H, G \xleftarrow{\$} \Omega_G$, $\delta^* \xleftarrow{\$} \mathsf{MSP}^{asy}, (pk, sk) \leftarrow Gen^{asy}(1^n)$, and $b \xleftarrow{\$} \{0,1\}$.

In the next game, we replace the decryption algorithm Dec^{hy} with Dec^* where Dec^* on (e, c, d) does the following:

1. If e^* is defined and $e = e^*$: abort and return $\bot$.
2. Else do:
   - Compute $\hat{\delta} := Dec_{sk}^{asy}(e)$.
   - If $\hat{\delta} = \bot$: query $H'(\delta^* \oplus 1)$,[1] abort and output $\bot$.
   - Otherwise set $\hat{h} := H(\hat{\delta}\|c)$.
   - If $e \neq Enc_{pk}^{asy}(\hat{\delta}; \hat{h})$: query $H'(\delta^* \oplus 1)$, (see Footnote 1) abort and output $\bot$.

---

[1] This extra query is needed later to prove that Game 4 and Game 5 are identical.

- Else if $d = H'(\hat{\delta})$: compute $\hat{a} := G(\hat{\delta})$ and output $Dec_{\hat{a}}^{sy}(c)$.
- Else: output $\perp$.

Therefore, Game 1 is as follows:

Game 1:

---

let $H' \stackrel{\$}{\leftarrow} \Omega_{H'}$

let $m_0, m_1 \leftarrow A_{hy}^{H,G,H',Dec^*}(pk)$

let $c^* \leftarrow Enc_{G(\delta^*)}^{sy}(m_b)$, $e^* \leftarrow Enc_{pk}^{asy}(\delta^*; H(\delta^*\|c^*))$

let $b' \leftarrow A_{hy}^{H,G,H',Dec^*}(e^*, c^*, H'(\delta^*))$

return $[b = b']$

We prove that the probabilities of success in Game 0 and Game 1 have negligible difference. We can conclude the result by the fact that the asymmetric encryption scheme is well-spread. We present the proof of the following lemma in Sect. 4.

Lemma 4. *If the asymmetric encryption scheme Π^{asy} is well-spread, then*

$$\left| \Pr[1 \leftarrow Game\ 0] - \Pr[1 \leftarrow Game\ 1] \right| \leq O\left(\frac{(q_H + q_{dec} + 1)^{9/5}}{2^{\omega(\log(n))/5}} \right) =: \ell(n).$$

It is clear that $\ell(n)$ is a negligible function and as a result Game 0 and Game 1 have negligible difference.

We replace $G(\delta^*)$ and $H'(\delta^*)$ with random elements in the next game.

Game 2:

---

let $H' \stackrel{\$}{\leftarrow} \Omega_{H'}$, $a^* \stackrel{\$}{\leftarrow} \text{KSP}^{sy}$, $d^* \stackrel{\$}{\leftarrow} \text{MSP}^{asy}$

let $m_0, m_1 \leftarrow A_{hy}^{H,G,H',Dec^*}(pk)$

let $c^* \leftarrow Enc_{a^*}^{sy}(m_b)$, $e^* \leftarrow Enc_{pk}^{asy}(\delta^*; H(\delta^*\|c^*))$

let $b' \leftarrow A_{hy}^{H,G,H',Dec^*}(e^*, c^*, d^*)$

return $[b = b']$

Now, we can prove that $\Pr[1 \leftarrow Game\ 2] = 1/2 + \mathsf{negl}(\mathsf{n})^{sy}$. This follows from the one-time security assumption of the symmetric encryption scheme. We postpone the detailed proof of the following lemma to Sect. 4 in favor of having a simple proof.

Lemma 5. *If the symmetric encryption scheme Π^{sy} is one-time secure, then* $\Pr[1 \leftarrow Game\ 2] = 1/2 + \mathbf{negl(n)}^{sy}$.

By using Lemma 5, we only need to show that the difference between the success probabilities of Game 1 and Game 2 is negligible.

Note that if we were in the classical random oracle setting, we could define the **bad** event to be querying G or H' on input δ^* and argue that the two games are indistinguishable until the bad event happens. However, there is no well-defined concept for the bad event when the adversary A can query G and H' in superposition and each quantum query can contain δ^* in some sense. Therefore, we use the O2H Lemma 1 to obtain an upper bound for $\left| \Pr[1 \leftarrow Game\ 1] - \Pr[1 \leftarrow Game\ 2] \right|$.

Let $A^{G \times H'}$ be an adversary that has quantum access to random oracle $G \times H' \Big(where\ (G \times H')(\delta) := \big(G(\delta), H'(\delta) \big) \Big)$. The adversary $A^{G \times H'}$ on input $\big(\delta^*, (a^*, d^*) \big)$ does the following:

The adversary $A^{G \times H'}\big(\delta^*, (a^*, d^*) \big)$:

---

let $H \xleftarrow{\$} \Omega_H, (pk, sk) \leftarrow Gen^{asy}(1^n), b \xleftarrow{\$} \{0, 1\}$
let $m_0, m_1 \leftarrow A_{hy}^{H, G, H', Dec^*}(pk)$
let $c^* \leftarrow Enc_{a^*}^{sy}(m_b),\ e^* \leftarrow Enc_{pk}^{asy}(\delta^*; H(\delta^* \| c^*))$
let $b' \leftarrow A_{hy}^{H, G, H', Dec^*}(e^*, c^*, d^*)$
return $[b = b']$

Note that the adversary $A^{G \times H'}$ makes at most $q_{o2h} := q_G + q_{H'} + 2q_{dec}$ queries to the random oracle $G \times H'$ in order to respond to the A_{hy}-queries.[2]

Let C be an oracle algorithm that on input δ^* does the following: pick $i \xleftarrow{\$} \{1, \ldots, q_{o2h}\}$ and $(a^*, d^*) \xleftarrow{\$} \mathsf{KSP}^{sy} \times \mathsf{MSP}^{asy}$, run $A^{G \times H'}\big(\delta^*, (a^*, d^*) \big)$ until (just before) the i-th query, measure the argument of the $G \times H'$-query in the computational basis, output the measurement outcome (when $A^{G \times H'}$ makes less than i queries, C outputs $\perp \notin \{0, 1\}^{n_1}$). Note that with this definition we have $P_A^1 = \Pr[1 \leftarrow Game\ 1]$ and $P_A^2 = \Pr[1 \leftarrow Game\ 2]$ where P_A^1 and P_A^2 are defined in O2H Lemma 1 for the adversary $A^{G \times H'}$. Therefore, we will define Game 3 such that $P_C = \Pr[1 \leftarrow Game\ 3]$ where P_C is defined in O2H Lemma 1 for the adversary $C^{G \times H'}$. Thus by O2H Lemma 1:

$$\left| \Pr[1 \leftarrow Game\ 1] - \Pr[1 \leftarrow Game\ 2] \right| \leq 2q_{o2h} \sqrt{\Pr[1 \leftarrow Game\ 3]}.$$

---

[2] For example, to respond to a query to the random oracle G with input register I and output register O, the adversary $A^{G \times H'}$ prepares an additional register T (for the output of H') in state $|+\rangle^{n_1}$ and invokes $U_{G \times H'}$ on I, O, T. It is easy to verify that this leaves T unchanged and applies U_G to I, O. (This idea was already used in [18] to ignore part of the output of an oracle.)

We define Game 3 as follows:

Game 3:

> let $H' \xleftarrow{\$} \Omega_{H'}$, $a^* \xleftarrow{\$} \mathsf{KSP}^{sy}$, $d^* \xleftarrow{\$} \mathsf{MSP}^{asy}$, $i \xleftarrow{\$} \{1, \ldots, q_{o2h}\}$
> **run until** *i-th query to oracle* $G \times H'$
> > **let** $m_0, m_1 \leftarrow A_{hy}^{H,G,H',Dec^*}(pk)$
> > **let** $c^* \leftarrow Enc_{a^*}^{sy}(m_b)$, $e^* \leftarrow Enc_{pk}^{asy}(\delta^*; H(\delta^*\|c^*))$
> > **let** $b' \leftarrow A_{hy}^{H,G,H',Dec^*}(e^*, c^*, d^*)$
>
> **measure** the argument $\tilde{\delta}$ of the *i*-th query to oracle $G \times H'$
> **return** $[\tilde{\delta} = \delta^*]$

In the next game, we replace the random oracle H' with a $2(q_{H'} + q_{dec})$-wise independent function. Random polynomials of degree $2(q_{H'} + q_{dec}) - 1$ over finite field $GF(2^{n_1})$ are $2(q_{H'} + q_{dec})$-wise independent. Let Ω_{wise} be the set of all such polynomials.

Game 4:

> let $H' \xleftarrow{\$} \Omega_{wise}$, $a^* \xleftarrow{\$} \mathsf{KSP}^{sy}$, $d^* \xleftarrow{\$} \mathsf{MSP}^{asy}$, $i \xleftarrow{\$} \{1, \ldots, q_{o2h}\}$
> **run until** *i-th query to oracle* $G \times H'$
> > **let** $m_0, m_1 \leftarrow A_{hy}^{H,G,H',Dec^*}(pk)$
> > **let** $c^* \leftarrow Enc_{a^*}^{sy}(m_b)$, $e^* \leftarrow Enc_{pk}^{asy}(\delta^*; H(\delta^*\|c^*))$
> > **let** $b' \leftarrow A_{hy}^{H,G,H',Dec^*}(e^*, c^*, d^*)$
>
> **measure** the argument $\tilde{\delta}$ of the *i*-th query to oracle $G \times H'$
> **return** $[\tilde{\delta} = \delta^*]$

Due to a result by Zhandry [17], a $2(q_{H'} + q_{dec})$-wise independent function H' is perfectly indistinguishable from a random function when the adversary makes at most $q_{H'} + q_{dec}$ queries to H'. Therefore, Game 3 and Game 4 are identical.

We replace the decryption algorithm Dec^* with a new decryption algorithm Dec^{**} in Game 5. Dec^{**} has access to the description (as a polynomial) of H'. Dec^{**} on input (e, c, d) works as follows:

1. If e^* is defined and $e = e^*$: output $\bot$.
2. Else do:
   - Calculate all roots of the polynomial $H' - d$. Let S be the set of those roots.
   - If there exists $\hat{\delta} \in S \setminus \{\delta^*\}$ such that $e = Enc_{pk}^{asy}(\hat{\delta}; H(\hat{\delta}\|c))$:
     - query H' on input $\hat{\delta}$.
     - compute $\hat{a} := G(\hat{\delta})$ and return $Dec_{\hat{a}}^{sy}(c)$.

– Else if $e = Enc_{pk}^{asy}(\delta^*; H(\delta^* \| c))$:

   • If $H'(\delta^*) = d$, then compute $\hat{a} := G(\delta^*)$ and return $Dec_{\hat{a}}^{sy}(c)$.

   • Else: return $\perp$.

– Else: query H' on random input $\delta \xleftarrow{\$} (\mathrm{MSP}^{asy} \setminus \{\delta^*\})$, and output $\perp$.

Note that Dec^{**} depends on the randomness used in choosing H'. This is formally unproblematic (it is comparable to Dec^{**} implicitly depending on secret key) and appears only in intermediate game within the proof. We emphasis that finding roots of polynomial $H' - d$ is possible in polynomial time [5] and it does not involve query to the polynomial H'. (We need that Dec^{**} as well as all other parts of our games run in polynomial time because we want to use the one-way security of the asymmetric encryption scheme in Lemma 6 below.)

Game 5:

> let $H' \xleftarrow{\$} \Omega_{wise}$, $a^* \xleftarrow{\$} \mathrm{KSP}^{sy}$, $d^* \xleftarrow{\$} \mathrm{MSP}^{asy}$, $i \xleftarrow{\$} \{1, \ldots, q_{o2h}\}$
> **run until** i-th query to *oracle* $G \times H'$
>> let $m_0, m_1 \leftarrow A_{hy}^{H,G,H',\boxed{Dec^{**}}}(pk)$
>> let $c^* \leftarrow Enc_{a^*}^{sy}(m_b)$, $e^* \leftarrow Enc_{pk}^{asy}(\delta^*; H(\delta^* \| c^*))$
>> let $b' \leftarrow A_{hy}^{H,G,H',\boxed{Dec^{**}}}(e^*, c^*, d^*)$
>
> **measure** the argument $\tilde{\delta}$ of the i-th query to oracle $G \times H'$
> **return** $[\tilde{\delta} = \delta^*]$

In order to show that Game 4 and Game 5 are identical, we need to prove that the two decryption algorithms Dec^* and Dec^{**} return the same output. Also, note that Game 4 and Game 5 succeed if they measure a query containing the argument δ^*. Therefore, we have to prove that the total number of queries submitted to the random oracles G and H' are equal in two decryption algorithms and the number of queries with argument δ^* are equal and appear at the same time.

Suppose the adversary submits a decryption query (e, c, d). Let $\hat{\delta} := Dec_{sk}^{asy}(e)$. We consider the following cases:

1. If $\hat{\delta} = \perp$: In this case, both decryption algorithms return $\perp$ and query the random oracle H', but not on input δ^*.
2. If $\hat{\delta} \neq \perp$, $\hat{\delta} \neq \delta^*$ and $H'(\hat{\delta}) \neq d$: Note that $\hat{\delta} \neq \delta^*$ implies that $e \neq e^*$ and $e \neq Enc_{pk}^{asy}(\delta^*; H(\delta^* \| c))$. Therefore, there are two subcases:

  (a) If $e \neq Enc_{pk}^{asy}(\hat{\delta}; H(\hat{\delta} \| c))$, then the decryption algorithm Dec^* queries the random oracle H' on input $\delta^* \oplus 1$ and the decryption algorithm Dec^{**} queries H' on a random element from $\mathrm{MSP}^{asy} \setminus \{\delta^*\}$ since $\hat{\delta} \notin S$. Both algorithms return $\perp$.

  (b) Else, the decryption algorithm Dec^* queries random oracle H' on input $\hat{\delta}$ and the decryption algorithm Dec^{**} queries H' on a random element from $\mathrm{MSP}^{asy} \setminus \{\delta^*\}$ since $\hat{\delta} \notin S$. Both algorithms return $\perp$.

3. If $\hat{\delta} \neq \perp, \hat{\delta} \neq \delta^*$ and $H'(\hat{\delta}) = d$: Note that $\hat{\delta} \neq \delta^*$ implies that $e \neq e^*$ and $e \neq Enc_{pk}^{asy}(\delta^*; H(\delta^* \| c))$. Therefore, there are two subcases:
   (a) If $e \neq Enc_{pk}^{asy}(\hat{\delta}; H(\hat{\delta} \| c))$, then the decryption algorithm Dec^* queries the random oracle H' on input $\delta^* \oplus 1$ and outputs $\perp$, and the decryption algorithm Dec^{**} queries H' on a random element from $\mathrm{MSP}^{asy} \setminus \{\delta^*\}$ and outputs $\perp$.
   (b) Else, both decryption algorithms query random oracles G and H' on input $\hat{\delta}$ and output $Dec_{G(\hat{\delta})}^{sy}$.

4. If $\hat{\delta} = \delta^*$ and $H'(\hat{\delta}) \neq d$: There are three subcases:
   (a) If e^* is defined and $e = e^*$: Then both decryption algorithms return $\perp$ without any query to the random oracles G and H'.
   (b) Else if $e \neq Enc_{pk}^{asy}(\delta^*; H(\delta^* \| c))$: Then the decryption algorithm Dec^* queries the random oracle H' on input $\delta^* \oplus 1$ and the decryption algorithm Dec^{**} queries H' on a random element from $\mathrm{MSP}^{asy} \setminus \{\delta^*\}$. Both decryption algorithms return $\perp$.
   (c) Else, both decryption algorithms query H' on input δ^* and output $\perp$.

5. If $\hat{\delta} = \delta^*$ and $H'(\hat{\delta}) = d$: There are three subcases:
   (a) If e^* is defined and $e = e^*$: Then both decryption algorithms return $\perp$ without any query to the random oracles G and H'.
   (b) Else if $e \neq Enc_{pk}^{asy}(\delta^*; H(\delta^* \| c))$: Then the decryption algorithm Dec^* queries the random oracle H' on input $\delta^* \oplus 1$ and decryption algorithm Dec^{**} queries H' on a random element from $\mathrm{MSP}^{asy} \setminus \{\delta^*\}$. Both decryption algorithms return $\perp$.
   (c) Else, both decryption algorithms query random oracles G and H' on input δ^* and output $Dec_{G(\delta^*)}^{sy}$.

Hence, $\Pr[1 \leftarrow Game\ 4] = \Pr[1 \leftarrow Game\ 5]$.

Note that Dec^{**} does not use the secret key of the asymmetric encryption scheme to decrypt the ciphertext. This will allow us below to make use of the one-way security of Π^{asy} (This is only possible if the secret key is never used).

The next step is to replace the random coins $H(\delta^* \| c^*)$ of the asymmetric encryption scheme by truly random coins from COIN^{asy}.

Game 6:

> let $H' \xleftarrow{\$} \Omega_{wise}$ $H' \xleftarrow{\$} \Omega_{H'}$, $a^* \xleftarrow{\$} \mathrm{KSP}^{sy}$, $d^* \xleftarrow{\$} \mathrm{MSP}^{asy}$, $i \xleftarrow{\$} \{1, \ldots, q_{o2h}\}$
> **run until** i-th query to oracle $G \times H'$
> > let $m_0, m_1 \leftarrow A_{hy}^{H,G,H',Dec^{**}}(pk)$
> > let $c^* \leftarrow Enc_{a^*}^{sy}(m_b), e^* \leftarrow Enc_{pk}^{asy}(\delta^* \quad)$
> > let $b' \leftarrow A_{hy}^{H,G,H',Dec^{**}}(e^*, c^*, d^*)$
> **measure** the argument $\tilde{\delta}$ of the i-th query to oracle $G \times H'$
> **return** $[\tilde{\delta} = \delta^*]$

Suppose that adversary A_{hy} makes $q_{0GH'}$ queries to the random oracle $G \times H'$ before the challenge query and $q_{1GH'}$ queries after the challenge query. In order to obtain an upper bound for $\left| \Pr[1 \leftarrow Game\ 5] - \Pr[1 \leftarrow Game\ 6] \right|$, we use O2HA Lemma 2. Let A_0^H be a quantum adversary that has oracle access to the random oracle H. The adversary A_0^H does the following:

The adversary A_0^H:

> **let** $G \xleftarrow{\$} \Omega_G$, $H' \xleftarrow{\$} \Omega_{wise}$, $(pk, sk) \leftarrow Gen^{asy}(1^n)$, $b \xleftarrow{\$} \{0,1\}$,
> $\quad a^* \xleftarrow{\$} KSP^{sy}$, $d^* \xleftarrow{\$} MSP^{asy}$, $i \xleftarrow{\$} \{1, \ldots, q_{o2h}\}$
> **run until** i-*th query to oracle* $G \times H'$
> $\quad$ **let** $m_0, m_1 \leftarrow A_{hy}^{H,G,H',Dec^{**}}(pk)$
> $\quad$ **let** $c^* \leftarrow Enc_{a^*}^{sy}(m_b)$
> **return** c^*

Let A_1^H be an adversary that has quantum access to the random oracle H and can use the final state of A_0^H. Therefore, he can access all the random variables that are chosen by A_0^H and also he can use the output of A_0^H. The adversary A_1^H on input (δ^*, h^*) does the following:

The adversary $A_1^H(\delta^*, h^*)$:

> **let** $\delta^* \xleftarrow{\$} MSP^{asy}$
> **if** $i > q_{0GH'}$ **then**
> $\quad$ **run until** $(i - q_{0GH'})$-*th query to oracle* $G \times H'$
> $\quad\quad$ **let** $e^* \leftarrow Enc_{pk}^{asy}(\delta^*; h^*)$
> $\quad\quad$ **let** $b' \leftarrow A_{hy}^{H,G,H',Dec^{**}}(e^*, c^*, d^*)$
> **measure** the argument $\tilde{\delta}$ of the i-th query to oracle $G \times H'$
> **return** $[\tilde{\delta} = \delta^*]$

Note that the adversary A_0^H may be stopped before receiving the challenge query (or when $i \leq q_{0GH'}$), in this case the adversary A_1^H measures the argument $\tilde{\delta}$ of i-th query to the random oracle $G \times H'$ and outputs $[\tilde{\delta} = \delta^*]$. If $i > q_{0GH'}$, then the adversary A_1^H continues to run the adversary A_{hy} till the $(i - q_{0GH'})$-th query to the random oracle $G \times H'$ and measures the argument $\tilde{\delta}$ of i-th query to the random oracle $G \times H'$ and outputs $[\tilde{\delta} = \delta^*]$. Note that with these definitions we have $P_A^1 = \Pr[1 \leftarrow Game\ 5]$ and $P_A^2 = \Pr[1 \leftarrow Game\ 6]$ where P_A^1 and P_A^2 are as in the O2HA Lemma 2 for the random oracle H.

A_0^H makes q_0 queries to the random oracle H, and A_1^H makes at most q_1 queries to the random oracle H. Let C be an oracle algorithm that on input δ^* does the following: pick $j \xleftarrow{\$} \{1, \ldots, q_1\}$ and $h^* \xleftarrow{\$} \{0,1\}^{n_2}$, run $A_1^H(\delta^*, h^*)$ until (just before) the j-th query to the random oracle H, measure the argument of that query in the computational basis, output the measurement outcome (when

A_1^H makes less than j queries, C outputs $\perp \notin \{0,1\}^n$). Now, we can introduce Game 7 such that by O2HA Lemma 2,

$$\left| \Pr[1 \leftarrow Game\ 5] - \Pr[1 \leftarrow Game\ 6] \right| \leq 2q_1\sqrt{Pr[1 \leftarrow Game\ 7]} + q_0 2^{-n_1/2+2}.$$

Game 7:

$$
\boxed{
\begin{aligned}
&\textbf{let } H' \xleftarrow{\$} \Omega_{wise},\ a^* \xleftarrow{\$} \mathsf{KSP}^{sy},\ d^* \xleftarrow{\$} \mathsf{MSP}^{asy},\ i \xleftarrow{\$} \{1,\ldots,q_{o2h}\} \\
&\textbf{run until } i\text{-th query to oracle } G \times H' \\
&\quad \Big|\ \textbf{let } m_0, m_1 \leftarrow A_{hy}^{H,G,H',Dec^{**}}(pk) \\
&\quad \Big|\ \textbf{let } c^* \leftarrow Enc_{a^*}^{sy}(m_b) \\
&\textbf{let } \delta^* \xleftarrow{\$} \mathsf{MSP}^{asy},\ j \xleftarrow{\$} \{1,\ldots,q_1\} \\
&\textbf{run until } j\text{-th query to oracle } H \\
&\quad \Big|\ \textbf{if } i > q_{0GH'} \textbf{ then} \\
&\quad \Big|\quad \Big|\ \textbf{run until } (i - q_{0GH'})\text{-th query to oracle } G \times H' \\
&\quad \Big|\quad \Big|\quad \Big|\ \textbf{let } e^* \leftarrow Enc_{pk}^{asy}(\delta^*; h^*) \\
&\quad \Big|\quad \Big|\quad \Big|\ \textbf{let } b' \leftarrow A_{hy}^{H,G,H',Dec^{**}}(e^*, c^*, d^*) \\
&\quad \Big|\ \textbf{measure the argument } \tilde{\delta} \text{ of the } i\text{-th query to oracle } G \times H' \\
&\textbf{measure the argument } \hat{\delta}\|\hat{c} \text{ of the } j\text{-th query to oracle } H \\
&\textbf{return } [\hat{\delta} = \delta^*] \wedge [\hat{c} = c^*]
\end{aligned}
}
$$

The next lemma shows that the success probabilities in Game 6 and Game 7 are negligible. We present the proof of the lemma in Sect. 4.

Lemma 6. *If the asymmetric scheme Π^{asy} is one-way secure then*

$$Pr[1 \leftarrow Game\ 6] \leq negl(n)^{asy} \ and \ Pr[1 \leftarrow Game\ 7] \leq negl(n)^{asy}.$$

Combining this with the bounds derived above we can conclude that

$$
Pr[1 \leftarrow Game\ 0] \leq \frac{1}{2} + negl(n)^{sy} + O\left(\frac{(q_H + q_{dec} + 1)^{9/5}}{2^{\omega(\log(n))/5}} \right)
$$
$$
+ 2q_{o2h}\sqrt{negl(n)^{asy}} + 2q_1\sqrt{negl(n)^{asy}} + q_0 2^{-n_1/2+2}. \qquad \square
$$

4 Deferred Proofs

4.1 Proof of Lemma 4

Proof. We list all the possibilities that the adversary can do to differentiate between the two games. Suppose that the adversary sends the ciphertext (e, c, d). Note that if $e \neq e^*$ or e^* is not defined, then two decryption algorithms Dec^{hy}

and Dec^* return the same output and nothing is left to show. Therefore we analyze the following cases where e^* is defined and $e = e^*$.

1. $(e = e^*, c = c^*, d \neq d^*)$ or $(e = e^*, c \neq c^*, d \neq d^*)$: In these two cases, the two decryption algorithms return $\bot$.
2. $(e = e^*, c \neq c^*, d = d^*)$: This means that $Enc_{pk}^{asy}(\delta^*; H(\delta^* \| c)) = Enc_{pk}^{asy}(\delta^*; H(\delta^* \| c^*))$. This is a collision in the sense of Lemma 3 since δ^* is chosen randomly and the $Enc_{pk}^{asy}(\delta^*; H(\delta^* \| \cdot))$ has min-entropy $\omega(\log(n))$.
   Therefore, it occurs with probability at most $O\left(\frac{(q_H + q_{dec} + 1)^{9/5}}{2^{\omega(\log(n))/5}}\right)$.
3. $(e = e^*, c = c^*, d = d^*)$. This query never occurs.

We can conclude that:

$$\left| \Pr[1 \leftarrow Game\ 0] - \Pr[1 \leftarrow Game\ 1] \right| \leq O\left(\frac{(q_H + q_{dec} + 1)^{9/5}}{2^{\omega(\log(n))/5}}\right).$$

$\square$

4.2   Proof of Lemma 5

Proof. Let $\varepsilon(n) := \Pr[1 \leftarrow Game\ 2]$. We construct the adversary A^{sy} such that:

$$\Pr[PriK_{A^{sy}, \Pi^{sy}}^{OT} = 1] = \varepsilon(n).$$

The adversary A^{sy} on input 1^n does the following:

1. Run $Gen^{asy}(1^n)$ to obtain (pk, sk).
2. Run the adversary $A_{hy}(pk)$.
3. Use a $2(q_H + q_{dec} + 1)$-wise independent function, a $2(q_G + q_{dec})$-wise independent function, and a $2(q_{H'} + q_{dec})$-wise independent function to answer the queries submitted to the random oracles H, G and H', respectively.
4. Whenever A_{hy} outputs challenge messages (m_0, m_1), do the following:
   - Select $b \xleftarrow{\$} \{0, 1\}$, $r \xleftarrow{\$} COIN^{sy}$, $\delta^* \xleftarrow{\$} MSP^{asy}$, $a^* \leftarrow KSP^{sy}$, $d^* \xleftarrow{\$} \{0, 1\}^{n_1}$.
   - Set $c^* := Enc_{a^*}^{sy}(m_b; r)$ and $e^* := Enc_{pk}^{asy}(\delta^*; H(\delta^*, c^*))$.
   - Send (e^*, c^*, d^*) to the adversary A_{hy}.
5. Answer the random oracle queries and decryption queries as before.
6. When A_{hy} returns bit b', output the same bit b'.

It is obvious that $\Pr[PriK_{A^{sy}, \Pi^{sy}}^{OT} = 1] = \varepsilon(n)$. Therefore, $\varepsilon(n) \leq 1/2 + \mathsf{negl}(n)^{sy}$.

$\square$

4.3   Proof of Lemma 6

As the proof for two games is similar we provide the instances for Game 7 in brackets $[\![\ldots]\!]$ wherever there is a difference.

Proof. Let $\varepsilon(n) := \Pr[1 \leftarrow Game\ 6]$ $[\![:= \Pr[1 \leftarrow Game\ 7]]\!]$. We construct an adversary A^{asy} such that:

$$\Pr[PubK^{OW}_{A^{asy}, \Pi^{asy}} = 1] = \varepsilon(n).$$

The adversary A^{asy} on input $(1^n, pk, y)$ does the following:

1. Run the adversary $A_{hy}(pk)$.
2. Use a $2(q_H + q_{dec})$-wise independent function, a $2(q_G + q_{dec})$-wise independent function, and a polynomial of degree $2(q_{H'} + q_{dec}) - 1$ to answer the queries submitted to random oracles H, G and H', respectively.
3. Answer the decryption queries using Dec^{**}.
4. Whenever A_{hy} outputs challenge messages (m_0, m_1), do the following:
   - Select $b \stackrel{\$}{\leftarrow} \{0,1\}$, $r \stackrel{\$}{\leftarrow} \mathtt{COIN}^{sy}$, $a^* \leftarrow \mathtt{KSP}^{sy}$, $d^* \stackrel{\$}{\leftarrow} \{0,1\}^{n_1}$.
   - Set $c^* := Enc^{sy}_{a^*}(m_b; r)$ and $e^* := y$.
   - Send (e^*, c^*, d^*) to the adversary A_{hy}.
5. Answer the random oracle queries as before and to the decryption queries using Dec^{**}.
6. When A_{hy} returns bit b' and halts, A^{asy} selects $i \stackrel{\$}{\leftarrow} \{1, \cdots, q_{o2h}\}$ $[\![i \stackrel{\$}{\leftarrow} \{1, \cdots, q_1\}]\!]$ and measures the argument $\hat{\delta}$ of i-th $[\![(i + q_0)$-th $]\!]$ query to the random oracle $G \times H'$ $[\![H]\!]$ and outputs $\hat{\delta}$ (When A_{hy} makes less than i queries output $\perp$).

It is obvious that $\Pr[PubK^{OW}_{A^{asy}, \Pi^{asy}} = 1] = \varepsilon(n)$. Therefore, $\varepsilon(n) \leq \mathsf{negl}(n)^{asy}$. $\square$

5   A Variant of OAEP

The following definitions are similar to the definitions presented in [9], except we define them in the presence of a **quantum** adversary.

Definition 6 (Quantum partial-domain one-way function). *We say a function $f : \{0,1\}^{n+k_1} \times \{0,1\}^{k_0} \rightarrow \{0,1\}^m$ is partial-domain one-way if for any polynomial time quantum adversary A,*

$$\Pr[\tilde{s} = s : s \stackrel{\$}{\leftarrow} \{0,1\}^{n+k_1},\ t \stackrel{\$}{\leftarrow} \{0,1\}^{k_0},\ \tilde{s} \leftarrow A(f(s,t))] \leq negl(n).$$

Definition 7. *Let $G : \{0,1\}^{k_0} \rightarrow \{0,1\}^{k-k_0}$, $H : \{0,1\}^{k-k_0} \rightarrow \{0,1\}^{k_0}$ and $H' : \{0,1\}^k \rightarrow \{0,1\}^k$ be random oracles. The Q-OAEP $= (Gen, Enc, Dec)$ encryption scheme is defined as:*

1. **Gen:** *Specifies an instance of the injective function f and its inverse f^{-1}. Therefore, the public key and secret key are f and f^{-1} respectively.*
2. **Enc:** *Given a message $m \in \{0,1\}^n$, the encryption algorithm computes*

$$s := m\|0^{k_1} \oplus G(r) \quad and \quad t := r \oplus H(s),$$

where $r \stackrel{\$}{\leftarrow} \{0,1\}^{k_0}$, and outputs the ciphertext $(c,d) := \left(f(s,t), H'(s\|t) \right)$.

3. **Dec:** *Given a ciphertext* (c, d), *the decryption algorithm does the following:*
   - *When* $c \notin \operatorname{Im} f$:
     (a) *If* c^* *is defined (where* c^* *is the challenge ciphertext), then query the random oracle* H' *on input* $(s^* \| t^*) \oplus 1$ *(where* $f(s^*, t^*) = c^*$*) and return* $\perp$.
     (b) *If* c^* *is not defined, then query the random oracle* H' *on a random input and return* $\perp$.
   - *When* $c \in \operatorname{Im} f$, *the decryption algorithm extracts* $(s, t) = f^{-1}(c)$. *If* $H'(s \| t) \neq d$ *it returns* $\perp$, *otherwise it does the following:*
     (a) *query the random oracle* H *on input* s *and compute* $r := t \oplus H(s)$.
     (b) *query the random oracle* G *on input* r *and compute* $M := s \oplus G(r)$.
     (c) *if the* k_1 *least significant bits of* M *are zero then return the* n *most significant bits of* M, *otherwise return* $\perp$.

Note that k_0 *and* k *depend on the security parameter* n.

Note that *Dec* contains several unnecessary oracle calls (after it already decided to output $\perp$). These obviously do not effect correctness or security, but make the proof a bit simple to formulate.

Theorem 2. *If the underlying injective function is quantum partial-domain one-way, then the Q-OAEP scheme is IND-CCA secure in the quantum random oracle model.*

Proof. Since the proof is similar and relatively easier compared to the proof of Fujisaki-Okamoto transform, we only present the main games in pseudocode and the intuition of the their negligibility. Let Ω_H, Ω_G, $\Omega_{H'}$ be the set of all function $H : \{0,1\}^{k-k_0} \to \{0,1\}^{k_0}$, $G : \{0,1\}^{k_0} \to \{0,1\}^{k-k_0}$ and $H' : \{0,1\}^k \to \{0,1\}^k$, respectively. Let A be a polynomial time quantum adversary that attacks the OAEP-cryptosystem in the sense of IND-CCA in the quantum random oracle model and makes at most q_H, q_G and $q_{H'}$ queries to the random oracles H, G and H' respectively and q_{dec} decryption queries.

Game 0:

---
let $H \xleftarrow{\$} \Omega_H, G \xleftarrow{\$} \Omega_G, H' \xleftarrow{\$} \Omega_{H'}, r \xleftarrow{\$} \{0,1\}^{k_0}, (pk, sk) \leftarrow Gen(1^n)$
let $m_0, m_1 \leftarrow A^{H,G,H',Dec}(pk)$
let $b \xleftarrow{\$} \{0,1\}, s^* := m_b \| 0^{k_1} \oplus G(r), t^* := r \oplus H(s^*), c^* := f(s^*, t^*),$
   $d^* := H'(s^* \| t^*)$
let $b' \leftarrow A^{H,G,H',Dec}(c^*, d^*)$
return $[b = b']$

Game 1:

> **let** $H \xleftarrow{\$} \Omega_H, G \xleftarrow{\$} \Omega_G, H' \xleftarrow{\$} \Omega_{H'}, r \xleftarrow{\$} \{0,1\}^{k_0}, (pk, sk) \leftarrow Gen(1^n)$,
> $\alpha^* \xleftarrow{\$} \{0,1\}^{k-k_0}$
> **let** $m_0, m_1 \leftarrow A^{H,G,H',Dec}(pk)$
> **let** $b \xleftarrow{\$} \{0,1\}, s^* = m_b || 0^{k_1} \oplus \alpha^*, t^* = r \oplus H(s^*), c^* = f(s^*, t^*)$,
> $d^* := H'(s^* || t^*)$
> **let** $b' \leftarrow A^{H,G,H',Dec}(c^*, d^*)$
> **return** $[b = b']$

The probability of success in Game 1 is $1/2$ for the reason that s^* is a random element and independent of the bit b.

Game 2:

> **let** $H \xleftarrow{\$} \Omega_H, G \xleftarrow{\$} \Omega_G, H' \xleftarrow{\$} \Omega_{H'}, r \xleftarrow{\$} \{0,1\}^{k_0}, (pk, sk) \leftarrow Gen(1^n)$,
> $\alpha^* \xleftarrow{\$} \{0,1\}^{k-k_0}, i \xleftarrow{\$} \{1,\ldots, q_G + q_{dec}\}$
> **run until** *i-th query to oracle G*
> > **let** $m_0, m_1 \leftarrow A^{H,G,H',Dec}(pk)$
> > **let** $b \xleftarrow{\$} \{0,1\}, s^* := m_b || 0^{k_1} \oplus \alpha^*, t^* := r \oplus H(s^*), c^* := f(s^*, t^*)$,
> > $d^* := H'(s^* || t^*)$
> > **let** $b' \leftarrow A^{H,G,H',Dec}(c^*, d^*)$
>
> **measure** the argument $\tilde{r}$ of the i-th query to oracle G
> **return** $[\tilde{r} = r]$ (When A makes less than i queries return $\perp$)

By O2H Lemma 1,

$$|\Pr[1 \leftarrow Game\ 0] - \Pr[1 \leftarrow Game\ 1]| \leq 2(q_G + q_{dec})\sqrt{\Pr[1 \leftarrow Game\ 2]}.$$

Game 3:

> **let** $H \xleftarrow{\$} \Omega_H, G \xleftarrow{\$} \Omega_G, H' \xleftarrow{\$} \Omega_{H'}, r \xleftarrow{\$} \{0,1\}^{k_0}, (pk, sk) \leftarrow Gen(1^n)$,
> $\alpha^* \xleftarrow{\$} \{0,1\}^{k-k_0}, i \xleftarrow{\$} \{1,\ldots, q_G + q_{dec}\}, \beta^* \xleftarrow{\$} \{0,1\}^{k_0}$
> **run until** *i-th query to oracle G*
> > **let** $m_0, m_1 \leftarrow A^{H,G,H',Dec}(pk)$
> > **let** $b \xleftarrow{\$} \{0,1\}, s^* := m_b || 0^{k_1} \oplus \alpha^*, t^* := r \oplus \beta^*, c^* := f(s^*, t^*)$,
> > $d^* := H'(s^* || t^*)$
> > **let** $b' \leftarrow A^{H,G,H',Dec}(c^*, d^*)$
>
> **measure** the argument $\tilde{r}$ of the i-th query to oracle G
> **return** $[\tilde{r} = r]$ (When A makes less than i queries return $\perp$)

Since t^* and s^* are random and independent of r, the probability of success in Game 3 is $\frac{1}{2^{k_0}}$.

Game 4:

> **let** $H \xleftarrow{\$} \Omega_H, G \xleftarrow{\$} \Omega_G, H' \xleftarrow{\$} \Omega_{H'}, r \xleftarrow{\$} \{0,1\}^{k_0}, (pk, sk) \leftarrow Gen(1^n),$
> $\alpha^* \xleftarrow{\$} \{0,1\}^{k-k_0}, i \xleftarrow{\$} \{1, \ldots, q_G + q_{dec}\}, \beta^* \xleftarrow{\$} \{0,1\}^{k_0},$
> $j \xleftarrow{\$} \{1, \ldots, q_H + q_{dec}\}$
> **run until** j-th query to oracle H
> > **run until** i-th query to oracle G
> > > **let** $m_0, m_1 \leftarrow A^{H,G,H',Dec}(pk)$
> > > **let** $b \xleftarrow{\$} \{0,1\}, s^* := m_b\|0^{k_1} \oplus \alpha^*, t^* := r \oplus \beta^*, c^* := f(s^*, t^*),$
> > > $d^* := H'(s^*\|t^*)$
> > > **let** $b' \leftarrow A^{H,G,H',Dec}(c^*, d^*)$
> > **measure** the argument $\tilde{r}$ of the i-th query to oracle G
> **measure** the argument $\tilde{s}$ of the j-th query to oracle H
> **return** $[\tilde{s} = s^*]$ (When A makes less than j queries return $\perp$)

By O2H Lemma 1,

$$|\Pr[1 \leftarrow Game\ 2] - \Pr[1 \leftarrow Game\ 3]| \leq 2(q_H + q_{dec})\sqrt{\Pr[1 \leftarrow Game\ 4]}.$$

Game 5:

> **let** $H \xleftarrow{\$} \Omega_H, G \xleftarrow{\$} \Omega_G, H' \xleftarrow{\$} \Omega_{H'}, r \xleftarrow{\$} \{0,1\}^{k_0}, (pk, sk) \leftarrow Gen(1^n),$
> $s^* \xleftarrow{\$} \{0,1\}^{k-k_0}, i \xleftarrow{\$} \{1, \ldots, q_G + q_{dec}\}, \beta^* \xleftarrow{\$} \{0,1\}^{k_0},$
> $j \xleftarrow{\$} \{1, \ldots, q_H + q_{dec}\}, d^* \xleftarrow{\$} \{0,1\}^{k}$
> **run until** j-th query to oracle H
> > **run until** i-th query to oracle G
> > > **let** $m_0, m_1 \leftarrow A^{H,G,H',Dec}(pk)$
> > > **let** $b \xleftarrow{\$} \{0,1\}, s^* := m_b\|0^{k_1} \oplus \alpha^*, t^* := r \oplus \beta^*, c^* := f(s^*, t^*),$
> > > **let** $b' \leftarrow A^{H,G,H',Dec}(c^*, d^*)$
> > **measure** the argument $\tilde{r}$ of the i-th query to oracle G
> **measure** the argument $\tilde{s}$ of the j-th query to oracle H
> **return** $[\tilde{s} = s^*]$ (When A makes less than j queries return $\perp$)

Game 6:

---

let $H \xleftarrow{\$} \Omega_H, G \xleftarrow{\$} \Omega_G, H' \xleftarrow{\$} \Omega_{H'}, r \xleftarrow{\$} \{0,1\}^{k_0}, (pk, sk) \leftarrow Gen(1^n)$,

$\alpha^* \xleftarrow{\$} \{0,1\}^{k-k_0}, i \xleftarrow{\$} \{1,\ldots,q_G + q_{dec}\}, \beta^* \xleftarrow{\$} \{0,1\}^{k_0}$,

$j \xleftarrow{\$} \{1,\ldots,q_H + q_{dec}\}, d^* \xleftarrow{\$} \{0,1\}^k, \ell \xleftarrow{\$} \{1,\ldots,q_{H'} + q_{dec}\}$

run until ℓ-th query to oracle H'

   run until j-th query to oracle H

      run until i-th query to oracle G

         let $m_0, m_1 \leftarrow A^{H,G,H',Dec}(pk)$

         let $b \xleftarrow{\$} \{0,1\}$, $s^* := m_b \| 0^{k_1} \oplus \alpha^*$, $t^* := r \oplus \beta^*$, $c^* = f(s^*, t^*)$

         let $b' \leftarrow A^{H,G,H',Dec}(c^*, d^*)$

      measure the argument $\tilde{r}$ of the i-th query to oracle G

   measure the argument $\tilde{s}$ of the j-th query to oracle H

measure the argument $(\tilde{s}, \tilde{t})$ of the ℓ-th query to oracle H'

return $[\tilde{s} = s^*] \wedge [\tilde{t} = t^*]$ (When A makes less than ℓ queries return $\bot$)

By O2H Lemma 1, $\cdot$

$$|\Pr[1 \leftarrow Game\ 4] - \Pr[1 \leftarrow Game\ 5]| \leq 2(q_{H'} + q_{dec})\sqrt{\Pr[1 \leftarrow Game\ 6]}.$$

Therefore, we only need to prove that the probability of success in Game 5 and Game 6 are negligible. Since a $2q$-wise independent function is indistinguishable from a random oracle provided the adversary makes at most q queries [17], we replace H' in Game 5 and Game 6 with a random polynomials of the proper degree. Let Ω_{wise} be the set of all such polynomials.

Game 5.b:

---

let $H \xleftarrow{\$} \Omega_H, G \xleftarrow{\$} \Omega_G, H' \xleftarrow{\$} \Omega_{wise}, r \xleftarrow{\$} \{0,1\}^{k_0}, (pk, sk) \leftarrow Gen(1^n)$,

$\alpha^* \xleftarrow{\$} \{0,1\}^{k-k_0}, i \xleftarrow{\$} \{1,\ldots,q_G + q_{dec}\}, \beta^* \xleftarrow{\$} \{0,1\}^{k_0}$,

$j \xleftarrow{\$} \{1,\ldots,q_H + q_{dec}\}, d^* \xleftarrow{\$} \{0,1\}^k$

run until j-th query to oracle H

   run until i-th query to oracle G

      let $m_0, m_1 \leftarrow A^{H,G,H',Dec}(pk)$

      let $b \xleftarrow{\$} \{0,1\}$, $s^* := m_b \| 0^{k_1} \oplus \alpha^*$, $t^* := r \oplus \beta^*$, $c^* = f(s^*, t^*)$

      let $b' \leftarrow A^{H,G,H',Dec}(c^*, d^*)$

   measure the argument $\tilde{r}$ of the i-th query to oracle G

measure the argument $\tilde{s}$ of the j-th query to oracle H

return $[\tilde{s} = s^*]$ (When A makes less than j queries return $\bot$)

By Zhandry's result [17]:

$$\Pr[1 \leftarrow Game\ 5] = \Pr[1 \leftarrow Game\ 5.b].$$

Now we define the decryption algorithm Dec^* that on input (c, d) does as follows:

1. It calculates the roots of polynomial $H' - d$. Let S be the set of all the roots.
2. If there exists $(s, t) \in S$ such that $f(s, t) = c$, then it outputs a message m using (s, t) and similar to the algorithm Dec. Otherwise it outputs $\bot$.

Game 5.c:

> let $H \xleftarrow{\$} \Omega_H, G \xleftarrow{\$} \Omega_G, H' \xleftarrow{\$} \Omega_{wise}, r \xleftarrow{\$} \{0,1\}^{k_0}, (pk, sk) \leftarrow Gen(1^n)$,
> $\alpha^* \xleftarrow{\$} \{0,1\}^{k-k_0}, i \xleftarrow{\$} \{1, \ldots, q_G + q_{dec}\}, \beta^* \xleftarrow{\$} \{0,1\}^{k_0}$,
> $j \xleftarrow{\$} \{1, \ldots, q_H + q_{dec}\}, d^* \xleftarrow{\$} \{0,1\}^k$
> **run until** j-th query to oracle H
> > **run until** i-th query to oracle G
> > > let $m_0, m_1 \leftarrow A^{H,G,H',Dec^*}(pk)$
> > > let $b \xleftarrow{\$} \{0,1\}, s^* := m_b\|0^{k_1} \oplus \alpha^*, t^* := r \oplus \beta^*, c^* = f(s^*, t^*)$
> > > let $b' \leftarrow A^{H,G,H',Dec^*}(c^*, d^*)$
> > **measure** the argument $\tilde{r}$ of the i-th query to oracle G
> **measure** the argument $\tilde{s}$ of the j-th query to oracle H
> **return** $[\tilde{s} = s^*]$ (When A makes less than j queries return $\bot$)

We show that two decryption algorithms Dec and Dec^* return the same output with the same number of queries to the random oracle H. For given ciphertext (c, d):

1. If $c \notin \mathrm{Im}\ f$, then both decryption algorithms return $\bot$ with no query to the random oracle H.
2. If $c \in \mathrm{Im}\ f$. Let $(\hat{s}, \hat{t}) := f^{-1}(c)$. There are two subcases:
   – If $H'(\hat{s}\|\hat{t}) \neq d$, then both algorithms return $\bot$ with no query to the random oracle H.
   – If $H'(\hat{s}\|\hat{t}) = d$, then both decryption algorithms return the same output and query H on input $\hat{s}$ for the reason that $(\hat{s}, \hat{t}) \in S$ and $f(\hat{s}, \hat{t}) = c$.

As a result:
$$\Pr[1 \leftarrow Game\ 5.b] = \Pr[1 \leftarrow Game\ 5.c].$$

Note that the decryption algorithm Dec^* does not use the secret key f^{-1}, therefore we can reduce the success probability of Game 5.c to the partial-domain one-wayness of function f.

We repeat a similar approach (define Game 6.b and Game 6.c as before) to prove the success probability of Game 6 is negligible. Note that the decryption algorithm Dec^{**} does as follows in the case of Game 6:

1. It calculates the roots of polynomial $H' - d$. Let S be the set of all the roots.
2. If there exists $(s, t) \in S$ such that $f(s, t) = c$, then it queries the random oracle H' on input $(s\|t)$ and outputs a message m using (s, t) and similar to the algorithm Dec.

3. Else:
   - If c^* is defined and $c = c^*$, then query H' on input $(s^* \| t^*)$ and return $\perp$.
   - If c^* is defined and $c \neq c^*$, then query H' on input $(s^* \| t^*) \oplus 1$ and return $\perp$.
   - If c^* is not defined then query H' on a random input and return $\perp$.

We show that two decryption algorithms Dec and Dec^{**} return the same output with the same number of queries to the random oracle H'. For given ciphertext (c, d):

1. If $c \notin \mathrm{Im}\ f$, then both decryption algorithms return $\perp$ and query the random oracle H' on a random input or on input $(s^* \| t^*) \oplus 1$.
2. If $c \in \mathrm{Im}\ f$ and c^* is defined. Let $(\hat{s}, \hat{t}) := f^{-1}(c)$. Then:
   - If $H'(\hat{s} \| \hat{t}) = d$, then both decryption algorithms return the same output and query H' on input $(\hat{s} \| \hat{t})$.
   - If $H'(\hat{s} \| \hat{t}) \neq d$ and $c \neq c^*$, then both algorithms return $\perp$ and query the random oracle H' on an input different from $(s^* \| t^*)$.
   - If $H'(\hat{s} \| \hat{t}) \neq d$ and $c = c^*$, then both algorithms return $\perp$ and query the random oracle H' on input $(s^* \| t^*)$.
3. If $c \in \mathrm{Im}\ f$ and c^* is not defined. Let $(\hat{s}, \hat{t}) := f^{-1}(c)$. Then:
   - If $H'(\hat{s} \| \hat{t}) \neq d$, then both algorithms return $\perp$ and query the random oracle H' on an input.
   - If $H'(\hat{s} \| \hat{t}) = d$, then both decryption algorithms return the same output and query H' on input $(\hat{s} \| \hat{t})$.

By combining all the inequalities from the proof, we can conclude that:

$$\Pr[1 \leftarrow Game\ 0] \leq 1/2 + \mathsf{negl}(\mathsf{n}).$$

Since our security proof does not depend on the bit padding, the message space can be extended to the set $\{0,1\}^{n+k_1}$. $\qquad\qquad\square$

Acknowledgments. This work was supported by the Estonian ICT program 2011-2015 (3.2.1201.13-0022), the European Union through the European Regional Development Fund through the sub-measure "Supporting the development of R&D of info and communication technology", by the European Social Fund's Doctoral Studies and Internationalisation Programme DoRa, by the Estonian Centre of Excellence in Computer Science, EXCS.

References

1. Alagic, G., Broadbent, A., Fefferman, B., Gagliardoni, T., Schaffner, C., Jules, M.S.: Computational security of quantum encryption. IACR ePrint 2016/424, April 2016
2. Ambainis, A., Rosmanis, A., Unruh, D.: Quantum attacks on classical proof systems (the hardness of quantum rewinding). In: FOCS 2014, pp. 474–483. IEEE, October 2014

3. Bellare, M., Rogaway, P.: Random oracles are practical: a paradigm for designing efficient protocols. In: Denning, D.E., Pyle, R., Ganesan, R., Sandhu, R.S., Ashby, V. (eds.) Proceedings of the 1st ACM Conference on Computer and Communications Security, CCS 1993, 3–5 November 1993, Fairfax, Virginia, USA, pp. 62–73. ACM (1993)

4. Bellare, M., Rogaway, P.: Optimal asymmetric encryption. In: Santis, A. (ed.) EUROCRYPT 1994. LNCS, vol. 950, pp. 92–111. Springer, Heidelberg (1995). doi:10.1007/BFb0053428

5. Ben-Or, M.: Probabilistic algorithms in finite fields. In: 22nd Annual Symposium on Foundations of Computer Science, 28–30 October 1981, Nashville, Tennessee, USA, pp. 394–398. IEEE Computer Society (1981)

6. Boneh, D., Dagdelen, Ö., Fischlin, M., Lehmann, A., Schaffner, C., Zhandry, M.: Random oracles in a quantum world. In: Lee, D.H., Wang, X. (eds.) ASIACRYPT 2011. LNCS, vol. 7073, pp. 41–69. Springer, Heidelberg (2011). doi:10. 1007/978-3-642-25385-0_3

7. Boneh, D., Zhandry, M.: Secure signatures and chosen ciphertext security in a quantum computing world. In: Canetti, R., Garay, J.A. (eds.) CRYPTO 2013. LNCS, vol. 8043, pp. 361–379. Springer, Heidelberg (2013). doi:10.1007/978-3-642-40084-1_21

8. Fujisaki, E., Okamoto, T.: Secure integration of asymmetric and symmetric encryption schemes. In: Wiener, M. (ed.) CRYPTO 1999. LNCS, vol. 1666, pp. 537–554. Springer, Heidelberg (1999). doi:10.1007/3-540-48405-1_34

9. Fujisaki, E., Okamoto, T., Pointcheval, D., Stern, J.: RSA-OAEP is secure under the RSA assumption. J. Cryptology 17(2), 81–104 (2004)

10. Shor, P.W.: Polynomial-time algorithms for prime factorization and discrete logarithms on a quantum computer. SIAM J. Comput. 26(5), 1484–1509 (1997)

11. Shoup, V.: OAEP reconsidered. In: Kilian, J. (ed.) CRYPTO 2001. LNCS, vol. 2139, pp. 239–259. Springer, Heidelberg (2001). doi:10.1007/3-540-44647-8_15

12. Targhi, E.E., Tabia, G.N., Unruh, D.: Quantum collision-resistance of non-uniformly distributed functions. In: Takagi, T. (ed.) PQCrypto 2016. LNCS, vol. 9606, pp. 79–85. Springer, Heidelberg (2016). doi:10.1007/978-3-319-29360-8_6

13. Unruh, D.: Quantum position verification in the random oracle model. In: Garay, J.A., Gennaro, R. (eds.) CRYPTO 2014. LNCS, vol. 8617, pp. 1–18. Springer, Heidelberg (2014). doi:10.1007/978-3-662-44381-1_1

14. Unruh, D.: Revocable quantum timed-release encryption. In: Nguyen, P.Q., Oswald, E. (eds.) EUROCRYPT 2014. LNCS, vol. 8441, pp. 129–146. Springer, Heidelberg (2014). doi:10.1007/978-3-642-55220-5_8

15. Unruh, D.: Non-interactive zero-knowledge proofs in the quantum random oracle model. In: Oswald, E., Fischlin, M. (eds.) EUROCRYPT 2015. LNCS, vol. 9057, pp. 755–784. Springer, Heidelberg (2015). doi:10.1007/978-3-662-46803-6_25

16. Yuen, H.: A quantum lower bound for distinguishing random functions from random permutations. Quantum Inf. Comput. 14(13–14), 1089–1097 (2014)

17. Zhandry, M.: Secure identity-based encryption in the quantum random oracle model. In: Safavi-Naini, R., Canetti, R. (eds.) CRYPTO 2012. LNCS, vol. 7417, pp. 758–775. Springer, Heidelberg (2012). doi:10.1007/978-3-642-32009-5_44

18. Zhandry, M.: A note on the quantum collision and set equality problems. Quantum Inf. Comput. 15(7&8), 557–567 (2015)

Multi-key FHE from LWE, Revisited

Chris Peikert[✉] and Sina Shiehian

Computer Science and Engineering, University of Michigan, Ann Arbor, USA
cpeikert@alum.mit.edu

Abstract. Traditional fully homomorphic encryption (FHE) schemes only allow computation on data encrypted under a *single* key. López-Alt, Tromer, and Vaikuntanathan (STOC 2012) proposed the notion of *multi-key* FHE, which allows homomorphic computation on ciphertexts encrypted under different keys, and also gave a construction based on a (somewhat nonstandard) assumption related to NTRU. More recently, Clear and McGoldrick (CRYPTO 2015), followed by Mukherjee and Wichs (EUROCRYPT 2016), proposed a multi-key FHE that builds upon the LWE-based FHE of Gentry, Sahai, and Waters (CRYPTO 2013). However, unlike the original construction of López-Alt *et al.*, these later LWE-based schemes have the somewhat undesirable property of being "single-hop for keys:" all relevant keys must be known at the start of the homomorphic computation, and the output cannot be usefully combined with ciphertexts encrypted under other keys (unless an expensive "bootstrapping" step is performed).

In this work we construct two multi-key FHE schemes, based on LWE assumptions, which are *multi-hop for keys*: the output of a homomorphic computation on ciphertexts encrypted under a set of keys can be used in further homomorphic computation involving *additional* keys, and so on. Moreover, incorporating ciphertexts associated with new keys is a relatively efficient "native" operation akin to homomorphic multiplication, and does not require bootstrapping (in contrast with all other LWE-based solutions). Our systems also have smaller ciphertexts than the previous LWE-based ones; in fact, ciphertexts in our second construction are simply GSW ciphertexts with no auxiliary data.

1 Introduction

Secure *multiparty computation* (MPC) is an important and well-studied problem in cryptography. In MPC, multiple users want to jointly perform a computation on their respective inputs via an interactive protocol. Informally, the goal is for the protocol to reveal nothing more than the *output* of the computation.

C. Peikert—This material is based upon work supported by the National Science Foundation under CAREER Award CCF-1054495 and CNS-1606362, and by the Alfred P. Sloan Foundation. The views expressed are those of the authors and do not necessarily reflect the official policy or position of the National Science Foundation or the Sloan Foundation.

M. Hirt and A. Smith (Eds.): TCC 2016-B, Part II, LNCS 9986, pp. 217–238, 2016.
DOI: 10.1007/978-3-662-53644-5_9

Fully homomorphic encryption (FHE) is a powerful tool for constructing secure MPC protocols. One approach suggested in Gentry's seminal work [9], and later optimized by Asharov *et al.* [4], is to have an initial phase in which all parties run a protocol to generate a sharing of an FHE secret key, then use the public key to encrypt their inputs and publish the ciphertexts. The parties then *locally* compute an encryption of the output using homomorphic operations. Finally, they run a protocol to decrypt the encrypted output, using their secret key shares. Overall, this approach requires the set of involved parties to be known in advance, and for them to run interactive protocols both *before* and *after* their local computation.

López-Alt *et al.* [15] (hereafter LTV) introduced the interesting notion of *on-the-fly MPC*, in which the set of parties who contribute inputs to the computation, and even the computation itself, need not be fixed in advance, and can even be chosen adaptively. In addition, there is no interaction among the parties at the outset: any user whose data might potentially be used simply uploads her encrypted input to a central server in advance, and can then go offline. The server then uses the uploaded data to compute (or continue computing) a desired function, and when finished, outputs an encrypted output. Finally, the parties whose inputs were used in the computation—and *only* those parties—run an interactive protocol to jointly decrypt the ciphertext and obtain the output.

Multi-key FHE. Traditional FHE schemes only allow computation on data encrypted under a *single* key, and therefore are not suitable for on-the-fly MPC, where users' inputs must be encrypted under different keys. As a tool for constructing on-the-fly protocols, LTV proposed a new type of FHE scheme, which they called *multi-key FHE* (MK-FHE). Such a scheme extends the FHE functionality to allow homomorphic computation on ciphertexts encrypted under different, independent keys. Decrypting the result of such a computation necessarily requires all of the corresponding secret keys.

In [15], LTV constructed an MK-FHE scheme based on a variant of the NTRU cryptosystem [13]. Its security was based on a new and somewhat non-standard assumption on polynomial rings, which, unlike the commonly used learning with errors (LWE) assumption [20] or its ring-based analogue [16], is not currently supported by a worst-case hardness theorem.[1] (LTV also constructed MK-FHE based on ring-LWE, but limited only to a *logarithmic* number of keys and circuit depth.) Subsequently, Clear and McGoldrick [8] gave an LWE-based construction for an *unlimited* number of keys, using a variant of the FHE scheme of Gentry *et al.* [11] (hereafter GSW). Later, Mukherjee and Wichs [18] provided another exposition of the Clear-McGoldrick scheme, and built a two-round (plain) MPC protocol upon it.

---

[1] Indeed, Albrecht *et al.* [1], and later Kirchner and Fouque [14], recently gave attacks on "overstretched" NTRU problems like those used in [15], where the running times range from slightly subexponential to even polynomial-time, depending on the parameterization.

Static Versus Dynamic. We observe that the LTV multi-key FHE, to extend the terminology of [10], is *"multi-hop* for keys," or, more concisely, "dynamic:" one can perform a homomorphic computation on a collection of ciphertexts encrypted under some set of keys, then use the resulting ciphertext as an input to further homomorphic computation on ciphertexts encrypted under *additional* keys, and so on. (Multi-hop homomorphic computation is naturally supported by essentially all known single-key FHE schemes as well.) The on-the-fly MPC protocol of [15] naturally inherits this dynamic flavor, which is very much in the spirit of "on the fly" computation, since it allows reusing encrypted results across different computations.

By contrast, it turns out that neither of the MK-FHE constructions from [8,18] appear to be dynamic, but are instead only *static* (i.e., single-hop for keys): once a homomorphic computation has been performed on a collection of ciphertexts encrypted under some set of keys, the output cannot easily be used in further computation involving additional keys. Instead, one must restart the whole computation from scratch (incorporating all the relevant keys from the very beginning), or perform an expensive "bootstrapping" step, which may be even more costly.[2] This rules out a dynamic computation, since all involved parties must be known before the computation begins. In summary, existing constructions of MK-FHE and on-the-fly MPC from standard (worst-case) lattice assumptions still lack basic functionality that has been obtained from more heuristic assumptions.

1.1 Our Results

In this work we construct two (leveled) multi-key FHE schemes, for any number of keys, from LWE assumptions. Like the original MK-FHE scheme of [15], and unlike those of [8,18], both of our schemes are dynamic (i.e., multi-hop for keys), and hence are suitable for dynamic on-the-fly MPC. Specifically, in our schemes one can homomorphically compute on ciphertexts encrypted under several keys, then use the result in further computation on ciphertexts under additional keys, and so on. Moreover, incorporating ciphertexts associated with new keys into the computation is a relatively efficient "native" operation, akin to GSW ciphertext multiplication, which does not require bootstrapping. In addition, our schemes are also naturally bootstrappable (as usual, under appropriate circular-security assumptions), and can therefore support unbounded homomorphic computations for any polynomial number of keys. We now describe our two systems in more detail, and discuss their different efficiency and security tradeoffs.

Scheme #1: Large Ciphertexts, Standard LWE. The security of our first scheme, which is described in Sect. 3, is based on the standard n-dimensional decision-LWE assumption (appropriately parameterized), but has rather large ciphertexts and correspondingly slow homomorphic operations. Actually, the ciphertexts are

---

[2] Indeed, a recent concurrent and independent work by Brakerski and Perlman [6] follows this bootstrapping approach; we provide a comparison to our work in Sect. 1.1.

	Public key	Ciphertext	Key Hops	Must Bootstrap?
[8, 18]	$\tilde{O}(nd^2)$	$\tilde{O}(n^4d^4) \to \tilde{O}(n^2k^2d^2)$	Single	No
[6]	$\tilde{O}(n^3)$	$\tilde{O}(nk)$	Multiple	Yes
Our Scheme #1	$\tilde{O}(n(K+d)^2)$	$\tilde{O}(n^3k(K+d)^4)$	Multiple	No
Our Scheme #2	$\tilde{O}(n^4(K+d)^4)$	$\tilde{O}(n^2k^2(K+d)^2)$	Multiple	No

Fig. 1. Properties of LWE-based MK-FHE schemes, where all sizes are in bits. Here k denotes the actual number of secret keys associated with the ciphertext, with a designed upper bound of K; d denotes the boolean circuit depth the scheme is designed to homomorphically evaluate (without bootstrapping); and n is the dimension of the underlying LWE problem used for security. (The $\tilde{O}$ notation hides logarithmic factors in these parameters.) The arrow $\to$ for [8, 18] denotes the change in size following the single "hop" from fresh ciphertexts (under single keys) to multi-key ciphertexts.

about an n factor *smaller* than fresh ciphertexts in the systems from [8, 18] (see Fig. 1), but unlike in those systems, our ciphertexts remain rather large even *after* multi-key homomorphic operations. Essentially, this is the price of being dynamic—indeed, it is possible at any point to "downgrade" our ciphertexts to ordinary GSW ciphertexts, by giving up the ability to extend ciphertexts to additional keys.

Scheme #2: Small Ciphertexts, Circular LWE. In our second scheme, which is described in Sect. 4, ciphertexts are simply GSW ciphertexts, and are therefore (relatively) small and admit correspondingly efficient homomorphic operations. This efficiency comes at the price of rather large *public keys* (which are comparable to fresh ciphertexts in the systems from [8, 18]) and a correspondingly slow algorithm for extending ciphertexts to additional keys. This efficiency profile seems preferable to our first scheme's, because applications of MK-FHE would typically involve many more homomorphic operations than extensions to new keys. Therefore, we consider this scheme to be our main contribution.

Interestingly, the security of our second scheme appears to require a natural *circular security* assumption for LWE. Despite some positive results for circular security of LWE-based encryption [3], we do not yet see a way to prove security under standard LWE. We point out, however, that our assumption is no stronger than the circular-security assumptions that are used to "bootstrap" FHE, because any circular-secure FHE is itself fully key-dependent message secure [9]. So in a context where our system is bootstrapped to obtain unbounded FHE, we actually incur no additional assumption.

Comparison with [6]. A concurrent and independent work by Brakerski and Perlman [6], which also constructs (unbounded) dynamic multi-key FHE from LWE, was posted to ePrint shortly after our original preprint appeared there. (Both works were submitted to CRYPTO'16, but only [6] was accepted.) The construction of Brakerski and Perlman follows the "bootstrapping" approach mentioned above, and is focused on minimizing the ciphertext size. Specifically, their multi-key ciphertexts grow only *linearly* in the number of secret keys associated with the ciphertext. In addition, they describe an "on-the-fly" bootstrapping algorithm that requires only a linear amount of "local" memory (even though the encrypted secret keys are much larger). However, all this comes at the cost of needing to perform an expensive bootstrapping operation whenever incorporating a ciphertext encrypted under a new key, and also for *every* homomorphic multiplication/NAND operation. (Essentially, this is because the linear-sized ciphertexts are ordinary LWE vectors, not GSW matrices.)

By contrast, our work gives (leveled) dynamic multi-key FHE schemes for which both homomorphic multiplication and incorporation of new keys are much more efficient "native" operations, requiring only a few standard GSW-style matrix operations. This comes at the cost of relatively larger ciphertexts, which naïvely grow at least quartically in the maximum number of keys (see Fig. 1). However, we point out that using ordinary bootstrapping, our constructions can also be made to support an unbounded number of keys, and with ciphertext sizes that grow only quadratically in the number of associated keys.

1.2   Technical Overview

For context, we start with a brief overview of the prior (single-hop for keys) MK-FHE constructions of [8,18], and the challenge in making them dynamic. In these systems, a fresh ciphertext that decrypts under secret key $\mathbf{t} \in \mathbb{Z}^n$ is a GSW ciphertext $\mathbf{C} \in \mathbb{Z}_q^{n \times m}$ encrypted to the corresponding public key $\mathbf{P}$, along with an encryption $\mathbf{D}$ of the *encryption randomness* used to produce $\mathbf{C}$ from $\mathbf{P}$. (Specifically, each entry of the randomness matrix is encrypted as a separate GSW ciphertext.)

To perform a homomorphic computation on fresh ciphertexts $(\mathbf{C}_i, \mathbf{D}_i)$ that are respectively encrypted under secret keys $\mathbf{t}_i$ for (say) $i = 1, 2$, we first *extend* each ciphertext to an ordinary GSW ciphertext

$$\hat{\mathbf{C}}_i = \begin{bmatrix} \mathbf{C}_i & \mathbf{X}_i \\ & \mathbf{C}_i \end{bmatrix} \in \mathbb{Z}_q^{2n \times 2m} \tag{1}$$

that decrypts to the same message under the *concatenated* key $(\mathbf{t}_1, \mathbf{t}_2)$, and then perform normal GSW homomorphic operations on these extended ciphertexts. Essentially, extending $\mathbf{C}_1$ is done by considering the extra "junk" term $(\mathbf{t}_2 - \mathbf{t}_1) \cdot \mathbf{C}_1$ that arises from decrypting $\mathbf{C}_1$ under the wrong secret key $\mathbf{t}_2$, and cancelling it out via a ciphertext $\mathbf{X}_1$ that "decrypts" under $\mathbf{t}_1$ to (the negation of) the same junk term. To produce $\mathbf{X}_1$ we use linearly homomorphic operations on $\mathbf{D}_1$ (the encryption of $\mathbf{C}_1$'s randomness relative to $\mathbf{P}_1$), along with some additional information about $\mathbf{t}_1$ relative to a shared public parameter.

We point out that in the above scheme, it is not clear how to obtain an encryption of $\hat{\mathbf{C}}_i$'s underlying encryption randomness—indeed, it is not even clear what composite *public key* $\hat{\mathbf{P}}$ the ciphertext $\hat{\mathbf{C}}_i$ would be relative to, nor whether valid encryption randomness for $\hat{\mathbf{C}}_i$ exists at all! (Indeed, for certain natural ways of combining the public keys $\mathbf{P}_i$, valid encryption randomness is not likely to exist.) This is what prevents the extended ciphertexts from satisfying the same invariant that fresh ciphertexts satisfy, which makes the scheme only single-hop for keys. Moreover, even if we could produce an encryption of the ciphertext randomness (assuming it exists), it is not clear whether we could later re-extend an arbitrary ciphertext $\mathbf{C} \in \mathbb{Z}_q^{2n \times 2m}$ that decrypts under $(\mathbf{t}_1, \mathbf{t}_2)$ to an additional key $\mathbf{t}_3$: the block upper-triangular structure from Eq. (1) would produce a $4n$-by-$4m$ matrix, which is too large.

Our Approach. To overcome the above difficulties, our ciphertexts and/or public keys consist of different information, whose invariants can be maintained after extension to additional keys. In particular, we forego maintaining *encryption randomness* relative to a *varying* public key, and instead only maintain *commitment randomness* relative to a *fixed* public parameter, along with an encryption of that randomness.[3] Concretely, this works in two different ways in our two schemes, as we now explain.

Scheme #1. In our first system (given in Sect. 3), a ciphertext under a secret key $\mathbf{t} \in \mathbb{Z}^{kn}$—which would typically be the concatenation of $k \geq 1$ individual secret keys—consists of three components:

1. a (symmetric-key) *GSW ciphertext* $\mathbf{C} \in \mathbb{Z}_q^{kn \times km}$ that decrypts under $\mathbf{t}$,
2. a GSW-style *homomorphic commitment* (à la [12]) $\mathbf{F} \in \mathbb{Z}_q^{n \times m}$ to the same message, relative to a public parameter, and
3. a special encryption $\mathbf{D}$ under $\mathbf{t}$ of the *commitment randomness* underlying $\mathbf{F}$.

To extend such a ciphertext to a new secret key $\mathbf{t}^* \in \mathbb{Z}^n$, we simply extend the GSW ciphertext $\mathbf{C}$ to some

$$\mathbf{C}' = \begin{bmatrix} \mathbf{C} & \mathbf{X} \\ & \mathbf{F} \end{bmatrix} \in \mathbb{Z}_q^{(k+1)n \times (k+1)m},$$

where $\mathbf{X}$ is produced from $\mathbf{D}$ (in much the same way as above) to cancel out the "junk" term that comes from "decrypting" $\mathbf{F}$ with $\mathbf{t}^*$. The commitment $\mathbf{F}$ and its encrypted randomness $\mathbf{D}$ remain unchanged, except that we need to pad $\mathbf{D}$ with zeros to make it valid under $(\mathbf{t}, \mathbf{t}^*)$.

Finally, it is not too hard to design homomorphic addition and multiplication operations for ciphertexts having the above form: as shown in [12], GSW commitments admit exactly the same homomorphic operations as GSW encryption, so we can maintain a proper commitment. The homomorphic operations also have a natural, predictable effect on the underlying commitment randomness, so we can use the encrypted randomness $\mathbf{D}_i$ along with the GSW ciphertexts $\mathbf{C}_i$ to maintain correct encrypted commitment randomness.

---

[3] We note that the previous constructions from [8,18] also require a public parameter, so we are not changing the model.

Scheme #2. Our second system (given in Sect. 4) works differently from all the previous ones. In it, ciphertexts are simply GSW ciphertexts, with no extra components, so they support the standard homomorphic operations. To support extending ciphertexts to additional keys, each *public key* contains a *commitment* to its secret key **t**, along with an appropriate encryption under **t** of the commitment randomness. (This cyclical relation between secret key and commitment randomness is what leads to our circular-security assumption.) We show how to combine two public keys to get a ciphertext, under the *concatenation* of their secret keys $\mathbf{t}_1, \mathbf{t}_2$, that encrypts the *tensor product* $\mathbf{t}_1 \otimes \mathbf{t}_2$ of those keys. By applying homomorphic operations, it is then fairly straightforward to extend a ciphertext that decrypts under one of the keys to a ciphertext that decrypts under their concatenation.

2   Preliminaries

In this work, vectors are denoted by lower-case bold letters (e.g., **a**), and are *row* vectors unless otherwise indicated. Matrices are denoted by upper-case bold letters (e.g., **A**). We define $[k] := \{1, \ldots, k\}$ for any non-negative integer k.

Approximations. As in many works in lattice cryptography, we work with "noisy equations" and must quantify the quality of the approximation. For this purpose we use the notation $\approx$ to indicate that the two sides are approximately equal up to some *additive* error, and we always include a bound on the magnitude of this error. For example,

$$x \approx y \qquad\qquad (\text{error } E)$$

means that $x = y + e$ for some $e \in [-E, E]$. In the case of vectors or matrices, the error bound applies to every entry of the error term, i.e., it is an ℓ_∞ bound.

For simplicity of analysis, in this work we use the following rather crude "expansion" bounds to quantify error growth. (Sharper bounds can be obtained using more sophisticated tools like subgaussian random variables.) Because $\|\mathbf{x} \cdot \mathbf{y}^t\|_\infty \leq \|\mathbf{x}\|_\infty \cdot \|\mathbf{y}\|_1$ and $\|\mathbf{y}\|_1 \leq \dim(\mathbf{y}) \cdot \|\mathbf{y}\|_\infty$, we have implications like

$$\mathbf{X} \approx \mathbf{Y} \qquad\qquad (\text{error } E)$$
$$\implies \mathbf{X} \cdot \mathbf{R} \approx \mathbf{Y} \cdot \mathbf{R}. \qquad (\text{error height } (\mathbf{R}) \cdot \|\mathbf{R}\|_\infty \cdot E)$$

for any $\mathbf{X}, \mathbf{Y}, \mathbf{R}$.

Tensor Products. The *tensor* (or *Kronecker*) product $\mathbf{A} \otimes \mathbf{B}$ of an m_1-by-n_1 matrix $\mathbf{A}$ with an m_2-by-n_2 matrix $\mathbf{B}$, both over a common ring $\mathcal{R}$, is the $m_1 m_2$-by-$n_1 n_2$ matrix consisting of m_2-by-n_2 blocks, whose (i,j)th block is $a_{i,j} \cdot \mathbf{B}$, where $a_{i,j}$ denotes the (i,j)th entry of $\mathbf{A}$.

It is clear that

$$r(\mathbf{A} \otimes \mathbf{B}) = (r\mathbf{A}) \otimes \mathbf{B} = \mathbf{A} \otimes (r\mathbf{B})$$

for any scalar $r \in \mathcal{R}$. We extensively use the *mixed-product property* of tensor products, which says that

$$(\mathbf{A} \otimes \mathbf{B}) \cdot (\mathbf{C} \otimes \mathbf{D}) = (\mathbf{AC}) \otimes (\mathbf{BD})$$

for any matrices $\mathbf{A}, \mathbf{B}, \mathbf{C}, \mathbf{D}$ of compatible dimensions. In particular,

$$(\mathbf{A} \otimes \mathbf{B}) = (\mathbf{A} \otimes \mathbf{I}_{\mathrm{height}(\mathbf{B})}) \cdot (\mathbf{I}_{\mathrm{width}(\mathbf{A})} \otimes \mathbf{B}) = (\mathbf{I}_{\mathrm{height}}(\mathbf{A}) \otimes \mathbf{B}) \cdot (\mathbf{A} \otimes \mathbf{I}_{\mathrm{width}}(\mathbf{B})).$$

2.1 Cryptographic Definitions

Definition 1. *A leveled multi-hop, multi-key FHE scheme is a tuple of efficient randomized algorithms* (Setup, Gen, Enc, Dec, EvalNAND) *having the following properties:*

- Setup($1^\lambda, 1^k, 1^d$), *given the security parameter λ, a bound k on the number of keys, and a bound d on the circuit depth, outputs a public parameter pp. (All the following algorithms implicitly take pp as an input.)*
- Gen() *outputs a public key pk and secret key sk.*
- Enc(pk, μ), *given a public key pk and a message $\mu \in \{0, 1\}$, outputs a ciphertext c. For convenience, we assume that c implicitly contains a reference to pk.*
- Dec($(sk_1, sk_2, \ldots, sk_t), c$), *given a tuple of secret keys $sk_1, \ldots, sk_t$ and a ciphertext c, outputs a bit.*
- EvalNAND(c_1, c_2), *given two ciphertexts c_1, c_2, outputs a ciphertext $\hat{c}$. For convenience, we assume that $\hat{c}$ implicitly contains a reference to each public key associated with either c_1 or c_2 (or both).*

These algorithms should satisfy correctness *and* compactness *functionality properties, as defined below.*

We now describe how to homomorphically evaluate a given boolean circuit composed of NAND gates and having one output wire, which is without loss of generality. The algorithm Eval($C, (c_1, \ldots, c_N)$), given a circuit C having N input wires, first associates c_i with the ith input wire for each $i = 1, \ldots, N$. Then for each gate (in some topological order) having input wires i, j and output wire k, it computes $c_k \leftarrow$ EvalNAND(c_i, c_j). Finally, it outputs the ciphertext associated with the output wire.

We stress that the above homomorphic evaluation process is qualitatively different from the ones defined in [15,18], because when homomorphically evaluating each gate we can only use the key(s) associated with the input ciphertexts *for that gate alone*; this is what makes the computation multi-hop. By contrast, homomorphic evaluation in [15,18] is given all the input ciphertexts and public keys from the start, so it can (and does, in the case of [18]) use this knowledge before evaluating any gates.

Definition 2 (Correctness). *A leveled multi-hop, multi-key FHE scheme is correct if for all positive integers λ, k, d, for every circuit C of depth at most d having N input wires, for every function $\pi \colon [N] \to [k]$ (which associates each* *input wire with a key pair), and for every $x \in \{0,1\}^N$, the following experiment succeeds with $1 - \text{negl}(\lambda)$ probability: generate a public parameter $pp \leftarrow$ Setup$(1^\lambda, 1^k, 1^d)$, generate key pairs $(pk_j, sk_j) \leftarrow$ Gen$()$ for each $j \in [k]$, generate ciphertexts $c_i \leftarrow$ Enc$(pk_{\pi(i)}, x_i)$ for each $i \in [N]$, let $\hat{c} \leftarrow$ Eval$(C, (c_1, \ldots, c_N))$, and finally test whether*

$$\text{Dec}((sk_j), \hat{c}) = C(x_1, \ldots, x_N),$$

where Dec *is given those secret keys sk_j corresponding to the public keys referenced by $\hat{c}$.*

Definition 3 (Compactness). *A leveled multi-hop, multi-key FHE scheme is compact if there exists a polynomial $p(\cdot, \cdot, \cdot)$ such that in the experiment from Definition 2, $|\hat{c}| \leq p(\lambda, k, d)$. In other words, the length of $\hat{c}$ is independent of C and N, but can depend polynomially on λ, k, and d.*

2.2   Learning with Errors

For a positive integer dimension n and modulus q, and an error distribution χ over $\mathbb{Z}$, the LWE distribution and decision problem are defined as follows. For an $\mathbf{s} \in \mathbb{Z}^n$, the LWE distribution $A_{\mathbf{s},\chi}$ is sampled by choosing a uniformly random $\mathbf{a} \leftarrow \mathbb{Z}_q^n$ and an error term $e \leftarrow \chi$, and outputting $(\mathbf{a}, b = \langle \mathbf{s}, \mathbf{a} \rangle + e) \in \mathbb{Z}_q^{n+1}$.

Definition 4. *The decision-LWE$_{n,q,\chi}$ problem is to distinguish, with non-negligible advantage, between any desired (but polynomially bounded) number of independent samples drawn from $A_{\mathbf{s},\chi}$ for a single $\mathbf{s} \leftarrow \chi^n$, and the same number of uniformly random and independent samples over $\mathbb{Z}_q^{n+1}$.[4]*

A standard instantiation of LWE is to let χ be a *discrete Gaussian* distribution (over $\mathbb{Z}$) with parameter $r = 2\sqrt{n}$. A sample drawn from this distribution has magnitude bounded by, say, $r\sqrt{n} = \Theta(n)$ except with probability at most 2^{-n}. For this parameterization, it is known that LWE is at least as hard as *quantumly* approximating certain "short vector" problems on n-dimensional lattices, in the worst case, to within $\tilde{O}(q\sqrt{n})$ factors [20]. Classical reductions are also known for different parameterizations [5, 19].

In this work it will be convenient to use a form of LWE that is somewhat syntactically different from, but computationally equivalent to, the one defined above. Letting $\mathbf{s} = (-\bar{\mathbf{s}}, 1) \in \mathbb{Z}^n$ where $\bar{\mathbf{s}} \leftarrow \chi^{n-1}$, notice that an LWE sample $\mathbf{b} = (\mathbf{a}, b = \langle \mathbf{s}, \mathbf{a} \rangle + e) \in \mathbb{Z}_q^n$ drawn from $A_{\bar{\mathbf{s}},\chi}$ is simply a uniformly random vector satisfying

$$\langle \mathbf{s}, \mathbf{b} \rangle = \mathbf{s} \cdot \mathbf{b}^t = e \approx 0. \tag{2}$$

Therefore, decision-LWE$_{n-1,q,\chi}$ is equivalent to the problem of distinguishing samples having the above form (and in particular, satisfying Eq. (2)) from uniformly random ones.

---

[4] Notice that in the above definition, the coordinates of $\mathbf{s}$ are drawn from the error distribution χ; as shown in [3], this form of the problem is equivalent to the one where $\mathbf{s} \leftarrow \mathbb{Z}_q^n$ is drawn uniformly at random.

More generally, for $\mathbf{s} \in \mathbb{Z}^n$ as above and some $t = \mathrm{poly}(n)$, we will need to generate uniformly random vectors $\mathbf{b} \in \mathbb{Z}_q^{tn}$ that satisfy

$$(\mathbf{I}_t \otimes \mathbf{s}) \cdot \mathbf{b} = \mathbf{e} \approx \mathbf{0},$$

for some $\mathbf{e} \leftarrow \chi^t$. This is easily done by concatenating t independent samples from $A_{\bar{\mathbf{s}},\chi}$; clearly, the result is indistinguishable from uniform assuming the hardness of decision-$\mathrm{LWE}_{n,q,\chi}$.

2.3   Gadgets and Decomposition

Here we recall the notion of a "gadget" [17], which is used for decomposing $\mathbb{Z}_q$-elements—or more generally, vectors or matrices over $\mathbb{Z}_q$—into short vectors or matrices over $\mathbb{Z}$. We also define some new notation that will be convenient for our application.

For simplicity, throughout this work we use the standard "powers of two" gadget vector

$$\mathbf{g} = (1, 2, 4, 8, \ldots, 2^{\ell-1}) \in \mathbb{Z}_q^\ell, \quad \text{where } \ell = \lceil \lg q \rceil.$$

The "bit decomposition" function $\mathbf{g}^{-1} \colon \mathbb{Z}_q \to \{0,1\}^\ell$ outputs a binary *column* vector (over $\mathbb{Z}$) consisting of the binary representation of (the canonical representative in $\{0, 1, \ldots, q-1\}$ of) its argument. As such, it satisfies the identity $\mathbf{g} \cdot \mathbf{g}^{-1}[a] = a$. (This identity explains the choice of notation $\mathbf{g}^{-1}$; we stress that $\mathbf{g}^{-1}$ is a *function*, not a vector itself.) Symmetrically, we define the notation

$$[a]\mathbf{g}^{-t} := \mathbf{g}^{-1}[a]^t,$$

which outputs a binary *row* vector and satisfies the identity $[a]\mathbf{g}^{-t} \cdot \mathbf{g}^t = a$. (This identity explains why we place the bracketed argument to the *left* of $\mathbf{g}^{-t}$.)

More generally, we define the operation denoted by $(\mathbf{I}_n \otimes \mathbf{g}^{-1})[\cdot]$, which applies $\mathbf{g}^{-1}$ entrywise to a height-n vector/matrix, and thereby produces a height-$n\ell$ binary output that satisfies the convenient identity

$$(\mathbf{I}_n \otimes \mathbf{g}) \cdot (\mathbf{I}_n \otimes \mathbf{g}^{-1})[\mathbf{A}] = \mathbf{A}.$$

Similarly, we define $[\cdot](\mathbf{I}_n \otimes \mathbf{g}^{-t})$ to apply $\mathbf{g}^{-t}$ entrywise to a width-n vector/matrix, thereby producing a width-$n\ell$ output that satisfies

$$[\mathbf{A}](\mathbf{I}_n \otimes \mathbf{g}^{-t}) \cdot (\mathbf{I}_n \otimes \mathbf{g}^t) = \mathbf{A}.$$

For the reader who is familiar with previous works that use gadget techniques, the matrix $\mathbf{I}_n \otimes \mathbf{g}$ is exactly the n-row gadget matrix $\mathbf{G}$, and $(\mathbf{I}_n \otimes \mathbf{g}^{-1})[\cdot]$ is exactly the bit-decomposition operation $\mathbf{G}^{-1}$ on height-n vectors/matrices. In this work we adopt the present notation because we use several different dimensions n, and because it interacts cleanly with tensor products of vectors and matrices, which we use extensively in what follows.

3   Large-Ciphertext Construction

In this section we describe our first construction of a multi-hop, multi-key FHE, which has small keys but rather large ciphertexts (although fresh ciphertexts are still smaller than in prior constructions). For simplicity, we describe the scheme in the symmetric-key setting, but then note how to obtain a public-key scheme using a standard transformation.

The system is parameterized by a dimension n, modulus q, and error distribution χ for the underlying LWE problem; we also let $m = \lceil 2n \log q \rceil$. For concreteness, we let χ be the standard discrete Gaussian error distribution with parameter $2\sqrt{n}$; to recall, the samples it produces have magnitudes bounded by some $E = \Theta(n)$ except with exponentially small $2^{-\Omega(n)}$ probability. The modulus q is instantiated in Sect. 3.3, based on a desired depth of homomorphic computation and number of distinct keys. The scheme is defined as follows.

- Setup: output a uniformly random $\mathbf{A} \in \mathbb{Z}_q^{n \times m}$.
- Gen($\mathbf{A}$): choose $\bar{\mathbf{t}} \leftarrow \chi^{n-1}$ and define $\mathbf{t} := (-\bar{\mathbf{t}}, 1) \in \mathbb{Z}^n$. Choose $\mathbf{e} \leftarrow \chi^m$ and define

$$\mathbf{b} := \mathbf{t}\mathbf{A} + \mathbf{e}$$
$$\approx \mathbf{t}\mathbf{A} \in \mathbb{Z}_q^m. \qquad\qquad (\text{error } E) \qquad (3)$$

Output $\mathbf{t}$ as the secret key and $\mathbf{b}$ as the associated public extension key.
- Enc($\mathbf{t}, \mu \in \{0, 1\}$): do the following, outputting $(\mathbf{C}, \mathbf{F}, \mathbf{D})$ as the ciphertext.
  1. As described in Sect. 2.2, choose an LWE matrix $\bar{\mathbf{C}} \in \mathbb{Z}_q^{n \times n\ell}$ that satisfies $\mathbf{t}\bar{\mathbf{C}} \approx \mathbf{0}$, and define

$$\mathbf{C} := \bar{\mathbf{C}} + \mu(\mathbf{I}_n \otimes \mathbf{g}) \in \mathbb{Z}_q^{n \times n\ell}.$$

Notice that $\mathbf{C}$ is simply a *GSW ciphertext* encrypting μ under secret key $\mathbf{t}$:

$$\mathbf{t}\mathbf{C} = \mathbf{t}\bar{\mathbf{C}} + \mu(\mathbf{t} \otimes 1) \cdot (\mathbf{I}_n \otimes \mathbf{g}) \approx \mu(\mathbf{t} \otimes \mathbf{g}). \qquad (\text{error } E_\mathbf{C}) \qquad (4)$$

  2. In addition, choose a uniformly random $\mathbf{R} \in \{0, 1\}^{m \times n\ell}$ and define

$$\mathbf{F} := \mathbf{A}\mathbf{R} + \mu(\mathbf{I}_n \otimes \mathbf{g}) \in \mathbb{Z}_q^{n \times n\ell}. \qquad (5)$$

We view $\mathbf{F}$ as a *commitment* to the message μ under randomness $\mathbf{R}$.
  3. Finally, choose (as described in Sect. 2.2) an LWE matrix $\bar{\mathbf{D}} \in \mathbb{Z}_q^{nm\ell \times n\ell}$ that satisfies

$$(\mathbf{I}_{m\ell} \otimes \mathbf{t}) \cdot \bar{\mathbf{D}} \approx \mathbf{0},$$

and define $\mathbf{D} := \bar{\mathbf{D}} + (\mathbf{R} \otimes \mathbf{g}^t \otimes \mathbf{e}_n^t)$, where $\mathbf{e}_n \in \mathbb{Z}^n$ is the nth standard basis vector (so $\mathbf{t} \cdot \mathbf{e}^t = 1$). We therefore have

$$(\mathbf{I}_{m\ell} \otimes \mathbf{t}) \cdot \mathbf{D} \approx \mathbf{R} \otimes \mathbf{g}^t. \qquad (\text{error } E_\mathbf{D}) \qquad (6)$$

We view $\mathbf{D}$ as a kind of *encryption* of the commitment randomness $\mathbf{R}$.

- Dec(t, (**C**, **F**, **D**)): this is standard GSW decryption of **C** under **t**, which works due to Eq. (4).

Remark 1. The above scheme is defined in the symmetric-key setting, i.e., Enc uses the secret key **t** to generate LWE samples. We can obtain a public-key scheme using a standard technique, namely, have the encryption algorithm rerandomize some public LWE samples to generate as many additional samples as needed. More formally, we define $\mathbf{B} := \mathbf{A} - \mathbf{e}_n^t \otimes \mathbf{b}$. Then because $\mathbf{t} \cdot \mathbf{e}_n^t = 1$, we have

$$\mathbf{tB} \approx \mathbf{0}. \qquad\qquad (\text{error } E)$$

The public-key encryption algorithm then constructs $\bar{\mathbf{C}}, \bar{\mathbf{D}}$ by generating fresh samples as $\mathbf{B} \cdot \mathbf{x}$ for fresh uniformly random $\mathbf{x} \in \{0,1\}^m$. It is easy to verify that $\mathbf{t}(\mathbf{Bx}) \approx 0$ with error $m \cdot E$. Security follows from a standard argument, using the LWE assumption to make **b** (and thereby **B**) uniformly random, and then the leftover hash lemma to argue that the distribution of the fresh samples is negligibly far from uniform.

Theorem 1. *The above scheme is IND-CPA secure assuming the hardness of the decision-LWE$_{n-1,q,\chi}$ problem.*

Proof. We prove that the view of an attacker in the real game is indistinguishable from its view in a game in which the public extension key and every ciphertext are uniformly random and independent of the message; this clearly suffices for IND-CPA security. We proceed by a considering the following sequence of hybrid experiments:

Game 0: This is the real IND-CPA game.

Game 1: In this game the public extension key and the **C**, **D** components of every ciphertext are uniformly random and independent (but **F** is constructed in the same way). More precisely:

1. Choose uniformly random public parameter **A** and extension key **b**, and give them to the adversary.
2. For each encryption query, choose uniformly random and independent $\mathbf{C} \in \mathbb{Z}_q^{n \times n\ell}$ and $\mathbf{D} \in \mathbb{Z}_q^{nm\ell \times n\ell}$, construct **F** exactly as in Enc, and give ciphertext (**C**, **F**, **D**) to the adversary.

Game 2: This is the ideal game; the only change from the previous game is that each **F** is chosen uniformly at random.

We claim that Games 0 and 1 are computationally indistinguishable under the LWE hypothesis. To prove this we describe a simulator $\mathcal{S}$ that is given an unbounded source of samples; when they are LWE samples it simulates Game 0, and when they are uniformly random samples it simulates Game 1. It works as follows:

- Draw m samples and form a matrix $\bar{\mathbf{A}} \in \mathbb{Z}_q^{n \times m}$ with the samples as its columns. Choose a uniformly random extension key $\mathbf{b} \in \mathbb{Z}_q^m$, and let the public parameter $\mathbf{A} = \bar{\mathbf{A}} + \mathbf{e}_n^t \otimes \mathbf{b}$.

– On encryption query μ, draw samples to construct matrices $\bar{\mathbf{C}}$ and $\bar{\mathbf{D}}$, and define $\mathbf{C}, \mathbf{D}$ from these as in Enc. Also construct $\mathbf{F}$ exactly as in Enc.

If the simulator's input distribution is $A_{\bar{\mathbf{t}}, \chi}$ for some $\bar{\mathbf{t}} \leftarrow \chi^{n-1}$, then the first $n - 1$ rows of $\bar{\mathbf{A}}$ are uniformly random, hence $\mathbf{A}$ is uniformly random by construction. Moreover, $\mathbf{b} \approx (-\bar{\mathbf{t}}, 1) \cdot \mathbf{A}$ has the same distribution as in the real game. Finally, $\bar{\mathbf{C}}$ and $\bar{\mathbf{D}}$ are constructed exactly as in the real game, so $\mathcal{S}$ perfectly simulates Game 0.

By contrast, if the simulator's input distribution is uniform, then $\mathbf{A}$ and $\mathbf{b}$ are uniformly random and independent. Similarly, because $\bar{\mathbf{C}}$ and $\bar{\mathbf{D}}$ are uniform and independent of everything else, so are $\mathbf{C}$ and $\mathbf{D}$. Therefore, $\mathcal{S}$ perfectly simulates Game 1. This proves the first claim.

Finally, we claim that Games 1 and 2 are statistically indistinguishable. This follows directly from the leftover hash lemma. This concludes the proof.

3.1 Extending Ciphertexts

We first describe how to *extend* a ciphertext to an additional secret key $\mathbf{t}^*$, using the associated public extension key $\mathbf{b}^* \approx \mathbf{t}^* \mathbf{A} \in \mathbb{Z}_q^m$. More precisely, suppose we have a ciphertext that encrypts μ under secret key $\mathbf{t} \in \mathbb{Z}^{n'}$. (Here the dimension n' can be arbitrary, but typically $n' = nk$ for some positive integer k, and $\mathbf{t}$ is the concatenation of k individual secret keys, each of dimension n.) The ciphertext therefore consists of component matrices

$$\mathbf{C} \in \mathbb{Z}_q^{n' \times n'\ell}, \quad \mathbf{F} \in \mathbb{Z}_q^{n \times n\ell}, \quad \mathbf{D} \in \mathbb{Z}_q^{n'm\ell \times n\ell}$$

that satisfy Eqs. (4)–(6) for some short commitment randomness $\mathbf{R} \in \mathbb{Z}^{m \times n\ell}$. (Notice that the dimensions of $\mathbf{F}$ and the width of $\mathbf{D}$ do not depend on n'.)

Our goal is to extend $(\mathbf{C}, \mathbf{F}, \mathbf{D})$ to a new ciphertext $(\mathbf{C}', \mathbf{F}', \mathbf{D}')$ that satisfies Eqs. (4)–(6) with respect to the concatenated secret key $\mathbf{t}' = (\mathbf{t}, \mathbf{t}^*) \in \mathbb{Z}^{n'+n}$ and some short commitment randomness $\mathbf{R}'$. We do so as follows.

– The commitment and its randomness are unchanged: we define $\mathbf{F}' := \mathbf{F}$ and $\mathbf{R}' := \mathbf{R}$. This clearly preserves Eq. (5).
– Similarly, the encrypted randomness also is essentially unchanged, up to some padding by zeros: we define

$$\mathbf{D}' := (\mathbf{I}_{m\ell} \otimes \begin{pmatrix} \mathbf{I}_{n'} \\ \mathbf{0}_{n \times n'} \end{pmatrix})) \cdot \mathbf{D} \in \mathbb{Z}_q^{(n'+n)m\ell \times n\ell}.$$

Then Eq. (6) is preserved: $(\mathbf{I}_{m\ell} \otimes \mathbf{t}') \cdot \mathbf{D}' = (\mathbf{I}_{m\ell} \otimes \mathbf{t}) \cdot \mathbf{D} \approx \mathbf{R} \otimes \mathbf{g}^t = \mathbf{R}' \otimes \mathbf{g}^t$.
– Lastly, we define

$$\mathbf{C}' := \begin{pmatrix} \mathbf{C} \ \mathbf{X} \\ \mathbf{F} \end{pmatrix} \in \mathbb{Z}_q^{(n'+n) \times (n'+n)\ell}$$

where $\mathbf{X}$ is defined as follows:

$$\mathbf{s} := [-\mathbf{b}^*](\mathbf{I}_m \otimes \mathbf{g}^{-t}) \in \{0, 1\}^{m\ell}, \tag{7}$$
$$\mathbf{X} := (\mathbf{s} \otimes \mathbf{I}_{n'}) \cdot \mathbf{D} \in \mathbb{Z}_q^{n' \times n\ell}.$$

We now do the error analysis for ciphertext extension. Notice that by construction,

$$\begin{aligned} \mathbf{tX} &= (1 \otimes \mathbf{t}) \cdot (\mathbf{s} \otimes \mathbf{I}_{n'}) \cdot \mathbf{D} \\ &= (\mathbf{s} \otimes 1) \cdot (\mathbf{I}_{m\ell} \otimes \mathbf{t}) \cdot \mathbf{D} \\ &\approx \mathbf{s} \cdot (\mathbf{R} \otimes \mathbf{g}^t) && (\text{Eq. (6), error } m\ell \cdot E_{\mathbf{D}}) \\ &= -\mathbf{b}^* \mathbf{R}. && (\text{Eq. (7)}) \end{aligned}$$

Putting everything together, we see that Eq. (4) is preserved:

$$\begin{aligned} \mathbf{t}'\mathbf{C}' &\approx \big(\mu(\mathbf{t} \otimes \mathbf{g})\, \mathbf{tX} + \mathbf{t}^* \mathbf{F}\big) && (\text{Eq. (4); error } E_{\mathbf{C}}) \\ &= \big(\mu(\mathbf{t} \otimes \mathbf{g})\, \mathbf{tX} + \mathbf{t}^* \mathbf{AR} + \mu(\mathbf{t}^* \otimes \mathbf{g})\big) && (\text{Eq. (5)}) \\ &\approx \big(\mu(\mathbf{t} \otimes \mathbf{g})\, \mathbf{tX} + \mathbf{b}^* \mathbf{R} + \mu(\mathbf{t}^* \otimes \mathbf{g})\big) && (\text{Eq. (3); error } m\|\mathbf{R}\|_\infty \cdot E) \\ &\approx \mu(\mathbf{t}' \otimes \mathbf{g}). && (\text{error } m\ell \cdot E_{\mathbf{D}}) \end{aligned}$$

In total, the error in the new ciphertext $\mathbf{C}'$ is

$$E_{\mathbf{C}'} = E_{\mathbf{C}} + m\|\mathbf{R}\|_\infty \cdot E + m\ell \cdot E_{\mathbf{D}}.$$

We remark that the error growth is merely additive, so we can extend to multiple new keys with only additive error growth per key. This is important for bootstrapping a multi-key ciphertext, where the first step is to extend the circularly encrypted secret keys to the keys that the ciphertext is encrypted under.

3.2 Homomorphic Operations

We now describe homomorphic addition and multiplication for the above cryptosystem. Suppose we have two ciphertexts $(\mathbf{C}_1, \mathbf{F}_1, \mathbf{D}_1)$ and $(\mathbf{C}_2, \mathbf{F}_2, \mathbf{D}_2)$ that respectively encrypt μ_1 and μ_2, with commitment randomness $\mathbf{R}_1$ and $\mathbf{R}_2$, under a common secret key $\mathbf{t} \in \mathbb{Z}^{n'}$. (As in the previous subsection, everything below works for arbitrary dimension n' and key $\mathbf{t}$, but typically $n' = nk$ for some positive integer k, and $\mathbf{t}$ is the concatenation of k individual secret keys.) Recall that the ciphertext components

$$\mathbf{C}_i \in \mathbb{Z}_q^{n' \times n'\ell}, \quad \mathbf{F}_i \in \mathbb{Z}_q^{n \times n\ell}, \quad \mathbf{D}_i \in \mathbb{Z}_q^{n'm\ell \times n\ell}$$

satisfy Eqs. (4)–(6) for some short commitment randomness $\mathbf{R}_i \in \mathbb{Z}^{m \times n\ell}$.

- **Negation and scalar addition.** (These are used to homomorphically compute $\text{NAND}(\mu_1, \mu_2) = 1 - \mu_1\mu_2$ for $\mu_i \in \{0, 1\}$.) To homomorphically negate a message for a ciphertext $(\mathbf{C}, \mathbf{F}, \mathbf{D})$, just negate each of the components. It is clear that this has the desired effect, and that the associated commitment randomness and error terms are also negated. To homomorphically add a constant $c \in \mathbb{Z}$ to a message, just add $c(\mathbf{I}_{n'} \otimes \mathbf{g})$ to both $\mathbf{C}$ and $\mathbf{F}$. It is clear that this has the desired effect, and leaves the commitment randomness and error terms unchanged.

- **Addition.** To homomorphically add, we simply add the corresponding matrices, outputting

$$(\mathbf{C}_{\mathrm{add}}, \mathbf{F}_{\mathrm{add}}, \mathbf{D}_{\mathrm{add}}) := (\mathbf{C}_1 + \mathbf{C}_2, \mathbf{F}_1 + \mathbf{F}_2, \mathbf{D}_1 + \mathbf{D}_2).$$

It is easy to verify that Eqs. (4)–(6) hold for the new ciphertext with message $\mu_{\mathrm{add}} = \mu_1 + \mu_2$ and commitment randomness $\mathbf{R}_{\mathrm{add}} = \mathbf{R}_1 + \mathbf{R}_2$, where the errors in the approximations are also added.

- **Multiplication.** To homomorphically multiply, we define the short matrices

$$\mathbf{S}_c := (\mathbf{I}_{n'} \otimes \mathbf{g}^{-1})[\mathbf{C}_2] \in \{0,1\}^{n'\ell \times n'\ell}, \tag{8}$$

$$\mathbf{S}_f := (\mathbf{I}_n \otimes \mathbf{g}^{-1})[\mathbf{F}_2] \in \{0,1\}^{n\ell \times n\ell}, \tag{9}$$

$$\mathbf{S}_d := (\mathbf{I}_{n'm\ell} \otimes \mathbf{g}^{-1})[\mathbf{D}_2] \in \{0,1\}^{n'm\ell^2 \times n\ell}, \tag{10}$$

and output the ciphertext consisting of

$$\mathbf{C}_{\mathrm{mul}} := \mathbf{C}_1 \cdot \mathbf{S}_c$$
$$\mathbf{F}_{\mathrm{mul}} := \mathbf{F}_1 \cdot \mathbf{S}_f$$
$$\mathbf{D}_{\mathrm{mul}} := \mathbf{D}_1 \cdot \mathbf{S}_f + (\mathbf{I}_{m\ell} \otimes \mathbf{C}_1) \cdot \mathbf{S}_d.$$

The associated commitment randomness is defined as

$$\mathbf{R}_{\mathrm{mul}} := \mathbf{R}_1 \cdot \mathbf{S}_f + \mu_1 \mathbf{R}_2.$$

We now show that the ciphertext output by homomorphic multiplication satisfies Eqs. (4)–(6) for key $\mathbf{t}$, message $\mu_{\mathrm{mul}} = \mu_1 \mu_2$, and commitment randomness $\mathbf{R}_{\mathrm{mul}}$. We already know that Eq. (4), the GSW ciphertext relation, is satisfied by construction of $\mathbf{C}_{\mathrm{mul}}$ as the homomorphic product of GSW ciphertexts $\mathbf{C}_1, \mathbf{C}_2$. Specifically:

$$
\begin{aligned}
\mathbf{t}\mathbf{C}_{\mathrm{mult}} &= \mathbf{t}\mathbf{C}_1 \cdot \mathbf{S}_c \\
&\approx \mu_1(\mathbf{t} \otimes \mathbf{g}) \cdot \mathbf{S}_c && (\text{error } n'\ell \cdot E_{\mathbf{C}_1}) \\
&= \mu_1 \mathbf{t}\mathbf{C}_2 && (\text{Eq. (8)}) \\
&\approx \mu_1 \mu_2 (\mathbf{t} \otimes \mathbf{g}). && (\text{error } \mu_1 E_{\mathbf{C}_2})
\end{aligned}
$$

Similarly, Eq. (5) is satisfied by construction of $\mathbf{F}_{\mathrm{mul}}$ as the homomorphic product of commitments $\mathbf{F}_1, \mathbf{F}_2$:

$$
\begin{aligned}
\mathbf{F}_{\mathrm{mul}} &= \mathbf{F}_1 \cdot \mathbf{S}_f \\
&= (\mathbf{A}\mathbf{R}_1 + \mu_1(\mathbf{I}_n \otimes \mathbf{g})) \cdot \mathbf{S}_f \\
&= \mathbf{A}\mathbf{R}_1 \cdot \mathbf{S}_f + \mu_1 \mathbf{F}_2 && (\text{Eq. (9)}) \\
&= \mathbf{A}\mathbf{R}_1 \cdot \mathbf{S}_f + \mu_1 \mathbf{A}\mathbf{R}_2 + \mu_1 \mu_2 (\mathbf{I}_n \otimes \mathbf{g}) \\
&= \mathbf{A}\mathbf{R}_{\mathrm{mult}} + \mu_1 \mu_2 (\mathbf{I}_n \otimes \mathbf{g}).
\end{aligned}
$$

Finally, to see that Eq. (6) holds for $\mathbf{D}_{\mathrm{mul}}$, first notice that

$$(\mathbf{I}_{m\ell} \otimes \mathbf{t}) \cdot \mathbf{D}_1 \cdot \mathbf{S}_f \approx (\mathbf{R}_1 \otimes \mathbf{g}^t) \cdot (\mathbf{S}_f \otimes 1) \qquad \text{(Eqs. (6); error } n\ell \cdot E_{\mathbf{D}_1})$$
$$= (\mathbf{R}_1 \cdot \mathbf{S}_f) \otimes \mathbf{g}^t. \qquad\qquad\qquad\qquad (11)$$

In addition,

$$(\mathbf{I}_{m\ell} \otimes \mathbf{t}) \cdot (\mathbf{I}_{m\ell} \otimes \mathbf{C}_1) \cdot \mathbf{S}_d = (\mathbf{I}_{m\ell} \otimes \mathbf{t}\mathbf{C}_1) \cdot \mathbf{S}_d$$
$$\approx \mu_1 (\mathbf{I}_{m\ell} \otimes \mathbf{t} \otimes \mathbf{g}) \cdot \mathbf{S}_d \quad \text{(Eq. (4); error } n'\ell \cdot E_{\mathbf{C}_1})$$
$$= \mu_1 (\mathbf{I}_{m\ell} \otimes \mathbf{t}) \cdot \mathbf{D}_2 \qquad\qquad \text{(Eq. (10))}$$
$$\approx (\mu_1 \mathbf{R}_2) \otimes \mathbf{g}^t \qquad\qquad \text{(Eq. (6); error } \mu_1 \cdot E_{\mathbf{D}_2})$$
$$\qquad\qquad\qquad\qquad\qquad\qquad\qquad\qquad (12)$$

Summing Eqs. (11) and (12) yields

$$(\mathbf{I}_{m\ell} \otimes \mathbf{t}) \cdot \mathbf{D}_{\mathrm{mul}} \approx \mathbf{R}_{\mathrm{mul}} \otimes \mathbf{g}^t$$

with error $n\ell \cdot E_{\mathbf{D}_1} + n'\ell \cdot E_{\mathbf{C}_1} + \mu_1 \cdot E_{\mathbf{D}_2}$ as desired.

3.3   Instantiating the Parameters

We now bound the worst-case error growth when homomorphically evaluating a depth-d circuit of NAND gates for up to k individual keys. As above, let $n' = nk$. For a ciphertext $(\mathbf{C}, \mathbf{F}, \mathbf{D})$ with commitment randomness $\mathbf{R}$, define the "max error"

$$E^* := \max(E_{\mathbf{C}}, E_{\mathbf{D}}, E \cdot \|\mathbf{R}\|_\infty).$$

By the bounds from the previous subsection, for two ciphertexts with max error at most E^*, their homomorphic NAND has max error at most $(n(k + 1)\ell + 1) \cdot E^* = \mathrm{poly}(n, k, \ell) \cdot E^*$. Similarly, when we extend a ciphertext with max error at most E^*, the result has max error at most $(m(\ell + 1) + 1) \cdot E^* = \mathrm{poly}(n, \ell) \cdot E^*$. Therefore, for any depth-d homomorphic computation on fresh ciphertexts encrypted under k keys, the result has max error at most

$$\mathrm{poly}(n, k, \ell)^{k+d}.$$

The GSW decryption algorithm works correctly on a ciphertext as long as its error is smaller than $q/4$, hence it suffices to choose a modulus q that exceeds the above quantity by a factor of four. Recalling that $\ell = \Theta(\log q) = \tilde{O}(k + d)$, this corresponds to a worst-case approximation factor of $\mathrm{poly}(n, k, d)^{k+d}$ for n-dimensional lattice problems.

We also remark that when bootstrapping a k-key ciphertext, we first extend the circularly encrypted secret keys to the k relevant keys, incurring only additive $\mathrm{poly}(n, k, \ell)$ error growth, then we run the bootstrapping algorithm. Using an algorithm from [2,7] that incurs only additive $\mathrm{poly}(n, k, \ell)$ error growth, we can use a modulus q that is as small as slightly super-polynomial $q = n^{\omega(1)}$ and still support any polynomial number of keys.

4   Smaller-Ciphertext Construction

In this section we describe a multi-hop, multi-key FHE having smaller cipher-texts and more efficient homomorphic operations than the one in Sect. 3. Indeed, ciphertexts in this system are simply GSW ciphertexts (with no additional information), which admit the usual homomorphic operations. These efficiency improvements come at the cost of larger public extension keys, as well as a circular-security assumption.

Recall that in the scheme from the previous section, a ciphertext includes a commitment to the message, along with a special encryption of the commitment randomness. By contrast, in the scheme described below, the *extension key* contains a commitment to the *secret key*, along with an encryption (under the secret key) of the commitment randomness. (Using the commitment randomness to hide the secret key, and using the secret key to hide the commitment randomness, is what leads to a circular-security assumption.) We show how to combine two extension keys to get an encryption, under the concatenation of the secret keys, of the *tensor product* of those keys; this in turn lets us extend a ciphertext encrypted under one of the keys to their concatenation. We now describe the construction.

As in the previous section, the scheme is parameterized by LWE parameters n and q, the standard error distribution χ (which is E-bounded for $E = \Theta(n)$), and $m = \lceil 2n \log q \rceil$. The system is defined as follows.

- Setup: output a uniformly random $\mathbf{A} \in \mathbb{Z}_q^{n \times m}$.
- Gen($\mathbf{A}$): do the following, outputting $\mathbf{t}$ as the secret key and $(\mathbf{b}, \mathbf{P}, \mathbf{D})$ as the public extension key.
  1. Choose $\bar{\mathbf{t}} \leftarrow \chi^{n-1}$ and define $\mathbf{t} := (-\bar{\mathbf{t}}, 1) \in \mathbb{Z}^n$. Choose $\mathbf{e} \leftarrow \chi^m$ and define

$$\mathbf{b} := \mathbf{t}\mathbf{A} + \mathbf{e}$$
$$\approx \mathbf{t}\mathbf{A} \in \mathbb{Z}_q^m. \qquad\qquad (\text{error } E)$$

  2. Choose a uniformly random $\mathbf{R} \leftarrow \{0, 1\}^{m \times n^2 \ell}$ and define

$$\mathbf{P} := \mathbf{A}\mathbf{R} + (\mathbf{I}_n \otimes \mathbf{t} \otimes \mathbf{g}) \in \mathbb{Z}_q^{n \times n^2 \ell}.$$

  3. As described in Sect. 2.2, choose an LWE matrix $\bar{\mathbf{D}} \in \mathbb{Z}_q^{nm\ell \times n^2\ell}$ that satisfies $(\mathbf{I}_{m\ell} \otimes \mathbf{t}) \cdot \bar{\mathbf{D}} \approx \mathbf{0}$ (with error E), and define $\mathbf{D} := \bar{\mathbf{D}} + (\mathbf{R} \otimes \mathbf{g}^t \otimes \mathbf{e}_n^t)$, where $\mathbf{e}_n \in \{0, 1\}^n$ denotes the nth standard basis vector. Notice that, because $\mathbf{t} \cdot \mathbf{e}_n^t = 1$, we have

$$(\mathbf{I}_{m\ell} \otimes \mathbf{t}) \cdot \mathbf{D} \approx \mathbf{R} \otimes \mathbf{g}^t. \qquad\qquad (\text{error } E)$$

- Enc($\mathbf{t}, \mu \in \{0, 1\}$): This is standard GSW encryption. Specifically, as described in Sect. 2.2, choose an LWE matrix $\bar{\mathbf{C}} \in \mathbb{Z}_q^{n \times n\ell}$ that satisfies $\mathbf{t}\bar{\mathbf{C}} \approx \mathbf{0}$, and output the ciphertext $\mathbf{C} := \bar{\mathbf{C}} + \mu(\mathbf{I}_n \otimes \mathbf{g})$. Notice that $\mathbf{t}, \mathbf{C}$ satisfy the GSW relation

$$\mathbf{t}\mathbf{C} = \mathbf{t}\bar{\mathbf{C}} + \mu(\mathbf{t} \otimes 1) \cdot (\mathbf{I}_n \otimes \mathbf{g}) \approx \mu(\mathbf{t} \otimes \mathbf{g}). \qquad (\text{error } E_\mathbf{C})$$

- Dec($\mathbf{t}, \mathbf{C}$): this is standard GSW decryption.

We again stress that ciphertexts in the above system are just GSW ciphertexts (with no auxiliary information), so homomorphic addition and multiplication work as usual (and as in Sect. 3). Therefore, we only need to show how to extend ciphertexts to new keys, which we do below in Sect. 4.1.

For security, we rely on the following circular hardness assumption: that LWE samples for secret $\bar{t} \leftarrow \chi^n$ are indistinguishable from uniform, even given $(\mathbf{b}, \mathbf{A}, \mathbf{P}, \mathbf{D})$ as constructed by Setup and Gen (using secret $\bar{t}$). We remark that this assumption is "circular" because $\mathbf{D}$ computationally hides (but statistically determines) $\mathbf{R}$ under $\bar{t}$, and $\mathbf{P}$ hides $\bar{t}$ using $\mathbf{R}$.

The proof of the following theorem follows immediately from the assumption.

Theorem 2. *The above scheme is IND-CPA secure under the above circular-security assumption.*

Proof. The proof follows immediately from the assumption: in the real IND-CPA game, the adversary gets the public information $(\mathbf{b}, \mathbf{A}, \mathbf{P}, \mathbf{D})$ along with ciphertexts generated from LWE samples with secret $\bar{t}$. In the ideal world, these samples are instead uniformly random, and hence perfectly hide the encrypted messages. Indistinguishability of the two worlds follows directly from the circular-security assumption.

4.1   Extending a Ciphertext to a New Key

We now show how to extend a (potentially multi-key) ciphertext to an additional key, so as to preserve the GSW relation for the concatenation of the secret keys. Specifically, suppose we have a ciphertext $\mathbf{C} \in \mathbb{Z}_q^{n' \times n'\ell}$ that encrypts μ under a key $\mathbf{t} \in \mathbb{Z}^{n'}$, i.e.,

$$\mathbf{t}\mathbf{C} \approx \mu(\mathbf{t} \otimes \mathbf{g}). \qquad\qquad (\text{error } E_{\mathbf{C}})$$

In this setting, $n' = nk$ for some positive integer $k \geq 1$, and $\mathbf{t} = (\mathbf{t}_1, \ldots, \mathbf{t}_k)$ is the concatenation of k individual secret keys $\mathbf{t}_i \in \mathbb{Z}^n$ for which we know the associated vector $\mathbf{b}_i \approx \mathbf{t}_i \mathbf{A} \in \mathbb{Z}_q^m$ (with error E) from the public extension key. (We will not need the extension key's other components $\mathbf{P}, \mathbf{D}$.)

We wish to extend $\mathbf{C}$ to an additional secret key $\mathbf{t}^*$ for which we know the associated matrices $\mathbf{P}^*, \mathbf{D}^*$ from the public extension key (we will not need the associated $\mathbf{b}^*$). More precisely, we want to generate a ciphertext $\mathbf{C}'$ that encrypts μ under $\mathbf{t}' = (\mathbf{t}, \mathbf{t}^*) \in \mathbb{Z}^{n(k+1)}$, i.e., we want

$$\mathbf{t}'\mathbf{C}' \approx \mu(\mathbf{t}' \otimes \mathbf{g}) = \mu\left(\mathbf{t} \otimes \mathbf{g} \; \mathbf{t}^* \otimes \mathbf{g}\right).$$

To do this, we output

$$\mathbf{C}' := \begin{pmatrix} \mathbf{C} & \mathbf{X} \\ & \mathbf{X}^* \end{pmatrix} \qquad\qquad (13)$$

where $\mathbf{X}' = \begin{pmatrix} \mathbf{X} \\ \mathbf{X}^* \end{pmatrix}$ is as defined below. Notice that by construction,

$$\mathbf{t}'\mathbf{C}' \approx \left(\mu(\mathbf{t} \otimes \mathbf{g}) \; \mathbf{t}'\mathbf{X}'\right). \qquad\qquad (\text{error } E_{\mathbf{C}})$$

Below we show how to satisfy

$$\mathbf{t'X'} = \mathbf{tX} + \mathbf{t^*X^*} \approx \mu(\mathbf{t^*} \otimes \mathbf{g}) \tag{14}$$

with error

$$E_{\mathbf{X'}} = (n^2 \cdot (k\ell + 1)^2 \cdot m + E_{\mathbf{C}}) \cdot E,$$

which yields $\mathbf{t'C'} \approx \mu(\mathbf{t'} \otimes \mathbf{g})$ with error $E_{\mathbf{C'}} = \max\{E_{\mathbf{C}}, E_{\mathbf{X'}}\} = E_{\mathbf{X'}}$, as desired.

Remark 2. While the error bound $E_{\mathbf{C'}} = \mathbf{E_C} \cdot E + \text{poly}(n, k, \ell)$ is multiplicative in the original error $E_{\mathbf{C}}$, we can still extend to multiple new keys while incurring just one factor-of-E increase in the error. This is important for bootstrapping a multi-key ciphertext, where the first step is to extend the circularly encrypted secret keys to all the keys that the ciphertext is encrypted under. The method works by naturally generalizing Eq. (13) to a matrix with blocks along the diagonal and top row only.

Constructing $\mathbf{X'}$. We construct $\mathbf{X'}$ in two steps:

1. Using just the $\mathbf{b}_i$ and $\mathbf{P^*}$, $\mathbf{D^*}$ (but not the ciphertext $\mathbf{C}$), we construct $\mathbf{Y'} = \begin{pmatrix} \mathbf{Y} \\ \mathbf{Y^*} \end{pmatrix}$ that satisfies

$$\mathbf{t'Y'} = \mathbf{tY} + \mathbf{t^*Y^*} \approx (\mathbf{t} \otimes \mathbf{t^*} \otimes \mathbf{g}) \tag{15}$$

   with error $E_{\mathbf{Y'}} = (k\ell + 1) \cdot m \cdot E$. This construction is described below.

2. We then obtain $\mathbf{X'}$ by multiplying $\mathbf{Y'}$ by a certain binary matrix that is derived from the ciphertext $\mathbf{C}$. Essentially, this step just replaces $\mathbf{t}$ with $\mu\mathbf{g}$ in the right-hand side of Eq. (15), while consuming the existing $\mathbf{g}$. Let $\bar{\mathbf{C}} := \mathbf{C} \cdot (\mathbf{e}_n^t \otimes \mathbf{I}_\ell) \in \mathbb{Z}_q^{nk \times \ell}$ consist of the last ℓ columns of $\mathbf{C}$, so that

$$\mathbf{t}\bar{\mathbf{C}} \approx \mu(\mathbf{t} \otimes \mathbf{g}) \cdot (\mathbf{e}_n^t \otimes \mathbf{I}_\ell) = \mu\mathbf{g}. \qquad (\text{error } E_{\mathbf{C}}) \tag{16}$$

Define the binary matrix

$$\mathbf{S} := (\mathbf{I}_{nk} \otimes \mathbf{I}_n \otimes \mathbf{g}^{-1})\left[\bar{\mathbf{C}} \otimes \mathbf{I}_n\right] \in \{0,1\}^{n^2 k\ell \times \ell}, \tag{17}$$

and observe that

$$
\begin{aligned}
\mathbf{t'Y'} \cdot \mathbf{S} &\approx (\mathbf{t} \otimes \mathbf{t^*} \otimes \mathbf{g}) \cdot \mathbf{S} && (\text{Eq. (15); error } n^2 k\ell \cdot E_{\mathbf{Y'}}) \\
&= (\mathbf{t} \otimes \mathbf{t^*}) \cdot (\bar{\mathbf{C}} \otimes \mathbf{I}_n) && (\text{Eq. (17)}) \\
&= (\mathbf{t}\bar{\mathbf{C}}) \otimes \mathbf{t^*} \\
&\approx \mu(\mathbf{g} \otimes \mathbf{t^*}). && (\text{Eq. (16)}, \|\mathbf{t^*}\|_\infty \leq E, \text{ so error } E_{\mathbf{C}} \cdot E) \tag{18}
\end{aligned}
$$

Notice that the right-hand side of Eq. (18) is exactly the desired right-hand side of Eq. (14), but permuted (because the arguments of the Kronecker product are swapped). So let $\boldsymbol{\Pi}$ be the permutation matrix for which $(\mathbf{g} \otimes \mathbf{t^*})\boldsymbol{\Pi} = (\mathbf{t^*} \otimes \mathbf{g})$ for any $\mathbf{t^*}$, and define

$$\mathbf{X'} := \mathbf{Y'} \cdot \mathbf{S} \cdot \boldsymbol{\Pi},$$

which by the above satisfies Eq. (14), as desired.

Constructing $\mathbf{Y}'$. We now describe the construction of $\mathbf{Y}' = \begin{pmatrix} \mathbf{Y} \\ \mathbf{Y}^* \end{pmatrix}$ to satisfy Eq. (15). To do this we use the public matrices $\mathbf{P}^*, \mathbf{D}^*$ associated with $\mathbf{t}^*$, which by construction satisfy

$$\mathbf{P}^* = \mathbf{A}\mathbf{R}^* + (\mathbf{I}_n \otimes \mathbf{t}^* \otimes \mathbf{g})$$

$$(\mathbf{I}_{m\ell} \otimes \mathbf{t}^*) \cdot \mathbf{D}^* \approx \mathbf{R}^* \otimes \mathbf{g}^t \qquad\qquad \text{(error } E) \qquad\qquad (19)$$

for some binary matrix $\mathbf{R}^* \in \{0,1\}^{m \times n^2\ell}$. Recalling that $\mathbf{t} \in \mathbb{Z}^{nk}$ is the concatenation of k individual secret keys $\mathbf{t}_i \in \mathbb{Z}^n$, we also define $\mathbf{b} \in \mathbb{Z}_q^{mk}$ to be the concatenation of the associated $\mathbf{b}_i \approx \mathbf{t}_i\mathbf{A} \in \mathbb{Z}_q^m$ (all with error E), so

$$\mathbf{b} \approx \mathbf{t} \cdot (\mathbf{I}_k \otimes \mathbf{A}). \qquad\qquad \text{(error } E) \qquad\qquad (20)$$

First, we define

$$\mathbf{Y} := \mathbf{I}_k \otimes \mathbf{P}^* = (\mathbf{I}_k \otimes \mathbf{A}\mathbf{R}^*) + (\mathbf{I}_{nk} \otimes \mathbf{t}^* \otimes \mathbf{g}).$$

Observe that

$$
\begin{aligned}
\mathbf{t}\mathbf{Y} &= \mathbf{t} \cdot (\mathbf{I}_k \otimes \mathbf{A}\mathbf{R}^*) + (\mathbf{t} \otimes 1 \otimes 1) \cdot (\mathbf{I}_{nk} \otimes \mathbf{t}^* \otimes \mathbf{g}) \\
&= \mathbf{t} \cdot (\mathbf{I}_k \otimes \mathbf{A}) \cdot (\mathbf{I}_k \otimes \mathbf{R}^*) + (\mathbf{t} \otimes \mathbf{t}^* \otimes \mathbf{g}) \\
&\approx \mathbf{b} \cdot (\mathbf{I}_k \otimes \mathbf{R}^*) + (\mathbf{t} \otimes \mathbf{t}^* \otimes \mathbf{g}). \qquad\qquad \text{(Eq. (20); error } m \cdot E.)
\end{aligned}
$$

Therefore, in order to satisfy Eq. (15), it suffices to construct $\mathbf{Y}^*$ to satisfy

$$\mathbf{t}^*\mathbf{Y}^* \approx -\mathbf{b} \cdot (\mathbf{I}_k \otimes \mathbf{R}^*).$$

with error $km\ell \cdot E$. To do this, we define

$$\mathbf{s} := -[\mathbf{b}](\mathbf{I}_k \otimes \mathbf{I}_m \otimes \mathbf{g}^{-t}) \in \{0,1\}^{km\ell} \qquad\qquad (21)$$

$$\mathbf{Y}^* := (\mathbf{s} \otimes \mathbf{I}_n) \cdot (\mathbf{I}_k \otimes \mathbf{D}^*).$$

Then observe that

$$
\begin{aligned}
\mathbf{t}^*\mathbf{Y}^* &= (1 \otimes \mathbf{t}^*) \cdot (\mathbf{s} \otimes \mathbf{I}_n) \cdot (\mathbf{I}_k \otimes \mathbf{D}^*) \\
&= (\mathbf{s} \otimes 1) \cdot (\mathbf{I}_{km\ell} \otimes \mathbf{t}^*) \cdot (\mathbf{I}_k \otimes \mathbf{D}^*) \\
&\approx \mathbf{s} \cdot (\mathbf{I}_k \otimes \mathbf{R}^* \otimes \mathbf{g}^t) \qquad\qquad \text{(Eq. (19); error } km\ell \cdot E) \\
&= -\mathbf{b} \cdot (\mathbf{I}_k \otimes \mathbf{R}^*) \qquad\qquad \text{(Eq. (21))}
\end{aligned}
$$

as desired. This completes the construction and analysis.

4.2   Instantiating the Parameters

We now bound the worst-case error growth when homomorphically evaluating a depth-d circuit of NAND gates for up to k individual keys. As above, let $n' = nk$. For two ciphertexts with error bounded by E^*, their homomorphic

NAND has error bounded by $(n'\ell + 1) \cdot E^* = \text{poly}(n, k, \ell) \cdot E^*$. Similarly, when we extend a ciphertext with error bounded by E^*, the result has error bounded by $(n^2 \cdot (k\ell+1)^2 \cdot m + E^*) \cdot E = \text{poly}(n, k, \ell) \cdot E^*$. Therefore, for any depth-d homomorphic computation on fresh ciphertexts encrypted under k keys, the result has error bounded by $\text{poly}(n, k, \ell)^{k+d}$. Therefore, it suffices to choose a modulus q that exceeds four times this bound. Recalling that $\ell = \Theta(\log q) = \tilde{O}(k + d)$, this corresponds to a worst-case approximation factor of $\text{poly}(n, k, d)^{k+d}$ for n-dimensional lattice problems.

We also remark that when bootstrapping a k-key ciphertext, we first extend the circularly encrypted secret keys to the k relevant keys, incurring only a single factor-of-E plus additive $\text{poly}(n, k, \ell)$ error growth, then we run the bootstrapping algorithm. Using an algorithm from [2,7] that incurs only additive $\text{poly}(n, k, \ell)$ error growth, we can use a modulus q that is as small as slightly super-polynomial $q = n^{\omega(1)}$ and still support any polynomial number of keys.

References

1. Albrecht, M., Bai, S., Ducas, L.: A subfield lattice attack on overstretched NTRU assumptions. In: Robshaw, M., Katz, J. (eds.) CRYPTO 2016. LNCS, vol. 9814, pp. 153–178. Springer, Heidelberg (2016). doi:10.1007/978-3-662-53018-4_6
2. Alperin-Sheriff, J., Peikert, C.: Faster bootstrapping with polynomial error. In: Garay, J.A., Gennaro, R. (eds.) CRYPTO 2014. LNCS, vol. 8616, pp. 297–314. Springer, Heidelberg (2014). doi:10.1007/978-3-662-44371-2_17
3. Applebaum, B., Cash, D., Peikert, C., Sahai, A.: Fast cryptographic primitives and circular-secure encryption based on hard learning problems. In: Halevi, S. (ed.) CRYPTO 2009. LNCS, vol. 5677, pp. 595–618. Springer, Heidelberg (2009). doi:10.1007/978-3-642-03356-8_35
4. Asharov, G., Jain, A., López-Alt, A., Tromer, E., Vaikuntanathan, V., Wichs, D.: Multiparty computation with low communication, computation and interaction via threshold FHE. In: Pointcheval, D., Johansson, T. (eds.) EUROCRYPT 2012. LNCS, vol. 7237, pp. 483–501. Springer, Heidelberg (2012). doi:10.1007/978-3-642-29011-4_29
5. Brakerski, Z., Langlois, A., Peikert, C., Regev, O., Stehlé, D.: Classical hardness of learning with errors. In: STOC, pp. 575–584 (2013)
6. Brakerski, Z., Perlman, R.: Lattice-based fully dynamic multi-key FHE with short ciphertexts. In: Robshaw, M., Katz, J. (eds.) CRYPTO 2016. LNCS, vol. 9814, pp. 190–213. Springer, Heidelberg (2016). doi:10.1007/978-3-662-53018-4_8
7. Brakerski, Z., Vaikuntanathan, V.: Lattice-based FHE as secure as PKE. In: ITCS, pp. 1–12 (2014)
8. Clear, M., McGoldrick, C.: Multi-identity and multi-key leveled FHE from learning with errors. In: Gennaro, R., Robshaw, M. (eds.) CRYPTO 2015. LNCS, vol. 9216, pp. 630–656. Springer, Heidelberg (2015). doi:10.1007/978-3-662-48000-7_31
9. Gentry, C.: A fully homomorphic encryption scheme. PhD thesis, Stanford University (2009). http://crypto.stanford.edu/craig
10. Gentry, C., Halevi, S., Vaikuntanathan, V.: i-hop homomorphic encryption and rerandomizable Yao circuits. In: Rabin, T. (ed.) CRYPTO 2010. LNCS, vol. 6223, pp. 155–172. Springer, Heidelberg (2010). doi:10.1007/978-3-642-14623-7_9

11. Gentry, C., Sahai, A., Waters, B.: Homomorphic encryption from learning with errors: conceptually-simpler, asymptotically-faster, attribute-based. In: Canetti, R., Garay, J.A. (eds.) CRYPTO 2013. LNCS, vol. 8042, pp. 75–92. Springer, Heidelberg (2013). doi:10.1007/978-3-642-40041-4_5

12. Gorbunov, S., Vaikuntanathan, V., Wichs, D.: Leveled fully homomorphic signatures from standard lattices. In: STOC, pp. 469–477 (2015)

13. Hoffstein, J., Pipher, J., Silverman, J.H.: NTRU: A ring-based public key cryptosystem. In: Buhler, J.P. (ed.) ANTS 1998. LNCS, vol. 1423, pp. 267–288. Springer, Heidelberg (1998). doi:10.1007/BFb0054868

14. Kirchner, P., Fouque, P.-A.: Comparison between subfield and straightforward attacks on NTRU. Cryptology ePrint Archive, Report 2016/717 (2016). http://eprint.iacr.org/2016/717

15. López-Alt, A., Tromer, E., Vaikuntanathan, V.: On-the-fly multiparty computation on the cloud via multikey fully homomorphic encryption. In: STOC, pp. 1219–1234 (2012)

16. Lyubashevsky, V., Peikert, C., Regev, O.: On ideal lattices, learning with errors over rings. J. ACM 60(6), 43:1–43:35 (2013). Preliminary version in Eurocrypt 2010

17. Micciancio, D., Peikert, C.: Trapdoors for lattices: simpler, tighter, faster, smaller. In: Pointcheval, D., Johansson, T. (eds.) EUROCRYPT 2012. LNCS, vol. 7237, pp. 700–718. Springer, Heidelberg (2012). doi:10.1007/978-3-642-29011-4_41

18. Mukherjee, P., Wichs, D.: Two round multiparty computation via multi-key FHE. In: Fischlin, M., Coron, J.-S. (eds.) EUROCRYPT 2016. LNCS, vol. 9666, pp. 735–763. Springer, Heidelberg (2016). doi:10.1007/978-3-662-49896-5_26

19. Peikert, C.: Public-key cryptosystems from the worst-case shortest vector problem. In: STOC, pp. 333–342 (2009)

20. Regev, O.: On lattices, learning with errors, random linear codes, and cryptography. J. ACM 56(6), 1–40 (2009). Preliminary version in STOC 2005

Obfuscation and Multilinear Maps

Secure Obfuscation in a Weak Multilinear Map Model

Sanjam Garg[1]([✉]), Eric Miles[2], Pratyay Mukherjee[1], Amit Sahai[2],
Akshayaram Srinivasan[1], and Mark Zhandry[3,4]

[1] University of California, Berkeley, USA
{sanjamg,pratyay85,akshayaram}@berkeley.edu
[2] UCLA and Center for Encrypted Functionalities, Los Angeles, USA
{enmiles,sahai}@cs.ucla.edu
[3] MIT, Cambridge, USA
[4] Princeton University, Princeton, USA
mzhandry@princeton.edu

Abstract. All known candidate indistinguishability obfuscation (iO) schemes rely on candidate multilinear maps. Until recently, the strongest proofs of security available for iO candidates were in a generic model that only allows "honest" use of the multilinear map. Most notably, in this model the zero-test procedure only reveals whether an encoded element is 0, and nothing more.

However, this model is inadequate: there have been several attacks on multilinear maps that exploit extra information revealed by the zero-test procedure. In particular, Miles, Sahai and Zhandry (Crypto'16) recently gave a polynomial-time attack on several iO candidates when instantiated with the multilinear maps of Garg, Gentry, and Halevi (Eurocrypt'13), and also proposed a new "weak multilinear map model" that captures all known polynomial-time attacks on GGH13.

In this work, we give a new iO candidate which can be seen as a small modification or generalization of the original candidate of Garg, Gentry,

This paper is a merged version of [GMS16, MSZ16b].

S. Garg, P. Mukherjee and A. Srinivasan—Research supported in part from a DARPA/ARL SAFEWARE award, AFOSR Award FA9550-15-1-0274, NSF CRII Award 1464397 and an Okawa Foundation Research Grant. The views expressed are those of the authors and do not reflect the official policy or position of the funding agencies.

E. Miles and A. Sahai—Research supported in part from a DARPA/ARL SAFEWARE award, NSF Frontier Award 1413955, NSF grants 1228984, 1136174, 1118096, and 1065276, a Xerox Faculty Research Award, a Google Faculty Research Award, an equipment grant from Intel, and an Okawa Foundation Research Grant. This material is based upon work supported by the Defense Advanced Research Projects Agency through the ARL under Contract W911NF-15-C-0205. The views expressed are those of the author and do not reflect the official policy or position of the Department of Defense, the National Science Foundation, or the U.S. Government.

M. Zhandry—Supported in part by the Defense Advanced Research Projects Agency (DARPA) and the U.S. Army Research Office under contract number W911NF-15-C-0226.

M. Hirt and A. Smith (Eds.): TCC 2016-B, Part II, LNCS 9986, pp. 241–268, 2016.
DOI: 10.1007/978-3-662-53644-5_10

Halevi, Raykova, Sahai, and Waters (FOCS'13). We prove its security in the weak multilinear map model, thus giving the first iO candidate that is provably secure against all known polynomial-time attacks on GGH13. The proof of security relies on a new assumption about the hardness of computing annihilating polynomials, and we show that this assumption is implied by the existence of pseudorandom functions in NC^1.

1   Introduction

Candidates for multilinear maps [GGH13a, CLT13, GGH15, CLT15, Hal15], also called graded encoding schemes, have formed the substrate for achieving the important goal of general-purpose indistinguishability obfuscation (iO) [BGI+01, BGI+12]. Several iO candidates have appeared in the literature starting with the work of [GGH+13b]. However, all known proofs of security for candidate obfuscation schemes have relied on assumptions that are justified only in a generic multilinear group model, where, informally speaking, the adversary is limited to using the multilinear map only in an honest manner. Most notably, this model allows the adversary to submit encodings for a zero test, and in the model the adversary only learns whether the encoding is an encoding of zero or not, and nothing more.

Unfortunately this last aspect of the modeling of multilinear maps has proven extremely elusive to achieve in multilinear map candidates: zero testing seems to reveal quite a bit more than just whether an encoded element is zero or not. Indeed, all candidate constructions of multilinear maps have been shown to suffer from "zeroizing" attacks [GGH13a, CHL+15, BWZ14, CGH+15, HJ15, BGH+15, Hal15, CLR15, MF15, MSZ16a] that show how to exploit the additional information leaked by zero testing to attack various schemes constructed on top of multilinear maps. In particular, a work by Miles, Sahai, and Zhandry [MSZ16a] gave the first polynomial-time attack on several candidate constructions of iO [BR14, BGK+14, PST14, AGIS14, MSW14, BMSZ16] when those constructions are instantiated using the original multilinear map candidate due to Garg, Gentry, and Halevi [GGH13a]. Thus, these attacks show that our modeling of multilinear map candidates is insufficient, even as a heuristic for arguing security.

The work of Badrinarayanan et al. [BMSZ16] explicitly addressed the question of whether a weaker model of security of multilinear maps can suffice for proving the security of iO. In particular, such a model of *weak multilinear maps* must take into account known attacks on the candidate multilinear map — that is, all known polynomial-time attacks must be allowable in the model. While there are several long-standing iO candidates that are not known to be broken (see, e.g., [AJN+16, Appendix A]), until recently there has not been any model for justifying their security. The work of [BMSZ16] gave the first such positive result, and showed that in one such weak multilinear map model, obfuscation for evasive functions can be proven secure with only minor modifications to existing iO candidates. [MSZ16a] posited another, more specific, weak multilinear map model that captured all known polynomial-time attacks in the context of the GGH13 multilinear map candidate. However, that work did not answer the question of whether one can construct an iO candidate for general programs that is provably secure in this model.

Our Contribution. In this work we answer this question in the affirmative, showing a new construction of an iO candidate, which can be seen as a small modification or generalization of the original iO candidate of [GGH+13b], and we prove its security in the weak multilinear map model of [MSZ16a].

We prove the security of our candidate under a new assumption about the hardness of computing annihilating polynomials (cf. Definition 4), and we show that this assumption is implied by the existence of pseudorandom functions (PRF) in NC^1. Interestingly, if our assumption is true because a PRF exists and can be computed by a matrix branching program of size $t(n)$, then our construction will only depend on this size bound $t(n)$, and not on any other details of the PRF! Indeed, our construction will just need to be padded to have size at least roughly $t(n)$, and no modification will be necessary at all if the program being obfuscated is already larger than $t(n)$.

Philosophically, this is reminiscent of the recent work on time-lock puzzles of [BGJ+15], where their construction of a puzzle needs to be padded to have the size of some program that computes a long non-parallelizable computation. Technically, however, our methods appear to be completely unrelated.

We now give an overview of the GGH13 multilinear map candidate. Following that, we describe an objective that is common to all known polynomial-time attacks on the GGH13 multilinear map, and use this to explain the weak multilinear map model of [MSZ16a]. We then present some starting intuition followed by an outline of the proof that our new candidate is secure against all known polynomial-time attacks on GGH13 (including [MSZ16a]).

1.1 Overview of GGH13

For GGH13 [GGH13a] with k levels of multilinearity, the plaintext space is a quotient ring $R_g = R/gR$ where R is the ring of integers in a number field and $g \in R$ is a "small element" in that ring. The space of encodings is $R_q = R/qR$ where q is a "big integer". An instance of the scheme relies on two secret elements, the generator g itself and a uniformly random denominator $z \in R_q$. A small plaintext element α is encoded "at level one" as $u = [e/z]_q$ where e is a "small element" in the coset of α, that is $e = \alpha + gr$ for some small $r \in R$.

Addition/subtraction of encodings at the same level is just addition in R_q, and it results in an encoding of the sum at the same level, so long as the numerators do not wrap around modulo q. Similarly multiplication of elements at levels i, i' is a multiplication in R_q, and as long as the numerators do not wrap around modulo q the result is an encoding of the product at level $i + i'$.

The scheme also includes a "zero-test parameter" in order to enable testing for zero at level k. Noting that a level-k encoding of zero is of the form $u = [gr/z^k]_q$, the zero-test parameter is an element of the form $\mathbf{p}_{zt} = [hz^k/g]_q$

for a "somewhat small element" $h \in R$. This lets us eliminate the z^k in the denominator and the g in the numerator by computing $[\mathbf{p}_{zt} \cdot u]_q = h \cdot r$, which is much smaller than q because both h, r are small. If u is an encoding of a non-zero α, however, then multiplying by $\mathbf{p}_{zt}$ leaves a term of $[h\alpha/g]_q$ which is not small. Testing for zero therefore consists of multiplying by the zero-test parameter modulo q and checking if the result is much smaller than q.

Note that above we describe the "symmetric" setting for multilinear maps where there is only one z, and its powers occur in the denominators of encodings. More generally, there is an "asymmetric" setting where there are multiple z_i.

1.2   Overview of the Model

To motivate our model (which is essentially that of [MSZ16a] with some clarifications), we note that all known polynomial-time attacks [GGH13a, HJ15, MSZ16a] on the GGH13 graded encoding scheme share a common property. As mentioned above, these attacks work by using information leaked during zero testing. More precisely, these attacks compute a set of top-level 0-encodings via algebraic manipulations on some set of initial encodings, then apply the zero test to each top level encoding, and then perform an algebraic computation on the *results* of the zero testing to obtain an element in the ideal $\langle g \rangle$. In particular, the latter computation is agnostic to the particular value of g and to the randomization values r chosen for each initial encoding.

After obtaining a set of elements from $\langle g \rangle$, the prior attacks then use these in various different ways to mount attacks on different cryptographic constructions built on top of GGH13. However, those details are not important to us. In our model (as suggested in [MSZ16a]), if the adversary succeeds in just generating an element in the ideal $\langle g \rangle$, we will say that the adversary has won.

Our model captures the type of attack described above as follows. Like the standard ideal graded encoding model, our model $\mathcal{M}$ is an oracle that maintains a table mapping generic representations called "handles" to encodings of elements $a_i \in \mathbb{Z}_p \simeq R/\langle g \rangle$. However, rather than just storing each value a_i (along with its level), we store the formal $\mathbb{Z}_p$-polynomial $a_i + g \cdot r_i$, where g is a formal variable common to all encodings and r_i is a "fresh" formal variable chosen for each a_i. Then, an adversary may use the handles to perform any set of level-respecting algebraic computations on the initial set of encodings. The result of any such computation is an encoding f which is represented as a $\mathbb{Z}_p$-polynomial in the variables g and $\{r_i\}$.

When the adversary submits a handle to a top-level encoding f for zero-testing, $\mathcal{M}$ checks whether f's constant term is 0 (which corresponds to a 0-encoding in the standard ideal model). If so, $\mathcal{M}$ returns a handle to the formal polynomial f/g (corresponding to the result of the GGH13 zero-testing procedure), and otherwise $\mathcal{M}$ responds "not zero."

Finally, the adversary may submit a post-zero-test polynomial Q of degree at most $2^{o(\lambda)}$, where throughout the paper λ is the security parameter. $\mathcal{M}$ checks whether Q, when evaluated on the set of zero-tested encodings $\{f/g\}$ the adversary has created, produces a non-zero polynomial in which every monomial is divisible by g; i.e., it checks whether Q produces a non-zero polynomial that *is* zero mod g. If so, $\mathcal{M}$ outputs "WIN", indicating that the adversary's attack was successful. Note that any such Q is an annihilating polynomial (Definition 4) for the set $\{f/g \mod g\}$.

On the Degree Bound. The bound $\deg(Q) \leq 2^{o(\lambda)}$ for efficient adversaries may seem somewhat artificial. Indeed, arithmetic circuits of size $\mathsf{poly}(\lambda)$ can have arbitrary exponential degree.

However, using the GGH13 graded encoding scheme, such high-degree polynomials appear difficult to compute in the non-idealized setting. This is because, in all known polynomial-time attacks on GGH13, the post-zero-test computations cannot be performed modulo the GGH13 parameter q while maintaining the correctness of the attack. Indeed, there is no modulus M known with respect to which the computations can be performed while still maintaining correctness of attacks, unless the modulus M is so large that working modulo M results in computations that are identical to the computations over $\mathbb{Z}$.

Let us explore the intuition behind why this seems to be the case. Let d be the dimension of the ring R over $\mathbb{Z}$. Recall that the goal of the attacker in our model is to recover an element of the ideal $\langle g \rangle$. In order to safely work modulo M, it needs to be the case that $M\mathbb{Z}^d$ is a sublattice of the ideal lattice $\langle g \rangle$. But g is a secret parameter of the GGH13 scheme. Until the adversary finds out something about g, it cannot be sure that any modulus M it chooses will be safe (and indeed if the computation overflows with respect to M, almost certainly any information relevant to g will be lost). But the only way we know to learn anything about g is to find an element in $\langle g \rangle$, which was the goal of the attack to begin with.

Therefore, multiplication of two elements potentially doubles the size of the elements, and an element of exponential degree will likely have exponential size. It seems difficult even to perform post-zero-test computations of *superpolynomial* degree.

At a technical level, we need to restrict to degree $2^{o(\lambda)}$ due to our use of the Schwartz-Zippel lemma, which ceases to give useful bounds when Q has larger degree.

1.3   Intuition: Obfuscation Using an Explicit NC^1 PRF

To build intuition, we first describe a construction assuming an explicit PRF in NC^1. Later we will show that simply the *existence* of an NC^1 PRF (in fact, a more general assumption that is implied by the existence of such PRF) suffices for our purpose.

Consider an obfuscator that, given a matrix branching program A, first turns each matrix $A_{i,b}$ into a block-diagonal matrix

$$P_{i,b} = \begin{pmatrix} A_{i,b} & \\ & R_{i,b}^K \end{pmatrix}$$

where the $R_{i,b}^K$ form an "auxiliary" branching program which, on input x, computes a value $\rho_x \cdot g$ where ρ_x is the output[1] of an NC^1 PRF on input x.

The $P_{i,b}$ matrices are then randomized as in previous works using Kilian randomization [Kil88] plus independent scalars for each matrix, and encoded as in previous works using [GGH13a] multilinear maps and the "straddling set" level structure from [BGK+14]. Thus, the only deviation from the "standard recipe" for obfuscation are the auxiliary matrices $R_{i,b}$ matrices described above. Note that an honest evaluation of P on input x results in roughly the following evaluation:

$$P(x) = A(x) + g \cdot \rho_x.$$

The proof of security for this obfuscator starts with the analysis of [BGK+14, BMSZ16], which decomposes each top-level 0-encoding produced by the adversary into a linear combination of "honest evaluation" polynomials $f_{x_1}, \ldots, f_{x_m}$ over the obfuscated branching program, for some $\mathrm{poly}(\lambda)$-size set of inputs $x_1, \ldots, x_m$ on which the BP evaluates to 0. Thus, we can view any post-zero-test polynomial Q (produced by the adversary) as a polynomial in $\{f_{x_j}/g\}_{j \in [m]}$.

For each x_j, we can write

$$f_{x_j} = f_{x_j}^{(0)} + g \cdot f_{x_j}^{(1)} + g^2 \cdot f_{x_j}^{(2)} + \cdots$$

where $f_{x_j}^{(0)}, f_{x_j}^{(1)}, \ldots$ are polynomials over just the randomness $\{r_i\}$ of the GGH13 graded encoding (i.e. they do not contain the variable g). Since the "main branching program" A evaluates to 0 on each x_j, we can show that $f_{x_j}^{(0)}$ is the 0 polynomial, which means that $f_{x_j}/g = f_{x_j}^{(1)} + g \cdot f_{x_j}^{(2)} + \cdots$. Thus by algebraic independence, if Q annihilates $\{f_{x_j}/g\}_{j \in [m]}$ mod g, it must in particular annihilate $\{f_{x_j}^{(1)}\}_{j \in [m]}$.

We can further decompose the structure of each such $f_{x_j}^{(1)}$ by writing it as

$$f_{x_j}^{(1)} = \widehat{f}_{x_j}^{(1)} + \rho_{x_j}$$

where ρ_{x_j} is the pseudorandom multiplier of g produced via the PRF computation $R^K(x_j)$ (which is independent of the $\{r_i\}$ values). Intuitively, if Q annihilates the polynomials $f_{x_j}^{(1)}$, then by algebraic independence it must annihilate $\{\rho_{x_1}, \rho_{x_2}, \ldots \rho_{x_m}\}$ as formal polynomials. However, since for a PPT attacker such variables are pseudorandom in a large field of size $p > 2^\lambda$, Q cannot exist except with negligible probability (as otherwise it could be used to efficiently distinguish between the PRF and a random function).

1.4   Overview of the Security Proof

We now give an overview of our proof of security, building on the above intuition. Given a branching program A, our obfuscator first transforms each matrix $A_{i,b}$ again into a block-diagonal matrix

---

[1] For simplicity we abuse notations of branching programs, in that it outputs a ring element instead of a bit. It is straightforward to embed multiple branching programs into one to achieve this effect.

$$\begin{pmatrix} A_{i,b} & \\ & B_{i,b} \end{pmatrix}$$

where, in contrast to the intuition presented above, each auxiliary $B_{i,b}$ is simply a uniform random matrix over $\mathbb{Z}_p$. (As mentioned above, this can be seen as a generalization of [GGH+13b], where this same block-diagonal structure was used but the $B_{i,b}$ matrix was a random diagonal matrix. Note that we choose $B_{i,b}$ to be completely random instead.) Note that this obfuscator does not hard-wire into it a branching program for a PRF, or for any other specific function aside from the branching program A that is being obfuscated.

The proof of security follows the argument presented above, up to the point of showing that a "successful" post-zero-test polynomial Q must in particular annihilate the polynomials $\{f_{x_j}^{(1)}\}_{j \in [m]}$. Unlike in the hardwired-PRF construction however, each $f_{x_j}^{(1)}$ now does not contain an explicit PRF output. Still, each can be viewed a polynomial in the entries of the original branching program A, the randomization values chosen by the obfuscator (including the $B_{i,b}$ matrices), and the randomization values r_i in the GGH13 encodings.

The core of our proof shows that if Q annihilates the set $\{f_{x_j}^{(1)}\}_{j \in [m]}$, then it must also annihilate a corresponding set of "generic BP evaluation polynomials"

$$e_{x_j} := \beta_0 \times \prod_{i=1}^{\ell} \beta_{i,(x_j)_{\mathsf{inp}(i)}} \times \beta_{\ell+1}$$

where $\{\beta_{i,b}\}_{j \in [\ell], b \in \{0,1\}}$ (resp. $\beta_0, \beta_{\ell+1}$) are matrices (resp. vectors) of independent variables, corresponding to the $B_{i,b}$ matrices. This uses the Schwartz-Zippel lemma, and two additional techniques. The first is that if Q annihilates a set of polynomials $\{p_i = p_i' + u \cdot p_i''\}_i$ where the variable u appears in no p_i', then by algebraic independence Q must also annihilate $\{p_i'\}_i$. The second is that if a set of polynomials $\{q_i\}_i$ can be obtained from another set of polynomials $\{p_i\}_i$ via a change of variables, and Q annihilates $\{p_i\}_i$, then Q also annihilates $\{q_i\}_i$.

Our main assumption (Assumption 1) states that annihilating a poly-size subset of $\{e_x\}_{x \in \{0,1\}^n}$ is not possible. We observe in Theorem 2 that, in particular, this assumption is implied by the existence of PRF in NC^1. However, we believe the above assumption to be quite plausible independent of the fact that a PRF in NC^1 would imply its validity.

1.5 Extensions

Single-Input vs Dual-Input Branching Programs. Our obfuscator, following Barak et al. [BGK+14], uses dual-input branching programs, which allows us to prove VBB security in the weak multilinear map model. The obfuscator of [BGK+14] can also be modified to use single-input branching programs, though then only iO security is proved in the plain generic model. Unfortunately, we are unable to prove iO security for a single-input variant of our construction. The problem is that a post-zero-test encoding can now consist of elements coming from exponentially many inputs. This means that an annihilating polynomial Q may annihilate an exponential set of "generic BP evaluation polynomials." This prevents us from embedding Assumption 1 into the security proof.

However, if the input domain of the obfuscated program is polynomial-sized instead of exponential, then there are only a polynomial number of possible BP evaluation polynomials. Thus, we are able to embed Assumption 1. Therefore, in the case of polynomial-sized domain, the single input version of our obfuscator achieves iO security.

Order Revealing Encryption. Our techniques can also be applied to the order-revealing encryption scheme of [BLR+15]. Order-revealing encryption is a symmetric encryption scheme that lets one publicly compare the order of plaintexts, but no information beyond the order of the plaintexts is revealed.

In the scheme of [BLR+15], ciphertexts are generated by encoding branching program matrices analagous to how they are encoded in obfuscation — Kilian randomize and multiply by a random scalar. The branching program arises from the state transition matrices of the finite automata for comparing two integers.

We note that their scheme was shown to be insecure in the weak multilinear map model by [MSZ16a]. To protect against these attacks, we similarly extend the branching program matrices into a block diagonal matrix with the new block being a random matrix, before applying Kilian randomization.

Security readily follows from our analysis, using a "base-B" version of Assumption 1, where B is the number of ciphertexts the adversary sees. That is, we can consider a version of our assumption where the matrix branching programs have inputs that are represented base B, and each layer of the branching program reads a single digit, selecting one of B matrices for that layer. Such a base-B assumption follows from the standard binary version of Assumption 1 by decomposing each digit into $\log B$ bits.

Model Variations. In Sect. 5, we consider a variant of our model that more closely reflects the GGH13 encodings. Here, the r_i used to encode are no longer treated as formal variables, but are instead treated as actual ring elements sampled from some distribution. In GGH13, the distribution on r_i depends on the ring element a_i — in our model, we therefore allow the r_i to have arbitrary correlations with the a_i, as long as the conditional min-entropy of r_i given a_i is high. This min-entropy requirement is satisfied by GGH13 encodings. We note that switching to r_i being ring elements makes the adversary's winning condition easier, as there are now fewer constraints on the post-zero-test polynomial Q.

We show that, with a small modification to the proof, our obfuscator is also secure in this variant model. If the r_i were uniformly random in some fixed subset of the ring, the Schwartz-Zippel lemma would suffice for adapting our original security proof to this setting. However, as we allow the r_i to be non-uniform and potentially come from different distributions, we need a new variant of the Schwartz-Zippel lemma for more general distributions. We prove this variant, which may be of independent interest, in Lemma 2.

Organization. In Sect. 2 we formally define our model. In Sect. 3 we give the details of our obfuscator, and in Sect. 4 we give the proof of security and discuss our assumption. In Sect. 5, we prove security in the alternative model discussed above.

2   The Model

In this section, we define our model for weak graded encoding schemes. The model is inspired by [CGH+15, Appendix A], and is essentially the same as the model given in [MSZ16a] except for some details that we clarify here.

Recall that in a graded encoding scheme, there is a universe set $\mathbb{U}$, and a *value* a can be encoded at a *level* $S \subseteq \mathbb{U}$, denoted by $[a]_S$. Addition, subtraction, and multiplication of encodings are defined provided that the levels satisfy certain restrictions, as follows.

- For any $S \subseteq \mathbb{U}$: $[a_1]_S \pm [a_2]_S := [a_1 \pm a_2]_S$.
- For any $S_1, S_2 \subseteq \mathbb{U}$ such that $S_1 \cap S_2 = \emptyset$: $[a_1]_{S_1} \cdot [a_2]_{S_2} := [a_1 \cdot a_2]_{S_1 \cup S_2}$.

Further, an encoding $[a]_{\mathbb{U}}$ at level $\mathbb{U}$ can be zero-tested, which checks whether $a = 0$.

In the standard ideal graded encoding model, a stateful oracle maintains a table that maps encodings to generic representations called *handles*. Each handle explicitly specifies the encoding's level, but is independent of the encoding's value. All parties have access to these handles, and can generate new handles by querying the oracle with arithmetic operations that satisfy the above restrictions. In addition, all parties may perform a zero-test query on any handle whose level is $\mathbb{U}$, which returns a bit indicating whether the corresponding value is 0.

Our model also implements these features, but adds new features to more closely capture the power that an adversary has in the non-idealized setting. The most important new feature is that a successful zero test returns a handle to a ring element that can further be manipulated, as opposed to just returning a bit.

We now formally describe the interfaces implemented by the oracle $\mathcal{M}$ that defines our model. For concreteness, we define $\mathcal{M}$ to explicitly work over the GGH13 ring $\mathcal{R} = \mathbb{Z}[X]/(X^\eta + 1)$ and the field $\mathbb{Z}_p \simeq \mathcal{R}/\langle g \rangle$ for an appropriate $g \in \mathcal{R}$.

Initialize Parameters. The first step in interacting with $\mathcal{M}$ is to initialize it with the security parameter $\lambda \in \mathbb{N}$. (Jumping ahead, this will be done by the obfuscator.) $\mathcal{M}$ defines the ring $\mathcal{R} = \mathbb{Z}[X]/(X^\eta + 1)$, where $\eta = \eta(\lambda)$ is chosen as in [GGH13a]. Then, $\mathcal{M}$ chooses $g \in \mathcal{R}$ according to the distribution in [GGH13a], and outputs the prime $p := |\mathcal{R}/\langle g \rangle| > 2^\lambda$. After initializing these parameters, $\mathcal{M}$ discards the value of g, and treats g as a *formal variable* in all subsequent steps.

Initialize Elements. After the parameters have been initialized, $\mathcal{M}$ is given a universe set $\mathbb{U}$ and a set of initial elements $\{[a_i]_{S_i}\}_i$ where $a_i \in \mathbb{Z}_p$ and $S_i \subseteq \mathbb{U}$ for each i. For each initial element $[a_i]_{S_i}$, $\mathcal{M}$ defines the formal polynomial

$f_i := a_i + g \cdot z_i$ over $\mathbb{Z}_p$. Here g is a formal variable that is common to all f_i, while z_i is a "fresh" formal variable[2] chosen for each f_i. Then $\mathcal{M}$ generates a handle h_i (whose representation explicitly specifies S_i but is independent of a_i), and stores the mapping "$h_i \rightarrow (f_i, S_i)$" in a table that we call the *pre-zero-test* table. Finally, $\mathcal{M}$ outputs the set of handles $\{h_i\}_i$.

Note that storing the formal polynomial f_i strictly generalizes the standard ideal model which just stores the value a_i. This is because a_i can always be recovered as the constant term of f_i, and this holds even for subsequent polynomials that are generated from the initial set via the algebraic operations defined next.

The above two initialization interfaces are each executed once, in the order listed; any attempt to execute them out of order or more than once will fail. $\mathcal{M}$ also implements the following algebraic interfaces.

Pre-zero-test Arithmetic. Given two input handles h_1, h_2 and an operation $\circ \in \{+, -, \cdot\}$, $\mathcal{M}$ first locates the corresponding polynomials f_1, f_2 and level sets S_1, S_2 in the pre-zero-test table. If h_1 and h_2 do not both appear in this table, the call to $\mathcal{M}$ fails. If the expression is undefined (i.e., $S_1 \neq S_2$ for $\circ \in \{+, -\}$, or $S_1 \cap S_2 \neq \emptyset$ for $\circ \in \{\cdot\}$), the call fails. Otherwise, $\mathcal{M}$ computes the formal polynomial $f := f_1 \circ f_2$ and the level set $S := S_1 \cup S_2$, generates a new handle h, and stores the mapping "$h \rightarrow (f, S)$" in the pre-zero-test table. Finally, $\mathcal{M}$ outputs h.

Zero-Testing. Given an input handle h, $\mathcal{M}$ first locates the corresponding polynomial f and level set S in the pre-zero-test table. If h does not appear in this table, or if $S \neq \mathbb{U}$, the call to $\mathcal{M}$ fails. If f's constant term is non-zero (recall that this term is an element of $\mathbb{Z}_p$), $\mathcal{M}$ outputs the string "non-zero". If instead f's constant term is 0, note that f must be divisible by the formal variable g, i.e. g appears in each of f's monomials. $\mathcal{M}$ computes the formal polynomial $f' := f/g$ over $\mathbb{Z}_p$, generates a new handle h', and stores the mapping "$h' \rightarrow f'$" in a table that we call the *post-zero-test* table. Finally, $\mathcal{M}$ outputs h'.

Post-zero-test Arithmetic. Given a set of input handles $h'_1, \ldots, h'_m$ and an m-variate polynomial Q over $\mathbb{Z}$ (represented as an arithmetic circuit), $\mathcal{M}$ first locates the corresponding polynomials $f'_1, \ldots, f'_m$ in the post-zero-test table. If any h'_i does not appear in this table, the call to $\mathcal{M}$ fails. Otherwise, $\mathcal{M}$ checks whether $Q(f'_1, \ldots, f'_m)$ is non-zero as a polynomial over $\mathbb{Z}_p$ which *is* zero modulo the variable g. In other words, $\mathcal{M}$ checks that $Q(f'_1, \ldots, f'_m)$ contains at least one monomial whose coefficient is not zero modulo p, and that g appears in all such non-zero monomials.[3] If this check passes, $\mathcal{M}$ outputs "WIN", otherwise it outputs $\perp$.

---

[2] Here and for the remainder of the paper, we use z_i rather than r_i to denote the randomization values in GGH13 encodings, to avoid conflicting with the random matrices R chosen by the obfuscator. We will not need to work with the GGH13 level denominators, which were previously denoted by z_i.

[3] Note that this corresponds to finding a non-trivial element in the ideal $\langle g \rangle$.

Definition 1. *A (possibly randomized) adversary interacting with the model $\mathcal{M}$ is efficient if it runs in time* $\mathsf{poly}(\lambda)$, *and if each Q submitted in a post-zero-test query has degree* $2^{o(\lambda)}$. *Such an adversary wins if it ever submits a post-zero-test query that causes $\mathcal{M}$ to output "WIN".*

3   The Obfuscator

Our obfuscator for matrix branching programs is closely related to that of Badrinarayanan et al. [BMSZ16]. The main difference is that, before randomizing and encoding, each matrix $A_{i,b}$ is first transformed into a block-diagonal matrix

$$\begin{pmatrix} A_{i,b} & \\ & B_{i,b} \end{pmatrix}$$

where each $B_{i,b}$ is uniformly random.

We now describe our obfuscator $\mathcal{O}$. $\mathcal{O}$ is instantiated with two parameters, $t = t(n,\lambda)$ and $s = s(n,\lambda)$, that correspond to those in Assumption 1.

Input. $\mathcal{O}$ takes as input a dual-input matrix branching program[4] BP of length m, width w, and input length n. Such a matrix branching program consists of an input-selection function $\mathsf{inp} : [m] \to [n] \times [n]$, $4m$ matrices $\{A_{i,b_1,b_2} \in \{0,1\}^{w \times w}\}_{i \in [m]; b_1, b_2 \in \{0,1\}}$, and two "bookend" vectors $A_0 \in \{0,1\}^{1 \times w}$ and $A_{m+1} \in \{0,1\}^{w \times 1}$. BP is evaluated on input $x \in \{0,1\}^n$ by checking whether

$$A_0 \times \prod_{i \in [m]} A_{i,x(i)} \times A_{m+1}$$

is zero or non-zero, where we abbreviate $x(i) := (x_{\mathsf{inp}(i)_1}, x_{\mathsf{inp}(i)_2})$. We make three requirements on BP (cf. [BMSZ16, Sect. 3]).

1. It is forward non-shortcutting, defined below.
2. For each $i \in [m] : \mathsf{inp}(i)_1 \neq \mathsf{inp}(i)_2$.
3. For each pair $j \neq k \in [n]$, there exists $i \in [m]$ such that $\mathsf{inp}(i) \in \{(j,k),(k,j)\}$.

Definition 2 ([BMSZ16]). *A branching program* $A_0, \{A_{i,b}\}_{i \in [\ell], b \in \{0,1\}}, A_{\ell+1}$ *is forward (resp. reverse) non-shortcutting if, for every input x, the vector*

$$A_0 \times \prod_{i \in [\ell]} A_{i,x(i)} \qquad \left(resp. \qquad \prod_{i \in [\ell]} A_{i,x(i)} \times A_{\ell+1} \right)$$

is non-zero. It is non-shortcutting if it is both forward and reverse non-shortcutting.

---

[4] These can be constructed from any NC^1 formula with $m = \mathsf{poly}(n)$ and $w = 5$ by Barrington's theorem [Bar86]. Obfuscating NC^1 formulas is sufficient to obfuscate all polynomial-size circuits [GGH+13b, BR14, App14].

Step 0: Initialize Model. $\mathcal{O}$ first sends the security parameter λ to the model $\mathcal{M}$, and receives back a prime p.

Step 1: Pad BP. $\mathcal{O}$'s first modification to BP is to pad it with identity matrices (if necessary) so that it contains a set of t layers $i_1 < \ldots < i_t$ such that $(\mathsf{inp}(i_1)_1, \ldots, \mathsf{inp}(i_t)_1)$ cycles t/n times through $[n]$. This choice of inp is specifically to allow a branching program of the form in Assumption 1 to be transformed into one with input selection function $\mathsf{inp}(\cdot)_1$. We use $\ell \le t + m$ to denote the length of the padded branching program.

Step 2: Extend Matrices. Next, $\mathcal{O}$ extends the matrices as mentioned above. To do this, it selects 4ℓ uniformly random matrices $\{B_{i,b_1,b_2} \in \mathbb{Z}_p^{s \times s}\}_{i \in [\ell]; b_1, b_2 \in \{0,1\}}$ and one uniformly random vector $B_{\ell+1} \in \mathbb{Z}_p^{s \times 1}$, and defines the following matrices and vectors.

$$A_0' := (A_0 \quad 0^s) \qquad A_{i,b_1,b_2}' := \begin{pmatrix} A_{i,b_1,b_2} & \\ & B_{i,b_1,b_2} \end{pmatrix} \qquad A_{\ell+1}' := \begin{pmatrix} A_{\ell+1} \\ B_{\ell+1} \end{pmatrix}$$

Note that this satisfies

$$A_0' \times \prod_{i \in [\ell]} A_{i,x(i)}' \times A_{\ell+1}' \;=\; A_0 \times \prod_{i \in [\ell]} A_{i,x(i)} \times A_{\ell+1}$$

for every input $x \in \{0,1\}^n$.

Step 3: Randomize. Next, $\mathcal{O}$ generates uniformly random non-singular matrices $\{R_i\}_{i \in [\ell+1]}$ and uniformly random non-zero scalars $\alpha_0, \{\alpha_{i,b_1,b_2}\}_{i \in [\ell]; b_1, b_2 \in \{0,1\}}$, $\alpha_{\ell+1}$. Then it computes the randomized branching program, denoted $\widehat{BP}$, as follows.

$$\widehat{A_0} := \alpha_0 A_0' \times R_1^{adj} \qquad \widehat{A_{i,b_1,b_2}} := \alpha_{i,b_1,b_2} R_i \times A_{i,b_1,b_2}' \times R_{i+1}^{adj}$$

$$\widehat{A_{\ell+1}} := \alpha_{\ell+1} R_{\ell+1} \times A_{\ell+1}'$$

Here R_i^{adj} denotes the adjugate matrix of R_i that satisfies $R_i^{adj} \times R_i = \det(R_i) \cdot I$. It is easy to see that $\widehat{BP}$ computes the same function as BP, i.e.

$$\widehat{A_0} \times \prod_{i \in [\ell]} \widehat{A_{i,x(i)}} \times \widehat{A_{\ell+1}} = 0 \quad \Leftrightarrow \quad A_0 \times \prod_{i \in [\ell]} A_{i,x(i)} \times A_{\ell+1} = 0$$

for every input $x \in \{0,1\}^n$.

Step 4: Encode. Finally, $\mathcal{O}$ initializes $\mathcal{M}$ with the elements of the $\widehat{A}$ matrices. To do this, it uses the level structure in [BGK+14] constructed from so-called *straddling sets*. We defer the details to Appendix A, but we remark that this level structure has the property that each "honest evaluation" $\widehat{BP}(x) = \widehat{A_0} \times \prod_i \widehat{A_{i,x(i)}} \times \widehat{A_{\ell+1}}$ results in an encoding at level $\mathbb{U}$. This, in combination with the zero-test procedure, allows the obfuscated program to be evaluated.

$\mathcal{M}$'s pre-zero-test table can now be viewed as containing the variables $Y_0, \{Y_{i,b_1,b_2}\}_{i\in[\ell];b_1,b_2\in\{0,1\}}, Y_{\ell+1}$ of the following form.

$$Y_0 = \widehat{A_0} + gZ_0 \qquad Y_{i,b_1,b_2} = \widehat{A_{i,b_1,b_2}} + gZ_{i,b_1,b_2} \qquad Y_{\ell+1} = \widehat{A_{\ell+1}} + gZ_{\ell+1}$$

Here g is a formal variable and each Z matrix is a matrix of formal variables, while the $\widehat{A}$ matrices contain $\mathbb{Z}_p$-elements.

The final branching program $\widehat{BP} = \mathcal{O}(BP)$ has length ℓ (satisfying $t \le \ell \le m + t$) and width $w + s$. In the proof of Theorem 3, we will use the fact that any branching program of the form in Assumption 1 can be transformed (by padding with identity matrices) into one with length ℓ whose input selection function is the same as $\mathsf{inp}(\cdot)_1$.

Definition 3. $\mathcal{O}$ *is secure in the model $\mathcal{M}$ of Sect. 2 if, for every BP matching $\mathcal{O}$'s input specification and every efficient adversary $\mathcal{A}$ interacting with $\mathcal{M}$, $\Pr[\mathcal{A}$ wins$] < \mathsf{negl}(\lambda)$ when $\mathcal{M}$ is initialized by $\mathcal{O}(BP)$. (Here the probability is over the randomness of $\mathcal{O}$ and $\mathcal{A}$.)*

4   Security of Our Obfuscator

We first state two definitions, and then state the assumption under which we will prove security. After that, we prove our security theorem.

Definition 4. *Let $f_1, \dots, f_m$ be a set of polynomials over some common set of variables. Then an m-variate polynomial Q annihilates $\{f_i\}_{i\in[m]}$ if $Q(f_1, \dots, f_m)$ is zero as a formal polynomial.*

Definition 5. *A matrix branching program BP is L-bounded for $L \in \mathbb{N}$ if every intermediate value computed when evaluating BP on any input is at most L. In particular all of BP's outputs and matrix entries are $\le L$.*

Our assumption essentially states that no efficiently computable polynomial can annihilate *every* branching program's evaluation polynomials on some efficiently computable set of inputs. (The assumption is parameterized by the length t and width s of the branching program.) In the assumption, we implicitly use a more general notion of how a branching program computes a function than was used in the previous section. Namely, the function computed can have range $[2^\lambda]$ (rather than $\{0,1\}$) by taking the output to be the value resulting from multiplying the appropriate vectors and matrices (rather than a bit indicating whether this value is 0).

Assumption 1 The (t,s)-**branching** **program** **un-annihilatability** **(BPUA) assumption.** *Let $t = \mathsf{poly}(n,\lambda)$ and $s = \mathsf{poly}(n,\lambda)$ be parameters. Let $\mathcal{A}$ denote a PPT that, on input $(1^n, 1^\lambda)$, outputs a $\mathsf{poly}(\lambda)$-size set $\mathcal{X} \subseteq \{0,1\}^n$ and a $\mathsf{poly}(\lambda)$-size, $2^{o(\lambda)}$-degree polynomial Q over $\mathbb{Z}$.*

For all n and for sufficiently large λ, all primes $2^\lambda < p \le 2^{\mathsf{poly}(\lambda)}$, and all such $\mathcal{A}$, there exists a (single-input) 2^λ-bounded matrix branching program

$BP : \{0,1\}^n \rightarrow [2^\lambda]$ of length t and width s, whose input selection function iterates over the n input bits t/n times, such that

$$\Pr\left[Q\left(\{BP(x)\}_{x \in \mathcal{X}}\right) = 0 \ (\mathrm{mod} \ p)\right] < \mathsf{negl}(\lambda)$$

where the probability is over $\mathcal{A}$'s randomness.

We observe that Assumption 1 is in particular implied by the existence of PRF in NC^1 secure against P/poly (with t, s related to the size of such PRF).

Theorem 2. Let t and s be as in Assumption 1. If there exists a PRF F_k : $\{0,1\}^n \rightarrow [2^\lambda]$ that

- is computable by a length-t/n, width-s, 2^λ-bounded matrix branching program BP_k, and
- is secure against non-uniform, polynomial-time adversaries (i.e. secure against P/poly)

then Assumption 1 holds.

Note that we take BP_k's matrix entries to be computed as a function of the PRF key k.

Proof. Assume that Assumption 1 is false, and fix a PPT $\mathcal{A}$ and a prime p such that

$$\Pr\left[Q\left(\{BP(x)\}_{x \in \mathcal{X}}\right) = 0 \ (\mathrm{mod} \ p)\right] \geq 1/\mathsf{poly}(\lambda)$$

for every BP of the form in Assumption 1. We give a PPT $\mathcal{A}'$ with oracle access to O that distinguishes with probability $\geq 1/\mathsf{poly}(\lambda)$ whether O implements BP_k for a uniform k or implements a uniform function $F : \{0,1\}^n \rightarrow [2^\lambda]$. We note that hardwiring p into $\mathcal{A}'$ is the only place where non-uniformity is needed.

$\mathcal{A}'$ simply runs $\mathcal{A}$ to get Q and $\mathcal{X}$, and computes $d := Q(O(x)_{x \in \mathcal{X}})$ (mod p). Note that $\mathcal{A}'$ runs in time $\mathsf{poly}(\lambda)$ because Q and p both have this size. If O implements BP_k, then $d = 0$ with probability $\geq 1/\mathsf{poly}(\lambda)$. To see this, note that BP_k can be transformed (by padding with identity matrices) into an equivalent branching program of the form in Assumption 1 due to the input selection function there.

On the other hand, if O implements a random function, then since $p > 2^\lambda$ and $\deg(Q) = 2^{o(\lambda)}$, $d = 0$ with probability $< \mathsf{negl}(\lambda)$ by the Schwartz-Zippel lemma.

For further discussion on our assumption, including the plausibility of PRF necessary for Theorem 2, see Sect. 4.2.

4.1   Our Main Theorem

Theorem 3. Let $\mathcal{O}$ be the obfuscator from Sect. 3 with parameters t and s. If the (t, s)-BPUA assumption holds, $\mathcal{O}$ is secure in the model $\mathcal{M}$ of Sect. 2.

We note that this theorem also implies that $\mathcal{O}$ achieves VBB security in the model from Sect. 2. To see this, first note that the initialization, pre-zero-test arithmetic, and zero-test interfaces can be simulated with error $\mathsf{negl}(\lambda)$ exactly as in the proof of [BMSZ16, Theorem 5.1]. Further, a simulator can simply respond to every post-zero-test query with $\perp$, and the additional error introduced by this is bounded by $\mathsf{negl}(\lambda)$ due to Theorem 3.

Proof. Fix a PPT adversary $\mathcal{A}$ and assume for contradiction that, with probability $\epsilon \geq 1/\mathsf{poly}(\lambda)$, $\mathcal{A}$ obtains a set of valid post-zero-test handles $h'_1, \ldots, h'_m$ and constructs a size-$\mathsf{poly}(\lambda)$, degree-$2^{o(\lambda)}$, m-variate polynomial Q over $\mathbb{Z}$ such that the post-zero-test query $(Q, h'_1, \ldots, h'_m)$ causes $\mathcal{M}$ to output "WIN". By the definition of $\mathcal{M}$, each handle h'_j must then correspond to a polynomial f'_j such that $f_j := g \cdot f'_j$ is a level-$\mathbb{U}$ polynomial in $\mathcal{M}$'s pre-zero-test table with constant term 0.

Recall that $\mathcal{M}$ is initialized with the set of $\mathbb{Z}_p$ values $\{a_i\}_i$ from the branching program $\widehat{BP}$ created by $\mathcal{O}(BP)$, and that for each such value $\mathcal{M}$ stores a polynomial $a_i + g \cdot z_i$ with formal variables g, z_i. Thus each f_j is a $\mathbb{Z}_p$-polynomial with variables $g, \{z_i\}_i$. In the following, we use $\overline{f_j}$ to denote the polynomial over the set of $\mathcal{M}$'s initial elements such that $\overline{f_j}(\{a_i + g \cdot z_i\}_i) = f_j$.

Decomposing $\overline{f_j}$. For any input x, let $\overline{f_x}$ denote the matrix product polynomial that corresponds to evaluating $\widehat{BP}(x)$, and note that $\overline{f_x}(\{a_i\}_i) = 0 \pmod{p} \Leftrightarrow \widehat{BP}(x) = 0 \Leftrightarrow BP(x) = 0$. The results of [BGK+14, BMSZ16] (summarized in Lemma 1 following this proof) show that, with probability $1 - \mathsf{negl}(\lambda)$ over the randomness of $\mathcal{O}$, for each $j \in [m]$ there is a $\mathsf{poly}(\lambda)$-size set $\mathcal{X}_j$ such that: (1) $\overline{f_j}$ is a linear combination of the polynomials $\{\overline{f_x}\}_{x \in \mathcal{X}_j}$, and (2) $BP(x) = 0$ for every $x \in \mathcal{X}_j$. (Note that the conditions of the lemma are satisfied, as we can assume wlog that the post-zero-test query we are analyzing is the first to which $\mathcal{M}$ has responded with "WIN".)

The set $\mathcal{X}_j$ and the coefficients in the linear combination depend only on the *structure* of $\overline{f_j}$, and not on $\mathcal{O}$'s randomness. So, more precisely, Lemma 1 says that if $\overline{f_j}$ is *not* a linear combination of $\{\overline{f_x}\}_{x \in \mathcal{X}_j}$ for some $\mathcal{X}_j$ that satisfies $\bigwedge_{x \in \mathcal{X}_j}(BP(x) = 0)$, then $f_j = \overline{f_j}(\{a_i + g \cdot z_i\}_i)$ has constant term 0 with probability $< \mathsf{negl}(\lambda)$ over the randomness of $\mathcal{O}$. Thus, we condition on the event that each $\overline{f_j}$ is decomposable in this way, which has probability $1 - \mathsf{negl}(\lambda)$.

Structure of $\overline{f_x}$. Let $\mathcal{X} := \bigcup_{j \in [m]} \mathcal{X}_j$, and consider the polynomial $f_x := \overline{f_x}(\{a_i + g \cdot z_i\}_i)$ for any $x \in \mathcal{X}$. This is a $\mathbb{Z}_p$-polynomial with variables $g, \{z_i\}_i$, so we can "stratify" by g, writing

$$f_x = f_x^{(0)} + g \cdot f_x^{(1)} + g^2 \cdot f_x^{(2)} \tag{1}$$

where g does not appear in the polynomials $f_x^{(0)}$ and $f_x^{(1)}$, i.e. they are polynomials in just the variables $\{z_i\}_i$. From the analysis above, we know that $f_x^{(0)}$ is the identically 0 polynomial; if not, we would not have $\overline{f_x}(\{a_i\}_i) = 0 \pmod{p}$, and thus would not have $BP(x) = 0$. So, we can write

$$f_x/g = f_x^{(1)} + g \cdot f_x^{(2)}. \tag{2}$$

The fact that the post-zero-test query $(Q, h_1', \ldots, h_m')$ causes $\mathcal{M}$ to output "WIN" implies that $Q(f_1', \ldots, f_m') = Q(f_1/g, \ldots, f_m/g)$ is not identically zero as a polynomial in variables g and $\{z_i\}_i$, but is identically zero modulo the variable g. Let L_j denote the linear polynomial such that $\overline{f_j} = L_j(\{\overline{f_x}\}_{x \in \mathcal{X}_j})$. Then for each $j \in [m]$, we can write

$$f_j = \overline{f_j}(\{a_i + g \cdot z_i\}_i) = L_j\left(\left\{\overline{f_x}(\{a_i + g \cdot z_i\})_i)\right\}_{x \in \mathcal{X}_j}\right) = L_j\left(\{f_x\}_{x \in \mathcal{X}_j}\right).$$

Since each L_j is linear, we then obtain an $|\mathcal{X}|$-variate polynomial Q', with $\deg(Q') = \deg(Q)$, such that $Q'(\{f_x/g\}_{x \in \mathcal{X}}) = Q(\{f_j/g\}_{j \in [m]})$. Then, using (2) and the fact that $Q(\{f_j/g\}_{j \in [m]})$ is identically zero modulo the variable g, we must have that $Q'(\{f_x^{(1)}\}_{x \in \mathcal{X}})$ is the identically zero polynomial. In other words, Q' annihilates the set of polynomials $\{f_x^{(1)}\}_{x \in \mathcal{X}}$.

We now analyze the structure of the $f_x^{(1)}$ to show that such a Q' violates the (t, s)-BPUA assumption, which will complete the proof.

Structure of $f_x^{(1)}$. Recalling the notation from Sect. 3, each $\overline{f_x}$ is a polynomial in the entries of $Y_0, \{Y_{i,b_1,b_2}\}_{i \in [\ell]; b_1, b_2 \in \{0,1\}}, Y_{\ell+1}$. Specifically, it is the polynomial

$$\overline{f_x} = Y_0 \times \prod_{i \in [\ell]} Y_{i,x(i)} \times Y_{\ell+1}$$

where we abbreviate $x(i) := (x_{\mathsf{inp}(i)_1}, x_{\mathsf{inp}(i)_2})$. Notice that f_x is the polynomial obtained from $\overline{f_x}$ after making the following substitution.

$$Y_0 = \widehat{A_0} + gZ_0 \qquad Y_{i,b_1,b_2} = \widehat{A_{i,b_1,b_2}} + gZ_{i,b_1,b_2} \qquad Y_{\ell+1} = \widehat{A_{\ell+1}} + gZ_{\ell+1}$$

Then, because $f_x^{(1)}$ is the coefficient of g in f_x (see (1)) and the $\widehat{A}$ matrices are of the form

$$\widehat{A_0} = \alpha_0 A_0' \times R_1^{adj} \qquad \widehat{A_{i,b_1,b_2}} = \alpha_{i,b_1,b_2} R_i \times A_{i,b_1,b_2}' \times R_{i+1}^{adj}$$

$$\widehat{A_{\ell+1}} = \alpha_{\ell+1} R_{\ell+1} \times A_{\ell+1}'$$

we can expand the $\widehat{A}$ matrices to write $f_x^{(1)} = d_x + \alpha_0 \cdot d_x'$, where

$$d_x := Z_0 R_1 \left(\prod_{i=1}^{\ell} \alpha_{i,x(i)} A_{i,x(i)}'\right) \alpha_{\ell+1} A_{\ell+1}' \rho_0$$

and d_x' is another polynomial. Here we denote $\rho_0 := \prod_{i=2}^{\ell+1} \det(R_i)$, which arises from the fact that $R_i \times R_i^{adj} = \det(R_i) \cdot I$. Below, we will use the fact that α_0 does not appear in d_x.

Now recall that the A' matrices are constructed as

$$A'_0 := (A_0 \quad 0^s) \qquad A'_{i,b_1,b_2} := \begin{pmatrix} A_{i,b_1,b_2} & \\ & B_{i,b_1,b_2} \end{pmatrix} \qquad A'_{\ell+1} := \begin{pmatrix} A_{\ell+1} \\ B_{\ell+1} \end{pmatrix}$$

where the A matrices are the original branching program input to $\mathcal{O}$. We consider two cases: either

- Q' annihilates $\left\{ f_x^{(1)} \right\}_{x \in \mathcal{X}}$ when considered as polynomials in variables Z, R, B, and α (i.e. when only the A matrices are taken to be $\mathbb{Z}_p$-values), or
- it does not, but with probability $\epsilon \geq 1/\mathsf{poly}(\lambda)$ over the distribution on R, B, and α, Q' annihilates the set $\left\{ f_x^{(1)} \right\}_{x \in \mathcal{X}}$ when considered as polynomials in variables Z.

Here and throughout the remainder of the proof, we use the phrase "variables Z" to refer to the set of all variables arising from the Z matrices, and similarly for R, B, and α.

We now show that the first case contradicts the (t, s)-BPUA assumption, while the second case is ruled out by the Schwartz-Zippel lemma.

Case 1: Q' annihilates $\left\{ f_x^{(1)} \right\}_{x \in \mathcal{X}}$ as polynomials in variables Z, R, B, and α.
Because we can write $f_x^{(1)} = d_x + \alpha_0 \cdot d'_x$, where d_x does not contain the variable α_0, if Q' annihilates $\left\{ f_x^{(1)} \right\}_{x \in \mathcal{X}}$ as polynomials in variables Z, R, B, and α, it must also annihilate $\{d_x\}_{x \in \mathcal{X}}$.

Next, we perform the following change of variables: we set each R matrix to be the identity matrix (which in particular induces $\rho_0 = 1$), we set each α scalar to 1, and we set $Z_0 = (uV \quad B_0)$ for a new variable u and new vectors of variables V, B_0 which have lengths w and s respectively (recall that the A and B matrices have dimensions w and s respectively). Applying this change of variables to d_x, we obtain the polynomial $e_x + u \cdot e'_x$, where

$$e_x := B_0 \times \prod_{i \in [\ell]} B_{i,x(i)} \times B_{\ell+1}$$

and e'_x is another polynomial. Because $e_x + u \cdot e'_x$ was obtained from d_x via a change of variables, if Q' annihilates $\{d_x\}_{x \in \mathcal{X}}$ then it must also annihilate $\{e_x + u \cdot e'_x\}_{x \in \mathcal{X}}$. Further, since the variable u does not appear in e_x, Q' must also annihilate $\{e_x\}_{x \in \mathcal{X}}$.

However, this contradicts the (t, s)-BPUA assumption: by construction of inp and ℓ in Sect. 3, any branching program of the form in Assumption 1 can be embedded into the B matrices, and thus there is an efficiently computable distribution on degree-$2^{o(\lambda)}$ polynomials that annihilates all such branching programs with probability $\geq 1/\mathsf{poly}(\lambda)$.

Case 2: $\Pr_{R,B,\alpha}\left[Q'\textit{annihilates}\ \left\{f_x^{(1)}\right\}_{x\in\mathcal{X}}\ \textit{as polynomials in variables}\ Z\right] \geq$ $1/\text{poly}(\lambda)$. If Case 1 does not hold, then $Q'(\{f_x^{(1)}\}_{x\in\mathcal{X}})$ must contain some non-zero monomial. View this monomial as being over the variables g and Z, whose coefficient is a non-zero polynomial γ of degree $2^{o(\lambda)}$ in variables R, B, and α. (The degree bound on γ comes from the fact that Q' has degree $2^{o(\lambda)}$ and each $f_x^{(1)}$ has degree $\text{poly}(\lambda)$.)

If Case 2 holds, we must have $\Pr_{R,B,\alpha}[\gamma(R,B,\alpha)=0] \geq 1/\text{poly}(\lambda)$. However, this contradicts the Schwartz-Zippel lemma, because we are working over the field $\mathbb{Z}_p$ with $p > 2^\lambda$, and the distribution on the variables R, B, α is $2^{-\Omega(\lambda)}$-close to each being uniform and independent. Indeed, the distributions on the B variables are uniform over $\mathbb{Z}_p$, the distributions on the α variables are uniform over $\mathbb{Z}_p\backslash\{0\}$, and the distributions on the R variables are uniform over $\mathbb{Z}_p$ conditioned on each matrix R_i being non-singular.

We now prove the lemma that was used in the proof of Theorem 3. We will need the following result from [BMSZ16]. Recall that $\overline{f_x}$ denotes the matrix product polynomial that corresponds to evaluating $\widehat{BP}(x)$.

Theorem 4 ([BMSZ16]). *Fix $x \in \{0,1\}^n$, and consider the following matrices from Sect. 3: $A_i' := A_{i,x(i)}'$, $\widehat{A}_i := \widehat{A}_{i,x(i)}$, and R_i. Consider also a polynomial f in the entries of the $\widehat{A}$ matrices in which each monomial contains at most one variable from each $\widehat{A}_i$. Let f' be the polynomial derived from f after making the substitution $\widehat{A}_i = R_{i-1}^{adj} \times A_i' \times R_i$, and suppose that f' is identically 0 as a polynomial over the R_i.*

Then either f is identically zero as a polynomial over its formal variables (namely the $\widehat{A}_i$), or else f is a constant multiple of the matrix product polynomial $\overline{f_x} = \widehat{A}_0 \times \cdots \times \widehat{A}_{\ell+1}$.

We remark that the proof of this theorem requires that the A' matrices form a non-shortcutting branching program (see Definition 2), and that for us this is implied by the distribution on the B matrices and the fact that A is forward non-shortcutting.

Lemma 1. *Let BP be any forward-non-shortcutting branching program, and let the model $\mathcal{M}$ from Sect. 2 be initialized by the obfuscator $\mathcal{O}(BP)$ with parameters t, s as described in Sect. 3.*

Let $\mathcal{A}$ be an efficient adversary interacting with $\mathcal{M}$, and let $\{h_j\}_{j\in[m]}$ be the set of all handles $\mathcal{A}$ has received that map to a level-$\mathbb{U}$ polynomial with constant term 0 in $\mathcal{M}$'s pre-zero-test table; denote these polynomials by $\{f_j\}_{j\in[m]}$. Assume that $\mathcal{A}$ has not received "WIN" in response to any post-zero-test query.

Then with probability $1 - \text{negl}(\lambda)$ over the randomness of $\mathcal{O}$, there exist $\text{poly}(\lambda)$-size sets $\mathcal{X}_1, \ldots, \mathcal{X}_m$ such that: (1) for each $j \in [m]$, f_j is a linear combination of the polynomials $\{\overline{f_x}\}_{x\in\mathcal{X}_j}$, and (2) for each $j \in [m]$ and each $x \in \mathcal{X}_j$, $BP(x) = 0$.

Proof. The proof follows the analysis of [BMSZ16, Theorem 5.1], which builds on [BGK+14]. We assume that the lemma's conclusion holds for $f_1, \ldots, f_{m-1}$, and prove that it holds for f_m with probability $1 - \mathsf{negl}(\lambda)$. This inductively implies the lemma.

As in the proof of Theorem 3, let $\overline{f_m}$ be the polynomial over the set of $\mathcal{M}$'s initial elements such that $f_m = \overline{f_m}(\{a_i + g \cdot z_i\})$. Because $\overline{f_m}$ is at level $\mathbb{U}$, we can use the procedure given by [BGK+14, Sect. 6] (cf. [BMSZ16, Lemma 5.3]) to decompose it as

$$\overline{f_m} = \sum_{x \in \mathcal{X}_m} f_{m,x}$$

with equality as formal polynomials, where $\mathcal{X}_m$ is a $\mathsf{poly}(\lambda)$-size set given by the decomposition, and each $f_{m,x}$ is a non-identically-zero polynomial at level $\mathbb{U}$ that only has variables from matrices in $\widehat{BP}$ that correspond to input x.

Notice that f_m has constant term 0 iff $\overline{f_m}(\{a_i\}_i) = 0$. Then following the [BGK+14, Sect. 6] analysis, the independence of the α_{i,b_1,b_2} randomization variables along with the fact that $\overline{f_m}(\{a_i\}_i) = 0$ implies $\Pr[\exists x \in \mathcal{X}_m : f_{m,x}(\{a_i\}_i) \neq 0] < \mathsf{negl}(\lambda)$, where the probability is over $\mathcal{O}$'s randomness. Assume for the remainder that $f_{m,x}(\{a_i\}_i) = 0$ for all $x \in \mathcal{X}_m$, which occurs with probability $1 - \mathsf{negl}(\lambda)$.

Consider the moment just before $\mathcal{A}$ submits the handle h_m (corresponding to f_m) for zero-testing. At this point, since we assume the lemma's conclusion holds for $f_1, \ldots, f_{m-1}$ and that $\mathcal{A}$ has never received "WIN" in response to any post-zero-test query, $\mathcal{A}$'s view can be completely derived from the set $\{BP(x) \mid x \in \bigcup_{j \in [m-1]} \mathcal{X}_j\}$. In particular, $\mathcal{A}$'s view is independent of the randomness generated by $\mathcal{O}$.

Now fix some $x \in \mathcal{X}_m$. The values $\{a_i\}_i$ are generated by $\mathcal{O}$ from the original branching program BP by choosing the randomization matrices R and the other randomization values α, B, and performing the computation described in Sect. 3. We can thus view $f_{m,x}$ as a polynomial $f'_{m,x}$ over the R variables whose coefficients are polynomials in the variables α, B. Then because $f_{m,x}$ only has variables from matrices corresponding to input x and is not identically zero, Theorem 4 implies that either $f_{m,x}$ is a constant multiple of $\overline{f_x}$, or else $f'_{m,x}$ is not the identically zero polynomial.

Because we assume $f_{m,x}(\{a_i\}_i) = 0$ for the particular sample of $\{a_i\}_i$ generated by $\mathcal{O}$, if $f'_{m,x}$ is not identically zero, then one of two things must have occurred. Either every coefficient of $f'_{m,x}$ became 0 after the choice of α, B, or some choice of α, B yields a fixed $\mathbb{Z}_p$-polynomial that evaluated to 0 on the choice of the R matrices. However, both of these events have probability $1 - \mathsf{negl}(\lambda)$ by the Schwartz-Zippel lemma. Thus, since $\mathcal{A}$'s view (and in particular $f'_{m,x}$) is independent of $\mathcal{O}$'s randomness, we conclude that with probability $1 - \mathsf{negl}(\lambda)$, $f_{m,x}$ is a constant multiple of $\overline{f_x}$.

Finally, note that if $f_{m,x}$ is a (non-zero) constant multiple of $\overline{f_x}$ and if $f_{m,x}(\{a_i\}_i) = 0$, then $\overline{f_x}(\{a_i\}_i) = 0$, which is equivalent to $BP(x) = 0$.

4.2   Further Discussion of Our Assumption

We first note that PRFs such as those in the statement of Theorem 2 can be constructed from any boolean NC^1 PRF, provided $s \geq 5\lambda$ and t is a sufficiently large polynomial. The idea is to take λ copies of a width-5, length-t boolean PRF (constructed via [Bar86]), scale the ith copy by 2^i for $i = 0, \ldots, \lambda - 1$, and put them into a block-diagonal BP of width 5λ with appropriate bookend vectors to sum the scaled copies.

We note that for complicated programs whose length is already larger than t, the overhead for protecting against zeroing attacks is mainly due to increasing the width by s. The multiplicative overhead is thus $(w + s)^2/w^2$ where w is the original width of the branching program. Thus, for many applications, it is likely best to minimize s, potentially at the expense of a slightly larger t. Next, we describe how to modify the above idea to obtain a branching program of *constant* width.

Making the PRF Computation have Constant Width. We now explain that the width s can actually be taken to be a constant. There are many ways to accomplish this. Perhaps the simplest is the following. Ben Or and Cleve [Cle88] show how to convert any arithmetic formula into a matrix branching program consisting of 3×3 matrices, where the matrix product gives

$$\begin{pmatrix} 1 & f(x) & 0 \\ 0 & 1 & 0 \\ 0 & 0 & 1 \end{pmatrix}$$

Then the output $f(x)$ can be selected by multiplying by the appropriate bookend vectors.

For any invertible constant c in the ring, by left- and right- multiplying the branching program by the constant matrices

$$\begin{pmatrix} c & 0 & 0 \\ 0 & 1 & 0 \\ 0 & 0 & 1 \end{pmatrix} \text{ and } \begin{pmatrix} c^{-1} & 0 & 0 \\ 0 & 1 & 0 \\ 0 & 0 & 1 \end{pmatrix},$$

the product of the branching program matrices becomes

$$\begin{pmatrix} 1 & cf(x) & 0 \\ 0 & 1 & 0 \\ 0 & 0 & 1 \end{pmatrix}$$

Next, by concatenating the branching programs for f_1 and f_2, the result of the matrix product is

$$\begin{pmatrix} 1 & f_1(x) + f_2(x) & 0 \\ 0 & 1 & 0 \\ 0 & 0 & 1 \end{pmatrix}$$

Let $f_0, \ldots, f_{\lambda-1}$ be independent formulas for computing a pseudorandom bit. It is therefore possible to construct a matrix branching program whose matrix product is

$$\begin{pmatrix} 1 & \sum_{i=1}^{\lambda-1} 2^i f_i(x) & 0 \\ 0 & 1 & 0 \\ 0 & 0 & 1 \end{pmatrix}$$

By multiplying by the appropriate bookend vectors, the result is $\sum_{i=1}^{\lambda-1} 2^i f_i(x)$. By the pseudorandomness of the f_i, this is a pseudorandom value in $[0, 2^\lambda - 1]$.

Varying the Assumption Strength. We also note that, based on whether we wish t, s to be polynomial, logarithmic, or constant, we can obtain assumptions of varying strength. For example, we can have the following.

Assumption 5 (The poly/poly-BPUA Assumption). *There exist polynomials t, s such that the (t, s)-BPUA assumption holds.*

Assumption 6 (The poly/const-BPUA Assumption). *There exists polynomial t and* constant s *such that the (t, s)-BPUA assumption holds.*

Assumption 7 (The polylog/const-BPUA Assumption). *There exists polylogarithmic t and constant s such that the (t, s)-BPUA assumption holds.*

We can thus get a trade-off between efficiency and assumption strength - stronger assumptions (those with smaller s and t) very naturally correspond to more efficient obfuscators.

Dual Input Assumptions. We could have similarly made dual-input versions of the above assumptions. However, we observe that the single input and dual input variants are equivalent, up to constant factors in t and s.

In particular, any single input branching program of length t and width s can be turned into a dual input program of length $t/2$ and width s by premultiplying branching program matrices. That is, set $A'_{i,b_0,b_1} = A_{2i-1,b_0} \cdot A_{2i,b_1}$ and $\mathsf{inp}_b(i) = \mathsf{inp}(2i - b)$.

Moreover, any dual input branching program of length t and width s can be converted into a single input branching program of length $2t$ and width $2s$ via the following transformation:

$$A'_{2i-1,b} = \begin{pmatrix} A_{i,b,0} & A_{i,b,1} \\ 0^{s\times s} & 0^{s\times s} \end{pmatrix} \quad A'_{2i,b} = \begin{pmatrix} (1-b)I_s & 0^{s\times s} \\ bI_s & 0^{s\times s} \end{pmatrix}$$

$$\mathsf{inp}(i) = \begin{cases} \mathsf{inp}_0((i+1)/2) & \text{if } i \text{ is odd} \\ \mathsf{inp}_1(i/2) & \text{if } i \text{ is even} \end{cases}$$

Notice that $A'_{2i-1,b_0} \cdot A'_{2i,b_1} = \begin{pmatrix} A_{i,b_0,b_1} & 0^{s\times s} \\ 0^{s\times s} & 0^{s\times s} \end{pmatrix}$.

5   Security in an Alternative Model

In this section, we define a second model for weak multilinear maps, and we show that the proof of Theorem 3 can be modified to give security in this model as well. The main difference as compared to the model in Sect. 2 is that this model no longer treats the z_i as formal variables, but instead considers z_i sampled in some fashion by the encoding procedure.

We now formally describe the interfaces implemented by the oracle $\mathcal{M}$ that defines our model. For concreteness, we define $\mathcal{M}$ to explicitly work over the GGH13 ring $\mathcal{R} = \mathbb{Z}[X]/(X^\eta + 1)$ and the field $\mathbb{Z}_p \simeq \mathcal{R}/\langle g \rangle$ for an appropriate $g \in \mathcal{R}$.

$\mathcal{M}$ is parameterized by a family of distributions $\{D_{p,\{a_i\}_{i \in [n]}}\}$ for prime p and sets of integers $\{a_i\}_{i \in [n]} \subseteq \mathbb{Z}_p$ of size n. Each $D_{p,\{a_i\}_{i \in [n]}}$ is a product distribution $D_1 \times \cdots \times D_n$ where the D_i are distributions over $\mathbb{Z}_p$.

Initialize Parameters. This is identical to the model of Sect. 2. The first step in interacting with $\mathcal{M}$ is to initialize it with the security parameter $\lambda \in \mathbb{N}$. (Jumping ahead, this will be done by the obfuscator.) $\mathcal{M}$ defines the ring $\mathcal{R} = \mathbb{Z}[X]/(X^\eta + 1)$, where $\eta = \eta(\lambda)$ is chosen as in [GGH13a]. Then, $\mathcal{M}$ chooses $g \in \mathcal{R}$ according to the distribution in [GGH13a], and outputs the prime $p := |\mathcal{R}/\langle g \rangle| > 2^\lambda$. After initializing these parameters, $\mathcal{M}$ discards the value of g, and treats g as a *formal variable* in all subsequent steps.

Initialize Elements. After the parameters have been initialized, $\mathcal{M}$ is given a universe set $\mathbb{U}$ and a set of initial elements $\{[a_i]_{S_i}\}_i$ where $a_i \in \mathbb{Z}_p$ and $S_i \subseteq \mathbb{U}$ for each i. $\mathcal{M}$ then samples a set of ring elements $\{z_i\}$ from $D_{p,\{a_i\}}$.

$\mathcal{M}$ defines the formal polynomial $f_i := a_i + g \cdot z_i$ over $\mathbb{Z}_p$. Here g is a formal variable that is common to all f_i. Then $\mathcal{M}$ generates a handle h_i (whose representation explicitly specifies S_i but is independent of a_i), and stores the mapping "$h_i \to (f_i, S_i)$" in a table that we call the *pre-zero-test* table. Finally, $\mathcal{M}$ outputs the set of handles $\{h_i\}_i$.

The above two initialization interfaces are each executed once, in the order listed; any attempt to execute them out of order or more than once will fail. The only difference with the model in Sect. 2 is that the z_i are no longer formal variables, but are now actual ring elements.

$\mathcal{M}$ also implements the following algebraic interfaces.

Pre-zero-test Arithmetic. Given two input handles h_1, h_2 and an operation $\circ \in \{+, -, \cdot\}$, $\mathcal{M}$ first locates the corresponding polynomials f_1, f_2 and level sets S_1, S_2 in the pre-zero-test table. If h_1 and h_2 do not both appear in this table, the call to $\mathcal{M}$ fails. If the expression is undefined (i.e., $S_1 \neq S_2$ for $\circ \in \{+, -\}$, or $S_1 \cap S_2 \neq \emptyset$ for $\circ \in \{\cdot\}$), the call fails. Otherwise, $\mathcal{M}$ computes the formal polynomial $f := f_1 \circ f_2$ and the level set $S := S_1 \cup S_2$, generates a new handle h, and stores the mapping "$h \to (f, S)$" in the pre-zero-test table. Finally, $\mathcal{M}$ outputs h.

Zero-Testing. Given an input handle h, $\mathcal{M}$ first locates the corresponding polynomial f and level set S in the pre-zero-test table. If h does not appear in this table, or if $S \neq \mathbb{U}$, the call to $\mathcal{M}$ fails. If f's constant term is non-zero (recall that this term is an element of $\mathbb{Z}_p$), $\mathcal{M}$ outputs the string "non-zero". If instead f's constant term is 0, note that f must be divisible by the formal variable g, i.e. g appears in each of f's monomials. $\mathcal{M}$ computes the formal polynomial $f' := f/g$ over $\mathbb{Z}_p$, generates a new handle h', and stores the mapping "$h' \to f'$" in a table that we call the *post-zero-test* table. Finally, $\mathcal{M}$ outputs h'.

Post-zero-test Arithmetic. Given a set of input handles $h'_1, \ldots, h'_m$ and an m-variate polynomial Q over $\mathbb{Z}$ (represented as an arithmetic circuit), $\mathcal{M}$ first locates the corresponding polynomials $f'_1, \ldots, f'_m$ in the post-zero-test table. If any h'_i does not appear in this table, the call to $\mathcal{M}$ fails. Otherwise, $\mathcal{M}$ checks whether $Q(f'_1, \ldots, f'_m)$ is non-zero as a polynomial over $\mathbb{Z}_p$ which *is* zero modulo the variable g. In other words, $\mathcal{M}$ checks that the constant term of $Q(f'_1, \ldots, f'_m)$ is 0, but that some other coefficient is non-zero. If this check passes, $\mathcal{M}$ outputs "WIN", otherwise it outputs $\bot$.

Definition 6. *A (possibly randomized) adversary interacting with the model $\mathcal{M}$ is* efficient *if it runs in time* $\mathsf{poly}(\lambda)$, *and if each Q submitted in a post-zero-test query has degree* $2^{o(\lambda)}$. *Such an adversary* wins *if it ever submits a post-zero-test query that causes $\mathcal{M}$ to output "WIN".*

Definition 7. *Let $O = \{O_p\}$ be a (family of) distributions over initial elements $\{[a_i]_{S_i}\}_{i \in [n]}$. Consider model $\mathcal{M}$ parameterized by distribution family $\{D_{p,\{a_i\}_{i \in [n]}}\}$. $\mathcal{M}$ satisfies the* unpredictability probability *relative to O if the following holds. For each $i \in [n]$, the expected guessing probability of z_i drawn from $D_{p,\{a_i\}_{i \in [n]}}$ (where the expectation is over the choice of $\{a_i\}_{i \in [n]}$) is at most $2^{-\Omega(\lambda)}$.*

The above definition captures the fact that in GGH13 encodings, the z_i elements are chosen with min-entropy at least $\Omega(\lambda)$, yielding a guessing probability of $2^{-\Omega(\lambda)}$. This holds even in the "low noise" variants, due to the large dimensional space that the z_i are drawn from. As required by GGH13, our definition allows the z_i to depend on a_i; however we allow for an even more general condition where the z_i can depend on *all* of the $\{a_j\}$. Moreover, we only require the guessing probability to be small *on average*.

5.1   A New Variant of the Schwartz-Zippel Lemma

We now prove a generalization of the Schwartz-Zippel lemma, which will allow us to prove security in the alternative model described above. The standard Schwartz-Zippel lemma applies to variables chosen independently and uniformly from some (possibly restricted) set. Here, we instead allow the variables to be chosen from arbitrary distributions with sufficient min-entropy, and we even allow some correlations among the variables.

Let $\mathbb{F}$ be a finite field, and let $P \in \mathbb{F}[x_1, \ldots, x_n]$ be an arbitrary polynomial of degree at most d. Let $X_1, \ldots, X_n$ be potentially correlated random variables over $\mathbb{F}$. Let $p_i(x_1, \ldots, x_{i-1})$ be the guessing probability of X_i conditioned on $X_j = x_j$ for each $j < i$. That is,

$$p_i(x_1, \ldots, x_{i-1}) = \max_{x_i \in \mathbb{F}} \Pr[X_i = x_i | X_j = x_j \forall j < i]$$

Let p_i be the expectation of $p_i(x_1, \ldots, x_{i-1})$ when x_j are drawn from X_j: $p_i = \mathbb{E}[p_i(X_1, \ldots, X_{i-1})]$. Let $p_{\max} = \max_i p_i$ be the maximum of the p_i.

Lemma 2. *Let $\mathbb{F}, d, n, P, X_1, \ldots, X_n, p_{\max}$ be as above. Then*

$$\Pr_{X_1, \ldots, X_n}[P(X_1, \ldots, X_n) = 0] \le d \cdot p_{\max}.$$

Proof. The proof will be by induction on n. The case $n = 1$ follows from the fact that a degree d polynomial has at most d roots. Assume the lemma holds up to $n - 1$. Let d_n be the maximum degree of x_n in P. Consider first sampling $X_1, \ldots, X_{n-1}$. Plugging into P, we get a polynomial $P_{X_1, \ldots, X_{n-1}}(x_n) = P(X_1, \ldots, X_{n-1}, x_n)$ in x_n of degree at most d_n. Then consider sampling X_n conditioned on the outcome of $X_1, \ldots, X_{n-1}$. P gives zero if and only if one of two conditions are met:

- $P_{X_1, \ldots, X_{n-1}}$ is identically zero. Let e_0 be the probability of this event. Let $e_{\ne 0} = 1 - e_0$ be the probability that $P_{X_1, \ldots, X_{n-1}}$ is not identically zero
- $P_{X_1, \ldots, X_{n-1}}$ is not identically zero, and X_n is a root of $P_{X_1, \ldots, X_{n-1}}$.

Let q_0 be the expectation of $p_n(X_1, \ldots, X_{n-1})$ conditioned on $P_{X_1, \ldots, X_{n-1}}$ being identically 0, and let $q_{\ne 0}$ be the expectation conditioned on $P_{X_1, \ldots, X_{n-1}}$ *not* being identically 0. Note that $p_n = e_0 q_0 + e_{\ne 0} q_{\ne 0}$. Also, not that $q_0, q_{\ne 0} \ge 0$. Therefore, $e_{\ne 0} q_{\ne 0} \le p_n$.

The coefficient of $x_n^{d_n}$ in $P_{X_1, \ldots, X_{n-1}}$ is a polynomial of total degree at most $d - d_n$ in $X_1, \ldots, X_{n-1}$. If $d_n = d$, the coefficient is actually a constant and must be non-zero (with probability 1). In the case $d_n < d$, we can apply the inductive hypothesis to bound the probability that this coefficient is 0 by $(d - d_n)p_{\max}$. Thus, in either case, the probability e_0 that $P_{X_1, \ldots, X_{n-1}}$ is identically 0 is at most $(d - d_n)p_{\max}$.

We now bound the probability that $P_{X_1, \ldots, X_{n-1}}$ is not identically zero, and X_n is a root of $P_{X_1, \ldots, X_{n-1}}$. Since $P_{X_1, \ldots, X_{n-1}}$ is not identically 0 and has degree at most d_n, there are at most d_n roots. Thus, the probability that X_n is a root is at most $d_n p_n(X_1, \ldots, X_{n-1})$. Taking the expectation conditioned on $P_{X_1, \ldots, X_{n-1}}$ being not identically 0, we get a bound of $d_n q_{\ne 0}$ on the probability that $P = 0$ conditioned on $P_{X_1, \ldots, X_{n-1}}$ being identically 0. The joint probability is therefore at most $e_{\ne 0} d_n q_{\ne 0} \le d_n p_n \le d_n p_{\max}$.

Putting everything together, the probability that $P = 0$ is at most $(d - d_n)p_{\max} + d_n p_{\max} = d p_{\max}$.

5.2   Security in the Alternative Model

Security in the alternative model is given by the following theorem. We note that, analagously to Sect. 4, this theorem also implies that $\mathcal{O}$ achieves VBB security in the alternative model.

Theorem 8. *Let $\mathcal{O}$ be the obfuscator from Sect. 3 with parameters t and s. Let $\mathcal{M}$ be the model defined above, parameterized by some distribution family $\{D_{p,\{a_i\}_{i\in[n]}}\}$. If the (t,s)-BPUA assumption holds, and if $\mathcal{M}$ satisfies the unpredictability property relative to the elements outputted by $\mathcal{O}$, then $\mathcal{O}$ is secure in the model $\mathcal{M}$.*

The proof of Theorem 8 follows the proof of Theorem 3 almost exactly, with the only difference being that, when analyzing Case 2, we apply Lemma 2 instead of the standard Schwartz-Zippel lemma. We omit further details.

A   Straddling Set Level Structure

Here we describe the level structure for the graded encoding scheme that is used by the obfuscator $\mathcal{O}$ when initializing the model $\mathcal{M}$ with the values of $\widehat{BP}$ (see Sect. 3). This construction is due to Barak et al. [BGK+14], and was used in several subsequent works. It relies on the following notion of a *straddling set system.*[5]

Definition 8. *A straddling set system with n entries is a universe set $\mathbb{U}$ and a collection of subsets $\mathbb{S} = \{S_{i,b} \subseteq \mathbb{U}\}_{i\in[n],b\in\{0,1\}}$ such that*

1. $\bigcup_{i\in[n]} S_{i,0} = \bigcup_{i\in[n]} S_{i,1} = \mathbb{U}$, *and*
2. *for any distinct $C, D \subseteq \mathbb{S}$ such that $\bigcup_{S\in C} S = \bigcup_{S\in D} S$, there exists $b \in \{0,1\}$ such that $C = \{S_{i,b}\}_{i\in[n]}$ and $D = \{S_{i,1-b}\}_{i\in[n]}$.*

For any n, the following is a straddling set system with n entries over the universe $\mathbb{U} = \{1,\ldots,2n-1\}$ (for a proof see [BGK+14, Appendix A]).

$$S_{1,0} = \{1\}, S_{2,0} = \{2,3\}, \ldots, S_{i,0} = \{2i-2, 2i-1\}, \ldots, S_{n,0} = \{2n-2, 2n-1\}$$

$$S_{1,1} = \{1,2\}, \ldots, S_{i,1} = \{2i-1, 2i\}, \ldots, S_{n-1,1} = \{2n-3, 2n-2\}, S_{n,1} = \{2n-1\}$$

We now describe the level structure that is used to encode $\widehat{BP}$. For each input index $i \in [n]$, let r_i denote the number of layers in which bit i is read, and create a straddling set system with r_i entries. We denote the universe set of this straddling set system by $\mathbb{U}^{(i)}$, and its subsets by $\{S_{j,b}^{(i)}\}_{j\in[r_i],b\in\{0,1\}}$. The overall universe set is then $\mathbb{U} := \bigcup_{i\in[n]} \mathbb{U}^{(i)} \cup \{L, R\}$, where we assume that the $\mathbb{U}^{(i)}$ are pairwise disjoint, and L and R are new symbols that don't appear in any $\mathbb{U}^{(i)}$.

---

[5] For the analysis that we borrow from [BGK+14,BMSZ16], namely Lemma 1, we will not need the *strong* straddling set systems due to [MSW14].

Then, for each matrix $\widehat{A_{j,b_1,b_2}}$ in $\widehat{BP}$, each entry of this matrix is encoded at level

$$S_{k_1,b_1}^{(\mathsf{inp}(j)_1)} \cup S_{k_2,b_2}^{(\mathsf{inp}(j)_2)}$$

where k_1, k_2 are defined such that layer j is the k_1-th layer in which input bit $\mathsf{inp}(j)_1$ is read and the k_2-th layer in which input bit $\mathsf{inp}(j)_2$ is read. Finally, each entry of $\widehat{A_0}$ is encoded at level $\{L\}$, and each entry of $\widehat{A_{\ell+1}}$ is encoded at level $\{R\}$.

References

[AGIS14] Ananth, P., Gupta, D., Ishai, Y., Sahai, A.: Optimizing obfuscation: avoiding Barrington's theorem. In: Proceedings of the 2014 ACM SIGSAC Conference on Computer and Communications Security, pp. 646–658 (2014)

[AJN+16] Ananth, P., Jain, A., Naor, M., Sahai, A., Yogev, E.: Universal constructions and robust combiners for indistinguishability obfuscation and witness encryption. In: Robshaw, M., Katz, J. (eds.) CRYPTO 2016. LNCS, vol. 9815, pp. 491–520. Springer, Heidelberg (2016). doi:10.1007/978-3-662-53008-5_17

[App14] Applebaum, B.: Bootstrapping obfuscators via fast pseudorandom functions. In: Sarkar, P., Iwata, T. (eds.) ASIACRYPT 2014. LNCS, vol. 8874, pp. 162–172. Springer, Heidelberg (2014). doi:10.1007/978-3-662-45608-8_9

[Bar86] Mix Barrington, D.A.: Bounded-width polynomial-size branching programs recognize exactly those languages in NC1. In: STOC (1986)

[BGH+15] Brakerski, Z., Gentry, C., Halevi, S., Lepoint, T., Sahai, A., Tibouchi, M.: Cryptanalysis of the quadratic zero-testing of GGH. Cryptology ePrint Archive, Report 2015/845 (2015). http://eprint.iacr.org/

[BGI+01] Barak, B., Goldreich, O., Impagliazzo, R., Rudich, S., Sahai, A., Vadhan, S., Yang, K.: On the (im)possibility of obfuscating programs. In: Kilian, J. (ed.) CRYPTO 2001. LNCS, vol. 2139, pp. 1–18. Springer, Heidelberg (2001). doi:10.1007/3-540-44647-8_1

[BGI+12] Barak, B., Goldreich, O., Impagliazzo, R., Rudich, S., Sahai, A., Vadhan, S.P., Yang, K.: On the (im)possibility of obfuscating programs. J. ACM **59**(2), 6 (2012)

[BGJ+15] Bitansky, N., Goldwasser, S., Jain, A., Paneth, O., Vaikuntanathan, V., Waters, B.: Time-lock puzzles from randomized encodings. Cryptology ePrint Archive, Report 2015/514 (2015). http://eprint.iacr.org/

[BGK+14] Barak, B., Garg, S., Kalai, Y.T., Paneth, O., Sahai, A.: Protecting obfuscation against algebraic attacks. In: Nguyen, P.Q., Oswald, E. (eds.) EUROCRYPT 2014. LNCS, vol. 8441, pp. 221–238. Springer, Heidelberg (2014). doi:10.1007/978-3-642-55220-5_13

[BLR+15] Boneh, D., Lewi, K., Raykova, M., Sahai, A., Zhandry, M., Zimmerman, J.: Semantically secure order-revealing encryption: multi-input functional encryption without obfuscation. In: Oswald, E., Fischlin, M. (eds.) EUROCRYPT 2015. LNCS, vol. 9057, pp. 563–594. Springer, Heidelberg (2015). doi:10.1007/978-3-662-46803-6_19

[BMSZ16] Badrinarayanan, S., Miles, E., Sahai, A., Zhandry, M.: Post-zeroizing obfuscation: new mathematical tools, and the case of evasive circuits. In: Fischlin, M., Coron, J.-S. (eds.) EUROCRYPT 2016. LNCS, vol. 9666, pp. 764–791. Springer, Heidelberg (2016). doi:10.1007/978-3-662-49896-5_27

[BR14] Brakerski, Z., Rothblum, G.N.: Virtual black-box obfuscation for all circuits via generic graded encoding. In: Lindell, Y. (ed.) TCC 2014. LNCS, vol. 8349, pp. 1–25. Springer, Heidelberg (2014). doi:10.1007/978-3-642-54242-8_1

[BWZ14] Boneh, D., Wu, D.J., Zimmerman, J.: Immunizing multilinear maps against zeroing attacks. Cryptology ePrint Archive, Report 2014/930 (2014). http://eprint.iacr.org/

[CGH+15] Coron, J.-S., et al.: Zeroizing without low-level zeroes: new MMAP attacks and their limitations. In: Gennaro, R., Robshaw, M. (eds.) CRYPTO 2015. LNCS, vol. 9215, pp. 247–266. Springer, Heidelberg (2015). doi:10.1007/978-3-662-47989-6_12

[CHL+15] Cheon, J.H., Han, K., Lee, C., Ryu, H., Stehlé, D.: Cryptanalysis of the multilinear map over the integers. In: Oswald, E., Fischlin, M. (eds.) EUROCRYPT 2015. LNCS, vol. 9056, pp. 3–12. Springer, Heidelberg (2015). doi:10.1007/978-3-662-46800-5_1

[Cle88] Cleve, R.: Computing algebraic formulas with a constant number of registers. In: Proceedings of the Twentieth Annual ACM Symposium on Theory of Computing, STOC 1988, pp. 254–257. ACM, New York (1988)

[CLR15] Cheon, J.H., Lee, C., Ryu, H.: Cryptanalysis of the new CLT multilinear maps. Cryptology ePrint Archive, Report 2015/934 (2015). http://eprint.iacr.org/

[CLT13] Coron, J.-S., Lepoint, T., Tibouchi, M.: Practical multilinear maps over the integers. In: Canetti, R., Garay, J.A. (eds.) CRYPTO 2013. LNCS, vol. 8042, pp. 476–493. Springer, Heidelberg (2013). doi:10.1007/978-3-642-40041-4_26

[CLT15] Coron, J.-S., Lepoint, T., Tibouchi, M.: New multilinear maps over the integers. In: Gennaro, R., Robshaw, M. (eds.) CRYPTO 2015. LNCS, vol. 9215, pp. 267–286. Springer, Heidelberg (2015). doi:10.1007/978-3-662-47989-6_13

[GGH13a] Garg, S., Gentry, C., Halevi, S.: Candidate multilinear maps from ideal lattices. In: Johansson, T., Nguyen, P.Q. (eds.) EUROCRYPT 2013. LNCS, vol. 7881, pp. 1–17. Springer, Heidelberg (2013). doi:10.1007/978-3-642-38348-9_1

[GGH+13b] Garg, S., Gentry, C., Halevi, S., Raykova, M., Sahai, A., Waters, B.: Candidate indistinguishability obfuscation and functional encryption for all circuits. In: FOCS, pp. 40–49 (2013)

[GGH15] Gentry, C., Gorbunov, S., Halevi, S.: Graph-induced multilinear maps from lattices. In: Dodis, Y., Nielsen, J.B. (eds.) TCC 2015. LNCS, vol. 9015, pp. 498–527. Springer, Heidelberg (2015). doi:10.1007/978-3-662-46497-7_20

[GMS16] Garg, S., Mukherjee, P., Srinivasan, A.: Obfuscation without the vulnerabilities of multilinear maps. Cryptology ePrint Archive, Report 2016/390 (2016). http://eprint.iacr.org/

[Hal15] Halevi, S.: Graded encoding, variations on a scheme. IACR Cryptology ePrint Archive, 2015:866 (2015)

[HJ15] Hu, Y., Jia, H.: Cryptanalysis of GGH map. IACR Cryptology ePrint Archive, 2015:301 (2015)

[Kil88] Kilian, J.: Founding cryptography on oblivious transfer. In: STOC, pp. 20–31 (1988)

[MF15] Minaud, B., Fouque, P.-A.: Cryptanalysis of the new multilinear map over the integers. Cryptology ePrint Archive, Report 2015/941 (2015). http:// eprint.iacr.org/

[MSW14] Miles, E., Sahai, A., Weiss, M.: Protecting obfuscation against arithmetic attacks. IACR Cryptology ePrint Archive, 2014:878 (2014)

[MSZ16a] Miles, E., Sahai, A., Zhandry, M.: Annihilation attacks for multilinear maps: cryptanalysis of indistinguishability obfuscation over GGH13. In: Robshaw, M., Katz, J. (eds.) CRYPTO 2016. LNCS, vol. 9815, pp. 629–658. Springer, Heidelberg (2016). doi:10.1007/978-3-662-53008-5_22

[MSZ16b] Miles, E., Sahai, A., Zhandry, M.: Secure obfuscation in a weak multilinear map model: a simple construction secure against all known attacks. Cryptology ePrint Archive, Report 2016/588 (2016). http://eprint.iacr. org/2016/588

[PST14] Pass, R., Seth, K., Telang, S.: Indistinguishability obfuscation from semantically-secure multilinear encodings. In: Garay, J.A., Gennaro, R. (eds.) CRYPTO 2014. LNCS, vol. 8616, pp. 500–517. Springer, Heidelberg (2014). doi:10.1007/978-3-662-44371-2_28

Virtual Grey-Boxes Beyond Obfuscation: A Statistical Security Notion for Cryptographic Agents

Shashank Agrawal[1]([⊠]), Manoj Prabhakaran[2], and Ching-Hua Yu[2]

[1] University of Texas at Austin, Austin, USA
sagrawal@cs.utexas.edu
[2] University of Illinois at Urbana-Champaign, Champaign, USA
{mmp,cyu17}@illinois.edu

Abstract. We extend the simulation-based definition of Virtual Grey Box (VGB) security – originally proposed for obfuscation (Bitansky and Canetti 2010) – to a broad class of cryptographic primitives. These include functional encryption, graded encoding schemes, bi-linear maps (with über assumptions), as well as unexplored ones like homomorphic functional encryption.

Our main result is a characterization of VGB security, in all these cases, in terms of an *indistinguishability-preserving* notion of security, called Γ^*-s-IND-PRE security, formulated using an extension of the recently proposed *Cryptographic Agents* framework (Agrawal et al. 2015). We further show that this definition is equivalent to an indistinguishability based security definition that is restricted to "concentrated" distributions (wherein the outcome of any computation on encrypted data is essentially known ahead of the computation).

A result of Bitansky et al. (2014), who showed that VGB obfuscation is equivalent to strong indistinguishability obfuscation (SIO), is obtained by specializing our result to obfuscation. Our proof, while sharing various elements from the proof of Bitansky et al., is simpler and significantly more general, as it uses Γ^*-s-IND-PRE security as an intermediate notion. Our characterization also shows that the semantic security for graded encoding schemes (Pass et al. 2014), is in fact an instance of this same definition.

We also present a composition theorem for Γ^*-s-IND-PRE security. We can then recover the result of Bitansky et al. (2014) regarding the existence of VGB obfuscation for all NC^1 circuits, simply by instantiating this composition theorem with a reduction from obfuscation of NC^1 circuits to graded encoding schemas (Barak et al. 2014) and the assumption that there exists an Γ^*-s-IND-PRE secure scheme for the graded encoding schema (Pass et al. 2014).

1 Introduction

Many recent advances in theoretical cryptography deal with obfuscation, multi-linear maps, various forms of functional encryption and more generally, tools that

© International Association for Cryptologic Research 2016
M. Hirt and A. Smith (Eds.): TCC 2016-B, Part II, LNCS 9986, pp. 269–296, 2016.
DOI: 10.1007/978-3-662-53644-5_11

enable computation on encrypted data. These tools are relatively new (compared to say, encryption, signatures and secure multi-party computation): for instance, the first formal definitions of obfuscation appeared only at the turn of the century [4,16]. As such our understanding of these tools and their security properties is relatively limited, and continues to generate steady interest within the field.

In this paper, we further push the boundaries of what we know regarding the security notions for these emerging cryptographic objects. To illustrate our findings, consider defining a new primitive, called *Homomorphic Functional Encryption* (HFE): HFE requires a private-key for encryption and decryption, but allows public homomorphic operations — for concreteness, addition — on ciphertexts, and also lets one use the private-key to generate function-keys that can be used to securely evaluate functions on ciphertexts (the function-key may reveal the function associated with it). Note that this allows a user with a collection ciphertexts $(c_1, \cdots, c_n)$ and a key for a function f, to evaluate $f(\sum_{i \in S} x_i)$, where x_i is the plaintext of c_i and $S \subseteq [n]$. We study two possible security notions for HFE, stated roughly below:

- A simulation-based security definition s-SIM,[1] in which a set of ciphertexts and function-keys can be simulated by a computationally unbounded simulator which is allowed to query $f(\sum_{i \in S} x_i)$ for only polynomially many subsets S.
- An indistinguishability based definition IND-CON, in which it is enough that, given a key for a function f, the ciphertexts for two "concentrated" distributions over plaintexts are indistinguishable. A pair of plaintext distributions $(\mathcal{D}_0, \mathcal{D}_1)$ is said to be concentrated for f if there is a function F such that for all $S \subseteq [n]$, $f(\sum_{i \in S} x_i) = F(S)$, with high probability over $(x_1, \cdots, x_n) \leftarrow \mathcal{D}_b$ for both $b = 0$ and $b = 1$ (i.e., the outcome is predictable just from the subset).

IND-CON is a fairly basic requirement: if the plaintext distribution is promised to be such that the function reveals virtually no information about the plaintexts (as the outcome of every function evaluation is known *a priori*), then the ciphertexts and function keys should hide which exact distribution the plaintexts were drawn from. On the other hand, the simulation-based definition requires the security to hold irrespective of the input distribution. The simulator needs to fool only an adversary who makes polynomially many queries, but no matter which of the *exponentially many subset queries* the adversary evaluates using the simulated ciphertexts and function keys, the outcome should match what the actual evaluation would have given. Remarkably,

> *our result implies that these two definitions are equivalent to each other.*

This is a significant generalization of a similar surprising result by Bitansky et al. [9], who studied the problem of obfuscation of circuits with boolean outputs. There it was shown that virtual grey-box (VGB) obfuscation and strong-indistinguishability obfuscation (SIO) are equivalent. In this work, we abstract out the fundamental properties underlying this equivalence and show that it covers a much wider spectrum of primitives beyond obfuscation, including HFE,

---

[1] *s* stands for statistical, indicating that the simulator is computationally unbounded.

(function-hiding) functional encryption, graded encoding schemes (with semantic security [19]), bi-linear maps (with über assumptions similar to the ones in [2]), etc.

Our main tool for establishing this equivalence is an intermediate security definition, which we cast in the recently formulated framework of *Cryptographic Agents* [2]. The Cryptographic Agents framework unifies several disparate cryptographic *objects*, akin to how the universal composition framework [11] unifies the study of *protocols* like oblivious-transfer, commitment and zero-knowledge proofs. Perhaps more significantly, it provides a definitional framework that, unlike the universal composition framework and the constructive cryptography framework [17], is based on indistinguishability-preservation (IND-PRE). We extend this security property, as well as introduce a new "test family" (which specifies the nature of the environment in which the security property should hold) as follows:

- we introduce the notion of *statistical* indistinguishability preserving (*s*-IND-PRE) security;
- we formulate a non-interactive test family Γ^*, which provides arbitrary auxiliary information about the objects being encoded, but — being non-interactive — prevents the adversary from adaptively influencing their choice.

We show that the resulting security definition of Γ^*-*s*-IND-PRE is equivalent to both the *s*-SIM and IND-CON definitions sketched above. These two definitions are formulated to apply to all primitives in the framework: when applied to obfuscation they yield the same definitions as in the equivalence result of [9], namely VGB obfuscation and SIO, respectively, thereby recovering the main result of [9] as a corollary.

We emphasize that our result is not about a particular primitive like obfuscation or HFE, *but about the framework itself.* Thus, for any primitive which can be modeled in the Cryptographic Agents framework, this equivalence holds.[2] For example, we observe that the "semantic-security" notion for graded encoding schemes introduced by Pass et al. [19] (or more precisely, its strengthening, as used in [9]) corresponds to Γ^*-*s*-IND-PRE security, and hence is also equivalent to corresponding *s*-SIM and IND-CON security definitions.

A New Composition Theorem for Cryptographic Agents. Another important component in our extension of the agents framework is a composition theorem. Given that our security definition involves a computationally unbounded adversary in the ideal world, the original composition theorem of [2] breaks down. However, we present a new information-theoretic variant of

---

[2] We point out that for certain primitives, like simple functional encryption and fully-homomorphic encryption, for which the number of ideal computations that a user can make — given a set of (evaluation or decryption) keys and ciphertexts — is only polynomially large, this equivalence is easier to establish. This is because, then a simulator can make *all possible ideal queries* that the user can ever make, and use plaintexts consistent with their results to generate the simulated ciphertexts.

the notion of reduction — statistical reduction — between two schemas, to re-establish a composition theorem. Specifically, we show that

> *a statistical reduction from a schema Σ to another schema Σ^* can be combined with a secure scheme for Σ^*, to obtain a secure scheme for Σ,*

where security refers to Γ^*-s-IND-PRE security.

An illustrative application of this composition theorem is to recover another result of [9] regarding the existence of VGB obfuscation for all NC^1 circuits. Indeed, once cast in our framework, this result is natural and immediate: [5] gave (using a different terminology) a reduction from obfuscation of NC^1 circuits to graded encoding schemas, and [9,19] put forth the assumption that there exists a Γ^*-s-IND-PRE secure scheme for the graded encoding schema. Under this assumption, our composition theorem immediately yields the result that VGB obfuscation exists for all NC^1 circuits.

Our Contributions. Below we summarize the contributions discussed above:

- We extend the cryptographic agents framework [2] to include the notion of statistical hiding and a new security definition called s-IND-PRE. Specifically, we consider Γ^*-s-IND-PRE security, where Γ^* is a family of computationally unbounded tests, which do not accept messages from the user. We also present two security definitions, IND-CON (indistinguishability for concentrated distributions) and s-SIM (statistical simulation security) for all schemas, which generalize the notions of SIO and VGB obfuscation to all schemas.
- Our main result is that all the above definitions are equivalent (for any schema). For the case of obfuscation, this result was proven in [9].
- We define a notion of *statistical reductions* and prove a *composition theorem* for Γ^*-s-IND-PRE security and statistical reductions. In particular, this can be used to reprove the existence of Γ^*-s-IND-PRE secure obfuscation for all of NC^1, assuming "strong-sampler semantically-secure" graded encoding schemes [9,19], and relying on an interpretation of a construction in [5] as a statistical reduction from the obfuscation schema to the graded encoding schema.

The above results clarify and significantly generalize the results in a small but influential collection of recent works on the foundations of security definitions for cryptographic objects [2,8,9,19]. Specifically,

> *our results generalize the notion of "Virtual Grey-Box security" beyond the realm of obfuscation.*

In particular, they help us better understand the security notions for graded encoding schemes. Also, they give concrete ways to prove VGB security for *future* constructions of homomorphic functional encryption, function-hiding functional encryption, etc.: a composition theorem that can be directly used if the construction uses VGB secure components, and an equivalence with IND-CON security, which would typically be easier to prove from scratch.

Finally, our results also enrich the nascent framework of Cryptographic Agents. We consider this an important contribution, as this framework can play a significant role in developing our understanding of the definitional aspects of emerging cryptographic primitives. Indeed, our result itself illustrates the usefulness of this framework, as it allowed us to extend a non-trivial result about obfuscation to a general result about unbounded simulation.

1.1 Technical Overview

We outline the definitional aspects first, and then present a high-level sketch of the proof of our main theorem (IND-CON $\Leftrightarrow$ Γ^*-s-IND-PRE $\Leftrightarrow$ Γ^*-s-SIM), and the composition theorem.

Security Definitions. Cryptographic agents and IND-PRE security were introduced as a means to define security for a large class of modern cryptographic primitives — including obfuscation, functional encryption, fully homomorphic-encryption and graded encoding schemes — avoiding the notion of simulation [2].

A scheme Π (consisting of two algorithms $\mathcal{O}$ and $\mathcal{E}$, analogous to the obfuscation and evaluation algorithms, in the case of obfuscation), is said to be IND-PRE secure for a schema Σ (which is defined by a family of idealized "agents" to which a user will only have black-box access in an ideal world) if every *test* in the ideal world that hides a challenge bit continues to hide the challenge bit in the real world. A cryptographic primitive is fully defined by the schema Σ as well as the test family Γ for which the indistinguishability preservation property holds.

We extend this notion naturally to consider *statistical hiding* in the ideal world. In s-IND-PRE security, a test in Γ is required to be hiding in the real world only if it is statistically hiding in the ideal world — i.e., hiding against computationally unbounded adversaries (who are still limited to making polynomial number of accesses to the agents uploaded by the test). Further, we introduce a sharper *quantitative notion* of s-IND-PRE security, which makes explicit the (polynomial) gap permitted between the extent of ideal world hiding and real world hiding.[3]

We also introduce a new test family denoted by Γ^*, which consists of computationally unbounded tests, which do not accept any messages from the adversary. Alternately, a test in Γ^* can be considered as sampling a collection of agents to upload, and a string of bits to communicate to the adversary (taking only a challenge bit as input in the experiments).

Combined, the above two elements fully define Γ^*-s-IND-PRE. Next, we turn our attention to giving two security definitions which are not of the indistinguishability-preserving genre. Firstly, s-SIM is a statistical simulation based security notion, which, on the face of it, is a stronger definition than s-IND-PRE. In s-SIM security, it is required that for every real world adversary

---

[3] In IND-PRE security as defined in [2], it is only required that a negligible distinguishing probability in the ideal world translates to a negligible distinguishing probability in the real world. The security notion here is tighter in that it requires indistinguishability to be preserved up to a polynomial loss, even if the original distinguishing probability in the ideal world is not negligible.

Adv, there is an ideal world simulator $\mathcal{S}$, which has a similar distinguishing probability as Adv has in the real world experiment. To be a strong security guarantee, we require that the simulator cannot depend on the test (but it can depend on Adv). We instantiate s-SIM security against the test-family Γ^*. This generalizes the notion of VGB security for obfuscation (see Proposition 1 in Sect. 5).

The other security definition we introduce, called IND-CON (for indistinguishability of concentrated distributions) generalizes the notion of SIO introduced by [9] for obfuscation, to all schemas. Here indistinguishability is required only against tests which upload agents from two distributions which are not only indistinguishable in the ideal world, but in fact "concentrated" — with high probability, the outcome of any query strategy[4] is already determined.

Equivalence of Security Notions. It is easy to see that Γ^*-s-SIM $\Rightarrow$ Γ^*-s-IND-PRE $\Rightarrow$ IND-CON.[5] Our main result is a proof that the reverse implications hold as well, and hence the three notions are identical.

Our proof could be seen as a simplification and significant generalization of the proof in [9] that SIO implies VGB obfuscation. We briefly overview the proof of [9] before explaining our version. There it is shown how to construct a computationally unbounded simulator which receives access to a single circuit computing a binary function, makes only polynomially many queries to the circuit, and learns a sufficiently accurate approximation of the circuit so that it can simulate it to the given adversary, provided that the obfuscation scheme is SIO secure. The simulator iteratively narrows down the set of possibilities for the circuit it is given access to, by making carefully chosen queries. Firstly, the simulator narrows down the possibilities to a set of circuits R such that a uniform distribution over R is a concentrated distribution (this is called the *concentration step* of the proof). However, the adversary may behave differently on certain circuits within this set; the computationally unbounded simulator can identify this subset D.[6] To determine if the circuit is from D using a small number of queries, the simulator relies on SIO security: since the adversary can

---

[4] As opposed to the case of obfuscation, for general schemas, a query can typically depend on previous queries. For example, in a graded encoding schema, it may be the case that a "zero-test" can be performed only after performing a sequence of operations on encodings provided by the test. A query-strategy is a polynomially deep (but exponentially large) tree which fully specifies a (deterministic) choice of ideal world queries based on the outcomes of the previous queries, and potentially using the agents generated by those queries.

[5] In this chain, we may insert a weaker version of s-SIM, which allows the simulator to depend on the test as well as the adversary (but not on the challenge bit given to the test), between Γ^*-s-SIM and Γ^*-s-IND-PRE security. Since all these notions turn out to be the same, in this paper we avoid defining the weaker simulation. However, for more general test families, or without the requirement of statistical security, this notion of a simulation could be of independent interest.

[6] More precisely there are two parts of D, corresponding to positive and negative distinguishing advantage. For simplicity, here we assume that only one such part is non-empty.

distinguish the obfuscation of each of the circuits in D from the obfuscation of a random circuit in R (with distinguishing advantage of the same sign), it follows that it can distinguish the obfuscation of a random circuit in D from a random circuit in R. Hence, by SIO security, it must be the case that the uniform distribution over D is not concentrated around the same majority outcome as R is (and possibly, not concentrated at all). This is exploited to argue that a small set of queries can be found to check if the circuit is in D or not (this is called the *majority-separation step*). If after making these queries, the simulator determines that the circuit is not in D, it can obfuscate a random circuit from R and present it to the adversary. On the other hand, if it is in D, this allows the simulator to make significant progress, because as D is not concentrated, it must be a significantly small fraction of R. The simulator *iterates the concentration and majority-separation steps alternately* until it determines that the circuit is not in D. To complete the proof, it is argued that the number of iterations (and the number of queries within each iteration) is logarithmic in the size of the space of circuits being obfuscated.

In our proofs, the simulation is required only in showing that Γ^*-s-IND-PRE security implies Γ^*-s-SIM security. Here, the simulator can rely on the "stronger" s-IND-PRE security guarantee, and obtain a "separating query" more directly, without relying on R being concentrated: indeed, if D is distinguishable from R in the real world, then s-IND-PRE security guarantees that there is a (small-depth) query strategy that separates the two. Performing this query strategy either allows D to be significantly shrunk, or allows R to be significantly shrunk (since otherwise, it will not be a sufficiently separating query strategy). If R shrinks, then D is redefined with respect to the new R (and may become as large as the new R). Iterating this procedure makes D empty, with the number of iterations being logarithmic in the size of the space of agents.

Roughly, the above argument corresponds to the majority-separation step in the proof of [9]. An analogue of the concentration step appears in the proof that IND-CON security implies Γ^*-s-IND-PRE security, described below.

A potentially difficult part in proving IND-PREsecurity in general is that it requires one to show that *every* ideal-hiding test is real-hiding, and it is not clear which tests are ideal-hiding. Our proof can in fact be viewed as a characterization of tests in Γ^* that are statistically ideal-hiding. A test in Γ^* can be identified with a pair of distributions $\mathcal{D}_0$ and $\mathcal{D}_1$, corresponding to the collection of agents (and auxiliary information) it generates when the challenge bit is 0 and 1 respectively. For a test to be ideal hiding, the outcome of any (polynomial depth) query-strategy must have essentially the same distribution for both $\mathcal{D}_0$ and $\mathcal{D}_1$, but the distributions are not necessarily concentrated (which requires the outcome of any query strategy to be essentially deterministic). We give a simple combinatorial lemma which shows that

for any distribution $\mathcal{D}$ over agents and auxiliary information, there is a polynomial-depth query strategy that breaks down $\mathcal{D}$ into concentrated distributions (plus negligible mass on an unconcentrated distribution).

The query strategy reveals which constituent concentrated distribution a collection of agents come from. Hence, if $\mathcal{D}_0$ and $\mathcal{D}_1$ are ideal-hiding, then both of them should have essentially the same distribution over concentrated distributions. Now, for each concentrated distribution, IND-CON security guarantees that the two distributions are real-hiding too.

Simplification and Generalization. We highlight two contributions of our result, given the prior work of [9]. Technically, it simplifies the proof by changing a nested iterative construction (used in the simulator), into two separate constructions, each with a simple iterative procedure. At a more conceptual level, seemingly technical details in the proof of [9] — namely, the concentration step and the majority-separation step — are reflected in two separate concrete concepts (namely, IND-CON $\Rightarrow \Gamma^*$-s-IND-PRE and Γ^*-s-IND-PRE $\Rightarrow \Gamma^*$-s-SIM).

But more importantly our result also ties these results to the new framework of cryptographic agents. While the development of the notions of VGB obfuscation and SIO were important contributions to our understanding of obfuscation, our result shows that their equivalence has more to do with certain structural properties of the security definition (captured in Γ^*-s-IND-PRE security) rather than obfuscation itself. Indeed, we show that the same security definition, applied to the graded encoding schema captures the independently developed notion of "semantic-security" for graded encoding [19].[7] More broadly, Γ^*-s-IND-PRE security can be used to model über assumptions for a variety of cryptographic encoding schemes (e.g., groups, groups with bi-linear pairings etc.). Our result shows that *in all these cases*, there is an equivalent simulation based security notion as well as a low-level security notion for concentrated distributions.

Composition Theorem. In [2] a notion of reduction was defined and it was shown that IND-PRE security composes with reductions: if Σ reduces to Σ^*, and Σ^* has an IND-PRE secure scheme, then so does Σ. However, this composition theorem breaks down in the case of s-IND-PRE security, since it involves an ideal-world adversary who is computationally unbounded. However, if the reduction is a *statistical reduction* — i.e., Σ can be information-theoretically securely constructed based on Σ^*— then we show that the composition theorem holds. Further, the composition theorem holds even if we restrict to the test family Γ^*.

A consequence of this composition theorem is that we can readily obtain the result that, if a strong-sampler semantically-secure graded encoding scheme exists, then there exists a VGB obfuscation scheme for NC^1 circuits. We point out that in obtaining this result, we do not rely on the IND-CON security definition at all. While [9] crucially used the notion of SIO for obtaining this result, the notion of Γ^*-s-IND-PRE is sufficient: the proof relies on the fact that Γ^*-s-IND-PRE is equivalent to VGB security for obfuscation and to strong-sampler semantic

---

[7] The original notion in [19] essentially corresponds to s-IND-PRE security for a test family which requires the tests to be efficient. Without this requirement, the security notion is termed strong-sampler semantic-security [9].

security for graded encoding schemes and on the composition theorem for Γ^*-s-IND-PRE (as well as the existence of a statistical reduction from obfuscation for NC^1 to graded encoding schemes).

1.2 Related Work

A formal study of obfuscation was initiated in the works of Hada [16] and Barak et al. [4] only about a decade and a half ago. The latter proposed several notions of obfuscation: virtual black-box (VBB), differing-inputs obfuscation (diO), indistinguishability obfuscation (iO), etc., with VBB being the strongest. Further definitions appeared later [8,14,15]. In particular, Bitansky and Canetti proposed the definition of Virtual Grey-Box (VGB) obfuscation [8].

Much work has appeared on the definitional front for other primitives like functional encryption as well [1,3,6,7,10,12,18]. The recent framework of Cryptographic Agents [2] unified many of the concepts underlying the definitions of obfuscation, functional encryption and other cryptographic objects. Our results are formulated in this new framework, and hence extends to all primitives that can be expressed as cryptographic agent schemas.

Recently, Bitansky et al. [9] gave a surprising characterization of VGB obfuscation as being equivalent to a seemingly simpler definition of obfuscation, called *strong indistinguishability obfuscation* (SIO). Further, based on this, they showed that under a variant of a semantic-security assumption on graded encoding schemes (a.k.a. multi-linear maps) [19], any NC^1 circuit can be VGB-obfuscated. Both these results can be obtained as corollaries of our result.

2 Preliminaries

We use κ to denote the security parameter. For two functions f and g, we write $f(g)$ to denote the function $f \circ g$, so that $f(g)(x) = f(g(x))$. If $X = \{X_\lambda\}_{\lambda \in \mathbb{N}}$ and $Y = \{Y_\lambda\}_{\lambda \in \mathbb{N}}$ are distribution ensembles over $\{0,1\}$, we write $X \approx Y$ if there is a negligible function negl such that $|\Pr[X_\lambda = 1] - \Pr[Y_\lambda = 1]| \leq \mathsf{negl}(\lambda)$.

We work with the same framework of cryptographic agents as was originally proposed by Agrawal et al. [2], except that we consider *statistical* hiding in the ideal world and focus on a new family of tests which are computationally unbounded and do not receive messages from adversaries. We summarize the salient features of the framework here, and provide further details in Appendix A for the sake of self-containment.

Agents and Sessions. Agents are used to model idealizations of entities like ciphertexts, keys, encodings and obfuscations. An agent is an interactive Turing Machine, derived from a family of agents all of whose programs are identical, but may have different contents in a read-only parameter tape (e.g., message in a ciphertext, the function in a functional-encryption key, or the program in an obfuscation). Agents may interact with each other (e.g., a ciphertext agent and a key agent) to produce outputs that a user can access. This is modeled by *sessions*. A session consists of a finite ordered set of agents which can interact with each other according to their programs (e.g., a ciphertext agent can send its message to a key agent), and result in updated states for the agents as well as outputs from each agent in the session. Updated state may be the same as the original state, and the outputs may be empty.

Ideal world. The ideal system for a schema $\Sigma = (\mathcal{P}_{\mathsf{auth}}, \mathcal{P}_{\mathsf{user}})$, where $\mathcal{P}_{\mathsf{auth}}$ and $\mathcal{P}_{\mathsf{user}}$ are agent families, consists of two parties Test and User and a fixed third party $\mathcal{B}[\Sigma]$ (for "black-box"). Test receives a "secret bit" b as input and User produces an output bit b'. Test and User can, at any point, choose an agent and **upload** it to $\mathcal{B}[\Sigma]$. Test is allowed to upload agents from $\mathcal{P}_{\mathsf{test}} := \mathcal{P}_{\mathsf{auth}} \cup \mathcal{P}_{\mathsf{user}}$ and User agents from $\mathcal{P}_{\mathsf{user}}$. Whenever an agent is uploaded, $\mathcal{B}[\Sigma]$ sends a unique handle for that agent to User.

A **query** is a request for session execution. At any point in time, User may request an execution of a session, by sending an ordered tuple of handles $(h_1, \ldots, h_t)$ along with their inputs. $\mathcal{B}[\Sigma]$ reports back the outputs from the session, and also gives new handles corresponding to the configurations of the agents when the session terminated. (Note that after a session, the old handles for the agents are not invalidated.)

We define the random variable $\mathrm{IDEAL}\langle \mathsf{Test}(b) \mid \Sigma \mid \mathsf{User} \rangle$ to be the output of User in an execution of the above system, when Test gets b as input. We write $\mathrm{IDEAL}\langle \mathsf{Test} \mid \Sigma \mid \mathsf{User} \rangle$ in the case when the input to Test is a uniformly random bit. We also define $\mathrm{TIME}\langle \mathsf{Test} \mid \Sigma \mid \mathsf{User} \rangle$ as the maximum number of steps taken by Test (with a random input), $\mathcal{B}[\Sigma]$ and User in total.

Definition 1 ((Statistical) Ideal world hiding). *A* Test *is* η-*s-hiding w.r.t. a schema* Σ *if, for all unbounded users* User *who make at most* η *queries,*

$$\left| \Pr[\mathrm{IDEAL}\langle \mathsf{Test}(0) \mid \Sigma \mid \mathsf{User} \rangle = 1] - \Pr[\mathrm{IDEAL}\langle \mathsf{Test}(1) \mid \Sigma \mid \mathsf{User} \rangle = 1] \right| \leq \frac{1}{\eta}.$$

Real World. A *cryptographic scheme* consists of programs $\mathcal{O}$ and $\mathcal{E}$, where $\mathcal{O}$ is an encoding (or objectification) procedure for agents in $\mathcal{P}_{\mathsf{test}}$ and $\mathcal{E}$ is an execution procedure. The real world execution for a scheme $(\mathcal{O}, \mathcal{E})$ consists of Test, a user that we shall generally denote as Adv, and the encoder $\mathcal{O}$. ($\mathcal{E}$ features as part of an honest user.) Test uploads agents to the encoder $\mathcal{O}$, who encodes them and sends the resulting cryptographic agents to Adv. $(\mathcal{O}, \mathcal{E})$ are generally memory-less from one invocation to the next, except that $\mathcal{E}$ has access to a list of all objects it ever received. For certain schemes, it is important to let $\mathcal{O}$ and $\mathcal{E}$ have access to persistent keys generated during a set-up phase, which is also incorporated into the model via a public-secret key pair $(\mathsf{MPK}, \mathsf{MSK})$ (for details see Appendix A).

We define the random variable $\mathrm{REAL}\langle \mathsf{Test}(b) \mid \mathcal{O} \mid \mathsf{Adv} \rangle$ to be the output of Adv in an execution of the above system, when Test gets b as input; as before, we omit b from the notation to indicate a random bit. Also, as before, $\mathrm{TIME}\langle \mathsf{Test} \mid \mathcal{O} \mid \mathsf{User} \rangle$ is the maximum number of steps taken by Test (with a random input), $\mathcal{O}$ and User in total.

Definition 2 (Real world hiding). *A* Test *is η-hiding w.r.t. $\mathcal{O}$ if for all adversaries* Adv *who run for at most η time,*

$$|\Pr[\text{REAL}\langle\text{Test}(0) \mid \mathcal{O} \mid \text{Adv}\rangle = 1] - \Pr[\text{REAL}\langle\text{Test}(1) \mid \mathcal{O} \mid \text{Adv}\rangle = 1]| \leq \frac{1}{\eta}.$$

Definition 3 (Admissibility of schemes). *A cryptographic agent scheme* $\Pi = (\mathcal{O}, \mathcal{E})$ *is said to be an* admissible scheme *for a schema Σ if the following conditions hold.*

– *Correctness.* $\forall$ PPT User *and* $\forall$ Test,

$$\text{IDEAL}\langle\text{Test} \mid \Sigma \mid \text{User}\rangle \approx \text{REAL}\langle\text{Test} \mid \mathcal{O} \mid \mathcal{E} \circ \text{User}\rangle.$$

If the difference is 0, $(\mathcal{O}, \mathcal{E})$ is said to have perfect correctness.
– *Efficiency. There exists a polynomial* poly *such that,* $\forall$ PPT User, $\forall$ Test,

$$\text{TIME}\langle\text{Test} \mid \mathcal{O} \mid \mathcal{E} \circ \text{User}\rangle \leq \text{poly}(\text{TIME}\langle\text{Test} \mid \Sigma \mid \text{User}\rangle, \kappa).$$

Γ^* **Test Family.** This family consists of computationally unbounded tests which do not accept any messages from the user/adversary. Without loss of generality, such a test is fully characterized by a distribution over $\{0,1\}^* \times \mathcal{P}_{\text{test}}^*$.[8]

The first part of a $\boldsymbol{P} \in \{0,1\}^* \times \mathcal{P}_{\text{test}}^*$, which we denote as $\boldsymbol{P}_0 \in \{0,1\}^*$, is a message from test to the user/adversary; the remaining components of the vector $\boldsymbol{P}$ denote a (possibly empty) collection of agents from $\mathcal{P}_{\text{test}}$.

We write $\mathcal{O}(\boldsymbol{P})$ to denote a random encoding of $\boldsymbol{P}$ which consists of $(\boldsymbol{P}_0, \mathcal{O}(\boldsymbol{P}_1), \cdots, \mathcal{O}(\boldsymbol{P}_i))$ (as well as the public-key MPK if $\mathcal{O}$ involves a set-up). We write $\text{Adv}(\mathcal{O}(\boldsymbol{P}))$ to denote the random variable corresponding to the bit output by Adv when given $\mathcal{O}(\boldsymbol{P})$.

Definition 4 (IND-PRE security). *An admissible cryptographic agent scheme* $\Pi = (\mathcal{O}, \mathcal{E})$ *is said to be a p-Γ^*-s-INDPRE-secure scheme for a schema Σ if for all κ, all* Test $\in \Gamma^*$, *and every polynomial η, if* Test *is $p(\eta(\kappa))$-s-hiding w.r.t. Σ, then it is $\eta(\kappa)$-hiding w.r.t. $\mathcal{O}$.*

If $\Pi = (\mathcal{O}, \mathcal{E})$ *is p-Γ^*-s-INDPRE-secure for some polynomial p, then we simply refer to it as Γ^*-s-INDPRE -secure scheme.*

We also define a simulation-based security notion in the agents framework.

Definition 5 (Simulation-based security). *An admissible cryptographic agent scheme* $\Pi = (\mathcal{O}, \mathcal{E})$ *is said to be a p-Γ^*-s-SIM-secure scheme for a schema Σ if for all κ, all polynomials ℓ, η, and any adversary* Adv *which runs in time at most $\ell(\kappa)$, there exists a computationally unbounded simulator $\mathcal{S}$ that makes at most $p(\eta(\kappa), \ell(\kappa))$ queries, such that for all* Test $\in \Gamma^*$,

$$|\Pr[\text{IDEAL}\langle\text{Test} \mid \Sigma \mid \mathcal{S}\rangle = 1] - \Pr[\text{REAL}\langle\text{Test} \mid \mathcal{O} \mid \text{Adv}\rangle = 1]| \leq \frac{1}{\eta(\kappa)}.$$

---

[8] In proving our results, we can assume an upper-bound on the number of bits communicated by the test, as there will be a bound on the running time of an adversary that it interacts with.

A *cryptographic agent scheme* $\Pi = (\mathcal{O}, \mathcal{E})$ *is said to be a* Γ^*-*s*-SIM-*secure scheme if it is a* p-Γ^*-*sSIM*-*secure scheme for some (bivariate) polynomial* p.

We remark that one can consider a weaker notion of simulation where $\mathcal{S}$ can depend on Test. As we shall see, for Γ^*, this weaker notion is no different from the notion defined above.

2.1   Concentrated Distributions

Recall that in the ideal world, User can make queries — i.e., requests to run sessions — to $\mathcal{B}[\Sigma]$ and obtain the outcome of the session (and handles for the updated configurations of the agents involved in the session). User can carry this out repeatedly, and adaptively. The following definition captures this procedure (for a deterministic User).

Definition 6 (Query Strategy). *A* d-*query-strategy is a tree of depth at most* d *where each internal node* u *is labeled with a query* q_u *and each outgoing edge from* u *is labeled with a different possible outcome of* q_u. *The execution of a query strategy* Q *on a* $\boldsymbol{P} \in \{0,1\}^* \times \mathcal{P}_{\text{test}}^*$ *is a path in this tree starting from the root node, such that an edge from node* u, *labeled with an answer* ans, *is present in the path if and only if the outcome of running a session on (the updated configurations of)* $\boldsymbol{P}$ *with the query* q_u *is* ans. *The outcome of the entire execution, denoted by* $\boldsymbol{P}(Q)$ *is the (concatenated) outcomes of all the queries in the path. We use the convention that the first query in* Q *is an empty query and its answer is the auxiliary information* $\boldsymbol{P}_0 \in \{0,1\}^*$.

We now define concentrated distributions over collections of agents and indistinguishability between them.

Definition 7 (Concentrated distributions). *A distribution ensemble* $\mathcal{D}$ *over* $\{0,1\}^{\ell(\kappa)} \times \bigcup_{i=0}^{\ell(\kappa)} \mathcal{P}_{\text{test}}^i$ *is said to be* η-*concentrated if for all* κ *there exists a function* A *(called an answer function) which maps query strategies to answers, such that for all depth* $\eta(\kappa)$ *query strategy* Q,

$$\Pr_{\boldsymbol{P} \leftarrow \mathcal{D}(\kappa)}[\boldsymbol{P}(Q) \neq A(Q)] \leq \frac{1}{\eta(\kappa)}.$$

A pair of distribution ensembles $(\mathcal{D}_0, \mathcal{D}_1)$ *is said to be* η-*concentrated if they are both* η-*concentrated with the same answer function.*

Definition 8 (Indistinguishability of concentrated distributions). *An admissible scheme* $\Pi = (\mathcal{O}, \mathcal{E})$ *is* q-IND-CON *secure for* $\Sigma = (\mathcal{P}_{\text{auth}}, \mathcal{P}_{\text{user}})$ *if for all* κ, *every polynomial* η, *and any pair of distribution ensembles* $(\mathcal{D}_0, \mathcal{D}_1)$ *over* $\{0,1\}^{\ell(\kappa)} \times \bigcup_{i=0}^{\ell(\kappa)} \mathcal{P}_{\text{test}}^i$ *which are* $q(\eta(\kappa))$-*concentrated, we have that for any* PPT *adversary* Adv *with running time at most* $\eta(\kappa)$,

$$\left| \Pr_{\boldsymbol{P} \leftarrow \mathcal{D}_0(\kappa)}[\mathsf{Adv}(\mathcal{O}(\boldsymbol{P})) = 1] - \Pr_{\boldsymbol{P} \leftarrow \mathcal{D}_1(\kappa)}[\mathsf{Adv}(\mathcal{O}(\boldsymbol{P})) = 1] \right| \leq \frac{1}{\eta(\kappa)}.$$

A scheme $\Pi = (\mathcal{O}, \mathcal{E})$ *is* IND-CON *secure if it is* q-IND-CON *secure for some polynomial* q.

A Probability Lemma. The following is a simple lemma which can be used to relate two distributions with a small statistical difference to a single common distribution; further, the lemma allows the common distribution to avoid a subset S of the sample space, provided the given distributions have low mass on it. Below, $\Delta\left(\cdot,\cdot\right)$ denotes the statistical difference between two distributions.

Lemma 1. *For any two probability distributions $\mathcal{A}_0$, $\mathcal{A}_1$ over the same sample space, and any subset S of the sample space, there exists $\epsilon \leq \Delta\left(\mathcal{A}_0,\mathcal{A}_1\right) + \min\{\Pr_{a \leftarrow \mathcal{A}_0}[a \in S], \Pr_{a \leftarrow \mathcal{A}_1}[a \in S]\}$, a distribution $\mathcal{A}_{\overline{S}}$ over $\overline{S}$, and two distributions $\mathcal{A}_0', \mathcal{A}_1'$ such that for each $b \in \{0,1\}$, $\mathcal{A}_b$ is equal to the distribution of a in the following experiment:*

$$\alpha \sim \text{Bernoulli}(\epsilon);\ \textit{if}\ \alpha = 0, a \leftarrow \mathcal{A}_{\overline{S}},\ \textit{else}\ a \leftarrow \mathcal{A}_b'.$$

Proof. Given distributions $\mathcal{A}_0, \mathcal{A}_1$ over a sample space T and a set $S \subseteq T$, the goal is to construct a distribution $\mathcal{A}_{\overline{S}}$ over $\overline{S}$ such that sampling according to $\mathcal{A}_0$ (resp. $\mathcal{A}_1$) is the same as sampling according to $\mathcal{A}_{\overline{S}}$ with probability $1 - \epsilon$ and according to another distribution $\mathcal{A}_0'$ (resp. $\mathcal{A}_1'$) with probability ϵ. Intuitively, $\mathcal{A}_{\overline{S}}$ is the "intersection" of $\mathcal{A}_0$ and $\mathcal{A}_1$ over $\overline{S}$, and $\mathcal{A}_0'$ (resp. $\mathcal{A}_1'$) is the "remaining distribution" after $\mathcal{A}_{\overline{S}}$ is cut out from $\mathcal{A}_0$ (resp. $\mathcal{A}_1$).

More formally, define weight functions $f, f_0, f_1 : T \to [0,1]$ as follows:

$$f(a) = \begin{cases} \min\{\mathcal{A}_0(a), \mathcal{A}_1(a)\} & \text{if } a \in \overline{S} \\ 0 & \text{if } a \in S \end{cases} \quad \text{and} \quad \begin{aligned} f_0(a) &= \mathcal{A}_0(a) - f(a) \\ f_1(a) &= \mathcal{A}_1(a) - f(a) \end{aligned}$$

where $\mathcal{A}_b(a)$ denotes the probability mass on a according to the distribution $\mathcal{A}_b$. Furthermore, set $\epsilon = 1 - \sum_a f(a) = \sum_a f_0(a) = \sum_a f_1(a)$. Then, we define the distributions $\mathcal{A}_{\overline{S}}, \mathcal{A}_0', \mathcal{A}_1'$ as follows:

$$\mathcal{A}_{\overline{S}}(a) = f(a)/(1 - \epsilon), \qquad \mathcal{A}_0(a) = f_0(a)/\epsilon, \qquad \mathcal{A}_1(a) = f_1(a)/\epsilon.$$

(If $\epsilon = 1$, we let $\mathcal{A}_{\overline{S}}$ be an arbitrary probability distribution; similarly if $\epsilon = 0$, $\mathcal{A}_0, \mathcal{A}_1$ are arbitrary.) Then for $b \in \{0,1\}$, for all $a \in T$, $\mathcal{A}_b(a) = (1-\epsilon)\mathcal{A}_{\overline{S}}(a) + \epsilon\mathcal{A}_b'(a)$, as required by the lemma.

It remains to prove the claimed upper bound on ϵ. Let $g(a) = \min\{\mathcal{A}_0(a), \mathcal{A}_1(a)\}$ for all a. Note that $\sum_a \mathcal{A}_b(a) - g(a) = \Delta\left(\mathcal{A}_0, \mathcal{A}_1\right)$ for $b \in \{0,1\}$ and $\sum_a g(a) - f(a) \leq \min\{\Pr_{a \leftarrow \mathcal{A}_0}[a \in S], \Pr_{a \leftarrow \mathcal{A}_1}[a \in S]\}$. Hence $\epsilon = \sum_a f_0(a) = \sum_a \mathcal{A}_0(a) - g(a) + \sum_a g(a) - f(a) \leq \Delta\left(\mathcal{A}_0, \mathcal{A}_1\right) + \min\{\Pr_{a \leftarrow \mathcal{A}_0}[a \in S], \Pr_{a \leftarrow \mathcal{A}_1}[a \in S]\}$. $\square$

3   Equivalence of Definitions

In this section we prove our main results (Theorems 1 and 2).

Theorem 1 (Equivalence of IND-CON and s-IND-PRE). *A cryptographic agent scheme $\Pi = (\mathcal{O}, \mathcal{E})$ is a Γ^*-s-IND-PRE-secure scheme for a schema Σ if and only if it is IND-CON secure for Σ.*

To prove Theorem 1, or more specifically, that IND-CON $\Rightarrow$ Γ^*-s-IND-PRE, we rely on the following lemma, which gives a query strategy that can be used to narrow down a distribution over agents to a concentrated distribution (except with negligible probability over the choice of the agents). As sketched in Sect. 1.1, this lemma gives a characterization of hiding tests in terms of concentrated distributions and is at the heart of proving Theorem 1.

Below, for a distribution $\mathcal{D}$ over agent vectors and a query strategy Q, $\mathcal{D}|_{Q \to \mathsf{ans}}$ denotes the distribution obtained by restricting $\mathcal{D}$ to the subset $\{P | P(Q) = \mathsf{ans}\}$. Below, when we say that a distribution $\mathcal{D}|_{Q \to \mathsf{ans}}$ is ρ-concentrated, we consider concentration against depth ρ query-strategies which can optionally use the handles resulting from the query-strategy Q, as well as the original handles (this is relevant only for schemas with stateful agents).

Lemma 2. *Let $\mathcal{P}_{\mathsf{test}}$ be a set of agents with polynomially long representation. Then, for any polynomial ρ, there exists a polynomial π such that for any polynomial η, any function $\varepsilon > 0$, and any distribution $\mathcal{D}$ over $\mathcal{R}^\eta = \{0,1\}^\eta \times \mathcal{P}_{\mathsf{test}}^\eta$, there is a $\pi(\eta \cdot \log \frac{1}{\varepsilon})$-query strategy Q^* such that*

$$\Pr_{P \leftarrow \mathcal{D}}[\mathcal{D}|_{Q^* \to P(Q^*)} \text{ not } \rho(\eta)\text{-concentrated}] \leq \varepsilon.$$

Proof. The query strategy can be defined as repeatedly, conditioned on the previous queries and answers, identifying and carrying out a query strategy whose answer is not concentrated (i.e., no one answer has probability more than $1 - \rho(\eta)$) until the remaining distribution is $\rho(\eta)$-concentrated, or the budget on the number of queries (depth of the strategy) has been exhausted. We shall show that this leads to the mass in unconcentrated leaves of the query strategy tree to be at most ε.

More formally, consider a tree in T which each node v is associated with a subset $R_v \subseteq \mathcal{R}^\eta$ and (unless it is a leaf node) with a query strategy Q_v. The set at the root of T is the entire set $\mathcal{R}^\eta$. For $R \subseteq \mathcal{R}^\eta$, let $\mathcal{D}|_R$ denote the distribution $\mathcal{D}$ restricted to the set R. A node v in T is a leaf node either if the distribution $\mathcal{D}|_{R_v}$ is $\sigma := \rho(\eta)$-concentrated or if v is at a depth σ. For every internal node v, Q_v is a query strategy of depth at most σ such that for all ans, $\Pr_{P \leftarrow \mathcal{D}|_{R_v}}[P(Q_v) = \mathsf{ans}] \leq 1 - \frac{1}{\sigma}$. Note that such a Q_v exists since $\mathcal{D}|_{R_v}$ is not σ-concentrated (v being an internal node). For each possible answer ans to Q_v, v has a child v_{ans} such that $R_{v_{\mathsf{ans}}} = \{P \in R_v \mid P(Q_v) = \mathsf{ans}\}$.

Let L_ℓ be the set of all nodes at depth ℓ in T. Note that for each $v \in L_\ell$, $|R_v| \geq 1$, whereas $\sum_{v \in L_\ell} |R_v| \leq |\mathcal{R}^\eta|$. Therefore, $|L_\ell| \leq |\mathcal{R}^\eta|$. On the other hand, note that if u is a child of v in T, then $\Pr_{P \leftarrow \mathcal{D}}[P \in R_u \mid P \in R_v] \leq 1 - \frac{1}{\sigma}$. Thus for all $v \in L_\ell$, $\Pr_{P \leftarrow \mathcal{D}}[P \in R_v] \leq (1 - \frac{1}{\sigma})^\ell$. Hence, $\Pr_{P \leftarrow \mathcal{D}}[P \in \bigcup_{v \in L_\ell} R_v] \leq (1 - \frac{1}{\sigma})^\ell \cdot |\mathcal{R}^\eta|$.

If we choose $\ell = \Omega(\sigma \cdot \log(|\mathcal{R}^\eta|/\varepsilon))$ then $\Pr_{P \leftarrow \mathcal{D}}[P \in \bigcup_{v \in L_\ell} R_v] \leq \varepsilon$. Note that $|\mathcal{R}^\eta| = \zeta^\eta$ for some polynomial ζ (determined by the size of $\mathcal{P}_{\mathsf{test}}$) and $\sigma = \mathrm{poly}(\eta)$, so that ℓ is polynomial in $\eta \cdot \log \frac{1}{\varepsilon}$. Our query strategy Q^* is obtained from T by executing the first ℓ query strategies in it. The depth of Q^* is $\ell \cdot \sigma$, again a polynomial in $\eta \cdot \log \frac{1}{\varepsilon}$. $\qquad \square$

We prove the two directions of Theorem 1 separately. Intuitively, IND-CON security is a "weaker" notion, and hence the first direction below is easier to see. The second direction relies on Lemma 2.

Γ^*-s-IND-PRE $\Rightarrow$ IND-CON: Suppose that for some polynomial q, $\Pi = (\mathcal{O}, \mathcal{E})$ is a q-Γ^*-s-IND-PRE secure scheme for a schema Σ. We shall show that Π is q-IND-CON secure for Σ.

Let η be a polynomial, and $(\mathcal{D}_0, \mathcal{D}_1)$ be a pair of distribution ensembles which are $q(\eta)$-concentrated. Let A denote the answer function that maps depth $q(\eta)$ query strategies to answers, so that for any such query strategy Q, for both $b \in \{0,1\}$, we have $\Pr_{P \leftarrow \mathcal{D}_b}[P(Q) \neq A(Q)] \leq \frac{1}{q(\eta)}$.

Consider the test Test which on input $b \in \{0,1\}$, uploads a sample from the distribution $\mathcal{D}_b$. Observe that Test $\in \Gamma^*$. Consider any unbounded ideal-world user User that makes at most $q(\eta)$ queries. For each setting of the random-tape of User, its behavior can be identified with a query strategy of depth at most $q(\eta)$. For any such strategy Q, irrespective of the bit b, with probability at least $1 - 1/q(\eta)$ User receives the answer $A(Q)$. Thus, for any User which makes at most $q(\eta)$ queries $|\Pr[\text{IDEAL}\langle \text{Test}(0) \mid \Sigma \mid \text{User}\rangle = 1] - \Pr[\text{IDEAL}\langle \text{Test}(1) \mid \Sigma \mid \text{User}\rangle = 1]| \leq 1/q(\eta)$. That is, Test is $q(\eta)$-s-hiding w.r.t. Σ.

Then, since Π is a q-Γ^*-s-IND-PRE secure scheme for Σ, we have that Test is η-hiding w.r.t. $\mathcal{O}$. That is, for any adversary Adv with running time at most η, $|\Pr[\text{REAL}\langle \text{Test}(0) \mid \Sigma \mid \text{User}\rangle = 1] - \Pr[\text{REAL}\langle \text{Test}(1) \mid \Sigma \mid \text{User}\rangle = 1]| \leq 1/\eta$. But $\Pr[\text{REAL}\langle \text{Test}(b) \mid \Sigma \mid \text{User}\rangle = 1]$ is simply $\Pr_{P \leftarrow \mathcal{D}_b}[\text{Adv}(\mathcal{O}(P)) = 1]$.

Hence, by the definition of IND-CON security, Π is q-IND-CON secure for Σ.

IND-CON $\Rightarrow$ Γ^*-s-IND-PRE: Suppose Π is an IND-CON secure scheme for Σ. Then, there is a polynomial q such that it is q-IND-CON secure. We shall show that Π is p-Γ^*-s-IND-PRE secure, for some polynomial p.

Let Test be an arbitrary test in Γ^*, that is η^*-hiding w.r.t. Σ. We shall show that Test is η-hiding w.r.t. Π, where $\eta^* = p(\eta)$ (for a polynomial p to be determined).

We consider the space $\mathcal{R}^\eta$ of all possible agents vector produced by tests, i.e., $\mathcal{R}^\eta = \{0,1\}^\eta \times \mathcal{P}_{\text{test}}^\eta$.[9] Let $\mathcal{D}_0$ and $\mathcal{D}_1$ be the distributions over $\mathcal{R}^\eta$, produced by Test on input $b = 0$ and $b = 1$ respectively. Now, we apply Lemma 2 to the distribution $\mathcal{D}_0$, with η as above, $\rho(\eta) := 2q(\eta/2)$, and (say) $\varepsilon = 2^{-\eta}$. Let Q be the query strategy guaranteed by the lemma. Also, let $\mu = \rho(\eta)/2$.

Recall that each root-to-leaf path in a query strategy is labeled by a sequence of responses, ans. We define two subsets of leaves B and C which correspond to answers that can potentially differentiate between $\mathcal{D}_0$ and $\mathcal{D}_1$. Let $B = \{\text{ans} \mid \mathcal{D}_0|_{Q \rightarrow \text{ans}}$ is not 2μ-concentrated$\}$. Also let $C = \{\text{ans} \mid \mathcal{D}_0|_{Q \rightarrow \text{ans}}$ is 2μ-concentrated around some answer function A, but $\mathcal{D}_1|_{Q \rightarrow \text{ans}}$ is not μ-concentrated around $A\}$. For ans $\notin B \cup C$, the pair of distributions $(\mathcal{D}_0|_{Q \rightarrow \text{ans}}, \mathcal{D}_1|_{Q \rightarrow \text{ans}})$ is μ-concentrated.

---

[9] Note that we truncate the auxiliary information to $\eta(\kappa)$ bits, and the number of agents uploaded by the test to $\eta(\kappa)$. This is because, to show that Test is η-hiding w.r.t. Π, it is enough to consider adversaries who read at most η bits of the messages from Test.

We argue, relying on the fact that Test is η^*-hiding, that the mass of $B \cup C$ under $\mathcal{D}_0$ is $O(\mu/\eta^*)$. Firstly, mass of B under $\mathcal{D}_0$ is bounded by Lemma 2 to at most ε. Next, for each ans $\in C$, let A_{ans} be the answer function that $\mathcal{D}_0|_{Q \to \mathsf{ans}}$ is 2μ-concentrated around. Since $\mathcal{D}_1|_{Q \to \mathsf{ans}}$ is not μ-concentrated around A_{ans}, there is some query strategy Q_{ans} with depth at most μ, such that $\Pr_{P \leftarrow \mathcal{D}_1|_{Q \to \mathsf{ans}}}[P(Q_{\mathsf{ans}}) \neq A_{\mathsf{ans}}(Q_{\mathsf{ans}})] > 1/\mu$. But since $\mathcal{D}_0|_{Q \to \mathsf{ans}}$ is 2μ-concentrated around A_{ans} and Q_{ans} has depth less than 2μ, $\Pr_{P \leftarrow \mathcal{D}_0|_{Q \to \mathsf{ans}}}[P(Q_{\mathsf{ans}}) \neq A_{\mathsf{ans}}(Q_{\mathsf{ans}})] \leq 1/(2\mu)$. Now, consider a 2-phase query strategy Q' that in the first phase carries out Q and at the end of it, if ans $\in C$ is obtained, then follows up with the query strategy Q_{ans}. Q' is of depth at most $\pi(\eta^2) + \mu$ (which we shall arrange to be less than η^*). We may write the answer $P(Q')$ as $\mathsf{ans}_1 || \mathsf{ans}_2$, where ans_1 and ans_2 are the answers to the first and second phases of queries, respectively (if $\mathsf{ans}_1 \notin C$, then ans_2 will be empty). Then,

$$\Pr_{P \leftarrow \mathcal{D}_0}[P(Q') = \mathsf{ans}||A_{\mathsf{ans}}(Q_{\mathsf{ans}}) \text{ for ans} \in C] \geq \Pr_{P \leftarrow \mathcal{D}_0}[P(Q) \in C] \cdot \left(1 - \frac{1}{2\mu}\right)$$

$$\Pr_{P \leftarrow \mathcal{D}_1}[P(Q') = \mathsf{ans}||A_{\mathsf{ans}}(Q_{\mathsf{ans}}) \text{ for ans} \in C]$$

$$< \Pr_{P \leftarrow \mathcal{D}_1}[P(Q) \in C] \cdot \left(1 - \frac{1}{\mu}\right)$$

$$\leq \left(\Pr_{P \leftarrow \mathcal{D}_0}[P(Q) \in C] + 1/\eta^*\right) \cdot \left(1 - \frac{1}{\mu}\right)$$

The difference between these two probabilities is more than $\Pr_{P \leftarrow \mathcal{D}_0}[P(Q) \in C] \cdot \frac{1}{2\mu} - \frac{1}{\eta^*}$. But as the depth of Q' is less than η^* (as we ensure below), and Test is η^*-hiding, this difference is upper-bounded by $\frac{1}{\eta^*}$. Hence $\Pr_{P \leftarrow \mathcal{D}_0}[P(Q) \in C] \leq \frac{4\mu}{\eta^*}$.

Now, we view the test, on each input b, as sampling its agents vector P by first sampling the answer $P(Q)$, and then sampling P conditioned on this answer. $P(Q)$ itself is sampled from the distribution $\mathcal{A}_b = \{P(Q)\}_{P \leftarrow \mathcal{D}_b}$. Now, we invoke Lemma 1 on the distributions $\mathcal{A}_0$ and $\mathcal{A}_1$ with the set $S = B \cup C$. This results in $\epsilon = O(\frac{\mu}{\eta^*})$, given the above bound (and since $\Delta(\mathcal{A}_0, \mathcal{A}_1) \leq 1/\eta^*$). Thus, the test, with probability $1 - \epsilon$ samples ans $\notin B \cup C$ (from a distribution independent of b) and then samples $P \leftarrow \mathcal{D}_b|_{Q \to \mathsf{ans}}$. (With the remaining ϵ probability, it samples P depending on b as appropriate.) Recall that, for ans $\notin B \cup C$, we have that $(\mathcal{D}_0|_{Q \to \mathsf{ans}}, \mathcal{D}_1|_{Q \to \mathsf{ans}})$ is μ-concentrated, where $\mu = q(\eta/2)$. Hence we can apply the q-IND-CON security to conclude that no adversary can distinguish between $b = 0$ and $b = 1$ in the real experiment with advantage more than $\epsilon + (1 - \epsilon)\eta/2$. We shall set $\epsilon < \eta/2$ so that this advantage is less than η, as we need to prove.

To finish the proof we need to ensure that $\eta^* > \pi(\eta^2) + \mu$ and $\epsilon < \eta/2$. This is satisfied by setting, say, $\eta^* > \pi(\eta^2) + q(\eta/2)$. Thus, we can set p to be, say, the polynomial $p(\eta) := \pi(\eta^2) + q(\eta/2) + 1$.

Theorem 2 (Equivalence of s-IND-PRE and s-SIM). *A cryptographic agent scheme $\Pi = (\mathcal{O}, \mathcal{E})$ is a Γ^*-s-IND-PRE-secure scheme for a schema Σ if and only if it is Γ^*-s-SIM-secure for the same schema.*

Proof. Intuitively, Γ^*-s-IND-PRE security is "weaker" than Γ^*-s-SIM security, and hence the first direction below is easier to see.

$\underline{\Gamma^*\text{-}s\text{-SIM} \Rightarrow \Gamma^*\text{-}s\text{-IND-PRE}}$: Suppose $\Sigma = (\mathcal{O}, \mathcal{E})$ is a p-Γ^*-s-SIM secure scheme for Σ, for some (bivariate) polynomial p. We shall show that Σ is a q-Γ^*-s-IND-PRE schema for a polynomial q to be determined.

For a Test $\in \Gamma^*$ and η, suppose there exists a PPT adversary Adv which runs in at most η time but can distinguish between Test with bit 0 and 1 with probability at least $1/\eta$. That is,

$$\left| \Pr[\text{REAL}\langle \text{Test}(0) \mid \mathcal{O} \mid \text{Adv}\rangle = 1] - \Pr[\text{REAL}\langle \text{Test}(1) \mid \mathcal{O} \mid \text{Adv}\rangle = 1] \right| > 1/\eta.$$

We need to show that there is an ideal world user User, which makes at most $q(\eta)$ queries and achieves a distinguishing advantage of at least $1/q(\eta)$.

Since Π is p-Γ^*-s-SIM secure, given Adv which runs in time at most η, there exists an unbounded simulator $\mathcal{S}$ making at most $p(3\eta, \eta)$ queries, such that for all tests (and in particular, for Test) and $b \in \{0, 1\}$:

$$\left| \Pr[\text{IDEAL}\langle \text{Test}(b) \mid \Sigma \mid \mathcal{S}\rangle = 1] - \Pr[\text{REAL}\langle \text{Test}(b) \mid \mathcal{O} \mid \text{Adv}\rangle = 1] \right| \leq \frac{1}{3\eta}.$$

And therefore,

$$\left| \Pr[\text{IDEAL}\langle \text{Test}(0) \mid \Sigma \mid \mathcal{S}\rangle = 1] - \Pr[\text{IDEAL}\langle \text{Test}(1) \mid \Sigma \mid \mathcal{S}\rangle = 1] \right| >$$
$$\frac{1}{\eta} - \frac{2}{3\eta} = \frac{1}{3\eta}.$$

We set q such that $q(\eta) \geq p(3\eta, \eta)$ and $\frac{1}{3\eta} \geq \frac{1}{q(\eta)}$. For instance, we can set $q(x) = p(3x, x) + 3x$.

Note that in the above proof, we could allow $\mathcal{S}$ to depend on Test, and therefore, even the weaker notion of simulation mentioned after Definition 5 implies IND-PRE security.

$\underline{\Gamma^*\text{-}s\text{-IND-PRE} \Rightarrow \Gamma^*\text{-}s\text{-SIM}}$: Suppose $\Pi = (\mathcal{O}, \mathcal{E})$ is q-Γ^*-s-IND-PRE secure for a schema Σ. Fix a polynomial η and a PPT adversary Adv whose running time is upper-bounded by a polynomial ℓ. We shall construct a simulator $\mathcal{S}$ for Adv in the ideal world, which makes at most $p(\eta, \ell)$ queries for some polynomial p, and suffers a simulation error of at most $1/\eta$. Below, we write η to mean $\max(\eta, \ell)$, so that we may assume that $\eta \geq \ell$.

In the ideal world, when a test Test $\in \Gamma^*$ uploads a $\widehat{\boldsymbol{P}} \in \{0,1\}^* \times \mathcal{P}^*_{\text{test}}$, $\mathcal{S}$ attempts to learn a sufficiently accurate approximation $\boldsymbol{P}^\dagger$ using a polynomial depth query strategy, and then faithfully simulates $\mathcal{O}(\boldsymbol{P}^\dagger)$ to Adv. Note that since Adv's running time is upper-bounded by the polynomial ℓ, w.l.o.g, the simulator considers $\widehat{\boldsymbol{P}}$ to be in $\{0,1\}^\ell \times \mathcal{P}^{\ell'}_{\text{test}}$, where ℓ' is the lesser of ℓ and the actual number of agents uploaded by Test.

$\mathcal{S}$ defines $R_i \subseteq \{0,1\}^\ell \times \mathcal{P}^{\ell'}_{\text{test}}$ and $D_i \subseteq R_i$ inductively as follows, for integers $i \geq 0$, up till $i = i^*$ such that $D_{i^*} = \emptyset$. It then samples $\boldsymbol{P}^\dagger \leftarrow R_{i^*}$ and uses it to complete the simulation.

Below, we write $\mathsf{Adv}(\mathcal{O}(R_i))$ to denote the random variable corresponding to the output of Adv when a random $\boldsymbol{P} \leftarrow R_i$ is encoded using $\mathcal{O}$ and given to Adv; also, recall that $\mathsf{Adv}(\mathcal{O}(\boldsymbol{P}))$ denotes the similar random variable when the fixed agent vector $\boldsymbol{P}$ is encoded and given to Adv.

1. Firstly, for each i, we define D_i^* in terms of R_i, as follows. $D_i^* = D_{i,0}^* \cup D_{i,1}^*$, where

$$D_{i,b}^* = \left\{ \boldsymbol{P} \in R_i \mid (-1)^b (\Pr[\mathsf{Adv}(\mathcal{O}(\boldsymbol{P})) = 1] - \Pr[\mathsf{Adv}(\mathcal{O}(R_i)) = 1]) > \frac{1}{\eta} \right\}.$$

   Below, we shall iteratively define sets $D_{i,0}$ and $D_{i,1}$, and let $D_i := D_{i,0} \cup D_{i,1}$. We shall maintain the invariant that, for all $i \geq 0$, $D_{i,\beta} \subseteq D_{i,\beta}^*$, and the uploaded agent vector $\widehat{\boldsymbol{P}} \in R_i \backslash (D_i^* \backslash D_i)$ (i.e., $\boldsymbol{P} \in R_i$, and if $\boldsymbol{P} \in D_i^*$ then $\boldsymbol{P} \in D_i$).
2. $R_0 = \{0,1\}^\ell \times \mathcal{P}_{\text{test}}^{\ell'}$, $D_{0,0} = D_{0,0}^*$, and $D_{0,1} = D_{0,1}^*$.
3. If $D_i \neq \emptyset$, we define R_{i+1} and D_{i+1} as follows.
   Suppose $D_{i,\beta} \neq \emptyset$. Then, consider the test $\mathsf{Test}_{i,\beta} \in \Gamma^*$, which on input $b = 0$ uploads $\boldsymbol{P} \leftarrow D_{i,\beta}$, and on input $b = 1$, uploads $\boldsymbol{P} \leftarrow R_i.^{10}$ Since $D_{i,\beta}$ is not empty, we have

$$|\Pr[\mathrm{REAL}\langle \mathsf{Test}_{i,\beta}(0) \mid \mathcal{O} \mid \mathsf{Adv}\rangle = 1] - \Pr[\mathrm{REAL}\langle \mathsf{Test}_{i,\beta}(1) \mid \mathcal{O} \mid \mathsf{Adv}\rangle = 1]|$$

$$= (-1)^\beta \frac{1}{|D_{i,\beta}|} \sum_{\boldsymbol{P} \in D_{i,\beta}} (\Pr[\mathsf{Adv}(\mathcal{O}(\boldsymbol{P})) = 1] - \Pr[\mathsf{Adv}(\mathcal{O}(R_i)) = 1]) > \frac{1}{\eta}$$

   because for each $\boldsymbol{P} \in D_{i,\beta} \subseteq D_{i,\beta}^*$, we have $(-1)^\beta (\Pr[\mathsf{Adv}(\mathcal{O}(\boldsymbol{P})) = 1] - \Pr[\mathsf{Adv}(\mathcal{O}(R_i)) = 1]) > \frac{1}{\eta}$. That is, $\mathsf{Test}_{i,\beta}$ is not η-hiding (against Adv, which runs for less than $\ell \leq \eta$ time). Since the scheme $\Pi = (\mathcal{O}, \mathcal{E})$ is Γ^*-s-IND-PRE-secure, there must exist an ideal world adversary, or equivalently, a query strategy $Q_{i,\beta}$ of depth at most $q(\eta)$ which has advantage of more than $\sigma := 1/q(\eta)$ in distinguishing $\mathsf{Test}_{i,\beta}(0)$ and $\mathsf{Test}_{i,\beta}(1)$.
   If $D_{i,\beta} = \emptyset$, $Q_{i,\beta}$ is taken as the empty query strategy. For each $\beta \in \{0,1\}$, $\mathcal{S}$ executes the query strategy $Q_{i,\beta}$ to obtain an answer $\mathsf{ans}_{i,\beta}$. It defines $R_i' = \{\boldsymbol{P} \in R_i \mid \boldsymbol{P}(Q_{i,0}) = \mathsf{ans}_{i,0}, \boldsymbol{P}(Q_{i,1}) = \mathsf{ans}_{i,1}\}$, and $D_{i,\beta}' = \{\boldsymbol{P} \in D_{i,\beta} \mid \boldsymbol{P}(Q_{i,\beta}) = \mathsf{ans}_{i,\beta}\}$. If $|R_i'| \leq (1 - \sigma)|R_i|$, then set $R_{i+1} = R_i'$ and $D_{i+1,\beta} = D_{i+1,\beta}^*$. Otherwise, set $R_{i+1} = R_i$ (and hence $D_{i+1,\beta}^* = D_{i,\beta}^*$) and $D_{i+1,\beta} = D_{i,\beta}'$.

The above iteration terminates for the least i such that $D_i = \emptyset$. Then we have the property that the uploaded agent $\widehat{\boldsymbol{P}} \in R_i \backslash D_i^*$, which means that

$$\left| \Pr[\mathsf{Adv}(\mathcal{O}(\widehat{\boldsymbol{P}})) = 1] - \Pr[\mathsf{Adv}(\mathcal{O}(R_i)) = 1] \right| \leq \frac{1}{\eta}.$$

Thus $\mathcal{S}$ completes the simulation by sampling $\boldsymbol{P}^\dagger \leftarrow R_i$ and giving $\mathcal{O}(\boldsymbol{P}^\dagger)$ to Adv.

---

[10] Note that $\mathsf{Test}_{i,\beta}$ may be computationally inefficient. This is the only reason we are not able to prove analogous results for a test-family that is like Γ^* but restricted to PPT tests.

Note that if $|R'_i| > (1 - \sigma)|R_i|$ then $|D'_{i,\beta}| \leq (1 - \sigma)|D_{i,\beta}|$, because otherwise $Q_{i,\beta}$ cannot distinguish $\mathsf{Test}_{i,\beta}$ with advantage σ (as, for $b = 0$ and $b = 1$, it receives an answer other than $\mathsf{ans}_{i,\beta}$ with probability less than σ). Therefore, we make progress in each iteration: either $|R_{i+1}| \leq (1 - \sigma)|R_i|$ (in which case $|D_{i+1}| \leq |R_{i+1}|$), or $|R_{i+1}| = |R_i|$ and $|D_{i+1,\beta}| \leq (1 - \sigma)|D_{i,\beta}|$. Hence, for $i^* \leq \log^2_{1-\sigma}|R_0|$ we have $D_i = \emptyset$.

The total number of queries made by the simulator above is bounded by $q(\eta) \cdot \log^2_{1-\sigma}|R_0|$. Note that $\log_2|R_0| \leq \ell + n_\Sigma \cdot \ell$, where n_Σ is a (polynomial) upper-bound on the number of bits required to represent an agent in the schema Σ. Also, $\left|\frac{1}{\log_2(1-\sigma)}\right| = O(q(\eta))$, so that $\log^2_{1-\sigma}|R_0| = O((n_\Sigma \cdot \ell \cdot q(\eta))^2)$. Hence, we can set $p(\eta, \ell)$ to be $q(\eta)$ times this polynomial.   □

3.1 Extensions: Limited Agent-Space and Resettable Tests

Firstly, in the above results we can use a test-family which is a subset of Γ^* as follows. Note that the tests in Γ^* may upload any number of agents and send messages of any length (i.e., we considered agents in $\{0,1\}^* \times \mathcal{P}^*_{\mathsf{test}}$). But our proofs go through unchanged if we restrict to a subset of Γ^* which uses an arbitrary subset of $\{0,1\}^* \times \mathcal{P}^*_{\mathsf{test}}$. (In this case, IND-CON is suitably modified to use the same subset.) In particular, we may restrict to the test-family $\Gamma^*_1 \subseteq \Gamma^*$ which uploads a single agent and does not give any auxiliary information. Thus, every test in this family is fully characterized by a distribution over $\mathcal{P}^*_{\mathsf{test}}$. A variant of IND-CON, say $\mathsf{IND-CON}_1$, can be defined where distribution ensembles only over $\mathcal{P}_{\mathsf{test}}$ are considered.

Secondly, we consider the possibility of using a test-family that is larger than Γ^*. Above, the restriction to Γ^* was crucial in allowing the construction of a composite query strategy by grafting a query strategy onto the leaves of another query strategy. However, if the test allowed itself to be treated as an agent — i.e., allowing a User to access Test from any state in its history — then the above equivalences would carry over. Thus, we may define a test-family Γ_{reset} consisting of tests which are allowed to accept messages from the user/adversary and react to them, but also allows the user/adversary to reset it to the beginning (without changing its random tape). Then the above proofs extend to show that IND-CON $\Leftrightarrow \Gamma_{\mathsf{reset}}$-$s$-IND-PRE $\Leftrightarrow \Gamma_{\mathsf{reset}}$-$s$-SIM, for all schemas. Note that tests in Γ^* are effectively resettable and hence $\Gamma_{\mathsf{reset}} \supseteq \Gamma^*$. We defer a formal definition of Γ_{reset} to the final version.

4   Reductions and Compositions

A *hybrid scheme* $(\mathcal{O}, \mathcal{E})^{\Sigma^*}$ is a cryptographic agent scheme in which $\mathcal{O}$ and $\mathcal{E}$ have access to $\mathcal{B}[\Sigma^*]$, as shown in Fig. 1 (in the middle), where $\Sigma^* = (\mathcal{P}^*_{\mathsf{auth}}, \mathcal{P}^*_{\mathsf{user}})$.[11] In general, the honest user would be replaced by an adversarial user Adv. Note that

---

[11] If $\mathcal{O}$ has a setup phase (see Appendix A), we require that $\mathcal{O}_{\mathsf{user}}$ uploads agents only in $\mathcal{P}^*_{\mathsf{user}}$ but $\mathcal{O}_{\mathsf{auth}}$ can upload any agent in $\mathcal{P}^*_{\mathsf{auth}} \cup \mathcal{P}^*_{\mathsf{user}}$.

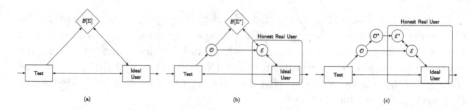

Fig. 1. $(\mathcal{O}, \mathcal{E})$ in (b) is a reduction from schema Σ to Σ^*. The security requirement is that no adversary Adv in the system (a) can distinguish that execution from an execution of the system in (b) (with Adv taking the place of honest real user). The correctness requirement is that the ideal User in (b) behaves the same as the ideal User interacting directly with $\mathcal{B}[\Sigma]$ (as in Fig. 2(a)). (c) shows the composition of the hybrid scheme $(\mathcal{O}, \mathcal{E})^{\Sigma^*}$ with a scheme $(\mathcal{O}^*, \mathcal{E}^*)$ that s-IND-PRE-securely implements Σ^*.

the output bit of Adv in such a system is given by the random variable IDEAL$\langle$Test $\circ$ $\mathcal{O} \mid \Sigma^* \mid$ Adv$\rangle$, where Test $\circ$ $\mathcal{O}$ denotes the combination of Test and $\mathcal{O}$.

We introduce a new *information-theoretic* notion of reduction between schemata which would allow for composition of Γ^*-s-IND-PRE secure schemes. When compared to [2], the main difference is that we require the hybrid world to be secure against *unbounded* adversaries (who make a polynomial number of queries). Further, the simulator is allowed to depend on the adversary.

Definition 9 (Statistical Reduction). *We say that a (hybrid) cryptographic agent scheme* $\Pi = (\mathcal{O}, \mathcal{E})$ *statistically reduces* Σ *to* Σ^* *with respect to* $\widetilde{\Gamma}$, *if there exists a polynomial* p *such that for all* unbounded User *who make at most* $\eta(\kappa)$ *queries for some polynomial* η,

1. *Correctness:* $\forall$ Test, IDEAL$\langle$Test $\mid \Sigma \mid$ User$\rangle \approx$ IDEAL$\langle$Test $\circ \mathcal{O} \mid \Sigma^* \mid \mathcal{E} \circ$ User$\rangle$.
2. *Simulation:* $\exists$ *a simulator* $\mathcal{S}_{\mathsf{User}}$ *which makes at most* $p(\eta(\kappa))$ *queries s.t.*
   $\forall$ Test $\in \widetilde{\Gamma}$, IDEAL$\langle$Test $\mid \Sigma \mid \mathcal{S}_{\mathsf{User}}\rangle \approx$ IDEAL$\langle$Test $\circ \mathcal{O} \mid \Sigma^* \mid$ User$\rangle$.

If there exists a scheme that reduces Σ *to* Σ^*, *then we say* Σ *reduces to* Σ^*. *(Note that correctness is required for all tests, not just those in* $\widetilde{\Gamma}$.)

Figure 1 illustrates a reduction. It also shows how such a reduction can be composed with an IND-PRE-secure scheme for Σ^*. We now prove the main result of this section, in a manner very similar to that of Agrawal et al. [2].

Theorem 3 (Composition). *For any two schemata,* Σ *and* Σ^*, *if* $(\mathcal{O}, \mathcal{E})$ *reduces* Σ *to* Σ^* *with respect to* Γ^* *and* $(\mathcal{O}^*, \mathcal{E}^*)$ *is a* Γ^*-s-IND-PRE *secure scheme for* Σ^*, *then* $(\mathcal{O} \circ \mathcal{O}^*, \mathcal{E}^* \circ \mathcal{E})$ *is a* Γ^*-s-IND-PRE *secure scheme for* Σ.

Proof. Let $(\mathcal{O}', \mathcal{E}') = (\mathcal{O} \circ \mathcal{O}^*, \mathcal{E}^* \circ \mathcal{E})$. Also, let Test$' = $ Test $\circ \mathcal{O}$ and User$' = \mathcal{E} \circ$ User. We first show that for all Test and PPT User, $(\mathcal{O}', \mathcal{E}')$ is a correct agent scheme for

Σ. We have

$$\text{REAL}\langle \text{Test} \mid \mathcal{O}' \mid \mathcal{E}' \circ \text{User} \rangle = \text{REAL}\langle \text{Test}' \mid \mathcal{O}^* \mid \mathcal{E}^* \circ \text{User}' \rangle$$

$$\overset{(a)}{\approx} \text{IDEAL}\langle \text{Test}' \mid \Sigma^* \mid \text{User}' \rangle$$

$$= \text{IDEAL}\langle \text{Test} \circ \mathcal{O} \mid \Sigma^* \mid \mathcal{E} \circ \text{User} \rangle$$

$$\overset{(b)}{\approx} \text{IDEAL}\langle \text{Test} \mid \Sigma \mid \text{User} \rangle$$

where (a) follows from the correctness guarantee of IND-PRE security of $(\mathcal{O}^*, \mathcal{E}^*)$ (Definition 3), and (b) follows from the correctness guarantee of $(\mathcal{O}, \mathcal{E})$ being a reduction of Σ to Σ^* (Definition 9). (Both (a) and (b) hold for all tests.) The other equalities are by regrouping the components in the system.

It remains to prove that there exists a polynomial p such that for all large enough κ, all Test $\in \Gamma^*$, and every polynomial η, if Test is $p(\eta(\kappa))$-s-hiding w.r.t. Σ then Test is $\eta(\kappa)$-hiding w.r.t. $\mathcal{O}'$.

Suppose that for some polynomial p', $(\mathcal{O}^*, \mathcal{E}^*)$ is a p'-Γ^*-s-IND-PRE secure scheme for Σ^*. We know that since $(\mathcal{O}, \mathcal{E})$ is a statistical reduction of Σ to Σ^* w.r.t. Γ^*, there exists a polynomial p^* such that for all unbounded User who make at most $\mu(\kappa)$ queries (for some polynomial μ), there exists a simulator $\mathcal{S}_{\text{User}}$ which makes at most $p^*(\mu(\kappa))$ queries such that for all Test $\in \Gamma^*$,

$$| \Pr[\text{IDEAL}\langle \text{Test} \mid \Sigma \mid \mathcal{S}_{\text{User}} \rangle = 1] -$$
$$\Pr[\text{IDEAL}\langle \text{Test} \circ \mathcal{O} \mid \Sigma^* \mid \text{User} \rangle = 1]| \leq \text{negl}(\kappa). \quad (1)$$

So let p be a polynomial such that $1/p(x) \leq \max\{1/p'(x) - 2 \cdot \text{negl}(x), 1/p^*(p'(x))\}$ for all $x \geq 0$.

Let Test be an arbitrary test in Γ^*, η be any polynomial, and $\overline{\text{User}}$ be any unbounded user who makes at most $p'(\eta(\kappa))$ queries. We can apply Eq. 1 on Test(b) and $\overline{\text{User}}$ to get

$$| \Pr[\text{IDEAL}\langle \text{Test}(b) \mid \Sigma \mid \mathcal{S}_{\overline{\text{User}}} \rangle = 1] -$$
$$\Pr[\text{IDEAL}\langle \text{Test}(b) \circ \mathcal{O} \mid \Sigma^* \mid \overline{\text{User}} \rangle = 1]| \leq \text{negl}(\kappa) \quad (2)$$

for $b \in \{0, 1\}$. Here the simulator $\mathcal{S}_{\overline{\text{User}}}$ makes at most $p^*(p'(\eta(\kappa))) \leq p(\eta(\kappa))$ queries.

If Test is $p(\eta(\kappa))$-s-hiding w.r.t. Σ, then for all unbounded User$'$ who make at most $p(\eta(\kappa))$ queries,

$$| \Pr[\text{IDEAL}\langle \text{Test}(0) \mid \Sigma \mid \text{User}' \rangle = 1] -$$
$$\Pr[\text{IDEAL}\langle \text{Test}(1) \mid \Sigma \mid \text{User}' \rangle = 1]| \leq \frac{1}{p(\eta(\kappa))}. \quad (3)$$

Recall that $\mathsf{Test}' = \mathsf{Test} \circ \mathcal{O}$ and if $\mathsf{Test} \in \Gamma^*$ then $\mathsf{Test}' \in \Gamma^*$ too. Now by using Eqs. 2 and 3 with User' set to $\mathcal{S}_{\overline{\mathsf{User}}}$, we get

$$|\Pr[\text{IDEAL}\langle \mathsf{Test}'(0) \mid \Sigma^* \mid \overline{\mathsf{User}} \rangle = 1] - \Pr[\text{IDEAL}\langle \mathsf{Test}'(1) \mid \Sigma^* \mid \overline{\mathsf{User}} \rangle = 1]|$$
$$\leq \frac{1}{p(\eta(\kappa))} + 2 \cdot \mathsf{negl}(\kappa) \leq \frac{1}{p'(\eta(\kappa))}.$$

Thus Test' is $p'(\eta(\kappa))$-s-hiding w.r.t. Σ^*. This implies that Test' is $\eta(\kappa)$-hiding w.r.t. $\mathcal{O}^*$, and by regrouping the components, we have that Test is $\eta(\kappa)$-hiding w.r.t. $\mathcal{O}'$. $\qquad\square$

We also have the following result regarding transitivity of reduction.

Theorem 4 (Transitivity of Reduction). *For any three schemata, Σ_1, Σ_2, Σ_3, if Σ_1 statistically reduces to Σ_2 and Σ_2 statistically reduces to Σ_3, then Σ_1 statistically reduces to Σ_3.*

Proof. If $\Pi_1 = (\mathcal{O}_1, \mathcal{E}_1)$ and $\Pi_2 = (\mathcal{O}_2, \mathcal{E}_2)$ are schemes that carry out the statistical reduction of Σ_1 to Σ_2 and that of Σ_2 to Σ_3, respectively, we claim that the scheme $\Pi = (\mathcal{O}_1 \circ \mathcal{O}_2, \mathcal{E}_2 \circ \mathcal{E}_1)$ is a statistical reduction of Σ_1 to Σ_3. The correctness of this reduction follows from the correctness of the given reductions. Further, if $\mathcal{S}_1$ and $\mathcal{S}_2$ are the simulators associated with the two reductions, we can define a simulator $\mathcal{S}$ for the composed reduction as $\mathcal{S}_2 \circ \mathcal{S}_1$. $\qquad\square$

5 Applications

In this section we briefly summarize how the above results can be instantiated to rederive the main results of [9]. We start off by defining the obfuscation schema.

Obfuscation Schema. If $\mathcal{F}$ is a family of circuits, we define

$$\Sigma_{\text{OBF}(\mathcal{F})} := (\emptyset, \mathcal{F}).$$

That is, in the ideal execution User obtains handles for agents which simple compute $\mathcal{F}$ on their inputs and write the result on to their output tapes. We shall consider setup-free, IND-PRE secure implementations $(\mathcal{O}, \mathcal{E})$ of $\Sigma_{\text{OBF}(\mathcal{F})}$.

The following propositions which easily follow from the definitions. Below we refer to the test-family Γ_1^* from Sect. 3.1.

Proposition 1. *For a function family $\mathcal{F}$, a Γ_1^*-s-SIM secure scheme for $\Sigma_{\text{OBF}(\mathcal{F})}$ is a VGB obfuscation scheme for $\mathcal{F}$, and vice-versa.*

With the modification to IND-CON also to distributions over a single agent (circuit), which we called IND-CON$_1$ in Sect. 3.1, we have the following proposition.

Proposition 2. *For a function family $\mathcal{F}$, an INDCON$_1$ secure scheme for $\Sigma_{\text{OBF}(\mathcal{F})}$ is an SIO scheme for $\mathcal{F}$ and vice versa.*

These propositions, combined with Theorems 1 and 2 (as extended in Sect. 3.1), yields the following result of [9] as a corollary.

Corollary 5 *An obfuscation scheme is a VGB obfuscation for a function family $\mathcal{F}$ if and only if it is an SIO for $\mathcal{F}$.*

Next we describe how the security of the VGB obfuscation construction given in [9] follows as a corollary of our composition theorem.

Graded Encoding Schema. Following "set-based" graded encoding [5,9,13, 19], we define the graded encoding schema $\Sigma_{GE} = (\emptyset, \mathcal{P}_{\mathsf{user}}^{GE})$, where $\mathcal{P}_{\mathsf{user}}^{GE}$ contains a single type of agent. The schema is specified by a ring $\mathcal{R}(+, \times)$ and a subset $\mathfrak{S}$ of $2^{[k]}$ for a level $k \in \mathbb{N}$ (where $[k] = \{1, 2, \ldots, n\}$). The persistent state of an agent $P \in \mathcal{P}_{\mathsf{user}}^{GE}$ is a pair (x, S) where $x \in \mathcal{R}$ and $S \in \mathfrak{S}$, which it maintains on its work-tape (initially copied from its parameter tape). When invoked without an input, it sends (x, S) to a peer agent in the session. When invoked with an input *Oper* on its input tape, it operates as follows (before entering a blocking state):

- *Oper* $= +$ (resp. $-$): It reads a message (x', S') from its incoming communication tape. If $S = S'$, it updates its work-tape with $(x + x', S)$ (resp. $(x - x', S)$); otherwise, it writes $\perp$ on its output tape.
- *Oper* $= \times$: It reads a message (x', S') from its incoming communication tape. If $S' \in \mathfrak{S}$ and $S \cap S' = \emptyset$, it updates its work-tape with $(x \times x', S \cup S')$; otherwise, it writes $\perp$ on its output tape.
- *Oper* $=$ Zero$-$Test: It first checks whether S is the universe set $[k]$. If not, it writes $\perp$ on its output tape. Otherwise, if $x = 0$ it writes 1; otherwise, 0.

The following proposition is an immediate consequence of the definition of strong-sampler semantic security [9].

Proposition 3. *A graded encoding scheme is strong-sampler semantically secure if and only if it is a Γ^*-sIND-PRE secure scheme for the schema Σ_{GE}.*

The following is a restatement of a result in [5] (that [9] relies on), formalized as a statistical reduction.

Proposition 4. *For any function family $\mathcal{F} \in$ NC^1, there exists a statistical reduction from $\Sigma_{\mathrm{OBF}(\mathcal{F})}$ to Σ_{GE}.*

The following result of [9] is then an immediate corollary of the above two propositions and the composition theorem (Theorem 3) as well as the fact that a Γ^*-s-IND-PRE secure scheme for $\Sigma_{\mathrm{OBF}(\mathcal{F})}$ is a VGB obfuscation (from Theorem 2 and Proposition 1).

Corollary 6. *If there exists a strong-sampler semantically-secure graded encoding scheme, then there exists a VGB obfuscation scheme for any function family $\mathcal{F} \in$ NC^1.*

Acknowledgments. This work was supported in part by NSF grant 12-28856. Part of this work was carried out while the authors were visiting the Simons Institute for Theoretical Computer Science, supported by the Simons Foundation and by the DIMACS/Simons Collaboration in Cryptography through NSF grant CNS 15-23467.

Part of this work was done when the first author was at the University of Illinois at Urbana-Champaign. At University of Texas at Austin, he is supported by NSF CNS-1228599, CNS-1414082 and DARPA SafeWare.

A Preliminaries

The following description of the Cryptographic Agents model is adapted from [2], and follows it closely.

A.1 Agents

Definition 10 (Agents and Family of Agents). *An agent is an interactive Turing Machine, with the following modifications:*

- *There is a special read-only parameter tape, which always consists of a security parameter κ, and possibly other parameters.*
- *There is an a priori restriction on the size of all the tapes other than the randomness tape (including input, communication and work tapes), as a function of the security parameter.*
- *There is a special* blocking state *such that if the machine enters such a state, it remains there if the input tape is empty. Similarly, there are blocking states which let the machine block if any combination of the communication tape and the input tape is empty.*

An agent family is a maximal set of agents with the same program (i.e., state space and transition functions), but possibly different contents in their parameter tapes. We also allow an agent family to be the empty set $\emptyset$.

Note that an agent who enters a blocking state can move out of it if its configuration is changed by adding a message to its input tape and/or communication tape. However, if the agent enters a halting state, it will not move out of that state. An agent who never enters a blocking state is called a *non-reactive agent*. An agent who never reads or writes from a communication tape is called a *non-interactive agent*.

Definition 11 (Session). *A session maps a finite ordered set of agents, their configurations and inputs, to outputs and (updated) configurations of the same agents, as follows. The agents are initialized with the given inputs on their input tapes, and then executed together until they are deadlocked.*[12] *The result of applying the session is defined as the collection of outputs and configurations of the agents when the session terminates (if it terminates; if not, the result is left undefined).*

---

[12] More precisely, the first agent is executed till it enters a blocking or halting state, and then the second and so forth, in a round-robin fashion, until all the agents remain in blocking or halting states for a full round. After each execution of an

We shall be restricting ourselves to collections of agents such that sessions involving them are guaranteed to terminate. Note that we have defined a session to have only an initial set of inputs, so that the outcome of a session is well-defined (without the need to specify how further inputs would be chosen).

Definition 12 (Ideal Agent Schema). *A (well-behaved) ideal agent schema $\Sigma = (\mathcal{P}_{\mathsf{auth}}, \mathcal{P}_{\mathsf{user}})$, or simply schema, is a pair of agent families, such that there is a polynomial poly such that for any session of agents belonging to $\mathcal{P}_{\mathsf{auth}} \cup \mathcal{P}_{\mathsf{user}}$ (with any inputs and any configurations, with the same security parameter κ), the session terminates within $\mathrm{poly}(\kappa, t)$ steps, where t is the number of agents in the session.*

A.2   Security Definitions

We define what it means for a cryptographic agent scheme to securely implement a given ideal agent schema. Intuitively, the security notion is of *indistinguishability preservation*: if two executions using an ideal schema are indistinguishable, we require them to remain indistinguishable when implemented using a cryptographic agent scheme.

Ideal World. The ideal system for a schema Σ consists of two parties Test and User and a fixed third party $\mathcal{B}[\Sigma]$ (for "black-box"). All three parties have a security parameter κ built-in. We shall explicitly refer to their random-tapes as r, s and t. Test receives a "secret bit" b as input and User produces an output bit b'. The interaction between User, Test and $\mathcal{B}[\Sigma]$ can be summarized as follows:

- **Uploading agents.** Let $\Sigma = (\mathcal{P}_{\mathsf{auth}}, \mathcal{P}_{\mathsf{user}})$ where we associate $\mathcal{P}_{\mathsf{test}} := \mathcal{P}_{\mathsf{auth}} \cup \mathcal{P}_{\mathsf{user}}$ with Test and $\mathcal{P}_{\mathsf{user}}$ with User. Test and User can, at any point, choose an agent from its agent family and send it to $\mathcal{B}[\Sigma]$. More precisely, User can send a string to $\mathcal{B}[\Sigma]$, and $\mathcal{B}[\Sigma]$ will instantiate an agent $\mathcal{P}_{\mathsf{user}}$, with the given string (along with its own security parameter) as the contents of the parameter tape, and all other tapes being empty. Similarly, Test can send a string and a bit indicating whether it is a parameter for $\mathcal{P}_{\mathsf{auth}}$ or $\mathcal{P}_{\mathsf{user}}$, and it is used to instantiate an agent $\mathcal{P}_{\mathsf{auth}}$ or $\mathcal{P}_{\mathsf{user}}$, accordingly.[13] Whenever an agent is instantiated, $\mathcal{B}[\Sigma]$ sends a unique handle (a serial number) for that agent to User; the handle also indicates whether the agent belongs to $\mathcal{P}_{\mathsf{auth}}$ or $\mathcal{P}_{\mathsf{user}}$.
- **Query.** A query is a request for session execution. At any point in time, User may request an execution of a session, by sending an ordered tuple of handles $(h_1, \ldots, h_t)$ (from among all the handles obtained thus far from $\mathcal{B}[\Sigma]$) to specify the configurations of the agents in the session, along with their

---

agent, the contents of its outgoing communication tape are interpreted as an ordered sequence of messages to each of the other agents in the session (some or all of them possibly being empty messages), and copied over to the respective agents' incoming communication tapes.

[13] In fact, for convenience, we allow Test and User to specify multiple agents in a single message to $\mathcal{B}[\Sigma]$.

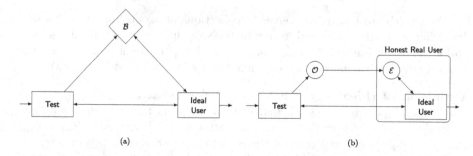

Fig. 2. The ideal world (on the left) and the real world with an honest user.

inputs. $\mathcal{B}[\Sigma]$ reports back the outputs from the session, and also gives new handles corresponding to the configurations of the agents when the session terminated.[14] If an agent halts in a session, no new handle is given for that agent.

Observe that only User receives any output from $\mathcal{B}[\Sigma]$; the communication between Test and $\mathcal{B}[\Sigma]$ is one-way. (See Fig. 2.)

Real World. A *cryptographic scheme* (or simply scheme) consists of a pair of (possibly stateful and randomized) programs $(\mathcal{O}, \mathcal{E})$, where $\mathcal{O}$ is an encoding procedure for agents in $\mathcal{P}_{\text{test}}$ and $\mathcal{E}$ is an execution procedure. The real world execution for a scheme $(\mathcal{O}, \mathcal{E})$ consists of Test, a user that we shall generally denote as Adv and the encoder $\mathcal{O}$. ($\mathcal{E}$ features as part of an honest user in the real world execution: see Fig. 2.) Test remains the same as in the ideal world, except that instead of sending an agent to $\mathcal{B}[\Sigma]$, it sends it to the encoder $\mathcal{O}$. In turn, $\mathcal{O}$ encodes this agent and sends the resulting cryptographic agent to Adv.

Syntactic Requirements on $(\mathcal{O}, \mathcal{E})$. $(\mathcal{O}, \mathcal{E})$ may or may not use a "setup" phase. In the latter case we call it a *setup-free cryptographic agent scheme*, and $\mathcal{O}$ is required to be a memory-less program that takes an agent $P \in \mathcal{P}_{\text{test}}$ as input and outputs a cryptographic agent that is sent to Adv. If the scheme has a setup phase, $\mathcal{O}$ consists of a triplet of memory-less programs $(\mathcal{O}_{\text{setup}}, \mathcal{O}_{\text{auth}}, \mathcal{O}_{\text{user}})$: in the real world execution, first $\mathcal{O}_{\text{setup}}$ is run to generate a secret-public key pair (MSK, MPK); MPK is sent to Adv. Subsequently, when $\mathcal{O}$ receives an agent $P \in \mathcal{P}_{\text{auth}}$ it will invoke $\mathcal{O}_{\text{auth}}(P, \text{MSK})$, and when it receives an agent $P \in \mathcal{P}_{\text{user}}$, it will invoke $\mathcal{O}_{\text{user}}(P, \text{MPK})$, to obtain a cryptographic agent that is then sent to Adv.

$\mathcal{E}$ is required to be memoryless as well, except that when it gives a handle to a User, it can record a string against that handle, and later when User requests a

---

[14] Note that if the same handle appears more than once in the tuple $(h_1, \ldots, h_t)$, it is interpreted as multiple agents with the same configuration (but possibly different inputs). Also note that after a session, the old handles for the agents are not invalidated; so a User can access a configuration of an agent any number of times, by using the same handle.

session execution, $\mathcal{E}$ can access the string recorded for each handle in the session. There is a *compactness requirement* that the size of this string is *a priori* bounded (note that the state space of the ideal agents are also *a priori* bounded). If there is a setup phase, $\mathcal{E}$ can also access MPK each time it is invoked.

References

1. Agrawal, S., Agrawal, S., Badrinarayanan, S., Kumarasubramanian, A., Prabhakaran, M., Sahai, A.: On the practical security of inner product functional encryption. In: Katz, J. (ed.) PKC 2015. LNCS, vol. 9020, pp. 777–798. Springer, Heidelberg (2015). doi:10.1007/978-3-662-46447-2_35
2. Agrawal, S., Agrawal, S., Prabhakaran, M.: Cryptographic agents: towards a unified theory of computing on encrypted data. In: Oswald, E., Fischlin, M. (eds.) EUROCRYPT 2015. LNCS, vol. 9057, pp. 501–531. Springer, Heidelberg (2015). doi:10.1007/978-3-662-46803-6_17
3. Agrawal, S., Gorbunov, S., Vaikuntanathan, V., Wee, H.: Functional encryption: new perspectives and lower bounds. In: Canetti, R., Garay, J.A. (eds.) CRYPTO 2013. LNCS, vol. 8043, pp. 500–518. Springer, Heidelberg (2013). doi:10.1007/978-3-642-40084-1_28
4. Barak, B., Goldreich, O., Impagliazzo, R., Rudich, S., Sahai, A., Vadhan, S., Yang, K.: On the (im)possibility of obfuscating programs. In: Kilian, J. (ed.) CRYPTO 2001. LNCS, vol. 2139, pp. 1–18. Springer, Heidelberg (2001). doi:10.1007/3-540-44647-8_1
5. Barak, B., Garg, S., Kalai, Y.T., Paneth, O., Sahai, A.: Protecting obfuscation against algebraic attacks. In: Nguyen, P.Q., Oswald, E. (eds.) EUROCRYPT 2014. LNCS, vol. 8441, pp. 221–238. Springer, Heidelberg (2014). doi:10.1007/978-3-642-55220-5_13
6. Barbosa, M., Farshim, P.: On the semantic security of functional encryption schemes. In: Kurosawa, K., Hanaoka, G. (eds.) PKC 2013. LNCS, vol. 7778, pp. 143–161. Springer, Heidelberg (2013). doi:10.1007/978-3-642-36362-7_10
7. Bellare, M., O'Neill, A.: Semantically-secure functional encryption: possibility results, impossibility results and the quest for a general definition. In: Abdalla, M., Nita-Rotaru, C., Dahab, R. (eds.) CANS 2013. LNCS, vol. 8257, pp. 218–234. Springer, Heidelberg (2013)
8. Bitansky, N., Canetti, R.: On strong simulation and composable point obfuscation. In: Rabin, T. (ed.) CRYPTO 2010. LNCS, vol. 6223, pp. 520–537. Springer, Heidelberg (2010). doi:10.1007/978-3-642-14623-7_28
9. Bitansky, N., Canetti, R., Kalai, Y.T., Paneth, O.: On virtual grey box obfuscation for general circuits. In: Garay, J.A., Gennaro, R. (eds.) CRYPTO 2014. LNCS, vol. 8617, pp. 108–125. Springer, Heidelberg (2014). doi:10.1007/978-3-662-44381-1_7
10. Boneh, D., Sahai, A., Waters, B.: Functional encryption: definitions and challenges. In: Ishai, Y. (ed.) TCC 2011. LNCS, vol. 6597, pp. 253–273. Springer, Heidelberg (2011). doi:10.1007/978-3-642-19571-6_16
11. Canetti, R.: Universally composable security: a new paradigm for cryptographic protocols. In: Proceedings of the 42nd IEEE Symposium on Foundations of Computer Science. FOCS 2001 (2001)
12. De Caro, A., Iovino, V., Jain, A., O'Neill, A., Paneth, O., Persiano, G.: On the achievability of simulation-based security for functional encryption. In: Canetti, R., Garay, J.A. (eds.) CRYPTO 2013, Part II. LNCS, vol. 8043, pp. 519–535. Springer, Heidelberg (2013)

13. Garg, S., Gentry, C., Halevi, S., Raykova, M., Sahai, A., Waters, B.: Candidate indistinguishability obfuscation and functional encryption for all circuits. In: FOCS (2013). http://eprint.iacr.org/

14. Goldwasser, S., Kalai, Y.T.: On the impossibility of obfuscation with auxiliary input. In: Proceedings of the 46th Annual IEEE Symposium on Foundations of Computer Science. FOCS 2005 (2005)

15. Goldwasser, S., Rothblum, G.N.: On best-possible obfuscation. In: Vadhan, S.P. (ed.) TCC 2007. LNCS, vol. 4392, pp. 194–213. Springer, Heidelberg (2007)

16. Hada, S.: Zero-knowledge and code obfuscation. In: Okamoto, T. (ed.) ASI-ACRYPT 2000. LNCS, vol. 1976, pp. 443–457. Springer, Heidelberg (2000). doi:10.1007/3-540-44448-3_34

17. Maurer, U.: Constructive cryptography - a new paradigm for security definitions and proofs. In: Theory of Security and Applications - Joint Workshop, TOSCA 2011, Saarbrücken, Germany, 31 March–1 April 2011, Revised Selected Papers, pp. 33–56 (2011). http://dx.doi.org/10.1007/978-3-642-27375-9_3

18. O'Neill, A.: Definitional issues in functional encryption. Cryptology ePrint Archive, Report 2010/556 (2010). http://eprint.iacr.org/

19. Pass, R., Seth, K., Telang, S.: Indistinguishability obfuscation from semantically-secure multilinear encodings. In: Garay, J.A., Gennaro, R. (eds.) CRYPTO 2014. LNCS, vol. 8616, pp. 500–517. Springer, Heidelberg (2014). doi:10.1007/978-3-662-44371-2_28

Attribute-Based Encryption

Deniable Attribute Based Encryption
for Branching Programs from LWE

Daniel Apon[1($\boxtimes$)], Xiong Fan[2], and Feng-Hao Liu[3]

[1] University of Maryland, College Park, USA
dapon@cs.umd.edu
[2] Cornell University, Ithaca, USA
xfan@cs.cornell.edu
[3] Florida Atlantic University, Boca Raton, USA
fenghao.liu@fau.edu

Abstract. Deniable encryption (Canetti et al. CRYPTO '97) is an intriguing primitive that provides a security guarantee against not only eavesdropping attacks as required by semantic security, but also stronger coercion attacks performed after the fact. The concept of deniability has later demonstrated useful and powerful in many other contexts, such as leakage resilience, adaptive security of protocols, and security against selective opening attacks. Despite its conceptual usefulness, our understanding of how to construct deniable primitives under standard assumptions is restricted.

In particular from standard lattice assumptions, i.e. Learning with Errors (LWE), we have only flexibly and non-negligible advantage deniable public-key encryption schemes, whereas with the much stronger assumption of indistinguishable obfuscation, we can obtain at least fully sender-deniable PKE and computation. How to achieve deniability for other more advanced encryption schemes under standard assumptions remains an interesting open question.

In this work, we construct a flexibly bi-deniable Attribute-Based Encryption (ABE) scheme for all polynomial-size Branching Programs from LWE. Our techniques involve new ways of manipulating Gaussian noise that may be of independent interest, and lead to a significantly sharper analysis of noise growth in Dual Regev type encryption schemes. We hope these ideas give insight into achieving deniability and related properties for further, advanced cryptographic systems from lattice assumptions.

1 Introduction

Deniable encryption, introduced by Canetti et al. [14] at CRYPTO 1997, is an intriguing primitive that allows Alice to privately communicate with Bob in a way that resists not only eavesdropping attacks as required by semantic security, but also stronger *coercion attacks* performed after the fact. An eavesdropper Eve stages a cocercion attack by additionally approaching Alice (or Bob, or both) *after* a ciphertext is transmitted and demanding to see all secret information:

© International Association for Cryptologic Research 2016
M. Hirt and A. Smith (Eds.): TCC 2016-B, Part II, LNCS 9986, pp. 299–329, 2016.
DOI: 10.1007/978-3-662-53644-5_12

the plaintext, the random coins used by Alice for encryption, and any private keys held by Bob (or Alice) related to the ciphertext. In particular, Eve can use this information to "fully unroll" the *exact transcript* of some deterministic decryption procedure purportedly computed by Bob, as well as verify that the exact coins and decrypted plaintext in fact produce the coerced ciphertext. A secure deniable encryption scheme should maintain privacy of the sensitive data originally communicated between Alice and Bob under the coerced ciphertext (instead substituting a benign yet *convincing* plaintext in the view of Eve), even in the face of such a revealing attack and even if Alice and Bob may not interact during the coercion phase.

Historically, deniable encryption schemes have been challenging to construct. Under standard assumptions, Canetti et al. [14] constructed a sender-deniable[1] PKE where the distinguishing advantage between real and fake openings is an inverse polynomial depending on the public key size. But it was not until 2011 that O'Neill, Peikert, and Waters [35] proposed the first constructions of bi-deniable PKE with *negligible* deniability distinguishing advantage: from simulatable PKE generically, as well as from Learning with Errors (LWE [36]) directly.

Concurrently, Bendlin et al. [8] showed an inherent limitation: any non-interactive public-key encryption scheme may be receiver-deniable (resp. bi-deniable) only with *non-negligible* $\Omega(1/\mathsf{size}(\mathsf{pk}))$ distinguishing advantage in the deniability experiment. Indeed, O'Neill et al. [35] bypass the impossibility result of [8] by working in the so-called *flexible*[2] model of deniability. In the flexible of deniability, private keys sk are distributed by a central key authority. In the event that Bob is coerced to reveal a key sk that decrypts chosen ciphertext ct^*, the key authority distributes a *faking key* fk to Bob, which Bob can use to generate a fake key sk^* (designed to behave identically to sk except on ciphertext ct^*). If this step is allowed, then O'Neill et al. demonstrate that for their constructions, Eve has at most negligible advantage in distinguishing whether Bob revealed an honest sk or fake sk^*.

A major breakthrough in deniable encryption arrived with the work of Sahai and Waters [38], who proposed the first sender-deniable PKE with negligible distinguishing advantage from indistinguishability obfuscation ($i\mathcal{O}$) for P/poly [22]. The concept of deniability has been demonstrated useful in the contexts of leakage resilience [20], adaptive security for protocols, and as well as deniable computation (or algorithms) [16,19,23]. In addition to coercion resistance, a bi-deniable encryption scheme is a non-committing encryption scheme [15], as well as a scheme secure under selective opening (SOA) attacks [7], which are of independent theoretical interest.

Very recently, De Caro, Iovino, and O'Neill [17] gave various constructions of deniable *functional* encryption. First, they show a generic transformation of any

---

[1] We differentiate between sender-, receiver-, and bi-deniable schemes. A bi-deniable scheme is both sender- and receiver-deniable.

[2] We borrow the name "flexible" from Boneh, Lewi, and Wu [10] as the original term "multi-distributional" of O'Neill et al. [35] is used to define a slightly different security property in the recent work by De Caro et al. [17] than we achieve here.

IND-secure FE scheme for circuits into a flexibly receiver-deniable FE for circuits. Second, they give a direct construction of receiver-deniable FE for Boolean formulae from bilinear maps. Further, in the stronger *multi-distributional* model of deniable functional encryption – where there are special "deniable" set-up and encryption algorithms in addition to the plain ones, and where under coercion, it may non-interactively be made to seem as only the normal algorithms were used – De Caro et al. [17] construct receiver-deniable FE for circuits under the additional (powerful) assumption of different-inputs obfuscation (*diO*).

De Caro et al. [17] also show (loosely speaking) that any receiver-deniable FE implies SIM-secure FE for the same functionality. Following [17], we also emphasize that deniability for functional encryption is a **strictly stronger** property than SIM security, since fixed coerced ciphertexts must decrypt correctly and benignly *in the real world*. Finally, we mention that in concurrent work, Apon, Fan, and Liu, in an unpublished work [5], construct flexibly bi-deniable inner product encryption from standard *lattice* assumptions. This work generalizes and thus subsumes the prior results of [5].

Despite the apparent theoretical utility in understanding the extent to which cryptographic constructions are deniable, our current knowledge of constructing such schemes from standard lattice assumptions is still limited. From LWE, we have only flexible and non-negligible advantage deniable encryption schemes (or IPE from [5]), whereas with the much more powerful assumption of indistinguishability obfuscation (*iO*), we can obtain at least fully-secure sender-deniable PKE and computation [16,19,23], or as mentioned above even a multi-distributional receiver-deniable FE for all circuits from the even stronger assumption of *diO*.

1.1 Our Contributions

In this work, we further narrow this gap by investigating a richer primitive – attribute-based encryption (ABE) [9,28,30,31] – *without* the use of obfuscation as a black box primitive. We hope that the techniques developed in this work can further shed light on deniability for even richer schemes such as functional encryption [9,12,22,29] under standard assumptions.

- Our main contribution is the construction of a flexibly bi-deniable ABE for poly-sized branching programs (which can compute NC1 via Barrington's theorem [6]) from the standard Learning with Errors assumption [36].

Theorem 1.1. *Under the standard* LWE *assumption, there is a flexibly bi-deniable attribute-based encryption scheme for all poly-size branching programs.*

Recall that in an attribute-based encryption (ABE) scheme for a family of functions $\mathcal{F} : \mathcal{X} \to \mathcal{Y}$, every secret key sk_f is associated with a predicate $f \in \mathcal{F}$, and every ciphertext ct_x is associated with an attribute $x \in \mathcal{X}$. A ciphertext ct_x can be decrypted by a given secret key sk_f to its payload message m only when $f(x) = 0 \in \mathcal{Y}$. Informally, the typical security notion for an ABE scheme is *collusion resistance*, which means no collection of keys can provide information on a ciphertext's message, if the individual keys are not authorized to decrypt the ciphertext in the first place. Intuitively, a bi-deniable ABE must provide both collusion and coercion resistance.

Other contributions of this work can be summarized as:

- A new form of the Extended Learning with Errors (eLWE) assumption [2,13, 35], which is convenient in the context of Dual Regev type ABE/FE schemes that apply the Leftover Hash Lemma [21] in their security proofs.
- An explicit, tightened noise growth analysis for lattice-based ABE for branching programs. Prior work used the loose l_∞ norm to give a rough upper bound, which is technically insufficient to achieve deniability using our proof techniques. (We require matching upper *and* lower bounds on post-evaluation noise sizes.)

The eLWE assumption above is roughly the standard LWE assumption, but where the distinguisher also receives "hints" on the LWE sample's noise vector e in the form of inner products, i.e. distributions $\{ \mathbf{A}, \boldsymbol{b} = \mathbf{A}^T \boldsymbol{s} + \boldsymbol{e}, \boldsymbol{z}, \langle \boldsymbol{z}, \boldsymbol{e} \rangle \}$ where (intuitively) $\boldsymbol{z}$ is a decryption key in the real system (which are denoted r elsewhere). Our contribution here is a new reduction from the standard LWE assumption to our correlated variant of extended-LWE, eLWE$^+$, where the adversary requests arbitrary correlations (expressed as a matrix $\mathbf{R}$) between the hints, in the case of a prime poly-size modulus with noise-less hints. We show this by extending the LWE to eLWE reduction of Alperin-Sheriff and Peikert [2] to our setting.

1.2 Our Approach

At a high level, our work begins with the ABE for branching programs of Gorbunov and Vinayagamurthy [30]. We will augment the basic ABE-BP = (Setup, Keygen, Enc, Dec) with an additional suite of algorithms (DenSetup, DenEnc, SendFake, RecFake) to form our flexibly bi-deniable ABE-BP. Doing so requires careful attention to the setting of parameters, as we explain in the sequel.

We remark now that – due to reasons related to the delicateness of our parameter setting – the ABE scheme of [30] is *particularly suited* to being made bi-deniable, as compared to similar schemes such as the ABE for arithmetic circuits of Boneh et al. [9]. We will explain this in what follows as well.

Intuition for Our New Deniability Mechanism. As in the work of O'Neill et al. [35], our approach to bi-deniability relies primarily on a curious property of Dual Regev type [24] secret keys: by correctness of any such scheme, each key r is guaranteed to behave as intended for some $1 - \mathsf{negl}(n)$ fraction of the possible random coins used to encrypt, but system parameters may be set so that each key is also guaranteed to be *faulty* (i.e. fail to decrypt) on some $\mathsf{negl}(n)$ fraction of the possible encryption randomness. More concretely, each secret key vector r in lattice-based schemes is sampled from an m-dimensional Gaussian distribution, as is the error term e (for LWE public key $\mathbf{A} \in \mathbb{Z}_q^{n \times m}$). For every fixed r, with overwhelming probability over the choice of e, the vectors $r, e \in \mathbb{Z}_q^m$

will point in highly uncorrelated directions in m-space. However, if the vector r and e happen to point in similar directions, the error magnitude will be (loosely) *squared* during decryption.

Our scheme is based around the idea that a receiver, coerced on honest key-ciphertext pair (r, ct^*), can use the key authority's *faking key* fk to learn the precise error vector e^* used to construct ct^*. Given e^*, r, and fk, the receiver re-samples a fresh secret key r^* that is functionally-equivalent to the honest key r, except that r^* is strongly correlated with the vector e^* in ct^*. When the coercer then attempts to decrypt the challenge ciphertext ct^* using r^*, the magnitude of decryption error will artificially grow and cause the decryption to output the value we want to deny to. Yet, when the coercer attempts to decrypt any other independently-sampled ciphertext ct, decryption will succeed with overwhelming probability under r^* if it would have under r.

We emphasize that to properly show coercion resistance (when extending this intuition to the case of Dual Regev ABE instead of Dual Regev PKE), this behavior of r^* should hold *even when* ct *and* ct^* *embed the same attribute* x. (Indeed, the majority of our effort is devoted to ensuring this simple geometric intuition allows a valid instantiation of the denying algorithms (DenSetup, DenEnc, SendFake, RecFake) without "damaging" the basic operation of (Setup, Keygen, Enc, Dec) in the underlying ABE scheme.)

Then, given the ability to "artificially blow-up" the decryption procedure of a specific key on a ciphertext-by-ciphertext basis, we can employ an idea originally due to Canetti et al. [14] of *translucent sets*, but generalized to the setting of ABE instead of PKE, to construct our new, flexibly bi-deniable ABE-BP scheme out of the framework provided by the "plain" SIM-secure ABE-BP scheme of [30].

Highlights of the Gorbunov-Vinayagamurthy Scheme. In the ABE for (width 5) branching programs of [30], bits a are "LWE-encoded" by the vector

$$\psi_{\mathbf{A}, s, a} = s^T(\mathbf{A} + a \cdot \mathbf{G}) + e \in \mathbb{Z}_q^m$$

where $\mathbf{G}$ is the gadget matrix [34].

The ciphertext ct encrypting message μ under BP-input x is given by

$$\mathsf{ct} = (\psi_0, \psi^c, \{\psi_i\}_{i \in [\ell]}, \{\psi_{0,i}\}_{i \in [5]}, c),$$

and is composed of a Dual Regev ct-pair of vectors (ψ_0, c) encrypting the ciphertext's message μ, an encoding ψ^c representing the (freshly randomized) encoding of the constant 1, five encodings $\{\psi_{0,i}\}_{i \in [5]}$ representing a (freshly randomized) encoding of the initial state of a width-5, length-ℓ branching program BP, and ℓ encodings $\{\psi_i\}_{i \in [\ell]}$ – one for each step of the branching program's evaluation, storing a constant-sized *permutation matrix* associated with the i-th level of BP. Note that each "LWE encoding" ψ is performed under a distinct public key matrix $\mathbf{A}, \mathbf{A}^c, \{\mathbf{A}_i\}$, or $\{\mathbf{A}_{0,i}\}$ respectively.

The (key-homomorphic) evaluation procedure takes as input a ciphertext $\mathsf{ct} = (\psi_0, \psi^c, \{\psi_i\}, \{\psi_{0,i}\}, c)$ and the public key $\mathsf{pk} = (\mathbf{A}, \mathbf{A}^c, \{\mathbf{A}_i\}, \{\mathbf{A}_{0,i}\})$, as well as the cleartext branching program description BP and the BP-input x.

It produces the evaluated public key $\mathbf{V}_{\mathsf{BP}}$ and the evaluated encoding $\psi_{\mathsf{BP}(\boldsymbol{x})}$. Given a short secret key *vector* $\boldsymbol{r} \in \mathbb{Z}^{2m}$ matching (some public coset $\boldsymbol{u}$ of) the lattice generated by $[\mathbf{A}|\mathbf{V}_{\mathsf{BP}}] \in \mathbb{Z}^{n \times 2m}$, the encoding *vector* $\psi_{\mathsf{BP}(\boldsymbol{x})}$ (whose Dual Regev encoding-components (ψ_0, c) also match coset $\boldsymbol{u}$) can be decrypted to the message μ if and only if $\mathsf{BP}(\boldsymbol{x}) = \mathsf{accept} = \mathbf{0}$.

On the Necessity of Exact Noise Control. In order to push the intuition for our deniability mechanism through for an ABE of the above form, we must overcome a number of technical hurdles.

The major challenge is an implicit technical requirement to *very tightly control* the precise noise magnitude of evaluated ciphertexts. In previous functional (and homomorphic) encryption schemes from lattices, the emphasis is placed on upper bounding evaluated noise terms, to ensure that they do not grow too large and cause decryption to fail. Moreover, security (typically) holds for any ciphertext noise level at or above the starting ciphertexts' noises. In short, noise growth during evaluation is nearly always undesirable.

As with previous schemes, we too must upper bound the noise growth of evaluated ciphertexts in order to ensure basic correctness of our ABE. But unlike previous schemes, we must take the step of also (carefully) *lower bounding* the noise growth during the branching program evaluation (which technically motivates deviating from the l_∞ norm of prior analyses). This is due to the fact, highlighted above, that producing directional alignment between a key and error term can at most *square* the noise present during decryption. Since coercion resistance requires that it must always be possible to deny any ciphertext originally intended for any honest key, it must be that, with overwhelming probability, every honest key and every honest ciphertext produce evaluated error that is no less than the square root of the maximum noise threshold tolerated.

In a little more detail – as we will later demonstrate in Sect. 4 – in dimension m there is precisely an expected $\mathsf{poly}(m)$ *gap* in magnitude between the inner products of **(i)** two relatively *orthogonal* key/error vectors $\boldsymbol{r}, \boldsymbol{e}_{\mathsf{BP}(\boldsymbol{x})}$, and **(ii)** two highly *correlated* key/error vectors $\boldsymbol{r}^*, \boldsymbol{e}_{\mathsf{BP}(\boldsymbol{x})}$. The ability to deny is based around our ability to design $\boldsymbol{r}^*$ that are statistically indistinguishable from $\boldsymbol{r}$ in the attacker's view, but where $\boldsymbol{r}^*$ "punctures out" decryptions of ciphertexts with error vectors pointing in the *direction* of $\boldsymbol{e}_{\mathsf{BP}(\boldsymbol{x})}$ in m-space (error-vector directions are unique to each honest ct with overwhelming probability).

Crucially, this approach **generically forces** the use of a *polynomial-sized modulus* q in the scheme.[3] In particular, when error vectors $\boldsymbol{e}$ may (potentially) grow to be some superpolynomial magnitude in the dimension m of the public/secret keys, we totally lose any efficiently testable notion of "error vector orientation in m-space" for the purposes of Dual Regev type decryption.

Further, in order to "correctly trace and distinguish" different orientations throughout the computation of an arbitrary branching program BP, we are

---

[3] One consequence of a poly-size modulus requirement is that the fully key-homomorphic scheme of Boneh et al. [9], taken verbatim, can only be denied for up to NC0 functions using our approach. Past this, attempts to produce fake keys in an identical manner to this work may be detected by a statistical test under coercion.

required to make careful use of *multi-dimensional Gaussian* distributions. These are sampled using covariance matrices $\mathbf{Q} \in \mathbb{Z}^{m \times m}$ that allow us to succinctly describe the underlying, geometric *randomized rotation* action on error vector *orientations* in m-space with each arithmetic operation of the BP evaluation in the overall ABE-BP scheme. (We use the geometrically-inspired term "rotation matrix" to describe our low-norm matrices $\mathbf{R}$ for this reason.)

An additional subtlety in our new noise analysis is that we require the individual multiplications of the ct evaluation procedure to have *independently sampled* error vectors in each operand-encoding – and thus be "independently oriented" – in order for the overall analysis to go through correctly. (While there could in principle be some way around this technical obstacle in the analysis, we were unable to find one.) This appears to a priori exclude a straightforward denying procedure for *all circuits* [9], where a gate's input wires' preceding sub-circuits may have cross-wires between them. But it naturally permits denying *branching program computations*, where at the i-th time-step, an i-th *independently generated* ct-component is merged into an accumulated BP state, as with [30].

Finally, we mention that an inherent limitation in the techniques of Apon et al. [5], used to construct (the weaker notion of) flexibly bi-deniable inner product encryption from LWE, is bypassed in the current work at the cost of supporting only BP computations of an a-priori bounded length ℓ. Namely, it was the case in [5] that the *length* ℓ of the attribute vector $\boldsymbol{w}$ had to be "traded off" against the dimension m of the public/secret keys. We suppress the details, other than to point out that this issue can be resolved by artificially boosting the magnitude of the low-norm matrices used to generate error terms in fresh ciphertexts from $\{-1, 1\}$ up to $\{-\Theta(m\ell), \Theta(m\ell)\}$-valued matrices. This, of course, requires knowing the length ℓ of the branching program up front. (Intuitively, this technical change as compared to [5] allows for a sharp *inductive lower bound* on the *minimum* noise growth across all possible function-input pairs that might be evaluated in a given instance of our bi-deniable ABE-BP scheme.)

1.3   Future Directions

The next, most natural question is whether bi-deniable functional encryption can be built out of similar techniques (from only LWE), perhaps by leveraging our bi-deniable ABE for NC1 computations as a building block. We briefly sketch one possible approach and the obstacles encountered. Recall that Goldwasser et al. [27] show to transform the combination of (i) any ABE for a circuit family $\mathcal{C}$, (ii) fully homomorphic encryption, and (iii) a randomized encoding scheme (such as Yao's garbled circuits) into a 1-key (resp. *bounded collusion*) SIM-secure functional encryption scheme for $\mathcal{C}$.

If we instantiate the Goldwasser et al. transformation with our deniable ABE, we get a functional encryption scheme for NC1. We can then boost functional encryption for shallow circuits to functional encryption for all circuits using the "trojan method" of Ananth et al. [4]. As it turns out, it is easy to directly prove *flexible receiver-deniability* of the final scheme, independently of but matching the generic results of De Caro et al. [17] for receiver-deniable FE.

Unfortunately, we do not know how to prove (even, flexible) *sender-deniability* of this final scheme. Roughly speaking, the problem is that each ciphertext's attribute in such a scheme contains an FHE ciphertext ct_{FHE} for its attribute, and this attribute leaks to the attacker (resp. cocercer) on decryptions that succeed. In particular, there is nothing stopping the coercer from demanding that the sender also provide randomness r_S that *opens the attribute's FHE ciphertext*.

We speculate that a possible way around this obstacle would be to use an *adaptively-secure* homomorphic encryption scheme for NC1 computations. Note that adaptively-secure FHE is known to be impossible for circuits with $\omega(\log(n))$ depth due to a counting argument lower bound by Katz, Thiruvengadam, and Zhou [32], but this leaves open the possibility of an NC1-homomorphic encryption scheme with the necessary properties to re-obtain (flexible) sender deniability for lattice-based FE. We leave this as an intriguing open problem for future work.

2 Preliminaries

Notations. Let PPT denote probabilistic polynomial time. We use bold upper-case letters to denote matrices, and bold lowercase letters to denote vectors, where vectors are by default column vectors throughout the paper. We let λ be the security parameter, $[n]$ denote the set $\{1, ..., n\}$, and $|t|$ denote the number of bits in a string or vector t. We denote the i-th bit value of a string s by $s[i]$. We use $[\cdot|\cdot]$ to denote the concatenation of vectors or matrices, and $||\cdot||$ to denote the norm of vectors or matrices respectively. We use the ℓ_2 norm for all vectors unless explicitly stated otherwise.

We present necessary background knowledge of branching programs and lattices (such as the LWE assumption and lattice sampling algorithms) in full version.

Randomness Extraction. We will use the following lemma to argue the indistinghishability of two different distributions, which is a generalization of the leftover hash lemma proposed by Dodis et al. [21].

Lemma 2.1 ([1]). *Suppose that $m > (n + 1)\log q + w(\log n)$. Let $\mathbf{R} \in \{-1, 1\}^{m \times k}$ be chosen uniformly at random for some polynomial $k = k(n)$. Let $\mathbf{A}, \mathbf{B}$ be matrix chosen randomly from $\mathbb{Z}_q^{n \times m}, \mathbb{Z}_q^{n \times k}$ respectively. Then, for all vectors $\mathbf{w} \in \mathbb{Z}^m$, the two following distributions are statistically close:*

$$(\mathbf{A}, \mathbf{AR}, \mathbf{w}^T\mathbf{R}) \approx (\mathbf{A}, \mathbf{B}, \mathbf{w}^T\mathbf{R})$$

Learning With Errors. The LWE problem was introduced by Regev [36], who showed that solving it *on the average* is as hard as (quantumly) solving several standard lattice problems *in the worst case*.

Definition 2.2 (LWE). *For an integer $q = q(n) \geq 2$, and an error distribution $\chi = \chi(n)$ over $\mathbb{Z}_q$, the learning with errors problem $\mathsf{LWE}_{n,m,q,\chi}$ is to distinguish between the following pairs of distributions:*

$$\{\mathbf{A}, \boldsymbol{b} = \mathbf{A}^T \boldsymbol{s} + \boldsymbol{e}\} \ and \ \{\mathbf{A}, \boldsymbol{u}\}$$

where $\mathbf{A} \xleftarrow{\$} \mathbb{Z}_q^{n \times m}$, $\boldsymbol{s} \xleftarrow{\$} \mathbb{Z}_q^n$, $\boldsymbol{u} \xleftarrow{\$} \mathbb{Z}_q^m$, *and* $\boldsymbol{e} \xleftarrow{\$} \chi^m$.

Trapdoors and Sampling Algorithms. We will use the algorithms TrapGen, SampleLeft, SampleRight, ExtBasis, Invert first proposed in [1,18,24,34] to sample short vectors from specified lattices. For details of these sampling algorithms, please refer to the full version.

3 New Definitions and Tools

In this section, we first describe our new notion of flexibly bi-deniable ABE, which is a natural generalization of the flexibly bi-deniable PKE of [35]. Then we define the notion of a flexibly attribute-based bi-translucent set (AB-BTS), which generalizes the idea of bi-translucent set (BTS) in the work [35]. Using a similar argument as in the work [35], we can show that an AB-BTS suffices to construct bi-deniable ABE. In the last part of this section, we define a new assumption called Extended LWE Plus, and show its hardness by giving a reduction from the standard LWE problem.

3.1 Flexibly Bi-Deniable ABE: Syntax and Deniability Definition

A flexibly bi-deniable key-policy attribute based encryption for a class of Boolean circuits $\mathcal{C} : \{0,1\}^\ell \rightarrow \{0,1\}$ consists a tuple of PPT algorithms $\Pi =$ (Setup, Keygen, Enc, Dec, DenSetup, DenEnc, SendFake, RecFake). We describe them in detail as follows:

Setup(1^λ): On input the security parameter λ, the setup algorithm outputs public parameters pp and master secret key msk.

Keygen(msk, f): On input the master secret key msk and a function $f \in \mathcal{C}$, it outputs a secret key sk_f.

Enc(pp, $\boldsymbol{x}, \mu; r_S$): On input the public parameter pp, an attribute/message pair $(\boldsymbol{x}, \mu)$ and randomness r_S, it outputs a ciphertext $c_{\boldsymbol{x}}$.

Dec($\mathsf{sk}_f, c_{\boldsymbol{x}}$): On input the secret key sk_f and a ciphertext $c_{\boldsymbol{x}}$, it outputs the corresponding plaintext μ if $f(\boldsymbol{x}) = 0$; otherwise, it outputs $\bot$.

DenSetup(1^λ): On input the security parameter λ, the deniable setup algorithm outputs pubic parameters pp, master secret key msk and faking key fk.

DenEnc(pp, $\boldsymbol{x}, \mu; r_S$): On input the public parameter pp, an attribute/message pair $(\boldsymbol{x}, \mu)$ and randomness r_S, it outputs a ciphertext $c_{\boldsymbol{x}}$.

SendFake(pp, r_S, μ, μ'): On input public parameters pp, original random coins r_S, message μ of DenEnc and desired message μ', it outputs a faked random coin r_S'.

RecFake(pp, fk, $c_{\boldsymbol{x}}, f, \mu'$): On input public parameters pp, faking key fk, a ciphertext $c_{\boldsymbol{x}}$, a function $f \in \mathcal{C}$, and desired message μ', the receiver faking algorithm outputs a faked secret key sk_f'.

Correctness. We say the flexibly bi-deniable ABE scheme described above is correct, if for any $(\mathsf{msk}, \mathsf{pp}) \leftarrow \mathsf{S}(1^\lambda)$, where $\mathsf{S} \in \{\mathsf{Setup}, \mathsf{DenSetup}\}$, any message μ, function $f \in \mathcal{C}$, and any attribute vector $\boldsymbol{x}$ where $f(\boldsymbol{x}) = 0$, we have $\mathsf{Dec}(\mathsf{sk}_f, c_{\boldsymbol{x}}) = \mu$, where $\mathsf{sk}_f \leftarrow \mathsf{Keygen}(\mathsf{msk}, f)$ and $c_{\boldsymbol{x}} \leftarrow \mathsf{E}(\mathsf{pp}, \boldsymbol{x}, \mu; r_S)$ where $\mathsf{E} \in (\mathsf{Enc}, \mathsf{DenEnc})$.

Bi-Deniability Definition. Let μ, μ' be two arbitrary messages, not necessarily different. We propose the bi-deniability definition by describing real experiment $\mathbf{Expt}_{\mathcal{A},\mu,\mu'}^{\mathsf{Real}}(1^\lambda)$ and faking experiment $\mathbf{Expt}_{\mathcal{A},\mu,\mu'}^{\mathsf{Fake}}(1^\lambda)$ regarding adversary $\mathcal{A} = (\mathcal{A}_1, \mathcal{A}_2, \mathcal{A}_3)$ as shown in Fig. 1:

where $\mathsf{KG}(\mathsf{msk}, \boldsymbol{x}^*, \cdot)$ returns a secret key $\mathsf{sk}_f \leftarrow \mathsf{Keygen}(\mathsf{msk}, f)$ if $f(\boldsymbol{x}^*) \neq 0$ and $\perp$ otherwise.

1. $(\boldsymbol{x}^*, \mathsf{state}_1) \leftarrow \mathcal{A}_1(\lambda)$
2. $(\mathsf{pp}, \mathsf{msk}) \leftarrow \mathsf{Setup}(1^\lambda)$
3. $c'_{\boldsymbol{x}^*} \leftarrow \mathsf{Enc}(\mathsf{pp}, \boldsymbol{x}^*, \mu; r_S)$
4. $(f^*, \mathsf{state}_2) \leftarrow \mathcal{A}_2^{\mathsf{KG}(\mathsf{msk}, \boldsymbol{x}^*, \cdot)}(\mathsf{pp}, \mathsf{state}_1, c'_{\boldsymbol{x}^*})$
5. $\mathsf{sk}_{f^*} \leftarrow \mathsf{Keygen}(\mathsf{msk}, f^*)$
6. $b \leftarrow \mathcal{A}_3^{\mathsf{KG}(\mathsf{msk}, \boldsymbol{x}^*, \cdot)}(\mathsf{sk}_{f^*}, \mathsf{state}_2, r_S)$
7. Output $b \in \{0, 1\}$

(a) $\mathbf{Expt}_{\mathcal{A}}^{\mathsf{Real}}(1^\lambda)$

1. $(\boldsymbol{x}^*, \mathsf{state}_1) \leftarrow \mathcal{A}_1(\lambda)$
2. $(\mathsf{pp}, \mathsf{msk}, \mathsf{fk}) \leftarrow \mathsf{DenSetup}(1^\lambda)$
3. $c'_{\boldsymbol{x}^*} \leftarrow \mathsf{DenEnc}(\mathsf{pp}, \boldsymbol{x}^*, \mu'; r_S)$
4. $(f^*, \mathsf{state}_2) \leftarrow \mathcal{A}_2^{\mathsf{KG}(\mathsf{msk}, \boldsymbol{x}^*, \cdot)}(\mathsf{pp}, \mathsf{state}_1, c'_{\boldsymbol{x}^*})$
5. $r'_S \leftarrow \mathsf{SendFake}(\mathsf{pp}, \mu, \mu', r_S)$
6. $\mathsf{sk}_{f^*} \leftarrow \mathsf{RecFake}(\mathsf{pp}, \mathsf{fk}, c'_{\boldsymbol{x}}, \boldsymbol{v}^*, \mu')$
7. $b \leftarrow \mathcal{A}_3^{\mathsf{KG}(\mathsf{msk}, \boldsymbol{x}^*, \cdot)}(\mathsf{sk}_{f^*}, \mathsf{state}_2, r'_S)$
8. Output $b \in \{0, 1\}$

(b) $\mathbf{Expt}_{\mathcal{A}}^{\mathsf{Fake}}(1^\lambda)$

Fig. 1. Security experiments for bi-deniable ABE

Definition 3.1 (Flexibly Bi-Deniable ABE). *An ABE scheme Π is bi-deniable if for any two messages μ, μ', any probabilistic polynomial-time adversaries $\mathcal{A}$ where $\mathcal{A} = (\mathcal{A}_1, \mathcal{A}_2, \mathcal{A}_3)$, there is a negligible function $\mathsf{negl}(\lambda)$ such that*

$$\mathbf{Adv}_{\mathcal{A},\mu,\mu'}^{\Pi}(1^\lambda) = |\mathbf{Pr}[\mathbf{Expt}_{\mathcal{A},\mu,\mu'}^{\mathsf{Real}}(1^\lambda) = 1] - \mathbf{Pr}[\mathbf{Expt}_{\mathcal{A},\mu,\mu'}^{\mathsf{Fake}}(1^\lambda) = 1]| \leq \mathsf{negl}(\lambda)$$

3.2   Attribute Based Bitranslucent Set Scheme

In this section, we define the notion of a *Attribute Based Bitranslucent Set* (AB-BTS), which is an extension of bitranslucent sets (BTS) as defined by O'Neill et al. in [35]. Our new notion permits a more fine-grained degree of access control, where pseudorandom samples and secret keys are associated with attributes $\boldsymbol{x}$, and the testing algorithm can successfully distinguish a pseudorandom sample from a truly random one if and only if the attribute of the sample is accepted under a given secret key's policy f – i.e. when $f(\boldsymbol{x}) = \boldsymbol{0}$. This concept is reminiscent of *attribute-based encryption* (ABE), and in fact, we will show in the sequel how to construct a flexibly bi-deniable ABE from an AB-BTS. This is analogous to the construction of a flexibly bi-deniable PKE from O'Neill et al.'s BTS. We present the formal definition below.

Let $\mathcal{F}$ be some family of functions. An attribute based bitranslucent set (AB-BTS) scheme for $\mathcal{F}$ consists of the following algorithms:

Setup(1^λ): On input the security parameter, the normal setup algorithm outputs a public parameter pp and master secret key msk.

DenSetup(1^λ): On input the security parameter, the deniable setup algorithm outputs a public parameter pp, master secret key msk and faking key fk.

Keygen(msk, f): On input the master secret key msk and a function $f \in \mathcal{F}$, the key generation algorithm outputs a secret key sk_f.

P - and U-samplers SampleP(pp, $\boldsymbol{x}$; r_S) and SampleU(pp, $\boldsymbol{x}$; r_S) output some $\boldsymbol{c}$.

TestP(sk_f, $\boldsymbol{c_x}$): On input a secret key sk_f and a ciphertext $\boldsymbol{c_x}$, the P-tester algorithm outputs 1 (accepts) or 0 (rejects).

FakeSCoins(pp, r_S): On input a public parameters pp and randomness r_S, the sender-faker algorithm outputs randomness r_S^*.

FakeRCoins(pp, fk, $\boldsymbol{c_x}$, f): On input a public parameters pp, the faking key fk, a ciphertext $\boldsymbol{c_x}$ and a function $f \in \mathcal{F}$, the receiver-faker algorithm outputs a faked secret key sk_f'.

Definition 3.2 (AB-BTS). *We say a scheme* $\Pi = $ (Setup, DenSetup, Keygen, SampleP, SampleU, TestP, FakeSCoins, FakeRCoins) *is an AB-BTS scheme for a function family* $\mathcal{F}$ *if it satisfies:*

1. *(Correctness.) The following experiments accept or respectively reject with overwhelming probability over the randomness.*
   - *Let* (pp, msk) $\leftarrow$ Setup(1^λ), $f \in \mathcal{F}$, $\mathsf{sk}_f \leftarrow$ Keygen(msk, f). *If* $f(\boldsymbol{x}) = 0$ *and* $\boldsymbol{c_x} \leftarrow$ SampleP(pp, $\boldsymbol{x}$; r_S), *then* TestP(sk_f, $\boldsymbol{c_x}$) $= 1$; *otherwise,* TestP(sk_f, $\boldsymbol{c_x}$) $= 0$.
   - *Let* (pp, msk) $\leftarrow$ Setup(1^λ), $f \in \mathcal{F}$, $\mathsf{sk}_f \leftarrow$ Keygen(msk, f), $\boldsymbol{c} \leftarrow$ SampleU(pp; r_S). *Then* TestP(sk_f, $\boldsymbol{c}$) $= 0$.
2. *(Indistinguishable public parameters.) The public parameters* pp *generated by the two setup algorithms* (pp, msk) $\leftarrow$ Setup(1^λ) *and* (pp, msk, fk) $\leftarrow$ DenSetup(1^λ) *should be indistinguishable.*
3. *(Selective bi-deniability.) Let* $\mathcal{F}$ *be a family of functions. We define the following two experiments: the real experiment* $\mathbf{Expt}_{\mathcal{A},\mathcal{F}}^{\mathsf{Real}}(1^\lambda)$ *and the faking experiment* $\mathbf{Expt}_{\mathcal{A},\mathcal{F}}^{\mathsf{Fake}}(1^\lambda)$ *regarding an adversary* $\mathcal{A} = (\mathcal{A}_1, \mathcal{A}_2, \mathcal{A}_3)$ *as shown in Fig. 2:*
   where KG(msk, $\boldsymbol{x}^*$, $\cdot$) *returns a secret key* $\mathsf{sk}_f \leftarrow$ Keygen(msk, f) *if* $f \in \mathcal{F}$ *and* $f(\boldsymbol{x}^*) \neq 0$; *it returns* $\bot$ *otherwise. We also require that* $f^* \in \mathcal{F}$.
   We say the scheme is selectively bi-deniable for $\mathcal{F}$, *if for any probabilistic polynomial-time adversaries* $\mathcal{A} = (\mathcal{A}_1, \mathcal{A}_2, \mathcal{A}_3)$, *there is a negligible function* negl(λ) *such that*

$$\mathbf{Adv}_{\mathcal{A}}^{\Pi}(1^\lambda) = |\mathbf{Pr}[\mathbf{Expt}_{\mathcal{A},\mathcal{F}}^{\mathsf{Real}}(1^\lambda) = 1] - \mathbf{Pr}[\mathbf{Expt}_{\mathcal{A},\mathcal{F}}^{\mathsf{Fake}}(1^\lambda) = 1]| \leq \mathsf{negl}(\lambda)$$

Remark 3.3. *Correctness for the faking algorithms is implied by the bi-deniability property. In particular, with overwhelming probability over the overall randomness, the following holds: let* (pp, msk, fk) $\leftarrow$ DenSetup(1^λ), $f \in \mathcal{F}$, $\mathsf{sk}_f \leftarrow$ Keygen(msk, f), $\boldsymbol{x}$ *be a string and* $\boldsymbol{c_x} \leftarrow$ SampleP(pp, x; r_S), *then*

(a) $(f^*, x^*, \text{state}_1) \leftarrow \mathcal{A}_1(\lambda)$
(b) $(\text{pp}, \text{msk}, \text{fk}) \leftarrow \text{DenSetup}(1^\lambda)$
(c) $c \leftarrow \text{SampleU}(\text{pp}, x^*; r_S)$
(d) $\text{state}_2 \leftarrow \mathcal{A}_2^{\text{KG}(\text{msk}, x^*, \cdot)}(\text{pp}, \text{state}_1, c)$
(e) $\text{sk}_{f^*} \leftarrow \text{Keygen}(\text{msk}, f^*)$
(f) $b \leftarrow \mathcal{A}_3^{\text{KG}(\text{msk}, x^*, \cdot)}(\text{sk}_{f^*}, c, \text{state}_2, r_S)$
(g) Output $b \in \{0, 1\}$

(a) $\mathbf{Expt}_{\mathcal{A}}^{\text{Real}}(1^\lambda)$

(a) $(f^*, x^*, \text{state}_1) \leftarrow \mathcal{A}_1(\lambda)$
(b) $(\text{pp}, \text{msk}, \text{fk}) \leftarrow \text{DenSetup}(1^\lambda)$
(c) $c \leftarrow \text{SampleP}(\text{pp}, x^*; r_S)$
(d) $\text{state}_2 \leftarrow \mathcal{A}_2^{\text{KG}(\text{msk}, x^*, \cdot)}(\text{pp}, \text{state}_1, c)$
(e) $r_S' \leftarrow \text{FakeSCoins}(\text{pp}, r_S)$
(f) $\text{sk}_{f^*} \leftarrow \text{FakeRCoins}(\text{pp}, \text{fk}, c, f^*)$
(g) $b \leftarrow \mathcal{A}_3^{\text{KG}(\text{msk}, x^*, \cdot)}(\text{sk}_{f^*}, c, \text{state}_2, r_S')$
(h) Output $b \in \{0, 1\}$

(b) $\mathbf{Expt}_{\mathcal{A}}^{\text{Fake}}(1^\lambda)$

Fig. 2. Security experiments for AB-BTS

- $\text{SampleU}(\text{pp}, x; \text{FakeSCoins}(\text{pp}, r_S)) = c_x,$
- $\text{TestP}(\text{FakeRCoins}(\text{pp}, \text{fk}, c_x, f), c_x) = 0$
- *For any other x', let $c' \leftarrow \text{SampleP}(\text{pp}, x'; r_S')$, then (with overwhelming probability) we have*

$$\text{TestP}(\text{FakeRCoins}(\text{pp}, \text{fk}, c_x, f), c') = \text{TestP}(\text{sk}_f, c').$$

It is not hard to see that if one of these does not hold, then one can easily distinguish the real experiment from the faking experiment.

Remark 3.4. *Canetti et al. [14] gave a simple encoding technique to construct a sender-deniable encryption scheme from a translucent set. O'Neill, Peikert, and Waters [35] used a similar method to construct a flexibly bi-deniable encryption from a bi-translucent set scheme. Here we further observe that the same method as well allows us to construct a flexibly bi-deniable ABE scheme from bi-deniable AB-BTS. We present the construction in Sect. 4.4.*

3.3   Extended LWE and Our New Variant

O'Neill et al. [35] introduced the Extended LWE problem, which allows a "hint" on the error vector x to leak in form of a noisy inner product. They observe a trivial "blurring" argument shows that LWE reduces to eLWE when the hint-noise βq is superpolynomially larger than the magnitude of samples from χ, and also allows for unboundedly many *independent* hint vectors $\langle z, x_i \rangle$ while retaining LWE-hardness.

Definition 3.5 (Extended LWE). *For an integer $q = q(n) \geq 2$, and an error distribution $\chi = \chi(n)$ over $\mathbb{Z}_q$, the extended learning with errors problem $\text{eLWE}_{n,m,q,\chi,\beta}$ is to distinguish between the following pairs of distributions:*

$$\{\mathbf{A}, \mathbf{b} = \mathbf{A}^T \mathbf{s} + \mathbf{e}, \mathbf{z}, \langle \mathbf{z}, \mathbf{b} - \mathbf{e} \rangle + e'\} \text{ and } \{\mathbf{A}, \mathbf{u}, \mathbf{z}, \langle \mathbf{z}, \mathbf{u} - \mathbf{x} \rangle + e'\}$$

where $\mathbf{A} \xleftarrow{\$} \mathbb{Z}_q^{n \times m}$, $\mathbf{s} \xleftarrow{\$} \mathbb{Z}_q^n$, $\mathbf{u} \xleftarrow{\$} \mathbb{Z}_q^m$, $\mathbf{e}, \mathbf{z} \xleftarrow{\$} \chi^m$ and $e' \xleftarrow{\$} \mathcal{D}_{\beta q}$.

Further, Alperin-Sheriff and Peikert [2] show that LWE reduces to eLWE with a polynomial modulus and no hint-noise (i.e. $\beta = 0$), even in the case of a bounded number of *independent* hints.

We introduce the following new form of extended-LWE, called eLWE$^+$, which considers leaking a pair of *correlated hints* on the same noise vector. Our security proof of the AB-BTS construction relies on this new assumption.

Definition 3.6 (Extended LWE Plus). *For integer $q = q(n) \geq 2, m = m(n)$, an error distribution $\chi = \chi(n)$ over $\mathbb{Z}_q$, and a matrix $\mathbf{R} \in \mathbb{Z}_q^{m \times m}$, the extended learning with errors problem $\mathsf{eLWE}^+_{n,m,q,\chi,\beta,\mathbf{R}}$ is to distinguish between the following pairs of distributions:*

$$\{\mathbf{A}, \mathbf{b} = \mathbf{A}^T \mathbf{s} + \mathbf{e}, \mathbf{z}_0, \mathbf{z}_1, \langle \mathbf{z}_0, \mathbf{b} - \mathbf{e} \rangle + e, \langle \mathbf{R}\mathbf{z}_1, \mathbf{b} - \mathbf{e} \rangle + e'\} \text{ and}$$

$$\{\mathbf{A}, \mathbf{u}, \mathbf{z}_0, \mathbf{z}_1, \langle \mathbf{z}_0, \mathbf{u} - \mathbf{e} \rangle + e, \langle \mathbf{R}\mathbf{z}_1, \mathbf{u} - \mathbf{e} \rangle + e'\}$$

where $\mathbf{A} \xleftarrow{\$} \mathbb{Z}_q^{n \times m}$, $\mathbf{s} \xleftarrow{\$} \mathbb{Z}_q^n$, $\mathbf{u} \xleftarrow{\$} \mathbb{Z}_q^m$, $\mathbf{e}, \mathbf{z}_0, \mathbf{z}_1 \xleftarrow{\$} \chi^m$ and $e, e' \xleftarrow{\$} \mathcal{D}_{\beta q}$.

Hardness of Extended-LWE$^+$. A simple observation, following prior work, is that when χ is poly(n)-bounded and the hint noise βq (and thus, modulus q) is superpolynomial in n, then $\mathsf{LWE}_{n,m,q,\chi}$ trivially reduces to $\mathsf{eLWE}^+_{n,m,q,\chi,\beta,\mathbf{R}}$ for every $\mathbf{R} \in \mathbb{Z}_q^{m \times m}$ so that $\mathbf{R}\mathbf{z}_1$ has poly(n)-bounded norm. This is because, for any $r = \omega(\sqrt{\log n}), c \in \mathbb{Z}$, the statistical distance between $\mathcal{D}_{\mathbb{Z},r}$ and $c + \mathcal{D}_{\mathbb{Z},r}$ is at most $O(|c|/r)$.

However, our cryptosystem will require a polynomial-size modulus q. So, we next consider the case of *prime* modulus q of poly(n) size and no noise on the hints (i.e. $\beta = 0$). Following [2][4], it will be convenient to swap to the "knapsack" form of LWE, which is: given $\mathbf{H} \leftarrow \mathbb{Z}_q^{(m-n) \times m}$ and $\mathbf{c} \in \mathbb{Z}_q^{m-n}$, where either $\mathbf{c} = \mathbf{H}\mathbf{e}$ for $\mathbf{e} \leftarrow \chi^m$ or $\mathbf{c}$ uniformly random and independent of $\mathbf{H}$, determine which is the case (with non-negligible advantage). The "extended-plus" form of the knapsack problem also reveals a pair of hints $(\mathbf{z}_0, \mathbf{z}_1, \langle \mathbf{z}_0, \mathbf{e} \rangle, \langle \mathbf{R}\mathbf{z}_1, \mathbf{e} \rangle)$. Note the equivalence between LWE and knapsack-LWE is proven in [33] for $m \geq n + \omega(\log n)$.

Theorem 3.7. *For $m \geq n + \omega(\log n)$, for every prime $q = $ poly(n), for every $\mathbf{R} \in \mathbb{Z}_q^{m \times m}$, and for every $\beta \geq 0$, $\mathbf{Adv}_{\mathcal{B}\mathcal{A}}^{\mathsf{LWE}_{n,m,q,\chi}}(1^\lambda) \geq (1/q^2)\mathbf{Adv}_{\mathcal{A}}^{\mathsf{eLWE}^+_{n,m,q,\chi,\beta,\mathbf{R}}}(1^\lambda)$.*

---

[4] We note that a higher quality reduction from LWE to eLWE is given in [13] in the case of binary secret keys. However for our cryptosystem, it will be more convenient to have secret key coordinates in $\mathbb{Z}_q$, so we extend the reduction of [2] to eLWE$^+$ instead.

Proof. We construct an LWE to eLWE$^+$ reduction $\mathcal{B}$ as follows. $\mathcal{B}$ receives a knapsack-LWE instance $\mathbf{H} \in \mathbb{Z}_q^{(m-n)\times m}, \mathbf{c} \in \mathbb{Z}_q^{m-n}$. It samples $e', z_0, z_1 \leftarrow \chi^m$ and uniform $v_0, v_1 \leftarrow \mathbb{Z}_q^{m-n}$. It chooses any $\mathbf{R} \in \mathbb{Z}_q^{m\times m}$, then sets

$$\mathbf{H}' := \mathbf{H} - v_0 z_0^T - v_1 (\mathbf{R}z_1)^T \in \mathbb{Z}_q^{(m-n)\times m},$$
$$\mathbf{c}' := \mathbf{c} - v_0 \cdot \langle z_0, e' \rangle - v_1 \cdot \langle \mathbf{R}z_1, e' \rangle \in \mathbb{Z}_q^{m-n}.$$

It sends $(\mathbf{H}', \mathbf{c}', z_0, z_1, \langle z_0, e' \rangle, \langle \mathbf{R}z_1, e' \rangle)$ to the knapsack-eLWE$^+$ adversary $\mathcal{A}$, and outputs what $\mathcal{A}$ outputs.

Notice that when $\mathbf{H}, \mathbf{c}$ are independent and uniform, so are $\mathbf{H}', \mathbf{c}'$, in which case $\mathcal{B}$'s simulation is perfect.

Now, consider the case when $\mathbf{H}, \mathbf{c}$ are drawn from the knapsack-LWE distribution, with $\mathbf{c} = \mathbf{H}x$ for $e \leftarrow \chi^m$. In this case, $\mathbf{H}'$ is uniformly random over the choice of $\mathbf{H}$, and we have

$$c' = \mathbf{H}x - v_0 \cdot \langle z_0, e' \rangle - v_1 \cdot \langle \mathbf{R}z_1, e' \rangle$$
$$= \left(\mathbf{H}' + v_0 z_0^T + v_1 (\mathbf{R}z_1)^T \right) e - v_0 \cdot \langle z_0, e' \rangle - v_1 \cdot \langle \mathbf{R}z_1, e' \rangle$$
$$= \mathbf{H}'e + v_0 \cdot \langle z_0, e - e' \rangle + v_1 \cdot \langle \mathbf{R}z_1, e - e' \rangle.$$

Define the event $E = [E_0 \wedge E_1]$ as

$$E_0 \stackrel{\text{def}}{=} [\langle z_0, e \rangle = \langle z_0, e' \rangle],$$
$$E_1 \stackrel{\text{def}}{=} [\langle \mathbf{R}z_1, e \rangle = \langle \mathbf{R}z_1, e' \rangle].$$

If event E occurs, then the reduction $\mathcal{B}$ perfectly simulates a pseudorandom instance of knapsack-eLWE$^+$ to $\mathcal{A}$, as then $v_0 \cdot \langle z_0, e - e' \rangle + v_1 \cdot \langle \mathbf{R}z_1, e - e' \rangle$ vanishes, leaving $c' = \mathbf{H}'e$ for $\mathbf{H}' \leftarrow \mathbb{Z}_q^{(m-n)\times m}$ and $e \leftarrow \chi^m$ as required. Otherwise since q is prime, the reduction $\mathcal{B}$ (incorrectly) simulates an independent and uniform instance of knapsack-eLWE$^+$ to $\mathcal{A}$, as then either one of $v_0 \cdot \langle z_0, e - e' \rangle$ or $v_1 \cdot \langle \mathbf{R}z_1, e - e' \rangle$ does not vanish, implying that c' is uniform in $\mathbb{Z}_q^{m-n}$ over the choice of v_0 (resp. v_1) alone, independent of the choices of $\mathbf{H}'$ and x.

It remains to analyze the probability that event E occurs. Because e and e' are i.i.d., we may define the random variable $\mathcal{Z}_0$ that takes values $\langle z_0, e^* \rangle \in \mathbb{Z}_q$ and the random variable $\mathcal{Z}_1$ that takes values $\langle \mathbf{R}z_1, e^* \rangle \in \mathbb{Z}_q$ jointly over choice of $e^* \leftarrow \chi^m$, and analyze their collision probabilities independently. Since the collision probability of *any* random variable $\mathcal{Z}$ is at least $1/|\mathsf{Supp}(\mathcal{Z})|$, we have that $\Pr[E] \geq \min CP[\mathcal{Z}_0] \cdot \min CP[\mathcal{Z}_1] = 1/q^2 = 1/\mathsf{poly}(n)$, and the theorem follows. $\qquad\square$

4   Flexibly Bi-Deniable Attribute-Based Encryption (ABE) for Branching Programs

In this section, we present our flexibly bi-deniable ABE for bounded-length Branching Program. We organize our approach into the following three steps:

(1) first, we recall the encoding scheme proposed in the SIM-secure ABE-BP of [30]; (2) Then, we present our flexibly bi-deniable attribute bi-translucent set (AB-BTS) scheme, as was defined in Definition 3.2. Our AB-BTS construction uses the ideas of Gorbunov and Vinayagamurthy [30], with essential modifications that allow us to tightly upper and lower bound evaluated noise terms. As discussed in the Introduction, this tighter analysis plays a key role in proving bi-deniability. (3) Finally, we show how to obtain the desired bi-deniable ABE scheme from our AB-BTS. As pointed out by Canetti et al. [14] and O'Neill et al. [35], a bitranslucent set scheme implies flexibly bi-deniable PKE. We observe that the same idea generalizes to the case of an AB-BTS scheme and flexibly bi-deniable ABE in a straightforward manner.

4.1   Encoding Schemes for Branching Programs

Basic Homomorphic Encoding. Before proceeding to the public key evaluation algorithm, we first described basic homomorphic addition and multiplication over public keys and encoded ciphertexts based on the techniques in [3,9,25].

Definition 4.1 (LWE Encoding). *For any matrix* $\mathbf{A} \leftarrow \mathbb{Z}_q^{n \times m}$, *we define an LWE encoding of a bit* $a \in \{0,1\}$ *with respect to a public key* $\mathbf{A}$ *and randomness* $s \leftarrow \mathbb{Z}_q^n$ *as*

$$\psi_{\mathbf{A},s,a} = s^T(\mathbf{A} + a \cdot \mathbf{G}) + e \in \mathbb{Z}_q^m$$

for error vector $e \leftarrow \chi^m$ *and the gadget matrix* $\mathbf{G} \in \mathbb{Z}_q^{n \times m}$.

In our construction, all LWE encodings will be encoded using the same LWE secret s, thus for simplicity, we will simply refer to such an encoding as $\psi_{\mathbf{A},a}$.

For homomorphic addition, the addition algorithm takes as input two encodings $\psi_{\mathbf{A},a}, \psi_{\mathbf{A}',a'}$, and outputs the sum of them. Let $\mathbf{A}^+ = \mathbf{A} + \mathbf{A}'$ and $a^+ = a + a'$

$$\mathsf{Add}(\psi_{\mathbf{A},a}, \psi_{\mathbf{A}',a'}) = \psi_{\mathbf{A},a} + \psi_{\mathbf{A}',a'} = \psi_{\mathbf{A}^+,a^+}$$

For homomorphic multiplication, the multiplication algorithm takes as input two encodings $\psi_{\mathbf{A},a}, \psi_{\mathbf{A}',a}$, and outputs an encoding $\psi_{\mathbf{A}^\times,a^\times}$, where $\mathbf{A}^\times = -\mathbf{A}\mathbf{G}^{-1}(\mathbf{A}')$ and $a^\times = aa'$.

$$\mathsf{Mult}(\psi_{\mathbf{A},a}, \psi_{\mathbf{A}',a'}) = -\psi \cdot \mathbf{G}^{-1}(\mathbf{A}') + a \cdot \psi' = \psi_{\mathbf{A}^\times,a^\times}$$

Public Key Evaluation Algorithm. Following the notation in [30], we define a public evaluation algorithm $\mathsf{Eval}_{\mathsf{pk}}$. The algorithm takes as input a description of the branching program BP, a collection of public keys $\{\mathbf{A}_i\}_{i \in [\ell]}$ (one for each attribute bit x_i), a collection of public keys $\mathbf{V}_{0,i}$ for initial state vector and an auxiliary matrix $\mathbf{A}^c$, and outputs an evaluated public key corresponding to the branching program BP.

$$\mathbf{V}_{\mathsf{BP}} \leftarrow \mathsf{Eval}_{\mathsf{pk}}(\mathsf{BP}, \{\mathbf{A}_i\}_{i \in [\ell]}, \{\mathbf{V}_{0,i}\}_{i \in [5]}, \mathbf{A}^c)$$

where the auxiliary matrix $\mathbf{A}^c$ are used to encoded constant 1 for each input wire. We also define matrix $\mathbf{A}'_i = \mathbf{A}^c - \mathbf{A}_i$ as a public key used to encode

$1 - x_i$. By the definition of branching programs, the output $\mathbf{V}_{\mathsf{BP}} \in \mathbb{Z}_q^{n \times m}$ is the homomorphically generated public key $\mathbf{V}_{L,1}$ at position 1 of the state vector for the L-th step of the branching program evaluation.

Recall that in the definition of branching programs, BP is represented by the tuple $\{\mathsf{var}(t), \{\gamma_{t,i,0}, \gamma_{t,i,1}\}_{i \in [5]}\}$ for $t \in [L]$, and the initial state vector is set to be $\mathbf{v}_0 = (1, 0, 0, 0, 0)$. Further, for $t \in [L]$, the computation is performed as $\mathbf{v}_t[i] = \mathbf{v}_{t-1}[\gamma_{t,i,0}](1 - x_{\mathsf{var}(t)}) + \mathbf{v}_{t-1}[\gamma_{t,i,1}] \cdot x_{\mathsf{var}(t)}$. It is important for the security proof (among other reasons) that the evaluated state vector in each step is independent of the attribute vector.

Encoding Evaluation Algorithm. We define an encoding evaluation algorithm $\mathsf{Eval}_{\mathsf{ct}}$ that takes as input the description of a branching program BP, an attribute vector $\mathbf{x}$, a set of encodings for the attribute $\{\mathbf{A}_i, \psi_i := \psi_{\mathbf{A}_i, x_i}\}_{i \in [\ell]}$, encodings of the initial state vector $\{\mathbf{V}_{0,i}, \psi_{0,i} := \psi_{\mathbf{V}_{0,i}\mathbf{v}_0[i]}\}_{i \in [5]}$ and an encoding of a constant 1, i.e. $\psi^c := \psi_{\mathbf{A}^c, 1}$. The algorithm $\mathsf{Eval}_{\mathsf{ct}}$ outputs an encoding of the result $y := \mathsf{BP}(\mathbf{x})$ with respect to the homomorphically derived public key $\mathbf{V}_{\mathsf{BP}} := \mathbf{V}_{L,1}$

$$\psi_{\mathsf{BP}} \leftarrow \mathsf{Eval}_{\mathsf{ct}}(\mathsf{BP}, \mathbf{x}, \{\mathbf{A}_i, \psi_i\}_{i \in [\ell]}, \{\mathbf{V}_{0,i}, \psi_{0,i}\}_{i \in [5]}, \{\mathbf{A}^c, \psi^c\})$$

As mentioned above, in branching program computation, for $t \in [L]$, we have for all $i \in [5]$

$$\mathbf{v}_t[i] = \mathbf{v}_{t-1}[\gamma_{t,i,0}](1 - x_{\mathsf{var}(t)}) + \mathbf{v}_{t-1}[\gamma_{t,i,1}] \cdot x_{\mathsf{var}(t)}$$

The evaluation algorithm proceeds inductively to update the encoding of the state vector for each step of the branching program. Next, we need to instantiate this inductive computation using the homomorphic operations described above, i.e. Add, Mult. Following the notation used in [30], we define $\psi'_i := \psi_{\mathbf{A}'_i, (1-x_i)} = \mathbf{s}^T(\mathbf{A}'_i + (1-x_i)\mathbf{G}) + \mathbf{e}'_i$, where $\mathbf{A}'_i = \mathbf{A}^c - \mathbf{A}_i$, to denote the encoding of $1 - x_i$. This encoding can be computed using $\mathsf{Add}(\psi_{\mathbf{A}^c, 1}, -\psi_{\mathbf{A}_i, x_i})$. Then assuming at time $t - 1 \in [L]$ we hold encodings of the state vector $\{\psi_{\mathbf{V}_{t-1,i}, \mathbf{v}_{t-1}[i]}\}_{i \in [5]}$. For $i \in [5]$, we compute the encodings of new state values as

$$\psi_{i,t} = \mathsf{Add}(\mathsf{Mult}(\psi'_{\mathsf{var}(t)}, \psi_{t-1, \gamma_0}), \mathsf{Mult}(\psi_{\mathsf{var}(t)}, \psi_{t-1, \gamma_1}))$$

where $\gamma_0 := \gamma_{t,i,0}$ and $\gamma_1 := \gamma_{t,i,1}$. We omit the correctness proof of the encoding here, which is presented in [30].

The above algorithms suffice for us to describe our construction. To analyze the scheme, we need to encode a simulated public key and evaluate the simulated key. Due to lack of space, we present details on these simulated encodings and evaluations, plus some useful lemmas to bound the norm of errors in the simulation, in full version. As mentioned in the Introduction, these new bounds play a critical, technical role in our formal proof of flexible bi-deniability.

4.2 Construction of Flexibly Bi-Deniable ABE for Branching Programs

In this part, we present our flexibly bi-deniable AB-BTS scheme for bounded-length Branching Programs. We use a semantically-secure public key encryption $\Pi = (\mathsf{Gen}', \mathsf{Enc}', \mathsf{Dec}')$ with message space $\mathcal{M}_\Pi = \mathbb{Z}_q^{m \times m}$ and ciphertext space $\mathcal{C}_\Pi$. For a family of branching programs of length bounded by L and input space $\{0,1\}^\ell$, the description of $\mathsf{BiDenAB\text{-}BTS} = (\mathsf{Setup}, \mathsf{DenSetup}, \mathsf{Keygen}, \mathsf{SampleP}, \mathsf{SampleU}, \mathsf{TestP}, \mathsf{FakeRCoins}, \mathsf{FakeSCoins})$ are as follows:

- $\mathsf{Setup}(1^\lambda, 1^L, 1^\ell)$: On input the security parameter λ, the length of the branching program L and length of the attribute vector ℓ,
    1. Set the LWE dimension be $n = n(\lambda)$, modulus $q = q(n, L)$. Choose Gaussian distribution parameter $s = s(n)$. Let $\mathsf{params} = (n, q, m, s)$.
    2. Sample one random matrix associated with its trapdoor as

$$(\mathbf{A}, \mathbf{T_A}) \leftarrow \mathsf{TrapGen}(q, n, m)$$

    3. Choose $\ell + 6$ random matrices $\{\mathbf{A}_i\}_{i \in [\ell]}, \{\mathbf{V}_{0,i}\}_{i \in [5]}, \mathbf{A}^c$ from $\mathbb{Z}_q^{n \times m}$.
    4. Choose a random vector $\boldsymbol{u} \in \mathbb{Z}_q^n$.
    5. Compute a public/secret key pair $(\mathsf{pk}', \mathsf{sk}')$ for a semantically secure public key encryption $(\mathsf{pk}', \mathsf{sk}') \leftarrow \mathsf{Gen}'(1^\lambda)$
    6. Output the public parameter pp and master secret key msk as

$$\mathsf{pp} = (\mathsf{params}, \mathbf{A}, \{\mathbf{A}_i\}_{i \in [\ell]}, \{\mathbf{V}_{0,i}\}_{i \in [5]}, \mathbf{A}^c, \boldsymbol{u}, \mathsf{pk}'), \quad \mathsf{msk} = (\mathbf{T_A}, \mathsf{sk}')$$

- $\mathsf{DenSetup}(1^\lambda, 1^L, 1^\ell)$: On input the security parameter λ, the length of branching program L and length of attribute vector ℓ, the deniable setup algorithm runs the same computation as setup algorithm, and outputs

$$\mathsf{pp} = (\mathsf{params}, \mathbf{A}, \{\mathbf{A}_i\}_{i \in [\ell]}, \{\mathbf{V}_{0,i}\}_{i \in [5]}, \mathbf{A}^c, \boldsymbol{u}, \mathsf{pk}'), \quad \mathsf{msk} = (\mathbf{T_A}, \mathsf{sk}') \quad \mathsf{fk} = (\mathbf{T_A}, \mathsf{sk}')$$

- $\mathsf{Keygen}(\mathsf{msk}, \mathsf{BP})$: On input the master secret key msk and the description of a branching program BP, $\mathsf{BP} = (v_0, \{\mathsf{var}(t), \{\gamma_{t,i,0}, \gamma_{t,i,1}\}_{i \in [5]}\}_{t \in [L]})$.
    1. Homomorphically compute a public matrix with respect to the branching program BP: $\mathbf{V}_{\mathsf{BP}} \leftarrow \mathsf{Eval}_{\mathsf{pk}}(\mathsf{BP}, \{\mathbf{A}_i\}_{i \in [\ell]}, \{\mathbf{V}_{0,i}\}_{i \in [5]}, \mathbf{A}^c)$.
    2. Sample a low norm vector $\boldsymbol{r}_{\mathsf{BP}} \in \mathbb{Z}_q^{2m}$, using

$$\boldsymbol{r}_{\mathsf{BP}} \leftarrow \mathsf{SampleLeft}(\mathbf{A}, \mathbf{T_A}, (\mathbf{V}_{\mathsf{BP}} + \mathbf{G}), \boldsymbol{u}, sq)$$

such that $\boldsymbol{r}_{\mathsf{BP}}^T \cdot [\mathbf{A} | \mathbf{V}_{\mathsf{BP}} + \mathbf{G}] = \boldsymbol{u}$.
    3. Output the secret key $\mathsf{sk}_{\mathsf{BP}}$ for branching program as $\mathsf{sk}_{\mathsf{BP}} = (\boldsymbol{r}_{\mathsf{BP}}, \mathsf{BP})$.
- $\mathsf{SampleP}(\mathsf{pp}, \boldsymbol{x})$: On input public parameters pp and attribute $\boldsymbol{x}$,
    1. Choose an LWE secret $\boldsymbol{s} \in \mathbb{Z}_q^n$ uniformly at random.
    2. Choose noise vector $\boldsymbol{e} \leftarrow \mathcal{D}_{\mathbb{Z}_q^m, \alpha}$, and compute $\psi_0 = \boldsymbol{s}^T \mathbf{A} + \boldsymbol{e}$.
    3. Choose one random matrices $\mathbf{R}^c \leftarrow \{-1, 1\}^{m \times m}$, and let $\boldsymbol{e}^c = \boldsymbol{e}^T \mathbf{R}^c$. Compute an encoding of constant 1: $\psi^c = \boldsymbol{s}^T(\mathbf{A}^c + \mathbf{G}) + \boldsymbol{e}^c$.
    4. Encode each bit $i \in [\ell]$ of the attribute vector:

(a) Choose a random matrix $\mathbf{R}_i \leftarrow \{-1,1\}^{m \times m}$, and let $e_i = e^T \mathbf{R}_i$.

(b) Compute $\psi_i = s^T(\mathbf{A}_i + x_i \mathbf{G}) + e_i$.

5. Encode the initial state vector $v_0 = (1,0,0,0,0)$, for $i \in [5]$

   (a) Choose a random matrix $\mathbf{R}'_{0,i} \leftarrow \{-1,1\}^{m \times m}$, and let $\mathbf{R}_{0,i} = \eta \mathbf{R}'_{0,i}, e_{0,i} = e^T \mathbf{R}_{0,i}$, where the noise scaling parameter η is set in Sect. 4.3.

   (b) Compute $\psi_{0,i} = s^T(\mathbf{V}_{0,i} + v_0[i]\mathbf{G}) + e_{0,i}$.

6. Compute $c = s^T u + e$, where $e \leftarrow \mathcal{D}_{\mathbb{Z}_q,s}$

7. Use PKE to encrypt randomly chosen matrices $\mathbf{R}^c, \{\mathbf{R}_i\}_{i \in [\ell]}$ and $\{\mathbf{R}_{0,i}\}_{i \in [5]}$:

$$\mathbf{T}_i \leftarrow \mathsf{Enc}'(\mathsf{pk}', \mathbf{R}_i), \mathbf{T}^c \leftarrow \mathsf{Enc}'(\mathsf{pk}', \mathbf{R}^c), \mathbf{T}_{0,i} \leftarrow \mathsf{Enc}'(\mathsf{pk}', \mathbf{R}_{0,i})$$

8. Output the ciphertext

$$\mathsf{ct}_x = (x, \psi_0, \{\psi_i\}_{i \in [\ell]}, \psi^c, \{\psi_{0,i}\}_{i \in [5]}, c, \{\mathbf{T}_i\}_{i \in [\ell]}, \mathbf{T}^c, \{\mathbf{T}_{0,i}\}_{i \in [5]})$$

- $\mathsf{SampleU}(\mathsf{pp}, x)$: Output a uniformly random vector $\mathsf{ct} \in \mathbb{Z}_q^m \times \mathbb{Z}_q^{\ell m} \times \mathbb{Z}_q^{\ell m} \times \mathbb{Z}_q^{5m} \times \mathbb{Z}_q \times \mathcal{C}_\Pi^\ell \times \mathcal{C}_\Pi \times \mathcal{C}_\Pi^5$.
- $\mathsf{TestP}(\mathsf{sk_{BP}}, \mathsf{ct}_x)$: On input the secret key $\mathsf{sk_{BP}}$ for a branching program BP and a ciphertext associated with attribute x, if $\mathsf{BP}(x) = 0$, output $\perp$, otherwise,
  1. Homomorphically compute the evaluated ciphertext of result $\mathsf{BP}(x)$

  $$\psi_{\mathsf{BP}} \leftarrow \mathsf{Eval}_{\mathsf{ct}}(\mathsf{BP}, x, \{\mathbf{A}_i, \psi_i\}_{i \in [\ell]}, \{\mathbf{V}_{0,i}, \psi_{0,i}\}_{i \in [5]}, \{\mathbf{A}_i^c, \psi_i^c\}_{i \in [\ell]})$$

  2. Then compute $\phi = [\psi_0 | \psi_{\mathsf{BP}}]^T \cdot r_{\mathsf{BP}}$. Accept ct_x as a P-sample if $|c - \phi| < 1/4$, otherwise reject.
- $\mathsf{FakeSCoins}(r_S)$: Simply output the P-sample c as the randomness r_S^* that would cause $\mathsf{SampleU}$ to output c_x.
- $\mathsf{FakeRCoins}(\mathsf{pp}, \mathsf{fk}, \mathsf{ct}_x, \mathsf{BP})$: On input the public parameters pp, the faking key fk, a ciphertext ct_x and description of a branching program BP
  1. If $\mathsf{BP}(x) \neq 0$, then output $\mathsf{sk}_f \leftarrow \mathsf{Keygen}(\mathsf{fk}, \mathsf{BP})$.
  2. Otherwise, parse ciphertext ct_x as

  $$\mathsf{ct}_x = (x, \psi_0, \{\psi_i\}_{i \in [\ell]}, \psi^c, \{\psi_{0,i}\}_{i \in [5]}, c, \{\mathbf{T}_i\}_{i \in [\ell]}, \mathbf{T}^c, \{\mathbf{T}_{0,i}\}_{i \in [5]})$$

  Compute $e \leftarrow \mathsf{Invert}(\mathbf{A}, \mathbf{T_A}, \psi_0)$. Then decrypt $(\{\mathbf{T}_i\}_{i \in [\ell]}, \mathbf{T}^c, \{\mathbf{T}_{0,i}\}_{i \in [5]})$ respectively using $\mathsf{Dec}(\mathsf{sk}', \cdot)$ to obtain $\{\mathbf{R}_i\}_{i \in [\ell]}, \mathbf{R}^c, \{\mathbf{R}_{0,i}\}_{i \in [5]}$. Compute evaluated error

  $$e_{\mathsf{BP}} \leftarrow \mathsf{Eval}_{\mathsf{ct}}(\mathsf{BP}, x, \{\mathbf{A}_i, e^T \mathbf{R}_i\}_{i \in [\ell]}, \{\mathbf{V}_{0,i}, e^T \mathbf{R}_{0,i}\}_{i \in [5]}, \{\mathbf{A}^c, e^T \mathbf{R}^c\})$$

  such that $e_{\mathsf{BP}} = e^T \mathbf{R}_{\mathsf{BP}}$.
  3. Homomorphically compute a public matrix with respect to the branching program BP: $\mathbf{V}_{\mathsf{BP}} \leftarrow \mathsf{Eval}_{\mathsf{pk}}(\mathsf{BP}, \{\mathbf{A}_i\}_{i \in [\ell]}, \{\mathbf{V}_{0,i}\}_{i \in [5]}, \{\mathbf{A}_i^c\}_{i \in [\ell]})$. Then sample a properly distributed secret key $r_{\mathsf{BP}} \in \mathbb{Z}_q^{2m}$, using

  $$r_{\mathsf{BP}} \leftarrow \mathsf{SampleLeft}(\mathbf{A}, \mathbf{T_A}, (\mathbf{V}_{\mathsf{BP}} + \mathbf{G}), u, s)$$

4. Sample correlation vector $\mathbf{y}_0 \leftarrow \mathcal{D}_{\mathbb{Z}_q^m, \beta^2 q^2 \mathbf{I}_{m \times m}}$. Then sample correlation coefficient $\mu \leftarrow \mathcal{D}_\gamma$, and set vector $\mathbf{y}_1 = (\mu \mathbf{e}_{\mathsf{BP}} + \mathcal{D}_{\mathbf{Z}^m, \mathbf{Q}})q$, where

$$\mathbf{Q} = \beta^2 \mathbf{I}_{m \times m} - \gamma^2 \alpha^2 \mathbf{R}_{\mathsf{BP}}^T \mathbf{R}_{\mathsf{BP}} \tag{1}$$

5. Let $\mathbf{y} = (\mathbf{y}_0 | \mathbf{y}_1)$, then sample and output the faked secret key $\mathsf{sk}_{\mathsf{BP}}^* = \mathbf{r}_{\mathsf{BP}}^*$ as $\mathbf{r}_{\mathsf{BP}}^* \leftarrow \mathbf{y} + \mathcal{D}_{\Lambda + \mathbf{r}_{\mathsf{BP}} - \mathbf{y}, \sqrt{s^2 - \beta^2}}$, using $\mathsf{SampleD}(\mathsf{ExtBasis}(\mathbf{A}, \mathbf{T_A}, \mathbf{V}_{\mathsf{BP}} + \mathbf{G}), \mathbf{r}_{\mathsf{BP}} - \mathbf{y}, \sqrt{s^2 - \beta^2})$, where $\Lambda = \Lambda^\perp([\mathbf{A} | \mathbf{V}_{\mathsf{BP}} + \mathbf{G}])$.

The SampleP algorithm is similar to the ABE ciphertexts in the work [30], except that we add another scaling factor η to the rotation matrices $\mathbf{R}_{0,i}$'s. This allows us to both upper and lower bound the noise growth, which is essential to achieve bi-deniability. Detailed analysis can be found in full version. As we discussed in the introduction, the FakeRCoins embeds the evaluated noise into the secret key, so that it will change the decrypted value of the targeted ciphertext, but not others. Next we present the theorem we achieve and a high level ideas of the proof. We defer the formal analysis after the proof intuition.

Theorem 4.2. *Assuming the hardness of extended-LWE$_{q,\beta'}$, the above algorithms form a secure attribute-based bitranslucent set schemem, as in Definition 3.2.*

Overview of Our Security Proof. At a high level, our security proof begins at the Fake experiment (cf. Definition 3.1 for a formal description), where first a ciphertext ct^* and its associated noise terms $\mathbf{e}^*$ are sampled, then a fake key $\mathbf{r}^*$ is generated that "artificially" fails to decrypt any ciphertext with noise vector (oriented close to) $\mathbf{e}^*$. In the end, we will arrive at the Real experiment, where an honest key $\mathbf{r}$ is generated that "genuinely" fails to decrypt the honestly generated, coerced ciphertext ct^*. (Multi-ct coercion security follows by a standard hybrid argument that repeatedly modifies respective $\mathbf{r}^*$ to $\mathbf{r}$ for each coerced ct^* in order.) In order to transition from Fake to Real, we move through a sequence of computationally- or statistically-indistinguishable hybrid experiments.

The first set of intermediate experiments (represented by H_1 and H_2 in our formal proof) embeds the attribute $\boldsymbol{x}$ of the challenge ciphertext ct^* in the public parameters, in a similar fashion to the beginning of every SIM-secure proof of lattice-based ABE. Indistinguishability follows via the Leftover Hash Lemma [21]. (Note that the additional hybrid in our proof is used to ensure that the random rotation matrices $\mathbf{R}$ employed by the LHL for public key embedding of $\boldsymbol{x}$ are the *exact same* matrices $\mathbf{R}$ as used to generate the noise terms of the coerced ct^*, and uses the security of any semantically-secure PKE for computational indistinguishability.)

The next set of intermediate experiments (given by $\mathsf{H}_3, \mathsf{H}_4$, and H_5 in our formal proof) perform the "main, new work" of our security proof. Specifically, they "swap the order" of the generation of the pk matrices $\{\mathbf{A}\}$, the public coset $\boldsymbol{u}$ (in the public parameters and in the coerced ciphertext), and the error vector(s) $\mathbf{e}$ in the coerced ciphertext components. (An additional hybrid is used to toggle the order of a "correlation vector" y – a random, planted vector used to allow for a more modular analysis of these steps.) In each case, we give a *statistical* argument that the adversary's view in adjacent hybrids is indistinguishable or identical, using elementary properties of multi-dimensional Gaussians.

In the next step (given by H_6), we apply the eLWE$^+$ assumption to (roughly) change every component of the coerced ciphertext ct* to uniform – except for the final c^* component used to blind the message μ.

In the final step (given by H_7), we transition to the Real experiment by changing the c^* component to uniform (in the presence of Dual Regev decryption under honest z), using our sharper noise analysis as described above to show statistical indistinguishability of the final decryption output of z on ct*.

Lemma 4.3. *For parameters set in Sect. 4.3, the AB-BTS defined above satisfies the correctness property in Definition 3.2.*

Proof. As we mentioned in Remark 3.3, the correctness of faking algorithms is implied by the bi-deniability property. Therefore, we only need to prove the correctness of normal decryption algorithm. For branching program BP and input x, such that $\mathsf{BP}(x) = 1$, we compute $\psi_{t,i}$ for $t \in [\ell]$ as

$$\psi_{t,i} = \mathsf{Add}(\mathsf{Mult}(\psi'_{\mathsf{var}(t)}, \psi_{t-1,\gamma_0}), \mathsf{Mult}(\psi_{\mathsf{var}(t)}, \psi_{t-1,\gamma_1}))$$

$$= \mathsf{Add}\Big([s^T(-\mathbf{A}'_{\mathsf{var}(t)}\mathbf{G}^{-1}(\mathbf{V}_{t-1,\gamma_0}) + (v_t[\gamma_0] \cdot (1 - x_{\mathsf{var}(t)})) \cdot \mathbf{G}) + e_1],$$

$$([s^T(-\mathbf{A}'_{\mathsf{var}(t)}\mathbf{G}^{-1}(\mathbf{V}_{t-1,\gamma_1}) + (v_t[\gamma_1] \cdot x_{\mathsf{var}(t)}) \cdot \mathbf{G}) + e_2]\Big)$$

$$= s^T\Big[\underbrace{(-\mathbf{A}'_{\mathsf{var}(t)}\mathbf{G}^{-1}(\mathbf{V}_{t-1,\gamma_0}) - \mathbf{A}'_{\mathsf{var}(t)}\mathbf{G}^{-1}(\mathbf{V}_{t-1,\gamma_1}))}_{\mathbf{V}_{t,i}}$$

$$+ \underbrace{(v_t[\gamma_0] \cdot (1 - x_{\mathsf{var}(t)}) + v_t[\gamma_1] \cdot x_{\mathsf{var}(t)})}_{v_t[i]} \cdot \mathbf{G}\Big] + e_{t,i}$$

At the end of the ciphertext evaluation, since $\mathsf{BP}(x) = 1$, we can obtain $\psi_{\mathsf{BP}} = s^T(\mathbf{V}_{\mathsf{BP}} + \mathbf{G}) + e_{\mathsf{BP}}$, where $e_{\mathsf{BP}} = e^T\mathbf{R}_{\mathsf{BP}}$. Recall that the secret key $\mathsf{sk} = r_{\mathsf{BP}}$ satisfying $[\mathbf{A}|\mathbf{V}_{\mathsf{BP}} + \mathbf{G}] \cdot r_{\mathsf{BP}} = u$. Then for $c - [\psi_0|\psi_{\mathsf{BP}}] \cdot r_{\mathsf{BP}}$, it holds that

$$c - [\psi_0|\psi_{\mathsf{BP}}]^T \cdot r_{\mathsf{BP}} = e - e^T\mathbf{R}_{\mathsf{BP}} \cdot r_{\mathsf{BP}}$$

Now we need to compute a bound for the final noise term. By applying analysis of norm in full version, we obtain that

$$||e^T|| \cdot ||\mathbf{R}_{\mathsf{BP}}|| + 2m^{1.5}\ell||e|| \le (2m^{1.5}\ell + \eta\sqrt{m})||e|| \le \alpha\sqrt{m}(2m^{1.5}\ell + \eta\sqrt{m}) \cdot sq\sqrt{m} \le \frac{1}{4}$$

So by setting the parameters appropriately, as in Sect. 4.3, we have that

$$|c - [\psi_0|\psi_{\mathsf{BP}}]^T \cdot r_{\mathsf{BP}}| \le 1/4$$

and the lemma follows. $\square$

Lemma 4.4. *Assuming the hardness of extended-*$\mathsf{LWE}_{q,\beta'}$*, the AB-BTS scheme described above is bi-deniable as defined in Definition 3.2.*

Proof. First, we notice that because $\mathsf{SampleU}$ simply outputs its random coins as a uniformly random ct, we can use ct itself as the coins.

We prove the bi-deniability property by a sequence of hybrids H_i with details as follows:

Hybrid H_0: Hybrid H_0 is the same as the view of adversary $\mathcal{A}$ in the right-hand faking experiment in the definition of bi-deniability. We use the fact that algorithm Invert successfully recovers e from ct with overwhelming probability over all randomness in the experiment.

Hybrid H_1: In hybrid H_2, we switch the encryptions of matrices $(\{\mathbf{R}_i\}_{i\in[\ell]}, \{\mathbf{R}_{0,i}\}_{i\in[5]}, \mathbf{R}^c)$ in the ciphertext to encryptions of zero. Recall that in hybrid H_0, we encrypt the randomness matrices $(\{\mathbf{R}_i\}_{i\in[\ell]}, \{\mathbf{R}_{0,i}\}_{i\in[5]}, \mathbf{R}^c)$ using semantically secure PKE Π, i.e.

$$\mathbf{T}_i \leftarrow \mathsf{Enc}'(\mathsf{pk}', \mathbf{R}_i), \quad \mathbf{T}^c \leftarrow \mathsf{Enc}'(\mathsf{pk}', \mathbf{R}^c), \quad \mathbf{T}_{0,i} \leftarrow \mathsf{Enc}'(\mathsf{pk}', \mathbf{R}_{0,i})$$

In hybrid H_1, we just set

$$\mathbf{T}_i \leftarrow \mathsf{Enc}'(\mathsf{pk}', \mathbf{0}), \quad \mathbf{T}^c \leftarrow \mathsf{Enc}'(\mathsf{pk}', \mathbf{0}), \quad \mathbf{T}_{0,i} \leftarrow \mathsf{Enc}'(\mathsf{pk}', \mathbf{0})$$

to be encryptions of $\mathbf{0} \in \mathbb{Z}^{m\times m}$ to replace encryptions of matrices $(\{\mathbf{R}_i\}_{i\in[\ell]}, \{\mathbf{R}_{0,i}\}_{i\in[5]}, \mathbf{R}^c)$.

Hybrid H_2: In hybrid H_2, we embed random matrices $(\{\mathbf{R}_i\}_{i\in[\ell]}, \{\mathbf{R}_{0,i}\}_{i\in[5]}, \mathbf{R}^c)$ and challenge attribute x^* in the public parameters pp. Recall that in hybrid H_1 the matrices $(\{\mathbf{A}_i\}_{i\in[\ell]}, \{\mathbf{V}_{0,i}\}_{i\in[5]}, \mathbf{A}^c)$ are sampled at random. In hybrid H_2, we slightly change how these matrices are generated. Let $x^* = (x_1^*, ..., x_\ell^*)$ be the challenge attribute that the adversary $\mathcal{A}$ intends to attack. We sample matrices $(\{\mathbf{R}_i\}_{i\in[\ell]}, \{\mathbf{R}'_{0,i}\}_{i\in[5]}, \mathbf{R}^c)$ uniformly random from $\{-1,1\}^{m\times m}$ and set $\mathbf{R}_{0,i} = \eta\mathbf{R}'_{0,i}$, which would be used both in the generation of public parameters and challenge ciphertext. We set $(\{\mathbf{A}_i\}_{i\in[\ell]}, \{\mathbf{V}_{0,i}\}_{i\in[5]}, \mathbf{A}^c)$ respectively as

$$\mathbf{A}_i = \mathbf{A}\mathbf{R}_i - x_i^*\mathbf{G}, \quad \mathbf{V}_{0,i} = \mathbf{A}\mathbf{R}_{0,i} - v_0[i]\mathbf{G}, \quad \mathbf{A}^c = \mathbf{A}\mathbf{R}^c - \mathbf{G}$$

where $v_0 = [1,0,0,0,0]$. The rest of the hybrid remains unchanged.

Hybrid H_3: In hybrid H_3, we change the generation of matrix $\mathbf{A}$ and vector u in public parameters pp.

Let $\mathbf{A}$ be a random matrix in $\mathbb{Z}_q^{n\times m}$. The construction of matrices $(\{\mathbf{A}_i\}_{i\in[\ell]}, \{\mathbf{V}_{0,i}\}_{i\in[5]}, \mathbf{A}^c)$ remains the same, as in hybrid H_2. Sample error vectors e that would be used in algorithm $\mathsf{SampleP}$ later. Then compute the error vector

$$e_{\mathsf{BP}^*} \leftarrow \mathsf{Eval}_{\mathsf{ct}}(\mathsf{BP}, x, \{\mathbf{A}_i, e^T\mathbf{R}_i\}_{i\in[\ell]}, \{\mathbf{V}_{0,i}, e^T\mathbf{R}_{0,i}\}_{i\in[5]}, \{\mathbf{A}^c, e^T\mathbf{R}^c\})$$

and choose a correlation coefficient $\mu \leftarrow \mathcal{D}_\gamma$, and set vector $y_1 = (\mu e_{\mathsf{BP}^*} + \mathcal{D}_{\mathbb{Z}^m, \mathbf{Q}})q$, where

$$\mathbf{Q} = \beta^2\mathbf{I}_{m\times m} - \gamma^2\alpha^2\mathbf{R}_{\mathsf{BP}^*}^T\mathbf{R}_{\mathsf{BP}^*}$$

Then let $\boldsymbol{y} = (\boldsymbol{y}_0 | \boldsymbol{y}_1)$, where $\boldsymbol{y}_0 \leftarrow \mathcal{D}_{\mathbf{Z}_q^m, \beta^2 q^2 \mathbf{I}_{m \times m}}$. Sample vector $\mathbf{r}_{\mathsf{BP}^*} \leftarrow \boldsymbol{y} + \mathcal{D}_{\mathbb{Z}^{2m} - \boldsymbol{y}, (s^2 - \beta^2) q^2 \mathbf{I}_{2m \times 2m}}$, and compute matrix

$$\mathbf{V}_{\mathsf{BP}^*} \leftarrow \mathsf{Eval}_{\mathsf{pk}}(\mathsf{BP}, \{\mathbf{A}_i\}_{i \in [\ell]}, \{\mathbf{V}_{0,i}\}_{i \in [5]}, \mathbf{A}^c)$$

Set vector $\boldsymbol{u}$ in public parameters pp as $\boldsymbol{u} = [\mathbf{A} | \mathbf{V}_{\mathsf{BP}^*}] \cdot \mathbf{r}_{\mathsf{BP}^*}$. Since $\mathbf{A}$ is a random matrix without trapdoor $\mathbf{T_A}$ to answer key queries, we will use trapdoor $\mathbf{T_G}$ to answer queries as follows. Consider a secret key query for branching program BP such that $\mathsf{BP}(\boldsymbol{x}^*) = 0$. To respond, we do the following computations:

1. First, we compute

$$\mathbf{R}_{\mathsf{BP}} \leftarrow \mathsf{Eval}_{\mathsf{Sim}}(\mathsf{BP}, \boldsymbol{x}, \{\mathbf{R}_i\}_{i \in [\ell]}, \{\mathbf{R}_{0,i}\}_{i \in [5]}, \mathbf{R}^c, \mathbf{A})$$

to obtain a low-norm matrix $\mathbf{R}_{\mathsf{BP}} \in \mathbb{Z}_q^{m \times m}$ satisfying $\mathbf{A}\mathbf{R}_{\mathsf{BP}} - \mathsf{BP}(\boldsymbol{x}^*)\mathbf{G} = \mathbf{V}_{\mathsf{BP}}$.

2. Then, we sample $\mathbf{r}_{\mathsf{BP}}$ using

$$\mathbf{r}_{\mathsf{BP}} \leftarrow \mathsf{SampleRight}(\mathbf{A}, \mathbf{G}, \mathbf{R}_{\mathsf{BP}}, \mathbf{T_G}, \boldsymbol{u}, sq)$$

such that

$$\mathbf{r}_{\mathsf{BP}}^T \cdot [\mathbf{A} | \mathbf{V}_{\mathsf{BP}} + \mathbf{G}] = \boldsymbol{u}$$

By property of algorithm SampleRight, vector $\mathbf{r}_{\mathsf{BP}}$ is distributed as required.

The computation of answering P-sampler query, SampleP is the same as hybrid H_1 with error vectors $\boldsymbol{e}$, For faking receiver coins, FakeRCoins, simply output the vector $\mathbf{r}_{\mathsf{BP}^*}$ pre-sampled in the generation of vector $\boldsymbol{u}$ before.

Hybrid H_4: In hybrid H_4, we change the generation order of vector $\boldsymbol{y}$ and error vector $\boldsymbol{e}$.

First sample vector $\boldsymbol{y} = (\boldsymbol{y}_0 | \boldsymbol{y}_1) \leftarrow \mathcal{D}_{\mathbb{Z}^{2m}, \beta^2 q^2 \mathbf{I}_{2m \times 2m}}$ and compute $\mathbf{r}_{\mathsf{BP}^*}$ from $\boldsymbol{y}$ as in previous hybrid. Next, we compute error term $\boldsymbol{e}$ as $\boldsymbol{e} = \nu \boldsymbol{y}_1^T \mathbf{R}_{\mathsf{BP}^*} / q + \mathcal{D}_{\mathbb{Z}^m, \mathbf{Q}'}$, where $\nu \leftarrow \mathcal{D}_\tau, \tau = \gamma \alpha^2 / \beta^2$, and $\mathcal{D}_{\mathbb{Z}^m, \mathbf{Q}'}$ is sampled as $\mathbf{L}' \mathcal{D}_{\mathbf{Z}_1^m, \mathbf{I}_{m \times m}}$ for

$$\mathbf{Q}' = \mathbf{L}' \mathbf{L}'^T = \alpha^2 \mathbf{I} - \tau^2 \beta^2 \mathbf{R}_{\mathsf{BP}^*}^T \cdot \mathbf{R}_{\mathsf{BP}^*} \tag{2}$$

Additionally, we modify the challenge ciphertext to be

$$\psi_0^* = \boldsymbol{s}^T \mathbf{A} / q + \boldsymbol{e}, \quad \psi_i^* = \psi_0^{*T} \mathbf{R}_i / q, \quad \psi_{0,i}^* = \psi_0^{*T} \mathbf{R}_{0,i} / q, \quad \psi^{*c} = \psi_0^{*T} \mathbf{R}^c / q$$

and $c^* = \boldsymbol{s}^T \boldsymbol{u} + \mathcal{D}_{\mathbf{Z}^m, \alpha \mathbf{I}_{m \times m}}$.

Hybrid H_5: In hybrid H_5, we change the generation order of secret key $\mathbf{r}_{\mathsf{BP}^*}$ and vector $\boldsymbol{y}$.

We first sample matrix $\mathbf{r}_{\mathsf{BP}^*}$ from discrete Gaussian distribution $\mathcal{D}_{\mathbb{Z}^{2m}, s^2 q^2 \mathbf{I}_{2m \times 2m}}$, and set vector $\boldsymbol{u}$ in public parameters pp to be $\boldsymbol{u} = [\mathbf{A} | \mathbf{V}_{\mathsf{BP}^*}] \cdot \mathbf{r}_{\mathsf{BP}^*}$, where

$$\mathbf{V}_{\mathsf{BP}^*} \leftarrow \mathsf{Eval}_{\mathsf{pk}}(\mathsf{BP}, \{\mathbf{A}_i\}_{i \in [\ell]}, \{\mathbf{V}_{0,i}\}_{i \in [5]}, \{\mathbf{A}_i^c\}_{i \in [\ell]})$$

Then set $\boldsymbol{y} = (\boldsymbol{y}_0|\boldsymbol{y}_1) = r_{\mathsf{BP}^*}/2 + \mathcal{D}_{\mathbb{Z}^{2m},(\beta^2-s^2/4)q^2\mathbf{I}_{2m\times 2m}}$. The remainder of the hybrid remains roughly the same. In particular, the challenge ciphertext ct* is generated in the same manner as Hybrid H_4. We break the noise term e into two terms $e = e_0^{(1)} + e_0^{(2)} + \nu \boldsymbol{y}_1^T \mathbf{R}_{\mathsf{BP}^*}/q$, where $e_0^{(1)} \leftarrow \mathcal{D}_{\mathbb{Z}^m,\beta'\mathbf{I}_{m\times m}}$, $e_0^{(2)} \leftarrow \mathcal{D}_{\mathbb{Z}^m,\mathbf{Q}'-\beta'^2\mathbf{I}_{m\times m}}$ and $\beta' = \alpha/2$.

Hybrid H_6: In hybrid H_6, we change how the challenge ciphertext is generated by using the Extended-LWE$^+$ instance.

First sample uniformly random vector $\boldsymbol{b} \in \mathbb{Z}^m$ and set the challenge ciphertext as

$$\psi_0^* = \boldsymbol{b}/q + e_0^{(2)}, \quad \psi_i^* = \psi_0^{*T}\mathbf{R}_i/q, \quad \psi_{0,i}^* = \psi_0^{*T}\mathbf{R}_{0,i}/q, \quad \psi^{*c} = \psi_0^{*T}\mathbf{R}^c/q$$

and $c^* = r_{\mathsf{BP}^*}^T[\mathbf{I}_{m\times m}|\mathbf{R}_{\mathsf{BP}^*}](\boldsymbol{b}/q - e_0^{(1)}) + \mathcal{D}_{\mathbb{Z}^m,\alpha\mathbf{I}_{m\times m}}$.

Hybrid H_7: In hybrid H_7, we change the challenge ciphertext to be uniformly random.

In algorithm SampleP, sample uniformly random vectors $\mathsf{ct} \in \mathbb{Z}_q^m \times \mathbb{Z}_q^{\ell m} \times \mathbb{Z}_q^m \times \mathbb{Z}_q^{5m} \times \mathbb{Z}_q$ and outputs ct.

Claim 4.5. *Assuming the semantic security of PKE* $\Pi = (\mathsf{Gen}', \mathsf{Enc}', \mathsf{Dec}')$, *hybrid* H_0 *and* H_1 *are computationally indistinguishable.*

Proof. Observe there is only one difference between hybrids H_0 and H_1 occurs in the challenge ciphertext, i.e. the encryption (under PKE Π) of the random matrices $\mathbf{S}_i$ are replaced by encryption of 0. If a PPT adversary $\mathcal{A}$ distinguishes between the H_0-encryptions of $(\{\mathbf{R}_i\}_{i\in[\ell]}, \{\mathbf{R}_{0,i}\}_{i\in[5]}, \{\mathbf{R}^c\})$ and the H_1-encryptions of **0** with non-negligible probability, then we can construct an efficient reduction $\mathcal{B}$ that uses $\mathcal{A}$ to break the semantic security of PKE Π with similar probability. $\square$

Claim 4.6. *Hybrids* H_1 *and* H_2 *are statistically indistinguishable.*

Proof. Observe the only difference between hybrids H_1 and H_2 is the generation of matrices

$$(\{\mathbf{A}_i\}_{i\in[\ell]}, \{\mathbf{V}_{0,i}\}_{i\in[5]}, \{\mathbf{A}_i^c\}_{i\in[\ell]})$$

The random matrices $(\{\mathbf{R}_i\}_{i\in[\ell]}, \{\mathbf{R}_{0,i}\}_{i\in[5]}, \{\mathbf{R}_i^c\}_{i\in[\ell]})$ are used in the generation of public parameters pp:

$$\mathbf{A}_i = \mathbf{A}\mathbf{R}_i - x_i^*\mathbf{G}, \quad \mathbf{V}_{0,i} = \mathbf{A}\mathbf{R}_{0,i} - v_0[i]\mathbf{G}, \quad \mathbf{A}^c = \mathbf{A}\mathbf{R}^c - \mathbf{G}$$

and the construction of errors in challenge ciphertext

$$e_i = e^T\mathbf{R}_i, \quad e^c = e^T\mathbf{R}^c, \quad e_{0,i} = e^T\mathbf{R}_{0,i}$$

Then by Leftover Hash Lemma 2.1, the following two distributions are statistically indistinguishable

$$(\mathbf{A}, \{\mathbf{A}\mathbf{R}_i\}_{i\in[\ell]}, \{\mathbf{A}\mathbf{R}_{0,i}\}_{i\in[5]}, \{\mathbf{A}\mathbf{R}^c\}, \tilde{e}) \approx (\mathbf{A}, \{\mathbf{A}_i\}_{i\in[\ell]}, \{\mathbf{V}_{0,i}\}_{i\in[5]}, \{\mathbf{A}^c\}, \tilde{e})$$

where $\tilde{e} = (\{e_i\}_{i\in[\ell]}, \{e_{0,i}\}_{i\in[5]}, \{e^c\})$. Hence, hybrid H_0 and H_1 are statistically indistinguishable. $\square$

Claim 4.7. *Hybrids H_2 and H_3 are statistically indistinguishable.*

Proof. Observe there are three differences between hybrid H_2 and H_3: The generation of matrix $\mathbf{A}$ and vector $\boldsymbol{u}$ in pp, challenge secret key sk_{BP^*} and the computation methods to answer secret key queries. By the property of algorithm $\mathsf{TrapGen}(q, n, m)$, the distribution of matrix $\mathbf{A}$ in hybrid H_2 is statistically close to uniform distribution, from which matrix $\mathbf{A}$ in hybrid H_3 is sampled.

For secret key queries regarding branching program BP, in hybrid H_2, we sample vector r_{BP}, using

$$r_{\mathsf{BP}} \leftarrow \mathsf{SampleLeft}(\mathbf{A}, \mathbf{T_A}, (\mathbf{V}_{\mathsf{BP}} + \mathbf{G}), \boldsymbol{u}, s)$$

While in hybrid H_3, we sample vector r_{BP}, using

$$r_{\mathsf{BP}} \leftarrow \mathsf{SampleRight}(\mathbf{A}, \mathbf{G}, \mathbf{R}_{\mathsf{BP}}, \mathbf{T_G}, \boldsymbol{u}, sq)$$

By setting the parameters appropriately as specified in Sect. 4.3, and the properties of algorithms SampleLeft and SampleRight, the answers to secret key queries are statistically close.

By Leftover Hash Lemma 2.1, the distribution $([\mathbf{A}|\mathbf{V}_{\mathsf{BP}^*}], [\mathbf{A}|\mathbf{V}_{\mathsf{BP}^*}] \cdot r_{\mathsf{BP}^*})$ and $([\mathbf{A}|\mathbf{V}_{\mathsf{BP}^*}], \boldsymbol{u})$ are statistically close. Hence, hybrid H_2 and H_3 are statistically indistinguishable. □

Claim 4.8. *Hybrids H_3 and H_4 are statistically indistinguishable.*

Proof. The only difference between the two experiments is in the choice of $\boldsymbol{y}$ and e, specifically, the choice of the $\boldsymbol{y}_1$ component of $\boldsymbol{y} = (\boldsymbol{y}_0|\boldsymbol{y}_1)$. We will show that the joint distribution of $(e, \boldsymbol{y}_1)$ is identically distributed in these two hybrids:

In hybrid H_3, $\boldsymbol{y}_1$ is set as $\boldsymbol{y}_1 = (\mu e_{\mathsf{BP}^*} + \mathcal{D}_{\mathbb{Z}^m,\mathbf{Q}})q$, where $\mathbf{Q} = \beta^2 \mathbf{I}_{m \times m} - \gamma^2 \alpha^2 \mathbf{R}_{\mathsf{BP}^*}^T \cdot \mathbf{R}_{\mathsf{BP}^*}$ with $e \leftarrow \mathcal{D}_{\mathbb{Z}^m, \alpha^2 \mathbf{I}_{m \times m}}$ and

$$e_{\mathsf{BP}^*} \leftarrow \mathsf{Eval}_{\mathsf{ct}}(\mathsf{BP}, \boldsymbol{x}, \{\mathbf{A}_i, e^T \mathbf{R}_i\}_{i \in [\ell]}, \{\mathbf{V}_{0,i}, e^T \mathbf{R}_{0,i}\}_{i \in [5]}, \{\mathbf{A}^c, e^T \mathbf{R}^c\})$$

Therefore, in hybrid H_3, we may write the joint distribution of $(e, \boldsymbol{y}_1)$ as $\mathbf{T}_1 \cdot \mathcal{D}_{\mathbb{Z}^{2m}, \mathbf{I}_{2m \times 2m}}$, where $\mathbf{T}_1 \stackrel{\text{def}}{=} \begin{pmatrix} \alpha \mathbf{I}_{m \times m} & \mathbf{0}_{m \times m} \\ \gamma \alpha q \mathbf{R}_{\mathsf{BP}^*}^T & q\mathbf{L} \end{pmatrix}$ for $\mathbf{Q} = \mathbf{L}\mathbf{L}^T \in \mathbb{Z}^{m \times m}$ via the Cholesky decomposition Lemma in full version.

In hybrid H_4, vector $\boldsymbol{y} = (\boldsymbol{y}_0|\boldsymbol{y}_1)$ is sampled as $\boldsymbol{y} = (\boldsymbol{y}_0|\boldsymbol{y}_1) \leftarrow \mathcal{D}_{\mathbb{Z}^{2m}, \beta^2 q^2 \mathbf{I}_{2m \times 2m}}$. Then e is computed as $e = \nu \boldsymbol{y}_1^T \mathbf{R}_{\mathsf{BP}^*}/q + \mathcal{D}_{\mathbb{Z}^m, \mathbf{Q}'}$, where $\nu \leftarrow \mathcal{D}_{\tau}, \tau = \gamma \alpha^2/\beta^2$, and $\mathbf{Q}' = \alpha^2 \mathbf{I} - \tau^2 \beta^2 \mathbf{R}_{\mathsf{BP}^*}^T \mathbf{R}_{\mathsf{BP}^*}$. Then in hybrid H_4, we may write the joint distribution of $(e, \boldsymbol{y}_1)$ as $\mathbf{T}_2 \cdot \mathcal{D}_{\mathbb{Z}^{2m}, \mathbf{I}_{2m \times 2m}}$, where $\mathbf{T}_2 \stackrel{\text{def}}{=} \begin{pmatrix} \mathbf{L}' & \tau \beta \mathbf{R}_{\mathsf{BP}^*} \\ \mathbf{0}_{m \times m} & \beta q \mathbf{I}_{m \times m} \end{pmatrix}$ for $\mathbf{Q}' = \mathbf{L}'\mathbf{L}'^T \in \mathbb{Z}^{m \times m}$ via the Cholesky decomposition Lemma in full version.

We claim equality of the following systems of equations:

$$\mathbf{T}_1 \mathbf{T}_1^T = \begin{pmatrix} \alpha^2 \mathbf{I}_{m \times m} & \gamma \alpha^2 q \mathbf{R}_{\mathsf{BP}^*} \\ \gamma \alpha^2 q \mathbf{R}_{\mathsf{BP}^*}^T & \gamma^2 \alpha^2 q^2 \mathbf{R}_{\mathsf{BP}^*}^T \cdot \mathbf{R}_{\mathsf{BP}^*} + q^2 \mathbf{L}\mathbf{L}^T \end{pmatrix}$$

$$= \begin{pmatrix} \mathbf{L}'\mathbf{L}'^T + \tau^2 \beta^2 \mathbf{R}_{\mathsf{BP}^*} \cdot \mathbf{R}_{\mathsf{BP}^*}^T & \tau \beta^2 q \mathbf{R}_{\mathsf{BP}^*} \\ \tau \beta^2 q \mathbf{R}_{\mathsf{BP}^*}^T & \beta^2 q^2 \mathbf{I}_{m \times m} \end{pmatrix} = \mathbf{T}_2 \mathbf{T}_2^T.$$

This fact may be seen quadrant-wise by our choice of $\tau = \gamma \alpha^2 / \beta^2$ and the settings of $\mathbf{Q} = \mathbf{L}\mathbf{L}^T$ and $\mathbf{Q}' = \mathbf{L}'\mathbf{L}'^T$ in Eqs. (1) and (2). It then follows that $(\mathbf{T}_2^{-1}\mathbf{T}_1)(\mathbf{T}_2^{-1}\mathbf{T}_1)^T = \mathbf{I}_{2m \times 2m}$, implying $\mathbf{T}_1 = \mathbf{T}_2\mathbf{Q}^*$ for some orthogonal matrix $\mathbf{Q}^*$. Because the spherical Gaussian $\mathcal{D}_{\mathbb{Z}^{2m}, \mathbf{I}_{2m \times 2m}}$ is invariant under rigid transformations, we have $\mathbf{T}_1 \cdot \mathcal{D}_{\mathbb{Z}^{2m}, \mathbf{I}_{2m \times 2m}} = \mathbf{T}_2\mathbf{Q}^* \cdot \mathcal{D}_{\mathbb{Z}^{2m}, \mathbf{I}_{2m \times 2m}} = \mathbf{T}_2 \cdot \mathcal{D}_{\mathbb{Z}^{2m}, \mathbf{I}_{2m \times 2m}}$, and the claim follows. $\qquad \square$

Claim 4.9. *Hybrids* H_4 *and* H_5 *are statistically indistinguishable.*

Proof. Observe the main difference between hybrids H_4 and H_5 is the order of generation of vectors $\boldsymbol{y}$ and $\boldsymbol{r}_{\mathsf{BP}^*}$: In hybrid H_4, we first sample $\boldsymbol{y} = (\boldsymbol{y}_0 | \boldsymbol{y}_1) \leftarrow \mathcal{D}_{\mathbb{Z}^{2m}, \beta^2 q^2 \mathbf{I}_{2m \times 2m}}$ and set $\boldsymbol{r}_{\mathsf{BP}^*} \leftarrow \boldsymbol{y} + \mathcal{D}_{\mathbb{Z}^{2m} - \boldsymbol{y}, q^2(s^2 - \beta^2) \mathbf{I}_{2m \times 2m}}$, while in hybrid H_5, we first sample $\boldsymbol{r}_{\mathsf{BP}^*} \leftarrow \mathcal{D}_{\mathbb{Z}^{2m}, s^2 q^2 \mathbf{I}_{2m \times 2m}}$ and set $\boldsymbol{y} = (\boldsymbol{y}_0 | \boldsymbol{y}_1) \leftarrow \boldsymbol{r}_{\mathsf{BP}^*}/2 + \mathcal{D}_{\mathbb{Z}^{2m}, (\beta^2 - s^2/4) q^2 \mathbf{I}_{2m \times 2m}}$. By setting parameters appropriately as in Sect. 4.3, these two distributions are statistically close. $\qquad \square$

Claim 4.10. *Assuming the hardness of* extended-$\mathsf{LWE}^+_{n,m,q,\mathcal{D}_{\mathbb{Z}^m, \beta'}, \mathbf{R}}$ *for any adversarially chosen distribution over matrices* $\mathbf{R} \in \mathbb{Z}_q^{m \times m}$, *then hybrids* H_5 *and* H_6 *are computationally indistinguishable.*

Proof. Suppose $\mathcal{A}$ has non-negligible advantage in distinguishing hybrid H_5 and H_6, then we use $\mathcal{A}$ to construct an extended-LWE^+ algorithm $\mathcal{B}$ as follows:

Invocation. $\mathcal{B}$ invokes adversary $\mathcal{A}$ to commit to a challenge attribute vector $\boldsymbol{x}^* = (x_1^*, ..., x_\ell^*)$ and challenge branching program BP^*. Then $\mathcal{B}$ generates $\mathbf{R}_{\mathsf{BP}^*}$ by first sampling $(\{\mathbf{R}_i\}_{i \in [\ell]}, \{\mathbf{R}_{0,i}\}_{i \in [5]}, \{\mathbf{R}^c\})$ as in the hybrid, and computes

$$\mathbf{R}_{\mathsf{BP}} \leftarrow \mathsf{Eval}_{\mathsf{Sim}}(\mathsf{BP}, \boldsymbol{x}, \{\mathbf{R}_i\}_{i \in [\ell]}, \{\mathbf{R}_{0,i}\}_{i \in [5]}, \{\mathbf{R}^c\}, \mathbf{A})$$

Then it receives an extended-LWE^+ instance for the matrix $\mathbf{R} = \mathbf{R}_{\mathsf{BP}^*}$ as follows:

$$\{\mathbf{A}, \boldsymbol{b} = \boldsymbol{s}^T \mathbf{A} + \boldsymbol{e}, \boldsymbol{z}_0, \boldsymbol{z}_1, \langle \boldsymbol{z}_0, \boldsymbol{b} - \boldsymbol{e} \rangle + e, \langle \boldsymbol{z}_1^T \mathbf{R}, \boldsymbol{b} - \boldsymbol{e} \rangle + e'\}$$

where $\mathbf{A} \xleftarrow{\$} \mathbb{Z}_q^{n \times m}$, $\boldsymbol{s} \xleftarrow{\$} \mathbb{Z}_q^n$, $\boldsymbol{u} \xleftarrow{\$} \mathbb{Z}_q^m$, $\boldsymbol{e}, \boldsymbol{z}_0, \boldsymbol{z}_1 \xleftarrow{\$} \chi^n$ and $e, e' \xleftarrow{\$} \chi$. Algorithm $\mathcal{B}$ aims to leverage adversary $\mathcal{A}$'s output to solve the extended-LWE^+ assumption.

Setup. $\mathcal{B}$ generates matrices $(\{\mathbf{A}_i\}_{i \in [\ell]}, \{\mathbf{V}_{0,i}\}_{i \in [5]}, \{\mathbf{A}^c\})$ as specified in hybrid H_1. Then, $\mathcal{B}$ sets challenge secret key $\mathsf{sk}_{\mathsf{BP}^*} = \boldsymbol{r}_{\mathsf{BP}^*} = (\boldsymbol{r}_0^* | \boldsymbol{r}_1^*) = (\boldsymbol{z}_0 | \boldsymbol{z}_1)$ from extended-LWE^+ instance and computes vector $\boldsymbol{u}$ as in hybrid H_5.

Secret key queries. $\mathcal{B}$ answers adversary $\mathcal{A}$'s secret key queries as in hybrid H_2.

Challenge ciphertext. $\mathcal{B}$ answers adversary $\mathcal{A}$'s P-sample query by setting

$$\psi_0^* = \boldsymbol{b}/q + e_0^{(2)} + \nu \boldsymbol{y}_1^T \mathbf{R}_{\mathsf{BP}^*}/q, \quad \psi_i^* = \psi_0^{*T} \mathbf{R}_i/q, \quad \psi_{0,i}^* = \psi_0^{*T} \mathbf{R}_{0,i}/q, \quad \psi^{*c} = \psi_0^{*T} \mathbf{R}^c/q$$

and $c^* = \boldsymbol{r}_{\mathsf{BP}^*}^T [\mathbf{I}_{m \times m} | \mathbf{R}_{\mathsf{BP}^*}](\boldsymbol{b}/q - \boldsymbol{e}^{(1)}) + \mathcal{D}_{\mathbb{Z}^{2m}, \alpha \mathbf{I}_{m \times m}}$.

Faking receiver coin query. $\mathcal{B}$ answers adversary $\mathcal{A}$'s faking receiver coin query by outputting the extended-LWE instance's vector $\mathsf{sk}_{\mathsf{BP}^*} = r_{\mathsf{BP}^*}$.

Output. $\mathcal{B}$ outputs whatever $\mathcal{A}$ outputs.

We can rewrite the expression of $c^{*'}$ to be

$$
\begin{aligned}
c^{*'} &= ([\mathbf{A}^*|\mathbf{A}^*\mathbf{R}_{\mathsf{BP}^*}](\begin{smallmatrix} z_0 \\ z_1 \end{smallmatrix}))^T s/q + \mathcal{D}_{\mathbb{Z}_1, \alpha} \\
&= ((z_0|z_1)(\begin{smallmatrix} \mathbf{A}^{*T} \\ \mathbf{R}_{\mathsf{BP}^*}^T \mathbf{A}^{*T} \end{smallmatrix}))s/q + \mathcal{D}_{\mathbb{Z}_1, \alpha} = z_0 \mathbf{A}^{*T} s/q + z_1 \mathbf{R}_{\mathsf{BP}^*}^T \mathbf{A}^{*T} s/q + \mathcal{D}_{\mathbb{Z}_1, \alpha} \\
&= \langle z_0, b/q - e^{(1)} \rangle + \langle z_1^T \mathbf{R}_{\mathsf{BP}^*}, b/q - e^{(1)} \rangle + \mathcal{D}_{\mathbb{Z}_1, \alpha}
\end{aligned}
$$

We can see that if the eLWE^+ instance's vector b is pseudorandom, then the distribution simulated by $\mathcal{B}$ is exactly the same as H_5. If b is truly random and independent, then the distribution simulated by $\mathcal{B}$ is exactly the same as H_6. Therefore, if $\mathcal{A}$ can distinguish H_5 from H_6 with non-negligible probability, then $\mathcal{B}$ can break the $\mathsf{eLWE}^+_{n,m,q,\mathcal{D}_{(\alpha/2)q},\alpha',\mathbf{S}_{f^*}}$ problem for some $\alpha' \geq 0$ with non-negligible probability. $\qquad\square$

Claim 4.11. *Hybrids* H_6 *and* H_7 *are statistically indistinguishable.*

Proof. Recall the only difference between hybrids H_6 and H_7 is the generation of challenge ciphertext. In hybrid H_7, we observe if ψ_0^* is chosen from uniform distribution, then by Leftover Hash Lemma 2.1, it holds

$$
\psi_i^* = \psi_0^{*T} \mathbf{R}_i / q, \quad \psi_{0,i}^* = \psi_0^{*T} \mathbf{R}_{0,i} / q, \quad \psi^{*c} = \psi_0^{*T} \mathbf{R}^c / q
$$

is also uniformly random (in their marginal distribution). Therefore, it remains to show that c^* is still uniformly random even conditioned on fixed samples of $(\psi_0^*, \{\psi_i^*\}_i, \{\psi_{0,i}^*\}_i, \{\psi^c\})$.

As calculated above, we can unfold the expression of c^* as

$$
c^* = \langle z_0, b/q - x^{(1)} \rangle + \langle z_1^T \mathbf{R}_{\mathsf{BP}^*}, b/q - x^{(1)} \rangle + \mathcal{D}_{\mathbb{Z}_1, \alpha}
$$

We note that $b/q - x^{(1)} = \psi_0^* - x^{(1)} - x^{(2)} - \nu \mathbf{R}_{\mathsf{BP}^*} y_1 / q$, thus if we show that

$$
\langle \mathbf{R}_{\mathsf{BP}^*} z_1, \nu \mathbf{R}_{\mathsf{BP}^*} y_1 / q \rangle
$$

is close to uniform distribution (modulo 1), then c^* will also be close to the uniform distribution (modulo 1), as c^* is masked by this uniformly random number. Recall in hybrids, we set $y_1 = z_1/2 + (\text{shift})$, so it is sufficient to analyze

$$
\langle \mathbf{R}_{\mathsf{BP}^*} z_1, \nu \mathbf{R}_{\mathsf{BP}^*} y_1 / q \rangle = \nu \langle \mathbf{R}_{\mathsf{BP}^*} z_1, \mathbf{R}_{\mathsf{BP}^*} z_1 / q \rangle = \nu \| \mathbf{R}_{\mathsf{BP}^*}^* z_1 \|^2 / q
$$

By applying analysis of norm inductively on matrix $\mathbf{R}_{\mathsf{BP}^*}$, we can obtain that

$$
\| \mathbf{R}_{\mathsf{BP}^*}^* z_1 \|^2 / q \geq \frac{(\| \mathbf{R}_{0,j} z_1 \| - \Theta(m^{1.5}) \ell \| z_1 \|)^2}{q}
$$

Table 1. Parameter description and simple example setting

Parameters	Description	Setting
n, m	Lattice dimension	$n = \lambda, m = n^2 \log n$
ℓ	Length of input to branching program	$\ell = n$
q	Modulus (resp. bit-precision)	Smallest prime $\geq n^{1.5} m^{2.5} \omega(\log n)$
α	Sampling error terms e, e	$\frac{1}{n^{2.5} \log^3 n}$
β	Sampling correlation vector y	$\alpha/2$
γ	Sampling correlation coefficient μ	$\frac{1}{n \log^{1.5} n}$
s	Sampling secret key r	$3\beta/2$
η	Scaling parameter for $\mathbf{R}_{0,j}$	$\Theta(m\ell)$

where $\mathbf{R}_{0,j} \in \{-1, 1\}^{m \times m}$. Since vector z_1 is sampled from Gaussian with width sq, so its two-norm is at least $\sqrt{m}(sq)$ with overwhelming probability. Then by analysis of norm, the distribution $\nu \| \mathbf{R}_{\mathsf{BP}^*}^* z_1 \|^2 / q$ is a Gaussian distribution with width at least

$$d = \tau \frac{(\eta sqm - \Theta(m^2\ell)sq)^2}{q} = \frac{\gamma\alpha^2(\eta sqm - \Theta(m^2\ell)sq)^2}{\beta^2 q}$$

We recall again that ν was sampled from a Gaussian with parameter $\tau = \gamma\alpha^2/\beta^2$. By our setting of parameters, we have $d/\omega(\log(n)) \geq 1$. A Gaussian with such width is statistically close to uniform in the domain $\mathbb{Z}_1$. This completes the proof of Lemma 4.4. Further, Theorem 4.2 follows from Lemmas 4.3 and 4.4. A (flexibly) bi-deniable ABE from LWE then follows.   □

4.3   Parameter Setting

The parameters in Table 1 are selected in order to satisfy the following constraints:

- To ensure correctness in Lemma 4.3, we have $\alpha sqm(\eta\sqrt{m} + 2m^{1.5}\ell) \leq 1/4$.
- To ensure deniability in Hybrid H_7, we have $d/\omega(\log(n)) > \frac{\gamma\alpha^2(\eta sqm - \Theta(m^2\ell sq))^2}{\beta^2 q\omega(\log(n))} > 1$.
- To ensure large enough LWE noise, we need $\alpha \geq (\sqrt{n} \log^{1+\delta} n)/q$.
- To apply the leftover hash lemma, we need $m \geq 2n \log(q)$.
- To ensure that the matrix $\mathbf{Q}$ in FakeRCoins is positive definite, we have $\beta \geq \alpha\gamma\sqrt{\eta\sqrt{m} + 2m^{1.5}\ell}$; To ensure that the matrix $\mathbf{Q}'$ in the security proof is positive definite, we have $\alpha \geq \tau\beta\sqrt{\eta\sqrt{m} + 2m^{1.5}\ell}$. This constraint will also imply that in the security proof, both $\mathbf{Q}'$ and $\mathbf{Q}' - \beta'\mathbf{I}_{m \times m}$ are positive definite (note $\beta' = \alpha/2$).
- To ensure hybrids H_3 and H_5 are well-defined, we have $s > \beta$ and $\beta > s/2$. Let $s := (3/2)\beta$.

Regev [36] showed that for $q > \sqrt{m}/\beta'$, an efficient algorithm for $\mathsf{LWE}_{n,m,q,\chi}$ for $\chi = \mathcal{D}_{\beta'q}$ (and $\beta'q \geq \sqrt{n}\omega(\log(n))$) implies an efficient quantum algorithm for approximating the SIVP and GapSVP problems, to within $\tilde{O}(n/\beta')$ approximation factors in the worst case. Our example parameter setting yields a bi-deniable AB-BTS based on the (quantum) hardness of solving $\mathsf{SIVP}_{\tilde{O}(n^{9.5})}$, respectively $\mathsf{GapSVP}_{\tilde{O}(n^{9.5})}$. (We write this term to additionally absorb the $(1/q^2)$ loss from our LWE to eLWE^+ reduction.) We leave further optimizing the lattice problem approximation factor to future work.

4.4 From AB-BTS to Flexible Bi-Deniable ABE

We present the instantiation of a flexible bi-deniable ABE using our AB-BTS scheme described above. We let $\Sigma' = (\mathsf{Setup'}, \mathsf{DenSetup'}, \mathsf{Keygen'}, \mathsf{SampleP'}, \mathsf{SampleU'}, \mathsf{TestP'}, \mathsf{FakeRCoins'}, \mathsf{FakeSCoins'})$ be an AB-BTS scheme. Then the flexible bi-deniable ABE $\Sigma = (\mathsf{Setup}, \mathsf{DenSetup}, \mathsf{Keygen}, \mathsf{Enc}, \mathsf{DenEnc}, \mathsf{Dec}, \mathsf{SendFake}, \mathsf{RecFake})$ is:

- $\mathsf{Setup}(1^\lambda)$: Run algorithm $(\mathsf{pp'}, \mathsf{msk'}) \leftarrow \mathsf{Setup'}(1^\lambda)$ in AB-BTS and set $\mathsf{pp} = \mathsf{pp'}, \mathsf{msk} = \mathsf{msk'}$.
- $\mathsf{DenSetup}(1^\lambda)$: Run algorithm $(\mathsf{pp'}, \mathsf{msk'}, \mathsf{fk'}) \leftarrow \mathsf{DenSetup'}(1^\lambda)$ in AB-BTS and set $\mathsf{pp} = \mathsf{pp'}, \mathsf{msk} = \mathsf{msk'}, \mathsf{fk} = (\mathsf{fk'}, \mathsf{msk'})$.
- $\mathsf{Keygen}(\mathsf{msk}, f)$: Run algorithm $\mathsf{sk}_f \leftarrow \mathsf{Keygen'}(\mathsf{msk}, f)$ in AB-BTS and set $\mathsf{sk}_f = \mathsf{sk}'_f$.
- $\mathsf{Enc}(\mathsf{pp}, \boldsymbol{x}, \mu; (r_S^{(1)}, r_S^{(2)}))$: On input the message $\mu \in \{0,1\}$, if $\mu = 0$, then run $c_i \leftarrow \mathsf{SampleU'}(\mathsf{pp}, \boldsymbol{x}; r_S^{(i)})$ for $i = 1,2$, otherwise, $\mu = 1$, run $c_1 \leftarrow \mathsf{SampleU'}(\mathsf{pp}, \boldsymbol{x}; r_S^{(1)})$ and $c_2 \leftarrow \mathsf{SampleP'}(\mathsf{pp}, \boldsymbol{x}; r_S^{(2)})$. Output $\mathsf{ct}_{\boldsymbol{x}} = (c_1, c_2)$.
- $\mathsf{DenEnc}(\mathsf{pp}, \boldsymbol{x}, \mu; (r_S^{(1)}, r_S^{(2)}))$: On input the message $\mu \in \{0,1\}$, then run $c_i \leftarrow \mathsf{SampleP'}(\mathsf{pp}, \boldsymbol{x}; r_S^{(i)})$ for $i = 1,2$, otherwise, $\mu = 1$, run $c_1 \leftarrow \mathsf{SampleU'}(\mathsf{pp}, \boldsymbol{x}; r_S^{(1)})$ and $c_2 \leftarrow \mathsf{SampleP'}(\mathsf{pp}, \boldsymbol{x}; r_S^{(2)})$. Output $\mathsf{ct}_{\boldsymbol{x}} = (c_1, c_2)$.
- $\mathsf{Dec}(\mathsf{ct}_{\boldsymbol{x}}, \mathsf{sk}_f)$: If $f(\boldsymbol{x}) \neq 0$, then output $\bot$. Otherwise, parse $\mathsf{ct}_{\boldsymbol{x}} = (c_1, c_2)$ and run $b_i \leftarrow \mathsf{TestP'}(\mathsf{sk}_f, c_i)$ for $i = 1,2$. Output 0 if the $b_1 = b_2$ and 1 if $b_1 \neq b_2$.
- $\mathsf{SendFake}(\mathsf{pp}, r_S, \mu, \mu')$: If $\mu = \mu'$, return r_S. If $(\mu, \mu') = (0,1)$, then run $r_S^{*(2)} \leftarrow \mathsf{FakeSCoins'}(\mathsf{pp}, r_S^{(2)})$ and return $(r_S^{(1)}, r_S^{*(2)})$. Else if $(\mu, \mu') = (1,0)$, run $r_S^{*(1)} \leftarrow \mathsf{FakeSCoins'}(\mathsf{pp}, r_S^{(1)})$ and return $(r_S^{*(1)}, r_S^{(2)})$.
- $\mathsf{RecFake}(\mathsf{pp}, \mathsf{fk}, \mathsf{ct}_{\boldsymbol{x}}, f, \mu')$: Parse $\mathsf{ct}_{\boldsymbol{x}} = (c_1, c_2)$ and use fk to decrypt the ciphertext $\mathsf{ct}_{\boldsymbol{x}}$ then obtain the plaintext μ. If $\mu = \mu'$, then run the honest key generation of the BTS scheme, i.e. $\mathsf{sk}'_f \leftarrow \mathsf{Keygen'}(\mathsf{msk'}, f)$. Otherwise, run $\mathsf{sk}'_f \leftarrow \mathsf{FakeRCoins'}(\mathsf{pp}, \mathsf{fk}, c_{\mu+1}, f)$. Return sk'_f.

Similar to the work by Canetti et al. [14] and O'Neil et al. [35], the following desired theorem can be proven in a straightforward manner.

Theorem 4.12. *Assume that Σ' is a flexible bi-deniable AB-BTS, as in Definition 3.2. Then Σ is a flexibly bi-deniable ABE, as in Definition 3.1.*

Acknowledgments. We thank anonymous reviewers for their insightful comments. This work was performed in part under financial assistance award 70NANB15H328 from the U.S. Department of Commerce, National Institute of Standards and Technology, and was additionally supported in part by NSF award #1223623, NSF grants CNS-1314857, CNS-1453634, CNS-1518765, CNS-1514261, a Packard Fellowship, a Sloan Fellowship, two Google Faculty Research Awards, and a VMWare Research Award.

References

1. Agrawal, S., Boneh, D., Boyen, X.: Efficient lattice (H)IBE in the standard model. In: Gilbert, H. (ed.) EUROCRYPT 2010. LNCS, vol. 6110, pp. 553–572. Springer, Heidelberg (2010)
2. Alperin-Sheriff, J., Peikert, C.: Circular and KDM security for identity-based encryption. In: Fischlin, M., Buchmann, J., Manulis, M. (eds.) PKC 2012. LNCS, vol. 7293, pp. 334–352. Springer, Heidelberg (2012)
3. Alperin-Sheriff, J., Peikert, C.: Faster bootstrapping with polynomial error. In: Garay, J.A., Gennaro, R. (eds.) CRYPTO 2014, Part I. LNCS, vol. 8616, pp. 297–314. Springer, Heidelberg (2014)
4. Ananth, P., Brakerski, Z., Segev, G., Vaikuntanathan, V.: From selective to adaptive security in functional encryption. In: Gennaro, R., Robshaw, M. (eds.) CRYPTO 2015. LNCS, vol. 9216, pp. 657–677. Springer, Heidelberg (2015)
5. Apon, D., Fan, X., Liu, F.-H.: Bi-deniable inner product encryption from LWE. IACR Cryptology ePrint Archive, 2015:993 (2015)
6. Barrington, D.A.M.: Bounded-width polynomial-size branching programs recognize exactly those languages in nc^1. J. Comput. Syst. Sci. **38**(1), 150–164 (1989)
7. Bellare, M., Hofheinz, D., Yilek, S.: Possibility and impossibility results for encryption and commitment secure under selective opening. In: Joux, A. (ed.) EUROCRYPT 2009. LNCS, vol. 5479, pp. 1–35. Springer, Heidelberg (2009)
8. Bendlin, R., Nielsen, J.B., Nordholt, P.S., Orlandi, C.: Lower and upper bounds for deniable public-key encryption. In: Lee, D.H., Wang, X. (eds.) ASIACRYPT 2011. LNCS, vol. 7073, pp. 125–142. Springer, Heidelberg (2011)
9. Boneh, D., Gentry, C., Gorbunov, S., Halevi, S., Nikolaenko, V., Segev, G., Vaikuntanathan, V., Vinayagamurthy, D.: Fully key-homomorphic encryption, arithmetic circuit ABE and compact garbled circuits. In: Nguyen, P.Q., Oswald, E. (eds.) EUROCRYPT 2014. LNCS, vol. 8441, pp. 533–556. Springer, Heidelberg (2014)
10. Boneh, D., Lewi, K., David, J.W.: Constraining pseudorandom functions privately. IACR Cryptology ePrint Archive **2015**, 1167 (2015)
11. Boneh, D., Roughgarden, T., Feigenbaum, J. (eds.): 45th ACM STOC. ACM Press, June 2013
12. Boneh, D., Sahai, A., Waters, B.: Functional encryption: definitions and challenges. In: Ishai, Y. (ed.) TCC 2011. LNCS, vol. 6597, pp. 253–273. Springer, Heidelberg (2011)
13. Brakerski, Z., Langlois, A., Peikert, C., Regev, O., Stehlé, D.: Classical hardness of learning with errors. In: Boneh et al. [11], pp. 575–584 (2013)
14. Canetti, R., Dwork, C., Naor, M., Ostrovsky, R.: Deniable encryption. In: Kaliski Jr., B.S. (ed.) CRYPTO 1997. LNCS, vol. 1294, pp. 90–104. Springer, Heidelberg (1997)

15. Canetti, R., Feige, U., Goldreich, O., Naor, M.: Adaptively secure multi-party computation. In: 28th ACM STOC, pp. 639–648. ACM Press, May 1996
16. Canetti, R., Goldwasser, S., Poburinnaya, O.: Adaptively secure two-party computation from indistinguishability obfuscation. In: Dodis, Y., Nielsen, J.B. (eds.) TCC 2015, Part II. LNCS, vol. 9015, pp. 557–585. Springer, Heidelberg (2015)
17. Angelo De Caro, Vincenzo Iovino, and Adam O'Neill. Deniable functional encryption. In Public-Key Cryptography - PKC –19th IACR International Conference on Practice and Theory in Public-Key Cryptography, Taipei, Taiwan, March 6–9, Proceedings, Part I, pp. 196–222, (2016)
18. Cash, D., Hofheinz, D., Kiltz, E., Peikert, C.: Bonsai trees, or how to delegate a lattice basis. In: Gilbert, H. (ed.) EUROCRYPT 2010. LNCS, vol. 6110, pp. 523–552. Springer, Heidelberg (2010)
19. Dachman-Soled, D., Katz, J., Rao, V.: Adaptively secure, universally composable, multiparty computation in constant rounds. In: Dodis, Y., Nielsen, J.B. (eds.) TCC 2015, Part II. LNCS, vol. 9015, pp. 586–613. Springer, Heidelberg (2015)
20. Dachman-Soled, D., Liu, F.-H., Zhou, H.-S.: Leakage-resilient circuits revisited - optimal number of computing components without leak-free hardware, pp. 131–158 (2015)
21. Dodis, Y., Reyzin, L., Smith, A.: Fuzzy extractors: how to generate strong keys from biometrics and other noisy data. In: Cachin, C., Camenisch, J.L. (eds.) EUROCRYPT 2004. LNCS, vol. 3027, pp. 523–540. Springer, Heidelberg (2004)
22. Garg, S., Gentry, C., Halevi, S., Raykova, M., Sahai, A., Waters, B.: Candidate indistinguishability obfuscation and functional encryption for all circuits. In: 54th FOCS, pp. 40–49. IEEE Computer Society Press, October 2013
23. Garg, S., Polychroniadou, A.: Two-round adaptively secure MPC from indistinguishability obfuscation. In: Dodis, Y., Nielsen, J.B. (eds.) TCC 2015, Part II. LNCS, vol. 9015, pp. 614–637. Springer, Heidelberg (2015)
24. Gentry, C., Peikert, C., Vaikuntanathan, V.: Trapdoors for hard lattices and new cryptographic constructions. In: 40th ACM STOC, pp. 197–206. ACM Press, May 2008
25. Gentry, C., Sahai, A., Waters, B.: Homomorphic encryption from learning with errors: conceptually-simpler, asymptotically-faster, attribute-based. In: Canetti, R., Garay, J.A. (eds.) CRYPTO 2013, Part I. LNCS, vol. 8042, pp. 75–92. Springer, Heidelberg (2013)
26. Gilbert, H. (ed.): EUROCRYPT 2010. LNCS, vol. 6110. Springer, Heidelberg (2010)
27. Goldwasser, S., Kalai, Y.T., Popa, R.A., Vaikuntanathan, V., Zeldovich, N.: Reusable garbled circuits and succinct functional encryption. In: Boneh et al. [11], pp. 555–564 (2013)
28. Gorbunov, S., Vaikuntanathan, V., Wee, H.: Attribute-based encryption for circuits. In: Boneh et al. [11], pp. 545–554 (2013)
29. Gorbunov, S., Vaikuntanathan, V., Wee, H.: Predicate encryption for circuits from LWE. In: Gennaro, R., Robshaw, M. (eds.) CRYPTO 2015. LNCS, vol. 9216, pp. 503–523. Springer, Heidelberg (2015)
30. Gorbunov, S., Vinayagamurthy, D.: Riding on asymmetry: efficient ABE for branching programs. In: Iwata, T., et al. (eds.) ASIACRYPT 2015. LNCS, vol. 9452, pp. 550–574. Springer, Heidelberg (2015). doi:10.1007/978-3-662-48797-6_23
31. Goyal, V., Pandey, O., Sahai, A., Waters, B.: Attribute-based encryption for fine-grained access control of encrypted data. In: ACM CCS 2006, pp. 89–98. ACM Press, October/November 2006. Available as Cryptology ePrint Archive Report 2006/309

32. Katz, J., Thiruvengadam, A., Zhou, H.-S.: Feasibility and infeasibility of adaptively secure fully homomorphic encryption. In: Kurosawa, K., Hanaoka, G. (eds.) PKC 2013. LNCS, vol. 7778, pp. 14–31. Springer, Heidelberg (2013)
33. Micciancio, D., Mol, P.: Pseudorandom knapsacks and the sample complexity of LWE search-to-decision reductions. In: Rogaway [37], pp. 465–484
34. Micciancio, D., Peikert, C.: Trapdoors for lattices: simpler, tighter, faster, smaller. In: Pointcheval, D., Johansson, T. (eds.) EUROCRYPT 2012. LNCS, vol. 7237, pp. 700–718. Springer, Heidelberg (2012)
35. O'Neill, A., Peikert, C., Waters, B.: Bi-deniable public-key encryption. In: Rogaway [37], pp. 525–542 (2011)
36. Regev, O.: On lattices, learning with errors, random linear codes, and cryptography. In: 37th ACM STOC, pp. 84–93. ACM Press, May 2005
37. Rogaway, P. (ed.): CRYPTO 2011. LNCS, vol. 6841. Springer, Heidelberg (2011)
38. Sahai, A., Waters, B.: How to use indistinguishability obfuscation: deniable encryption, and more. In: Shmoys, D.B. (eds.) 46th ACM STOC, pp. 475–484. ACM Press, May/June 2014

Targeted Homomorphic
Attribute-Based Encryption

Zvika Brakerski[1]([⊠]), David Cash[2], Rotem Tsabary[1], and Hoeteck Wee[3]

[1] Weizmann Institute of Science, Rehovot, Israel
{zvika.brakerski,rotem.tsabary}@weizmann.ac.il
[2] Rutgers University, New Brunswick, USA
david.cash@cs.rutgers.edu
[3] ENS, CNRS and Columbia University, Paris, France
wee@di.ens.fr

Abstract. In (key-policy) attribute-based encryption (ABE), messages are encrypted respective to attributes x, and keys are generated respective to policy functions f. The ciphertext is decryptable by a key only if $f(x) = 0$. Adding homomorphic capabilities to ABE is a long standing open problem, with current techniques only allowing compact homomorphic evaluation on ciphertext respective to the same x. Recent advances in the study of multi-key FHE also allow cross-attribute homomorphism with ciphertext size growing (quadratically) with the number of input ciphertexts.

We present an ABE scheme where homomorphic operations can be performed compactly across attributes. Of course, decrypting the resulting ciphertext needs to be done with a key respective to a policy f with $f(x_i) = 0$ for *all* attributes involved in the computation. In our scheme, the *target policy* f needs to be known to the evaluator, we call this *targeted homomorphism*. Our scheme is secure under the polynomial hardness of learning with errors (LWE) with sub-exponential modulus-to-noise ratio.

We present a second scheme where there needs not be a single target policy. Instead, the decryptor only needs a set of keys representing policies f_j s.t. for any attribute x_i there exists f_j with $f_j(x_i) = 0$. In this scheme, the ciphertext size grows (quadratically) with the size of the *set of policies* (and is still independent of the number of inputs or attributes). Again, the target set of policies needs to be known at evaluation time. This latter scheme is secure in the random oracle model under the polynomial hardness of LWE with sub-exponential noise ratio.

For the full and most up-to-date version of this work, see Cryptology ePrint Archive http://eprint.iacr.org/2016/691.

Z. Brakerski and R. Tsabary—Supported by the Israel Science Foundation (Grant No. 468/14), the Alon Young Faculty Fellowship, Binational Science Foundation (Grant No. 712307) and Google Faculty Research Award.

H. Wee—Supported by ERC Project aSCEND (H2020 639554) and NSF Award CNS-1445424.

M. Hirt and A. Smith (Eds.): TCC 2016-B, Part II, LNCS 9986, pp. 330–360, 2016.
DOI: 10.1007/978-3-662-53644-5_13

1   Introduction

Consider a situation where a large number of data items $\mu_1, \mu_2, \ldots$ is stored on a remote cloud server. For privacy purposes, the data items are encrypted. The user, who holds the decryption key, can retrieve the encrypted data and decrypt it locally. Using fully homomorphic encryption (FHE) [20,34], it can also ask the server to evaluate a function g on the encrypted data, and produce an encryption of $g(\mu_1, \mu_2, \ldots)$ which can be sent back for decryption, all without compromising privacy. The state of the art homomorphic encryption schemes can be based on the hardness of the learning with errors (LWE) problem, and of particular importance to us is the scheme of Gentry et al. [22]. However, one could consider a case where the data belongs to a big organization, where different position holders have different access permissions to the data. That is, every user can only access some fraction of the encrypted items. A trivial solution would be to duplicate each data item, and encrypt each copy using the public keys of all permitted users. However, this might be unsatisfactory in many cases.

Attribute-based encryption (ABE) [26,35] is a special type of public-key encryption scheme that allows to implement access control.[1] A (master) public key mpk is used for encryption, and users are associated to secret keys sk_f corresponding to policy functions $f : \{0,1\}^\ell \to \{0,1\}$. The encryption of a message μ is labeled with a public attribute $x \in \{0,1\}^\ell$, and can be decrypted using sk_f if and only if $f(x) = 0$.[2] The security guarantee of ABE is collusion resistance: a coalition of users learns nothing about the plaintext message μ if none of their individual keys are authorized to decrypt the ciphertext. Goyal et al. [26] used bilinear maps to construct ABE for log-depth circuits. Gorbunov et al. [23] showed the first ABE scheme where the policies can be arbitrary (a-priori bounded) polynomial circuits, based on LWE. A scheme with improved parameters was presented by Boneh et al. [5].

Using ABE for encrypting our remote data, a user with access permission to a certain data item can retrieve and decrypt it, but what about private processing on the server side? This would require *homomorphic attribute-based encryption* (HABE). Intuitively, we would like a way for a user to specify a set of data items for which it has permission, as well as a function g to be applied, such that the server can evaluate g on those data items. We would like this procedure to be private, i.e. the server learns nothing about the contents, and compact, i.e. the size of the evaluated response is independent of the number of inputs and the complexity of g.

Gentry et al. [22] showed how to achieve this goal in the case where all items of interest have the same attribute x, but cannot allow any homomorphism across attributes, even if the decryptor is allowed to access all of them. It is possible to compose a standard ABE scheme together with *multi-key* FHE [16,27,31] to achieve HABE, at the cost of blowing up the ciphertext size with the number of inputs to the homomorphic function. We provide a proof for this fact in Appendix A.

---

[1] Throughout this work we will consider the flavor known as "key-policy" ABE.
[2] In the original formulation, the convention was opposite: that $f(x) = 1$ allows to decrypt. However in this work we use $f(x) = 0$ throughout.

1.1 Our Results

We show that under a proper relaxed formulation of the problem, there is a solution that allows cross-attribute evaluation, with the resulting ciphertext size not depending on the number of attributes at all. In the motivating example above, if the remote server holds various encrypted items under various attributes, then the client must specify which of these ciphertexts are allowed to participate in the computation. In our formulation, this is done by providing the server with the *policy* f associated with the user's decryption key (note that this is public information that does not compromise data privacy). The policy is a compact representation that indicates which attributes are accessible by the user and which are not, so the server can tell which ciphertexts are to be included. We call our notion targeted HABE (T-HABE) since the evaluator needs to know the *target policy* which will be used to decrypt the homomorphically evaluated ciphertext. We believe that our formulation can be useful in some situations, as illustrated by the motivating example above.

So far we discussed the case where the decryptor only has one secret key corresponding to a single policy, we call this *single target* (or single policy) HABE (ST-HABE). We extend this notion and consider *multi target* (or multi policy) HABE (MT-HABE), where the decryptor is defined not just by a single policy f, but rather by a collection of policies F. This means that the decryptor holds all $\{sk_f : f \in F\}$ and is thus allowed to decrypt ciphertexts with attribute x s.t. there exists $f \in F$ with $f(x) = 0$. This can be thought of as a single user with multiple keys, or as a collection of users who wish to perform homomorphic computation on the union of their permitted data items. In this setting, target homomorphism requires F to be known to the homomorphic evaluator. This notion trivially degenerates to the single-policy variant if F is a singleton set. A formal definition of T-HABE appears in Sect. 2.

We construct new ST-HABE and MT-HABE schemes as follows. In the single target setting, our scheme relies on the same hardness assumptions as previous (standard) ABE candidates [5,23], namely the polynomial hardness of learning with errors (LWE) with sub-exponential modulus-to-noise ratio. Our scheme is leveled both for policies and for homomorphic evaluation, which means that at setup time one can specify arbitrary depth bounds, and once they are set, all policies f and homomorphicly evaluated functions g must adhere by these bounds. We note that in terms of assumptions and functionality, our scheme performs as well as any known ABE for circuits and as well as any known FHE scheme (without bootstrapping). In fact, using the composition theorem in [17], we can get non-leveled full homomorphism. However, this requires a non-leveled MK-FHE as a building block, which is only known to exist under a circular security assumption (see e.g. [10]). We note that whereas the [17] result is stated for non-targeted HABE, it applies readily in this setting as well. See an outline of our construction in Sect. 1.2 below, and the full scheme in Sect. 4.

Our MT-HABE scheme relies on the same assumption but in the *random oracle model*, and furthermore the ciphertext size grows quadratically with the cardinality of the set F (i.e. if more policies are involved, more communication is needed),[3] however the ciphertext size is independent of the number of attributes and the complexity of g. Interestingly, we use the random oracle in order to generate a part of the *secret key*, and we show that security is still maintained. See an outline of our construction in Sect. 1.2 below, and the full scheme in Sect. 5.

1.2 Our Techniques

Previous works [11,16,22] observed that known LWE-based ABE schemes have the following structure. Given the public parameters pp and an attribute x, it is possible to derive a "designated public key" pk_x, which has the same structure as a public-key for Regev's famous encryption scheme [33] (more precisely "dual-Regev", introduced by Gentry et al. [21]), and indeed the encryption process is also identical to the dual-Regev scheme. Therefore, since the FHE scheme of Gentry, Sahai and Waters [22] (henceforth GSW) has the same key distribution as dual-Regev, one can just substitute the encryption procedure from dual-Regev to GSW, and single attribute homomorphism follows. To show that the evaluated ciphertext can be decrypted, GSW notice that the decryption procedure of the [23] ABE scheme can be seen as a two step process: first sk_f is preprocessed together with x to obtained $\mathsf{sk}_{f,x}$ which is a valid dual-Regev secret key for pk_x, and this key is used for standard dual-Regev decryption. This means that this key can also be used to decrypt GSW evaluated ciphertexts. This observation also carries over to the later ABE scheme of Boneh et al. [5]. A similar approach was used by Clear and McGoldrick [16] in conjunction with their multi-key homomorphism to achieve a homomorphic IBE (ABE where the policies are only point functions) where the ciphertext size grows with the number of attributes.

Our starting point is to consider a "dual" two-step decryption process for the [5] ABE, where given a ciphertext c_x relative to an attribute x, it is first pre-processed together with f to obtain $c_{x,f}$ which can then be decrypted by sk_f as a standard dual-Regev ciphertext. This is not a new perspective, in fact this is the original way [5,23] described their decryption process. We would hope, therefore, to apply targeted homomorphism by first preprocessing all input ciphertexts to make them correspond to the same sk_f, and then apply homomorphic evaluation. However, applied naively, preprocessing a GSW ciphertext destroys its homomorphic features. This is the reason GSW needed to reinterpret the decryption process in order for their approach to work even in the single input setting. We show how to modify the encryption procedure so as to allow preprocessing of a ciphertext *for any policy function f* without compromising its homomorphic features, which will allow to achieve targeted homomorphism for single policy (ST-HABE).

Our multi-target solution relies on the multi-key FHE scheme of [16], and in particular we use the simplified variant of Mukherjee and Wichs [31]. Recall that we have a set F of policies, where each attribute x in the computation has at least

---

[3] As in previous works, part of the ciphertext is redundant for decryption and can be truncated post-evaluation, which will lead to only linear dependence on $|F|$.

one policy $f \in F$ that can decrypt it. The basic idea is to group the ciphertexts according to the f's, preprocess them so all ciphertexts that correspond to a given f are now respective to the same (unknown) secret key sk_f. After preprocessing, the situation is equivalent to multi-key FHE with $|F|$ many users, each with their own key, so it would appear that we are in the clear. However, known LWE-based multi-key FHE schemes require *common public parameters*. In particular, all public keys are matrices which are identical except the last row, all secret keys are vectors with the last element being equal to 1. However, our preprocessing does not produce ciphertexts that conform with this requirement. In particular, our ciphertexts correspond to public keys that all share a prefix, but they differ in much more than a single row. We show that the [16,31] scheme can be generalized to the aforementioned case, however a fraction of the secret key needs to be known at homomorphic evaluation time. Whereas revealing this fraction of the key does not compromise security, it is generated independently for each policy f using the master secret key, and there appears to be no compact way to provide the key fractions for *all* policies in the public parameters. We resolve this using the random oracle heuristic, namely we show that we can generate a fraction of the secret key using the random oracle, which allows the homomorphic evaluator to learn the allowable part of all relevant keys and perform the multi-key homomorphism.

1.3   A More Formal Overview

Syntax. As mentioned earlier, in an ABE, ciphertexts are associated with an attribute x and a message μ, and decryption is possible using sk_f iff $f(x) = 0$. In a single-attribute homomorphic ABE, an evaluator given encryptions of $\mu_1, \mu_2, \ldots$, under the same attribute x and any circuit g, can compute an encryption of $g(\mu_1, \mu_2, \ldots)$ under the same attribute x. In a ST-HABE, an evaluator given encryptions of $\mu_1, \mu_2, \ldots$ under different attributes $x_1, x_2, \ldots$, any circuit g and a "target" f for which $f(x_1) = f(x_2) = \cdots = 0$, outputs a ciphertext that decrypts to $g(\mu_1, \mu_2, \ldots)$ under sk_f.

Prior ABE. We recall that in the [5] ABE, the public parameters contain a matrix $\mathbf{A}$, a vector $\mathbf{v}$ and a set of matrices $\mathbf{B}_1, \ldots, \mathbf{B}_\ell$, where ℓ is the supported attribute length. For all $x \in \{0,1\}^\ell$, we can define $\mathbf{B}_x = [\mathbf{B}_1 - x_1 \mathbf{G} \| \cdots \| \mathbf{B}_\ell - x_\ell \mathbf{G}]$, where $\mathbf{G}$ is the special "gadget" matrix, and use dual-Regev to encrypt messages w.r.t $[\mathbf{A} \| \mathbf{B}_x], \mathbf{v}$. Namely the ciphertexts are of the form $\mathbf{c} \approx [\mathbf{A} \| \mathbf{B}_x \| \mathbf{v}]^T \mathbf{s} + \mathbf{y}_\mu$, where $\mathbf{y}_\mu$ is some vector that encodes the message. Furthermore, given f, $\mathbf{B}_1, \ldots, \mathbf{B}_\ell$ can be preprocessed to obtain a matrix $\mathbf{B}_f$, and for all f, x s.t. $f(x) = 0$, there exists a publicly computable low-norm matrix $\mathbf{H} = \mathbf{H}_{f,x,\mathbf{B}_x}$ s.t. $\mathbf{B}_f = \mathbf{B}_x \mathbf{H}$. The secret key is a row vector $\mathsf{sk}_f = \mathbf{r}_f$ s.t. $\mathbf{r}_f [\mathbf{A} \| \mathbf{B}_f]^T = -\mathbf{v}^T$. Decryption proceeds by using $\widehat{\mathbf{H}} = \mathrm{diag}(\mathbf{I}, \mathbf{H}, 1)$ (i.e. a diagonal block matrix whose blocks are $\mathbf{I}, \mathbf{H}, 1$) to compute $\mathbf{c}_f = \widehat{\mathbf{H}}^T \mathbf{c}$ so that $\mathbf{c}_f \approx [\mathbf{A} \| \mathbf{B}_f \| \mathbf{v}]^T \mathbf{s} + \widehat{\mathbf{H}} \mathbf{y}_\mu$, and then using $\mathbf{r}_f$ to decrypt.

Warm-Up. Recall that in GSW style FHE, an encryption of μ under a secret key $\mathbf{r}$ is a matrix $\mathbf{D} + \mu\mathbf{G}$ where $\mathbf{rD} \approx \mathbf{0}^T$, where $\mathbf{G}$ is a gadget matrix of appropriate dimensions. As a warm-up, suppose we encrypt μ as

$$\mathbf{C} \approx [\mathbf{A}\|\mathbf{B}_x\|\mathbf{v}]^T\mathbf{S} + \mu\mathbf{G}.$$

That is, each column in the new ciphertext is essentially a ciphertext of the aforementioned ABE scheme (with different $\mathbf{y}$ in each column). Observe that $[\mathbf{r}_f\|1][\mathbf{A}\|\mathbf{B}_x\|\mathbf{v}]^T\mathbf{S} \approx \mathbf{0}^T$, so $\mathbf{C}$ is indeed a GSW style encryption of μ under the secret key $[\mathbf{r}_f\|1]$.

In order to achieve cross-attribute homomorphism, we would like to replace the matrix $[\mathbf{A}\|\mathbf{B}_x\|\mathbf{v}]^T\mathbf{S}$ in $\mathbf{C}$ with one that depends only on f and not x. Towards this goal, observe that

$$\widehat{\mathbf{H}}^T\mathbf{C} \approx [\mathbf{A}\|\mathbf{B}_f\|\mathbf{v}]^T\mathbf{S} + \mu\widehat{\mathbf{H}}^T\mathbf{G}.$$

Unfortunately, this is not a GSW style FHE ciphertext as described above because we have $\widehat{\mathbf{H}}^T\mathbf{G}$ instead of $\mathbf{G}$. In fact, GSW style homomorphic evaluation can be still made to work if we can ensure that $\widehat{\mathbf{H}}^T\mathbf{G}$ behaves like a gadget matrix (e.g. if the matrix $\widehat{\mathbf{H}}^T$ has a low-norm inverse, which is not true for a general $\mathbf{H}_{f,x,\mathbf{B}_x}$); instead, we provide a simpler fix that also yields shorter ciphertexts.

Our ST-HABE Scheme. Our ST-HABE ciphertext has two components. The first one is independent of x: $\mathbf{C} \approx [\mathbf{A}\|\mathbf{B}_0\|\mathbf{v}]^T\mathbf{S} + \mu\mathbf{G}$, where $\mathbf{B}_0$ is another matrix, like the other $\mathbf{B}_i$'s, which is added to the public parameters. The second one is similar to an ABE encryption of 0, with the same $\mathbf{S}$: $\mathbf{C}_x \approx \mathbf{B}_x^T\mathbf{S}$. Now, observe that

$$\mathbf{C}_f := \mathbf{C} + [\mathbf{0}\|\mathbf{H}^T\mathbf{C}_x\|\mathbf{0}] \approx [\mathbf{A}\|\mathbf{B}_0 + \mathbf{B}_f\|\mathbf{v}]^T\mathbf{S} + \mu\mathbf{G},$$

since $\mathbf{H}^T\mathbf{B}_x^T = \mathbf{B}_f^T$. Note that $\mathbf{C}_f$ is now indeed a GSW FHE ciphertext under the key $[\mathbf{r}_f\|1]$, where $\mathbf{r}_f$ is the modified ABE secret key satisfying

$$\mathbf{r}_f[\mathbf{A}\|\mathbf{B}_0 + \mathbf{B}_f]^T = \mathbf{v}^T.$$

The proof of security for the modified ABE scheme is very similar to that of [5] (in the simulation, we program $\mathbf{B}_0$ as $\mathbf{AR}_0$). See Sect. 4 for more details.

Our MT-HABE Scheme. For the multi-policy setting, assume for simplicity that we only have two attributes x, x' and two policies f, f' s.t. $f(x) = 0$, $f'(x') = 0$ (generalization is straightforward). After applying the transformation as above, we have $\mathbf{C}_f \approx [\mathbf{A}\|\mathbf{B}_0 + \mathbf{B}_f\|\mathbf{v}]^T\mathbf{S} + \mu\mathbf{G}$ and likewise for f'. In the background there are the secret keys $\mathbf{r}_f, \mathbf{r}_{f'}$. Let us partition $\mathbf{r}_f = [\mathbf{r}_1, \mathbf{r}_2]$, s.t. $\mathbf{r}_1\mathbf{A}^T + \mathbf{r}_2(\mathbf{B}_0 + \mathbf{B}_f)^T = -\mathbf{v}^T$. Likewise $\mathbf{r}_{f'} = [\mathbf{r}'_1, \mathbf{r}'_2]$. We show that the methods of [16,31] for achieving multi-key homomorphism generalize fairly straightforwardly whenever the value of the cross multiplication $\mathbf{r}_f[\mathbf{A}\|\mathbf{B}_0 + \mathbf{B}_{f'}\|\mathbf{v}]^T$ is publicly computable (note that the secret key for f is multiplied by the public key for f', and vice versa). One can verify that if the $\mathbf{r}_2$ components of the two keys are known, then this is indeed the case. Our approach is therefore to achieve multi-policy homomorphism by releasing the $\mathbf{r}_2$ components of the keys. This approach might seem risky, since information about the secret key is revealed. To see why this is not a problem, we recall that the key $\mathbf{r}_f$ is generated using a trapdoor for $\mathbf{A}$ such that $\mathbf{r}_f$ is distributed like a discrete Gaussian, conditioned on $\mathbf{r}_1\mathbf{A}^T + \mathbf{r}_2(\mathbf{B}_0 + \mathbf{B}_f)^T = -\mathbf{v}^T$. One can verify that the marginal distribution of $\mathbf{r}_2$ is Gaussian and completely independent of f (this fact had been utilized in [1,14]). Therefore there seems to be hope that releasing it might not hurt security. Another serious problem is that $\mathbf{r}_2$ is generated using secret information, and is not known to the evaluator. Unfortunately, we are only able to resolve this difficulty in the random oracle model, by generating $\mathbf{r}_2$ using the random oracle. Specifically, we apply the random oracle to $(\mathbf{A}, f)$ to obtain $\mathbf{r}_2$ for f. In a nutshell, producing $\mathbf{r}_2$ using a random oracle is secure since the security reduction can always program the response of the random oracle: if the call is on a function f s.t. $f(x^*) = 1$ (where x^* is the challenge attribute) then returning $\mathbf{r}_2$ is similar to answering a key generation query, and if $f(x^*) = 0$ then a random value can be returned, since a key generation query to f will never be issued and therefore no consistency issues arise. However, as described so far, this solution requires a special random oracle: one that samples from a discrete Gaussian distribution. We would like to rely on the standard binary random oracle. To this end, we will set $\mathbf{r}_f = [\mathbf{r}_1, \mathbf{r}_2]$ such that $\mathbf{r}_1$ is Gaussian and $\mathbf{r}_2$ is *binary*, conditioned on $\mathbf{r}_1\mathbf{A}^T + \mathbf{r}_2(\mathbf{B}_0 + \mathbf{B}_f)^T = -\mathbf{v}^T$. This will allow us to use a standard binary random oracle for the generation of $\mathbf{r}_2$.[4] In the proof of security, we use the discrete Gaussian sampler of Lyubashevsky and Wichs [28] instead of the Gaussian sampler of [2,30]. This sampler, which is based on rejection sampling, allows to sample from "partially Gaussian" distributions which is exactly what we need in order for the proof of security to go through. See Sect. 5 for more details. We note that for the sake of consistency, we also use this distribution of $\mathbf{r}_f$ in our single target scheme.

1.4   Other Related Work

Other works on homomorphic ABE include the works of Clear and McGoldrick [15,17]. In the former, program obfuscation is used to enhance the homomorphic ABE of [22] to support evaluating circuits of arbitrary depth. Still, cross-attribute homomorphism is not addressed. In the latter, it is shown how to use bootstrapping to leverage cross-attribute homomorphism into evaluating circuits without a depth bound. This result can be used in conjunction with our construction from Appendix A to achieve a non-compact solution, or in conjunction with our targeted scheme as explained above.

---

[4] Alternatively we could have shown that the Gaussian random oracle model is implied by the standard random oracle model. However this requires a fairly involved argument that we chose to avoid.

Brakerski and Vaikuntanathan [13] show how to extend the [5] ABE scheme to support attributes of unbounded polynomial length, and to provide semi-adaptive security guarantee. This was generalized by Goyal et al. [25] to a generic transformation that does not rely on the specific properties of the ABE scheme. Whereas the semi-adaptive transformation appears to be applicable here, it is not clear whether we can support unbounded attribute length using their methods and still maintain homomorphism. We leave this avenue of research for future work.

2   Targeted Homomorphic ABE

In this work, we define a notion of homomorphic ABE where the homomorphic evaluation process depends on the policy (or policies) that are used to decrypt the resulting ciphertext, we refer to such schemes as Targeted Homomorphic ABE (T-HABE). We start by defining the syntax of a T-HABE scheme, and proceed with definitions of correctness and security.

Definition 1 (Targeted Homomorphic ABE). *A* Targeted Homomorphic Attribute Based Encryption *(T-HABE) scheme is a tuple of* PPT *algorithms* THABE = (Setup, Enc, Keygen, TEval, Dec) *with the following syntax:*

- THABE.Setup(1^λ) *takes as input the security parameter (and possibly in addition some specification of the class of policies and class of homomorphic operations supported). It outputs a master secret key* msk *and a set of public parameters* pp.
- THABE.Enc$_{pp}(\mu, x)$ *uses the public parameters* pp *and takes as input a message* $\mu \in \{0, 1\}$ *and an attribute* $x \in \{0, 1\}^*$. *It outputs a ciphertext* ct.
- THABE.Keygen$_{msk}(f)$ *uses the master secret key* msk *and takes as input a policy* $f \in \mathcal{F}$. *It outputs a secret key* sk$_f$.
- THABE.TEval$_{pp}(F, ct^{(1)}, \dots, ct^{(k)}, g)$ *uses the public parameters* pp *and takes as input a set* F *of target policies,* k *ciphertexts* $ct^{(1)}, \dots, ct^{(k)}$ *and a function* $g \in \mathcal{G}$. *It outputs a ciphertext* ct^g.
- THABE.Dec(sk_F, ct^g) *takes as input a set of secret keys* sk$_F$ *for a set of policies* F, *with* sk$_F$ = $\{sk_f : f \in F\}$, *and a ciphertext* ct^g. *It outputs a message* $\mu \in \{0, 1\}$.

We will also consider a restriction of the above definition to the single-target setting, where the set F is only allowed to contain a single function. We call this Single Target HABE (ST-HABE). Explicit reference to the multi target setting is denoted MT-HABE.

Our correctness guarantee is that given the set of keys for the policy set F, an evaluated ciphertext decrypts correctly to the intended value.

Definition 2 (Correctness of T-HABE). *Let* $\{\mathcal{F}_\lambda\}_{\lambda \in \mathbb{N}}$ *be a class of policy functions and* $\{\mathcal{G}_\lambda\}_{\lambda \in \mathbb{N}}$ *be a class of evaluation functions. We say that* THABE = (Setup, Enc, Keygen, TEval, Dec) *is correct w.r.t* $\mathcal{F}, \mathcal{G}$ *if the following holds.*

Let $(\mathsf{msk}, \mathsf{pp}) = \mathsf{THABE.Setup}(1^\lambda)$. *Consider a set of functions* $F \subseteq \mathcal{F}_\lambda$ *of* $\mathrm{poly}(\lambda)$ *cardinality, and its matching set of secret keys* $\mathsf{sk}_F = \{\mathsf{sk}_f = \mathsf{THABE.Keygen}_{\mathsf{msk}}(f) : f \in F\}$, *a sequence of* $k \geq 1$ *messages and attributes* $\{(\mu^{(i)} \in \{0,1\}, x^{(i)} \in \{0,1\}^*)\}_{i \in [k]}$ *such that* $\forall x^{(i)}. \exists f \in F. f(x^{(i)}) = 0$, *and the sequence of their encryptions* $\{\mathsf{ct}^{(i)} = \mathsf{THABE.Enc}_{\mathsf{pp}}(\mu^{(i)}, x^{(i)})\}_{i \in [k]}$.

Then letting $g \in \mathcal{G}$ *for some* $g \in \{0,1\}^k \to \{0,1\}$, *and computing* $\mathsf{ct}^g = \mathsf{THABE.TEval}(F, \mathsf{ct}^{(1)}, \ldots, \mathsf{ct}^{(k)}, g)$, *it holds that*

$$\Pr[\mathsf{THABE.Dec}(\mathsf{sk}_F, \mathsf{ct}^g) \neq g(\mu^{(1)}, \ldots, \mu^{(k)})] = \mathrm{negl}(\lambda),$$

where the probability is taken over all of the randomness in the experiment.

We note that similarly to the definition of correctness of homomorphic encryption, we do not require correctness for ciphertexts that did not undergo homomorphic evaluation. However, this can be assumed w.l.o.g since the class $\mathcal{G}$ will always contain the identity function which will allow decryption by first evaluating identity and then decrypting.

Security is defined using the exact same experiment as standard ABE.

Definition 3 (Security for ABE/T-HABE). *Let* THABE *be an* T-HABE *scheme as above, and consider the following game between the challenger and adversary.*

1. *The adversary sends an attribute* x^* *to the challenger.*
2. *The challenger generates* $(\mathsf{msk}, \mathsf{pp}) = \mathsf{THABE.Setup}(1^\lambda)$ *and sends* pp *to the adversary.*
3. *The adversary makes arbitrarily many key queries by sending functions* f_i *(represented as circuits) to the challenger. Upon receiving such function, the challenger creates a key* $\mathsf{sk}_i = \mathsf{THABE.Keygen}_{\mathsf{msk}}(f_i)$ *and sends* sk_i *to the adversary.*
4. *The adversary sends a pair of messages* μ_0, μ_1 *to the challenger. The challenger samples* $b \in \{0,1\}$ *and computes* $\mathsf{ct}^* = \mathsf{THABE.Enc}_{\mathsf{pp}}(\mu_b, x^*)$. *It sends* ct^* *to the adversary.*
5. *The adversary makes arbitrarily many key queries as in Step 3 above.*
6. *The adversary outputs* $\tilde{b} \in \{0,1\}$.
7. *Let* legal *denote the event where all key queries of the adversary are such that* $f_i(x^*) = 1$. *If* legal, *the output of the game is* $b' = \tilde{b}$, *otherwise the output* b' *is a uniformly random bit.*

The advantage of an adversary $\mathcal{A}$ *is* $|\Pr[b' = b] - 1/2|$, *where* b, b' *are generated in the game played between the challenger and the adversary* $\mathcal{A}(1^\lambda)$.

The game above is called the selective *security game, because the adversary sends* x^* *before Step 2. The scheme* THABE *is* selectively secure *if any* PPT *adversary* $\mathcal{A}$ *only has negligible advantage in the selective security game.*

Stronger notions of security include semi-adaptive *security where step 1 only happens after step 2, and* adaptive *(or full) security where step 1 only happens after step 3.*

We note that the adversary has no benefit in making key queries for policies for which $f(x^*) = 0$ and therefore we can assume w.l.o.g that such queries are not made (this is obvious for selective and semi-adaptive security and slightly less obvious for adaptive security).

Negated Policies. We note again that as in previous lattice based ABE constructions, we allow decryption when $f(x) = 0$ and require that in the security game all queries are such that $f(x^*) = 1$.

3 Preliminaries

We denote vectors by lower-case bold letters (e.g. $\mathbf{v}$) and matrices by upper-case bold letters (e.g. $\mathbf{A}$). The i'th component of a vector $\mathbf{v}$ is denoted by v_i. The component in the ith row and the jth column of a matrix $\mathbf{A}$ is denoted by $\mathbf{A}[i, j]$. We denote the security parameter by λ and let $\mathrm{negl}(\lambda)$ denote a negligible function. Sets and distributions are usually denoted in plain uppercase. If S is a set, then we also use S to denote the uniform distribution over this set. The distinction will be clear from the context.

Elements of $\mathbb{Z}_q$ are represented by the integers in $(-q/2, q/2]$. In particular the absolute value of $x \in \mathbb{Z}_q$ is defined as $|x| = \min\{|y| : y \in \mathbb{Z}, y = x \pmod{q}\}$.

As in many previous works relying on the LWE assumption, we rely on distributions that are supported over a bounded domain. A distribution χ over $\mathbb{Z}$ is said to be *B-bounded* if it is supported only over $[-B, B]$. The infinity norm of a matrix $\mathbf{A}$ is defined as $\|\mathbf{A}\|_\infty = \max_{i,j} |\mathbf{A}[i, j]|$, and we write

$$\mathbf{A} \approx \mathbf{B} \quad (\text{err: } B)$$

to denote that $\|\mathbf{A} - \mathbf{B}\|_\infty \leq B$.

3.1 Learning with Errors (LWE)

The *Learning with Errors* (LWE) problem was introduced by Regev [33]. Our scheme relies on the hardness of its decisional version.

Definition 4 (Decisional LWE(DLWE) [33]). *Let λ be the security parameter, $n = n(\lambda)$ and $q = q(\lambda)$ be integers and let $\chi = \chi(\lambda)$ be a probability distribution over $\mathbb{Z}$. The $\mathrm{DLWE}_{n,q,\chi}$ problem states that for all $m = \mathrm{poly}(n)$, letting $\mathbf{A} \leftarrow \mathbb{Z}_q^{n \times m}$, $\mathbf{s} \leftarrow \mathbb{Z}_q^n$, $\mathbf{e} \leftarrow \chi^m$, and $\mathbf{u} \leftarrow \mathbb{Z}_q^m$, it holds that $(\mathbf{A}, \mathbf{s}^T \mathbf{A} + \mathbf{e}^T)$ and $(\mathbf{A}, \mathbf{u}^T)$ are computationally indistinguishable.*

In this work we only consider the case where $q \leq 2^n$. Recall that GapSVP_γ is the (promise) problem of distinguishing, given a basis for a lattice and a parameter d, between the case where the lattice has a vector shorter than d, and the case where the lattice doesn't have any vector shorter than $\gamma \cdot d$. SIVP is the search problem of finding a set of "short" vectors. The best known algorithms for GapSVP_γ ([36]) require at least $2^{\widetilde{\Omega}(n/\log \gamma)}$ time. We refer the reader to [32,33] for more information.

There are known reductions between $\mathsf{DLWE}_{n,q,\chi}$ and those problems, which allows us to appropriately choose the LWE parameters for our scheme. We summarize in the following corollary (which addresses the regime of sub-exponential modulus-to-noise ratio).

Corollary 1 ([29,30,32,33]). *For all $\epsilon > 0$ there exist functions $q = q(n) \leq 2^n, \chi = \chi(n)$ and $B = B(n)$ such that χ is B-bounded, $q/B \geq 2^{n^\epsilon}$ and such that $\mathsf{DLWE}_{n,q,\chi}$ is at least as hard as the classical hardness of GapSVP_γ and the quantum hardness of SIVP_γ for $\gamma = 2^{\Omega(n^\epsilon)}$.*

3.2 The Gadget Matrix

Let $\mathbf{g} = (1, 2, 4, \ldots, 2^{\lceil \log q \rceil - 1}) \in \mathbb{Z}_q^{\lceil \log q \rceil}$ and let $N = n \cdot \lceil \log q \rceil$. The *gadget matrix* $\mathbf{G}_n$ is defined as the diagonal concatenation of $\mathbf{g}$ n times. Formally, $\mathbf{G}_n = \mathbf{g} \otimes \mathbf{I}_n \in \mathbb{Z}_q^{n \times N}$. We omit the n when the dimension is clear from the context.

We define the inverse function $\mathbf{G}_n^{-1} : \mathbb{Z}_q^{n \times m} \to \{0, 1\}^{N \times m}$ which expands each entry $a \in \mathbb{Z}_q$ of the input matrix into a column of size $\lceil \log q \rceil$ consisting of the bits of the binary representation of a. We have the property that for any matrix $\mathbf{A} \in \mathbb{Z}_q^{n \times m}$, it holds that $\mathbf{G} \cdot \mathbf{G}^{-1}(\mathbf{A}) = \mathbf{A}$.

3.3 Trapdoors and Discrete Gaussians

Let $n, m, q \in \mathbb{N}$ and consider a matrix $\mathbf{A} \in \mathbb{Z}_q^{n \times m}$. For all $\mathbf{V} \in \mathbb{Z}_q^{n \times m'}$ and for any distribution P over $\mathbb{Z}^m$, we let $\mathbf{A}_P^{-1}(\mathbf{V})$ denote the random variable whose distribution is P conditioned on $\mathbf{A} \cdot \mathbf{A}_P^{-1}(\mathbf{V}) = \mathbf{V}$. A P-trapdoor for $\mathbf{A}$ is a procedure that can sample from a distribution within 2^{-n} statistical distance of $\mathbf{A}_P^{-1}(\mathbf{V})$ in time $\mathrm{poly}(n, m, m', \log q)$, for any $\mathbf{V}$. We slightly overload notation and denote a P-trapdoor for $\mathbf{A}$ by $\mathbf{A}_P^{-1}$.

The (centered) discrete Gaussian distribution over $\mathbb{Z}^m$ with parameter τ, denoted $D_{\mathbb{Z}^m, \tau}$, is the distribution over $\mathbb{Z}^m$ where for all $\mathbf{x}$, $\Pr[\mathbf{x}] \propto e^{-\pi \|\mathbf{x}\|^2 / \tau^2}$. When P is the Discrete Gaussian $D_{\mathbb{Z}^m, \tau}$, we denote $\mathbf{A}_P^{-1} = \mathbf{A}_\tau^{-1}$.

It had been established in a long sequence of works that it is possible to generate an almost uniform $\mathbf{A}$ together with a trapdoor as formalized below (the parameters are taken from [30] together with the Gaussian sampler of [9,21]).

Corollary 2 (Trapdoor Generation). *There exists an efficient procedure* $\mathsf{TrapGen}(1^n, q, m)$ *that outputs* $(\mathbf{A}, \mathbf{A}_{\tau_0}^{-1})$, *where* $\mathbf{A} \in \mathbb{Z}_q^{n \times m}$ *for all* $m \geq m_0$ *for* $m_0 = O(n \log q)$, $\mathbf{A}$ *is* 2^{-n}-*uniform and* $\tau_0 = O(\sqrt{n \log q \log n})$. *Furthermore, given* $\mathbf{A}_{\tau_0}^{-1}$, *one can obtain* $\mathbf{A}_\tau^{-1}$ *for any* $\tau \geq \tau_0$.

We will also use the "mixed" Gaussian-Binary sampler of Lyubashevsky and Wichs [28]. The following corollary is a consequence of example 2 in [28, Sect. 3.2], by adjusting the analysis for general $\mathbf{R}$ instead of random $\{-1, 0, 1\}$ entries.

Corollary 3 (Gaussian-Binary Sampler). *Let n, m, q be such that $m \geq n\lceil \log q \rceil$. With all but $O(2^{-n})$ probability over the choice of $\mathbf{A} \xleftarrow{\$} \mathbb{Z}_q^{n \times m}$, for all $\mathbf{R} \in \mathbb{Z}^{m \times N}$ with $N = n\lceil \log q \rceil$, one can obtain $[\mathbf{A} \| \mathbf{AR} + \mathbf{G}_n]_P^{-1}$ for $P = D_{\mathbb{Z}^m, \tau} \times \{0,1\}^N$ with $\tau = O(N\sqrt{mn} \cdot \|\mathbf{R}\|_\infty)$. Furthermore, for all $\mathbf{v}$, it holds that the marginal distribution of the last N coordinates of $[\mathbf{A} \| \mathbf{AR} + \mathbf{G}_n]_P^{-1}(\mathbf{v})$ are $O(2^{-n})$-uniform in $\{0,1\}^N$.*

3.4 Homomorphic Evaluation

We define the basic procedure that will be used for homomorphic evaluation of FHE ciphertexts and also in the ABE scheme [5,22,24].

Definition 5. *Let $n, q \in \mathbb{N}$. Consider $\mathbf{B}_1, \ldots, \mathbf{B}_\ell \in \mathbb{Z}_q^{n \times N}$ where $N = n\lceil \log q \rceil$, and denote $\vec{\mathbf{B}} = [\mathbf{B}_1 \| \cdots \| \mathbf{B}_\ell]$. Let f be a boolean circuit of depth d computing a function $\{0,1\}^\ell \to \{0,1\}$, and assume that f contains only NAND gates. We define $\mathbf{B}_f = \mathsf{Eval}(f, \vec{\mathbf{B}})$ recursively: associate $\mathbf{B}_1, \ldots, \mathbf{B}_\ell$ with the input wires of the circuit. For every wire w in f, let u, v be its predecessors and define*

$$\mathbf{B}_w = \mathbf{G} - \mathbf{B}_u \mathbf{G}^{-1}(\mathbf{B}_v). \tag{1}$$

Finally $\mathbf{B}_f$ is the matrix associated with the output wire.

The properties of Eval are summarized in the following facts.

Fact 1. *Consider $\mathbf{B}_1, \ldots, \mathbf{B}_\ell \in \mathbb{Z}_q^{n \times N}$ and $x \in \{0,1\}^\ell$. Denoting $\vec{\mathbf{B}} = [\mathbf{B}_1 \| \cdots \| \mathbf{B}_\ell]$ and $x\vec{\mathbf{G}} = [x_1 \mathbf{G} \| \cdots \| x_\ell \mathbf{G}]$, it holds that there exists an polynomial time algorithm $\mathsf{EvRelation}$ s.t. if $\mathbf{H} = \mathbf{H}_{f,x,\vec{\mathbf{B}}} = \mathsf{EvRelation}(f, x, \vec{\mathbf{B}})$ then $\|\mathbf{H}\|_\infty \leq (N+1)^d$ and furthermore*

$$(\mathbf{B}_f - f(x)\mathbf{G})^T = \mathbf{H}^T \cdot [\vec{\mathbf{B}} - x\vec{\mathbf{G}}]^T$$

where $\mathbf{B}_f = \mathsf{Eval}(f, \vec{\mathbf{B}})$.

In particular, if $\mathbf{B}_i = \mathbf{AR}_i + x_i \mathbf{G}$, i.e. $\vec{\mathbf{B}} = \mathbf{A}\vec{\mathbf{R}} + x\vec{\mathbf{G}}$ for $\vec{\mathbf{R}} = [\mathbf{R}_1 \| \cdots \| \mathbf{R}_\ell]$, then $\mathbf{B}_f = \mathbf{AR}_f + f(x)\mathbf{G}$ for $\mathbf{R}_f = \vec{\mathbf{R}} \cdot \mathbf{H}_{f,x,\vec{\mathbf{B}}}$.

To see why the fact holds, note that for the NAND evaluation in Eq. (1), one can verify that

$$\mathsf{EvRelation}(\text{NAND}, [x_u, x_v], [\mathbf{B}_u \| \mathbf{B}_v]) = \begin{bmatrix} -\mathbf{G}^{-1}(\mathbf{B}_v) \\ -x_u \mathbf{I} \end{bmatrix}.$$

Recursive application implies the general statement.

Fact 2. *Let $\mathbf{r} \in \mathbb{Z}_q^n$, $\mathbf{C}^{(1)}, \ldots, \mathbf{C}^{(k)} \in \mathbb{Z}_q^{n \times N}$ and $\mu^{(1)}, \ldots, \mu^{(k)} \in \{0,1\}$, be such that*

$$\mathbf{r}^T \mathbf{C}^{(i)} \approx \mu^{(i)} \mathbf{r}^T \mathbf{G} \qquad (\text{err} : B).$$

Let g be a boolean circuit of depth d computing a function $\{0,1\}^k \to \{0,1\}$, and assume that g contains only NAND gates. Let $\mathbf{C}_g = \mathsf{Eval}(g, \vec{\mathbf{C}})$, then

$$\mathbf{r}^T \mathbf{C}^{(i)} \approx g(\mu^{(1)}, \ldots, \mu^{(k)}) \mathbf{r}^T \mathbf{G} \qquad (\text{err} : B \cdot (N+1)^d).$$

3.5   Pseudorandom Functions

A pseudorandom function family is a pair of PPT algorithms PRF = (PRF.Gen, PRF.Eval), such that the key generation algorithm PRF.Gen(1^λ) takes as input the security parameter and outputs a seed $\sigma \in \{0,1\}^\lambda$. The evaluation algorithm PRF.Eval(σ, x) takes a seed $\sigma \in \{0,1\}^\lambda$ and an input $x \in \{0,1\}^*$, and returns a bit $y \in \{0,1\}$.

Definition 6. *A family* PRF *as above is secure if for every polynomial time adversary* $\mathcal{A}$ *it holds that*

$$\left| \Pr[\mathcal{A}^{\mathsf{PRF.Eval}(\sigma,\cdot)}(1^\lambda) = 1] - \Pr[\mathcal{A}^{\mathcal{O}(\cdot)}(1^\lambda) = 1] \right| = \mathrm{negl}(\lambda),$$

where $\sigma = \mathsf{PRF.Gen}(1^\lambda)$ *and* $\mathcal{O}$ *is a random oracle. The probabilities are taken over all of the randomness of the experiment.*

4   A Single Target Homomorphic ABE Scheme

In this section we present our construction of LWE-based Single Target HABE. As in previous works, a constant $\epsilon \in (0,1)$ determines the tradeoff between the hardness of the DLWE problem on which security is based, and the efficiency of the scheme.

The scheme supports any class of policies $\mathcal{F}_{\ell,d_{\mathcal{F}}} \subseteq \{0,1\}^\ell \to \{0,1\}$, and any class of operations $\mathcal{G}_{d_{\mathcal{G}}} \subseteq \{0,1\}^* \to \{0,1\}$, where $d_{\mathcal{F}}, d_{\mathcal{G}}$ is the bound on the depth of the circuit representation of each function in the set $\mathcal{F}, \mathcal{G}$, respectively. Out scheme works for any $\ell, d_{\mathcal{F}}, d_{\mathcal{G}} = \mathrm{poly}(\lambda)$.

- STHABE.Setup($1^\lambda, 1^\ell, 1^{d_{\mathcal{F}}}, 1^{d_{\mathcal{G}}}$). Choose n, q, B, χ, m as described in Sect. 4.1 below. Let $m = \max\{m_0, (n+1)\lceil \log q \rceil + 2\lambda\}$ (where m_0 is as in Corollary 2), $N = n\lceil \log q \rceil$ and $M = (m + N + 1)\lceil \log q \rceil$.
  Generate a matrix-trapdoor pair $(\mathbf{A}, \mathbf{A}_{\tau_0}^{-1}) = \mathsf{TrapGen}(1^n, q, m)$ (see Corollary 2), where $\mathbf{A} \in \mathbb{Z}_q^{n \times m}$. Generate matrices $\mathbf{B}_0, \mathbf{B}_1, \ldots, \mathbf{B}_\ell \xleftarrow{\$} \mathbb{Z}_q^{n \times N}$ and denote $\vec{\mathbf{B}} = [\mathbf{B}_1 \| \ldots \| \mathbf{B}_\ell]$. Generate a vector $\mathbf{v} \xleftarrow{\$} \mathbb{Z}_q^n$.
  Set $\mathsf{msk} = \mathbf{A}_{\tau_0}^{-1}$ and $\mathsf{pp} = (\mathbf{A}, \mathbf{B}_0, \vec{\mathbf{B}}, \mathbf{v})$.
- STHABE.Enc$_{\mathsf{pp}}(\mu, x)$, where $\mathsf{pp} = (\mathbf{A}, \mathbf{B}_0, \vec{\mathbf{B}}, \mathbf{v})$, $\mu \in \{0,1\}$ (however, this procedure is well defined for any $\mu \in \mathbb{Z}_q$ which will be useful for our next scheme) and $x \in \{0,1\}^\ell$.
  Sample a random matrix $\mathbf{S} \xleftarrow{\$} \mathbb{Z}_q^{n \times M}$, an error matrix $\mathbf{E}_A \xleftarrow{\$} \chi^{m \times M}$ and an error row vector $\mathbf{e}_v \xleftarrow{\$} \chi^M$.
  Generate $\ell + 1$ more error matrices as follows: For all $i \in [\ell]$ and $j \in [M]$, sample $\mathbf{R}_{i,j} \xleftarrow{\$} \{0,1\}^{m \times N}$. Let $\mathbf{E}_0, \ldots, \mathbf{E}_\ell$ be matrices of dimension $N \times M$ defined by $\mathbf{E}_i[j] = \mathbf{R}_{i,j}^T \mathbf{E}_A[j]$, where $\mathbf{E}_i[j]$, $\mathbf{E}_A[j]$ denotes the jth column of $\mathbf{E}_i, \mathbf{E}_A$ respectively. Let

$$\begin{bmatrix} \mathbf{C}_A \\ \mathbf{C}_0 \\ \mathbf{c}_v \end{bmatrix} = \begin{bmatrix} \mathbf{A}^T \\ \mathbf{B}_0^T \\ \mathbf{v}^T \end{bmatrix} \cdot \mathbf{S} + \begin{bmatrix} \mathbf{E}_A \\ \mathbf{E}_0 \\ \mathbf{e}_v \end{bmatrix} + \mu \mathbf{G}_{m+N+1}.$$

The rest of the ciphertext contains auxiliary information that will allow to decrypt given a proper functional secret key. For all $i \in [\ell]$ let

$$\mathbf{C}_i = [\mathbf{B}_i - x_i \mathbf{G}_n]^T \cdot \mathbf{S} + \mathbf{E}_i.$$

Denote $\mathbf{C}_x = \begin{bmatrix} \mathbf{C}_1 \\ \vdots \\ \mathbf{C}_\ell \end{bmatrix}$ and $\mathbf{E}_x = \begin{bmatrix} \mathbf{E}_1 \\ \vdots \\ \mathbf{E}_\ell \end{bmatrix}$.

The final ciphertext is $\mathsf{ct} = (x, \mathbf{C}_A, \mathbf{C}_0, \mathbf{c}_v, \mathbf{C}_x)$.

- STHABE.Keygen$_{\mathsf{msk}}(f)$. Given a circuit f computing a function $\{0,1\}^\ell \to \{0,1\}$, the key is generated as follows. Set $\mathbf{B}_f = \mathsf{Eval}(f, \vec{\mathbf{B}})$ (where Eval is as defined in Sect. 3.4).
Generate a random vector $\mathbf{r}_f' \overset{\$}{\leftarrow} \{0,1\}^N$. Let $\mathbf{r}_f^T = \mathbf{A}_\tau^{-1}(-(\mathbf{B}_0 + \mathbf{B}_f) \mathbf{r}_f'^T - \mathbf{v}^T)$, where $\tau = O(m \cdot N\ell \cdot (N+1)^{d_{\mathcal{F}}}) \geq \tau_0$ (the enlargement of τ is needed for the security proof to work). Note that

$$[\mathbf{r}_f \| \mathbf{r}_f' \| 1] \cdot [\mathbf{A} \| \mathbf{B}_0 + \mathbf{B}_f \| \mathbf{v}]^T = \mathbf{0}^T.$$

Output $\mathsf{sk}_f = [\mathbf{r}_f \| \mathbf{r}_f']$.
- STHABE.ApplyF$_{\mathsf{pp}}(\mathsf{ct}, f)$. This is an auxiliary function that is used for homomorphic evaluation below. It uses the public parameters pp and takes as input a ciphertext $\mathsf{ct} = (x, \mathbf{C}_A, \mathbf{C}_0, \mathbf{c}_v, \mathbf{C}_x)$ and a policy $f \in \mathcal{F}$, such that $f(x) = 0$. It computes and outputs a "functioned" ciphertext ct_f as follows.
Compute the matrix $\mathbf{H} = \mathbf{H}_{f,x,\vec{\mathbf{B}}} \in \mathbb{Z}_q^{\ell N \times N}$ as $\mathbf{H} = \mathsf{EvRelation}(f, x, \vec{\mathbf{B}})$ (see Fact 1), define $\mathbf{C}_f = \mathbf{H}^T \mathbf{C}_x$ and finally set

$$\widehat{\mathbf{C}}_f = \begin{bmatrix} \mathbf{C}_A \\ \mathbf{C}_0 + \mathbf{C}_f \\ \mathbf{c}_v \end{bmatrix}.$$

The "functioned" ciphertext is $\mathsf{ct}_f = \widehat{\mathbf{C}}_f$.
- STHABE.TEval$_{\mathsf{pp}}(f, \mathsf{ct}^{(1)}, \ldots, \mathsf{ct}^{(k)}, g)$. Given a policy $f \in \mathcal{F}$, k ciphertexts $\mathsf{ct}^{(1)}, \ldots, \mathsf{ct}^{(k)}$ and a function $g \in G$. $g \in \{0,1\}^k \to \{0,1\}$, for each $i \in [k]$ compute the matrix

$$\widehat{\mathbf{C}}_f^{(i)} = \mathsf{STHABE.ApplyF}_{\mathsf{pp}}(\mathsf{ct}^{(i)}, f).$$

Set $\mathsf{ct}^g = \mathbf{C}_f^g = \mathsf{Eval}(g, \widehat{\mathbf{C}}_f^{(1)}, \ldots, \widehat{\mathbf{C}}_f^{(k)})$ (see Definition 5 in Sect. 3.4).
- STHABE.Dec$(\mathsf{sk}_f, \mathsf{ct}^g)$. Given $\mathsf{sk}_f = [\mathbf{r}_f \| \mathbf{r}_f']$ and $\mathsf{ct}^g = \mathbf{C}_f^g$, compute the vector $\mathbf{c} = [\mathbf{r}_f \| \mathbf{r}_f' \| 1] \cdot \mathbf{C}_f^g$. Let $\mathbf{u}^T = (0, \ldots, 0, \lfloor q/2 \rfloor) \in \mathbb{Z}_q^{(m+N+1)}$. Compute $\tilde{\mu} = \mathbf{c}\mathbf{G}^{-1}(\mathbf{u})$. Output $\mu' = 0$ if $|\tilde{\mu}| \leq q/4$ and $\mu' = 1$ otherwise.

4.1   Choice of Parameters

The DLWE parameters n, q, B, χ are chosen according to constraints from the correctness and security analyses that follow. We require that $n \geq \lambda$, $q \leq 2^n$ and recall that $\ell = \text{poly}(\lambda) \leq 2^\lambda$. We recall that $m \geq m_0$ where $m_0 = O(n \log q)$, $N = n\lceil \log q \rceil$ and $M = (m + N + 1)\lceil \log q \rceil$, and we require that

$$2^{n^\epsilon} \geq 8 \cdot (N+1)^{2d_\mathcal{F}} \cdot (M+1)^{d_\mathcal{G}} \cdot \ell^2 \cdot P(m, N, M, \lceil \log q \rceil)$$

for $P(m, N, M, \lceil \log q \rceil) = \text{poly}(m, N, M, \lceil \log q \rceil) = n^{O(1)}$ defined in the correctness analysis below. These constraints can be met by setting $n = \tilde{O}(\lambda + d_\mathcal{F} + d_\mathcal{G})^{1/\epsilon}$.

We then choose q, χ, B accordingly based on Corollary 1, and note that it guarantees that indeed $q \leq 2^n$. Furthermore, this choice guarantees that

$$q/B \geq 2^{n^\epsilon} \geq 8 \cdot (N+1)^{2d_\mathcal{F}} \cdot (M+1)^{d_\mathcal{G}} \cdot \ell^2 \cdot P(m, N, M, \lceil \log q \rceil).$$

4.2   Correctness

Lemma 1. *The scheme* STHABE *with parameters* $\ell, d_\mathcal{F}, d_\mathcal{G}$ *is correct with respect to policy class* $\mathcal{F}_{\ell, d_\mathcal{F}}$ *and homomorphism class* $\mathcal{G}_{d_\mathcal{G}}$.

Proof. Let $(\mathsf{msk}, \mathsf{pp}) = \mathsf{STHABE.Setup}(1^\lambda, 1^\ell, 1^{d_\mathcal{F}}, 1^{d_\mathcal{G}})$. Consider a function $f \in \mathcal{F}$ and a matching secret key $\mathsf{sk}_f = \mathsf{STHABE.Keygen}_{\mathsf{msk}}(f)$, a sequence of $k \geq 1$ messages and attributes $\{(\mu^{(i)} \in \{0,1\}, x^{(i)} \in \{0,1\}^\ell)\}_{i \in [k]}$ such that $\{f(x^{(i)}) = 0\}_{i \in [k]}$, and the sequence of their encryptions respectively $\{\mathsf{ct}^{(i)} = \mathsf{STHABE.Enc}_{\mathsf{pp}}(\mu^{(i)}, x^{(i)})\}_{i \in [k]}$. For each ciphertext it holds that

$$\mathbf{C}_x \approx [\vec{\mathbf{B}} - x\vec{\mathbf{G}}] \qquad (\text{err: } mB)$$

Consider a function $g \in \mathcal{G}$ such that $g \in \{0,1\}^k \to \{0,1\}$, and let $\mathsf{ct}^g = \mathsf{STHABE.TEval}(f, \mathsf{ct}^{(1)}, \dots, \mathsf{ct}^{(k)}, g)$. Recall that for each ciphertext, during the execution of $\mathsf{STHABE.ApplyF}(\mathsf{ct}, f)$ we compute the matrix $\mathbf{C}_f^{(i)} = \mathbf{H}^{(i)} \mathbf{C}_x^{(i)}$.

By the properties stated at Fact 1, and since for all $i \in [k]$ $\|\mathbf{H}^{(i)}\|_\infty \leq (N+1)^{d_\mathcal{F}}$ and $f(x^{(i)}) = 0$, for each ciphertext it holds that

$$\mathbf{C}_f = \mathbf{H}^T \mathbf{C}_x \approx [\mathbf{B}_f - f(x)\mathbf{G}]^T \cdot \mathbf{S} = \mathbf{B}_f^T \cdot \mathbf{S} \qquad (\text{err: } (N+1)^{d_\mathcal{F}} \cdot \ell N \cdot mB)$$

and hence

$$\widehat{\mathbf{C}}_f \approx [\mathbf{A} \| \mathbf{B}_0 + \mathbf{B}_f \| \mathbf{v}]^T \cdot \mathbf{S} + \mu \mathbf{G} \qquad (\text{err: } mB \cdot (1 + (N+1)^{d_\mathcal{F}} \cdot \ell N)) \qquad (2)$$

(Note that Eq. (2) also holds when $\mu \in \mathbb{Z}_q$ instead of $\mu \in \{0,1\}$).

It therefore follows that

$$[\mathbf{r}_f \| \mathbf{r}_f' \| 1] \cdot \widehat{\mathbf{C}}_f \approx \mu \cdot [\mathbf{r}_f \| \mathbf{r}_f' \| 1] \cdot \mathbf{G}$$
$$(\text{err: } \left\| [\mathbf{r}_f \| \mathbf{r}_f' \| 1] \right\|_\infty \cdot mB \cdot (1 + (N+1)^{d_\mathcal{F}} \cdot \ell N) \cdot (m + N + 1))$$

Now consider a function $g \in \mathcal{G}$ such that $g \in \{0,1\}^k \to \{0,1\}$, and consider the execution of $\mathsf{STHABE.Dec}_{\mathsf{pp}}(\mathsf{sk}_f, \mathsf{ct}^g)$ where $\mathsf{ct}^g = \mathsf{STHABE.TEval}(f, \mathsf{ct}^{(1)}, \dots, \mathsf{ct}^{(k)}, g)$.

By Fact 2, denoting $\mu^g = g(\mu^{(1)}, \ldots, \mu^{(k)})$, we get

$$\mathbf{c} = [\mathbf{r}_f \| \mathbf{r}'_f \| 1] \cdot \mathbf{C}^g_f \approx \mu^g \cdot [\mathbf{r}_f \| \mathbf{r}'_f \| 1] \cdot \mathbf{G}$$

$$\left(\text{err:} \ \left\| [\mathbf{r}_f \| \mathbf{r}'_f \| 1] \right\|_\infty \cdot mB \cdot (1 + (N+1)^{d_{\mathcal{F}}} \cdot \ell N) \cdot (m + N + 1) \cdot (M+1)^{d_{\mathcal{G}}} \right)$$

and therefore

$$\tilde{\mu} = \mathbf{c} \mathbf{G}^{-1}(\mathbf{u}) \approx \mu^g \lfloor q/2 \rfloor \tag{3}$$

$$\left(\text{err:} \ \left\| [\mathbf{r}_f \| \mathbf{r}'_f \| 1] \right\|_\infty \cdot mB \cdot (1 + (N+1)^{d_{\mathcal{F}}} \ell N) \cdot (m + N + 1) \cdot (M+1)^{d_{\mathcal{G}}} \cdot \lceil \log q \rceil \right)$$

We conclude that we get correct decryption as long as the error in Eq. (3) is bounded away from $q/4$. We recall that by the properties of discrete Gaussians and since $\mathbf{r}'_f \in \{0,1\}^N$, it holds that $\left\| [\mathbf{r}_f \| \mathbf{r}'_f] \right\|_\infty \le \max\{\|\mathbf{r}_f\|_\infty, 1\} \le \tau \sqrt{m}$ with all but $2^{-(m)} = \text{negl}(\lambda)$ probability, where $\tau = O(\sqrt{mn} \cdot N^2 \ell \cdot (N+1)^{d_{\mathcal{F}}})$. Therefore, with all but negligible probability, the error is at most

$$\left\| [\mathbf{r}_f \| \mathbf{r}'_f \| 1] \right\|_\infty \cdot mB \cdot (1 + (N+1)^{d_{\mathcal{F}}} \ell N) \cdot (m + N + 1) \cdot (M+1)^{d_{\mathcal{G}}} \cdot \lceil \log q \rceil$$
$$\le O(\sqrt{mn} \cdot N^2 \ell \cdot (N+1)^{d_{\mathcal{F}}}) \sqrt{m} \cdot mB \cdot (1 + (N+1)^{d_{\mathcal{F}}} \ell N)$$
$$\cdot (m + N + 1) \cdot (M+1)^{d_{\mathcal{G}}} \cdot \lceil \log q \rceil$$
$$= B \cdot (N+1)^{2d_{\mathcal{F}}} \cdot (M+1)^{d_{\mathcal{G}}} \cdot \ell^2 \cdot P(m, N, M, \lceil \log q \rceil).$$

Since we set $B \le q / \left(8 \cdot (N+1)^{2d_{\mathcal{F}}} \cdot (M+1)^{d_{\mathcal{G}}} \cdot \ell^2 \cdot P(m, N, M, \lceil \log q \rceil) \right)$, it holds that the error is less than $q/4$. Hence,

$$\Pr[\text{STHABE.Dec}_{\text{pp}}(\text{sk}_f, \hat{\text{ct}}_f) \ne g(\mu^{(1)}, \ldots, \mu^{(k)})] = \text{negl}(\lambda).$$

4.3   Security

Lemma 2. *Under the* $\text{DLWE}_{n,q,\chi}$ *assumption, the scheme* STHABE *is selectively secure for the function classes* $\mathcal{F}, \mathcal{G}$. *Moreover, under this assumptions the scheme has* pseudorandom ciphertexts: *no polynomial time adversary can distinguish between the* $\mathbf{C}_A, \mathbf{C}_0, \mathbf{c}_v, \mathbf{C}_x$ *components of* ct* *and a set of uniform matrices of the same dimension. Furthermore, this is true even if the encryption algorithm is applied to an arbitrary* $\mu \in \mathbb{Z}_q$, *and not necessarily* $\mu \in \{0,1\}$.

The security proof is a straightforward extension of the proof of [5]. In fact, our setup and key generation procedure are identical to the [5] scheme, the only difference is the setting of the LWE parameters and the sampling of $\mathbf{r}'_f$ from the binary distribution rather than Gaussian. The latter issue only requires a minor change in the proof, namely replacing the [2,30] Gaussian sampler for matrices of the form $[\mathbf{A} \| \mathbf{A} \mathbf{R} + \mathbf{G}_n]$ with the [28] sampler which allows to sample from a part Gaussian part binary distribution for matrices of this form.

As for our ciphertexts, they are of the form $\widetilde{\mathbf{A}}^T \mathbf{S} + \widetilde{\mathbf{E}} + \mathbf{Y}_\mu$, where $\widetilde{\mathbf{A}}$ is derived from the public parameters, $\widetilde{\mathbf{E}}$ is noise, and $\mathbf{Y}_\mu$ is a matrix that is determined by the message μ. In [5], the ciphertext is of the form $\tilde{\mathbf{A}}^T \mathbf{s} + \tilde{\mathbf{e}} + \mathbf{y}'_\mu$. That is, we can almost think about our ciphertext as a matrix whose every column is a [5] ciphertext. The difference is that the encoding of the message $\mathbf{y}'_\mu$ is different from our $\mathbf{Y}_\mu$. However, [5] prove that their ciphertexts are *pseudorandom* and this means that they can mask $\mathbf{Y}_\mu$ regardless of its specific definition. The security of our scheme thus follows. The full proof follows.

Proof. Consider the selective security game as per Definition 3. Recall that in our scheme, an encryption of a message can be expressed as $\mathsf{ct} = (x, \mathbf{C})$, where

$$
\mathbf{C} = \tilde{\mathbf{A}}_x^T \mathbf{S} + \tilde{\mathbf{E}} + \tilde{\mathbf{Y}}_\mu = \begin{bmatrix} \mathbf{A}^T \\ \mathbf{B}_0^T \\ \mathbf{v}^T \\ (\vec{\mathbf{B}} - x\vec{\mathbf{G}})^T \end{bmatrix} \mathbf{S} + \begin{bmatrix} \mathbf{E}_A \\ \mathbf{E}_0 \\ \mathbf{e}_v \\ \mathbf{E}_x \end{bmatrix} + \begin{bmatrix} \mu \mathbf{G} \\ \mathbf{0} \end{bmatrix}.
$$

We note that all columns of $\mathbf{S}$ are identically and independently distributed and the same holds for all columns of $\tilde{\mathbf{E}}$. We intend to prove security using a hybrid on the columns of $\mathbf{C}$. That is, we will consider a modified game which is identical to the selective security game, except for the challenge phase, where the adversary gets either $\mathbf{c} = \tilde{\mathbf{A}}_{x^*}^T \mathbf{s} + \tilde{\mathbf{e}}$ or a completely uniform vector, and needs to distinguish the two cases. Specifically, $\mathbf{s}$ is a uniform vector, and $\tilde{\mathbf{e}} = [\mathbf{e}_A^T \| \mathbf{e}_0^T \| e_v \| \mathbf{e}_1^T \| \cdots \| \mathbf{e}_\ell^T]^T$, where the entries of $\mathbf{e}_A$ and e_v are sampled from χ and $\mathbf{e}_i = \mathbf{R}_i^T \mathbf{e}_A$ for $\mathbf{R}_0, \dots, \mathbf{R}_\ell$ which are uniform in $\{0,1\}^{m \times N}$ (recall that we choose the matrices $\mathbf{R}_i$ independently for each column of the ciphertext). We will refer to this game as the *column game* and denote the advantage of an adversary $\mathcal{A}'$ in this game as $|\Pr[b' = 1 | \mathbf{c}] - \Pr[b' = 1 | \text{uniform}]|$.

We start by showing that under the DLWE assumption, no polynomial time adversary can have noticeable advantage against the column game. Afterwards we will show that this implies the security of the scheme.

Consider an adversary $\mathcal{A}'$ for the column game discussed above, and let $\mathrm{Adv}[\mathcal{A}']$ denote its advantage in the column game. The proof will proceed by a sequence of hybrids, denote by $\mathrm{Adv}_{\mathcal{H}}[\mathcal{A}']$ the advantage of $\mathcal{A}'$ in the experiment described in hybrid $\mathcal{H}$.

Hybrid $\mathcal{H}_0$. This is the column game. By definition $\mathrm{Adv}[\mathcal{A}'] = \mathrm{Adv}_{\mathcal{H}_0}[\mathcal{A}']$.

Hybrid $\mathcal{H}_1$. We now change the way the matrices $\mathbf{B}_0$ and $\vec{\mathbf{B}}$ are generated. Recall that $\tilde{\mathbf{e}} = [\mathbf{e}_A^T \| \mathbf{e}_0^T \| e_v \| \mathbf{e}_1^T \| \cdots \| \mathbf{e}_\ell^T]^T$, where there exist $\mathbf{R}_0, \dots, \mathbf{R}_\ell$ which are uniform in $\{0,1\}^{m \times N}$ s.t. $\mathbf{e}_i = \mathbf{R}_i^T \mathbf{e}_A$. In this hybrid, we set $\mathbf{B}_i = \mathbf{A}\mathbf{R}_i + x_i \mathbf{G}_n$ instead of generating the $\mathbf{B}_i$ matrices uniformly.

Indistinguishability will follow from the extended leftover hash lemma as in [1, Lemma 13] (also used in [5]), since $m \geq (n+1)\lceil \log q \rceil + 2\lambda$.[5] We point out

---

[5] We note that they stated their lemma only for prime q, but in fact any q works for us since $\mathbf{R}_i$ have $\{0,1\}$ entries and since ± 1 are units over any ring $\mathbb{Z}_q$. Therefore matrix multiplication is a universal hash function for any distribution of binary vectors.

that the lemma can be used even though $\mathbf{A}$ is not uniform but only statistically close to uniform, since the argument here is information theoretic.

$$|\mathrm{Adv}_{\mathcal{H}_1}[\mathcal{A}'] - \mathrm{Adv}_{\mathcal{H}_0}[\mathcal{A}']| = \mathrm{negl}(\lambda).$$

We notice that in this hybrid, we now have that $\vec{\mathbf{B}} = \mathbf{A}\vec{\mathbf{R}} + x\vec{\mathbf{G}}$, where $\vec{\mathbf{R}} = [\mathbf{R}_1 \| \cdots \| \mathbf{R}_\ell]$.

Hybrid $\mathcal{H}_2$. In this hybrid we switch from generating sk_f using $\mathbf{A}_{\tau_0}^{-1}$ to generating them using $\mathbf{R}_0$ and $\vec{\mathbf{R}}$. We recall that we are only required to generate keys for f s.t. $f(x^*) = 1$, otherwise the adversary loses in the selective security game.

We recall that by definition, $\mathsf{sk}_f = [\mathbf{r}_f \| \mathbf{r}_f']$ where $\mathbf{r}_f' \xleftarrow{\$} \{0,1\}^N$ and $\mathbf{r}_f = \mathbf{A}_\tau^{-1}(-\mathbf{v} - (\mathbf{B}_0 + \mathbf{B}_f)\mathbf{r}_f'^T)$. Corollary 3 asserts that this is equivalent to sampling $[\mathbf{r}_f \| \mathbf{r}_f'] \xleftarrow{\$} [\mathbf{A}\|\mathbf{B}_0 + \mathbf{B}_f]_P^{-1}(-\mathbf{v})$ for $P = D_{\mathbb{Z}^m, \tau} \times \{0,1\}^N$, since the marginal distribution of $\mathbf{r}_f'$ is uniform binary, and the conditional distribution of $\mathbf{r}_f$ given $\mathbf{r}_f'$ is therefore the discrete Gaussian over the appropriate coset of the integer lattice. Denoting $\mathbf{H} = \mathbf{H}_{f, x^*, \vec{\mathbf{B}}}$, it holds that

$$\mathbf{B}_f - f(x^*)\mathbf{G}_n = \left(\vec{\mathbf{B}} - x^*\vec{\mathbf{G}} \right) \mathbf{H}.$$

Since $f(x^*) = 1$, we get that

$$\mathbf{B}_f = \mathbf{A}\vec{\mathbf{R}}\mathbf{H} + \mathbf{G}_n.$$

It also holds that

$$\mathbf{A}\mathbf{R}_0 + \mathbf{A}\vec{\mathbf{R}}\mathbf{H} = \mathbf{A}(\mathbf{R}_0 + \vec{\mathbf{R}}\mathbf{H})$$

Therefore, $[\mathbf{A}\|\mathbf{B}_0 + \mathbf{B}_f] = [\mathbf{A}\|\mathbf{A}\mathbf{R}_0 + \mathbf{A}\vec{\mathbf{R}}\mathbf{H} + \mathbf{G}_n] = [\mathbf{A}\|\mathbf{A}(\mathbf{R}_0 + \vec{\mathbf{R}}\mathbf{H}) + \mathbf{G}_n]$. By Corollary 3, given $\mathbf{R}_0, \vec{\mathbf{R}}$ and the computable matrix $\mathbf{H}$, we can sample from $[\mathbf{A}\|\mathbf{B}_0 + \mathbf{B}_f]_P^{-1}$, with $P = D_{\mathbb{Z}^m, \tau} \times \{0,1\}^N$ for all values of $\tau \geq \tau'$ for $\tau' = O\left(\sqrt{mn}N \cdot \left\|(\mathbf{R}_0 + \vec{\mathbf{R}}\mathbf{H})\right\|_\infty\right)$. This is true for all but $O(2^{-n})$ probability for random $\mathbf{A}$ and therefore, since TrapGen produces a distribution on $\mathbf{A}$ that is 2^{-n} uniform, it also holds for such matrices with all but $O(2^{-n})$ probability. Plugging in the bounds $\|\mathbf{H}\|_\infty \leq (N+1)^{d_{\mathcal{F}}}$, $\|\mathbf{R}_i\|_\infty = 1$, we get that $\left\|\mathbf{R}_0 + \vec{\mathbf{R}}\mathbf{H}\right\|_\infty \leq N\ell \cdot (N+1)^{d_{\mathcal{F}}}$ and therefore

$$\tau' = O(\sqrt{mn} \cdot N^2\ell \cdot (N+1)^{d_{\mathcal{F}}}).$$

Recall that we need to sample with $\tau = O(\sqrt{mn} \cdot N^2\ell \cdot (N+1)^{d_{\mathcal{F}}})$ and therefore, by appropriately setting τ, we can sample from $[\mathbf{A}\|\mathbf{B}_0 + \mathbf{B}_f]_P^{-1}$ up to $O(2^{-n})$ statistical distance.

It follows that after changing our method of sampling sk_f, the view of the adversary remains unchanged up to statistical distance of $\mathrm{poly}(\lambda) \cdot 2^{-n} = \mathrm{negl}(\lambda)$, since with all but $O(2^{-n})$ probability, our alternative sampler outputs a proper sample from a distribution that is within $O(2^{-n})$ statistical distance of $[\mathbf{A}\|\mathbf{B}_0 +$

$\mathbf{B}_f]_P^{-1}(-\mathbf{v})$. Since the number of key queries is at most $\text{poly}(\lambda)$, the result follows. We conclude that

$$|\text{Adv}_{\mathcal{H}_2}[\mathcal{A}'] - \text{Adv}_{\mathcal{H}_1}[\mathcal{A}']| = \text{negl}(\lambda).$$

We notice that in this hybrid, the challenger does not require $\mathbf{A}_{\tau_0}^{-1}$ at all.

Hybrid $\mathcal{H}_3$. In this hybrid, we change the distribution of $\mathbf{A}$ and sample it uniformly from $\mathbb{Z}_q^{n \times m}$ rather than via TrapGen. Since TrapGen samples $\mathbf{A}$ which is statistically indistinguishable from uniform, we conclude that the distribution produced in the two hybrids are statistically indistinguishable as well.

$$|\text{Adv}_{\mathcal{H}_3}[\mathcal{A}'] - \text{Adv}_{\mathcal{H}_2}[\mathcal{A}']| = \text{negl}(\lambda).$$

Hybrid $\mathcal{H}_4$. We change the contents of the challenge ciphertext as follows. We generate $\mathbf{s}, \mathbf{e}_A, e_v$ as before, and set $\mathbf{d} = \mathbf{A}^T \mathbf{s} + \mathbf{e}_A$, $d_v = \mathbf{v}^T \mathbf{s} + e_v$. The components of the vector $\mathbf{c}$ can now be expressed in terms of $\mathbf{d}, d_v$ since $\mathbf{c}^T = [\mathbf{d}^T \| \mathbf{d}^T \mathbf{R}_0 \| d_v \| \mathbf{d}^T \mathbf{R}_1 \| \cdots \| \mathbf{d}^T \mathbf{R}_\ell]$. This hybrid is in fact identical to the previous one, only notation had been changed.

$$\text{Adv}_{\mathcal{H}_4}[\mathcal{A}'] = \text{Adv}_{\mathcal{H}_3}[\mathcal{A}'].$$

We note that in this hybrid, given $\mathbf{d}, d_v$, the challenger does not need to know the values of $\mathbf{s}, \mathbf{e}_A, e_v$ since they are not used directly.

Hybrid $\mathcal{H}_5$. We change the distribution of $\mathbf{d}, d_v$ to be uniform in $\mathbb{Z}_q^m, \mathbb{Z}_q$. Indistinguishability follows by definition from the $\text{DLWE}_{n,q,\chi}$ assumption. We have

$$|\text{Adv}_{\mathcal{H}_5}[\mathcal{A}'] - \text{Adv}_{\mathcal{H}_4}[\mathcal{A}']| = \text{negl}(\lambda).$$

Hybrid $\mathcal{H}_6$. Finally, we change the distribution of $\mathbf{c}$ to uniform. By the leftover hash lemma, for all i it holds that $(\mathbf{A}, \mathbf{d}^T, \mathbf{A}\mathbf{R}_i, \mathbf{d}^T \mathbf{R}_i)$ are statistically close to uniform. Therefore this hybrid is statistically indistinguishable from the previous. We have that

$$|\text{Adv}_{\mathcal{H}_6}[\mathcal{A}] - \text{Adv}_{\mathcal{H}_5}[\mathcal{A}]| = \text{negl}(\lambda).$$

Clearly, in this hybrid the adversary has no advantage in the column game since $\mathbf{c}$ itself is uniform, so there is no difference between the two cases. It follows therefore that

$$\text{Adv}_{\mathcal{H}_6}[\mathcal{A}'] = 0,$$

and therefore

$$\text{Adv}[\mathcal{A}'] = \text{negl}(\lambda).$$

Having established the hardness of the column game, a straightforward hybrid argument over the columns of the ciphertext shows that no polynomial time adversary can have non-negligible advantage in a game that is identical to the selective security game, except $\widetilde{\mathbf{A}}_{x^*}^T \mathbf{S} + \widetilde{\mathbf{E}}$ in the generation of ct^* is replaced with a uniform matrix. Pseudorandomness of the ciphertext, and thus selective security, follows.

5   A Multi Target Homomorphic ABE Scheme

Using the multi-key FHE technique presented in [16,31], we generalize the single-target HABE scheme of the previous section to support homomorphic evaluations targeted to a *set of policies* instead of just one. In this variant, homomorphic evaluation is performed with respect to a set of policy functions $F = \{f_1, \ldots, f_d\}$ that "covers" all of the participating attributes. That is, any participating ciphertext's attribute zeros at least one function in F. The resulting ciphertext can be decrypted only with the set of keys corresponding to the set F.

We start in Sect. 5.1 by presenting a generalization to the [16,31] scheme that will be useful for our construction. Section 5.2 contains a description of the scheme, and the choice of parameters is in Sect. 5.3. Correctness and security analyses appear in Sects. 5.4 and 5.5.

5.1   A Generalized Multi-key FHE

We start with a describing a generalized version of the [16,31] MK-FHE scheme. Consider matrices $\mathbf{A} \in \mathbb{Z}_q^{n \times m}, \mathbf{B}_1, \ldots, \mathbf{B}_d \in \mathbb{Z}_q^{n \times N}$ and a vector $\mathbf{v} \in \mathbb{Z}_q^n$. For all $j \in [d]$ let $\mathbf{r}_j, \mathbf{r}_j'$ be vectors of dimensions m, N respectively, such that $[\mathbf{r}_j \| \mathbf{r}_j' \| 1] \cdot [\mathbf{A} \| \mathbf{B}_j \| \mathbf{v}]^T = \mathbf{0}$ and $\left\| [\mathbf{r}_j \| \mathbf{r}_j' \| 1] \right\|_\infty \leq B'$.

Let $\mathbf{C}^{(1)}, \ldots, \mathbf{C}^{(k)} \in \mathbb{Z}_q^{(m+N+1) \times M}$ be GSW-style encryptions of $\mu^{(1)}, \cdots, \mu^{(k)} \in \{0, 1\}$. That is, for all $i \in [k]$ there exists and index $j \in [d]$ and a matrix $\mathbf{S}^{(i)} \in \mathbb{Z}_q^{n \times M}$ for which

$$\mathbf{C}^{(i)} \approx [\mathbf{A} \| \mathbf{B}_j \| \mathbf{v}]^T \mathbf{S}^{(i)} + \mu^{(i)} \mathbf{G} \qquad \text{(err: } B) \tag{4}$$

(recall that $M = (m + N + 1) \lceil \log q \rceil$).

For all $i \in [k]$ let $\mathcal{X}^{(i)} = \{\mathbf{X}_{1,1}, \ldots, \mathbf{X}_{n,M}\}$ be a set of GSW-style encryptions of the entries of $\mathbf{S}^{(i)}$ under the same public key $[\mathbf{A} \| \mathbf{B}_j \| \mathbf{v}]^T$. So for all $\mathbf{X}_{a,b} \in \mathcal{X}^{(i)}$ we have

$$\mathbf{X}_{a,b} \approx [\mathbf{A} \| \mathbf{B}_j \| \mathbf{v}]^T \widetilde{\mathbf{S}}_{a,b}^{(i)} + \mathbf{S}^{(i)}[a, b] \mathbf{G} \qquad \text{(err: } B)$$

for some matrix $\widetilde{\mathbf{S}}_{a,b}^{(i)} \in \mathbb{Z}_q^{n \times M}$. Therefore,

$$[\mathbf{r}_j \| \mathbf{r}_j' \| 1] \cdot \mathbf{X}_{a,b} \approx \mathbf{S}^{(i)}[a, b] \cdot [\mathbf{r}_j \| \mathbf{r}_j' \| 1] \cdot \mathbf{G} \qquad \text{(err: } B' \cdot B \cdot (m + N + 1))$$

Let $\mathsf{LComb}(\mathcal{X}, \mathbf{u})$ be an algorithm that takes as input $\mathcal{X} = (\mathbf{X}_{1,1}, \ldots, \mathbf{X}_{n,M})$ as defined above and a vector $\mathbf{u} \in \mathbb{Z}_q^n$, and outputs a matrix $\mathbf{X} \in \mathbb{Z}_q^{(m+N+1) \times M}$ computed as follows:

For each $a \in [n], b \in [M]$ define a matrix $\mathbf{Z}_{a,b} \in \mathbb{Z}_q^{(m+N+1) \times M}$ consisting of zeros, where the only non-zero entry is $\mathbf{Z}_{a,b}[m + N + 1, b] = \mathbf{u}[a]$. Compute and output

$$\mathbf{X} = \sum_{a,b}^{n,M} \mathbf{X}_{a,b} \mathbf{G}^{-1}(\mathbf{Z}_{a,b}).$$

Lemma 3. *Consider the properties states above and let* $\mathbf{X}^{(i)} = \mathsf{LComb}$
$\left(\mathcal{X}^{(i)}, \mathbf{u}\right)$ *for some vector* $\mathbf{u} \in \mathbb{Z}_q^n$. *Then for all* $i \in [k]$, *it holds that*

$$[\mathbf{r}_j\|\mathbf{r}_j'\|1] \cdot \mathbf{X}^{(i)} \approx \mathbf{uS}^{(i)} \qquad (\text{err} : B' \cdot B \cdot (m + N + 1) \cdot nM \cdot \lceil \log q \rceil)$$

Proof. For all $i \in [k]$ It holds that

$$[\mathbf{r}_j\|\mathbf{r}_j'\|1] \cdot \mathbf{X}^{(i)} = [\mathbf{r}_j\|\mathbf{r}_j'\|1] \cdot \sum_{a,b}^{n,N} \mathbf{X}_{a,b} \mathbf{G}^{-1}(\mathbf{Z}_{a,b})$$

$$\approx \sum_{a,b}^{n,M} \mathbf{S}[a,b] \cdot [\mathbf{r}_j\|\mathbf{r}_j'\|1] \cdot \mathbf{GG}^{-1}(\mathbf{Z}_{a,b})$$

$$= \sum_{a,b}^{n,M} \mathbf{S}[a,b] \cdot (0, \ldots, 0, \mathbf{u}[a], 0, \ldots, 0) \quad (\text{Where } \mathbf{u}[a] \text{ is in the } b\text{th position}).$$

$$= \mathbf{uS}^{(i)} \qquad (\text{err} : B' \cdot B \cdot (m + N + 1) \cdot nM \cdot \lceil \log q \rceil)$$

Denoting $\mathsf{params} = (\mathbf{A}, \mathbf{B}_1, \ldots, \mathbf{B}_d, \mathbf{v})$, consider the following algorithm:
$\mathsf{Expand}_{\mathsf{params}}(\mathbf{C}, \mathcal{X}, (\mathbf{r}_1', \ldots, \mathbf{r}_d'), j)$ uses the parameters params and gets as input
a ciphertext $\mathbf{C}$ together with its auxiliary data $\mathcal{X}$ (as defined above), a sequence
of vectors $\mathbf{r}_1', \ldots, \mathbf{r}_d'$ of dimension N and an index $j \in [d]$. It computes and
outputs an "expanded" ciphertext $\widehat{\mathbf{C}}$ as follows:
    For all $t \in [d]\setminus\{j\}$ compute $\mathbf{X}_t = \mathsf{LComb}\left(\mathcal{X}, \mathbf{r}_t'(\mathbf{B}_t - \mathbf{B}_j)^T\right)$. Construct
and output the expanded matrix $\widehat{\mathbf{C}}$ as a $d \times d$ block matrix, where each block
$\widehat{\mathbf{C}}_{a,b} \in \mathbb{Z}_q^{(m+N+1)\times M}$ for $a, b \in [d]$ is defined as:

$$\widehat{\mathbf{C}}_{a,b} = \begin{cases} \mathbf{C} & a = b \\ \mathbf{X}_b & a = j, b \neq j \\ 0 & o.w. \end{cases}$$

Fact 3. *Consider the properties stated above. For all* $i \in [k]$ *let* $j \in [d]$ *such
that Eq. (4) holds and let* $\widehat{\mathbf{C}}^{(i)} = \mathsf{Expand}_{\mathsf{params}}(\mathbf{C}^{(i)}, \mathcal{X}^{(i)}, (\mathbf{r}_1', \ldots, \mathbf{r}_d'), j)$. *Let* $g \in
\{0,1\}^k \to \{0,1\}$ *be a circuit consisting of* NAND *gates of depth at most* d_g, *and
let* $\widehat{\mathbf{C}}^g = \mathsf{Eval}(g, \widehat{\mathbf{C}}^{(1)}, \ldots, \widehat{\mathbf{C}}^{(k)})$. *Then denoting* $\mathbf{r} = [\mathbf{r}_1\|\mathbf{r}_1'\|1\|\cdots\|\mathbf{r}_d\|\mathbf{r}_d'\|1]$ *and*
$\mu^g = g(\mu^{(1)}, \ldots, \mu^{(k)})$, *it holds that*

$$\mathbf{r} \cdot \widehat{\mathbf{C}}^g \approx \mu^g \cdot \mathbf{r} \cdot \mathbf{G}_{d(m+N+1)}$$

$$(\text{err} : B' \cdot B \cdot (m + N + 1)^2 \cdot (1 + nM \cdot \lceil \log q \rceil) \cdot kdM \cdot (dM + 1)^{d_g})$$

Proof. For all $i \in [k]$ it holds that

$$\widehat{\mathbf{C}}^{(i)} \approx \mathbf{I}_d \otimes \left([\mathbf{A}\|\mathbf{B}_j\|\mathbf{v}]^T \mathbf{S}^{(i)}\right) + \begin{bmatrix} 0 & & \cdots & & 0 \\ \mathbf{X}_1 \cdots \mathbf{X}_{j-1} & 0 & \mathbf{X}_{j+1} & \cdots & \mathbf{X}_d \\ 0 & & \cdots & & 0 \end{bmatrix}^{(i)}$$

$$+ \mu^{(i)} \mathbf{G}_{d(m+N+1)} \qquad (\text{err} : B)$$

where for all $t \in [d] \backslash \{j\}$, by Lemma 3 we have

$$[\mathbf{r}_j \| \mathbf{r}'_j \| 1] \cdot \mathbf{X}_t^{(i)} \approx \mathbf{r}'_t (\mathbf{B}_t - \mathbf{B}_j)^T \cdot \mathbf{S}^{(i)} \qquad (\text{err: } B' \cdot B \cdot (m + N + 1) \cdot nM \cdot \lceil \log q \rceil)$$

and therefore

$$[\mathbf{r}_t \| \mathbf{r}'_t \| 1] \cdot \mathbf{C}^{(i)} + [\mathbf{r}_j \| \mathbf{r}'_j \| 1] \cdot \mathbf{X}_t^{(i)} \approx \mu^{(i)} \cdot [\mathbf{r}_t \| \mathbf{r}'_t \| 1] \cdot \mathbf{G}$$
$$(\text{err: } B' \cdot B \cdot (m + N + 1) \cdot (1 + nM \cdot \lceil \log q \rceil))$$

from which it follows that

$$\mathbf{r} \cdot \widehat{\mathbf{C}}^{(i)} \approx \left[\mu^{(i)} \cdot [\mathbf{r}_1 \| \mathbf{r}'_1 \| 1] \cdot \mathbf{G} \| \ldots \| \mu^{(i)} \cdot [\mathbf{r}_d \| \mathbf{r}'_d \| 1] \cdot \mathbf{G} \right]$$
$$= \mu^{(i)} \cdot \mathbf{r} \cdot \mathbf{G}_{d(m+N+1)}$$
$$(\text{err: } B' \cdot B \cdot (m + N + 1) \cdot (1 + nM \cdot \lceil \log q \rceil))$$

Now let $g \in \{0,1\}^k \to \{0,1\}$, where g is of depth $d_{\mathcal{G}}$, and let $\widehat{\mathbf{C}}^g = \mathsf{Eval}(g, \widehat{\mathbf{C}})$. By Fact 2 we get

$$\mathbf{r} \cdot \widehat{\mathbf{C}}^g \approx \mu^g \cdot \mathbf{r} \cdot \mathbf{G}$$
$$(\text{err: } B' \cdot B \cdot (m + N + 1)^2 \cdot (1 + nM \cdot \lceil \log q \rceil) \cdot kdM \cdot (dM + 1)^{d_{\mathcal{G}}})$$

which completes the proof.

5.2 Our Scheme

Our Random Oracle. We consider a uniform random oracle $\mathcal{O}$. Namely, for every input $x \in \{0,1\}^*$, the value $\mathcal{O}(x)$ is a random variable that is uniformly distributed over $\{0,1\}^N$. The dimension of the vector N will be specified in the description of the scheme.

The Scheme. As in the STHABE construction, the scheme is parameterized with a security vs. efficiency trade-off constant $\epsilon \in (0,1)$, and supports a policies class $\mathcal{F}_{\ell,d_{\mathcal{F}}} \subseteq \{0,1\}^\ell \to \{0,1\}$ and homomorphism class $\mathcal{G}_{d_{\mathcal{G}}} \subseteq \{0,1\}^* \to \{0,1\}$. The scheme works for any $\ell, d_{\mathcal{F}}, d_{\mathcal{G}} = \mathrm{poly}(\lambda)$. We consider a family of pseudorandom functions PRF with seed length λ.

- THABE.Setup$(1^\lambda, 1^\ell, 1^{d_{\mathcal{F}}}, 1^{d_{\mathcal{G}}})$. Choose n, q, B, χ as described in Sect. 5.3 below, and generate $\mathbf{A}_{\tau_0}^{-1}$ and $\mathbf{A}, \mathbf{B}_0, \vec{\mathbf{B}}, \mathbf{v}$ as in STHABE.Setup. Generate a PRF seed $\sigma = \mathsf{PRF.Gen}(1^\lambda)$.
  Set $\mathsf{msk} = (\mathbf{A}_{\tau_0}^{-1}, \sigma)$ and $\mathsf{pp} = (\mathbf{A}, \mathbf{B}_0, \vec{\mathbf{B}}, \mathbf{v})$.
- THABE.Enc$_{\mathsf{pp}}(\mu, x)$. Let $(\mathbf{C}_A, \mathbf{C}_0, \mathbf{c}_v, \mathbf{C}_x) \leftarrow \mathsf{STHABE.Enc}_{\mathsf{pp}}(\mu, x)$ and denote $\mathbf{S} \in \mathbb{Z}_q^{n \times M}$ the randomness matrix that was generated in the encryption process.
  We now add an ABE-encryption of each entry of the matrix $\mathbf{S}$, respective to the attribute x. For all $a \in [n], b \in [M]$, let

$$\mathbf{X}_{a,b} \leftarrow \mathsf{STHABE.Enc}_{\mathsf{pp}}(\mathbf{S}[a,b], x)$$

As pointed out above, $\mathsf{STHABE.Enc_{pp}}$ is well defined and has some provable features even for $\mu \notin \{0, 1\}$, and indeed here we use it with $\mathbf{S}[a, b] \in \mathbb{Z}_q$. The final ciphertext is $\mathsf{ct} = (\mathbf{C}_A, \mathbf{C}_0, \mathbf{c}_v, \mathbf{C}_x, \mathcal{X} = (\mathbf{X}_{1,1}, \ldots, \mathbf{X}_{n,M}))$.

- $\mathsf{THABE.Keygen_{msk}}(f)$. Set $\mathbf{B}_f = \mathsf{Eval}(f, \vec{\mathbf{B}})$ and query the random oracle $\mathbf{r}'_f = \mathcal{O}(\mathbf{A}, f) \in \{0, 1\}^N$.

Let $\mathbf{r}_f^T = \mathbf{A}_\tau^{-1} \left(-(\mathbf{B}_0 + \mathbf{B}_f) {\mathbf{r}'_f}^T - \mathbf{v}^T \right)$, where $\tau = O(\sqrt{mn} \cdot N^2 \ell \cdot (N+1)^{d_{\mathcal{F}}}) \geq \tau_0$ (the enlargement of τ is needed for the security proof to work), such that the trapdoor function uses $\mathsf{PRF.Gen}(\sigma, f)$ as its randomness. Note that

$$[\mathbf{r}_f \| \mathbf{r}'_f \| 1] \cdot [\mathbf{A} \| \mathbf{B}_0 + \mathbf{B}_f \| \mathbf{v}]^T = \mathbf{0}^T.$$

Output $\mathsf{sk}_f = \mathbf{r}_f$.

- $\mathsf{THABE.Eval}(F, \mathsf{ct}^{(1)}, \ldots, \mathsf{ct}^{(k)}, g)$. Denoting $F = \{f_1, \ldots, f_d\}$, for every $i \in [k]$ let $j \in [d]$ be an index for which $f_j(x^{(i)}) = 0$. Compute

$$\widehat{\mathbf{C}}_f^{(i)} = \mathsf{STHABE.ApplyF_{pp}}(\mathsf{ct}^{(i)}, f_j),$$
$$\mathcal{X}_f^{(i)} = \{\mathsf{STHABE.ApplyF_{pp}}(\mathbf{X}, f_j) : \mathbf{X} \in \mathcal{X}^{(i)}\}$$

and for all $t \in [d]$ let $\mathbf{B}_{f_t} = \mathsf{Eval}(f_t, \vec{\mathbf{B}})$ and $\mathbf{r}'_t = \mathcal{O}(\mathbf{A}, f_t)$. Now compute

$$\mathbf{C}_F^{(i)} = \mathsf{Expand_{params}}(\widehat{\mathbf{C}}_f^{(i)}, \mathcal{X}_f^{(i)}, (\mathbf{r}'_1, \ldots, \mathbf{r}'_d), j)$$

where

$$\mathsf{params} = (\mathbf{A}, (\mathbf{B}_0 + \mathbf{B}_{f_1}), \ldots, (\mathbf{B}_0 + \mathbf{B}_{f_d}), \mathbf{v}).$$

Finally, set $\mathsf{ct}^g = \mathbf{C}_F^g = \mathsf{Eval}(g, \mathbf{C}_F)$.

- $\mathsf{THABE.Dec}(\mathsf{sk}_{f_1}, \ldots, \mathsf{sk}_{f_d}, \mathsf{ct}^g)$. For all $j \in [d]$ sample $\mathbf{r}'_{f_j} = \mathcal{O}(\mathbf{A}, f_j)$. Construct the concatenated key $\mathbf{r}_F = [\mathbf{r}_{f_1} \| \mathbf{r}'_{f_1} \| 1 \| \cdots \| \mathbf{r}_{f_d} \| \mathbf{r}'_{f_d} \| 1]$ and compute the vector $\mathbf{c} = \mathbf{r}_F \cdot \mathbf{C}_f^g$.

Let $\mathbf{u}^T = (0, \ldots, 0, \lfloor q/2 \rfloor) \in \mathbb{Z}_q^{d(m+N+1)}$. Compute $\tilde{\mu} = \mathbf{c}\mathbf{G}^{-1}(\mathbf{u})$. Output $\mu' = 0$ if $|\tilde{\mu}| \leq q/4$ and $\mu' = 1$ otherwise.

5.3 Choice of Parameters

The DLWE parameters n, q, B, χ are chosen according to constraints from the correctness and security analyses that follow. We require that $n \geq \lambda$, $q \leq 2^n$ and recall that $\ell, d = \mathsf{poly}(\lambda) \leq 2^\lambda$. We recall that $m = O(n \log q)$, $N = n \lceil \log q \rceil$ and $M = (m + N + 1) \lceil \log q \rceil$, and we require that

$$2^{n^\epsilon} \geq 8 \cdot (N+1)^{2 d_{\mathcal{F}}} \cdot (dM+1)^{d_g} \cdot d^{1.5} \cdot \ell^2 \cdot P(n, m, N, M, \lceil \log q \rceil)$$

for $P(n, m, N, M, \lceil \log q \rceil) = \mathsf{poly}(n, m, N, M, \lceil \log q \rceil) = n^{O(1)}$ defined in the correctness analysis below. These constraints can be met by setting $n = \tilde{O}(d_{\mathcal{F}} + \lambda d_g)^{1/\epsilon}$. We then choose q, χ, B accordingly based on Corollary 1. This choice guarantees that

$$q/B \geq 2^{n^\epsilon} \geq 8 \cdot (N+1)^{2 d_{\mathcal{F}}} \cdot (dM+1)^{d_g} \cdot d^{1.5} \cdot \ell^2 \cdot P(n, m, N, M, \lceil \log q \rceil).$$

5.4 Correctness

Lemma 4. *The scheme* THABE *with parameters* $\ell, d_{\mathcal{F}}, d_{\mathcal{G}}$ *is correct with respect to policy class* $\mathcal{F}_{\ell, d_{\mathcal{F}}}$ *and homomorphism class* $\mathcal{G}_{d_{\mathcal{G}}}$.

Proof. Let $(\mathsf{msk}, \mathsf{pp}) = \mathsf{THABE.Setup}(1^\lambda, 1^\ell, 1^{d_{\mathcal{F}}}, 1^{d_{\mathcal{G}}})$. Consider a set of $d \geq 1$ functions $F = \{f_1, \ldots, f_d \in\} \subseteq \mathcal{F}$ along with matching secret keys $\{\mathsf{sk}_f = \mathsf{THABE.Keygen}_{\mathsf{msk}}(f)\}_{f \in F}$. Consider a sequence of $k \geq 1$ messages and attributes $\{(\mu^{(i)} \in \{0,1\}, x^{(i)} \in \{0,1\}^\ell)\}_{i \in [k]}$, such that

$$\forall i \in [k] \ \exists j \in [d]: \ f_j(x^{(i)}) = 0,$$

and the sequence of their encryptions $\{\mathsf{ct}^{(i)} = \mathsf{THABE.Enc}_{\mathsf{pp}}(\mu^{(i)}, x^{(i)})\}_{i \in [k]}$. Let $g \in \mathcal{G}$ and consider the execution of $\mathsf{THABE.Eval}(F, \mathsf{ct}^{(1)}, \ldots, \mathsf{ct}^{(k)}, g)$. By Eq. (2), for all $i \in [k]$ the following holds:

- $\widehat{\mathbf{C}}_f^{(i)} \approx [\mathbf{A} \| \mathbf{B}_0 + \mathbf{B}_{f_j} \| \mathbf{v}]^T \mathbf{S}^{(i)} + \mu^{(i)} \mathbf{G}$ (err: $mB \cdot (1 + (N+1)^{d_{\mathcal{F}}} \cdot \ell N)$).
- $\forall \ \mathbf{X}_{a,b} \in \mathcal{X}_f^{(i)}$,
  $\mathbf{X}_{a,b} \approx [\mathbf{A} \| \mathbf{B}_0 + \mathbf{B}_{f_j} \| \mathbf{v}]^T \widetilde{\mathbf{S}}_{a,b}^{(i)} + \mathbf{S}^{(i)}[a,b] \mathbf{G}$ (err: $mB \cdot (1 + (N+1)^{d_{\mathcal{F}}} \cdot \ell N)$)
  for some $\widetilde{\mathbf{S}}_{a,b}^{(i)}$.
- $[\mathbf{r}_{f_j} \| \mathbf{r}'_{f_j} \| 1] \cdot [\mathbf{A} \| \mathbf{B}_0 + \mathbf{B}_{f_j} \| \mathbf{v}]^T = \mathbf{0}$

Therefore, considering $\mathsf{THABE.Dec}(\mathsf{sk}_{f_1}, \ldots, \mathsf{sk}_{f_d}, \mathsf{ct}^g)$, by Fact 3 it holds that

$$\mathbf{c} = \mathbf{r}_F \cdot \mathbf{C}_F^g \approx \mu^g \cdot \mathbf{r}_F \cdot \mathbf{G}$$
$$(\text{err: } \|\mathbf{r}_F\|_\infty \cdot mB(1 + (N+1)^{d_{\mathcal{F}}} \cdot \ell N) \cdot (m + N + 1)^2 \cdot (1 + nM \cdot \lceil \log q \rceil) \cdot$$
$$(dM+1)^{d_{\mathcal{G}}})$$

and therefore

$$\tilde{\mu} = \mathbf{c} \mathbf{G}^{-1}(\mathbf{u}) \approx \mu^g \lfloor q/2 \rceil \tag{5}$$
$$(\text{err: } \|\mathbf{r}_F\|_\infty \cdot mB(1 + (N+1)^{d_{\mathcal{F}}} \cdot \ell N) \cdot (m + N + 1)^2 \cdot (1 + nM \cdot \lceil \log q \rceil) \cdot$$
$$(dM+1)^{d_{\mathcal{G}}} \lceil \log q \rceil)$$

We conclude that we get correct decryption as long as the error in Eq. (5) is bounded away from $q/4$. We recall that by the properties of discrete Gaussians, it holds that $\|\mathbf{r}_F\|_\infty \leq \tau \sqrt{dm}$ with all but $2^{-dm} = \mathrm{negl}(\lambda)$ probability, where $\tau = O(\sqrt{mn} \cdot N^2 \ell \cdot (N+1)^{d_{\mathcal{F}}})$. Therefore, with all but negligible probability, the error is at most

$$\|\mathbf{r}_F\|_\infty \cdot mB(1 + (N+1)^{d_{\mathcal{F}}} \cdot \ell N) \cdot (m + N + 1)^2 \cdot (1 + nM \cdot \lceil \log q \rceil) \cdot$$
$$(dM+1)^{d_{\mathcal{G}}} \cdot \lceil \log q \rceil$$
$$\leq O(\sqrt{mn} \cdot N^2 \ell \cdot (N+1)^{d_{\mathcal{F}}}) \sqrt{dm} \cdot mB(1 + (N+1)^{d_{\mathcal{F}}} \cdot \ell N) \cdot (m + N + 1)^2 \cdot$$
$$(1 + nM \cdot \lceil \log q \rceil) \cdot (dM+1)^{d_{\mathcal{G}}} \cdot \lceil \log q \rceil$$
$$= B \cdot (N+1)^{2d_{\mathcal{F}}} \cdot (dM+1)^{d_{\mathcal{G}}} \cdot d^{1.5} \cdot \ell^2 \cdot P(n, m, N, M, \lceil \log q \rceil).$$

Since we set (see Sect. 5.3)

$$B \leq q/\left(8 \cdot (N+1)^{2d_{\mathcal{F}}} \cdot (dM+1)^{d_{\mathcal{G}}} \cdot d^{1.5} \cdot \ell^2 \cdot P(n, m, N, M, \lceil \log q \rceil)\right),$$

it holds that the error is less than $q/4$. Hence,

$$\Pr[\mathsf{THABE}.\mathsf{Dec}_{\mathsf{pp}}(\mathsf{sk}_F, \mathsf{ct}^g) \neq g(\mu^{(1)}, \ldots, \mu^{(k)})] = \mathsf{negl}(\lambda).$$

5.5 Security

Lemma 5. *In the random oracle model, under the* $\mathrm{DLWE}_{n,q,\chi}$ *assumption the scheme* THABE *is selectively secure for the function classes* $\mathcal{F}, \mathcal{G}$.

Proof. Consider the selective security game as per Definition 1 and let $\mathcal{A}$ be an adversary with advantage $\mathrm{Adv}[\mathcal{A}]$ in the selective security game. We start with a claim on random oracle queries that will be useful down the line. We classify oracle queries as follows. A query is *blind* if it is made before x^* is sent to the challenger. A query is *valid* if it is of the form $(\mathbf{D}, f)$ with $\mathbf{D} = \mathbf{A}$ and $f(x^*) = 1$ (for the matrix $\mathbf{A}$ in the public parameters). Let η be the probability that a blind and valid oracle query is made throughout the experiment. Clearly, since blind queries are made by the adversary before any information on $\mathbf{A}$ is given to him, the probability of any blind query has $\mathbf{D} = \mathbf{A}$ is at most $q^{-nm} = \mathsf{negl}(\lambda)$. Since the total number of queries is $\mathrm{poly}(\lambda)$ it holds that $\eta = \mathsf{negl}(\lambda)$.

The proof proceeds by a sequence of hybrids. Recall that in the random oracle model, the challenger needs to also be able to answer oracle queries at all steps of the security game.

Hybrid $\mathcal{H}_0$. In this hybrid, the challenger executes the selective security game as prescribed. Oracle queries are answered "on the fly": if the query is made for the first time, a fresh $\mathbf{r}$ is sampled uniformly from $\{0, 1\}^N$, and if the query had been made before then a consistent response is returned. By definition $\mathrm{Adv}[\mathcal{A}] = \mathrm{Adv}_{\mathcal{H}_0}[\mathcal{A}]$.

Hybrid $\mathcal{H}_1$. In this hybrid, the challenger, upon receiving x^*, checks whether any of the previous oracle calls had been blind and valid. If any such query had been made, the challenger aborts. Since this happens with negligible probability as analyzed above, the view of the adversary is statistically indistinguishable from the previous hybrid.

$$|\mathrm{Adv}_{\mathcal{H}_1}[\mathcal{A}] - \mathrm{Adv}_{\mathcal{H}_0}[\mathcal{A}]| = \mathsf{negl}(\lambda).$$

Hybrid $\mathcal{H}_2$. In this hybrid, we no longer use the PRF to generate randomness for the Gaussian sampler in Keygen queries. Instead, the challenger will keep track of all Keygen queries made so far. Given a Keygen query on a function f that was made before, it will answer consistently. When a new query is made, a new

random string is generated and used for the Gaussian sampling. The pseudorandomness property of the PRF guarantees that this hybrid is indistinguishable
from the previous one.

$$|\text{Adv}_{\mathcal{H}_2}[\mathcal{A}] - \text{Adv}_{\mathcal{H}_1}[\mathcal{A}]| = \text{negl}(\lambda).$$

From this point and on, we assume that a Keygen query is not made with the
same f more than once.

Hybrid $\mathcal{H}_3$. We now change the way non-blind and valid oracle queries, as well
as Keygen queries, are answered. First, we assume w.l.o.g that any non-blind and
valid oracle query is preceded by a Keygen query to the same function f (this
is allowed since $f(x^*) = 1$ by definition of a valid query). The Keygen query
itself is answered by using $\mathbf{A}_{\tau_0}^{-1}$ to sample $[\mathbf{r}_f\|\mathbf{r}_f'] = [\mathbf{A}\|\mathbf{B}_0 + \mathbf{B}_f]_P^{-1}(-\mathbf{v})$ where
$P = D_{\mathbb{Z}^m,\tau} \times \{0,1\}^N$. It then stores $\mathbf{r}_f'$ as the answer to the oracle query $(\mathbf{A}, f)$
(which at this point had necessarily not yet been made), and returns $\mathbf{r}_f$ as the
response to the Keygen(f) query.

Since Corollary 3 implies that the marginal distribution of the $\mathbf{r}'$ compo-
nent of $[\mathbf{A}\|\mathbf{B}_0 + \mathbf{B}_f]_\tau^{-1}(-\mathbf{v})$ is statistically indistinguishable from uniform over
$\{0,1\}^N$, it follows that the view of the adversary in this experiment is statisti-
cally close to the previous hybrid.

$$|\text{Adv}_{\mathcal{H}_3}[\mathcal{A}] - \text{Adv}_{\mathcal{H}_2}[\mathcal{A}]| = \text{negl}(\lambda).$$

Hybrid $\mathcal{H}_4$. At this point, we notice that the challenger in $\mathcal{H}_3$ can be simulated
via black box access to the challenger of our single key scheme described in
Sect. 4. This is because valid and non blind oracle queries are translated into
key generation queries, and all other queries are answered randomly. Since in
the proof of Lemma 2 we show that the encryption is secure even for non binary
messages, we can replace the encryptions of $\mathbf{S}$ in the challenge ciphertext with
encryptions of all 0, and asserts that this is indistinguishable to the adversary.

$$|\text{Adv}_{\mathcal{H}_4}[\mathcal{A}] - \text{Adv}_{\mathcal{H}_3}[\mathcal{A}]| = \text{negl}(\lambda).$$

Hybrid $\mathcal{H}_5$. Now that $\mathbf{S}$ is only used for generating the encryption of the message
bit μ, we can again use Lemma 2 to replace this part of the challenge ciphertext
with an encryption of 0.

$$|\text{Adv}_{\mathcal{H}_5}[\mathcal{A}] - \text{Adv}_{\mathcal{H}_4}[\mathcal{A}]| = \text{negl}(\lambda).$$

Clearly in this hybrid the adversary has no advantage since its view is independent of μ_b. Therefore $\text{Adv}_{\mathcal{H}_5}[\mathcal{A}] = 1/2$ and it follows that

$$|\text{Adv}[\mathcal{A}] - 1/2| = \text{negl}(\lambda),$$

which completes the proof of security.

Acknowledgments. We thank Vadim Lyubashevsky for numerous insightful discussions.

A   A Generic (Non-compact) Homomorphic ABE Construction

We show how to construct a non-targeted homomorphic ABE (HABE) given any ABE scheme and Multi-Key FHE scheme as building blocks. The main disadvantage of this construction is that the ciphertext's size grows at least linearly with the number of participants in the homomorphic evaluation. Interestingly, our method is very similar to the one presented in [17], despite the difference in the scheme's goal. Their construction relies on a leveled homomorphic ABE and uses it to create a non-leveled HABE scheme.

Below are definitions of ABE, MFHE and HABE, followed by our HABE construction and a brief proof of its correctness and security.

Definition 7 (ABE). *An* Attribute Based Encryption *(ABE) scheme is a tuple of* PPT *algorithms* ABE = (Setup, Enc, Keygen, Dec) *with the following syntax:*

- ABE.Setup(1^λ) *takes as input the security parameter and outputs a master secret key* msk *and a set of public parameters* pp.
- ABE.Enc$_{pp}(\mu, x)$ *uses the public parameters* pp *and takes as input a message* $\mu \in \{0, 1\}$ *and an attribute* $x \in \{0, 1\}^\ell$. *It outputs a ciphertext* ct.
- ABE.Keygen$_{msk}(f)$ *uses the master secret key* msk *and takes as input a function* $f \in \mathcal{F}$. *It outputs a secret key* sk$_f$.
- ABE.Dec(sk$_f$, ct) *takes as input a secret key* sk$_f$ *for a policy* f, *and a ciphertext* ct. *It outputs a message* $\mu \in \{0, 1\}$.

Definition 8 (MK-FHE). *A* Multi-Key Fully Homomorphic Encryption *(MK-FHE) scheme is a tuple of* PPT *algorithms* MFHE = (Setup, Enc, Keygen, Eval, Dec) *with the following syntax:*

- MFHE.Setup(1^λ) *takes as input the security parameter and generates public parameters* pp.
- MFHE.Keygen$_{pp}(1^\lambda)$ *uses the public parameters* pp *and outputs a pair of public key and secret key* (pk, sk).
- MFHE.Enc$_{pp}$(pk, μ) *uses the public parameters* pp *and takes as input a message* $\mu \in \{0, 1\}$ *and a public key* pk. *It outputs a ciphertext* ct.
- MFHE.Eval$_{pp}((ct^{(1)}, \ldots, ct^{(k)}), (pk^{(1)}, \ldots, pk^{(k)}), g)$ *uses the public parameters* pp *and takes as input* k *ciphertexts along with their respective public keys* $(pk^{(1)}, \ldots, pk^{(k)})$ *and a function* g. *It outputs a ciphertext* ct$_g$.
- MFHE.Dec$_{pp}$(sk$^{(1)}, \ldots,$ sk$^{(k)}$, ct$_g$) *uses the public parameters and takes as input a sequence of* k *secret keys* sk$^{(1)}, \ldots,$ sk$^{(k)}$ *and a ciphertext* ct$_g$. *It outputs a message* $\mu \in \{0, 1\}$.

Definition 9 (HABE). *An* Homomorphic ABE *(HABE) scheme is a tuple of* PPT *algorithms* HABE = (Setup, Enc, Keygen, Eval, Dec) *with the following syntax:*

- HABE.Setup(1^λ) *takes as input the security parameter and outputs a master secret key* msk *and a set of public parameters* pp.

- HABE.Enc$_{pp}(\mu, x)$ *uses the public parameters* pp *and takes as input a message* $\mu \in \{0, 1\}$ *and an attribute* $x \in \{0, 1\}^{\ell}$. *It outputs a ciphertext* ct.
- HABE.Keygen$_{msk}(f)$ *uses the master secret key* msk *and takes as input a function* $f \in \mathcal{F}$. *It outputs a secret key* sk$_f$.
- HABE.Eval(ct$^{(1)}, \ldots,$ ct$^{(k)}, g)$ *takes as input* k *ciphertexts* ct$^{(1)}, \ldots,$ ct$^{(k)}$ *and a function* $g \in \mathcal{G}$. *It outputs a ciphertext* ct^g.
- HABE.Dec(sk$_F$, ct^g) *takes as input a set of secret keys* sk$_F$ *for a set of policies* F, *with* sk$_F = \{$sk$_f : f \in F\}$, *and a ciphertext* ct^g. *It outputs a message* $\mu \in \{0, 1\}$.

Correctness. The correctness guarantee is that given a set of keys for a policy set F and a ciphertext that was evaluated from ciphertexts respective to attributes covered by F, the ciphertext decrypts correctly to the intended value.

Security. Security is defined using the same experiment as standard ABE (see Definition 3).

Construction of HABE. Consider an ABE black box and a MFHE black box. The construction works as follows:

- HABE.Setup(1^{λ})
  Let (pp$_{ABE}$, msk$_{ABE}$) $\leftarrow$ ABE.Setup(1^{λ}) and pp$_{MFHE}$ $\leftarrow$ MFHE.Setup(1^{λ}). Output pp $=$ (pp$_{ABE}$, pp$_{MFHE}$), msk $=$ msk$_{ABE}$.
- HABE.Enc$_{pp}(\mu, x)$. Let (pk, sk) $\leftarrow$ MFHE.Keygen$_{pp}$, where sk $\in \{0, 1\}^t$. Compute ct$_\mu$ $\leftarrow$ MFHE.Enc$_{pp}$(pk, μ) and ct$_{sk} = \{$ct$_{sk_i} =$ ABE.Enc$_{pp}($sk$_i, x)\}_{i \in [t]}$. Output ct $=$ (ct$_\mu$, ct$_{sk}$, pk, x).
- HABE.Keygen$_{msk}(f)$. Output ABEk$_f$ $\leftarrow$ ABE.Keygen$_{msk}(f)$.
- HABE.Eval(ct$^{(1)}, \ldots,$ ct$^{(k)}, g)$
  Let ct$_g$ $\leftarrow$ MFHE.Eval$_{pp}((\text{ct}_\mu^{(1)}, \ldots, \text{ct}_\mu^{(k)}), (\text{pk}^{(1)}, \ldots, \text{pk}^{(k)}), g)$. Output ct$^g =$ (ct$_g$, ct$_{sk}^{(1)}, \ldots,$ ct$_{sk}^{(k)}$).
- HABE.Dec(ABEk$_F$, ct^g).
  For all $i \in [k], j \in [t]$ compute sk$_j^{(i)} =$ ABE.Dec$_{pp}($ct$_{sk_j}^{(i)}$, ABEk$_f$), where $f \in F$ such that $f(x^{(i)}) = 0$. Compute and Output MFHE.Dec$_{pp}($sk$^{(1)}, \ldots,$ sk$^{(k)}$, ct$_g$).

Correctness Proof Sketch. Consider the execution of HABE.Dec(ABEk$_F$, ct^g). By the correctness of the ABE scheme we get correct decryptions of $\{$sk$^{(i)}\}_{i \in [k]}$, and by the correctness of the MFHE scheme we get a correct decryption of $g(\mu^{(1)}, \ldots, \mu^{(k)})$.

Security Proof Sketch. Consider the ABE selective security game, and assume that in HABE.Enc$_{pp}$ the challenger generates ABE encryptions of 0s instead of ABE encryptions of the bits of sk. By the security of the ABE scheme this change is indistinguishable to the adversarys, therefore in this case the ciphertext gives no information other than the MFHE encryption of the message μ. Hence by the security of the MFHE scheme the security of our construction follows.

References

1. Agrawal, S., Boneh, D., Boyen, X.: Efficient lattice (H)IBE in the standard model. In: Gilbert, H. (ed.) EUROCRYPT 2010. LNCS, vol. 6110, pp. 553–572. Springer, Heidelberg (2010)
2. Agrawal, S., Boneh, D., Boyen, X.: Lattice basis delegation in fixed dimension and shorter-ciphertext hierarchical IBE. In: Rabin, T. (ed.) CRYPTO 2010. LNCS, vol. 6223, pp. 98–115. Springer, Heidelberg (2010)
3. Ajtai, M.: Generating hard instances of lattice problems (extended abstract). In: Miller, G.L. (ed.) Proceedings of the Twenty-Eighth Annual ACM Symposium on the Theory of Computing, Philadelphia, Pennsylvania, USA, 22–24 May 1996, pp. 99–108. ACM (1996)
4. Alperin-Sheriff, J., Peikert, C.: Faster bootstrapping with polynomial error. In: Garay, J.A., Gennaro, R. (eds.) CRYPTO 2014, Part I. LNCS, vol. 8616, pp. 297–314. Springer, Heidelberg (2014)
5. Boneh, D., Gentry, C., Gorbunov, S., Halevi, S., Nikolaenko, V., Segev, G., Vaikuntanathan, V., Vinayagamurthy, D.: Fully key-homomorphic encryption, arithmetic circuit ABE and compact garbled circuits. In: Advances in Cryptology - EUROCRYPT –33rd Annual International Conference on the Theory and Applications of Cryptographic Techniques, Copenhagen, Denmark, 11–15 May 2014, Proceedings, pp. 533–556 (2014)
6. Boneh, D., Roughgarden, T., Feigenbaum, J. (eds.): Symposium on Theory of Computing Conference, STOC 2013, Palo Alto, CA, USA, 1–4 June 2013. ACM (2013)
7. Brakerski, Z.: Fully homomorphic encryption without modulus switching from classical GapSVP. In: Safavi-Naini, R., Canetti, R. (eds.) CRYPTO 2012. LNCS, vol. 7417, pp. 868–886. Springer, Heidelberg (2012)
8. Brakerski, Z., Gentry, C., Vaikuntanathan, V.: (Leveled) fully homomorphic encryption without bootstrapping. In: ITCS (2012)
9. Brakerski, Z., Langlois, A., Peikert, C., Regev, O., Stehlé, D.: Classical hardness of learning with errors. In: Boneh, D., et al. (eds.) [6], pp. 575–584
10. Brakerski, Z., Perlman, R.: Lattice-based fully dynamic multi-key FHE withshort ciphertexts. IACR Cryptology ePrint Archive, 2016:339 (2016, to appear)
11. Brakerski, Z., Vaikuntanathan, V.: Efficient fully homomorphic encryption from (standard) LWE. In: FOCS (2011)
12. Brakerski, Z., Vaikuntanathan, V.: Lattice-based FHE as secure as PKE. In: Naor, M. (ed.) Innovations in Theoretical Computer Science, ITCS 2014, Princeton, NJ, USA, 12–14 January 2014, pp. 1–12. ACM (2014)
13. Brakerski, Z., Vaikuntanathan, V.: Circuit-abe from LWE: unbounded attributesand semi-adaptive security. IACR Cryptology ePrint Archive, 2016:118 (2016, to appear)
14. Cash, D., Hofheinz, D., Kiltz, E., Peikert, C.: Bonsai trees, or how to delegate a lattice basis. J. Cryptology 25(4), 601–639 (2012)
15. Clear, M., McGoldrick, C.: Bootstrappable identity-based fully homomorphic encryption. In: Gritzalis, D., Kiayias, A., Askoxylakis, I. (eds.) CANS 2014. LNCS, vol. 8813, pp. 1–19. Springer, Heidelberg (2014)
16. Clear, M., McGoldrick, C.: Multi-identity and multi-key leveled FHE from learning with errors. In: Gennaro, R., Robshaw, M. (eds.) [19], pp. 630–656

17. Clear, M., McGoldrick, C.: Attribute-based fully homomorphic encryption with a bounded number of inputs. In: Pointcheval, D., Nitaj, A., Rachidi, T. (eds.) AFRICACRYPT 2016. LNCS, vol. 9646, pp. 307–324. Springer, Heidelberg (2016). doi:10.1007/978-3-319-31517-1_16
18. Fischlin, M., Coron, J.-S. (eds.): EUROCRYPT 2016. LNCS, vol. 9666. Springer, Heidelberg (2016)
19. Gennaro, R., Robshaw, M. (eds.): CRYPTO 2015. LNCS, vol. 9216. Springer, Heidelberg (2015)
20. Gentry, C.: A fully homomorphic encryption scheme. Ph.D. thesis, Stanford University (2009)
21. Gentry, C., Peikert, C., Vaikuntanathan, V.: Trapdoors for hard lattices and new cryptographic constructions. In: Dwork, C. (ed.) Proceedings of the 40th Annual ACM Symposium on Theory of Computing, Victoria, British Columbia, Canada, 17–20 May 2008, pp. 197–206. ACM (2008)
22. Gentry, C., Sahai, A., Waters, B.: Homomorphic encryption from learning with errors: conceptually-simpler, asymptotically-faster, attribute-based. In: Canetti, R., Garay, J.A. (eds.) CRYPTO 2013, Part I. LNCS, vol. 8042, pp. 75–92. Springer, Heidelberg (2013)
23. Gorbunov, S., Vaikuntanathan, V., Wee, H.: Attribute-based encryption for circuits. In: Boneh, D., et al. (eds.) [6], pp. 545–554
24. Gorbunov, S., Vaikuntanathan, V., Wichs, D.: Leveled fully homomorphic signatures from standard lattices. In: Servedio, R.A., Rubinfeld, R. (eds.) Proceedings of the Forty-Seventh Annual ACM on Symposium on Theory of Computing, STOC 2015, Portland, OR, USA, 14–17 June 2015, pp. 469–477. ACM (2015)
25. Goyal, R., Koppula, V., Waters, B.: Semi-adaptive security and bundling functionalities made generic and easy. Cryptology ePrint Archive, Report 2016/317 (2016). http://eprint.iacr.org/2016/317
26. Goyal, V., Pandey, O., Sahai, A., Waters, B.: Attribute-based encryption for fine-grained access control of encrypted data. In: Juels, A., Wright, R.N., di Vimercati, S.D.C. (eds.) Proceedings of the 13th ACM Conference on Computer and Communications Security, CCS 2006, Alexandria, VA, USA, 30 October–3 November 2006, pp. 89–98. ACM (2006)
27. López-Alt, A., Tromer, E., Vaikuntanathan, V.: On-the-fly multiparty computation on the cloud via multikey fully homomorphic encryption. In: Karloff, H.J., Pitassi, T. (eds.) Proceedings of the 44th Symposium on Theory of Computing Conference, STOC 2012, New York, NY, USA, 19–22 May 2012, pp. 1219–1234. ACM (2012)
28. Lyubashevsky, V., Wichs, D.: Simple lattice trapdoor sampling from a broad class of distributions. In: Katz, J. (ed.) PKC 2015. LNCS, vol. 9020, pp. 716–730. Springer, Heidelberg (2015)
29. Micciancio, D., Mol, P.: Pseudorandom knapsacks and the sample complexity of LWE search-to-decision reductions. In: Rogaway, P. (ed.) CRYPTO 2011. LNCS, vol. 6841, pp. 465–484. Springer, Heidelberg (2011)
30. Micciancio, D., Peikert, C.: Trapdoors for lattices: simpler, tighter, faster, smaller. In: Pointcheval, D., Johansson, T. (eds.) EUROCRYPT 2012. LNCS, vol. 7237, pp. 700–718. Springer, Heidelberg (2012)
31. Mukherjee, P., Wichs, D.: Two round multiparty computation via multi-key FHE. In: Fischlin, M., Coron, J. (eds.) [18], pp. 735–763
32. Peikert, C.: Public-key cryptosystems from the worst-case shortest vector problem: extended abstract. In: Proceedings of the 41st Annual ACM Symposium on Theory of Computing, STOC 2009, Bethesda, MD, USA, 31 May – 2 June 2009, pp. 333–342 (2009)

33. Regev, O.: On lattices, learning with errors, random linear codes, and cryptography. In: Proceedings of the 37th Annual ACM Symposium on Theory of Computing, Baltimore, MD, USA, 22–24 May 2005, pp. 84–93 (2005)
34. Rivest, R.L., Adleman, L., Dertouzos, M.L.: On data banks and privacy homomorphisms. Found. Secure Comput. (1978)
35. Sahai, A., Waters, B.: Fuzzy identity-based encryption. In: Cramer, R. (ed.) EUROCRYPT 2005. LNCS, vol. 3494, pp. 457–473. Springer, Heidelberg (2005)
36. Schnorr, C.: A hierarchy of polynomial time lattice basis reduction algorithms. Theor. Comput. Sci. **53**, 201–224 (1987)

Semi-adaptive Security and Bundling Functionalities Made Generic and Easy

Rishab Goyal$^{(\boxtimes)}$, Venkata Koppula, and Brent Waters

University of Texas at Austin, Austin, USA
{rgoyal,kvenkata,bwaters}@cs.utexas.edu

Abstract. Semi-adaptive security is a notion of security that lies between selective and adaptive security for Attribute-Based Encryption (ABE) and Functional Encryption (FE) systems. In the semi-adaptive model the attacker is forced to disclose the challenge messages before it makes any key queries, but is allowed to see the public parameters.

We show how to generically transform any selectively secure ABE or FE scheme into one that is semi-adaptively secure with the only additional assumption being public key encryption, which is already naturally included in almost any scheme of interest. Our technique utilizes a fairly simple application of garbled circuits where instead of encrypting directly, the encryptor creates a garbled circuit that takes as input the public parameters and outputs a ciphertext in the underlying selective scheme. Essentially, the encryption algorithm encrypts without knowing the 'real' public parameters. This allows one to delay giving out the underlying selective parameters until a private key is issued, which connects the semi-adaptive to selective security. The methods used to achieve this result suggest that the moral gap between selective and semi-adaptive security is in general much smaller than that between semi-adaptive and full security.

Finally, we show how to extend the above idea to generically bundle a family of functionalities under one set of public parameters. For example, suppose we had an inner product predicate encryption scheme where the length of the vectors was specified at setup and therefore fixed to the public parameters. Using our transformation one could create a system where for a single set of public parameters the vector length is not apriori bounded, but instead is specified by the encryption algorithm. The resulting ciphertext would be compatible with any private key generated to work on the same input length.

1 Introduction

Traditionally, in a public key encryption system a user will encrypt data m under a second user's public key to create a ciphertext. A receiver of the ciphertext can decrypt the data if they possess the corresponding secret key; otherwise,

B. Waters—Supported by NSF CNS-1228599 and CNS-1414082, DARPA SafeWare, Microsoft Faculty Fellowship, and Packard Foundation Fellowship.

M. Hirt and A. Smith (Eds.): TCC 2016-B, Part II, LNCS 9986, pp. 361–388, 2016.
DOI: 10.1007/978-3-662-53644-5_14

they will learn nothing. Over the last several years there has been a dramatic re-envisioning of the expressiveness of encryption systems with the introduction of Identity-Based Encryption (IBE) [12,19,34], Attribute-Based Encryption (ABE) [32] and culminating in Functional Encryption (FE) [33], which encompasses IBE and ABE.

In these systems a setup algorithm produces a master public/secret key pair, where the master public key is made public and the master secret key is retained by an authority. Any user can encrypt data m using the public parameters[1] to produce a ciphertext ct. In parallel the authority may issue (any number of times) to a user a secret key sk_f that allows the user to learn the output $f(m)$ of a ciphertext that encrypts data m. The message space $\mathcal{M}$ and function space $\mathcal{F}$ allowed depend on the expressiveness of the underlying cryptosystem.

The security of this class of systems is captured by an indistinguishability based security game between a challenger and an attacker.[2] In this game the challenger will first generate the master public key that it sends to the attacker. The attacker begins by entering the first key query phase where it will issue a polynomial number of key queries, each for a functionality $f \in \mathcal{F}$. For each query the attacker receives back a corresponding secret key sk_f. Next the attacker submits two challenge messages m_0, m_1 with the restriction that $f(m_0) = f(m_1)$ for all functions f queried on earlier. The challenger will flip a coin $b \in \{0,1\}$ and return a challenge ciphertext ct^* encrypting m_b. Next, the attacker will engage in a second set of private key queries with the same restrictions. Finally, it will output a guess $b' \in \{0,1\}$ and win if $b = b'$. For any secure scheme the probability of winning should be negligibly close to $\frac{1}{2}$.

The above game, called *full* or *adaptive* security game, captures our intuitive notion of what an indistinguishability based security game should look like. Namely, that an attacker cannot distinguish between two messages unless he receives keys that trivially allow him to — even if the attacker gets to adaptively choose what the keys and messages are. One issue faced by researchers is that when striving for a new functionality it is often difficult at first to achieve full security if we want to restrict ourselves to polynomial loss in the reductions and avoid relying on sub-exponential hardness assumptions. To ease the initial pathway people often consider security under a weaker notion of *selective* [17] security where the attacker is forced to submit the challenge messages m_0, m_1 *before* seeing the public parameters. After gaining this foothold, later work can circle back to move from selective to adaptive security.

Over the past decade there have been several examples of this process in achieving adaptive security for IBE, ABE and FE. The first such examples were the "partitioning" techniques developed by Boneh and Boyen [10] and Waters [35] in the context of achieving Identity-Based Encryption in the standard model, improving upon earlier selectively secure realizations [11,17]. While partitioning methods were helpful in realizing full security for IBE, they did not

---

[1] We use public parameters and master public key interchangeably.

[2] There also exists simulation-based notions of security [14,30], but these will not be a focus of this work.

generalize to more complex functionalities. To that end a new set of techniques were developed to move beyond partitioning which include those by Gentry and Halevi [21,22] and Waters' Dual System Encryption [36] methodology. The latter which spawned several other works within that methodology, e.g., [27,29,39].

More recently, Ananth et al. [2], building upon the bootstrapping concepts of [38], showed how to generically convert an FE scheme that supports arbitrary poly-sized circuits from selective security into one that achieves full security.[3] Their result, however, does not apply to the many ABE or FE schemes that fall below this threshold in functionality. Moreover meeting this bar might remain difficult as it has been shown [3,4,9] that achieving functional encryption for this level of functionality is as difficult as achieving indistinguishability obfuscation [7,20].

Delaying Parameters and Semi-Adaptive Security. One remarkable feature of almost all of the aforementioned works is that the security reductions treat the second key query phase identically to the first. Indeed papers will often simply describe the proof of Phase 2 key queries as being the same as Phase 1. Lewko and Waters [28] first departed from this paradigm in a proof where they gave an ABE scheme with a security reduction handled Phase 1 and Phase 2 keys differently. Central to their proof was what they called a "delayed parameters" technique that delayed the release of part of the public parameters in a way that gave a bridge for building adaptive security proofs utilizing selective type techniques. These ideas were extended and codified into a framework by Attrapadung [6].

Chen and Wee [18] introduced the definition of semi-adaptive security as a notion of security where an attacker discloses the challenge messages after it sees the public parameters, but before it makes any key queries. It is easy to see that this notion falls somewhere between selective and adaptive in terms of strength.

Most recently, Brakerski and Vaikuntanathan [16] gave an interesting circuit ABE scheme that was provably secure in the semi-adaptive model from the Learning with Error assumption [31]. Their cryptosystem and proof of security build upon the (arithmetic) circuit ABE scheme of Boneh et al. [13] and requires a somewhat elaborate two level application of these techniques integrated with a pseudorandom function (we note that some of the complexity is due to their parallel goal of bundling functionalities; we will return to this). Like the earlier work of [28], they also apply a "delayed parameter" concept, although its flavor and execution are significantly different.

1.1   Going from Selective to Semi-adaptive Security Generically

We now arrive at the first goal of this work.

Can we generically transform any selectively secure attribute-based encryption or functional encryption scheme into one that is semi-adaptively secure?

---

[3] We note that FE for poly-sized circuits is achievable by bootstrapping FE for **NC1** [23].

It turns out that this transformation is possible and moreover that the method to do so is quite simple. Here is our idea in a nutshell. Instead of encrypting the data outright, the encryptor will consider a circuit that fixes the message and randomness for encryption and takes the functional encryption scheme's public parameters as input. It then garbles this circuit and encrypts each pair of input wire values under pairs of standard PKE public keys provided by the authority. The garbled circuit plus pairs of encrypted wires comprise the ciphertext. In generating a secret key, the authority will output both the underlying functional encryption secret key as well as give one of the PKE secret keys for each pair corresponding to the underlying selectively secure FE public parameters. The decryption algorithm will first evaluate the garbled circuit to obtain the underlying ciphertext and then decrypt using the FE secret key. In this manner, the core FE parameters are literally not committed to until a key is given out.

We now elaborate our description. Let $\mathsf{FE}_{\mathsf{sel}} = (\mathsf{Setup}_{\mathsf{sel}}, \mathsf{Enc}_{\mathsf{sel}}, \mathsf{KeyGen}_{\mathsf{sel}}, \mathsf{Dec}_{\mathsf{sel}})$ be the underlying selectively secure FE scheme. Our semi-adaptively secure FE setup algorithm generates a master public/secret key pair $(\mathsf{mpk}_{\mathsf{sel}}, \mathsf{msk}_{\mathsf{sel}})$ using $\mathsf{Setup}_{\mathsf{sel}}$, and chooses 2ℓ public/secret key pairs $\{\mathsf{pk}_{i,b}, \mathsf{sk}_{i,b}\}$ for a semantically secure PKE scheme, where $\ell = |\mathsf{mpk}_{\mathsf{sel}}|$. The public key of $\mathsf{FE}_{\mathsf{sel}}$ consists of these 2ℓ PKE public keys $\{\mathsf{pk}_{i,b}\}$, but *not* the public key $\mathsf{mpk}_{\mathsf{sel}}$. To encrypt any message m, the encryptor constructs a circuit that takes as input an ℓ bit string str and outputs $\mathsf{Enc}_{\mathsf{sel}}(\mathsf{str}, m; r)$ – an encryption of m using str as the public key and r as randomness. The encryptor garbles this circuit and encrypts each of the 2ℓ garbled circuit input wire keys $w_{i,b}$ under the corresponding public key $\mathsf{pk}_{i,b}$. The ciphertext consists of the garbled circuit and the 2ℓ encrypted wire keys. The secret key for any function f consists of three parts — the master public key $\mathsf{mpk}_{\mathsf{sel}}$, ℓ PKE secret keys to decrypt half of the encrypted wire keys $w_{i,b}$ corresponding to $\mathsf{mpk}_{\mathsf{sel}}$, and $\mathsf{FE}_{\mathsf{sel}}$ secret key $\mathsf{sk}_{f,\mathsf{sel}}$ to decrypt the actual $\mathsf{FE}_{\mathsf{sel}}$ ciphertext. The key $\mathsf{sk}_{f,\mathsf{sel}}$ is simply generated using the $\mathsf{KeyGen}_{\mathsf{sel}}$ algorithm, and the ℓ PKE secret keys released correspond to the bits of $\mathsf{mpk}_{\mathsf{sel}}$. For decrypting any ciphertext, the decryptor first decrypts the encrypted input wire keys. Then, these wire keys are used to evaluate the garbled circuit. This evaluation results in an $\mathsf{FE}_{\mathsf{sel}}$ ciphertext under $\mathsf{mpk}_{\mathsf{sel}}$, which can be decrypted using $\mathsf{sk}_{f,\mathsf{sel}}$.

The crucial observation here is that the underling $\mathsf{FE}_{\mathsf{sel}}$ public key $\mathsf{mpk}_{\mathsf{sel}}$ is information theoretically hidden until any secret key is given out as the encryptor computes the ciphertext oblivious to the knowledge of $\mathsf{mpk}_{\mathsf{sel}}$. Therefore, the semi-adaptive security proof follows from a simple sequence of hybrids. In the first hybrid, we switch the ℓ encryptions of input wire keys (given out in the challenge ciphertext) which are never decrypted to encryptions of zeros. Next, in the following hybrid, we simulate the garbled circuit (given out in the challenge ciphertext) instead of constructing the actual encryption circuit and garbling it. After these two indistinguishable hybrid jumps, we could directly reduce the semi-adaptive security to selective security as the $\mathsf{FE}_{\mathsf{sel}}$ public key is hidden. Our construction and security proof is described in detail in Sect. 4.

The overhead associated with our transformation to semi-adaptive security is readily apparent. Instead of evaluating the underlying encryption algorithm, the transformed encryption algorithm will need to garble the encryption circuit. The ciphertext will grow proportionally to the size of this garbled circuit. Similarly, the decryption algorithm will first have to evaluate the garbled circuit before executing the core decryption. In the description above one will replace each bit of the original public parameters with a pair of PKE public keys. However, if one optimizes by using IBE instead of PKE for this step, the public parameters could actually become shorter than the original ones. In many cases our transformation will incur greater overhead than non-generic techniques designed with knowledge of the underlying scheme such as [18].

Interpreting Our Result. It is useful to step back and see what light our result can shed on the relationship between selective, semi-adaptive and adaptive security. Ideally, we would like to claim that semi-adaptive security gives us a half-way point between selective and adaptive where the next idea could take us all the way between the two endpoints. While this might turn out to be the case, the way in which we delay parameters seems primarily to exploit the closeness of selective and semi-adaptive security as opposed to crossing a great divide. To us this suggests that the moral gap between selective and semi-adaptive security is much smaller than that between semi-adaptive and full security (at least for functionalities that fall below the threshold needed by [2]). We view illuminating this relationship as one of the contributions of this paper.

1.2 Bundling Functionalities

We now turn to the second goal of our work. Before doing so, we describe a more general definition of functional encryption, which will later help us to explain our idea of bundling functionalities. Any functional encryption scheme is associated with a message space $\mathcal{M}$ and function space $\mathcal{F}$. In many scenarios, the function space $\mathcal{F}$ and message space $\mathcal{M}$ themselves consists of a sequence of function spaces $\{\mathcal{F}_n\}_n$ and message spaces $\{\mathcal{M}_n\}_n$ respectively, parameterized by the 'functionality index' n. In our definition of functional encryption, we assume that the setup algorithm takes two inputs - the security parameter λ and the functionality index n. This notation decouples the security of the scheme from the choice of functionality it provides. We note that such terminology has appeared in several prior works. For example, Goyal et al. [25] have a setup algorithm that takes as input the number of attributes along with the security parameter. Similarly, in the works of Boyen and Waters [15] and Agrawal et al. [1], the setup algorithms also take the length of vectors as an input. And other works [13,24] specify the maximum depth of a circuit in an ABE scheme during setup.

Using the above convention, $\mathsf{Setup}(1^\lambda, 1^n)$ creates a master public/secret key for message space $\mathcal{M}_n$ and function space $\mathcal{F}_n$. For example, in an inner product encryption scheme, the setup algorithm fixes the length of vectors to be encrypted once the master public key is fixed. However, one goal could be to allow more flexibility after the public key is published. In particular, would it be possible to have all message and function spaces available even after setup? Continuing our example, we might want an inner product encryption scheme where the encryptor/key generator are allowed to encrypt/generate keys for arbitrary length vectors after the public parameters have been fixed.

Looking more generally, a natural question to ask is — "Can we generically transform any (standard) functional encryption scheme into one where a *single* set of public parameters can support the union of underlying message/function spaces?" We answer this in the affirmative, and show a generic transformation using identity based encryption, pseudorandom functions and garbled circuits, all of which can be realized from standard assumptions. More formally, we show how to transform an FE scheme with message space $\{\mathcal{M}_n\}_n$ and function space $\{\mathcal{F}_n\}_n$ to an FE scheme for message space $\mathcal{M} = \cup_n \mathcal{M}_n$ and function space $\mathcal{F} = \cup_n \mathcal{F}_n$. The key for a function $f \in \mathcal{F}_n$ can be used with a ciphertext for message $m \in \mathcal{M}_n$ to compute $f(m)$. If f and m are not compatible (i.e. $f \in \mathcal{F}_n$ and $m \in \mathcal{M}_{n'}$), then the decryption fails.

As a simple instantiation, using our transformation, one can construct an inner product encryption scheme where the encryption algorithm and the key generation algorithm can both take arbitrary length vectors as input. However, given a secret key for vector v and an encryption of vector w, the decryption algorithm tests orthogonality only if v and w have same length; else the decryption algorithm fails. Similarly, our transformation can also capture the recent result of Brakerski and Vaikuntanathan [16]. They give a circuit ABE scheme where under a single set of parameters an encryptor can encrypt messages for an attribute of unbounded length. Later if a private key is given out and is tied to the same attribute length it can decrypt if the circuit matches. In our transformation we would start with a selective scheme for circuit ABE such as [24] where 1^n denotes the number of attributes and then apply our transformation. We observe that we could even choose to obtain more flexibility where we might allow both the attribute length and circuit depth to depend on 1^n.

Our Transformation for Bundling Functionalities. Our method for achieving such a transformation follows in a similar line to the selective to semi-adaptive transformation given above. In addition, it also amplifies to semi-adaptive security along the way for free. Recall, in the base scheme, $\mathsf{Setup}_{\mathsf{sel}}$ takes functionality index n as input and outputs master public/secret keys. Let $\ell(n)$ denote the bit-length of public keys output by $\mathsf{Setup}_{\mathsf{sel}}$. In our transformed scheme, the setup algorithm chooses IBE public/secret keys $(\mathsf{mpk}_{\mathsf{IBE}}, \mathsf{msk}_{\mathsf{IBE}})$ and sets $\mathsf{mpk}_{\mathsf{IBE}}$ as the public key. To encrypt a message $m \in \mathcal{M}_n$, the encryptor first chooses randomness r. It then constructs a circuit which takes a $\ell = \ell(n)$ bit input string str and outputs $\mathsf{Enc}_{\mathsf{sel}}(\mathsf{str}, m; r)$. The encryptor then garbles this circuit, and each wire key $w_{i,b}$ is encrypted for identity (n, i, b). The final ciphertext consists of the garbled circuit, together with encryptions of wire keys. Note that ℓ is not fixed during setup. It is defined (and used) during encryption, and depends on the functionality index of the message. The idea of using IBE to succinctly handle an unbounded number of public keys was also present in the work of [16].

The secret key for a function $f \in \mathcal{F}_t$ is computed as follows. First, the key generation algorithm chooses pseudorandom $\mathsf{FE}_{\mathsf{sel}}$ keys $(\mathsf{mpk}_t, \mathsf{msk}_t)$ using $\mathsf{Setup}_{\mathsf{sel}}(1^\lambda, 1^t)$. Next, it computes IBE secret keys for identities $(t, i, \mathsf{mpk}_t[i])$. Finally, it computes an $\mathsf{FE}_{\mathsf{sel}}$ secret key for the function f. The decryption procedure is similar to the one described in Sect. 1.1. Let $\mathsf{ct} = (C, \{\mathsf{ct}_{i,b}\})$ and $\mathsf{sk}_f = (\{\mathsf{sk}_i\}, \mathsf{sk}_{f,\mathsf{sel}})$. First, note that it is important that the message underlying the ciphertext, and the function underlying the secret key are compatible. If so, the decryptor first decrypts $\mathsf{ct}_{i,b}$ to compute the garbled circuit wire keys. Next, it evaluates the garbled circuit to get an $\mathsf{FE}_{\mathsf{sel}}$ ciphertext, which it then decrypts using $\mathsf{sk}_{f,\mathsf{sel}}$. The proof of security is along the lines of selective to semi-adaptive transformation proof.

The overhead involved in this transformation is similar to the overhead in going from selective to semi-adaptive security, except that the size of the garbled circuit, and the number of wire keys grows with the functionality index. Overhead comparisons between our approach and the non-generic approach of [16] are less clear, since their approach requires increasing the maximum depth of the circuit to accommodate a PRF evaluation before evaluating the main circuit.

Limits of Bunding Functionalities. One should be careful to point out the limits of such bundling. The main restriction is that in order for decryption to do anything useful the functionality index used to encrypt must match that of the private key; otherwise they simply are not compatible. Suppose FE is a functional encryption scheme with functionality class $\{\mathcal{F}_n\}$ and message class $\{\mathcal{M}_n\}$. Then, using our bundling approach, we get a functional encryption scheme for function space $\cup_n \mathcal{F}_n$ and message space $\cup_n \mathcal{M}_n$. However, the secret key for a function $f \in \mathcal{F}_n$ can only decrypt encryptions of messages in $\mathcal{M}_n$. So such a technique cannot be used to emulate a functionality such as ABE for DFAs [37] or Functional Encryption for Turning Machines [5] where the base private key is meant to operate on ciphertext corresponding to messages/attributes of unrestricted size. In general, our bundling approach cannot transform an FE scheme where secret keys decrypt bounded length encryptions to one where secrets keys can decrypt arbitrary length ciphertexts.

1.3   Encrypt Ahead Functional Encryption

We conclude by discussing a final potential application of our techniques that we call "Encrypt Ahead Functional Encryption". Our discussion is at an informal level and limited to this introduction.

Suppose that we would like to setup a functional encryption system and that a proper authority has already been identified. Furthermore suppose that several users have obtained data and are ready to encrypt. The only thing missing is the small detail that the algorithms comprising the cryptosystem have yet to be determined. Perhaps we are waiting on a security proof or maybe there is no reasonable candidate realization what so ever.

Normally, we would think that the lack of a encryption system would be a complete showstopper and that nothing could be done until it was in place.

However, as it turns out this need not be the case. Using a slight adaptation of our techniques an authority could publish the scheme's public parameters and user's could begin to encrypt data to create ciphertexts. The main idea is that the encryption algorithm will create a garbled circuit that takes a functional encryption scheme's public parameters (as before) *as well as* a description of the encryption algorithm itself. It then encrypts the corresponding input wires (for both the parameters and scheme description) under pairs of public keys in a similar manner to what was done above. Later when the actual cryptography is worked out the secret keys corresponding to the bits of the public parameters and scheme description can be given out as part of the functional encryption secret key and these are used to construct the ciphertext before decrypting. We call this concept "encrypt ahead" as encryption can occur prior to deciding on a scheme.

There are important caveats to encrypting ahead in this manner. While the setup and outer encryption algorithm need not know what the eventual core encryption algorithm is, one has to at least guess and then work with an upper bound on the core encryption algorithm's description and running time. If this guess turns out to be below the resources needed by the eventual scheme, the ciphertexts created will be unusable. Furthermore, until an actual scheme has been decided upon, the authority will be unable to create private keys and this aspect of the system will be stalled.

Paper Organization. We first introduce some preliminaries in Sect. 2. Next, in Sect. 3, we discuss functional encryption related preliminaries. In Sect. 4, we present our generic transformation from a selectively secure FE scheme to a semi-adaptively secure FE scheme. Finally in Sect. 5, we present our transformation for bundling functionalities.

2 Preliminaries

2.1 Garbled Circuits

Our definition of garbled circuits [40] is based upon the work of Bellare et al. [8]. Let $\{C_n\}_n$ be a family of circuits where each circuit in C_n takes n bit inputs. A garbling scheme GC for circuit family $\{C_n\}_n$ consists of polynomial-time algorithms Garble and Eval with the following syntax.

- Garble($C \in C_n, 1^\lambda$): The garbling algorithm takes as input the security parameter λ and a circuit $C \in C_n$. It outputs a garbled circuit G, together with $2n$ wire keys $\{w_{i,b}\}_{i \leq n, b \in \{0,1\}}$.
- Eval($G, \{w_i\}_{i \leq n}$): The evaluation algorithm takes as input a garbled circuit G and n wire keys $\{w_i\}_{i \leq n}$ and outputs $y \in \{0,1\}$.

Correctness: A garbling scheme GC for circuit family $\{C_n\}_n$ is said to be correct if for all λ, n, $x \in \{0,1\}^n$ and $C \in C_n$, Eval($G, \{w_{i,x_i}\}_{i \leq n}$) = $C(x)$, where $(G, \{w_{i,b}\}_{i \leq n, b \in \{0,1\}}) \leftarrow$ Garble($C, 1^\lambda$).

Security: Informally, a garbling scheme is said to be secure if for every circuit C and input x, the garbled circuit G together with input wires $\{w_{i,x_i}\}_{i \leq n}$ corresponding to some input x reveals only the output of the circuit $C(x)$, and nothing else about the circuit C or input x.

Definition 1. *A garbling scheme* GC $=$ (Garble, Eval) *for a class of circuits* $\mathcal{C} = \{\mathcal{C}_n\}_n$ *is said to be a secure garbling scheme if there exists a polynomial-time simulator* Sim *such that for all* λ, n, $C \in \mathcal{C}_n$ *and* $x \in \{0,1\}^n$, *the following holds:*

$$\left\{ \mathsf{Sim}\left(1^\lambda, 1^n, 1^{|C|}, C(x)\right) \right\} \approx_c \left\{ (G, \{w_{i,x_i}\}_{i \leq n}) : (G, \{w_{i,b}\}_{i \leq n, b \in \{0,1\}}) \leftarrow \mathsf{Garble}(C, 1^\lambda) \right\}.$$

While this definition is not as general as the definition in [8], it suffices for our construction.

2.2 Public Key Encryption

A Public Key Encryption (PKE) scheme PKE $=$ (Setup$_{\mathrm{PKE}}$, Enc$_{\mathrm{PKE}}$, Dec$_{\mathrm{PKE}}$) with message space $\mathcal{M} = \{\mathcal{M}_\lambda\}_\lambda$ consists of the following polynomial-time algorithms:

- Setup$_{\mathrm{PKE}}(1^\lambda) \rightarrow (\mathsf{pk}, \mathsf{sk})$: The setup algorithm is a randomized algorithm that takes security parameter λ as input and outputs a public-secret key pair $(\mathsf{pk}, \mathsf{sk})$.
- Enc$_{\mathrm{PKE}}(\mathsf{pk}, m \in \mathcal{M}_\lambda) \rightarrow \mathsf{ct}$: The encryption algorithm is a randomized algorithm that takes as inputs the public key pk, and a message m and outputs a ciphertext ct.
- Dec$_{\mathrm{PKE}}(\mathsf{sk}, \mathsf{ct}) \rightarrow \mathcal{M}_\lambda$: The decryption algorithm is a deterministic algorithm that takes as inputs the secret key sk, and a ciphertext ct and outputs a message m.

Correctness: For correctness, we require that for all $\lambda \in \mathbb{N}$, $m \in \mathcal{M}_\lambda$, and $(\mathsf{pk}, \mathsf{sk}) \leftarrow$ Setup$_{\mathrm{PKE}}(1^\lambda)$,

$$\Pr[\mathsf{Dec}_{\mathrm{PKE}}(\mathsf{sk}, \mathsf{Enc}_{\mathrm{PKE}}(\mathsf{pk}, m)) = m] = 1.$$

Security: For security, we require PKE to be semantically secure, i.e. the adversary must not be able to distinguish between encryptions of distinct messages of its own choosing even after receiving the public key. The notion of semantical security for PKE schemes is defined below.

Definition 2. *A PKE scheme* PKE $=$ (*Setup*$_{\mathrm{PKE}}$, *Enc*$_{\mathrm{PKE}}$, *Dec*$_{\mathrm{PKE}}$) *is said to be semantically secure if there exists* $\lambda_0 \in \mathbb{N}$ *such that for every PPT attacker* $\mathcal{A}$ *there exists a negligible function* negl($\cdot$) *such that for all* $\lambda \geq \lambda_0$, Adv$_{\mathcal{A}}^{\mathsf{PKE}}(\lambda) =$ $|\Pr[\mathsf{Exp\text{-}PKE}(\mathsf{PKE}, \lambda, \mathcal{A}) = 1] - 1/2| \leq$ negl(λ), *where* Exp-PKE *is defined in Fig. 1.*

$$
\begin{array}{l|l}
\textbf{Exp-PKE}(\mathsf{PKE}, \lambda, \mathcal{A}) & \textbf{Exp-IBE}(\mathsf{IBE}, \lambda, \mathcal{A}) \\
\hline
\quad (\mathsf{pk}, \mathsf{sk}) \leftarrow \mathsf{Setup}_{\mathrm{PKE}}(1^\lambda) & \quad (\mathsf{mpk}, \mathsf{msk}) \leftarrow \mathsf{Setup}_{\mathrm{IBE}}(1^\lambda) \\
\quad (m_0^*, m_1^*) \leftarrow \mathcal{A}(1^\lambda, \mathsf{pk}) & \quad (m_0^*, m_1^*, \mathsf{ID}^*) \leftarrow \mathcal{A}^{\mathsf{KeyGen}_{\mathrm{IBE}}(\mathsf{msk}, \cdot)}(1^\lambda, \mathsf{mpk}) \\
\quad b \leftarrow \{0, 1\}, \mathsf{ct}^* \leftarrow \mathsf{Enc}_{\mathrm{PKE}}(\mathsf{pk}, m_b^*) & \quad b \leftarrow \{0, 1\}, \mathsf{ct}^* \leftarrow \mathsf{Enc}_{\mathrm{IBE}}(\mathsf{mpk}, m_b^*, \mathsf{ID}^*) \\
\quad b' \leftarrow \mathcal{A}(\mathsf{ct}^*) & \quad b' \leftarrow \mathcal{A}^{\mathsf{KeyGen}_{\mathrm{IBE}}(\mathsf{msk}, \cdot)}(\mathsf{ct}^*) \\
\quad \text{Output } (b' \stackrel{?}{=} b) & \quad \text{Output } (b' \stackrel{?}{=} b)
\end{array}
$$

Fig. 1. The PKE and IBE security games. In both the games, we assume that the adversary $\mathcal{A}$ is stateful. And in the IBE security game, we also require that ID^* is not queried to the key generation oracle.

2.3 Identity-Based Encryption

An Identity-Based Encryption (IBE) scheme $\mathsf{IBE} = (\mathsf{Setup}_{\mathrm{IBE}}, \mathsf{KeyGen}_{\mathrm{IBE}}, \mathsf{Enc}_{\mathrm{IBE}}, \mathsf{Dec}_{\mathrm{IBE}})$ with message space $\mathcal{M} = \{\mathcal{M}_\lambda\}_\lambda$ and identity space $\mathcal{I} = \{\mathcal{I}_\lambda\}_\lambda$ consists of the following polynomial-time algorithms:

- $\mathsf{Setup}_{\mathrm{IBE}}(1^\lambda) \to (\mathsf{pp}, \mathsf{msk})$: The setup algorithm is a randomized algorithm that takes security parameter λ as input and outputs $(\mathsf{pp}, \mathsf{msk})$, where pp are public parameters and msk is the master secret key.
- $\mathsf{KeyGen}_{\mathrm{IBE}}(\mathsf{msk}, \mathsf{ID} \in \mathcal{I}_\lambda) \to \mathsf{sk}_{\mathsf{ID}}$: The key generation algorithm is a randomized algorithm that takes as inputs the master secret key msk, and an identity ID and outputs a secret key $\mathsf{sk}_{\mathsf{ID}}$.
- $\mathsf{Enc}_{\mathrm{IBE}}(\mathsf{pp}, m \in \mathcal{M}_\lambda, \mathsf{ID} \in \mathcal{I}_\lambda) \to \mathsf{ct}$: The encryption algorithm is a randomized algorithm that takes as inputs the public parameters pp, a message m, and an identity ID and outputs a ciphertext ct.
- $\mathsf{Dec}_{\mathrm{IBE}}(\mathsf{sk}_{\mathsf{ID}}, \mathsf{ct}) \to \mathcal{M}_\lambda \cup \{\bot\}$: The decryption algorithm is a deterministic algorithm that takes as inputs the secret key $\mathsf{sk}_{\mathsf{ID}}$, and a ciphertext ct and outputs a message m or $\bot$.

Correctness: For correctness, we require that for all $\lambda \in \mathbb{N}$, $m \in \mathcal{M}_\lambda$, $\mathsf{ID} \in \mathcal{I}_\lambda$, and $(\mathsf{pp}, \mathsf{msk}) \leftarrow \mathsf{Setup}_{\mathrm{IBE}}(1^\lambda)$,

$$
\Pr[\mathsf{Dec}_{\mathrm{IBE}}(\mathsf{KeyGen}_{\mathrm{IBE}}(\mathsf{msk}, \mathsf{ID}), \mathsf{Enc}_{\mathrm{IBE}}(\mathsf{pp}, m, \mathsf{ID})) = m] = 1.
$$

Security: For security, intuitively, we require that if an adversary has keys for identities $\{\mathsf{ID}_i\}_i$, and ct is a ciphertext for identity $\mathsf{ID}^* \neq \mathsf{ID}_i$ for all i, then the adversary must not be able to recover the underlying message. This is formally defined via the following security game between a challenger and an adversary.

Definition 3. *An IBE scheme* $\mathsf{IBE} = (\mathsf{Setup}_{\mathrm{IBE}}, \mathsf{KeyGen}_{\mathrm{IBE}}, \mathsf{Enc}_{\mathrm{IBE}}, \mathsf{Dec}_{\mathrm{IBE}})$ *is said to be fully secure if there exists* $\lambda_0 \in \mathbb{N}$ *such that for every PPT attacker* $\mathcal{A}$ *there exists a negligible function* $negl(\cdot)$ *such that for all* $\lambda \geq \lambda_0$, $\mathsf{Adv}_{\mathcal{A}}^{\mathsf{IBE}}(\lambda) = |\Pr[\mathsf{Exp\text{-}IBE}(\mathsf{IBE}, \lambda, \mathcal{A}) = 1] - 1/2| \leq negl(\lambda)$, *where* $\mathsf{Exp\text{-}IBE}$ *is defined in Fig. 1.*

3   Functional Encryption

The notion of functional encryption was formally defined in the works of Boneh et al. [14] and O'Neill [30]. A functional encryption scheme consists of a setup algorithm, an encryption algorithm, a key generation algorithm and a decryption algorithm. The setup algorithm takes the security parameter as input and outputs a public key and a master secret key. The encryption algorithm uses the public key to encrypt a message, while the key generation algorithm uses the master secret key to compute a secret key corresponding to a function. The decryption algorithm takes as input a ciphertext and a secret key, and outputs the function evaluation on the message.

The Functionality Index: Every functional encryption scheme is associated with a message space which defines the set of messages that can be encrypted, and a function space which defines the set of functions for which a secret key can be generated. In most schemes, the message space $\mathcal{M}$ and the function space $\mathcal{F}$ consists of a sequence of message spaces $\{\mathcal{M}_n\}_{n \in \mathbb{N}}$ and function spaces $\{\mathcal{F}_n\}_n$, both parameterized by the functionality index (the special case where $\mathcal{M}_n = \mathcal{M}$ and $\mathcal{F}_n = \mathcal{F}$ for all $n \in \mathbb{N}$ is discussed in Sect. 3.1).

The Choice of Functionality Index: A minor definitional issue that arises is with respect to the choice of functionality index. Some works use the security parameter itself to define a message space $\mathcal{M}_\lambda$ and function space $\mathcal{F}_\lambda$. For example, in the inner product FE scheme of Katz et al. [26], the message space and function space are set to be $\mathbb{Z}_q^\lambda$ during setup, where λ is the security parameter and q is an appropriately chosen modulus.

A more flexible approach is to decouple the security parameter and the *functionality index*, and allow the setup algorithm to take two inputs - a security parameter λ and a functionality index n. This additional parameter then defines the message space for the encryption algorithm and the function space for the key generation algorithm. Some existing works implicitly assume that the setup algorithm also receives such a parameter as input. For example, in the work of Goyal et al. [25], the universe $\mathcal{U} = \{1, 2, \ldots, n\}$ is defined as the universe of attributes for the ABE scheme. Other works, such as the inner product FE scheme of Agrawal et al. [1] explicitly mention this as an input to the setup algorithm. We will also use this approach in our formal definition of a functional encryption scheme.

Formal Definition: Let $\mathcal{M} = \{\mathcal{M}_n\}_{n \in \mathbb{N}}$, $\mathcal{R} = \{\mathcal{R}_n\}_{n \in \mathbb{N}}$ be families of sets, and $\mathcal{F} = \{\mathcal{F}_n\}$ a family of functions, where for all $n \in \mathbb{N}$ and $f \in \mathcal{F}_n$, $f : \mathcal{M}_n \to \mathcal{R}_n$. We will also assume that for all $n \in \mathbb{N}$, the set $\mathcal{F}_n$ contains an *empty function* $\epsilon_n : \mathcal{M}_n \to \mathcal{R}_n$. As in [14], the empty function is used to capture information that intentionally leaks from the ciphertext. For instance, in a PKE scheme, the length of the message could be revealed from the ciphertext. Similarly, in an attribute based encryption scheme, the ciphertext could reveal the attribute for which the message was encrypted.

A functional encryption scheme FE for function space $\{\mathcal{F}_n\}_{n \in \mathbb{N}}$ and message space $\{\mathcal{M}_n\}_{n \in \mathbb{N}}$ consists of four polynomial-time algorithms (Setup, Enc, KeyGen, Dec) with the following syntax.

- Setup($1^\lambda, 1^n$) → (mpk, msk): The setup algorithm is a randomized algorithm that takes as input the security parameter λ and the functionality index n, and outputs the master public/secret key pair (mpk, msk).
- Enc(mpk, $m \in \mathcal{M}_n$) → ct: The encryption algorithm is a randomized algorithm that takes as input the public key mpk and a message $m \in \mathcal{M}_n$ and outputs a ciphertext ct.
- KeyGen(msk, $f \in \mathcal{F}_n$) → sk_f: The key generation algorithm is a randomized algorithm that takes as input the master secret key msk and a function $f \in \mathcal{F}_n$ and outputs a secret key sk_f.
- Dec(sk_f, ct) → $\{0, 1, \perp\}$: The decryption algorithm is deterministic. It takes as input a ciphertext ct and a secret key sk_f and outputs $y \in \{0, 1, \perp\}$.

More General Definitions of Functional Encryption: It is possible to consider more general definitions for functional encryption. For example, one could consider a definition where the setup algorithm takes as input a security parameter λ, functionality index n and a depth-index d that bounds the circuit depth of $\mathcal{F}_n$. For simplicity of notation we avoid such extensions, although we believe that our results can be generalized for all such extensions.

Correctness: A functional encryption scheme FE = (Setup, Enc, KeyGen, Dec) is said to be correct if for all security parameter λ and functionality index n, functions $f \in \mathcal{F}_n$, messages $m \in \mathcal{M}_n$ such that (f, m) are compatible, and (mpk, msk) ← Setup($1^\lambda, 1^n$),

$$\Pr\left[\text{Dec}(\text{KeyGen}(\text{msk}, f), \text{Enc}(\text{mpk}, m)) = f(m)\right] = 1.$$

Security: Informally, a functional encryption scheme is said to be secure if an adversary having secret keys for functions $\{f_i\}_{i \leq k}$ and a ciphertext ct for message m learns only $\{f_i(m)\}_{i \leq k}$, and nothing else about the underlying message m. This can be formally captured via the following 'indistinguishability based' security definition.

Definition 4. *A functional encryption scheme* FE *is adaptively secure if there exists* $\lambda_0 \in \mathbb{N}$ *such that for all PPT adversaries* $\mathcal{A}$, *there exists a negligible function* negl($\cdot$) *such that for all* $\lambda > \lambda_0$, $n \in \mathbb{N}$, $|\Pr[\text{Exp-adaptive}(\text{FE}, \lambda, n, \mathcal{A}) = 1] - 1/2| \leq \text{negl}(\lambda)$, *where* Exp-adaptive *is defined in Fig. 2.*

A weaker notion of security is that of selective security, where the adversary must declare the challenge inputs before receiving the public parameters.

Exp-adaptive(FE, $\lambda, n, \mathcal{A}$)	Exp-selective(FE, $\lambda, n, \mathcal{A}$)
$(\mathsf{mpk}, \mathsf{msk}) \leftarrow \mathsf{Setup}(1^\lambda, 1^n)$	$(m_0^*, m_1^*) \leftarrow \mathcal{A}(1^\lambda)$
$(m_0^*, m_1^*) \leftarrow \mathcal{A}^{\mathsf{KeyGen}(\mathsf{msk}, \cdot)}(1^\lambda, \mathsf{mpk})$	$(\mathsf{mpk}, \mathsf{msk}) \leftarrow \mathsf{Setup}(1^\lambda)$
$b \leftarrow \{0, 1\}, \mathsf{ct}^* \leftarrow \mathsf{Enc}(\mathsf{mpk}, m_b^*)$	$b \leftarrow \{0, 1\}, \mathsf{ct}^* \leftarrow \mathsf{Enc}(\mathsf{mpk}, m_b^*)$
$b' \leftarrow \mathcal{A}^{\mathsf{KeyGen}(\mathsf{msk}, \cdot)}(\mathsf{ct}^*)$	$b' \leftarrow \mathcal{A}^{\mathsf{KeyGen}(\mathsf{msk}, \cdot)}(\mathsf{mpk}, \mathsf{ct}^*)$
Output ($b' \overset{?}{=} b$)	Output ($b' \overset{?}{=} b$)

$$\frac{\mathsf{Exp\text{-}semi\text{-}adp}(\mathsf{FE}, \lambda, n, \mathcal{A})}{}$$

$(\mathsf{mpk}, \mathsf{msk}) \leftarrow \mathsf{Setup}(1^\lambda, 1^n)$

$(m_0^*, m_1^*) \leftarrow \mathcal{A}(1^\lambda, \mathsf{mpk})$

$b \leftarrow \{0, 1\}, \mathsf{ct}^* \leftarrow \mathsf{Enc}(\mathsf{mpk}, m_b^*)$

$b' \leftarrow \mathcal{A}^{\mathsf{KeyGen}(\mathsf{msk}, \cdot)}(\mathsf{ct}^*)$

Output ($b' \overset{?}{=} b$)

Fig. 2. Experiments referred in Definitions 4, 5 and 6. We assume that the adversary $\mathcal{A}$ is stateful, $\epsilon_n(m_0^*) = \epsilon_n(m_1^*)$, and for all key queries f queried by $\mathcal{A}$ to KeyGen oracle, $f \in \mathcal{F}_n$ and $f(m_0^*) = f(m_1^*)$.

Definition 5. *A functional encryption scheme* FE *is selectively secure if there exists* $\lambda_0 \in \mathbb{N}$ *such that for all PPT adversaries* $\mathcal{A}$, *there exists a negligible function* $negl(\cdot)$ *such that for all* $\lambda > \lambda_0$, $n \in \mathbb{N}$, $|\Pr[\mathsf{Exp\text{-}selective}(\mathsf{FE}, \lambda, n, \mathcal{A}) = 1] - 1/2| \leq negl(\lambda)$, *where* Exp-selective *is defined in Fig. 2.*

Finally, we have an intermediate notion of security called semi-adaptive security, where the adversary must declare the challenge inputs before receiving any key queries.

Definition 6. *A functional encryption scheme* FE *is semi-adaptively secure if there exists* $\lambda_0 \in \mathbb{N}$ *such that for all PPT adversaries* $\mathcal{A}$, *there exists a negligible function* $negl(\cdot)$ *such that for all* $\lambda > \lambda_0$, $n \in \mathbb{N}$, $|\Pr[\mathsf{Exp\text{-}semi\text{-}adp}(\mathsf{FE}, \lambda, n, \mathcal{A}) = 1] - 1/2| \leq negl(\lambda)$, *where* Exp-semi-adp *is defined in Fig. 2.*

3.1 Functional Encryption with Uniform Function and Message Space

In the previous section, we saw a definition for functional encryption schemes where the setup algorithm takes the functionality index n as input, and outputs a master public/secret key pair specific to the functionality index n. As a result, the encryption algorithm, when using this public key, can only encrypt messages in the message space $\mathcal{M}_n$. Similarly, the key generation algorithm can only generate keys in the function space $\mathcal{F}_n$.

However, if there is exactly one message space, and exactly one function space (that is $\mathcal{M}_n = \mathcal{M}$ and $\mathcal{F}_n = \mathcal{F}$ for all n), then we can assume the setup algorithm takes only the security parameter as input. The remaining syntax, correctness and security definitions are same as before.

4   Selective to Semi-adaptive Security Generically

In this section, we show how to construct semi-adaptively secure functional encryption schemes from selectively secure functional encryption schemes, semantically secure public key encryption schemes, and secure garbled circuits. At a high-level, the idea is to delay the release of the base FE scheme's public parameters until the adversary makes first key query, and since in a semi-adaptive security game the adversary must submit its challenge before requesting any secret keys, therefore we could hope to invoke the selective security of the underlying FE scheme after receiving the challenge. However, the simulator needs to provide enough information to the adversary so that it could still perform encryptions before sending the challenge. To get around this problem, the encryption algorithm is modified to output a garbled circuit which takes as input the $\mathsf{FE_{sel}}$ public parameters and outputs the appropriate ciphertext. Essentially, the encryption algorithm encrypts without knowing the 'real' public parameters. The encryption algorithm would still need to hide the input wire keys such that a secret key reveals only half of them. Below we describe our approach in detail.

4.1   Construction

Let $\mathsf{FE_{sel}} = (\mathsf{Setup_{sel}}, \mathsf{KeyGen_{sel}}, \mathsf{Enc_{sel}}, \mathsf{Dec_{sel}})$ be a functional encryption scheme with function space $\{\mathcal{F}_n\}_n$ and message space $\{\mathcal{M}_n\}_n$. We use the polynomial $\ell(\lambda, n)$ to denote the size of the public key output by the $\mathsf{FE_{sel}}$ setup algorithm, where λ is the security parameter and n is the functionality index. We will simply write it as ℓ whenever clear from context.

Tools Required for Our Transformation: Let $\mathsf{GC} = (\mathsf{Garble}, \mathsf{Eval})$ be a garbling scheme for polynomial sized circuits, and $\mathsf{PKE} = (\mathsf{Setup_{PKE}}, \mathsf{Enc_{PKE}}, \mathsf{Dec_{PKE}})$ be a public key encryption scheme.

Our Transformation: We now describe our construction for semi-adaptively secure functional encryption scheme $\mathsf{FE} = (\mathsf{Setup}, \mathsf{Enc}, \mathsf{Dec}, \mathsf{KeyGen})$ with message space $\{\mathcal{M}_n\}_n$ and function space $\{\mathcal{F}_n\}_n$.

- $\mathsf{Setup}(1^\lambda, 1^n) \to (\mathsf{mpk}, \mathsf{msk})$: The setup algorithm first runs the PKE setup to compute 2ℓ public/secret key pairs $(\mathsf{pk}_{i,b}, \mathsf{sk}_{i,b})_{i \leq \ell, b \in \{0,1\}} \leftarrow \mathsf{Setup_{PKE}}(1^\lambda)$, independently and uniformly. It also runs $\mathsf{FE_{sel}}$ setup algorithm and generates master public/secret key pair $(\mathsf{mpk_{sel}}, \mathsf{msk_{sel}}) \leftarrow \mathsf{Setup_{sel}}(1^\lambda, 1^n)$. It sets $\mathsf{mpk} = \{\mathsf{pk}_{i,b}\}_{i \leq \ell, b \in \{0,1\}}$ and $\mathsf{msk} = \left(\mathsf{mpk_{sel}}, \mathsf{msk_{sel}}, \{\mathsf{sk}_{i,b}\}_{i \leq \ell, b \in \{0,1\}}\right)$.
- $\mathsf{Enc}(\mathsf{mpk}, m \in \mathcal{M}_n) \to \mathsf{ct}$: Let $\mathcal{C}\text{-}\mathsf{Enc\text{-}pk}_{m,r}^\ell$ be the canonical circuit which has message m and randomness r hardwired, takes an ℓ bit input x and computes $\mathsf{Enc_{sel}}(x, m; r)$; that is, it uses the input as a public key for the base FE scheme and encrypts message m using randomness r.
  The encryption algorithm constructs the circuit $\mathcal{C}\text{-}\mathsf{Enc\text{-}pk}_{m,r}^\ell$ using uniform randomness r, and it computes the garbled circuit as $(C, \{w_{i,b}\}_{i \leq \ell, b \in \{0,1\}}) \leftarrow$

Garble($\mathcal{C}$-Enc-pk$_{m,r}^{\ell}$, 1^{λ}). It then encrypts the garbled wire keys by computing ct$_{i,b}$ $\leftarrow$ Enc$_{\text{PKE}}$(pk$_{i,b}$, $w_{i,b}$) for $i \leq \ell$ and $b \in \{0,1\}$, where mpk $=$ $\{$pk$_{i,b}\}_{i \leq \ell, b \in \{0,1\}}$. Finally, it outputs a ciphertext ct which consists of the garbled circuit C and the 2ℓ ciphertexts $\{$ct$_{i,b}\}_{i \leq \ell, b \in \{0,1\}}$.

- KeyGen(msk, $f \in \mathcal{F}_n$) $\rightarrow$ sk$_f$: Let msk $=$ $\left(\text{mpk}_{\text{sel}}, \text{msk}_{\text{sel}}, \{\text{sk}_{i,b}\}_{i \leq \ell, b \in \{0,1\}}\right)$. The key generation algorithm first generates selective FE secret key corresponding to the function f by computing sk$_{f,\text{sel}}$ $\leftarrow$ KeyGen$_{\text{sel}}$(mpk$_{\text{sel}}$, f). It outputs sk$_f$ $=$ $\left(\text{mpk}_{\text{sel}}, \text{sk}_{f,\text{sel}}, \{\text{sk}_{i,\text{mpk}_{\text{sel}}[i]}\}_{i \leq \ell}\right)$ as the key for function f.

- Dec(sk$_f$, ct) $\rightarrow$ $\{0, 1, \bot\}$: Let sk$_f$ $=$ $\left(\text{mpk}_{\text{sel}}, \text{sk}_{f,\text{sel}}, \{\text{sk}_i\}_{i \leq \ell}\right)$ and ciphertext ct $=$ $\left(C, \{\text{ct}_{i,b}\}_{i \leq \ell, b \in \{0,1\}}\right)$. The decryption algorithm first decrypts the appropriate garbled circuit wires. Concretely, for $i \leq \ell$, it computes w_i $=$ Dec$_{\text{PKE}}$(sk$_i$, ct$_{i,\text{mpk}_{\text{sel}}[i]}$). It then uses these ℓ wire keys to evaluate the garbled circuit as $\widetilde{\text{ct}}$ $=$ Eval(C, $\{w_i\}_{i \leq \ell}$). Finally, it uses the secret key sk$_{f,\text{sel}}$ to decrypt the ciphertext $\widetilde{\text{ct}}$, and outputs Dec$_{\text{sel}}$(sk$_{f,\text{sel}}$, $\widetilde{\text{ct}}$).

Correctness. For all $\lambda, n \in \mathbb{N}$, message $m \in \mathcal{M}_n$, base FE keys (mpk$_{\text{sel}}$, msk$_{\text{sel}}$) $\leftarrow$ Setup$_{\text{sel}}$($1^{\lambda}, 1^n$), and 2ℓ PKE keys (pk$_{i,b}$, sk$_{i,b}$) $\leftarrow$ Setup$_{\text{PKE}}$(1^{λ}), the ciphertext corresponding to message m in our FE scheme is (C, $\{\text{ct}_{i,b}\}$), where (C, $\{w_{i,b}\}$) $\leftarrow$ Garble($\mathcal{C}$-Enc-pk$_{m,r}^{\ell}$, 1^{λ}) and ct$_{i,b}$ $\leftarrow$ Enc$_{\text{PKE}}$(pk$_{i,b}$, $w_{i,b}$).

For any function $f \in \mathcal{F}_n$, the corresponding secret key in our scheme consists of $\left(\text{mpk}_{\text{sel}}, \text{sk}_{f,\text{sel}}, \{\text{sk}_{i,\text{mpk}_{\text{sel}}[i]}\}\right)$, where sk$_{f,\text{sel}}$ $\leftarrow$ KeyGen$_{\text{sel}}$(msk$_{\text{sel}}$, f). The decryption algorithm first decrypts the encryptions of garbled circuit input wires corresponding to the public key mpk$_{\text{sel}}$ as $w_{i,\text{mpk}_{\text{sel}}[i]}$ $=$ Dec$_{\text{sel}}$(sk$_{i,\text{mpk}_{\text{sel}}[i]}$, ct$_{i,\text{mpk}_{\text{sel}}[i]}$). This follows from correctness of PKE scheme. Next, it computes ciphertext $\widetilde{\text{ct}}$ $=$ Eval(C, $\{w_{i,\text{mpk}_{\text{sel}}[i]}\}$) which is same as Enc$_{\text{sel}}$(mpk$_{\text{sel}}$, $m; r$) due to correctness of garbling scheme. Finally, the decryption algorithm computes Dec$_{\text{sel}}$(sk$_{f,\text{sel}}$, $\widetilde{\text{ct}}$) which is equal to $f(m)$ as the base FE scheme is also correct. Therefore, FE satisfies the functional encryption correctness condition.

Security. We will now show that the scheme described above is semi-adaptively secure.

Theorem 1. *Assuming* FE$_{\text{sel}}$ $=$ (Setup$_{\text{sel}}$, KeyGen$_{\text{sel}}$, Enc$_{\text{sel}}$, Dec$_{\text{sel}}$) *is a selectively-secure functional encryption scheme with* $\{\mathcal{F}_n\}_n$ *and* $\{\mathcal{M}_n\}_n$ *as function space and message space satisfying Definition 5,* GC $=$ (Garble, Eval) *is a secure garbling scheme for circuit family* $\mathcal{C} = \{\mathcal{C}_m\}_m$ *satisfying Definition 1, and* PKE $=$ (Setup$_{\text{PKE}}$, Enc$_{\text{PKE}}$, Dec$_{\text{PKE}}$) *is a semantically secure public key encryption scheme satisfying Definition 2, then* FE *forms a semi-adaptively secure functional encryption scheme satisfying Definition 6 for same function space and message space as the selective scheme.*

To formally prove our theorem, we describe the following sequence of games.

Game 1: This is the semi-adaptive security game described in Fig. 2.

1. (Setup Phase) The challenger first runs the PKE setup algorithm and $\mathsf{FE}_{\mathsf{sel}}$ setup algorithm to generate public/secret key pairs as $\left(\mathsf{pk}_{i,\beta}, \mathsf{sk}_{i,\beta}\right)_{i \leq \ell, \beta \in \{0,1\}}$ $\leftarrow \mathsf{Setup}_{\mathsf{PKE}}(1^\lambda)$ and $(\mathsf{mpk}_{\mathsf{sel}}, \mathsf{msk}_{\mathsf{sel}}) \leftarrow \mathsf{Setup}_{\mathsf{sel}}(1^\lambda, 1^n)$. It sets $\mathsf{mpk} = \{\mathsf{pk}_{i,\beta}\}$ and $\mathsf{msk} = \left(\mathsf{mpk}_{\mathsf{sel}}, \mathsf{msk}_{\mathsf{sel}}, \{\mathsf{sk}_{i,\beta}\}_{i \leq \ell, \beta \in \{0,1\}}\right)$, and sends mpk to $\mathcal{A}$.
2. (Challenge Phase)
   (a) $\mathcal{A}$ sends two challenge messages (m_0^*, m_1^*) to the challenger such that $\epsilon_n(m_0^*) = \epsilon_n(m_1^*)$, where $\epsilon_n(\cdot)$ is the empty function.
   (b) Challenger chooses a random bit $b \leftarrow \{0,1\}$, and computes the garbled circuit as $(C, \{w_{i,\beta}\}_{i \leq \ell, \beta \in \{0,1\}}) \leftarrow \mathsf{Garble}(\mathcal{C}\text{-}\mathsf{Enc\text{-}pk}_{m_b^*, r}^\ell, 1^\lambda)$.
   (c) It encrypts the wire keys $w_{i,\beta}$ as $\mathsf{ct}_{i,\beta}^* \leftarrow \mathsf{Enc}_{\mathsf{PKE}}(\mathsf{pk}_{i,\beta}, w_{i,\beta})$.
   (d) It sets challenge ciphertext as $\mathsf{ct}^* = \left(C, \{\mathsf{ct}_{i,\beta}^*\}_{i \leq \ell, \beta \in \{0,1\}}\right)$, and sends ct^* to $\mathcal{A}$.
3. (Key Query Phase)
   (a) $\mathcal{A}$ queries the challenger on polynomially many functions $f \in \mathcal{F}_n$ such that $f(m_0^*) = f(m_1^*)$.
   (a) For each queried function f, challenger generates the selective FE secret key as $\mathsf{sk}_{f,\mathsf{sel}} \leftarrow \mathsf{KeyGen}_{\mathsf{sel}}(\mathsf{msk}_{\mathsf{sel}}, f)$. It sets the secret key as $\mathsf{sk}_f = \left(\mathsf{mpk}_{\mathsf{sel}}, \mathsf{sk}_{f,\mathsf{sel}}, \{\mathsf{sk}_{i,\mathsf{mpk}_{\mathsf{sel}}[i]}\}_{i \leq \ell}\right)$, and sends sk_f to $\mathcal{A}$.
4. (Guess) Finally, $\mathcal{A}$ sends its guess b' and wins if $b = b'$.

Game 2: It is same as Game 1, except the way challenge ciphertext is created. In this game, while creating ct^*, challenger only encrypts garbled circuit wire keys corresponding to the bits of $\mathsf{mpk}_{\mathsf{sel}}$ and encrypts $\mathbf{0}$ at all other places.

1. (Setup Phase) The challenger first runs the PKE setup algorithm and $\mathsf{FE}_{\mathsf{sel}}$ setup algorithm to generate public/secret key pairs as $\left(\mathsf{pk}_{i,\beta}, \mathsf{sk}_{i,\beta}\right)_{i \leq \ell, \beta \in \{0,1\}}$ $\leftarrow \mathsf{Setup}_{\mathsf{PKE}}(1^\lambda)$ and $(\mathsf{mpk}_{\mathsf{sel}}, \mathsf{msk}_{\mathsf{sel}}) \leftarrow \mathsf{Setup}_{\mathsf{sel}}(1^\lambda, 1^n)$. It sets $\mathsf{mpk} = \{\mathsf{pk}_{i,\beta}\}_{i \leq \ell, \beta \in \{0,1\}}$ and $\mathsf{msk} = \left(\mathsf{mpk}_{\mathsf{sel}}, \mathsf{msk}_{\mathsf{sel}}, \{\mathsf{sk}_{i,\beta}\}_{i \leq \ell, \beta \in \{0,1\}}\right)$, and sends mpk to $\mathcal{A}$.
2. (Challenge Phase)
   (a) $\mathcal{A}$ sends two challenge messages (m_0^*, m_1^*) to the challenger such that $\epsilon_n(m_0^*) = \epsilon_n(m_1^*)$, where $\epsilon_n(\cdot)$ is the empty function.
   (b) Challenger chooses a random bit $b \leftarrow \{0,1\}$, and computes the garbled circuit as $(C, \{w_{i,\beta}\}_{i \leq \ell, \beta \in \{0,1\}}) \leftarrow \mathsf{Garble}(\mathcal{C}\text{-}\mathsf{Enc\text{-}pk}_{m_b^*, r}^\ell, 1^\lambda)$.
   (c) It then encrypts *half* of the 2ℓ wire keys as $\mathsf{ct}_{i,\beta}^* \leftarrow \mathsf{Enc}_{\mathsf{PKE}}(\mathsf{pk}_{i,\beta}, w_{i,\beta})$ if $\beta = \mathsf{mpk}_{\mathsf{sel}}[i]$, and $\mathsf{ct}_{i,\beta}^* \leftarrow \mathsf{Enc}_{\mathsf{PKE}}(\mathsf{pk}_{i,\beta}, \mathbf{0})$ otherwise.
   (d) It sets challenge ciphertext as $\mathsf{ct}^* = \left(C, \{\mathsf{ct}_{i,\beta}^*\}_{i \leq \ell, \beta \in \{0,1\}}\right)$, and sends ct^* to $\mathcal{A}$.
3. (Key Query Phase)

(a) $\mathcal{A}$ queries the challenger on polynomially many functions $f \in \mathcal{F}_n$ such that $f(m_0^*) = f(m_1^*)$.

(b) For each queried function f, challenger generates the selective FE secret key as $\mathsf{sk}_{f,\mathsf{sel}} \leftarrow \mathsf{KeyGen}_{\mathsf{sel}}(\mathsf{msk}_{\mathsf{sel}}, f)$. It sets the secret key as $\mathsf{sk}_f = \left(\mathsf{mpk}_{\mathsf{sel}}, \mathsf{sk}_{f,\mathsf{sel}}, \{\mathsf{sk}_{i,\mathsf{mpk}_{\mathsf{sel}}[i]}\}_{i \leq \ell}\right)$, and sends sk_f to $\mathcal{A}$.

4. (Guess) Finally, $\mathcal{A}$ sends its guess b' and wins if $b = b'$.

Game 3: It is same as Game 2, except the way challenge ciphertext is created. In this game, while creating ct^*, challenger simulates the garbled circuit instead of garbling the actual circuit.

1. (Setup Phase) The challenger first runs the PKE setup algorithm and $\mathsf{FE}_{\mathsf{sel}}$ setup algorithm to generate public/secret key pairs as $(\mathsf{pk}_{i,\beta}, \mathsf{sk}_{i,\beta})_{i \leq \ell, \beta \in \{0,1\}}$ $\leftarrow \mathsf{Setup}_{\mathrm{PKE}}(1^\lambda)$ and $(\mathsf{mpk}_{\mathsf{sel}}, \mathsf{msk}_{\mathsf{sel}}) \leftarrow \mathsf{Setup}_{\mathsf{sel}}(1^\lambda, 1^n)$. It sets $\mathsf{mpk} = \{\mathsf{pk}_{i,\beta}\}_{i \leq \ell, \beta \in \{0,1\}}$ and $\mathsf{msk} = \left(\mathsf{mpk}_{\mathsf{sel}}, \mathsf{msk}_{\mathsf{sel}}, \{\mathsf{sk}_{i,\beta}\}_{i \leq \ell, \beta \in \{0,1\}}\right)$, and sends mpk to $\mathcal{A}$.

2. (Challenge Phase)
   (a) $\mathcal{A}$ sends two challenge messages (m_0^*, m_1^*) to the challenger such that $\epsilon_n(m_0^*) = \epsilon_n(m_1^*)$, where $\epsilon_n(\cdot)$ is the empty function.
   (b) Challenger chooses a random bit $b \leftarrow \{0,1\}$.
   It computes $\widetilde{\mathsf{ct}}^* \leftarrow \mathsf{Enc}_{\mathsf{sel}}(\mathsf{mpk}_{\mathsf{sel}}, m_b^*)$ using uniform randomness.

   Next, it computes $(C, \{w_{i,\mathsf{mpk}_{\mathsf{sel}}[i]}\}) \leftarrow \mathsf{Sim}\left(1^\lambda, 1^\ell, 1^k, \widetilde{\mathsf{ct}}^*\right)$ (here, k is the size of the canonical circuit $\mathcal{C}\text{-}\mathsf{Enc}\text{-}\mathsf{pk}_{m,r}^\ell$).

   (c) It then encrypts half of the 2ℓ wire keys as $\mathsf{ct}_{i,\beta}^* \leftarrow \mathsf{Enc}_{\mathrm{PKE}}(\mathsf{pk}_{i,\beta}, w_{i,\beta})$ if $\beta = \mathsf{mpk}_{\mathsf{sel}}[i]$, and $\mathsf{ct}_{i,\beta}^* \leftarrow \mathsf{Enc}_{\mathrm{PKE}}(\mathsf{pk}_{i,\beta}, \mathbf{0})$ otherwise.
   (d) It sets challenge ciphertext as $\mathsf{ct}^* = \left(C, \{\mathsf{ct}_{i,\beta}^*\}_{i \leq \ell, \beta \in \{0,1\}}\right)$, and sends ct^* to $\mathcal{A}$.

3. (Key Query Phase)
   (a) $\mathcal{A}$ queries the challenger on polynomially many functions $f \in \mathcal{F}_n$ such that $f(m_0^*) = f(m_1^*)$.
   (b) For each queried function f, challenger generates the selective FE secret key as $\mathsf{sk}_{f,\mathsf{sel}} \leftarrow \mathsf{KeyGen}_{\mathsf{sel}}(\mathsf{msk}_{\mathsf{sel}}, f)$. It sets the secret key as $\mathsf{sk}_f = \left(\mathsf{mpk}_{\mathsf{sel}}, \mathsf{sk}_{f,\mathsf{sel}}, \{\mathsf{sk}_{i,\mathsf{mpk}_{\mathsf{sel}}[i]}\}_{i \leq \ell}\right)$, and sends sk_f to $\mathcal{A}$.

4. (Guess) Finally, $\mathcal{A}$ sends its guess b' and wins if $b = b'$.

Analysis. We now establish via a sequence of lemmas that no PPT adversary can distinguish between any two adjacent games with non-negligible advantage. To conclude, we also show that any PPT adversary that wins with non-negligible probability in the last game breaks the selective security of $\mathsf{FE}_{\mathsf{sel}}$ scheme.

Let $\mathcal{A}$ be any successful PPT adversary against our construction in the semi-adaptive security game (Fig. 2). In Game i, advantage of $\mathcal{A}$ is defined as $\mathsf{Adv}_{\mathcal{A}}^i = |\Pr[\mathcal{A} \text{ wins}] - 1/2|$. We then show via a sequence of claims that if $\mathcal{A}$'s advantage is non-negligible in Game i, then it has non-negligible advantage in Game $i + 1$ as well. Finally, in last game, we directly use $\mathcal{A}$ to attack the selective security of underlying FE scheme. Below we describe our hybrid games in more detail.

Lemma 1. *If* PKE *is a semantically secure public key encryption scheme, then for all PPT* $\mathcal{A}$, $|\mathsf{Adv}_{\mathcal{A}}^1 - \mathsf{Adv}_{\mathcal{A}}^2| \leq negl(\lambda)$ *for some negligible function* $negl(\cdot)$.

Proof. For proving indistinguishability of Games 1 and 2, we need to sketch ℓ intermediate hyrbrid games between these two, where ℓ is the length of master public key $\mathsf{mpk}_\mathsf{sel}$. Observe that in Game 1, ciphertexts $\mathsf{ct}_{i,\beta}^*$ are encryptions of garbled circuit input wire keys $w_{i,\beta}$ for both values of bit β; however, in Game 2, ciphertexts $\mathsf{ct}_{i,\beta}^*$ are encryptions of $w_{i,\beta}$ *if and only if* $\beta = \mathsf{mpk}_\mathsf{sel}[i]$, and they are encryptions of zeros otherwise. The high-level proof idea is to switch $\mathsf{ct}_{i,\beta}$ from encryptions of $w_{i,\beta}$ to encryptions of $\mathbf{0}$ one-at-a-time by using semantic security of PKE scheme. This could be done because the secret key $\mathsf{sk}_{i,\beta}$ is revealed only if $\beta = \mathsf{mpk}_\mathsf{sel}[i]$. Concretely, i^{th} intermediate hybrid between Game 1 and 2 proceeds same as Game 1 except that the first i ciphertexts $\mathsf{ct}_{j,\beta}^*$ is computed as $\mathsf{ct}_{j,\beta}^* \leftarrow \mathsf{Enc}_{\mathrm{PKE}}(\mathsf{pk}_{j,\beta}, \mathbf{0})$ if $\beta \neq \mathsf{mpk}_\mathsf{sel}[j]$, i.e. for $j \leq i$ and $\beta \neq \mathsf{mpk}_\mathsf{sel}[j]$, $\mathsf{ct}_{j,\beta}^*$ are encryptions of zero, and for $j > i$ or $\beta = \mathsf{mpk}_\mathsf{sel}[j]$, $\mathsf{ct}_{j,\beta}^*$ are encryptions of wire keys $w_{i,\beta}$. For the analysis, Game 1 is regarded as 0th intermediate hybrid, and Game 2 is regarded as ℓ^{th} intermediate hybrid. Below we show that $\mathcal{A}$'s advantage in distinguishing any pair of consecutive intermediate hybrid is negligibly small.

We describe a reduction algorithm $\mathcal{B}$ which breaks semantic security of the PKE scheme, if $\mathcal{A}$ distinguishes between intermediate hybrids $i - 1$ and i with non-negligible advantage. First, $\mathcal{B}$ receives the challenge public key pk^* from the PKE challenger. Next, $\mathcal{B}$ runs the Step 1 as in Game 1, except instead of running PKE Setup algorithm to compute public/secret key pair $(\mathsf{pk}_{i,\beta}, \mathsf{sk}_{i,\beta})$ when $\beta \neq \mathsf{mpk}_\mathsf{sel}[i]$, it sets $\mathsf{pk}_{i,1-\mathsf{mpk}_\mathsf{sel}[i]} = \mathsf{pk}^*$. After $\mathcal{A}$ submits its challenge messages (m_0^*, m_1^*) to $\mathcal{B}$, the reduction computes garbled circuit C and ciphertexts $\mathsf{ct}_{i,\beta}^*$ as in the $(i-1)^{th}$ intermediate hybrid, except to compute $\mathsf{ct}_{i,1-\mathsf{mpk}_\mathsf{sel}[i]}^*$, $\mathcal{B}$ sends $w_{i,1-\mathsf{mpk}_\mathsf{sel}[i]}$ and $\mathbf{0}$ as its challenge messages to the PKE challenger, and sets $\mathsf{ct}_{i,1-\mathsf{mpk}_\mathsf{sel}[i]}^*$ as the PKE challenge ciphertext. $\mathcal{B}$ runs the remaining game as in Game 1.[4] Finally, if $\mathcal{A}$ wins ($b = b'$), then $\mathcal{B}$ guesses 0 to indicate that $\mathsf{ct}_{i,1-\mathsf{mpk}_\mathsf{sel}[i]}^*$ was encryption of $w_{i,1-\mathsf{mpk}_\mathsf{sel}[i]}$, else it guesses 1 to indicate that it was encryption of zeros.

Note that when $w_{i,1-\mathsf{mpk}_\mathsf{sel}[i]}$ is encrypted by the PKE challenger, then $\mathcal{B}$ exactly simulates the view of intermediate hybrid $i - 1$ to $\mathcal{A}$. Otherwise if $\mathbf{0}$ is encrypted the view is of intermediate hybrid i. Therefore, $\mathcal{A}$'s advantage in any two consecutive intermediate hybrids is negligibly close as otherwise PKE scheme is not semantically secure. Hence, using ℓ intermediate hybrids we have

---

[4] It should be noted that $\mathcal{B}$ can still answer the secret key queries during the reduction because it only needs the secret keys $\mathsf{sk}_{i,\beta}$ corresponding to the public key $\mathsf{mpk}_\mathsf{sel}$ (i.e. $\beta = \mathsf{mpk}_\mathsf{sel}[i]$). Since $\mathcal{B}$ chooses all such secret keys, therefore it can answer $\mathcal{A}$'s secret key queries.

proved that switching $\mathsf{ct}_{i,\beta}$ from encryptions of $w_{i,\beta}$ to encryptions of $\mathbf{0}$ for $\beta \neq \mathsf{mpk}_\mathsf{sel}[i]$ causes at most negligible dip in $\mathcal{A}$'s advantage in Game 1. Therefore if $|\mathsf{Adv}^1_\mathcal{A} - \mathsf{Adv}^2_\mathcal{A}|$ is non-negligible, then the PKE scheme is not semantically secure.

Lemma 2. *If* GC *is a secure garbling scheme, then for all PPT* $\mathcal{A}$, $|\mathsf{Adv}^2_\mathcal{A} - \mathsf{Adv}^3_\mathcal{A}| \leq negl(\lambda)$ *for some negligible function* $negl(\cdot)$.

Proof. The proof of this lemma follows from the security of our garbling scheme. First, note that the simulation based definition of garbling security can be viewed as a game based definition between a challenger and an adversary. An adversary sends a circuit $C \in \mathcal{C}_m$ and input $x \in \{0,1\}^m$. The challenger then either honestly garbles the circuit, and sends the wire keys corresponding to x, or runs the simulator to compute the garbled circuit and the wire keys for x.

Suppose there exists an adversary $\mathcal{A}$ such that $\mathsf{Adv}^2_\mathcal{A} - \mathsf{Adv}^3_\mathcal{A}$ is non-negligible in λ. We will construct a reduction algorithm $\mathcal{B}$ that uses $\mathcal{A}$ to break the garbling security. $\mathcal{B}$ first chooses 2ℓ public/secret key pairs and sends $\{\mathsf{pk}_{i,\beta}\}$ to the $\mathcal{A}$. $\mathcal{B}$ also chooses the base FE scheme's master public/secret keys $(\mathsf{mpk}_\mathsf{sel}, \mathsf{msk}_\mathsf{sel})$. Next, $\mathcal{A}$ sends challenge messages m^*_0, m^*_1. The reduction algorithm chooses $b \leftarrow \{0,1\}$, randomness r and computes the circuit $\mathsf{ckt} = C\text{-}\mathsf{Enc\text{-}pk}^\ell_{m^*_b,r}$. It then sends circuit ckt and input $\mathsf{mpk}_\mathsf{sel}$ to the garbling challenger, and receives a garbled circuit C and ℓ wire keys $\{w_i\}$. The reduction algorithm then computes $\mathsf{ct}_{i,\beta} \leftarrow \mathsf{Enc_{PKE}}(\mathsf{pk}_{i,\beta}, w_i)$ if $\beta = \mathsf{mpk}[i]$, else $\mathsf{ct}_{i,\beta} \leftarrow \mathsf{Enc_{PKE}}(\mathsf{pk}_{i,\beta}, \mathbf{0})$. Finally $\mathcal{B}$ sends $(C, \{\mathsf{ct}_{i,\beta}\})$ to $\mathcal{A}$ as the challenge ciphertext. The key queries are identical in both Game 2 and Game 3. Finally, the adversary sends its guess b', and if $b = b'$, the reduction algorithm guesses that ckt was honestly garbled, else it guesses that ckt and wire keys were simulated.

Note that if the garbling challenger honestly garbled circuit ckt, then $\mathcal{B}$ exactly simulates the view of Game 2 to $\mathcal{A}$. Otherwise the view is of Game 3. As a result, if $\mathsf{Adv}^2_\mathcal{A} - \mathsf{Adv}^3_\mathcal{A}$ is non-negligible in λ, then $\mathcal{B}$ breaks the garbling scheme's security with non-negligible advantage.

Lemma 3. *If* FE_sel *is a selectively-secure functional encryption scheme, then for all PPT* $\mathcal{A}$, $\mathsf{Adv}^3_\mathcal{A} \leq negl(\lambda)$ *for some negligible function* $negl(\cdot)$.

Proof. We describe a reduction algorithm $\mathcal{B}$ which plays the selective indistinguishability based game with FE_sel challenger, and simulates Game 3 for adversary $\mathcal{A}$. $\mathcal{B}$ runs the Step 1 as in Game 3, except it does not choose FE_sel master public/secret key pair. It only generates 2ℓ PKE public/secret key pairs, sets $\mathsf{mpk} = \{\mathsf{pk}_{i,\beta}\}$, $\mathsf{msk} = \{\mathsf{sk}_{i,\beta}\}$, and sends mpk to $\mathcal{A}$. Next, $\mathcal{A}$ chooses two challenge messages (m^*_0, m^*_1), and sends those to $\mathcal{B}$. Reduction algorithm $\mathcal{B}$ forwards (m^*_0, m^*_1) to the FE_sel challenger as its challenge messages. Note that $\mathcal{B}$ is behaving as a selective adversary since it has not queried FE_sel challenger for a public key before sending its challenge messages. Now FE_sel challenger chooses a bit $b^* \leftarrow \{0,1\}$, runs the setup algorithm to compute key pair $(\mathsf{mpk}_\mathsf{sel}, \mathsf{msk}_\mathsf{sel})$, computes $\widetilde{\mathsf{ct}} \leftarrow \mathsf{Enc_{sel}}(\mathsf{mpk}_\mathsf{sel}, m^*_b)$, and sends public key $\mathsf{mpk}_\mathsf{sel}$ and ciphertext $\widetilde{\mathsf{ct}}^*$ to $\mathcal{B}$. $\mathcal{B}$ receives $\mathsf{mpk}_\mathsf{sel}$ and $\widetilde{\mathsf{ct}}^*$ from the challenger, and it simulates the garbled circuit $(C, \{w_i\}) \leftarrow \mathsf{Sim}\left(1^\lambda, 1^\ell, 1^k, \widetilde{\mathsf{ct}}^*\right)$. Next, it computes ciphertexts $\mathsf{ct}^*_{i,\mathsf{mpk_{sel}}[i]}$ as encryptions of w_i, and remaining ciphertexts as encryptions of $\mathbf{0}$. $\mathcal{B}$ sends the final challenge ciphertext ct^* as garbled circuit C and ciphertexts $\left\{\mathsf{ct}^*_{i,\beta}\right\}$ to $\mathcal{A}$. After receiving the challenge ciphertext, $\mathcal{A}$ is allowed to make polynomially many secret key queries sk_f for functions f, which $\mathcal{B}$ can answer by requesting corresponding secret keys $\mathsf{sk}_{f,\mathsf{sel}}$ from $\mathsf{FE_{sel}}$ challenger, and releasing $\left\{\mathsf{sk}_{i,\mathsf{mpk_{sel}}[i]}\right\}$ along with $\mathsf{sk}_{f,\mathsf{sel}}$. Finally, $\mathcal{A}$ sends its guess b' to $\mathcal{B}$, and $\mathcal{B}$ sends b' as its guess for $\mathsf{FE_{sel}}$ challenger's bit b^*.

Note that $\mathcal{B}$ exactly simulates the view of Game 3 to $\mathcal{A}$. Therefore, $\mathcal{A}$'s advantage in Game 3 is negligibly small as otherwise the underlying FE scheme is not selectively-secure. Thus if $\mathsf{Adv}^3_{\mathcal{A}}$ is non-negligible, then the $\mathsf{FE_{sel}}$ scheme is not selectively-secure.

5   Bundling Functionalities

In this section, we show how to transform a (standard) FE scheme to one where the public parameters can support the union of underlying message/function spaces. This transformation is similar to the one outlined in Sect. 4. The only difference is that instead of public key encryption, we need to use identity based encryption for encrypting the garbled circuit wire keys, and the underlying FE scheme's master public/secret keys are chosen pseudorandomly during the key generation phase.

5.1   Construction

Let $\mathsf{FE_{sel}} = (\mathsf{Setup_{sel}}, \mathsf{KeyGen_{sel}}, \mathsf{Enc_{sel}}, \mathsf{Dec_{sel}})$ be a functional encryption scheme with message space $\{\mathcal{M}_n\}_n$ and function space $\{\mathcal{F}_n\}_n$, where for each $n \in \mathbb{N}$, $f \in \mathcal{F}_n$, the domain of f is $\mathcal{M}_n$. Let $\ell\text{-pk}(\cdot,\cdot)$ denote the polynomial representing the size of the public key output by the setup algorithm, $\ell\text{-r}_s(\cdot,\cdot)$ the randomness required by $\mathsf{Setup_{sel}}$ and $\ell\text{-r}_e(\cdot,\cdot)$ the randomness used by $\mathsf{Enc_{sel}}$. Here, all the above polynomials take the security parameter as the first input and functionality index as the second input. For simplicity of notation, we will drop the dependence of these polynomials on the security parameter.

Tools Required for Our Transformation: Let $\mathsf{GC} = (\mathsf{Garble}, \mathsf{Eval})$ be a garbling scheme for circuit family $\mathcal{C} = \{\mathcal{C}_n\}_n$ such that the wire keys output by Garble have length $\ell\text{-w}(\lambda)$, where λ is the security parameter. Let F be a pseudo-random function family with key space $\{\mathcal{K}_\lambda\}_\lambda$, input space $\{\{0,1\}^{2\lambda}\}_\lambda$ and output space $\{0,1\}$. Finally, we also use an identity based encryption scheme $\mathsf{IBE} = (\mathsf{Setup_{IBE}}, \mathsf{Enc_{IBE}}, \mathsf{KeyGen_{IBE}}, \mathsf{Dec_{IBE}})$ with identity space $\{\{0,1\}^{2\lambda+1}\}_\lambda$ and message space $\{\{0,1\}^{\ell\text{-w}(\lambda)}\}_\lambda$.

Our Transformation: We will now describe our functional encryption scheme FE = (Setup, Enc, Dec, KeyGen) with message space $\mathcal{M} = \cup_n\{(n,m) : m \in \mathcal{M}_n\}$ and function space $\mathcal{F} = \cup_n\{(n,f) : f \in \mathcal{F}_n\} \cup \{\epsilon\}$. Hence, each message in $\mathcal{M}$ and function in $\mathcal{F}$ has two components - the first component reveals the functionality index, and the second component is the actual message/function. For each func $= (n, f) \in \mathcal{F}$ and msg $= (n', m) \in \mathcal{M}$, we define func(msg) $= f(m)$ if $n = n'$, $\perp$ otherwise. The empty function ϵ is defined as follows: for all messages msg $= (n, m) \in \mathcal{M}$, $\epsilon(\text{msg}) = (n, \epsilon_n(m))$ (recall $\epsilon_n(\cdot)$ is the empty function in $\mathcal{F}_n$).

- Setup(1^λ) $\rightarrow$ (mpk, msk): The setup algorithm first runs the IBE setup to compute $(\text{pp}_{\text{IBE}}, \text{msk}_{\text{IBE}}) \leftarrow \text{Setup}_{\text{IBE}}(1^\lambda)$. Next, it chooses a PRF key $K \leftarrow \mathcal{K}_\lambda$. It sets mpk $= \text{pp}_{\text{IBE}}$ and msk $= (\text{msk}_{\text{IBE}}, K)$.

- Enc(mpk, msg $\in \mathcal{M}$) $\rightarrow$ ct: Let msg $= (n, m)$, $t = \ell\text{-pk}(n)$, and $\mathcal{C}\text{-Enc-pk}_{m,r}^t$ be the canonical circuit which has message m, randomness r hardwired, takes a t bit input x and computes $\text{Enc}_{\text{sel}}(x, m; r)$; that is, it uses the input as a public key for the base FE scheme and encrypts message m using randomness r. The encryption algorithm first chooses randomness $r \leftarrow \{0,1\}^{\ell\text{-}r_e(n)}$. Next, it garbles the circuit $\mathcal{C}\text{-Enc-pk}_{m,r}^t$ by computing $(C, \{w_{i,b}\}_{i \leq t, b \in \{0,1\}}) \leftarrow \text{Garble}(\mathcal{C}\text{-Enc-pk}_{m,r}^t, 1^\lambda)$. It then encrypts the garbled wire keys by computing $\text{ct}_{i,b} \leftarrow \text{Enc}_{\text{IBE}}(\text{mpk}, w_{i,b}, (n, i, b))$. Note that both n and i can be represented as λ bit strings. The final ciphertext consists of the garbled circuit C and the $2t$ ciphertexts $\{\text{ct}_{i,b}\}_{i \leq t, b \in \{0,1\}}$.

- KeyGen(msk, func $\in \mathcal{F}$) $\rightarrow$ sk$_{\text{func}}$: Let func $= (n, f)$, msk $= (\text{msk}_{\text{IBE}}, K)$, $s = \ell\text{-}r_s(n)$ and $t = \ell\text{-pk}(n)$.
  The key generation algorithm computes an s bit pseudorandom string $r = (F(K, (n, 1)), \ldots, F(K, (n, s)))$. Next, it uses r as the randomness to generate the base FE keys $(\text{mpk}_n, \text{msk}_n) = \text{Setup}_{\text{sel}}(1^\lambda, 1^n; r)$. Note that the functionality index used for generating these keys is n, and therefore the size of mpk_n is $t = \ell\text{-pk}(n)$, and the amount of randomness required by $\text{Setup}_{\text{sel}}$ is $s = \ell\text{-}r_s(n)$. Next, it generates IBE secret keys corresponding to the identities $(n, i, \text{mpk}_n[i])$ for $i \leq t$. It computes t secret keys $\text{sk}_i \leftarrow \text{KeyGen}_{\text{IBE}}(\text{msk}_{\text{IBE}}, (n, i, \text{mpk}_n[i]))$. Finally, it generates an FE secret key corresponding to function f by computing $\text{sk}_{f,\text{sel}} \leftarrow \text{KeyGen}_{\text{sel}}(\text{msk}_n, f)$. It outputs $(\text{sk}_{f,\text{sel}}, \text{mpk}_n, \{\text{sk}_i\}_{i \leq t})$ as the key for function f.

- Dec(sk$_f$, ct) $\rightarrow \{0, 1, \perp\}$: Let $\text{sk}_f = (\text{sk}_{f,\text{sel}}, \text{mpk}_n, \{\text{sk}_i\}_{i \leq t})$ and ciphertext ct $= (C, \{\text{ct}_{i,b}\}_{i \leq t, b \in \{0,1\}})$. The decryption algorithm first decrypts the appropriate garbled circuit wires. For $i \leq t$, it computes $w_i = \text{Dec}_{\text{IBE}}(\text{sk}_i, \text{ct}_{i, \text{mpk}_n[i]})$. It then uses these t wire keys to evaluate the garbled circuit. It computes $\widetilde{\text{ct}} = \text{Eval}(C, \{w_i\}_{i \leq t})$. Finally, it uses the secret key $\text{sk}_{f,\text{sel}}$ to decrypt the ciphertext. The output is $\text{Dec}_{\text{sel}}(\text{sk}_{f,\text{sel}}, \widetilde{\text{ct}})$.

Correctness: Fix any λ, message msg $= (n, m) \in \mathcal{M}$, function func $= (n, f) \in \mathcal{F}$ and IBE keys $(\text{mpk}_{\text{IBE}}, \text{msk}_{\text{IBE}})$. Let $(G, \{w_{i,b}\}) \leftarrow \text{Garble}(\mathcal{C}\text{-Enc-pk}_{m,r}^{\ell\text{-pk}(n)}, 1^\lambda)$ and $\text{ct}_{i,b} \leftarrow \text{Enc}_{\text{IBE}}(\text{mpk}_{\text{IBE}}, w_{i,b}, (n, i, b))$. The ciphertext corresponding to message msg in our FE scheme is $(G, \{\text{ct}_{i,b}\})$. Now, let us consider the key for function func. Let $(\mathsf{mpk}_n, \mathsf{msk}_n)$ be the base FE scheme's keys as computed in the key generation phase. The secret key for f in our scheme consists of IBE keys $\{\mathsf{sk}_i \leftarrow \mathsf{KeyGen}_{\mathrm{IBE}}(\mathsf{msk}_{\mathrm{IBE}}, (n, i, \mathsf{mpk}_n[i]))\}$ and $\mathsf{FE}_{\mathrm{sel}}$ key $\mathsf{sk}_f \leftarrow \mathsf{KeyGen}_{\mathrm{sel}}(\mathsf{msk}_n, f)$.

The decryption algorithm first decrypts the IBE ciphertexts to recover the garbled circuit's wire keys $\{w_{i,\mathsf{mpk}_n[i]}\}$. Next, using $\mathsf{Eval}(G, \{w_{i,\mathsf{mpk}_n[i]}\})$, we can compute $\widetilde{\mathsf{ct}} = \mathsf{Enc}_{\mathrm{sel}}(\mathsf{mpk}_n, m; r)$. Finally, the decryption algorithm computes $\mathsf{Dec}_{\mathrm{sel}}(\mathsf{sk}_{f,\mathrm{sel}}, \widetilde{\mathsf{ct}}) = f(m)$.

5.2   Security Proof

We will now prove that the IBE scheme described above is semi-adaptive secure, as per Definition 6. Our proof consists of a sequence of hybrids. Let n^* denote the functionality index of the challenge inputs. The first hybrid corresponds to the semi-adaptive security game. In the second hybrid, the challenger uses a truly random function instead of a pseudorandom function. In the third hybrid, we use the security of the IBE scheme to modify the ciphertexts output as part of the challenge ciphertext. Instead of encrypting all the garbled circuit wire keys, the challenger encrypts $\mathbf{0}$ at positions that do not correspond to the base FE scheme's public key. Here, it is crucial that the challenger never outputs IBE keys corresponding to these 'off' positions. In the fourth hybrid, the garbled circuit is simulated using the challenge ciphertext of the base FE scheme. At this point, we can use the security of the base FE scheme to complete our argument.

Game 1: This is the semi-adaptive security game described in Fig. 2.

1. (Setup Phase) The challenger first runs the setup algorithm by choosing $(\mathsf{mpk}_{\mathrm{IBE}}, \mathsf{msk}_{\mathrm{IBE}}) \leftarrow \mathsf{Setup}_{\mathrm{IBE}}(1^\lambda)$ and $K \leftarrow \mathcal{K}_\lambda$. It sends $\mathsf{mpk}_{\mathrm{IBE}}$ to the adversary.
2. (Challenge Phase)
   (a) $\mathcal{A}$ sends two challenge messages $\mathsf{msg}_0 = (n^*, m_0), \mathsf{msg}_1 = (n^*, m_1)$ such that $\epsilon(\mathsf{msg}_0) = \epsilon(\mathsf{msg}_1)$.
   (b) The challenger chooses a random bit $b \leftarrow \{0, 1\}$, and computes the garbled circuit and its wire keys as $(C, \{w_{i,\beta}\}_{i \leq t^*, \beta \in \{0,1\}}) \leftarrow \mathsf{Garble}(\mathcal{C}\text{-}\mathsf{Enc}\text{-}\mathsf{pk}_{m_b,r}^{t^*}, 1^\lambda)$, where $t^* = |\ell\text{-}\mathsf{pk}(n^*)|$.
   (c) It then encrypts the wire keys as $\mathsf{ct}_{i,\beta} \leftarrow \mathsf{Enc}_{\mathrm{IBE}}(\mathsf{mpk}_{\mathrm{IBE}}, w_{i,\beta}, (n^*, i, \beta))$.
   (d) The challenger sets $\mathsf{ct} = (C, \{\mathsf{ct}_{i,\beta}\}_{i \leq t^*, \beta \in \{0,1\}})$ and sends ct to $\mathcal{A}$.
3. (Key Query Phase)
   (a) $\mathcal{A}$ queries the challenger on polynomially many functions $\mathsf{func} = (n, f) \in \mathcal{F}$ such that $\mathsf{func}(\mathsf{msg}_0) = \mathsf{func}(\mathsf{msg}_1)$. Let $s = \ell\text{-}r_s(n), t = \ell\text{-}\mathsf{pk}(n)$.
   (b) The challenger computes $r = (F(K, (n, 1)), \ldots, F(K, (n, s)))$ and $(\mathsf{mpk}_n, \mathsf{msk}_n) = \mathsf{Setup}_{\mathrm{sel}}(1^\lambda, 1^n; r)$.
   (c) It generates the IBE secret keys as $\mathsf{sk}_i \leftarrow \mathsf{KeyGen}_{\mathrm{IBE}}(\mathsf{msk}_{\mathrm{IBE}}, (n, i, \mathsf{mpk}_n[i]))$ and base FE scheme's secret key $\mathsf{sk}_{f,\mathrm{sel}} \leftarrow \mathsf{KeyGen}_{\mathrm{sel}}(\mathsf{msk}_n, f)$.
   (d) The challenger sets $\mathsf{sk}_{\mathsf{func}} = (\mathsf{sk}_{f,\mathrm{sel}}, \mathsf{mpk}_n, \{\mathsf{sk}_i\})$ and sends $\mathsf{sk}_{\mathsf{func}}$ to $\mathcal{A}$.
4. (Guess) Finally, $\mathcal{A}$ sends its guess b' and wins if $b = b'$.

Game 2: This game is identical to the previous one, except that the challenger uses a truly random function F_{rand} instead of the pseudorandom function F.

1. (Setup Phase) The challenger first runs the setup algorithm by choosing $(\mathsf{mpk}_{IBE}, \mathsf{msk}_{IBE}) \leftarrow \mathsf{Setup}_{IBE}(1^\lambda)$. It sends mpk_{IBE} to the adversary.
2. (Challenge Phase)
   (a) $\mathcal{A}$ sends two challenge messages $\mathsf{msg}_0 = (n^*, m_0), \mathsf{msg}_1 = (n^*, m_1)$ such that $\epsilon(\mathsf{msg}_0) = \epsilon(\mathsf{msg}_1)$.
   (b) The challenger chooses a random bit $b \leftarrow \{0,1\}$, and computes the garbled circuit and its wire keys as $(C, \{w_{i,\beta}\}_{i \leq t^*, \beta \in \{0,1\}}) \leftarrow$ $\mathsf{Garble}(\mathcal{C}\text{-}\mathsf{Enc}\text{-}\mathsf{pk}_{m_b,r}^{t^*}, 1^\lambda)$, where $t^* = |\ell\text{-}\mathsf{pk}(n^*)|$.
   (c) It then encrypts the wire keys as $\mathsf{ct}_{i,\beta} \leftarrow \mathsf{Enc}_{IBE}(\mathsf{mpk}_{IBE}, w_{i,\beta}, (n^*, i, \beta))$.
   (d) The challenger sets $\mathsf{ct} = (C, \{\mathsf{ct}_{i,\beta}\}_{i \leq t^*, \beta \in \{0,1\}})$ and sends ct to $\mathcal{A}$.
3. (Key Query Phase)
   (a) $\mathcal{A}$ queries the challenger on polynomially many functions $\mathsf{func} = (n, f) \in \mathcal{F}$ such that $\mathsf{func}(\mathsf{msg}_0) = \mathsf{func}(\mathsf{msg}_1)$. Let $s = \ell\text{-}r_s(n)$, $t = \ell\text{-}\mathsf{pk}(n)$.
   (b) The challenger computes $r = (\ F_{rand}(n,1), \ldots, F_{rand}(n,s)\ )$ and $(\mathsf{mpk}_n, \mathsf{msk}_n) = \mathsf{Setup}_{sel}(1^\lambda, 1^n; r)$.
   (c) It generates the IBE secret keys as $\mathsf{sk}_i \leftarrow \mathsf{KeyGen}_{IBE}(\mathsf{msk}_{IBE}, (n, i, \mathsf{mpk}_n[i]))$ and base FE scheme's secret key $\mathsf{sk}_{f,sel} \leftarrow \mathsf{KeyGen}_{sel}(\mathsf{msk}_n, f)$.
   (d) The challenger sets $\mathsf{sk}_{func} = (\mathsf{sk}_{f,sel}, \mathsf{mpk}_n, \{\mathsf{sk}_i\})$ and sends sk_{func} to $\mathcal{A}$.
4. (Guess) Finally, $\mathcal{A}$ sends its guess b' and wins if $b = b'$.

Game 3: This game is identical to the previous one. Here, we are introducing some syntactical changes. In this game, the challenger chooses the base FE scheme's keys $\mathsf{mpk}_{n^*}, \mathsf{msk}_{n^*}$ immediately after receiving the challenge messages.

1. (Setup Phase) The challenger first runs the setup algorithm by choosing $(\mathsf{mpk}_{IBE}, \mathsf{msk}_{IBE}) \leftarrow \mathsf{Setup}_{IBE}(1^\lambda)$. It sends mpk_{IBE} to the adversary.
2. (Challenge Phase)
   (a) $\mathcal{A}$ sends two challenge messages $\mathsf{msg}_0 = (n^*, m_0), \mathsf{msg}_1 = (n^*, m_1)$ such that $\epsilon(\mathsf{msg}_0) = \epsilon(\mathsf{msg}_1)$.
   (b) The challenger chooses $(\mathsf{mpk}_{n^*}, \mathsf{msk}_{n^*}) \leftarrow \mathsf{Setup}_{sel}(1^\lambda, 1^{n^*})$.
   (c) It chooses a random bit $b \leftarrow \{0,1\}$, and computes the garbled circuit and its wire keys as $(C, \{w_{i,\beta}\}_{i \leq t^*, \beta \in \{0,1\}}) \leftarrow \mathsf{Garble}(\mathcal{C}\text{-}\mathsf{Enc}\text{-}\mathsf{pk}_{m_b,r}^{t^*}, 1^\lambda)$, where $t^* = |\ell\text{-}\mathsf{pk}(n^*)|$.
   (d) It then encrypts the wire keys as $\mathsf{ct}_{i,\beta} \leftarrow \mathsf{Enc}_{IBE}(\mathsf{mpk}_{IBE}, w_{i,\beta}, (n^*, i, \beta))$.
   (e) The challenger sets $\mathsf{ct} = (C, \{\mathsf{ct}_{i,\beta}\}_{i \leq t^*, \beta \in \{0,1\}})$ and sends ct to $\mathcal{A}$.
3. (Key Query Phase)
   (a) $\mathcal{A}$ queries the challenger on polynomially many functions $\mathsf{func} = (n, f) \in \mathcal{F}$ such that $\mathsf{func}(\mathsf{msg}_0) = \mathsf{func}(\mathsf{msg}_1)$. Let $s = \ell\text{-}r_s(n)$, $t = \ell\text{-}\mathsf{pk}(n)$.
   (b) The challenger chooses $(\mathsf{mpk}_n, \mathsf{msk}_n) \leftarrow \mathsf{Setup}_{sel}(1^\lambda, 1^n)$ (if $\mathsf{msk}_n, \mathsf{mpk}_n$ have already been computed before, then it simply reuses those keys).
   (c) It generates the IBE secret keys as $\mathsf{sk}_i \leftarrow \mathsf{KeyGen}_{IBE}(\mathsf{msk}_{IBE}, (n, i, \mathsf{mpk}_n[i]))$ and base FE scheme's secret key $\mathsf{sk}_{f,sel} \leftarrow \mathsf{KeyGen}_{sel}(\mathsf{msk}_n, f)$.
   (d) The challenger sets $\mathsf{sk}_{func} = (\mathsf{sk}_{f,sel}, \mathsf{mpk}_n, \{\mathsf{sk}_i\})$ and sends sk_{func} to $\mathcal{A}$.
4. (Guess) Finally, $\mathcal{A}$ sends its guess b' and wins if $b = b'$.

Game 4: In this game, the challenger modifies the challenge ciphertext. Instead of encrypting the garbled circuit wire keys for all $i \leq t, \beta \in \{0,1\}$, the challenger encrypts zeroes at positions (i, β) if $\beta \neq \mathsf{mpk}_{n^*}[i]$.

1. (Setup Phase) The challenger first runs the setup algorithm by choosing $(\mathsf{mpk}_{\mathsf{IBE}}, \mathsf{msk}_{\mathsf{IBE}}) \leftarrow \mathsf{Setup}_{\mathsf{IBE}}(1^\lambda)$. It sends $\mathsf{mpk}_{\mathsf{IBE}}$ to the adversary.
2. (Challenge Phase)
   (a) $\mathcal{A}$ sends two challenge messages $\mathsf{msg}_0 = (n^*, m_0), \mathsf{msg}_1 = (n^*, m_1)$ such that $\epsilon(\mathsf{msg}_0) = \epsilon(\mathsf{msg}_1)$.
   (b) The challenger chooses $(\mathsf{mpk}_{n^*}, \mathsf{msk}_{n^*}) \leftarrow \mathsf{Setup}_{\mathsf{sel}}(1^\lambda, 1^{n^*})$.
   (c) It chooses a random bit $b \leftarrow \{0,1\}$, and computes the garbled circuit and its wire keys as $(C, \{w_{i,\beta}\}_{i \leq t^*, \beta \in \{0,1\}}) \leftarrow \mathsf{Garble}(\mathcal{C}\text{-}\mathsf{Enc}\text{-}\mathsf{pk}_{m_b, r}^{t^*}, 1^\lambda)$, where $t^* = |\ell\text{-}\mathsf{pk}(n^*)|$.
   (d) It encrypts wire keys at half the positions, and zeroes elsewhere.
       For each i, if $\beta = \mathsf{mpk}_{n^*}[i]$, $\mathsf{ct}_{i,\beta} \leftarrow \mathsf{Enc}_{\mathsf{IBE}}(\mathsf{mpk}_{\mathsf{IBE}}, w_{i,\beta}, (n^*, i, \beta))$, else $\mathsf{ct}_{i,\beta} \leftarrow \mathsf{Enc}_{\mathsf{IBE}}(\mathsf{mpk}_{\mathsf{IBE}}, \mathbf{0}, (n^*, i, \beta))$.
   (e) It then encrypts the wire keys as $\mathsf{ct}_{i,\beta} \leftarrow \mathsf{Enc}_{\mathsf{IBE}}(\mathsf{mpk}_{\mathsf{IBE}}, w_{i,\beta}, (n^*, i, \beta))$.
   (f) The challenger sets $\mathsf{ct} = (C, \{\mathsf{ct}_{i,\beta}\}_{i \leq t^*, \beta \in \{0,1\}})$ and sends ct to $\mathcal{A}$.
3. (Key Query Phase)
   (a) $\mathcal{A}$ queries the challenger on polynomially many functions $\mathsf{func} = (n, f) \in \mathcal{F}$ such that $\mathsf{func}(\mathsf{msg}_0) = \mathsf{func}(\mathsf{msg}_1)$. Let $s = \ell\text{-}\mathsf{r}_s(n), t = \ell\text{-}\mathsf{pk}(n)$.
   (b) The challenger chooses $(\mathsf{mpk}_n, \mathsf{msk}_n) \leftarrow \mathsf{Setup}_{\mathsf{sel}}(1^\lambda, 1^n)$ (if $\mathsf{msk}_n, \mathsf{mpk}_n$ have already been computed before, then it simply reuses those keys).
   (c) It generates the IBE secret keys as $\mathsf{sk}_i \leftarrow \mathsf{KeyGen}_{\mathsf{IBE}}(\mathsf{msk}_{\mathsf{IBE}}, (n, i, \mathsf{mpk}_n[i]))$ and base FE scheme's secret key $\mathsf{sk}_{f,\mathsf{sel}} \leftarrow \mathsf{KeyGen}_{\mathsf{sel}}(\mathsf{msk}_n, f)$.
   (d) The challenger sets $\mathsf{sk}_{\mathsf{func}} = (\mathsf{sk}_{f,\mathsf{sel}}, \mathsf{mpk}_n, \{\mathsf{sk}_i\})$ and sends $\mathsf{sk}_{\mathsf{func}}$ to $\mathcal{A}$.
4. (Guess) Finally, $\mathcal{A}$ sends its guess b' and wins if $b = b'$.

Game 5: In this game, the challenger simulates the garbled circuit when computing the challenge ciphertext.

1. (Setup Phase) The challenger first runs the setup algorithm by choosing $(\mathsf{mpk}_{\mathsf{IBE}}, \mathsf{msk}_{\mathsf{IBE}}) \leftarrow \mathsf{Setup}_{\mathsf{IBE}}(1^\lambda)$. It sends $\mathsf{mpk}_{\mathsf{IBE}}$ to the adversary.
2. (Challenge Phase)
   (a) $\mathcal{A}$ sends two challenge messages $\mathsf{msg}_0 = (n^*, m_0), \mathsf{msg}_1 = (n^*, m_1)$ such that $\epsilon(\mathsf{msg}_0) = \epsilon(\mathsf{msg}_1)$.
   (b) The challenger chooses $(\mathsf{mpk}_{n^*}, \mathsf{msk}_{n^*}) \leftarrow \mathsf{Setup}_{\mathsf{sel}}(1^\lambda, 1^{n^*})$.
   (c) It first chooses $b \leftarrow \{0,1\}$, computes $\widetilde{\mathsf{ct}} \leftarrow \mathsf{Enc}_{\mathsf{sel}}(\mathsf{mpk}_{n^*}, m_b)$.
   (d) It then uses $\widetilde{\mathsf{ct}}$ to simulate the garbled circuit.
       It computes $(\widetilde{C}, \{w_i\}) \leftarrow \mathsf{Sim}(1^\lambda, 1^{t^*}, 1^k, \widetilde{\mathsf{ct}})$, where $t^* = |\ell\text{-}\mathsf{pk}(n^*)|$ and k is the size of the circuit $\mathcal{C}\text{-}\mathsf{Enc}\text{-}\mathsf{pk}_{m,r}^{t^*}$.
   (e) It then encrypts the wire keys at half the positions, and zeroes at the remaining positions.
       For each i, if $\beta = \mathsf{mpk}_{n^*}[i]$, $\mathsf{ct}_{i,\beta} \leftarrow \mathsf{Enc}_{\mathsf{IBE}}(\mathsf{mpk}_{\mathsf{IBE}}, w_i, (n^*, i, \beta))$, else $\mathsf{ct}_{i,\beta} \leftarrow \mathsf{Enc}_{\mathsf{IBE}}(\mathsf{mpk}_{\mathsf{IBE}}, \mathbf{0}, (n^*, i, \beta))$.

(f) The challenger sets $\mathsf{ct} = (C, \{\mathsf{ct}_{i,\beta}\}_{i \leq t^*, \beta \in \{0,1\}})$ and sends ct to $\mathcal{A}$.

3. (Key Query Phase)

(a) $\mathcal{A}$ queries the challenger on polynomially many functions $\mathsf{func} = (n, f) \in \mathcal{F}$ such that $\mathsf{func}(\mathsf{msg}_0) = \mathsf{func}(\mathsf{msg}_1)$. Let $s = \ell\text{-}\mathsf{r}_s(n)$, $t = \ell\text{-}\mathsf{pk}(n)$.

(b) The challenger chooses $(\mathsf{mpk}_n, \mathsf{msk}_n) \leftarrow \mathsf{Setup}_{\mathsf{sel}}(1^\lambda, 1^n)$ (if $\mathsf{msk}_n, \mathsf{mpk}_n$ have already been computed before, then it simply reuses those keys).

(c) It generates the IBE secret keys as $\mathsf{sk}_i \leftarrow \mathsf{KeyGen}_{\mathrm{IBE}}(\mathsf{msk}_{\mathsf{IBE}}, (n, i, \mathsf{mpk}_n[i]))$ and base FE scheme's secret key $\mathsf{sk}_{f,\mathsf{sel}} \leftarrow \mathsf{KeyGen}_{\mathsf{sel}}(\mathsf{msk}_n, f)$.

(d) The challenger sets $\mathsf{sk}_{\mathsf{func}} = (\mathsf{sk}_{f,\mathsf{sel}}, \mathsf{mpk}_n, \{\mathsf{sk}_i\})$ and sends $\mathsf{sk}_{\mathsf{func}}$ to $\mathcal{A}$.

4. (Guess) Finally, $\mathcal{A}$ sends its guess b' and wins if $b = b'$.

Analysis. Let $\mathcal{A}$ be any PPT adversary against our construction in the semi-adaptive security game (Fig. 2) and $\mathsf{Adv}_{\mathcal{A}}^i$ denote the advantage of $\mathcal{A}$ in Game i. We will show that $\mathsf{Adv}_{\mathcal{A}}^i - \mathsf{Adv}_{\mathcal{A}}^{i+1}$ is negligible in λ for all i.

Lemma 4. *Assuming F is a secure pseudorandom function, for any PPT adversary $\mathcal{A}$, $|\mathsf{Adv}_{\mathcal{A}}^1 - \mathsf{Adv}_{\mathcal{A}}^2| \leq negl(\lambda)$.*

Proof. The proof of this lemma follows from a simple reduction to the security of PRF F. Suppose there exists an adversary $\mathcal{A}$ such that $|\mathsf{Adv}_{\mathcal{A}}^1 - \mathsf{Adv}_{\mathcal{A}}^2|$ is non-negligible. We will construct an algorithm $\mathcal{B}$ that uses $\mathcal{A}$ to break the PRF security. The reduction algorithm chooses an IBE master public/secret key pair $(\mathsf{mpk}_{\mathsf{IBE}}, \mathsf{msk}_{\mathsf{IBE}})$ and sends $\mathsf{mpk}_{\mathsf{IBE}}$ to the adversary. Next, it receives challenge messages $\mathsf{msg}_0, \mathsf{msg}_1$ with the restriction that $\epsilon(\mathsf{msg}_0) = \epsilon(\mathsf{msg}_1)$. It computes a challenge ciphertext and sends it to $\mathcal{A}$ (this step is identical in both Game 1 and Game 2). Next, the adversary queries for secret keys. For each queried function f, the reduction algorithm first computes the functionality index n and $s = \ell\text{-}\mathsf{r}_s(n)$. It then queries the PRF challenger for PRF evaluations at inputs (n, i) for $i \leq s$. It receives string r, which it uses as randomness to compute $\mathsf{FE}_{\mathsf{sel}}$ master keys $(\mathsf{mpk}_{\mathsf{sel}}, \mathsf{msk}_{\mathsf{sel}})$. The remaining steps (computing IBE secret keys and $\mathsf{sk}_{f,\mathsf{sel}}$) are identical in both Game 1 and Game 2. It sends $\mathsf{sk}_{\mathsf{func}}$ to $\mathcal{A}$, and $\mathcal{A}$ sends its guess b'. If $b = b'$, $\mathcal{B}$ outputs 1, indicating that the oracle was a pseudorandom function, else it outputs 0, indicating that the oracle was a truly random function. Clearly, if the PRF challenger used a pseudorandom function, then $\mathcal{A}$ participates in Game 1, else it participates in Game 2. This concludes our proof.

Lemma 5. *For any adversary $\mathcal{A}$, $\mathsf{Adv}_{\mathcal{A}}^2 = \mathsf{Adv}_{\mathcal{A}}^3$.*

Proof. The advantage of any adversary $\mathcal{A}$ is identical in Game 2 and Game 3. The only difference between the two games is that the challenger chooses $(\mathsf{mpk}_{n^*}, \mathsf{msk}_{n^*})$ immediately after receiving the challenge messages, instead of waiting for the first key query where the function is in $\mathcal{F}_{n^*}$. This does not affect the adversary's advantage.

Lemma 6. *Assuming IBE is a secure identity based encryption scheme (Definition 3), for any PPT adversary $\mathcal{A}$, $|\mathsf{Adv}_{\mathcal{A}}^3 - \mathsf{Adv}_{\mathcal{A}}^4| \leq negl(\lambda)$.*

Proof. Suppose there exists an adversary $\mathcal{A}$ such that $|\mathsf{Adv}_{\mathcal{A}}^3 - \mathsf{Adv}_{\mathcal{A}}^4|$ is non-negligible. We will construct a reduction algorithm $\mathcal{B}$ that uses $\mathcal{A}$ to break the security of IBE. First, $\mathcal{B}$ receives the IBE public key $\mathsf{mpk}_{\mathsf{IBE}}$, which it forwards to $\mathcal{A}$. The adversary then sends the challenge messages $\mathsf{msg}_0 = (n^*, m_0), \mathsf{msg}_1 = (n^*, m_1)$. Let $t^* = \ell\text{-pk}(n^*)$. $\mathcal{B}$ chooses $(\mathsf{mpk}_{n^*}, \mathsf{msk}_{n^*}) \leftarrow \mathsf{Setup}_{\mathsf{sel}}(1^\lambda, 1^{n^*})$. It then chooses $b \leftarrow \{0,1\}$ and computes garbled circuit C together with wire keys $\{w_{i,\beta}\}$ for message m_b. Next, it sends t^* challenge messages to the IBE challenger. For $i = 0$ to t, let $\beta_i' = 1 - \mathsf{mpk}_{n^*}[i]$. It sends challenge messages $(w_{i,\beta_i'}, \mathbf{0})$ and challenge identity (n^*, i, β_i'), and receives ciphertext $\mathsf{ct}_{i,\beta_i'}$. The reduction algorithm constructs the remaining ciphertexts by itself and sends $(C, \{\mathsf{ct}_{i,\beta}\})$ to $\mathcal{A}$.

Next, $\mathcal{A}$ sends key queries for functions in $\mathcal{F}$. Let $\mathsf{func} = (n, f) \in \mathcal{F}$ be such a function. The reduction algorithm needs to sends IBE secret keys as part of the secret key for func. If $n \neq n^*$, then $\mathcal{B}$ can simply query the IBE challenger for secret keys. If $n = n^*$, then the reduction algorithm needs to query the IBE challenger for keys corresponding to $(n^*, i, \mathsf{mpk}_{n^*}[i])$ only. In particular, the reduction does not need to query IBE keys for the challenge identities. After receiving the IBE secret keys $\{\mathsf{sk}_i\}$, $\mathcal{B}$ computes $\mathsf{sk}_{f,\mathsf{sel}} \leftarrow \mathsf{KeyGen}_{\mathsf{sel}}(\mathsf{msk}_n, f)$ and sends $\mathsf{sk}_{\mathsf{func}} = (\mathsf{sk}_{f,\mathsf{sel}}, \mathsf{mpk}_n, \{\mathsf{sk}_i\})$ to $\mathcal{A}$. Finally, $\mathcal{A}$ sends its guess b', and $\mathcal{B}$ forwards this guess to the IBE challenger.

Lemma 7. *Assuming* GC *is a secure garbling scheme (Definition 1), for any PPT adversary $\mathcal{A}$, $|\mathsf{Adv}_{\mathcal{A}}^4 - \mathsf{Adv}_{\mathcal{A}}^5| \leq \mathit{negl}(\lambda)$.*

The proof of this lemma is identical to the proof of Lemma 2.

Lemma 8. *Assuming* FE$_{\mathsf{sel}}$ *is a selectively secure functional encryption scheme for function space $\{\mathcal{F}_n\}_n$ (Definition 5), for any PPT adversary $\mathcal{A}$, $\mathsf{Adv}_{\mathcal{A}}^5 \leq \mathit{negl}(\lambda)$.*

The proof of this lemma is identical to the proof of Lemma 3.

References

1. Agrawal, S., Freeman, D.M., Vaikuntanathan, V.: Functional encryption for inner product predicates from learning with errors. In: Lee, D.H., Wang, X. (eds.) ASIACRYPT 2011. LNCS, vol. 7073, pp. 21–40. Springer, Heidelberg (2011)
2. Ananth, P., Brakerski, Z., Segev, G., Vaikuntanathan, V.: From selective to adaptive security in functional encryption. In: Gennaro, R., Robshaw, M. (eds.) CRYPTO 2015. LNCS, vol. 9216, pp. 657–677. Springer, Heidelberg (2015)
3. Ananth, P., Jain, A.: Indistinguishability obfuscation from compact functional encryption. In: Gennaro, R., Robshaw, M. (eds.) CRYPTO 2015. LNCS, vol. 9215, pp. 308–326. Springer, Heidelberg (2015)
4. Ananth, P., Jain, A., Sahai, A.: Achieving compactness generically: indistinguishability obfuscation from non-compact functional encryption. IACR Cryptology ePrint Archive (2015)

5. Ananth, P., Sahai, A.: Functional encryption for turing machines. In: Kushilevitz, E., et al. (eds.) TCC 2016-A. LNCS, vol. 9562, pp. 125–153. Springer, Heidelberg (2016). doi:10.1007/978-3-662-49096-9_6

6. Attrapadung, N.: Dual system encryption via doubly selective security: framework, fully secure functional encryption for regular languages, and more. In: Nguyen, P.Q., Oswald, E. (eds.) EUROCRYPT 2014. LNCS, vol. 8441, pp. 557–577. Springer, Heidelberg (2014)

7. Barak, B., Goldreich, O., Impagliazzo, R., Rudich, S., Sahai, A., Vadhan, S.P., Yang, K.: On the (Im)possibility of obfuscating programs. In: Kilian, J. (ed.) CRYPTO 2001. LNCS, vol. 2139, pp. 1–18. Springer, Heidelberg (2001)

8. Bellare, M., Hoang, V.T., Rogaway, P.: Foundations of garbled circuits. In: Proceedings of the 2012 ACM Conference on Computer and Communications Security, CCS 2012, pp. 784–796 (2012)

9. Bitansky, N., Vaikuntanathan, V.: Indistinguishability obfuscation from functional encryption. In: FOCS (2015)

10. Boneh, D., Boyen, X.: Secure identity based encryption without random oracles. In: Franklin, M. (ed.) CRYPTO 2004. LNCS, vol. 3152, pp. 443–459. Springer, Heidelberg (2004)

11. Boneh, D., Boyen, X.: Efficient selective-ID secure identity-based encryption without random oracles. In: Cachin, C., Camenisch, J.L. (eds.) EUROCRYPT 2004. LNCS, vol. 3027, pp. 223–238. Springer, Heidelberg (2004)

12. Boneh, D., Franklin, M.: Identity-based encryption from the Weil pairing. In: Kilian, J. (ed.) CRYPTO 2001. LNCS, vol. 2139, p. 213. Springer, Heidelberg (2001)

13. Boneh, D., Gentry, C., Gorbunov, S., Halevi, S., Nikolaenko, V., Segev, G., Vaikuntanathan, V., Vinayagamurthy, D.: Fully key-homomorphic encryption, arithmetic circuit ABE and compact garbled circuits. In: Nguyen, P.Q., Oswald, E. (eds.) EUROCRYPT 2014. LNCS, vol. 8441, pp. 533–556. Springer, Heidelberg (2014)

14. Boneh, D., Sahai, A., Waters, B.: Functional encryption: definitions and challenges. In: Ishai, Y. (ed.) TCC 2011. LNCS, vol. 6597, pp. 253–273. Springer, Heidelberg (2011)

15. Boyen, X., Waters, B.: Anonymous hierarchical identity-based encryption (without random oracles). In: Dwork, C. (ed.) CRYPTO 2006. LNCS, vol. 4117, pp. 290–307. Springer, Heidelberg (2006)

16. Brakerski, Z., Vaikuntanathan, V.: Circuit-abe from LWE: unbounded attributes and semi-adaptive security. IACR Cryptology ePrint Archive (2016)

17. Canetti, R., Halevi, S., Katz, J.: A forward-secure public-key encryption scheme. In: Biham, E. (ed.) EUROCRYPT 2003. LNCS, vol. 2656, pp. 255–271. Springer, Heidelberg (2003)

18. Chen, J., Wee, H.: Semi-adaptive attribute-based encryption and improved delegation for Boolean formula. In: Abdalla, M., De Prisco, R. (eds.) SCN 2014. LNCS, vol. 8642, pp. 277–297. Springer, Heidelberg (2014)

19. Cocks, C.: An identity based encryption scheme based on quadratic residues. In: Honary, B. (ed.) Cryptography and Coding 2001. LNCS, vol. 2260, pp. 360–363. Springer, Heidelberg (2001)

20. Garg, S., Gentry, C., Halevi, S., Raykova, M., Sahai, A., Waters, B.: Candidate indstinguishability obfuscation and functional encryption for all circuits. In: FOCS (2013)

21. Gentry, C.: Practical identity-based encryption without random oracles. In: Vaudenay, S. (ed.) EUROCRYPT 2006. LNCS, vol. 4004, pp. 445–464. Springer, Heidelberg (2006)

22. Gentry, C., Halevi, S.: Hierarchical identity based encryption with polynomially many levels. In: Reingold, O. (ed.) TCC 2009. LNCS, vol. 5444, pp. 437–456. Springer, Heidelberg (2009)
23. Gorbunov, S., Vaikuntanathan, V., Wee, H.: Functional encryption with bounded collusions via multi-party computation. In: Safavi-Naini, R., Canetti, R. (eds.) CRYPTO 2012. LNCS, vol. 7417, pp. 162–179. Springer, Heidelberg (2012)
24. Gorbunov, S., Vaikuntanathan, V., Wee, H.: Attribute-based encryption for circuits. In: STOC (2013)
25. Goyal, V., Pandey, O., Sahai, A., Waters, B.: Attribute-based encryption for fine-grained access control of encrypted data. In: Proceedings of the 13th ACM Conference on Computer and Communications Security, CCS 2006 (2006)
26. Katz, J., Sahai, A., Waters, B.: Predicate encryption supporting disjunctions, polynomial equations, and inner products. In: Smart, N.P. (ed.) EUROCRYPT 2008. LNCS, vol. 4965, pp. 146–162. Springer, Heidelberg (2008)
27. Lewko, A., Okamoto, T., Sahai, A., Takashima, K., Waters, B.: Fully secure functional encryption: attribute-based encryption and (hierarchical) inner product encryption. In: Gilbert, H. (ed.) EUROCRYPT 2010. LNCS, vol. 6110, pp. 62–91. Springer, Heidelberg (2010)
28. Lewko, A., Waters, B.: New proof methods for attribute-based encryption: achieving full security through selective techniques. In: Safavi-Naini, R., Canetti, R. (eds.) CRYPTO 2012. LNCS, vol. 7417, pp. 180–198. Springer, Heidelberg (2012)
29. Okamoto, T., Takashima, K.: Fully secure functional encryption with general relations from the decisional linear assumption. In: Rabin, T. (ed.) CRYPTO 2010. LNCS, vol. 6223, pp. 191–208. Springer, Heidelberg (2010)
30. O'Neill, A.: Definitional issues in functional encryption. Cryptology ePrint Archive, Report 2010/556 (2010)
31. Regev, O.: On lattices, learning with errors, random linear codes, and cryptography. In: STOC 2005
32. Sahai, A., Waters, B.: Fuzzy identity-based encryption. In: Cramer, R. (ed.) EUROCRYPT 2005. LNCS, vol. 3494, pp. 457–473. Springer, Heidelberg (2005)
33. Sahai, A., Waters, B.: Slides on functional encryption. PowerPoint presentation (2008). http://www.cs.utexas.edu/~bwaters/presentations/files/functional.ppt
34. Shamir, A.: Identity-based cryptosystems and signature schemes. In: Blakely, G.R., Chaum, D. (eds.) CRYPTO 1984. LNCS, vol. 196, pp. 47–53. Springer, Heidelberg (1985)
35. Waters, B.: Efficient identity-based encryption without random oracles. In: Cramer, R. (ed.) EUROCRYPT 2005. LNCS, vol. 3494, pp. 114–127. Springer, Heidelberg (2005)
36. Waters, B.: Dual system encryption: realizing fully secure ibe and hibe under simple assumptions. In: Halevi, S. (ed.) CRYPTO 2009. LNCS, vol. 5677, pp. 619–636. Springer, Heidelberg (2009)
37. Waters, B.: Functional encryption for regular languages. In: Safavi-Naini, R., Canetti, R. (eds.) CRYPTO 2012. LNCS, vol. 7417, pp. 218–235. Springer, Heidelberg (2012)
38. Waters, B.: A punctured programming approach to adaptively secure functional encryption. In: Gennaro, R., Robshaw, M. (eds.) CRYPTO 2015. LNCS, vol. 9216, pp. 678–697. Springer, Heidelberg (2015)
39. Wee, H.: Dual system encryption via predicate encodings. In: Lindell, Y. (ed.) TCC 2014. LNCS, vol. 8349, pp. 616–637. Springer, Heidelberg (2014)
40. Yao, A.: How to generate and exchange secrets. In: FOCS, pp. 162–167 (1986)

Functional Encryption

From Cryptomania to Obfustopia
Through Secret-Key Functional Encryption

Nir Bitansky[1]($\boxtimes$), Ryo Nishimaki[2], Alain Passelègue[3], and Daniel Wichs[4]

[1] MIT, Cambridge, USA
nirbitan@csail.mit.edu
[2] NTT, Secure Platform Laboratories, Tokyo, Japan
nishimaki.ryo@lab.ntt.co.jp
[3] ENS, Paris, France
alain.passelegue@ens.fr
[4] Northeastern University, Boston, USA
wichs@ccs.neu.edu

Abstract. Functional encryption lies at the frontiers of current research in cryptography; some variants have been shown sufficiently powerful to yield indistinguishability obfuscation (IO) while other variants have been constructed from standard assumptions such as LWE. Indeed, most variants have been classified as belonging to either the former or the latter category. However, one mystery that has remained is the case of *secret-key functional encryption* with an unbounded number of keys and ciphertexts. On the one hand, this primitive is not known to imply anything outside of minicrypt, the land of secret-key crypto, but on the other hand, we do no know how to construct it without the heavy hammers in obfustopia.

In this work, we show that (subexponentially secure) secret-key functional encryption is powerful enough to construct indistinguishability obfuscation if we additionally assume the existence of (subexponentially secure) plain public-key encryption. In other words, secret-key functional encryption provides a bridge from cryptomania to obfustopia.

On the technical side, our result relies on two main components. As our first contribution, we show how to use secret key functional encryption to get "exponentially-efficient indistinguishability obfuscation" (XIO), a notion recently introduced by Lin et al. (PKC '16) as a relaxation of IO. Lin et al. show how to use XIO and the LWE assumption to build IO. As our second contribution, we improve on this result by replacing its reliance on the LWE assumption with any plain public-key encryption scheme.

N. Bitansky—Supported by an IBM DARPA grant and an NJIT DARPA grant.
R. Nishimaki—This work was done in part while the author was visiting Northeastern University.
A. Passelègue—This work was done in part while the author was visiting Northeastern University. Supported in part by the *Direction Générale de l'Armement*.
D. Wichs—Supported in part by NSF grants CNS-1347350, CNS-1314722, CNS-1413964.

M. Hirt and A. Smith (Eds.): TCC 2016-B, Part II, LNCS 9986, pp. 391–418, 2016.
DOI: 10.1007/978-3-662-53644-5_15

Lastly, we ask whether secret-key functional encryption can be used to construct public-key encryption itself and therefore take us all the way from minicrypt to obfustopia. A result of Asharov and Segev (FOCS '15) shows that this is not the case under black-box constructions, even for exponentially secure functional encryption. We show, through a non-black box construction, that subexponentially secure-key functional encryption indeed leads to public-key encryption. The resulting public-key encryption scheme, however, is at most quasi-polynomially secure, which is insufficient to take us to obfustopia.

1   Introduction

The concept of *functional encryption* [17,45] extends that of traditional encryption by allowing the distribution of *functional decryption keys* that reveal specified functions of encrypted messages, but nothing beyond. This concept is one of the main frontiers in cryptography today. It offers tremendous flexibility in controlling and computing on encrypted data, is strongly connected to the holy grail of program obfuscation [3,14,44], and for many problems, may give superior solutions to obfuscation-based ones [28,29]. Accordingly, recent years have seen outstanding progress in the study of functional encryption, both in constructing functional encryption schemes and in exploring different notions, their power, and the relationship amongst them (see for instance, [1,2,4,5,7,11,15,16,20,21,24–26,32–36,40,42,47,49] and many more).

One striking question that has yet to be solved is *the gap between public-key and secret-key functional encryption schemes. In particular, does any secret-key scheme imply a public-key one?*

The answer to this question is nuanced and seems to depend on certain features of functional encryption schemes, such as the number of functional decryption keys and number of ciphertexts that can be released. For functional encryption schemes that only allow the release of an *a-priori bounded* number of functional keys (often referred to as *bounded collusion*), we know that the above gap is essentially the same as the gap between plain (rather than functional) secret-key encryption and public-key encryption, and should thus be as hard to bridge. Specifically, in the secret-key setting, such schemes supporting an unbounded number of ciphertexts can be constructed assuming low-depth pseudorandom generators (or just one-way functions in the single-key case) [34,47]. These secret-key constructions are then converted to public-key ones, relying on (plain) public-key encryption (and this is done quite directly by replacing invocations of a secret-key encryption scheme with invocations of a public-key one.) The same state of affairs holds when reversing the roles and considering a bounded number of ciphertexts and an unbounded number of keys [34,47]. In other words, in the terminology of Impagliazzo's complexity worlds [38], if the number of keys or ciphertexts is a-priori bounded, then symmetric-key functional encryption lies in *minicrypt*, the world of one-way functions, and public-key functional encryption lies in *cryptomania*, the world of public-key encryption.

For functional encryption schemes supporting an unbounded (polynomial) number of keys and unbounded number of ciphertexts, which will be the default notion throughout the rest of the paper, the question is far less understood. In the public-key setting, such functional encryption schemes with subexponential security are known to imply indistinguishability obfuscation [3,4,14]. In contrast, Bitansky and Vaikuntanathan [14] show that their construction of indistinguishability obfuscation using functional encryption may be insecure when instantiated with a secret-key functional encryption scheme. In fact, secret-key functional encryption schemes (even exponentially secure ones) are not known to imply any cryptographic primitive beyond those that follow from one-way functions. As far as we know the two notions of functional encryption may correspond to opposite extremes of the complexity spectrum: on one side, public-key schemes correspond to *obfustopia*, the world where indistinguishability obfuscation exists, and on the other side secret-key schemes may lie in minicrypt where there is even no (plain) public-key encryption.

One piece of evidence that may support such a view of the world is given by Asharov and Segev [6] who show that there do not exist *fully black-box* constructions of *plain* public-key encryption from secret-key functional encryption, even if the latter is exponentially secure. Still, while we may hope that such secret-key schemes could be constructed from significantly weaker assumptions than needed for public-key schemes, so far no such construction has been exhibited — all known constructions live in obfustopia.

1.1 Our Contributions

In this work, we shed new light on the question of secret-key vs public-key functional encryption (in the multi-key, multi-ciphertext setting). Our main result bridges the two notions based on (plain) public-key encryption.

Theorem 1 (Informal). *Assuming secret-key functional encryption and plain public-key encryption that are both subexponentially secure, there exists indistinguishability obfuscation, and in particular, also public-key functional encryption.*

In the terminology of Impagliazzo's complexity worlds: *secret-key functional encryption would turn cryptomania, the land of public-key encryption, into obfustopia.* This puts in new perspective the question of constructing such secret-key schemes from standard assumptions — any such construction would lead to indistinguishability obfuscation from standard assumptions.

The above result still does not settle the question of whether secret-key functional encryption on its own implies (plain) public-key encryption. Here we show that assuming subexponentially-secure secret-key functional encryption and (almost) exponentially-secure one-way functions, there exists (polynomially-secure) public-key encryption.

Theorem 2 (Informal). *Assuming subexponentially-secure secret-key functional encryption and $2^{n/\log\log n}$-secure one-way functions, there exists (polynomially-secure) public-key encryption.*

The resulting public-key encryption is not strong enough to take us to obfustopia. Concretely, the constructed scheme is not subexponentially secure as required by our first theorem — it can be *quasi-polynomially* broken. Nevertheless, the result does show that the black-box barrier shown by Asharov and Segev [6], which applies even if the underlying secret-key functional encryption scheme and one-way functions are *exponentially secure*, can be circumvented. Indeed, our construction uses the functional encryption scheme in a non-black-box way (see further details in the technical overview section below).

1.2   A Technical Overview

We now provide an overview of the main steps and ideas leading to our results.

Key Observation: From SKFE to (Strong) Exponentially-Efficient IO. Our first observation is that secret-key functional encryption (or SKFE in short) implies a weak form of indistinguishability obfuscators termed by Lin, Pass, Seth, and Telang [43] exponentially-efficient indistinguishability obfuscation (XIO). Like IO, this notion preserves the functionality of obfuscated circuits and guarantees that obfuscations of circuits of the same size and functionality are indistinguishable. However, in terms of efficiency the XIO notion only requires that an obfuscation $\widetilde{C}$ of a circuit $C : \{0,1\}^n \to \{0,1\}^m$ is just mildly smaller than its truth table, namely $|\widetilde{C}| \le 2^{\gamma n} \cdot \mathrm{poly}(|C|)$, for some compression factor $\gamma < 1$, and a fixed polynomial poly, rather than the usual requirement that *the time to obfuscate*, and in particular the size of $\widetilde{C}$, are polynomial in $|C|$. We show that SKFE implies a slightly stronger notion than XIO where the time to obfuscate C is bounded by $2^{\gamma n} \cdot \mathrm{poly}(|C|)$. We call this notion strong exponentially-efficient indistinguishability obfuscation (SXIO). (We note that, for either XIO or SXIO, we shall typically be interested in circuits over some polynomial size domain, which could be much larger than the circuit itself, e.g., $\{0,1\}^n$ where $n = 100 \log |C|$.)

Proposition 1 (Informal).

1. *For any constant $\gamma < 1$, there exists a transformation from SKFE to SXIO with compression factor γ.*
2. *For some subconstant $\gamma = o(1)$, there exists a transformation from subexponentially-secure SKFE to polynomially-secure SXIO with compression factor γ.*

We add more technical details regarding the proof of the above SXIO proposition later on. Both of our theorems stated above rely on the constructed SXIO as a main tool. We next explain, still at a high-level, how the first theorem is obtained. We then dive into further technical details about the proof of this theorem as well as the proof of the second theorem.

From SXIO to IO Through Public-key Encryption. Subexponentially-secure SXIO (or even XIO) schemes with a constant compression factor (as in Proposition 1) are already shown to be quite strong in [43] — assuming subexponential hardness of Learning with Errors (LWE) [46], they imply IO.

Corollary 1. (of Proposition 1 and [43]). *Assuming SKFE and LWE, both subexponentially secure, there exists IO.*

We go beyond the above corollary, showing that LWE can be replaced with a generic assumption — the existence of (plain) public-key encryption schemes. The transformation of [43] from LWE and XIO to IO, essentially relies on LWE to obtain a specific type of public-key functional encryption (PKFE) with certain succinctness properties. We show how to construct such PKFE from public-key encryption and SXIO. More details follow.

Concretely, the notion considered is of PKFE schemes that support a *single decryption key*. Furthermore, the time complexity of encryption is bounded by roughly $s^\beta \cdot d^{O(1)}$, where s and d are the size and depth of the circuit computing the function, and $\beta < 1$ is some compression factor. We call such schemes *weakly succinct PKFE schemes*. A weakly succinct PKFE for *boolean* functions (i.e., functions with a single output bit) is constructed by Goldwasser et al. [33] from (subexponentially-hard) LWE; in fact, the Goldwasser et al. construction has no dependence at all on the circuit size s (namely, $\beta = 0$).

Lin et al. [43] then show a transformation, relying on XIO, that extends the class of functions also to functions with a long output, rather than just boolean ones. (Their transformation is stated for the case that $\beta = 0$ assuming any constant XIO compression factor $\gamma < 1$, but can be extended to also work for any sufficiently small constant compression factor β for the PKFE.) Such weakly-succinct PKFE schemes can then be plugged in to the transformations of [3,14,44] to obtain full-fledged IO.[1]

We follow a similar blueprint. We first construct weakly-succinct PKFE for functions with a single output bit based on SXIO and PKE, rather than LWE (much of the technical effort in this work lies in this construction). We then bootstrap the construction to deal with multibit functions using (a slightly augmented version of) the transformation from [43].

Proposition 2 (Informal). *For any $\beta = \Omega(1)$, assuming PKE and SXIO with a small enough constant compression factor γ, there exists a single-key weakly-succinct PKFE scheme with compression factor β (for functions with long output).*

1.3 A Closer Look into the Techniques

We now provide further details regarding the proofs of the above Propositions 1 and 2 as well as the proof of Theorem 2.

SKFE to SXIO: The Basic Idea. To convey the basic idea behind the transformation, we first describe a construction of SXIO with compression

---

[1] The above is a slightly oversimplified account of [43]. They also rely on LWE to deduce the existence of puncturable PRFs in $\mathbf{NC}^1$ and show their transformation starting from weakly-succinct PKFE for functions in $\mathbf{NC}^1$. We avoid the reliance on puncturable PRFs in $\mathbf{NC}^1$ by constructing weakly-succinct PKFE for functions with no depth restriction, at the expense of allowing the complexity of encryption to scale polynomially in the depth. This is still sufficient for [14, Sect. 3.2].

$\gamma = 1/2$. We then explain how to extend it to obtain the more general form of Proposition 1.

Recall that in an SKFE scheme, first a master secret key MSK is generated, and can then be used to:

- encrypt (any number of) plaintext messages,
- derive (any number of) functional keys.

The constructed obfuscator sxi$\mathcal{O}$ is given a circuit C defined on domain $\{0,1\}^n$, where we shall assume for simplicity that the input length is even (this is not essential), and works as follows:

- For every $x \in \{0,1\}^{n/2}$, computes a ciphertext CT_x encrypting the circuit $C_x(\cdot)$ that given input $y \in \{0,1\}^{n/2}$, returns $C(x,y)$.
- For every $y \in \{0,1\}^{n/2}$, derives a functional decryption key SK_y for the function $U_y(\cdot)$ that given as input a circuit D of size at most $\max_x |C_x|$, returns $D(y)$.
- Outputs $\widetilde{C} = \left(\{\mathsf{CT}_x\}_{x \in \{0,1\}^{n/2}}, \{\mathsf{SK}_y\}_{y \in \{0,1\}^{n/2}} \right)$ as the obfuscation.

To evaluate $\widetilde{C}$ on input $(x,y) \in \{0,1\}^n$, simply decrypt

$$\mathsf{Dec}(\mathsf{SK}_y, \mathsf{CT}_x) = U_y(C_x) = C_x(y) = C(x,y).$$

Indeed, the required compression factor $\gamma = 1/2$ is achieved. Generating each ciphertext is proportional to the size of the message $|C_x| = \tilde{O}(|C|)$ and some fixed polynomial in the security parameter λ. Similarly the time to generate each functional key is proportional to the size of the circuit $|U_y| = \tilde{O}(|C|)$ and some fixed polynomial in the security parameter λ. Thus overall, the time to generate $\widetilde{C}$ is bounded by $2^{n/2} \cdot \mathrm{poly}(|C|, \lambda)$.

The indistinguishability guarantee follows easily from that of the underlying SKFE. Indeed, SKFE guarantees that for any two sequences $\boldsymbol{m} = \{m_i\}$ and $\boldsymbol{m'} = \{m_i'\}$ of messages to be encrypted and any sequence of functions $\{f_i\}$ for which keys are derived, encryptions of the $\boldsymbol{m}$ are indistinguishable from encryptions of the $\boldsymbol{m'}$, provided that the messages are not "separated by the functions", i.e. $f_j(m_i) = f_j(m_i')$ for every (i,j). In particular, any two circuits C and C' that have equal size and functionality will correspond to such two sequences of messages $\{C_x\}_{x \in \{0,1\}^{n/2}}$ and $\{C_x'\}_{x \in \{0,1\}^{n/2}}$, whereas $\{U_y\}_{y \in \{0,1\}^n}$ are indeed functions such that $U_y(C_x) = C(x,y) = C'(x,y) = U_y(C_x')$ for all (x,y). (The above argument works even given a very weak selective security definition where all messages and functions are chosen by the attacker ahead of time.)

As said, the above transformation achieves compression factor $\gamma = 1/2$. While such compression is sufficient for example to obtain IO based on LWE, it will not suffice for our two Theorems 1 and 2 (for the first we will need γ to be a smaller constant, and for the second we will need it to even be slightly subconstant). To prove Proposition 1 in its more general form, we rely on a result by Brakerski,

Komargodski, and Segev [20] that shows how to convert any SKFE into a t-input SKFE. A t-input scheme allows to encrypt a tuple of messages $(m_1, \ldots, m_t)$ each independently, and derive keys for t-input functions $f(m_1, \ldots, m_t)$. In their transformation, starting from a multi-key SKFE results in a multi-key t-input SKFE.

The general transformation then follows naturally. Rather than arranging the input space in a 2-dimensional cube $\{0,1\}^{n/2} \times \{0,1\}^{n/2}$ as we did before with a 1-input scheme, given a t-input scheme we can arrange it in a $(t+1)$-dimensional cube $\{0,1\}^{n/(t+1)} \times \cdots \times \{0,1\}^{n/(t+1)}$, and we will accordingly get compression $\gamma = 1/(t+1)$. The only caveat is that the BKS transformation incurs a security loss and blowup in the size of the scheme that can grow doubly exponentially in t. As long as t is constant the security loss and blowup are fixed polynomials. The transformation can also be invoked for slightly super-constant t (double logarithmic) assuming subexponential security of the underlying 1-input SKFE (giving rise to the second part of Proposition 1).

We remark that previously Goldwasser et al. [32] showed that t-input SKFE for polynomial t directly gives full-fledged IO. We demonstrate that even when t is small (even constant), t-input SKFE implies a meaningful obfuscation notion such as SXIO.

From SXIO and PKE to Weakly Succinct PKFE: Main Ideas. We now describe the main ideas behind our construction of a single-key weakly succinct PKFE. We shall focus on the main step of obtaining such a scheme for functions with a single output bit.[2]

Our starting point is the single-key PKFE scheme of Sahai and Seyalioglu [47] based on Yao's garbled circuit method [50]. Their scheme basically works as follows (we assume basic familiarity with the garbled circuit method):

- The master public key MPK consists of L pairs of public keys $\{\mathsf{PK}_i^0, \mathsf{PK}_i^1\}_{i \in L}$ for a (plain) public-key encryption scheme.
- A functional decryption key SK_f for a function (circuit) f of size L consists of the secret decryption keys $\{\mathsf{SK}_i^{f_i}\}_{i \in L}$ corresponding to the above public keys, according to the bits of f's description.
- To encrypt a message m, the encryptor generates a garbled circuit $\widehat{U}_m$ for the universal circuit U_m that given f, returns $f(m)$. It then encrypts the corresponding input labels $\{k_i^0, k_i^1\}_{i \in L}$ under the corresponding public keys.
- The decryptor in possession of SK_f can then decrypt to obtain the labels $\{k_i^{f_i}\}_{i \in L}$ and decode the garbled circuit to obtain $U_m(f) = f(m)$.

---

[2] Extending this to functions with multibit output is then done, based on SXIO, using a transformation of [43]. Concretely, given an m-bit output function $f(x)$ we consider a new single bit function $g_f(x, i)$ that returns the ith bit of $f(x)$. The function key is then derived for the boolean function g_f. The new encryption algorithm, for message x, produces an SXIO obfuscation of a circuit that given $i \in [m]$ uses the old encryption scheme to encrypt (m, i), deriving randomness using a puncturable PRF. The security of the construction is proven as in [43] based on a probabilistic IO argument [22]. (Mild) efficiency of the encryption then follows from the mild efficiency of the SXIO and PKFE with related (constant) compression factors.

Selective security of this scheme (where the function f and all messages are chosen ahead of time) follows from the semantic security of PKE and the garbled circuit guarantee which says that $\widehat{U}_m, \{k_i^{f_i}\}_{i \in L}$ can be simulated from $f(m)$.

The scheme is indeed *not* succinct in any way. The complexity of encryption and even the size of the ciphertext grows with the complexity of f. Nevertheless, it does seem that the encryption process has a much more succinct representation. In particular, computing a garbled circuit is a *decomposable* process — each garbled gate in $\widehat{U}_m$ depends on a single gate in the original circuit U_m and a small amount of randomness (for computing the labels corresponding to its wires). Furthermore, the universal circuit U_m itself is also decomposable — there exists a small (say, poly($|m|$, $\log L$)-sized) circuit that given i can output the i-th gate in U_m along with its neighbours. The derivation of randomness itself can also be made decomposable using a pseudorandom function. All in all, there exists a small (poly($|m|$, $\log L$, λ)-size, for security parameter λ), *decomposition circuit* $U_{m,K}^{\mathsf{de}}$ associated with a key $K \in \{0,1\}^\lambda$ for a pseudorandom function that can produce the ith garbled gate/input-label given input i.

Yet, the second part of the encryption process, where the input labels $\{k_i^0, k_i^1\}_{i \in L}$ are encrypted under the corresponding public keys $\{\mathsf{PK}_i^0, \mathsf{PK}_i^1\}_{i \in L}$, may not be decomposable at all. Indeed, in general, it is not clear how to even compress the representation of these $2L$ public-keys. In this high-level exposition, let us make the simplifying assumption that we have at our disposal a *succinct identity-based-encryption (IBE) scheme*. Such a scheme has a single public-key PK that allows to encrypt a message to an identity $\mathsf{id} \in \mathcal{ID}$ taken from an identity space $\mathcal{ID}$. Those in possession of a corresponding secret key $\mathsf{SK}_{\mathsf{id}}$ can decrypt and others learn nothing. Succinctness means that the complexity of encryption may only grow mildly in the size of the identity space. Concretely, by a factor of $|\mathcal{ID}|^\gamma$ for some small constant $\gamma < 1$. In the body, we show that such a scheme can be constructed from (plain) public-key encryption and SXIO (the construction relies on standard "puncturing techniques" and is pretty natural).

Equipped with such an IBE scheme, we can now augment the Sahai-Seyalioglu scheme to make sure that the entire encryption procedure is decomposable. Concretely, we will consider the identity space $\mathcal{ID} = [L] \times \{0,1\}$, augment the public key to only include the IBE's public key PK, and provide the decryptor with the identity keys $\{\mathsf{SK}_{(i,f_i)}\}_{i \in L}$. Encrypting the input labels $\{k_i^0, k_i^1\}_{i \in L}$ will now be done by simply encrypting to the corresponding identities $\{(i,0),(i,1)\}_{i \in L}$. This part of the encryption can now also be described by a small (say $L^\gamma \cdot$ poly(λ, $\log L$)-size) decomposition circuit $E_{K,K',\mathsf{PK}}^{\mathsf{de}}$ that has the PRF key K to derive input labels, the IBE public key PK, and another PRF key K' to derive randomness for encryption. Given an identity (i,b), it generates the corresponding encrypted input label.

At this point, a natural direction is to have the encryptor send a *compressed* version of the Sahai-Seyalioglu encryption, by first using SXIO to shield the two decomposition circuits $E_{K,K',\mathsf{PK}}^{\mathsf{de}}, U_{m,K}^{\mathsf{de}}$ and then sending the two obfuscations. Indeed, decryption can be done just as before by first reconstructing the expanded garbled circuit and input labels and then proceeding as before. Also, in terms of encryption complexity, provided that the IBE compression factor γ is a small enough constant, the entire encryption time will scale only sublinearly in the function's size $|f| = L$ (i.e., with L^β for some constant $\beta < 1$).

The only question is of course security. It is not too hard to see that if the decomposition circuits $E^{\mathsf{de}}_{K,K',\mathsf{PK}}, U^{\mathsf{de}}_{m,K}$ are given as black-boxes then security is guaranteed just as before. The challenge is to prove security relying only on the indistinguishability guarantee of SXIO. A somewhat similar challenge is encountered in the work of Bitansky et al. [12] when constructing *succinct randomized encodings*. In their setting, they obfuscate (using standard IO rather than SXIO) a decomposition circuit $C^{\mathsf{de}}_{x,K}$ (analogous to our $U^{\mathsf{de}}_{m,K}$) that computes the garbled gates of some succinctly represented long computation.

As already demonstrated in [12], proving the security of such a construction is rather delicate. As in the standard setting of garbled circuits, the goal is to gradually transition through a sequence of hybrids, from a real garbled circuit (that depends on the actual computation) to a simulated garbled circuit that depends just on the result of the computation. However, unlike the standard setting, here each of these hybrids should be generated by a *hybrid obfuscated decomposition circuit* and the attacker should not be able to tell them apart. As it turns out, "common IO gymnastics" are insufficient here, and we need to rely on the specific hybrid strategy used to transition between the different garbling modes is the proof of security for standard garbled circuits. One feature of the hybrid strategy which is dominant in this context is the amount of information that hybrid decomposition circuits need to maintain about the actual computation. Indeed, as the amount of this information grows so will the size of these decomposition circuits as will the size of the decomposition circuits in the actual construction (that will have to be equally padded to preserve indistinguishability).

Bitansky et al. show a hybrid strategy where the amount of information scales with the *space* of the computation (or *circuit width*). Whereas in their context this is meaningful (as the aim is to save comparing to the *time* of the computation), in our context this is clearly insufficient. Indeed, in our case the space of the computation given by the universal circuit U_m and the function f can be as large as f's description. Instead, we invoke a different hybrid strategy by Hemenway et al. [37] that scales only with the *circuit depth*. Indeed, this is the cause for the polynomial dependence on depth in our single-key PKFE construction. Below, we further elaborate on the Hemenway et al. hybrid strategy and how it is imported into our setting.

Decomposable Garbling and Pebbling. The work of Hemenway et al. [37] provided a useful abstraction for proving the security of Yao's garbled circuits via a sequence of hybrid games. The goal is to transition from a "real" garbled circuit, where each garbled gate is in "RealGate" mode consisting of four ciphertexts encrypting the two labels k^0_c, k^1_c of the output wire c under the labels of the input wires, to a "simulated" garbled circuit where each garbled gate is in SimGate mode consisting of four ciphertexts that all encrypt the same dummy label k^0_c. As an intermediate step, we can also create a garbled gate in CompDepSimGate mode consisting of four ciphertexts encrypting the same label

$k_c^{v(c)}$ where $v(c)$ is the value going over wire c during the computation $C(x)$ and therefore depends on the actual computation.

The transition from a real garbled circuit to a simulated garbled circuit proceeds via a sequence of hybrids where in each subsequent hybrid we can change one gate at a time from RealGate to CompDepSimGate (and vice versa) if all of its predecessors are in CompDepSimGate mode or it is an input gate, or change a gate from CompDepSimGate mode to SimGate mode (and vice versa) if all of its successors are in CompDepSimGate or SimGate modes. The goal of Hemenway et al. was to give a strategy using the least number of gates in CompDepSimGate mode as possible.[3] They abstracted this problem as a pebbling game and show that for circuits of depth d there exists a sequence of $2^{O(d)}$ hybrids with at most $O(d)$ gates in CompDepSimGate mode in any single hybrid.

In our case, we can give a decomposable circuit for each such hybrid game consisting of gates in RealGate, SimGate, CompDepSimGate modes. In particular, the decomposable circuit takes as input a gate index and outputs the garbled gate in the correct mode. We only need to remember which gate is in which mode, and for all gates in CompDepSimGate mode we need to remember the bit $v(c)$ going over the wire c during the computation $C(x)$. It turns out that the configuration of which mode each gate is in can be represented succinctly, and therefore the number of bits we need to remember is roughly proportional to the number of gates in CompDepSimGate mode in any given hybrid. Therefore, for circuits of depth d, the decomposable circuit is of size $O(d)$ and the number of hybrid steps is $2^{O(d)}$.

To ensure that the obfuscations of decomposable circuits corresponding to neighboring hybrids are indistinguishable we also need to rely on standard puncturing techniques. In particular, the gates are garbled using a punctured PRF and we show that in any transition between neighboring hybrids we can even give the adversary the PRF key punctured only on the surrounding of the gate whose mode is changed.

From SKFE to PKE: The Basic Idea. We end our technical exposition by explaining the basic idea behind the construction of public-key encryption (PKE) from SKFE. The construction is rather natural. Using subexponentially-secure SKFE and the second part of Proposition 1, we can obtain a poly(λ)-secure SXIO with a subconstant compression factor $\gamma = o(1)$; concretely, it can be for example $O(1/\log\log\lambda)$. We can now think about this obfuscator as a plain (efficient) indistinguishability obfuscator for circuits with input length at most $\log\lambda \cdot \log\log\lambda$.

Then, we take a construction of public-key encryption from IO and one-way functions where the input-size of obfuscated circuits can be scaled down at the expense of strengthening the one-way functions. For instance, following the basic *witness encryption paradigm* in [27], the public key can be a pseudorandom string $\mathsf{PK} = \mathsf{PRG}(s)$ for a $2^{n/\log\log n}$-secure length-doubling pseudorandom generator

---

[3] Their aim was proving adaptive security, which is completely orthogonal to our aim. However, for entirely different reasons, the above goal is useful in both their work and ours.

with seed length $n = \log \lambda \cdot \log \log \lambda$. Here the obfuscator is only invoked for a circuit with inputs in $\{0,1\}^n$. An encryption of m is simply an obfuscation of a circuit that has PK hardwired, and releases m only given a seed s such that PK $=$ PRG(s). Security follows essentially as in [27]. Note that in this construction, we cannot expect more than 2^n security, which is quasi-polynomial in the security parameter λ.

How Does the Construction Circumvent the Asharov-Segev Barrier?
As noted earlier, Asharov and Segev [6] show that even exponentially secure SKFE cannot lead to public-key encryption through a fully black-box construction (see their paper for details about the exact model). The reason that our construction does not fall under their criteria lies in the transformation from SKFE to SXIO with subconstant compression, and concretely in the Brakerski-Komargodski-Segev [20] transformation from SKFE to t-input SKFE that makes non-black-box use in the algorithms of the underlying SKFE scheme.

Organization. In Sect. 2, we provide preliminaries and basic definitions used throughout the paper. In Sect. 3, we introduce the definition of SXIO and present our construction based on SKFE schemes. In Sect. 4, we introduce a notion of decomposable garbling. In Sect. 5, we present our construction of IO from PKE and SXIO. In Sect. 6, we present a polynomially-secure PKE scheme from SKFE schemes.

2 Preliminaries

2.1 Standard Computational Concepts

We rely on the standard notions of Turing machines and Boolean circuits.

- We say that a (uniform) Turing machine is PPT if it is probabilistic and runs in polynomial time.
- A polynomial-size (or just polysize) circuit family $\mathcal{C}$ is a sequence of circuits $\mathcal{C} = \{C_\lambda\}_{\lambda \in \mathbb{N}}$, such that each circuit C_λ is of polynomial size $\lambda^{O(1)}$ and has $\lambda^{O(1)}$ input and output bits.
- We follow the standard habit of modeling any efficient adversary strategy as a family of polynomial-size circuits. For an adversary $\mathcal{A}$ corresponding to a family of polysize circuits $\{\mathcal{A}_\lambda\}_{\lambda \in \mathbb{N}}$, we often omit the subscript λ, when it is clear from the context.
- We say that a function $f : \mathbb{N} \to \mathbb{R}$ is negligible if for all constants $c > 0$, there exists $N \in \mathbb{N}$ such that for all $n > N$, $f(n) < n^{-c}$.
- If $\mathcal{X}^{(b)} = \{X_\lambda^{(b)}\}_{\lambda \in \mathbb{N}}$ for $b \in \{0,1\}$ are two ensembles of random variables indexed by $\lambda \in \mathbb{N}$, we say that $\mathcal{X}^{(0)}$ and $\mathcal{X}^{(1)}$ are computationally indistinguishable if for all polysize distinguishers $\mathcal{D}$, there exists a negligible function ν such that for all λ, $|\Pr[\mathcal{D}(X_\lambda^{(0)}) = 1] - \Pr[\mathcal{D}(X_\lambda^{(1)}) = 1]| \le \nu(\lambda)$.

2.2   Functional Encryption

Definition 1. (Multi-input secret-key functional encryption). *Let* $t(\lambda)$ *be a function,* $\overline{\mathcal{M}} = \{\overline{\mathcal{M}}_\lambda = \mathcal{M}_\lambda^{(1)} \times \cdots \times \mathcal{M}_\lambda^{(t(\lambda))}\}_{\lambda \in \mathbb{N}}$ *be a product message domain,* $\mathcal{Y} = \{\mathcal{Y}_\lambda\}_{\lambda \in \mathbb{N}}$ *a range, and* $\mathcal{F} = \{\mathcal{F}_\lambda\}_{\lambda \in \mathbb{N}}$ *a class of t-input functions* $f : \overline{\mathcal{M}}_\lambda \to \mathcal{Y}_\lambda$. *A t-input secret-key functional encryption (t-SKFE) scheme for* $\mathcal{M}, \mathcal{Y}, \mathcal{F}$ *is a tuple of algorithms* $\mathsf{SKFE}_t = (\mathsf{Setup}, \mathsf{KeyGen}, \mathsf{Enc}, \mathsf{Dec})$ *where:*

- $\mathsf{Setup}(1^\lambda)$ *takes as input the security parameter and outputs a master secret key* MPK.
- $\mathsf{KeyGen}(\mathsf{MSK}, f)$ *takes as input the master secret* MPK *and a function* $f \in \mathcal{F}$. *It outputs a secret key* SK_f *for* f.
- $\mathsf{Enc}(\mathsf{MSK}, m, i)$ *takes as input the master secret key* MPK, *a message* $m \in \mathcal{M}_\lambda^{(i)}$, *and an index* $i \in [t(\lambda)]$, *and outputs a ciphertext* CT_i.
- $\mathsf{Dec}(\mathsf{SK}_f, \mathsf{CT}_1, \dots, \mathsf{CT}_t)$ *takes as input the secret key* SK_f *for a function* $f \in \mathcal{F}$ *and ciphertexts* $\mathsf{CT}_1, \dots, \mathsf{CT}_t$, *and outputs some* $y \in \mathcal{Y}$, *or* $\perp$.

Correctness: For all tuples $\boldsymbol{m} = (m_1, \dots, m_t) \in \overline{\mathcal{M}}_\lambda$ *and any function* $f \in \mathcal{F}_\lambda$, *we have that*

$$\Pr\left[\mathsf{Dec}(\mathsf{SK}_f, \mathsf{CT}_1, \dots, \mathsf{CT}_t) = f(\boldsymbol{m}) : \begin{array}{l} \mathsf{MSK} \leftarrow \mathsf{Setup}(1^\lambda), \\ \mathsf{SK}_f \leftarrow \mathsf{KeyGen}(\mathsf{MSK}, f), \\ \forall i \ \mathsf{CT}_i \leftarrow \mathsf{Enc}(\mathsf{MSK}, m, i) \end{array}\right] = 1$$

Definition 2. (Selectively-secure multi-key t-SKFE). *We say that a tuple of algorithms* $\mathsf{SKFE}_t = (\mathsf{Setup}, \mathsf{KeyGen}, \mathsf{Enc}, \mathsf{Dec})$ *is a selectively-secure t-input secret-key functional encryption scheme for* $\overline{\mathcal{M}}, \mathcal{Y}, \mathcal{F}$, *if it satisfies the following requirement, formalized by the experiment* $\mathsf{Expt}_\mathcal{A}^{\mathsf{SKFE}_t}(1^\lambda, b)$ *between an adversary* $\mathcal{A}$ *and a challenger:*

1. *The adversary submits challenge message tuples* $\left\{(m_{i,1}^0, m_{i,1}^1, i)\right\}_{i \in [t]}, \dots,$ $\left\{(m_{i,q}^0, m_{i,q}^1, i)\right\}_{i \in [t]}$ *for all* $i \in [t]$ *to the challenger where* q *is an arbitrary polynomial in* λ.
2. *The challenger runs* $\mathsf{MSK} \leftarrow \mathsf{Setup}(1^\lambda)$
3. *The challenger generates ciphertexts* $\mathsf{CT}_{i,j} \leftarrow \mathsf{Enc}(\mathsf{MSK}, m_{i,j}^b, i)$ *for all* $i \in [t]$ *and* $j \in [q]$, *and gives* $\{\mathsf{CT}_{i,j}\}_{i \in [t], j \in [q]}$ *to* $\mathcal{A}$.
4. $\mathcal{A}$ *is allowed to make* q *function queries, where it sends a function* $f_j \in \mathcal{F}$ *to the challenger for* $j \in [q]$ *and* q *is an arbitrary polynomial in* λ. *The challenger responds with* $\mathsf{SK}_{f_j} \leftarrow \mathsf{KeyGen}(\mathsf{MSK}, f_j)$.
5. $\mathcal{A}$ *outputs a guess* b' *for* b.
6. *The output of the experiment is* b' *if the adversary's queries are valid:*

$$f_j(m_{1,j_1}^0, \dots, m_{t,j_t}^0) = f_j(m_{1,j_1}^1, \dots, m_{t,j_t}^1) \text{ for all } j_1, \dots, j_t, j \in [q].$$

Otherwise, the output of the experiment is set to be $\perp$.

We say that the functional encryption scheme is selectively-secure if, for all polysize adversaries $\mathcal{A}$, there exists a negligible function $\mu(\lambda)$, such that

$$\mathsf{Adv}_{\mathcal{A}}^{\mathsf{SKFE}_t} = \left| \Pr\left[\mathsf{Expt}_{\mathcal{A}}^{\mathsf{SKFE}_t}(1^\lambda, 0) = 1 \right] - \Pr\left[\mathsf{Expt}_{\mathcal{A}}^{\mathsf{SKFE}_t}(1^\lambda, 1) = 1 \right] \right| \leq \mu(\lambda).$$

We further say that SKFE_t is δ-selectively-secure, for some concrete negligible function $\delta(\cdot)$, if the above indistinguishability gap $\mu(\lambda)$ is smaller than $\delta(\lambda)^{\Omega(1)}$.

We recall the following theorem by Brakerski, Komargodski, and Segev, which states that one can build selectively-secure t-SKFE from any selectively-secure 1-SKFE. The transformation induces a significant blowup and security loss in the number of inputs t. This loss is polynomial as long as t is constant, but in general grows doubly-exponentially in t.

Theorem 3. [20]

1. *For $t = O(1)$, if there exists δ-selectively-secure single-input SKFE for P/poly, then there exists δ-selectively-secure t-input SKFE for P/poly.*
2. *There exists a constant $\varepsilon < 1$, such that for $t(\lambda) = \varepsilon \cdot \log\log(\lambda)$, $\tilde{\lambda} = 2^{(\log\lambda)^\varepsilon}$, $\delta(\tilde{\lambda}) = 2^{-\tilde{\lambda}^\varepsilon}$, if there exists δ-selectively-secure single-input SKFE for P/poly, then there exists polynomially-secure selectively-secure t-input SKFE for functions of size at most $2^{O((\log\lambda)^\varepsilon)}$. (Here $\tilde{\lambda}$ is the single-input SKFE security parameter and λ is the t-input SKFE security parameter.)*

Remark 1. (Dependence on circuit size in [20]). The [20] transformation incurs a $(s \cdot \tilde{\lambda})^{2^{O(t)}}$ blowup in parameters, where s is the size of maximal circuit size of supported functions, and $\tilde{\lambda}$ is the security parameter used in the underlying single-input SKFE. In the main setting of parameters considered there, $t = O(1)$, the security parameter λ of the t-SKFE scheme can be identified with $\tilde{\lambda}$ and s can be any polynomial in this security parameter. (Accordingly, the dependence on s is implicit there, and the blowup they address is $\lambda^{2^{O(t)}}$.)

For the second part of the theorem, to avoid superpolynomial blowup in λ, the security parameter $\tilde{\lambda}$ for the underlying SKFE and the maximal circuit size s should be set to $2^{O((\log\lambda)^\varepsilon)}$.

Definition 3. (Public-key functional encryption). *Let $\mathcal{M} = \{\mathcal{M}_\lambda\}_{\lambda \in \mathbb{N}}$ be a message domain, $\mathcal{Y} = \{\mathcal{Y}_\lambda\}_{\lambda \in \mathbb{N}}$ a range, and $\mathcal{F} = \{\mathcal{F}_\lambda\}_{\lambda \in \mathbb{N}}$ a class of functions $f : \mathcal{M} \to \mathcal{Y}$. A public-key functional encryption (PKFE) scheme for $\mathcal{M}, \mathcal{Y}, \mathcal{F}$ is a tuple of algorithms $\mathsf{PKFE} = (\mathsf{Setup}, \mathsf{KeyGen}, \mathsf{Enc}, \mathsf{Dec})$ where:*

- $\mathsf{Setup}(1^\lambda)$ *takes as input the security parameter and outputs a master secret key MSK and master public key MPK.*
- $\mathsf{KeyGen}(\mathsf{MSK}, f)$ *takes as input the master secret MSK and a function $f \in \mathcal{F}$. It outputs a secret key SK_f for f.*
- $\mathsf{Enc}(\mathsf{MPK}, m)$ *takes as input the master public key MPK and a message $m \in \mathcal{M}$, and outputs a ciphertext c.*
- $\mathsf{Dec}(\mathsf{SK}_f, c)$ *takes as input the secret key SK_f for a function $f \in \mathcal{F}$ and a ciphertext c, and outputs some $y \in \mathcal{Y}$, or $\bot$.*

Correctness: For any message $m \in \mathcal{M}$ and function $f \in \mathcal{F}$, we have that

$$\Pr \left[\mathsf{Dec}(\mathsf{SK}_f, c) = f(m) : \begin{array}{c} (\mathsf{MSK}, \mathsf{MPK}) \leftarrow \mathsf{Setup}(1^\lambda), \\ \mathsf{SK}_f \leftarrow \mathsf{KeyGen}(\mathsf{MSK}, f), \\ c \leftarrow \mathsf{Enc}(\mathsf{MPK}, m) \end{array} \right] = 1$$

Definition 4. (Selectively-secure single-key PKFE). *We say that a tuple of algorithm* $\mathsf{PKFE} = (\mathsf{Setup}, \mathsf{KeyGen}, \mathsf{Enc}, \mathsf{Dec})$ *is a selectively-secure single-key public-key functional encryption scheme for* $\mathcal{M}, \mathcal{Y}, \mathcal{F}$, *if it satisfies the following requirement, formalized by the experiment* $\mathsf{Expt}_\mathcal{A}^{\mathsf{PKFE}}(1^\lambda, b)$ *between an adversary* $\mathcal{A}$ *and a challenger:*

1. $\mathcal{A}$ *submits the message pair* $m_0^*, m_1^* \in \mathcal{M}$ *and a function* f *to the challenger.*
2. *The challenger runs* $(\mathsf{MSK}, \mathsf{MPK}) \leftarrow \mathsf{Setup}(1^\lambda)$, *generates ciphertext* $\mathsf{CT}^* \leftarrow \mathsf{Enc}(\mathsf{MPK}, m_b^*)$ *and a secret key* $\mathsf{SK}_f \leftarrow \mathsf{KeyGen}(\mathsf{MSK}, f)$. *The challenger gives* $(\mathsf{MPK}, \mathsf{CT}^*, sk_f)$ *to* $\mathcal{A}$.
3. $\mathcal{A}$ *outputs a guess* b' *for* b.
4. *The output of the experiment is* b' *if* $f(m_0^*) = f(m_1^*)$ *and* $\perp$ *otherwise.*

We say that the public-key functional encryption scheme is selectively-secure if, for all PPT adversaries $\mathcal{A}$, *there exists a negligible function* $\mu(\lambda)$, *such that*

$$\mathsf{Adv}_\mathcal{A}^{\mathsf{PKFE}} = \left| \Pr\left[\mathsf{Expt}_\mathcal{A}^{\mathsf{PKFE}}(1^\lambda, 0) = 1 \right] - \Pr\left[\mathsf{Expt}_\mathcal{A}^{\mathsf{PKFE}}(1^\lambda, 1) = 1 \right] \right| \leq \mu(\lambda).$$

We further say that PKFE *is* δ-*selectively secure, for some concrete negligible function* $\delta(\cdot)$, *if for all polysize distinguishers the above indistinguishability gap* $\mu(\lambda)$ *is smaller than* $\delta(\lambda)^{\Omega(1)}$.

We now further define a notion of succinctness for functional encryption schemes as above.

Definition 5. (Weakly Succinct functional encryption). *For a class of functions* $\mathcal{F} = \{\mathcal{F}_\lambda\}$ *over message domain* $\mathcal{M} = \{\mathcal{M}_\lambda\}$, *we let:*

- $n(\lambda)$ *be the input length of the functions in* $\mathcal{F}$,
- $s(\lambda) = \max_{f \in \mathcal{F}_\lambda} |f|$ *be a bound on the circuit size of functions in* $\mathcal{F}_\lambda$,
- $d(\lambda) = \max_{f \in \mathcal{F}_\lambda} \mathsf{depth}(f)$ *a bound on the depth, and*

A functional encryption scheme is

- *weakly succinct* [14] *if the size of the encryption circuit is bounded by* $s^\gamma \cdot \mathsf{poly}(n, \lambda, d)$, *where* poly *is a fixed polynomial, and* $\gamma < 1$ *is a constant. We call* γ *the compression factor.*

The following result from [14, Sect. 3.2] states that one can construct an indistinguishability obfuscator from any single-key weakly succinct public-key functional encryption scheme.

Theorem 4. ([14]). *If there exists a subexponentially secure single-key weakly succinct PKFE scheme, then there exists an indistinguishability obfuscator.*

2.3   Indistinguishability Obfuscation

Definition 6. (Indistinguishability obfuscator (IO) [8,9]). *A PPT machine* iO *is an* indistinguishability obfuscator *for a circuit class* $\{\mathcal{C}_\lambda\}_{\lambda \in \mathbb{N}}$ *if the following conditions are satisfied:*

- **Functionality:** *for all security parameters* $\lambda \in \mathbb{N}$, *for all* $C \in \mathcal{C}_\lambda$, *for all inputs* x, *we have that* $\Pr[C'(x) = C(x) : C' \leftarrow \mathsf{iO}(C)] = 1$.
- **Indistinguishability:** *for any polysize distinguisher* $\mathcal{D}$, *there exists a negligible function* $\mu(\cdot)$ *such that the following holds: for all security parameters* $\lambda \in \mathbb{N}$, *for all pairs of circuits* $C_0, C_1 \in \mathcal{C}_\lambda$ *of the same size and such that* $C_0(x) = C_1(x)$ *for all inputs* x, *then*

$$\left| \Pr\left[\mathcal{D}(\mathsf{iO}(C_0)) = 1 \right] - \Pr\left[\mathcal{D}(\mathsf{iO}(C_1)) = 1 \right] \right| \leq \mu(\lambda) \ .$$

We further say that iO *is* δ-*secure, for some concrete negligible function* $\delta(\cdot)$, *if for all polysize distinguishers the above indistinguishability gap* $\mu(\lambda)$ *is smaller than* $\delta(\lambda)^{\Omega(1)}$.

2.4   Succinct Identity-Based Encryption

We define identity-based encryption (IBE) [48] with a succinctness properties.

Definition 7. (Succinct IBE with γ-compression). *Let* $\mathcal{M}$ *be some message space and* $\mathcal{ID}$ *be an identity space. A* succint IBE *scheme with* γ-*compression for* $\mathcal{M}, \mathcal{ID}$ *is a tuple of algorithms* (Setup, KeyGen, Enc, Dec) *where:*

- Setup(1^λ) *is takes as input the security parameter and outputs a master secret key* MSK *and a master public key* MPK.
- KeyGen(MSK, id) *takes as input the master secret* MSK *and an identity* id $\in$ $\mathcal{ID}$. *It outputs a secret key* SK$_{\mathsf{id}}$ *for* id.
- Enc(MPK, id, m) *takes as input the public-parameter* MPK, *an identity* id $\in$ $\mathcal{ID}$, *and a message* $m \in \mathcal{M}$, *and outputs a ciphertext* c.
- Dec(SK$_{\mathsf{id}}$, c) *takes as input the secret key* SK$_{\mathsf{id}}$ *for an identity* id $\in \mathcal{ID}$ *and a ciphertext* c, *and outputs some* $m \in \mathcal{M}$, *or* $\bot$.

We require the following properties:

Correctness: *For any message* $m \in \mathcal{M}$ *and identity* id $\in \mathcal{ID}$, *we have that*

$$\Pr\left[\mathsf{Dec}(\mathsf{SK}_{\mathsf{id}}, c) = m : \begin{array}{l} (\mathsf{MSK}, \mathsf{MPK}) \leftarrow \mathsf{Setup}(1^\lambda), \\ \mathsf{SK}_{\mathsf{id}} \leftarrow \mathsf{KeyGen}(\mathsf{MSK}, \mathsf{id}), \\ c \leftarrow \mathsf{Enc}(\mathsf{MPK}, \mathsf{id}, m) \end{array} \right] = 1$$

Succinctness: *For any security parameter* $\lambda \in \mathbb{N}$, *identity space* $\mathcal{ID}$, *the size of the encryption circuit* Enc, *for messages of size* ℓ, *is at most* $|\mathcal{ID}|^\gamma \cdot \mathrm{poly}(\lambda, \ell)$.

In this work, we shall consider the following selective-security.

Definition 8. (Selectively-secure IBE). *A tuple of algorithms* IBE = (Setup, KeyGen, Enc, Dec) *is a* selectively-secure *IBE scheme for* $\mathcal{M}, \mathcal{ID}$ *if it satisfies the following requirement, formalized by the experiment* $\mathsf{Expt}_{\mathcal{A}}^{\mathsf{IBE}}(1^{\lambda}, b)$ *between an adversary* $\mathcal{A}$ *and a challenger:*

1. $\mathcal{A}$ *submits the challenge identity* $\mathsf{id}^* \in \mathcal{ID}$ *and the challenge messages* (m_0^*, m_1^*) *to the challenger.*
2. *The challenger runs* $(\mathsf{MSK}, \mathsf{MPK}) \leftarrow \mathsf{Setup}(1^{\lambda})$, *generates ciphertext* $\mathsf{CT}^* \leftarrow \mathsf{Enc}(\mathsf{MPK}, m_b^*)$ *and gives* $(\mathsf{MPK}, \mathsf{CT}^*)$ *to* $\mathcal{A}$.
3. $\mathcal{A}$ *is allowed to query (polynomially many) identities* $\mathsf{id} \in \mathcal{ID}$ *such that* $\mathsf{id} \neq \mathsf{id}^*$. *The challenger gives* $\mathsf{SK}_{\mathsf{id}} \leftarrow \mathsf{KeyGen}(1^{\lambda}, \mathsf{MSK}, \mathsf{id})$ *to the adversary.*
4. $\mathcal{A}$ *outputs a guess* b' *for* b. *The experiment outputs 1 if* $b' = b$, *0 otherwise.*

We say the IBE scheme is selectively-secure if, for all PPT adversaries $\mathcal{A}$, *there exists a negligible function* $\mu(\lambda)$, *it holds*

$$\mathsf{Adv}_{\mathcal{A}}^{\mathsf{IBE}} = \left| \Pr[\mathsf{Expt}_{\mathcal{A}}^{\mathsf{IBE}}(1^{\lambda}, 0) = 1] - \Pr[\mathsf{Expt}_{\mathcal{A}}^{\mathsf{IBE}}(1^{\lambda}, 1) = 1] \right| \leq \mu(\lambda).$$

We further say that IBE *is* δ-selectively secure, *for some concrete negligible function* $\delta(\cdot)$, *if for all polysize distinguishers the above indistinguishability gap* $\mu(\lambda)$ *is smaller than* $\delta(\lambda)^{\Omega(1)}$.

Theorem 5. *For any* $\beta < \gamma < 1$, *assuming there exists a* β-compressing *SXIO scheme for* P/poly *(defined in Sect. 3), a puncturable PRF, and a plain PKE scheme, there exists a succinct IBE scheme with* γ-compression. *Moreover, assuming the underlying primitives are* δ-secure *so is the resulting IBE scheme.*

We omit the proof of this theorem due to the limited space. See the full version of this paper [13].

We also omit the definition of puncturable PRF and (plain) PKE due to the limited space. Puncturable PRFs are constructed from OWFs [18,19,31,39]. See the full version of this paper [13] or references therein.

3   Strong Exponentially-Efficient Indistinguishability Obfuscation

Lin, Pass, Seth, and Telang [43] propose a variant of IO that has a weak (yet non-trivial) efficiency, which they call exponentially-efficient IO (XIO). All that this notion requires in terms of efficiency is that the size of an obfuscated circuit is sublinear in the size of the corresponding truth table. They also refer to a stronger notion that requires that also the time to obfuscate a given circuit is sublinear in the size of the truth table. This notion, which we call *strong* exponentially-efficient IO (SXIO), serves as one of the main abstractions in our work.

Definition 9 (Strong exponentially-efficient indistinguishability obfuscation (SXIO) [43]). *For a constant $\gamma < 1$, a machine $\mathsf{sxi}\mathcal{O}$ is a γ-compressing* strong exponentially-efficient indistinguishability obfuscator *(SXIO) for a circuit class $\{\mathcal{C}_\lambda\}_{\lambda \in \mathbb{N}}$ if it satisfies the functionality and indistinguishability in Definition 6 and the following efficiency requirements:*

Non-trivial Time Efficiency: *for any security parameter $\lambda \in \mathbb{N}$ and circuit $C \in \{\mathcal{C}_\lambda\}_{\lambda \in \mathbb{N}}$ with input length n, the running time of $\mathsf{sxi}\mathcal{O}$ on input $(1^\lambda, C)$ is at most $2^{n\gamma} \cdot \mathrm{poly}(\lambda, |C|)$.*

3.1   SXIO from Single-Input SKFE

In this section, we show that we can construct SXIO from any selectively-secure t-input SKFE scheme. We recall that such a t-SKFE scheme can be constructed from any selectively-secure 1-SKFE scheme, as stated in Theorem 3.

Theorem 6. *For any function $t(\lambda)$, if there exists δ-selectively-secure t-SKFE for $\mathsf{P/poly}$, then there exists $\frac{1}{t+1}$-compressing δ-secure SXIO for $\mathsf{P/poly}$.*

The idea of the construction of SXIO from SKFE is explained in the introduction. We immediately obtain the following corollary from Theorems 3 and 6.

Corollary 2. *1. If there exists δ-selectively-secure single-input SKFE for $\mathsf{P/poly}$, then there exists γ-compressing δ-secure SXIO for $\mathsf{P/poly}$ where $\gamma < 1$ is an arbitrary constant.*
2. Let $\varepsilon < 1$ be a constant and $\tilde{\lambda} = 2^{(\log \lambda)^\varepsilon}$. If there exists $2^{-\tilde{\lambda}^{\Omega(1)}}$-selectively-secure single-input SKFE for $\mathsf{P/poly}$, then there exists polynomially-secure SXIO with compression factor $\gamma(\lambda) = O(1/\log\log\lambda)$ for circuits of size at most $2^{O((\log\lambda)^\varepsilon)}$. (Here $\tilde{\lambda}$ is the single-input SKFE security parameter and λ is the SXIO security parameter.)

3.2   The Construction of SXIO

In what follows, given a circuit C, we identify its input space with $[N] = \{1, \ldots, N\}$ (so in particular, $N = 2^n$ if C takes n-bit strings as input). Let $\mathsf{SKFE}_t = (\mathsf{Setup}, \mathsf{KeyGen}, \mathsf{Enc}, \mathsf{Dec})$ be a selectively-secure t-input secret-key functional encryption scheme.

Construction. We construct an SXIO scheme $\mathsf{sxi}\mathcal{O}$ as follows.

$\mathsf{sxi}\mathcal{O}(1^\lambda, C)$: For every $j \in [N^{1/(t+1)}]$:

– let U_j be the t-input universal circuit that given $j_1, \ldots, j_{t-1} \in [N^{1/(t+1)}]$ and a t-input circuit D, returns $D(j_1, \ldots, j_{t-1}, j)$.
– let C_j be the t-input circuit that given $j_1, \ldots, j_t \in [N^{1/(t+1)}]$ returns $C(j_1, \ldots, j_t, j)$.

1. Generate MSK ← Setup(1^λ).
2. Generate $CT_{t,j}$ ← Enc(MSK, C_j, t) for $j \in [N^{1/(t+1)}]$.
3. Generate $CT_{i,j}$ ← Enc(MSK, j, i) for $i \in [t-1]$ and $j \in [N^{1/(t+1)}]$.
4. Generate SK_{U_j} ← KeyGen(MSK, U_j) for $j \in [N^{1/(t+1)}]$
5. sxi$\mathcal{O}(C) = (\{CT_{i,j}\}_{i \in [t], j \in [N^{1/(t+1)}]}, \{SK_{U_j}\}_{j \in [N^{1/(t+1)}]})$

Eval(sxi$\mathcal{O}, x$): To evaluate the obfuscated circuit, convert $x \in [N]$ into $(j_1, \ldots, j_t, j_{t+1}) \in [N^{1/(t+1)}]^{(t+1)}$ and output Dec($SK_{U_{j_{t+1}}}, CT_{1,j_1}, \ldots, CT_{t,j_t}$) .
   We omit the proof due to the limited space. See the full version [13].

Remark 2 (SXIO from succinct single-key SKFE). To get t-input SKFE as required above from 1-input SKFE, via the [20] transformation, the original SKFE indeed has to support an unbounded polynomial number of functional keys. We note that a similar SXIO construction is possible from a 1-input SKFE that supports a functional key for a single function f, but is *succinct* in the sense that encryption only grows mildly with the complexity of f, namely with $|f|^\beta$ for some constant $\beta < 1$.
   In more detail, assume a (1-input) single-key SKFE with succinctness as above, where the time to derive a key for a function f is bounded by $|f|^c \cdot \text{poly}(\lambda)$ for some constant $c \geq 1$. The SXIO will consist of a single key for the function f that given as input C_j, as defined above, returns $C_j(1), \ldots, C_j(N^{\frac{1}{c+1-\beta}})$, and encryptions of $C_1, \ldots, C_{N^{c-\beta/c+1-\beta}}$. Accordingly we still get SXIO with compression factor $\gamma = 1 - \frac{1-\beta}{c+1-\beta}$. This does not lead to arbitrary constant compression (in contrast with the theorem above), since $\frac{1}{2} \leq \gamma < 1$. Yet, it already suffices to obtain IO, when combined with LWE (as in Corollary 1).

4 Yao's Garbled Circuits Are Decomposable

In this section, we define the notion of decomposable garbled circuits. We can prove that the classical Yao's garbled circuit construction satisfies our definition of decomposability (in some parameter regime) though we omit the details about the proof due to the limited space. We use a decomposable garbling scheme as a building block to construct a PKFE scheme in Sect. 5.1.

4.1 Decomposable Garbling

Circuit garbling schemes [10,50] typically consist of algorithms (Gar.CirEn, Gar.InpEn, Gar.De). Gar.CirEn(C, K) is a circuit garbling algorithm that given a circuit C and secret key K, produces a garbled circuit $\widehat{C}$. Gar.InpEn(x, K) is an input garbling algorithm that takes an input x and the same secret key K, and produces a garbled input $\widehat{x}$. Gar.De($\widehat{C}, \widehat{x}$) is a decoder that given the garbled circuit and input decodes the result y.

In this work, we shall particularly be interested in garbling *decomposable circuits*. A decomposable circuit C can be represented by a smaller circuit C_{de} that can generate each of the gates in the circuit C (along with pointers to their neighbours). When garbling such circuits, we shall require that the garbling process will also be decomposable and will admit certain *decomposable security* properties. We next formally define the notion of decomposable circuits and decomposable garbling schemes.

Definition 10 (Decomposable Circuit). *Let $C : \{0,1\}^n \to \{0,1\}$ be a boolean circuit with L binary gates and W wires. Each gate $g \in [L]$ has an associated tuple (f, w_a, w_b, w_c) where $f : \{0,1\}^2 \to \{0,1\}$ is the binary function computed by the gate, $w_a, w_b \in [W]$ are the incoming wires, and $w_c \in [W]$ is the outgoing wire. A wire w_c can be the outgoing wire of at most a single gate, but can be used as an incoming wire to several different gates and therefore this models a circuit with fan-in 2 and unbounded fan-out. We define the predecessor gates of g to be the gates whose outgoing wires are w_a, w_b (at most 2 of them). We define the successor gates of g to be the gates that have w_c as an incoming wire. The gates are topologically ordered and labeled by $1, \ldots, L$ so that if j is a successor of i then $i < j$. A wire w is an input wire if it is not the outgoing wire of any gate. We assume that the wires $1, \ldots, n$ are the input wires. There is a unique output wire w which is not an incoming wire to any gate.*

We say that C is decomposable if there exists a smaller circuit C_{de}, called the decomposition circuit, that given a gate label $g \in [L]$ as input, outputs the associated tuple $C_{de}(g) = (f, w_a, w_b, w_c)$.

Definition 11 (Decomposable Garbling). *A decomposable garbling scheme consists of a tuple of three deterministic polynomial-time algorithms (Gar.CirEn, Gar.InpEn, Gar.De) that work as follows:*

- $\widehat{b}_i \leftarrow$ Gar.InpEn$(i, b; K)$: *takes as an input label $i \in [n]$, a bit $b \in \{0,1\}$, and secret key $K \in \{0,1\}^\lambda$, and outputs a garbled input bit $\widehat{b}_i$.*
- $\widehat{G}_g \leftarrow$ Gar.CirEn$(C_{de}, g; K)$: *takes as input a decomposition circuit $C_{de} : \{0,1\}^L \to \{0,1\}^*$, a gate label $g \in [L]$, and secret key $K \in \{0,1\}^\lambda$, and outputs a garbled gate $\widehat{G}_g$.*
- $y \leftarrow$ Gar.De$(\widehat{C}, \widehat{b})$: *takes as input garbled gates $\widehat{C} = \left\{\widehat{G}_g\right\}_{g \in [L]}$, and garbled input bits $\widehat{b} = \left\{\widehat{b}_i\right\}_{i \in [n]}$, and outputs $y \in \{0,1\}^m$.*

The scheme should satisfy the following requirements:

1. **Correctness:** *for every decomposable circuit C with decomposition circuit C_{de} and any input $b_1, \ldots, b_n \in \{0,1\}^n$, the decoding procedure Gar.De produces the correct output $y = C(b_1, \ldots, b_n)$.*
2. (σ, τ, δ)**-Decomposable Indistinguishability:** *There are functions $\sigma(\Phi, s, \lambda), \tau(\Phi) \in \mathbb{N}, \delta(\lambda) \leq 1$ such that for any security parameter λ, any input $x \in \{0,1\}^n$, and any two circuits (C, C') that:*

- *have the same topology Φ, and in particular the same size L and input-output lengths (n, m),*
- *have decomposition circuits $(C_{\mathsf{de}}, C'_{\mathsf{de}})$ of the same size s*
- *and agree on x: $C(x) = C'(x)$,*

there exist hybrid circuits $\left\{ \mathsf{Gar.HInpEn}^{(t)}, \mathsf{Gar.HCirEn}^{(t)} \mid t \in [\tau] \right\}$, each being of size at most σ, as well as (possibly inefficient) hybrid functions $\left\{ \mathsf{Gar.HPunc}^{(t)} \mid t \in [\tau] \right\}$ with the following syntax:

- $(K_{\mathsf{pun}}^{(t)}, g_{\mathsf{pun}}^{(t)}, i_{\mathsf{pun}}^{(t)}) \leftarrow \mathsf{Gar.HPunc}^{(t)}(K)$, *given a key $K \in \{0,1\}^{\lambda}$ and an index $t \in [\tau]$, outputs a punctured key $K_{\mathsf{pun}}^{(t)}$, a gate label $g_{\mathsf{pun}}^{(t)} \in [L]$, and an input label $i_{\mathsf{pun}}^{(t)} \in [n]$.*
- $\widehat{G}_g \leftarrow \mathsf{Gar.HCirEn}^{(t)}(g; K)$, *given a gate label $g \in [L]$, and a (possibly punctured) key K, outputs a fake garbled gate $\widehat{G}_g$.*
- $\widehat{b}_i \leftarrow \mathsf{Gar.HInpEn}^{(t)}(i, b; K)$, *given an input label $i \in [n]$, and a (possibly punctured) key K, outputs a fake garbled input bit $\widehat{b}_i$.*

We require that the following properties hold:

(a) **The hybrids transition from C to C':** *For any $K \in \{0,1\}^{\lambda}$, $g \in [L]$, $i \in [n]$, $b \in \{0,1\}$, we have:*

$\mathsf{Gar.CirEn}(C_{\mathsf{de}}, g; K) = \mathsf{Gar.HCirEn}^{(1)}(g; K),$
$\mathsf{Gar.InpEn}(i, b; K) = \mathsf{Gar.HInpEn}^{(1)}(i, b; K),$
$\mathsf{Gar.CirEn}(C'_{\mathsf{de}}, g; K) = \mathsf{Gar.HCirEn}^{(\tau)}(g; K),$
$\mathsf{Gar.InpEn}(i, b; K) = \mathsf{Gar.HInpEn}^{(\tau)}(i, b; K).$

(b) **Punctured keys preserve functionality:** *For any $K \in \{0,1\}^{\lambda}$, and $t \in [\tau - 1]$, and letting $(K_{\mathsf{pun}}^{(t)}, g_{\mathsf{pun}}^{(t)}, i_{\mathsf{pun}}^{(t)}) = \mathsf{Gar.HPunc}^{(t)}(K)$, it holds that, for any $g \neq g_{\mathsf{pun}}^{(t)}$, we have $\mathsf{Gar.HCirEn}^{(t)}(g; K) = \mathsf{Gar.HCirEn}^{(t)}(g; K_{\mathsf{pun}}^{(t)}) = \mathsf{Gar.HCirEn}^{(t+1)}(g, K)$, and for any $i \neq i_{\mathsf{pun}}^{(t)}$ and $b \in \{0,1\}$, we have $\mathsf{Gar.HInpEn}^{(t)}(i, b; K) = \mathsf{Gar.HInpEn}^{(t)}(i, b; K_{\mathsf{pun}}^{(t)}) = \mathsf{Gar.HInpEn}^{(t+1)}(i, b; K)$.*

(c) **Indistinguishability on punctured inputs:** *For any polysize distinguisher $\mathcal{D}$, security parameter $\lambda \in \mathbb{N}$, and circuits (C, C') as above,*

$$\left| \Pr\left[\mathcal{D}\left(\widehat{g}_{\mathsf{pun}}^{(t)}, \widehat{i}_{\mathsf{pun}}^{(t)}, \mathsf{Gar.HPunc}^{(t)}(K) \right) = 1 \right] - \right.$$
$$\left. \Pr\left[\mathcal{D}\left(\widehat{g}_{\mathsf{pun}}^{(t+1)}, \widehat{i}_{\mathsf{pun}}^{(t+1)}, \mathsf{Gar.HPunc}^{(t)}(K) \right) = 1 \right] \right| \leq \delta(\lambda) \ ,$$

where, for $t \geq 0$ we denote by $\widehat{g}_{\mathsf{pun}}^{(t)}$ the value $\mathsf{Gar.HCirEn}^{(t)}(g_{\mathsf{pun}}^{(t)}; K)$ and by $\widehat{i}_{\mathsf{pun}}^{(t)}$ the value $\mathsf{Gar.HInpEn}^{(t)}(i_{\mathsf{pun}}^{(t)}, x_{i_{\mathsf{pun}}^{(t)}}; K)$, with x being the input on which the two circuits C and C' agree on. The probability is over $K \leftarrow \{0,1\}^{\lambda}$, and $(K_{\mathsf{pun}}^{(t)}, g_{\mathsf{pun}}^{(t)}, i_{\mathsf{pun}}^{(t)}) = \mathsf{Gar.HPunc}^{(t)}(K)$.

We show that Yao's garbled circuit scheme, in fact, gives rise to a decomposable garbling scheme where the security loss and size of the hybrid circuits scales with the depth of the garbled circuits.

Theorem 7. *Let $\mathcal{C} = \{\mathcal{C}_\lambda\}_{\lambda \in \mathbb{N}}$ be a class of boolean circuits where each $C \in \mathcal{C}_\lambda$ has circuit size at most $L(\lambda)$, input size at most $n(\lambda)$, depth at most $d(\lambda)$, fan-out at most $\varphi(\lambda)$, and decomposition circuit of size at most $\Delta(\lambda)$. Then assuming the existence of δ-secure one-way functions, $\mathcal{C}$ has a decomposable garbling scheme with (σ, τ, δ)-decomposable indistinguishability where the bound on the size of hybrid circuits is $\sigma = \mathrm{poly}(\lambda, d, \log L, \varphi, \Delta)$, the number of hybrids is $\tau = L \cdot 2^{O(d)}$, and the indistinguishability gap is $\delta^{\Omega(1)}$.*

The proof is omitted due to the limited space. See the full version [13]. We rely heavily on the ideas of Hemenway et al. [37] which considered an orthogonal question of adaptively secure garbling schemes but (for entirely different reasons) developed ideas that are useful for decomposable garbling.

5 Single-Key Succinct PKFE from SXIO and PKE

This section consists of three subsections. The main part is constructing a weakly succinct PKFE scheme for boolean functions in Sect. 5.1. In Sect. 5.2, we present a transformation from weakly succinct PKFE schemes for boolean functions into ones for non-boolean functions. Lastly, we explain how the pieces come together to give IO from SKFE in Sect. 5.3.

5.1 Weakly Succinct PKFE for Boolean Functions

We now construct a single-key weakly succinct PKFE scheme for the class of boolean functions. The construction is based on succinct IBE, decomposable garbling, and SXIO.

Theorem 8. *Let $\mathcal{C} = \{\mathcal{C}_\lambda\}_{\lambda \in \mathbb{N}}$ be a family of circuits with a single output bit and let $n(\lambda), s(\lambda), d(\lambda)$ be bounds on their input length, size, and depth (respectively). For any constants β, γ such that $3\beta < \gamma < 1$, assuming a δ-secure, β-compressing SXIO for P/poly, there exists a constant α, such that given any δ-secure, α-compressing IBE, and δ-secure one-way functions, there exists a $2^d s \delta$-secure succinct PKFE for $\mathcal{C}$ with compression factor γ.*

Depth Preserving Universal Circuits. To prove the above theorem, we recall the existence of depth preserving universal circuits [23]. Concretely, any family of circuits $\mathcal{C}$ as considered in Theorem 8 has a uniform family of universal circuits $\{U_\lambda\}_{\lambda \in \mathbb{N}}$ with fan-out λ,[4] depth $O(d)$, and size $s^3 \cdot \mathrm{polylog}(s)$, for some fixed polynomial poly. Each such circuit takes as input a description $(f_1, \ldots, f_s)$ of a function in $\mathcal{C}$ and an input $(x_1, \ldots, x_n)$ and outputs $f(x)$. Furthermore, uniformity here means that each circuit has a decomposition circuit of size $\mathrm{polylog}(s)$.

---

[4] The restriction regarding fan-out is not stated explicitly in [23], but can always be achieved by blowing up the size and depth by a factor of at most $O(1)$.

Ingredients and Notation Used in the Construction

- We denote by $U^{(x)} : \{0,1\}^s \rightarrow \{0,1\}$ the universal circuit, with $x \in \{0,1\}^n$ being a hardwired bitstring, such that on input $(f_1, \ldots, f_s)$, the circuit $U^{(x)}$ outputs $f(x)$. This circuit has a decomposition circuit of size $\mathrm{poly}(n, \log(s))$, which we denote by $U_{\mathrm{de}}^{(x)}$. We also denote by L the number of gates in the circuit $U^{(x)}$.
- Let sxi$\mathcal{O}$ be a δ-secure, β-compressing SXIO scheme.
- Let IBE = (IBE.Setup, IBE.KeyGen, IBE.Enc, IBE.Dec) be δ-secure, succinct, IBE scheme with α-compression for the identity space being $\mathcal{ID} = [s] \times \{0,1\}$.
- Let (Gar.CirEn, Gar.InpEn, Gar.De) be a decomposable garbling scheme with (σ, τ, δ)-decomposable indistinguishability where $\tau = s2^{O(d)}$ and $\sigma = \mathrm{poly}(\lambda, n, d, \log(s))$. Such schemes are implied by δ-secure one-way functions (Theorem 7).
- Let $\mathcal{PPRF} = $ (PRF.Gen, PRF.Ev, PRF.Punc) be a δ-secure puncturable PRF. These are implied by δ-secure one-way functions [18,19,31,39].

Construction. The scheme consists of the following algorithms.

PKFE.Setup(1^λ):

- Run $(\mathsf{MSK}_{\mathsf{ibe}}, \mathsf{MPK}_{\mathsf{ibe}}) \leftarrow$ IBE.Setup(1^λ).
- Set $\mathsf{MSK} = \mathsf{MSK}_{\mathsf{ibe}}$, $\mathsf{MPK} = \mathsf{MPK}_{\mathsf{ibe}}$.

PKFE.Key(MSK, f):

- Compute $\mathsf{SK}_{i,f_i} \leftarrow$ IBE.KeyGen($\mathsf{MSK}_{\mathsf{ibe}}, (i, f_i)$) for $i \in [s]$, where $f = (f_1, \ldots, f_s)$.
- Return $\mathsf{SK}_f = \{\mathsf{SK}_{i,f_i}\}_{i \in [s]}$.

PKFE.Enc(MPK, x):

- Compute $U_{\mathrm{de}}^{(x)}$ and pick a garbling key $K \leftarrow \{0,1\}^\lambda$ and a punctured key $S \leftarrow$ PRF.Gen(1^λ);
- Generate an obfuscation $\widetilde{\mathsf{IGC}} = \mathsf{sxi}\mathcal{O}(1^\lambda, \mathsf{IGC}[K, S, \mathsf{MPK}])$ of the input garbling circuit defined in Fig. 1;
- Generate an obfuscation $\widetilde{\mathsf{GGC}} = \mathsf{sxi}\mathcal{O}(1^\lambda, \mathsf{GGC}[K, U_{\mathrm{de}}^{(x)}])$ of the gate garbling circuit defined in Fig. 2;
- Return $\mathsf{CT}_x = (\widetilde{\mathsf{IGC}}, \widetilde{\mathsf{GGC}})$.

PKFE.Dec($\mathsf{SK}_f, \mathsf{CT}_x$):

- For $i \in [s]$, run $\widetilde{\mathsf{IGC}}(i, f_i)$ to obtain an IBE ciphertext, and decrypt the output using SK_{i,f_i} to obtain $\hat{f}_i$.
- For all $g \in [L]$, run $\widetilde{\mathsf{GGC}}(g)$, in order to obtain the garbled gate $\widehat{G}_g$.
- Return $y \leftarrow$ Gar.De($\widehat{C}, \hat{f}$), with $\widehat{C} = \left\{\widehat{G}_g\right\}_{g \in [L]}$ and $\hat{f} = \left\{\hat{f}_i\right\}_{i \in [s]}$.

We omit the proof of correctness, succinctness, and security due to the limited space. See the full version for the complete proof of Theorem 8 [13].

Input Garbling Circuit $\mathsf{IGC}[K, S, \mathsf{MPK}]$

Hardwired: a garbling key K, a puncturable PRF key S, and $\mathsf{MPK} = \mathsf{MPK}_{\mathsf{ibe}}$.
Input: identity (i, b), consisting of an input label $i \in [s]$ and a bit $b \in \{0, 1\}$.
Padding: the circuit is padded to size $\mathsf{pad}_{\mathsf{IGC}}(s, d, n, \lambda)$, determined in the analysis.

1. Compute a corresponding garbled input bit $\widehat{b}_i = \mathsf{Gar.InpEn}(i, b; K)$.
2. Output an IBE encryption $\mathsf{IBE.Enc}(\mathsf{MPK}_{\mathsf{ibe}}, (i, b), \widehat{b}_i; \mathsf{PRF.Ev}_S(i, b))$.

Fig. 1. Circuit $\mathsf{IGC}[K, S, \mathsf{MPK}]$

Gate Garbling Circuit $\mathsf{GGC}[K, U_{\mathsf{de}}^{(x)}]$

Hardwired: a garbling key K and the decomposition circuit $U_{\mathsf{de}}^{(x)}$ of $U^{(x)}$.
Input: a gate label $g \in [L]$.
Padding: the circuit is padded to size $\mathsf{pad}_{\mathsf{GGC}}(s, d, n, \lambda)$, determined in the analysis.

Output $\widehat{G}_g = \mathsf{Gar.CirEn}(U_{\mathsf{de}}^{(x)}, g; K)$.

Fig. 2. Circuit $\mathsf{GGC}[K, U_{\mathsf{de}}^{(x)}]$

5.2 Weakly Succinct PKFE for Non-Boolean Functions

In this section, we give a transformation from weakly succinct PKFE schemes for boolean functions into ones for *non-boolean* functions.

Theorem 9. *Let* $\mathcal{C} = \{\mathcal{C}_\lambda\}_{\lambda \in \mathbb{N}}$ *be a family of circuits (with multiple output bits) and let* $n(\lambda), s(\lambda), d(\lambda)$ *be bounds on their input length, size, and depth (respectively). For any constants* $\beta < \gamma < 1$, *assuming a* β-*compressing SXIO for* P/poly, *there exists a constant* α, *such that given any* α-*compressing weakly succinct PKFE for boolean functions of size* $s \cdot \mathrm{polylog}(s)$ *and depth* $O(d)$, *and one-way functions, there exists a succinct PKFE for* $\mathcal{C}$ *with compression factor* γ. *If all primitives are* δ-*secure so is the resulting scheme.*

The transformation is essentially the same transformation presented in [43, Sect. 4], with the following differences:

- They use XIO rather than SXIO, which results in a PKFE scheme where only the size of ciphertexts is compressed, whereas the time to encrypt may be large. They then make an extra step, based on LWE, to make encryption efficient. Using SXIO directly as we do, allows avoiding this step.
- They start from weakly succinct PKFE for boolean functions where the size of ciphertexts is completely independent of the size s of the function class

considered. Due to this, they can start from XIO with any compression factor $\beta < 1$. In our notion of weakly succinct, there is dependence on s^α, for some $\alpha < 1$, and we need to make sure that β and α are appropriately chosen to account for this.

- As stated, their notion of weak succinctness for PKFE does not explicitly scale with the depth of the function class considered. Eventually, they apply their transformation to function classes in $\mathbf{NC}^1$, assuming puncturable PRFs in $\mathbf{NC}^1$ (which exist under LWE). Our succinctness notion allows polynomial dependence on the depth, which should be roughly preserved through the transformation.

The transformation and proof of security are almost identical to the ones in [43] and are omitted due to the limited space. See the full version [13].

5.3   Putting It All Together: From SKFE and PKE to IO

We obtain the following statements from the results proved in this section.

Theorem 10. *Let* $\mathcal{C} = \{\mathcal{C}_\lambda\}_{\lambda \in \mathbb{N}}$ *be a family of circuits (with multiple output bits) and let* $n(\lambda), s(\lambda), d(\lambda)$ *be bounds on their input length, size, and depth (respectively). Then, for any constant* $\gamma < 1$, *there exists a constant* β, *such that given any* δ-secure, β-compressing SXIO for P/poly, and δ-secure PKE, there exists $2^d s\delta$-secure, γ-compressing, weakly succinct PKFE for $\mathcal{C}$.

Combining the above theorem with the result from Sect. 3, we obtain the following corollary.

Corollary 3. *If there exist (1-input) SKFE for* P/poly *and PKE, both subexponentially-secure, then there exists IO for* P/poly.

Remark 3 (The security loss). In order, the known reductions [3,14] of IO to weakly-succinct PKFE incur a sub-exponential loss. Accordingly, reducing IO to SKFE based on our results incurs a similar loss. However, when restricting attention, to the transformation from SKFE to (weakly-succinct) PKFE, then the loss is $\mathrm{poly}(2^d, \lambda)$, for circuits of depth d. In particular, for $\mathbf{NC}^1$, our transformation incurs only polynomial security loss. Such a PKFE for $\mathbf{NC}^1$, can then be bootstrapped to all polynomial-size circuits using the transformation of [2], and assuming also weak PRFs in $\mathbf{NC}^1$.

In concurrent work [30,41], it is shown that weakly-succinct single-key PKFE can then be polynomially reduced to PKFE. In summary, SKFE and PRFs in $\mathbf{NC}^1$ can be polynomially reduced to PKFE for all polynomial-size circuits.

6   Polynomially-Secure PKE from Secret-Key FE

In this section, we construct PKE from SKFE. Our starting point is Corollary 2 that directly follows from Theorems 3 and 6.

We now show how to construct a PKE scheme from such SXIO.

The Construction. Let $\{\mathsf{PRG} : \{0,1\}^n \to \{0,1\}^{2n}\}_{n \in \mathbb{N}}$ be a length-doubling pseudorandom generator that is $2^{-n/\log\log n}$-secure. Let $\mathsf{sxi}\mathcal{O}$ be a SXIO with compression factor $\gamma(\lambda) = O(1/\log\log\lambda)$ (and poly(λ) security) for circuits of size at most $2^{O((\log\lambda)^{\epsilon})}$.

The scheme PKE = (KeyGen, Enc, Dec) is defined as follows:
KeyGen(1^{λ}):

- Sample a PRG seed $s \leftarrow \{0,1\}^{\log\lambda/\gamma(\lambda)}$.
- Output PK = PRG(s) and SK = s.

Enc(PK, x):

- Construct the circuit WE[x, PK] that takes $s' \in \{0,1\}^{\log\lambda/\gamma(\lambda)}$ as input and outputs x if PK = PRG(s') holds and $\perp$ otherwise.
- Output CT = $\mathsf{sxi}\mathcal{O}$(WE[x, PK])

Dec(SK, CT):

- Compute $x' = $ CT(SK).

Proposition 3. *PKE is a (polynomially-secure) public-key encryption scheme.*

We omit the proof due the limited space. See the full version [13].

Acknowledgements. We thank Vinod Vaikuntanathan and Hoeteck Wee for valuable discussions.

References

1. Agrawal, S., Gorbunov, S., Vaikuntanathan, V., Wee, H.: Functional encryption: new perspectives and lower bounds. In: Canetti, R., Garay, J.A. (eds.) CRYPTO 2013. LNCS, vol. 8043, pp. 500–518. Springer, Heidelberg (2013). doi:10.1007/978-3-642-40084-1_28

2. Ananth, P., Brakerski, Z., Segev, G., Vaikuntanathan, V.: From selective to adaptive security in functional encryption. In: Gennaro, R., Robshaw, M. (eds.) CRYPTO 2015. LNCS, vol. 9216, pp. 657–677. Springer, Heidelberg (2015). doi:10.1007/978-3-662-48000-7_32

3. Ananth, P., Jain, A.: Indistinguishability obfuscation from compact functional encryption. In: Gennaro, R., Robshaw, M. (eds.) CRYPTO 2015. LNCS, vol. 9215, pp. 308–326. Springer, Heidelberg (2015). doi:10.1007/978-3-662-47989-6_15

4. Ananth, P., Jain, A., Sahai, A.: Indistinguishability obfuscation from functional encryption forsimple functions. Cryptology ePrint Archive, Report 2015/730 (2015). http://eprint.iacr.org/2015/730

5. Ananth, P., Sahai, A.: Functional encryption for turing machines. In: Kushilevitz, E., Malkin, T. (eds.) TCC 2016. LNCS, vol. 9562, pp. 125–153. Springer, Heidelberg (2016). doi:10.1007/978-3-662-49096-9_6

6. Asharov, G., Segev, G.: Limits on the power of indistinguishability obfuscation and functional encryption. In: Guruswami, V. (ed.) 56th FOCS, pp. 191–209. IEEE Computer Society Press, October 2015

7. Badrinarayanan, S., Gupta, D., Jain, A., Sahai, A.: Multi-input functional encryption for unbounded arity functions. In: Iwata, T., Cheon, J.H. (eds.) ASIACRYPT 2015. LNCS, vol. 9452, pp. 27–51. Springer, Heidelberg (2015). doi:10.1007/978-3-662-48797-6_2

8. Barak, B., Goldreich, O., Impagliazzo, R., Rudich, S., Sahai, A., Vadhan, S., Yang, K.: On the (Im)possibility of obfuscating programs. In: Kilian, J. (ed.) CRYPTO 2001. LNCS, vol. 2139, pp. 1–18. Springer, Heidelberg (2001). doi:10.1007/3-540-44647-8_1

9. Barak, B., Goldreich, O., Impagliazzo, R., Rudich, S., Sahai, A., Vadhan, S.P., Yang, K.: On the (im)possibility of obfuscating programs. J. ACM **59**(2), 6 (2012)

10. Bellare, M., Hoang, V.T., Rogaway, P.: Foundations of garbled circuits. In: Yu, T., Danezis, G., Gligor, V.D. (eds.) ACM CCS 2012, pp. 784–796. ACM Press, October 2012

11. Bellare, M., O'Neill, A.: Semantically-secure functional encryption: possibility results, impossibility results and the quest for a general definition. In: Abdalla, M., Nita-Rotaru, C., Dahab, R. (eds.) CANS 2013. LNCS, vol. 8257, pp. 218–234. Springer, Heidelberg (2013). doi:10.1007/978-3-319-02937-5_12

12. Bitansky, N., Garg, S., Lin, H., Pass, R., Telang, S.: Succinct randomized encodings and their applications. In: Servedio, R.A., Rubinfeld, R. (eds.) 47th ACM STOC, pp. 439–448. ACM Press, June 2015

13. Bitansky, N., Nishimaki, R., Passelègue, A., Wichs, D.: From cryptomania to obfustopia through secret-key functional encryption. IACR Cryptology ePrint Archive 2016, 558 (2016)

14. Bitansky, N., Vaikuntanathan, V.: Indistinguishability obfuscation from functional encryption. In: Guruswami, V. (ed.) 56th FOCS, pp. 171–190. IEEE Computer Society Press, October 2015

15. Boneh, D., Gentry, C., Gorbunov, S., Halevi, S., Nikolaenko, V., Segev, G., Vaikuntanathan, V., Vinayagamurthy, D.: Fully key-homomorphic encryption, arithmetic circuit ABE and compact garbled circuits. In: Nguyen, P.Q., Oswald, E. (eds.) EUROCRYPT 2014. LNCS, vol. 8441, pp. 533–556. Springer, Heidelberg (2014). doi:10.1007/978-3-642-55220-5_30

16. Boneh, D., Lewi, K., Raykova, M., Sahai, A., Zhandry, M., Zimmerman, J.: Semantically secure order-revealing encryption: multi-input functional encryption without obfuscation. In: Oswald, E., Fischlin, M. (eds.) EUROCRYPT 2015. LNCS, vol. 9057, pp. 563–594. Springer, Heidelberg (2015). doi:10.1007/978-3-662-46803-6_19

17. Boneh, D., Sahai, A., Waters, B.: Functional encryption: definitions and challenges. In: Ishai, Y. (ed.) TCC 2011. LNCS, vol. 6597, pp. 253–273. Springer, Heidelberg (2011). doi:10.1007/978-3-642-19571-6_16

18. Boneh, D., Waters, B.: Constrained pseudorandom functions and their applications. In: Sako, K., Sarkar, P. (eds.) ASIACRYPT 2013. LNCS, vol. 8270, pp. 280–300. Springer, Heidelberg (2013). doi:10.1007/978-3-642-42045-0_15

19. Boyle, E., Goldwasser, S., Ivan, I.: Functional signatures and pseudorandom functions. In: Krawczyk, H. (ed.) PKC 2014. LNCS, vol. 8383, pp. 501–519. Springer, Heidelberg (2014). doi:10.1007/978-3-642-54631-0_29

20. Brakerski, Z., Komargodski, I., Segev, G.: Multi-input functional encryption in the private-key setting: stronger security from weaker assumptions. In: Fischlin, M., Coron, J.-S. (eds.) EUROCRYPT 2016. LNCS, vol. 9666, pp. 852–880. Springer, Heidelberg (2016). doi:10.1007/978-3-662-49896-5_30

21. Brakerski, Z., Segev, G.: Function-private functional encryption in the private-key setting. In: Dodis, Y., Nielsen, J.B. (eds.) TCC 2015. LNCS, vol. 9015, pp. 306–324. Springer, Heidelberg (2015). doi:10.1007/978-3-662-46497-7_12

22. Canetti, R., Lin, H., Tessaro, S., Vaikuntanathan, V.: Obfuscation of probabilistic circuits and applications. In: Dodis, Y., Nielsen, J.B. (eds.) TCC 2015. LNCS, vol. 9015, pp. 468–497. Springer, Heidelberg (2015). doi:10.1007/978-3-662-46497-7_19

23. Cook, S.A., Hoover, H.J.: A depth-universal circuit. SIAM J. Comput. **14**(4), 833–839 (1985)

24. Caro, A., Iovino, V., Jain, A., O'Neill, A., Paneth, O., Persiano, G.: On the achievability of simulation-based security for functional encryption. In: Canetti, R., Garay, J.A. (eds.) CRYPTO 2013. LNCS, vol. 8043, pp. 519–535. Springer, Heidelberg (2013). doi:10.1007/978-3-642-40084-1_29

25. Garg, S., Gentry, C., Halevi, S., Raykova, M., Sahai, A., Waters, B.: Candidate indistinguishability obfuscation and functional encryption for all circuits. In 54th FOCS, pp. 40–49. IEEE Computer Society Press, October 2013

26. Garg, S., Gentry, C., Halevi, S., Zhandry, M.: Functional encryption without obfuscation. In: Kushilevitz, E., Malkin, T. (eds.) TCC 2016. LNCS, vol. 9563, pp. 480–511. Springer, Heidelberg (2016). doi:10.1007/978-3-662-49099-0_18

27. Garg, S., Gentry, C., Sahai, A., Waters, B.: Witness encryption and its applications. In Boneh, D., Roughgarden, T., Feigenbaum, J. (eds.) 45th ACM STOC, pp. 467–476. ACM Press, June 2013

28. Garg, S., Pandey, O., Srinivasan, A.: Revisiting the cryptographic hardness of finding a nash equilibrium. In: Robshaw, M., Katz, J. (eds.) CRYPTO 2016. LNCS, vol. 9815, pp. 579–604. Springer, Heidelberg (2016). doi:10.1007/978-3-662-53008-5_20

29. Garg, S., Pandey, O., Srinivasan, A., Zhandry, M.: Breaking the sub-exponential barrier in obfustopia. Cryptology ePrint Archive, Report 2016/102 (2016). http://eprint.iacr.org/2016/102

30. Garg, S., Srinivasan, A.: Unifying security notions of functional encryption. IACR Cryptology ePrint Archive 2016:524 (2016)

31. Goldreich, O., Goldwasser, S., Micali, S.: How to construct random functions (extended abstract). In: 25th FOCS, pp. 464–479. IEEE Computer Society Press, October 1984

32. Goldwasser, S., Gordon, S.D., Goyal, V., Jain, A., Katz, J., Liu, F.-H., Sahai, A., Shi, E., Zhou, H.-S.: Multi-input functional encryption. In: Nguyen, P.Q., Oswald, E. (eds.) EUROCRYPT 2014. LNCS, vol. 8441, pp. 578–602. Springer, Heidelberg (2014). doi:10.1007/978-3-642-55220-5_32

33. Goldwasser, S., Kalai, Y.T., Popa, R.A., Vaikuntanathan, V., Zeldovich, N.: Reusable garbled circuits and succinct functional encryption. In: Boneh, D., Roughgarden, T., Feigenbaum, J. (eds.) 45th ACM STOC, pp. 555–564. ACM Press, June 2013

34. Gorbunov, S., Vaikuntanathan, V., Wee, H.: Functional encryption with bounded collusions via multi-party computation. In: Safavi-Naini, R., Canetti, R. (eds.) CRYPTO 2012. LNCS, vol. 7417, pp. 162–179. Springer, Heidelberg (2012). doi:10.1007/978-3-642-32009-5_11

35. Gorbunov, S., Vaikuntanathan, V., Wee, H.: Predicate encryption for circuits from LWE. In: Gennaro, R., Robshaw, M. (eds.) CRYPTO 2015. LNCS, vol. 9216, pp. 503–523. Springer, Heidelberg (2015). doi:10.1007/978-3-662-48000-7_25

36. Goyal, V., Jain, A., Koppula, V., Sahai, A.: Functional encryption for randomized functionalities. In: Dodis, Y., Nielsen, J.B. (eds.) TCC 2015. LNCS, vol. 9015, pp. 325–351. Springer, Heidelberg (2015). doi:10.1007/978-3-662-46497-7_13

37. Hemenway, B., Jafargholi, Z., Ostrovsky, R., Scafuro, A., Wichs, D.: Adaptively secure garbled circuits from one-way functions. In: Robshaw, M., Katz, J. (eds.) CRYPTO 2016. LNCS, vol. 9816, pp. 149–178. Springer, Heidelberg (2016). doi:10.1007/978-3-662-53015-3_6

38. Impagliazzo, R.: A personal view of average-case complexity. In: Proceedings of the Tenth Annual Structure in Complexity Theory Conference, Minneapolis, Minnesota, USA, June 19–22, 1995, pp. 134–147. IEEE Computer Society (1995)

39. Kiayias, A., Papadopoulos, S., Triandopoulos, N., Zacharias, T.: Delegatable pseudorandom functions and applications. In: Sadeghi, A.-R., Gligor, V.D., Yung, M. (eds.) ACM CCS 13, pp. 669–684. ACM Press, November 2013

40. Komargodski, I., Segev, G., Yogev, E.: Functional encryption for randomized functionalities in the private-key setting from minimal assumptions. In: Dodis, Y., Nielsen, J.B. (eds.) TCC 2015, Part II. LNCS, vol. 9015, pp. 352–377. Springer, Heidelberg (2015)

41. Li, B., Micciancio, D.: Compactness vs collusion resistance in functional encryption. IACR Cryptology ePrint Archive 2016:561 (2016)

42. Lin, H.: Indistinguishability obfuscation from constant-degree graded encoding schemes. In: Fischlin, M., Coron, J.-S. (eds.) EUROCRYPT 2016. LNCS, vol. 9665, pp. 28–57. Springer, Heidelberg (2016). doi:10.1007/978-3-662-49890-3_2

43. Lin, H., Pass, R., Seth, K., Telang, S.: Indistinguishability obfuscation with nontrivial efficiency. In: Cheng, C.-M., Chung, K.-M., Persiano, G., Yang, B.-Y. (eds.) PKC 2016. LNCS, vol. 9615, pp. 447–462. Springer, Heidelberg (2016). doi:10.1007/978-3-662-49387-8_17

44. Lin, H., Pass, R., Seth, K., Telang, S.: Output-compressing randomized encodings and applications. In: Kushilevitz, E., Malkin, T. (eds.) TCC 2016. LNCS, vol. 9562, pp. 96–124. Springer, Heidelberg (2016). doi:10.1007/978-3-662-49096-9_5

45. O'Neill, A.: Definitional issues in functional encryption. Cryptology ePrint Archive, Report 2010/556 (2010). http://eprint.iacr.org/2010/556

46. Regev, O.: On lattices, learning with errors, random linear codes, and cryptography. In: Gabow, H.N. Fagin, R. (eds.) 37th ACM STOC, pp. 84–93. ACM Press, May 2005

47. Sahai, A., Seyalioglu, H.: Worry-free encryption: functional encryption with public keys. In: Al-Shaer, E., Keromytis, A.D., Shmatikov, V. (eds.) ACM CCS 10, pp. 463–472. ACM Press, October 2010

48. Shamir, A.: Identity-based cryptosystems and signature schemes. In: Blakley, G.R., Chaum, D. (eds.) CRYPTO 1984. LNCS, vol. 196, pp. 47–53. Springer, Heidelberg (1985). doi:10.1007/3-540-39568-7_5

49. Waters, B.: A punctured programming approach to adaptively secure functional encryption. In: Gennaro, R., Robshaw, M. (eds.) CRYPTO 2015. LNCS, vol. 9216, pp. 678–697. Springer, Heidelberg (2015). doi:10.1007/978-3-662-48000-7_33

50. Yao, A.C.-C.: Protocols for secure computations (extended abstract). In: 23rd FOCS, pp. 160–164. IEEE Computer Society Press, November 1982

Single-Key to Multi-Key Functional Encryption with Polynomial Loss

Sanjam Garg$^{(\boxtimes)}$ and Akshayaram Srinivasan

University of California, Berkeley, USA
{sanjamg,akshayaram}@berkeley.edu

Abstract. Functional encryption (FE) enables fine-grained access to encrypted data. In a FE scheme, the holder of a secret key FSK_f (associated with a function f) and a ciphertext c (encrypting plaintext x) can learn $f(x)$ but nothing more.

An important parameter in the security model for FE is the number of secret keys that adversary has access to. In this work, we give a transformation from a FE scheme for which the adversary gets access to a single secret key (with ciphertext size sub-linear in the circuit for which this secret key is issued) to one that is secure even if adversary gets access to an unbounded number of secret keys. A novel feature of our transformation is that its security proof incurs only a *polynomial* loss.

1 Introduction

Functional encryption [SW05,BSW11,O'N10] generalizes the traditional notion of encryption by providing recipients fine-grained access to data. In a functional encryption (FE) system, holder of the master secret key MSK can derive secret key FSK_f for a circuit f. Given a ciphertext c (encrypting x) and the secret key FSK_f, one can learn $f(x)$ but nothing else about x is leaked. Functional encryption emerged as a generalization of several other cryptographic primitives like identity based encryption [Sha84,BF01,Coc01], attribute-based encryption [GPSW06,GVW13] and predicate encryption [KSW08,GVW15].

Single-Key vs Multi-Key. Results by Sahai and Seyalioglu [SS10] and Goldwasser, Kalai, Popa, Vaikuntanathan, and Zeldovich [GKP+13] provided FE scheme supporting all of P/poly circuits (based on standard assumptions). However, these constructions provide security only when the adversary is limited to obtaining a single functional secret key.[1] We call such a scheme as a *single-key* FE scheme. On the other hand, Garg, Gentry, Halevi, Raykova, Sahai and

---

This paper was jointly presented with the paper titled "Compactness vs Collusion Resistance in Functional Encryption" by Baiyu Li and Daniele Micciancio. Research supported in part from a DARPA/ARL SAFEWARE Award, AFOSR Award FA9550-15-1-0274, NSF CRII Award 1464397 and a research grant from the Okawa Foundation. The views expressed are those of the author and do not reflect the official policy or position of the funding agencies.

[1] These results could be generalized to support an a priori bounded number of functional secret keys.

© International Association for Cryptologic Research 2016
M. Hirt and A. Smith (Eds.): TCC 2016-B, Part II, LNCS 9986, pp. 419–442, 2016.
DOI: 10.1007/978-3-662-53644-5_16

Waters [GGH+13] construct an FE scheme for P/poly circuits and supporting security even when the adversary has access to an unbounded (polynomial) number of functional secret keys. We call such as scheme as a *multi-key* FE scheme. However, the work of Garg et al. assumes *indistinguishability obfuscation* (*iO*) [GGH+13].

A single-key FE scheme is said to have *weakly compact* ciphertexts if the size of the encryption circuit grows *sub-linearly* with the circuit for which secret key is given out. Ananth and Jain [AJ15] and Bitansky and Vaikuntanathan [BV15] showed that using single-key FE with weakly compact ciphertexts one can construct *iO* which can then be used to construct multi-key FE [GGH+13, Wat15]. However, this transformation incurs an exponential loss in security reduction. We ask:

Can we realize multi-key FE *from single-key* FE *with only a polynomial loss in the security reduction?*

1.1 Our Results

In this work, we answer the above question positively. More specifically, we give a *generic* transformation from single-key, compact FE to multi-key FE. Below, we highlight two additional features of our transformation:

1. Our transformation works even if the single key scheme we start with is *weakly selective* secure. The selective notion of security considered in literature restricts the adversary to commit to the challenge messages before seeing the public parameters but still allows functional secret key queries to be adaptively made (after seeing the challenge ciphertext and the public parameters). The weakly selective security (denoted by Sel*) restricts the adversary to commit to her challenge messages as well as make all the functional secret key queries before seeing the public parameters. Nonetheless, the multi-key scheme that we obtain is *selectively* secure.
2. For our transformation to work it is sufficient if the single-key scheme has *weakly compact* ciphertexts. However, the multi-key scheme resulting from our transformation has *fully compact* ciphertexts (independent of the circuit size).

Comparison with Concurrent and Independent Work. In a concurrent and independent work, Li and Micciancio [LM16] obtain a result similar to our, but using very different techniques. Their construction is based on two building blocks: SUM and PRODUCT constructions. The SUM and PRODUCT constructions take two FE schemes as input with security when q_1 and q_2 secret keys are given to the adversary, respectively. These constructions output a FE scheme with security when $q_1 + q_2$ and $q_1 \cdot q_2$ secret keys are provided to the adversary, respectively. Using these two building blocks, they present two constructions of multi-key FE with different security and efficiency tradeoffs. A nice feature of their result is that their construction just uses length doubling pseudorandom generator in addition to FE. However, their resultant multi-key FE scheme inherits the security and compactness property of the single-key scheme they start with. In particular, if the starting scheme in their transformation is weakly selectively secure (resp., weakly compact) then the resulting multi-key scheme is also weakly selectively secure (resp., weakly compact). On the other hand, our transformation always yields a selectively secure and fully compact scheme.

1.2 Obtaining Compactness and Adaptivity in FE

Using the transformation of Ananth, Brakerski, Segev and Vaikuntanathan [ABSV15] we can boost the security of our transformation from selectively to adaptive (while maintaining a polynomial loss). However, we note this transformation does not preserve compactness. In particular, even if the input to this transformation is a fully compact scheme, the resulting FE scheme is *non-compact* (where the ciphertext size can depend arbitrarily on the circuit size). In contrast, note that Ananth and Sahai [AS16] do provide an adaptively secure fully compact FE scheme based on $i\mathcal{O}$. Whether adaptive security with full compactness can be obtained from poly-hard FE is an interesting open problem. Partial progress on this question can be obtained using Hemenway et al. [HJO+15] who note that using the transformation of Ananth and Sahai [AS16] (starting with a fully compact selective FE, something that our transformation provides) along with adaptively secure garbled circuits [BHR12, HJO+15] yields an adaptively secure FE scheme whose ciphertext size grows with the on-line complexity of garbled circuits. The state of the art construction of adaptively secure garbled circuits [HJO+15] achieves an online-complexity that grows with the width of the circuit to be garbled. Hence, this yields a FE scheme with *width compact*

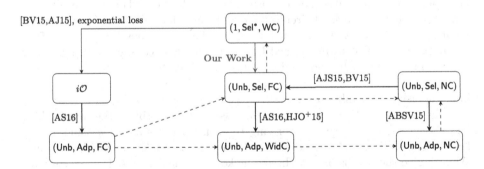

Fig. 1. Relationships between different notions of IND-FE parameterized by (xx, yy, zz). $xx \in \{1, \text{Unb}\}$ denotes the number of functional secret keys. $yy \in \{\text{Sel}^*, \text{Sel}, \text{Adp}\}$ denotes weakly selective, selective or adaptive security. $zz \in \{\text{NC}, \text{WC}, \text{FC}, \text{WidC}\}$ denotes the efficiency of the system: NC denotes non-compact ciphertexts, WC denotes weakly compact ciphertexts, FC denotes fully-compact ciphertexts and WidC denotes width-compact ciphertexts. Non-trivial relationships are given by solid arrows, and trivial relationships are given by dashed arrows.

ciphertexts (WidC); for which the size of the ciphertext grows with the *width* of circuits for which secret-keys are given out. We note that Ananth, Jain and Sahai [AJS15] and Bitansky and Vaikuntanthan [BV15] provide techniques for obtaining compactness in FE schemes. However, these results are limited to the selective security setting. Figure 1 shows known relationship between various notions of FE and the new relationships resulting from this work.

2   Our Techniques

We now give an overview of the techniques used in constructing multi-key, selective FE from single-key, weakly selective FE. We first give a description of a multi-key, selective FE scheme based on indistinguishability obfuscation ($i\mathcal{O}$). Though this result is not new, our construction is arguably different than the schemes of Garg et al. [GGH+13] and Waters [Wat15] and makes use of garbled circuits [Yao86]. Later, using techniques from works of Garg et al. [GPS15, GPSZ16] we obtain a FE scheme whose security can be based on *polynomially* hard single-key, weakly selective FE. The main novelty lies in designing a FE scheme from $i\mathcal{O}$ that is "amenable" to the techniques of Garg et al. [GPS15, GPSZ16] to avoid exponential loss in security.

$i\mathcal{O}$ **Based Construction.** Recall that a circuit garbling scheme (or randomized encoding in general) allows to encode an input x and a circuit C to obtain garbled input labels $\widetilde{x}$ and garbled circuit $\widetilde{C}$ respectively. Informally, the security of garbled circuits ensures that given $\widetilde{x}$ and $\widetilde{C}$, it is possible to learn $C(x)$ but nothing else. An additional feature of Yao's garbled circuits is that it is possible to encode the input x and the circuit C *separately* as long as the two encoding schemes share the same random tape.

At a high level, the ciphertext of our FE scheme corresponds to garbled input labels and the functional secret key corresponds to the garbled circuit. Intuitively, from the security of garbled circuits we can deduce that given the FE ciphertext c (encrypting x) and the functional secret key FSK_f it is possible to learn $f(x)$ but nothing else. But as mentioned before, to enable encoding the input x and the circuit C separately, the random coins used must be correlated in a certain way. The main crux of the construction is in achieving this correlation using indistinguishability obfuscation ($i\mathcal{O}$).

The correlation between the randomness used for garbling the input labels and the circuit is achieved by deriving the coins *pseudorandomly* using a PRF key S. This PRF key S also serves as the master secret key of our FE scheme. We now give the details of how the public key and the functional secret keys are derived from the master secret key S.

The public key of our FE scheme is an obfuscation of a program that takes as input some randomness r and outputs a "token" $t = \mathsf{PRG}(r)$ where PRG is a length doubling pseudorandom generator and a key $K = \mathsf{PRF}(S, t)$. The key K is used for deriving the input labels for the garbled circuit scheme say, that the two labels of the i-th input wire are given by $\{\mathsf{PRF}(K, i\|b)\}_{b \in \{0,1\}}$. The FE ciphertext encrypting a message m is given by the token t and the input labels

---

Input: r
Constants: PRF key S

1. Compute $t = \mathsf{PRG}(r)$.
2. Output $(t, K = \mathsf{PRF}(S, t))$

Fig. 2. Program implementing the public key

Input: Token t
Constants: PRF key S, PRF key S_f, Circuit C_f

1. Compute $K = \mathsf{PRF}(S, t)$.
2. Compute $\mathsf{L}_{i,b_i} = \mathsf{PRF}(K, i\|b_i)$ for all $i \in [n]$ and $b_i \in \{0, 1\}$.
3. Output the garbled circuit $\widetilde{C}_f$ with $\{\mathsf{L}_{i,b_i}\}_{i\in[n],b_i\in\{0,1\}}$ as the input labels and using $\mathsf{PRF}(S_f, t)$ as the random coins.

Fig. 3. Program implementing the functional secret key for a circuit C_f

corresponding to m i.e. $(t, \{\mathsf{PRF}(K, i\|m_i)\}_{i\in[n]})$. The description of the program implementing the public key is given in Fig. 2.

The functional secret key for a circuit C_f is an obfuscation of another program that takes as input the token t and first derives the key $K = \mathsf{PRF}(S, t)$. It then outputs a garbled circuit $\widetilde{C}_f$ where the garbled input labels are derived using key K. In particular, the input labels "encrypted" in the garbled evaluation table of $\widetilde{C}_f$ are given by $\{\mathsf{PRF}(K, i\|b)\}_{i\in[n],b\in\{0,1\}}$. The description of the program implementing the functional secret key is given in Fig. 3. The FE decryption corresponds to evaluation of this garbled circuit using the input labels given in the ciphertext. We now argue correctness and security.

The correctness of the above construction follows from having the "correct" input labels encrypted in the garbled evaluation tables in $\widetilde{C}_f$. It remains to show that the security holds when the obfuscation is instantiated with $i\mathcal{O}$. To achieve this, we use the punctured programming approach of Sahai and Waters [SW14].

We now give a high level overview of the security argument. The goal is to change from a hybrid where the adversary is given a challenge ciphertext encrypting message m_b for some $b \in \{0, 1\}$ to a hybrid where she is given a challenge ciphertext independent of the bit b. This is accomplished via a hybrid argument. In the first hybrid, we change the token t in the challenge ciphertext to an uniformly chosen random string t^* relying on the pseudorandomness property of the PRG. Next, we change the public key to be an obfuscation of a program that has the PRF key S punctured at t^* hardwired instead of S. The rest of the program is same as described in Fig. 2. Intuitively, the indistinguishability follows from $i\mathcal{O}$

Input: Token t
Constants: t^*, PRF key $S\{t^*\}$, PRF key $S_f\{t^*\}$, Circuit C_f, $\widetilde{C}_f^*$

- If $t \neq t^*$
  1. Compute $K = \mathsf{PRF}(S\{t^*\}, t)$.
  2. Compute $\mathsf{L}_{i,b_i} = \mathsf{PRF}(K, i\|b_i)$ for all $i \in [n]$ and $b_i \in \{0,1\}$.
  3. Output the garbled circuit $\widetilde{C}_f$ with $\{\mathsf{L}_{i,b_i}\}_{i \in [n], b_i \in \{0,1\}}$ as the input labels and using $\mathsf{PRF}(S_f\{t^*\}, t)$ as the random coins.
- Else, output $\widetilde{C}_f^*$.

Fig. 4. Program implementing the functional secret key for a circuit C_f in the security proof

security as the PRG has sparse images. In the next hybrid, the functional secret keys are generated as described in Fig. 4 where $\widetilde{C}_f^*$ hardwired in the program is exactly equal to garbled circuit $\widetilde{C}_f$ with $\{\mathsf{PRF}(K, i\|b_i)\}_{i \in [n], b_i \in \{0,1\}}$ (where $K = \mathsf{PRF}(S, t^*)$) as the input labels and generated using $\mathsf{PRF}(S_f, t^*)$ as the random coins. The indistinguishability of the two hybrids follows from $i\mathcal{O}$ security as the two programs described in Figs. 3 and 4 are functionally equivalent. Now, relying on the pseudorandomness at punctured point property of the PRF we change the input labels in the challenge ciphertext as well as the random coins used for generating $\widetilde{C}_f^*$ to uniformly chosen random strings. We can now change the challenge ciphertext to be independent of the bit b by relying on the security of garbled circuit. To be more precise, we change the input labels in the challenge ciphertext and $\widetilde{C}_f^*$ to be output of the garbled circuit simulator. Notice that we can still use the security of garbled circuits even if several garbled circuits share the same input labels. Thus, the above construction achieves security against unbounded collusions.

Construction from Poly Hard FE. The main idea behind our construction from polynomially hard, single-key, selectively secure FE is to simulate the effect of the obfuscation in the above construction using FE. To give a better insight into our construction we would first recall the FE to $i\mathcal{O}$ transformation of Ananth and Jain [AJ15] and Bitansky and Vaikuntanathan [BV15]. We note that this reduction suffers an exponential loss in security and we will be modifying this construction to achieve our goal of relying only on polynomially hard FE scheme. For this step, we rely on the techniques built by Garg et al. in [GPS15,GPSZ16] to avoid the exponential loss in security reduction. Parts of this section are adapted from [GPS15,GPSZ16].

FE to $i\mathcal{O}$ Transformation. We describe a modification of $i\mathcal{O}$ construction from FE of Bitansky and Vaikuntanathan [BV15] (Ananth and Jain [AJ15] take a slightly different route to achieve the same result). We note that the modified construction is not sufficient to obtain $i\mathcal{O}$ security but is "good enough" for our purposes.

The "obfuscation" of a circuit $C : \{0,1\}^\kappa \to \{0,1\}^\kappa$ consists of the following components: a FE ciphertext CT_ϕ and $\kappa + 1$ functional secret keys $\mathsf{FSK}_1, \cdots, \mathsf{FSK}_{\kappa+1}$ generated using independently sampled master secret keys $\mathsf{MSK}_1, \cdots, \mathsf{MSK}_{\kappa+1}$. CT_ϕ encrypts the empty string ϕ under the public key PK_1. The first κ functional secret keys $\mathsf{FSK}_1, \cdots, \mathsf{FSK}_\kappa$ implement the *bit-extension* functionality. To be more precise, FSK_i implements the function F_i that takes as input an $(i-1)$-bit string x and outputs two ciphertexts $\mathsf{CT}_{x\|0}$ and $\mathsf{CT}_{x\|1}$ encrypting $x\|0$ and $x\|1$ respectively under PK_{i+1}. The final function secret key $\mathsf{FSK}_{\kappa+1}$ implements the circuit C.

Let us discuss how to evaluate the "obfuscated" circuit on an input $x = x_1 \cdots x_\kappa$ where $x_i \in \{0,1\}$. The first step is to decrypt CT_ϕ using FSK_1 to obtain $\mathsf{CT}_0, \mathsf{CT}_1$. Depending on x_1 we choose either the left encryption (CT_0) or the right encryption (CT_1) and recursively decrypt the chosen ciphertext under FSK_2 and so on. After $\kappa + 1$ FE decryptions, we obtain the output of the circuit on input $x_1 \cdots x_\kappa$.

An alternate way to view this evaluation (which would be useful for this work) is as a traversal along a path from the root to a leaf node of a complete binary tree. The binary tree has the empty string at the root and traversal chooses either the left or the right child depending on the bits $x_1, x_2, \cdots, x_\kappa$ i.e. at level i, bit x_i is used to determine whether to go left or right. We would refer to this binary tree as the *evaluation binary tree*.

Our Construction. Recall that our main idea is to simulate the effect of obfuscation by appropriately modifying the above FE to $i\mathcal{O}$ transformation. We first explain the modifications to the "obfuscation" computing the master public key.

Let $C_{pk}[S]$ (having S hardwired) be the circuit that implements the public key of our $i\mathcal{O}$-based construction. Recall that this circuit takes as input some randomness r, expands it using the PRG to obtain the token t and outputs $(t, \mathsf{PRF}(S, t))$. The goal is to produce an "obfuscation" of this circuit using FE to $i\mathcal{O}$ transformation explained above. Recall that the FE to $i\mathcal{O}$ transformation has $\kappa + 1$ functional secret keys $\mathsf{FSK}_1, \cdots, \mathsf{FSK}_{\kappa+1}$ and an initial ciphertext CT_ϕ encrypting the empty string. The final functional secret key $\mathsf{FSK}_{\kappa+1}$ implements the circuit $C_{pk}[S]$. The first observation is that we cannot naively hardwire the PRF key in the circuit C_{pk}. This is because to achieve some "meaningful" mechanisms of hiding the PRF key (via puncturing) we need to go via the $i\mathcal{O}$ route that incurs an exponential loss in security. Therefore, the first modification is to change C_{pk} such that it takes the PRF key S as input instead of having it hardwired. We now include the PRF key S in the initial ciphertext CT_ϕ i.e. CT_ϕ is now an encryption of (ϕ, S). We run into the following problem: the initial ciphertext now contains the PRF key S whereas we actually need S to be given as input to the final circuit C_{pk} that is implemented in $\mathsf{FSK}_{\kappa+1}$. Therefore, we need a mechanism to make the PRF key S "available" to the final functional secret key $\mathsf{FSK}_{\kappa+1}$ so that it can compute PRF evaluation on the token. In other words, we need to "propagate" the PRF key S from the root to every leaf.

To propagate the PRF key, we make use of the "puncturing along the path" idea of Garg, Pandey and Srinivasan [GPS15]. This idea uses a primitive called as *prefix puncturable* PRF introduced in [GPS15]. Informally, prefix puncturable PRF allows to puncture the PRF key S at a specific prefix z to obtain S_z. The correctness guarantee is that given S_z, one can evaluate the PRF on any input x such that z is a prefix of x. The security guarantee is that as long as any adversary does not get access to S_z where z is a prefix of x, $\mathsf{PRF}(S, x)$ is indistinguishable from random string. An additional feature is that prefix puncturing can be done recursively i.e. given S_z one can obtain $S_{z\|0}$ and $S_{z\|1}$. Additionally, if we need to puncture the PRF key at an input x it is sufficient to change the distribution of FE ciphertexts only along the root to the leaf x in the evaluation binary tree. This gives us hope of basing security on polynomially hard FE. As a result, if we were to use this primitive, the problem reduces to the following: design a mechanism wherein the PRF key S prefix punctured at token t is available at the final functional secret key $\mathsf{FSK}_{\kappa+1}$ as this can then be used to derive $\mathsf{PRF}(S, t)$.

Recall that the circuit C_{pk} generates the token t as $\mathsf{PRG}(r)$ by taking r as input. If we naively try to combine this circuit with the "puncturing along the way" trick of Garg et al., we obtain S_r at the final functional secret key. It is not clear if there is a way of obtaining $S_{\mathsf{PRG}(r)}$ from S_r. Garg et al. [GPSZ16] faced a similar challenge in designing the sampler for trapdoor permutation and fortunately the solution they provide is applicable to our setting. The solution given in their work is to consider a different token generation mechanism. To be more precise, instead of generating the token as an output of a PRG on the input randomness r, the token now corresponds to a public key of a semantically secure encryption scheme. To give more details, the circuit C_{pk} now takes as input P which is a public key that also functions as the token. The circuit now computes $\mathsf{PRF}(S, P)$ and outputs a public key encryption of $\mathsf{PRF}(S, P)$ using P as the public key.[2] We combine this circuit with the "puncturing along the way" technique of Garg et al. to obtain the "obfuscation" of our public key.

The functional secret key for a function C_f (denoted by FSK_f) is constructed similarly to that of the public key. Recall that the functional secret key takes as input the token t (which is now given by the public key P) and computes $K = \mathsf{PRF}(S, t)$. It then uses the key K to derive the input garbled labels and outputs a garbled circuit $\widetilde{C}_f$. FSK_f also implements the "puncturing along the way" trick of Garg et al. to obtain S_P (which is the PRF key prefix punctured at P) which is used by the final circuit to derive the garbled input labels.

Proof Technique: "Tunneling." We now briefly explain the main proof technique called as the "tunneling" technique which is adapted from Garg et al.'s works [GPS15, GPSZ16]. Recall that the proof of our $i\mathcal{O}$ based construction relies on the punctured programming approach of Sahai and Waters [SW14]. We also follow a similar proof strategy. Let us explain how to "puncture" the master public key on the token P. At a high level, if we have punctured the PRF key at P then relying on the security guarantee of prefix punctured PRF to replace $\mathsf{PRF}(S, P)$ with a random string.

---

[2] Notice that if we know the secret key corresponding to the public key P, then we can recover $\mathsf{PRF}(S, P)$ which can then be used to derive the input garbled labels.

Recall that puncturing the PRF key S at a string P involves "removing" S_z for every z such that z is a strict prefix of P from the "obfuscation." To get better intuition on how the puncturing works it would be helpful to view the "obfuscation" in terms of the evaluation binary tree. As mentioned before, the crucial observation that helps us to base security on polynomially hard FE is that S_z where z is a prefix of P occurs only along the path from the root to the leaf node P in this tree. Hence, it is sufficient to change the distribution of the FE ciphertexts only along this path in such a manner that they don't contain S_z. To implement this change, we rely on the "Hidden trapdoor mechanism" (also called as the Trojan method) of Ananth et al. in [ABSV15]. To give more details, every functional secret key FSK_i implements a function F_i that has two "threads" of operation. In thread-1 or the normal mode of operation, it performs the bit-extension on input x and the prefix puncturing on input S_x. In thread-2 or the trapdoor mode, it does not perform any computation on the input (x, S_x) and instead outputs some fixed value that is hardwired. We change the FE ciphertexts in such a way that the trapdoor thread is invoked in every functional secret key when the "obfuscation" is run on input P. Metaphorically, we create a "tunnel" (i.e. a path from the root to a leaf where the trapdoor mode of operation is invoked in every intermediate node) from the root to the leaf labeled P in the complete binary tree corresponding to the obfuscation. Additionally, we change the FE ciphertexts along the path from root to leaf P such that they do not contain any prefix punctured keys. A consequence of our "tunneling" is that along the way we would have removed S_z for every z which is a strict prefix of P from the "obfuscation."

3   Preliminaries

λ denotes the security parameter. A function $\mu(\cdot) : \mathbb{N} \rightarrow \mathbb{R}^+$ is said to be negligible if for all polynomials $\mathsf{poly}(\cdot)$, $\mu(\lambda) < \frac{1}{\mathsf{poly}(\lambda)}$ for large enough λ. For a probabilistic algorithm $\mathcal{A}$, we denote $\mathcal{A}(x; r)$ to be the output of $\mathcal{A}$ on input x with the content of the random tape being r. We will omit r when it is implicit from the context. We denote $y \leftarrow \mathcal{A}(x)$ as the process of sampling y from the output distribution of $\mathcal{A}(x)$ with a uniform random tape. For a finite set S, we denote $x \leftarrow S$ as the process of sampling x uniformly from the set S. We model non-uniform adversaries $\mathcal{A} = \{\mathcal{A}_\lambda\}$ as circuits such that for all λ, $\mathcal{A}_\lambda$ is of size $p(\lambda)$ where $p(\cdot)$ is a polynomial. We will drop the subscript λ from the adversary's description when it is clear from the context. We will also assume that all algorithms are given the unary representation of security parameter 1^λ as input and will not mention this explicitly when it is clear from the context. We will use PPT to denote Probabilistic Polynomial Time algorithm. We denote $[\lambda]$ to be the set $\{1, \cdots, \lambda\}$. We will use $\mathsf{negl}(\cdot)$ to denote an unspecified negligible function and $\mathsf{poly}(\cdot)$ to denote an unspecified polynomial. We assume without loss of generality that all cryptographic randomized algorithms use λ-bits of randomness. If the algorithm needs more than λ-bit of randomness it can extend to arbitrary polynomial stretch using a pseudorandom generator (PRG).

A binary string $x \in \{0,1\}^\lambda$ is represented as $x_1 \cdots x_\lambda$. x_1 is the most significant (or the highest order bit) and x_λ is the least significant (or the lowest order bit). The i-bit prefix $x_1 \cdots x_i$ of the binary string x is denoted by $x_{[i]}$. We use $x \| y$ to denote concatenation of binary strings x and y. We say that a binary string y is a prefix of x if and only if there exists a string $z \in \{0,1\}^*$ such that $x = y \| z$.

Puncturable Pseudorandom Function. We recall the notion of puncturable pseudorandom function from [SW14]. The construction of pseudorandom function given in [GGM86] satisfies the following definition [BW13, KPTZ13, BGI14].

Definition 1. *A puncturable pseudorandom function* PRF *is a tuple of PPT algorithms* (KeyGen$_{\mathsf{PRF}}$, PRF, Punc) *with the following properties:*

- *Efficiently Computable: For all λ and for all $S \leftarrow$ KeyGen$_{\mathsf{PRF}}(1^\lambda)$, PRF$_S$: $\{0,1\}^{\mathsf{poly}(\lambda)} \to \{0,1\}^\lambda$ is polynomial time computable.*
- *Functionality is preserved under puncturing: For all λ, for all $y \in \{0,1\}^\lambda$ and $\forall x \neq y$,*

$$\Pr[\mathsf{PRF}_{S\{y\}}(x) = \mathsf{PRF}_S(x)] = 1$$

  where $S \leftarrow$ KeyGen$_{\mathsf{PRF}}(1^\lambda)$ and $S\{y\} \leftarrow$ Punc(S, y).
- *Pseudorandomness at punctured points: For all λ, for all $y \in \{0,1\}^\lambda$, and for all poly sized adversaries $\mathcal{A}$*

$$\left| \Pr[\mathcal{A}(\mathsf{PRF}_S(y), S\{y\}) = 1] - \Pr[\mathcal{A}(U_\lambda, S\{y\}) = 1] \right| \leq \mathsf{negl}(\lambda)$$

  where $S \leftarrow$ KeyGen$_{\mathsf{PRF}}(1^\lambda)$, $S\{y\} \leftarrow$ Punc(S, y) and U_λ denotes the uniform distribution over $\{0,1\}^\lambda$.

Symmetric Key Encryption. A Symmetric-Key Encryption scheme SKE is a tuple of algorithms (SK.KeyGen, SK.Enc, SK.Dec) with the following syntax:

- SK.KeyGen(1^λ): Takes as input an unary encoding of the security parameter λ and outputs a symmetric key SK.
- SK.Enc$_{SK}(m)$: Takes as input a message $m \in \{0,1\}^*$ and outputs an encryption C of the message m under the symmetric key SK.
- SK.Dec$_{SK}(C)$: Takes as input a ciphertext C and outputs a message m'.

We say that SKE is *correct* if for all λ and for all messages $m \in \{0,1\}^*$, $\Pr[\mathsf{SK.Dec}_{SK}(C) = m] = 1$ where $SK \leftarrow$ SK.KeyGen(1^λ) and $C \leftarrow$ SK.Enc$_{SK}(m)$.

Definition 2. *For all λ and for all polysized adversaries $\mathcal{A}$,*

$$\left| \Pr[\mathsf{Expt}_{1^\lambda, 0, \mathcal{A}} = 1] - \Pr[\mathsf{Expt}_{1^\lambda, 1, \mathcal{A}} = 1] \right| \leq \mathsf{negl}(\lambda)$$

where $\mathsf{Expt}_{1^\lambda, b, \mathcal{A}}$ *is defined below:*

- **Challenge Message Queries:** *The adversary* $\mathcal{A}$ *outputs two messages* m_0 *and* m_1 *such that* $|m_0| = |m_1|$ *for all* $i \in [n]$.
- *The challenger samples* $SK \leftarrow \mathsf{SK.KeyGen}(1^\lambda)$ *and generates the challenge ciphertext* C *where* $C \leftarrow \mathsf{SK.Enc}_{SK}(m_b)$. *It then sends* C *to* $\mathcal{A}$.
- *Output is* b' *which is the output of* $\mathcal{A}$.

Remark 1. We will denote range of a secret key FSK (denoted by $\mathsf{Range}_n(SK)$) to be $\{\mathsf{SK.Enc}(SK, x)\}_{x \in \{0,1\}^n}$ for a specific n. We will require that for any two secret keys SK_1, SK_2 where $SK_1 \neq SK_2$ we have $\mathsf{Range}_n(SK_1) \cap \mathsf{Range}_n(SK_2) = \phi$ with overwhelming probability. We will also require that the existence of an efficient procedure that checks if a given ciphertext c belongs to $\mathsf{Range}_n(SK)$ for a particular secret key SK. We call such a scheme to be symmetric key encryption with disjoint range. We note that symmetric key encryption with disjoint ranges can be obtained from one-way functions [LP09].

Public Key Encryption. A public-key Encryption scheme PKE is a tuple of algorithms $(\mathsf{PK.KeyGen}, \mathsf{PK.Enc}, \mathsf{PK.Dec})$ with the following syntax:

- $\mathsf{PK.KeyGen}(1^\lambda)$: Takes as input an unary encoding of the security parameter λ and outputs a public key, secret key pair (pk, sk).
- $\mathsf{PK.Enc}_{pk}(m)$: Takes as input a message $m \in \{0,1\}^*$ and outputs an encryption C of the message m under the public key pk.
- $\mathsf{PK.Dec}_{sk}(C)$: Takes as input a ciphertext C and outputs a message m'.

We say that PKE is *correct* if for all λ and for all messages $m \in \{0,1\}^*$, $\Pr[\mathsf{PK.Dec}_{sk}(C) = m] = 1$ where $(pk, sk) \leftarrow \mathsf{PK.KeyGen}(1^\lambda)$ and $C \leftarrow \mathsf{PK.Enc}_{pk}(m)$.

Definition 3. *For all* λ *and for all polysized adversaries* $\mathcal{A}$ *and for all messages* $m_0, m_1 \in \{0,1\}^*$ *such that* $|m_0| = |m_1|$,

$$|\Pr[\mathcal{A}(pk, \mathsf{PK.Enc}_{pk}(m_0)) = 1] - \Pr[\mathcal{A}(pk, \mathsf{PK.Enc}_{pk}(m_1)) = 1]| \leq \mathsf{negl}(\lambda)$$

where $(pk, sk) \leftarrow \mathsf{PK.KeyGen}(1^\lambda)$.

Prefix Puncturable Pseudorandom Functions. We now define the notion of prefix puncturable pseudorandom function PPRF. We note that the construction of the pseudorandom function in [GGM86] is prefix puncturable according to the following definition.

Definition 4. *A prefix puncturable pseudorandom function* PPRF *is a tuple of PPT algorithms* $(\mathsf{KeyGen}_{\mathsf{PPRF}}, \mathsf{PrefixPunc})$ *satisfying the following properties:*

- **Functionality is preserved under repeated puncturing:** *For all* λ, *for all* $y \in \cup_{k=0}^{\mathsf{poly}(\lambda)} \{0,1\}^k$ *and for all* $x \in \{0,1\}^{\mathsf{poly}(\lambda)}$ *such that there exists a* $z \in \{0,1\}^*$ *s.t.* $x = y \| z$,

$$\Pr[\mathsf{PrefixPunc}(\mathsf{PrefixPunc}(S, y), z) = \mathsf{PrefixPunc}(S, x)] = 1$$

where $S \leftarrow \mathsf{KeyGen}_{\mathsf{PPRF}}$.

– **Pseudorandomness at punctured prefix:** *For all λ, for all $x \in \{0,1\}^{\text{poly}(\lambda)}$, and for all poly sized adversaries $\mathcal{A}$*

$$|\Pr[\mathcal{A}(\text{PrefixPunc}(S, x), \text{Keys}) = 1] - \Pr[\mathcal{A}(U_\lambda, \text{Keys}) = 1]| \leq \text{negl}(\lambda)$$

where $S \leftarrow \text{KeyGen}_{\text{PRF}}(1^\lambda)$ and $\text{Keys} = \{\text{PrefixPunc}(S, x_{[i-1]} \| (1 - x_i))\}_{i \in [\text{poly}(\lambda)]}$ where $x_{[0]}$ denotes the empty string.

Remark 2. For brevity of notation, we will be denoting $\text{PrefixPunc}(S, y)$ by S_y.

Garbled Circuits. We now define the circuit garbling scheme of Yao [Yao86] and state the required properties.

Definition 5. *A circuit garbling scheme is a tuple of PPT algorithms given by $(\text{Garb.Circuit}, \text{Garb.Eval})$ with the following syntax:*

– *$\text{Garb.Circuit}(C)$: This is a randomized algorithm that takes in the circuit to be garbled and outputs garbled circuit and the set of garbled input labels: $\widetilde{C}, \{\text{Inp}_{i,b_i}\}_{i \in [\lambda], b_i \in \{0,1\}}$.*
– *$\text{Garb.Eval}(\widetilde{C}, \{\text{Inp}_{i,x_i}\}_{i \in [\lambda]})$: This is a deterministic algorithm that takes in $\{\text{Inp}_{i,x_i}\}_{i \in [\lambda]}$ and $\widetilde{C}$ as input and outputs a string y.*

Definition 6 (Correctness). *We say a circuit garbling scheme is correct if for all circuits C and for all inputs x:*

$$\Pr[\text{Garb.Eval}(\widetilde{C}, \{\text{Inp}_{i,x_i}\}_{i \in [\lambda]}) = C(x)] = 1$$

where $\widetilde{C}, \{\text{Inp}_{i,b_i}\}_{i \in [\lambda], b_i \in [\lambda]} \leftarrow \text{Garb.Circuit}(K, C)$.

Definition 7 (Security). *There exists a simulator Sim such that for all circuits C and input x:*

$$\{\widetilde{C}, \{\text{Inp}_{i,x_i}\}_{i \in [\lambda]}\} \stackrel{c}{\approx} \{\text{Sim}(1^\lambda, C, C(x))\}$$

Lemma 1 [Yao86,LP09]. *Assuming the existence of one-way functions there exists a circuit garbling scheme satisfying the security notion given in Definition 7.*

4   Functional Encryption: Security and Efficiency

We recall the syntax and security notions of functional encryption [BSW11, O'N10].

A functional encryption FE with the message space $\{0,1\}^*$ and function space $\mathcal{F}$ is a tuple of PPT algorithms (FE.Setup, FE.Enc, FE.KeyGen, FE.Dec) having the following syntax:

– FE.Setup(1^λ): Takes as input the unary encoding of the security parameter λ and outputs a public key PK and a master secret key MSK.

- FE.Enc(PK, m): Takes as input a message $m \in \{0,1\}^*$ and outputs an encryption c of m under the public key PK.
- FE.KeyGen(MSK, f): Takes as input the master secret key MSK and a function $f \in \mathcal{F}$ (given as a circuit) as input and outputs the function key FSK$_f$.
- FE.Dec(FSK$_f$, c): Takes as input the function key FSK$_f$ and the ciphertext c and outputs a string y.

Definition 8 (Correctness). *The functional encryption scheme* FE *is correct if for all λ and for all messages $m \in \{0,1\}^*$ and for all $f \in \mathcal{F}$,*

$$
\Pr \left[y = f(m) \,\middle|\, \begin{array}{l} (\text{PK}, \text{MSK}) \leftarrow \text{FE.Setup}(1^\lambda) \\ c \leftarrow \text{FE.Enc}(\text{PK}, m) \\ \text{FSK}_f \leftarrow \text{FE.KeyGen}(\text{MSK}, f) \\ y \leftarrow \text{FE.Dec}(\text{FSK}_f, c) \end{array} \right] = 1
$$

Security. We now give the formal definitions of the security notions. We start with the weakest notion of security namely *weakly selective security*.

Definition 9 (Weakly Selective Security). *The functional encryption scheme is said to be multi-key, weakly selective secure if for all λ and for all poly sized adversaries $\mathcal{A}$,*

$$
\left| \Pr[\text{Expt}_{\text{Sel}^*,1^\lambda,0,\mathcal{A}} = 1] - \Pr[\text{Expt}_{\text{Sel}^*,1^\lambda,1,\mathcal{A}} = 1] \right| \leq \text{negl}(\lambda)
$$

where $\text{Expt}_{\text{Sel},1^\lambda,b,\mathcal{A}}$ *is defined below:*

- **Challenge Message Queries:** *The adversary $\mathcal{A}$ outputs two messages m_0, m_1 such that $|m_0| = |m_1|$ and a set of functions $f_1, \cdots, f_q \in \mathcal{F}$ to the challenger. The parameter q is a priori unbounded.*
- *The challenger samples* (PK, MSK) $\leftarrow$ FE.Setup(1^λ) *and generates the challenge ciphertext $c \leftarrow$ FE.Enc(PK, m_b). The challenger also computes* FSK$_{f_i} \leftarrow$ FE.KeyGen(MSK, f_i) *for all $i \in [q]$. It then sends* (PK, c), $\{\text{FSK}_{f_i}\}_{i \in [q]}$ *to $\mathcal{A}$.*
- *If $\mathcal{A}$ makes a query f_j for some $j \in [q]$ to such that for any, $f_j(m_0) \neq f_j(m_1)$, output of the experiment is $\perp$. Otherwise, the output is b' which is the output of $\mathcal{A}$.*

Remark 3. We say that the functional encryption scheme FE is **single-key, weakly selectively secure** if the adversary $\mathcal{A}$ in $\text{Expt}_{\text{Sel}^*,1^\lambda,b,\mathcal{A}}$ is allowed to obtain the functional key for a single function f.

We now give the definition of selectively secure FE.

Definition 10 (Selective Security). *The functional encryption scheme is said to be multi-key, selectively secure FE if for all λ and for all poly sized adversaries $\mathcal{A}$,*

$$
\left| \Pr[\text{Expt}_{\text{Sel},1^\lambda,0,\mathcal{A}} = 1] - \Pr[\text{Expt}_{\text{Sel},1^\lambda,1,\mathcal{A}} = 1] \right| \leq \text{negl}(\lambda)
$$

where $\mathsf{Expt}_{\mathsf{Sel},1^\lambda,b,\mathcal{A}}$ *is defined below:*

- **Challenge Message Queries:** *The adversary* $\mathcal{A}$ *outputs two message vectors* m_0, m_1 *such that* $|m_0| = |m_1|$ *to the challenger.*
- *The challenger samples* $(\mathsf{PK}, \mathsf{MSK}) \leftarrow \mathsf{FE.Setup}(1^\lambda)$ *and generates the challenge ciphertext* $c \leftarrow \mathsf{FE.Enc}(\mathsf{PK}, m_b)$. *It then sends* (PK, c) *to* $\mathcal{A}$.
- **Function Queries:** $\mathcal{A}$ *adaptively chooses a function* $f \in \mathcal{F}$ *and sends it to the challenger. The challenger responds with* $\mathsf{FSK}_f \leftarrow \mathsf{FE.KeyGen}(\mathsf{MSK}, f)$. *The number of function queries made by the adversary is unbounded.*
- *If* $\mathcal{A}$ *makes a query* f *to functional key generation oracle such that,* $f(m_0) \neq f(m_1)$, *output of the experiment is* $\bot$. *Otherwise, the output is* b' *which is the output of* $\mathcal{A}$.

Remark 4. In the adaptive variant, the adversary is allowed to give challenge messages after seeing the public parameters and functional secret key queries.

Efficiency. We now define the efficiency requirements of a FE scheme.

Definition 11 (Fully Compact). *A functional encryption scheme* FE *is said to be fully compact if for all* $\lambda \in \mathbb{N}$ *and for all* $m \in \{0,1\}^*$ *the running time of the encryption algorithm* FE.Enc *is* $\mathsf{poly}(\lambda, |m|)$.

Definition 12 (Weakly Compact). *A functional encryption scheme is said to be weakly compact if the running time of the encryption algorithm* FE.Enc *is* $|\mathcal{F}|^{1-\epsilon}.\mathsf{poly}(\lambda, |m|)$ *for some* $\epsilon > 0$ *where* $|\mathcal{F}| = \max_{f \in \mathcal{F}} |C_f|$ *where* C_f *is the circuit implementing* f.

A functional encryption scheme is said to have *non-compact ciphertexts* if the running time of the encryption algorithm can depend arbitrarily on the maximum circuit size of the function family.

5   Our Transformation

In this section we describe our transformation from single-key, weakly selective secure functional encryption with fully compact ciphertexts to multi-key, selective secure functional encryption scheme. We later (in Sect. 6) show that it is sufficient for the single-key scheme to have weakly compact ciphertexts. We state the main theorem below.

Theorem 1. *Assuming the existence of single-key, weakly selective secure* FE *scheme with fully compact ciphertexts, there exists a multi-key, selective secure* FE *scheme with fully compact ciphertexts.*

The transformation from single-key, weakly selective secure FE scheme to multi-key, selective secure FE scheme uses the following primitives that are implied by single-key, weakly selective secure FE.

- A single-key, weakly selective FE scheme (FE.Setup, FE.KeyGen, FE.Enc, FE.Dec).
- A prefix puncturable PRF (PPRF, KeyGen$_\mathsf{PPRF}$, PrefixPunc).
- A Circuit garbling scheme (Garb.Circuit, Garb.Eval).
- A public key encryption scheme (PK.KeyGen, PK.Enc, PK.Dec).
- A symmetric key encryption scheme (SK.KeyGen, SK.Enc, SK.Dec) with disjoint range.

Notation. λ will denote our security parameter. Let the length of the secret key output by SK.KeyGen be λ_1, let length of the key output by KeyGen$_\mathsf{PPRF}$ be λ_2. We will denote length of public key output by PK.KeyGen to be κ. The message space is given by $\{0,1\}^\gamma$ and the function space is the set of all poly sized circuits taking γ-bit inputs.

The output of the transformation is a FE scheme (MKFE.Setup, MKFE.KeyGen, MKFE.Enc, MKFE.Dec). The formal description our construction appears in Fig. 5.

5.1 Correctness and Security

We first show correctness of our construction

Correctness. Recall that we need to show that if we decrypt a FE ciphertext encrypting m using a functional secret key for a function f then we obtain $f(m)$. We first argue that our FE ciphertext is distributed as $(pk, \{\mathsf{PRF}(S_{pk}, i\|m_i)\}_{i\in[\kappa]})$. From the correctness of FE decryption, we note that by iteratively decrypting CT_ϕ under $\mathsf{FSK}_1, \cdots, \mathsf{FSK}_{\kappa+1}$ using the bits of pk we obtain a public key encryption of S_{pk} under public key pk. From the correctness of public key decryption, we correctly recover S_{pk}. Hence our FE ciphertext is distributed as $(pk, \{\mathsf{PRF}(S_{pk}, i\|m_i)\}_{i\in[\kappa]})$.

We now look at the decryption procedure. We notice from the correctness of FE decryption procedure that by iteratively decrypting CT_ϕ^f under $\mathsf{FSK}_1^f, \cdots \mathsf{FSK}_{\kappa+1}^f$ using the bits of pk, we obtain $\widetilde{C_f}, \{c_{i,b_i}\}_{i\in[\gamma], b_i\in\{0,1\}}$ where $c_{i,b_i} \leftarrow \mathsf{SK.Enc}(\mathsf{PRF}(S_{pk}, i\|m_i), \mathsf{lnp}_{i,b_i})$ for every $i \in [\gamma]$ and $b_i \in \{0,1\}$. It follows from the correctness of SK.Dec and the fact that the symmetric key encryption we use has disjoint ranges, we correctly obtain $\{\mathsf{lnp}_{i,m_i}\}_{i\in[\kappa]}$. The correctness of our MKFE decryption now follows from the correctness of garbled circuit evaluation.

We note that length of the ciphertexts (and the size of the encryption circuit) in our MKFE scheme is independent of the circuit size of functions. Hence, the MKFE scheme has fully compact ciphertexts. We now state the main lemma for security.

Lemma 2. *Assuming single-key, weakly selective security of* FE, *semantic security of* SKE, *semantic security of* PKE, *and the security of prefix puncturable pseudorandom function* PPRF, *the* MKFE *construction described in Fig. 5 is multi-key, selectively secure.*

Before we describe the proof of Lemma 2, we first set up some notation.

- MKFE.Setup(1^λ) :
  1. Sample $S \leftarrow \mathsf{KeyGen}_{\mathsf{PPRF}}(1^\lambda)$ and $K_\phi \leftarrow \mathsf{KeyGen}_{\mathsf{PPRF}}(1^\lambda)$. Sample $sk \leftarrow \mathsf{SK.KeyGen}(1^\lambda)$ and compute $\Psi_i \leftarrow \mathsf{SK.Enc}(sk, 0^{\mathsf{len}_i(\lambda)})$ for all $i \in [\kappa + 1]$ where $\mathsf{len}_i(\cdot)$ is a length function that would be specified later.
  2. Sample $(\mathsf{PK}_i, \mathsf{MSK}_i) \leftarrow \mathsf{FE.Setup}(1^\lambda)$ for $i \in [\kappa + 1]$. Compute $\mathsf{FSK}_i \leftarrow \mathsf{FE.KeyGen}(\mathsf{MSK}_i, \mathsf{BitExt}_i[\Psi_i, \mathsf{PK}_{i+1}])$ for all $i \in [\kappa]$ and $\mathsf{FSK}_{\kappa+1} \leftarrow \mathsf{FE.KeyGen}(\mathsf{MSK}_{\kappa+1}, \mathsf{Output}_1[\Psi_{\kappa+1}])$ where $\mathsf{BitExt}_i[\cdot, \cdot]$ and $\mathsf{Output}_1[\cdot]$ are described in Figure 6.
  3. Compute $\mathsf{CT}_\phi \leftarrow \mathsf{FE.Enc}(\mathsf{PK}_1, (\phi, S, K_\phi, 0^{\lambda_1}, 0))$ where ϕ denotes a string of length 0 (a.k.a the empty string).
  4. Output the master public key PK to be $(\mathsf{CT}_\phi, \{\mathsf{FSK}_i\}_{i \in [\kappa]})$ and the master secret key $\mathsf{MSK} = S$.
- MKFE.Enc(PK, m) :
  1. Sample $(pk, tk) \leftarrow \mathsf{PK.KeyGen}(1^\lambda)$.
  2. For $i = 1, \cdots, \kappa$ compute: $(\mathsf{CT}_{pk_{[i-1]}\|0}, \mathsf{CT}_{pk_{[i-1]}\|1}) \leftarrow \mathsf{FE.Dec}(\mathsf{FSK}_i, \mathsf{CT}_{pk_{[i-1]}})$ where $\mathsf{CT}_{pk_{[0]}}$ is defined to be CT_ϕ.
  3. Compute $c := \mathsf{FE.Dec}(\mathsf{FSK}_{\kappa+1}, \mathsf{CT}_{pk})$ and recover $S_{pk} = \mathsf{PK.Dec}(tk, c)$.
  4. Compute $\{L_{i,m_i}\}_{i \in [\gamma]} \leftarrow \mathsf{PRF}(S_{pk}, i\|m_i)$ where m_i denotes the i-th bit of the message m.
  5. Output $(pk, \{L_{i,m_i}\}_{i \in [\gamma]})$.
- MKFE.KeyGen(MSK, f) :
  1. Sample $K_\phi^f \leftarrow \mathsf{KeyGen}_{\mathsf{PPRF}}(1^\lambda)$. Sample $sk^f \leftarrow \mathsf{SK.KeyGen}(1^\kappa)$ and compute $\Psi_i^f \leftarrow \mathsf{SK.Enc}(sk^f, 0^{\mathsf{len}_i^f(\kappa)})$ for all $i \in [\kappa+1]$ where $\mathsf{len}_i^f(\cdot)$ is a length function that would be specified later.
  2. Sample $(\mathsf{PK}_i^f, \mathsf{MSK}_i^f) \leftarrow \mathsf{FE.KeyGen}(1^\lambda)$ for $i \in [\kappa + 1]$.
  3. Compute $\mathsf{FSK}_i^f \leftarrow \mathsf{FE.KeyGen}(\mathsf{MSK}_i^f, \mathsf{BitExt}_i[\Psi_i^f, \mathsf{PK}_{i+1}^f])$ for all $i \in [\kappa]$ and $\mathsf{FSK}_{\kappa+1}^f \leftarrow \mathsf{FE.KeyGen}(\mathsf{MSK}_{\kappa+1}^f, \mathsf{Output}_2[\Psi_{\kappa+1}^f, C_f])$ where $\mathsf{BitExt}_i[\cdot, \cdot]$ and $\mathsf{Output}_2[\cdot, \cdot]$ are described in Figure 6 and C_f is the description of the circuit computing f.
  4. Compute $\mathsf{CT}_\phi^f \leftarrow \mathsf{FE.Enc}(\mathsf{PK}_1^f, (\phi, S, K_\phi^f, 0^{\lambda_1}, 0))$.
  5. Output $\mathsf{FSK}_f = (\mathsf{CT}_\phi^f, \{\mathsf{FSK}_i^f\}_{i \in [\kappa+1]})$.
- MKFE.Dec(FSK_f, CT) :
  1. Parse CT as $(pk, \{L_{i,m_i}\}_{i \in [\gamma]})$
  2. For $i = 1, \cdots, \kappa$ compute $(\mathsf{CT}_{(pk)_{[i-1]}\|0}^f, \mathsf{CT}_{(pk)_{[i-1]}\|1}^f) \leftarrow \mathsf{FE.Dec}(\mathsf{FSK}_i, \mathsf{CT}_{(pk)_{[i-1]}}^f)$ where $\mathsf{CT}_{(pk)_{[0]}}^f$ is defined to be CT_ϕ^f.
  3. Compute $\widetilde{C_f}, \{c_{i,b_i}\}_{i \in [\gamma], b_i \in \{0,1\}} \leftarrow \mathsf{FE.Dec}(\mathsf{FSK}_{\kappa+1}, \mathsf{CT}_{pk}^f)$. Decrypt c_{i,m_i} using L_{i,m_i} as the key and obtain Inp_{i,m_i} for every $i \in [\gamma]$ (to be more precise, first test if $c_{i,0}$ or $c_{i,1}$ is in the range of L_{i,m_i} and then decrypt $c_{i,b} \in \mathsf{Range}_\lambda(L_{i,m_i})\ \mathsf{Inp}_{i,m_i}$).
  4. Output $\mathsf{Garb.Eval}(\widetilde{C_f}, \{\mathsf{Inp}_{i,m_i}\}_{i \in [\gamma]})$.

Fig. 5. Transformation from single key to unbounded key secure

---

BitExt$_i[\Psi, \mathsf{PK}]$

Input. $x \in \{0,1\}^{i-1}$, S_x, K_x, sk, mode
Constants. Ψ, PK

- If mode $= 0$, compute $S_{x\|b} \leftarrow \mathsf{PrefixPunc}(S_x, b)$, $K_{x\|b} \leftarrow \mathsf{PrefixPunc}(K_x, b\|0)$ and $K'_{x\|b} \leftarrow \mathsf{PrefixPunc}(K_x, b\|1)$ for $b \in \{0,1\}$ and output $\{\mathsf{FE.Enc}(\mathsf{PK}, x\|b, S_{x\|b}, K_{x\|b}, sk, \text{mode}; K'_{x\|b})\}_{b\in\{0,1\}}$.
- Else, recover $(x\|0, \mathsf{CT}_{x\|0})$ and $(x\|1, \mathsf{CT}_{x\|1})$ from $\mathsf{SK.Dec}(sk, \Psi)$ and output $\{\mathsf{CT}_{x\|0}, \mathsf{CT}_{x\|1}\}$.

Output$_1[\Psi]$

Input. $x \in \{0,1\}^\kappa$, S_x, K_x, sk, mode
Constants. Ψ

- If mode $= 0$, output $\mathsf{PK.Enc}(x, S_x; K_x)$.
- Else, recover (x, Val_x) from $\mathsf{SK.Dec}(sk, \Psi)$ and output Val_x.

Output$_2[\Psi, C_f]$

Input. x where $x \in \{0,1\}^\kappa$, S_x, K_x^f, sk^f, mode
Constants. Ψ, C_f

- If mode $= 0$,
  1. Compute $(\widetilde{C_f}, \{\mathsf{Inp}_{i,b_i}\}_{i\in[\gamma], b_i\in\{0,1\}}) = \mathsf{Garb.Circuit}(C_f; K_{x\|0\|0}^f)$.
  2. Compute $\mathsf{L}_{i,b_i} \leftarrow \mathsf{PRF}(S_x, i\|b_i)$ for all $i \in [\gamma]$ and $b_i \in \{0,1\}$.
  3. Compute $c_{i,b_i} \leftarrow \mathsf{SK.Enc}(\mathsf{L}_{i,b_i}, \mathsf{Inp}_{i,b_i}; K_{x\|1\|i\|b_i}^f)$ for all $i \in [\gamma]$ and $b_i \in \{0,1\}$. For each bit $i \in [\gamma]$, permute $c_{i,0}$ and $c_{i,1}$ as per the bits of $K_{x\|0\|1}^f$.
  4. Output $\widetilde{C_f}, \{c_{i,b_i}\}_{i\in[\gamma], b_i\in\{0,1\}}$.
- Else, recover (x, Val_x) from $\mathsf{SK.Dec}(sk^f, \Psi)$ and output Val_x.

Fig. 6. Auxiliary circuits

Notation. Let $x \in \{0,1\}^\kappa$. Let $\mathsf{Prefixes}(x)$ denote the set of all prefixes (κ in number) of the string x. Formally,

$$\mathsf{Prefixes}(x) := \{x_{[i]}\}_{i\in[\kappa]}$$

Let $\mathsf{Siblings}(x)$ denote the set of siblings of all prefixes of x. Formally,

$$\mathsf{Siblings}(x) := \{y_{[i-1]}\|(1 - y_i) : \forall y \in \mathsf{Prefixes}(x), i \in [\kappa] \text{ where } |y| = i\}$$

Proof of Lemma 2. The proof proceeds via a hybrid argument.

- Hyb_0: In this hybrid, the adversary is given the challenge ciphertext encrypting the message m_b. To be more precise, the challenge ciphertext is given by $(pk^*, \{L_{i,(m_b)_i}\}_{i \in [\kappa]})$ where $(pk^*, sk^*) \leftarrow$ PK.KeyGen(1^κ) and $L_{i,(m_b)_i} \leftarrow$ PRF($S_{pk^*}, i \| (m_b)_i$) for all $i \in [\kappa]$. All key generation queries are generated as per the construction described in Fig. 5.

- Hyb_1: In this hybrid, we are going to "tunnel" through the path from root to the leaf node labeled pk^* in the master public key. This step is realized through a couple of intermediate hybrids.

  Let $P_1 := \mathsf{Prefixes}(pk^*)$ and $Q_1 = \mathsf{Siblings}(pk^*) \setminus P_1$. For every $z \in P_1 \cup Q_1$, let CT_z be the result of the iterated decryption procedure on the master public key with z as input.[3] Additionally, let Val_{pk^*} be the output of the decryption of CT_{pk^*} under $\mathsf{FSK}_{\kappa+1}$. Let

  $$\mathsf{str}_i = \|_{z \in P_1 \cup Q_1 \wedge |z|=i}(z, \mathsf{CT}_z)$$

  $$\mathsf{str}_{\lambda+1} = (pk^*, \mathsf{Val}^1_{pk^*})$$

  We set $\mathsf{len}_i(\lambda)$ to be the maximum length of str_i over all choices of pk^*. We pad str_i to this size.

- $\mathsf{Hyb}_{0,1}$: In this hybrid we are going to change how Ψ_i is generated. Instead of encrypting the all zeroes string of length $\mathsf{len}_i(\kappa)$, we encrypt str_i. Indistinguishability follows from the semantic security of the symmetric key encryption since the key sk is not needed to simulate Hyb_0 or $\mathsf{Hyb}_{0,1}$.

- $\mathsf{Hyb}_{0,2}$: In this hybrid we change how CT_ϕ is generated. Instead of generating CT_ϕ to be $\mathsf{FE.Enc}(\mathsf{PK}_1, (\phi, S, K_\phi, 0^{\lambda_1}, 0))$, we generate it as $\mathsf{FE.Enc}(\mathsf{PK}_1, (\phi, 0^{\lambda_2}, 0^{\lambda_2}, sk, 1))$. We now argue that $\mathsf{Hyb}_{0,2}$ is indistinguishable from $\mathsf{Hyb}_{0,1}$. Notice that output of $\mathsf{BitExt}_1[\Psi_1, \mathsf{PK}_2]$ is same on $(\phi, S, K_\phi, 0^{\lambda_1}, 0)$ and $(\phi, 0^{\lambda_2}, 0^{\lambda_2}, sk, 1)$. Also, the choice of the two messages and the functionality for which the secret key is obtained do not depend on the public parameters. Hence, it follows from the weakly selective security of FE scheme under PK_1 that $\mathsf{Hyb}_{0,1}$ and $\mathsf{Hyb}_{0,2}$ are indistinguishable.

- $\mathsf{Hyb}_{0,3}$: In this hybrid we are going to tunnel through the path from the root to the leaf labeled pk^*. To achieve this, we are going to change CT_z that is encrypted in Ψ_1 for every $z \in P_1$. We don't change the encryption when $z \in Q_1$. In particular, we change $\mathsf{CT}_z = \mathsf{FE.Enc}(\mathsf{PK}_{|z|+1}, (z, S_z, K_z, 0^{\lambda_1}, 0); K'_z)$ to $\mathsf{FE.Enc}(\mathsf{PK}_{|z|+1}, (z, 0^{\lambda_2}, 0^{\lambda_2}, sk, 1); r_z)$ where r_z is chosen uniformly at random. Notice that as a result S_z for every z that is a strict prefix of pk^* does not appear in the public key of our MKFE scheme.

We first introduce an ordering of strings in P_1. For every string $x, y \in P_1$ $x \prec y$ if and only if $|x| < |y|$. This induces a partial ordering of the strings in P_1. We let $\mathsf{Hyb}_{0,2,x}$ to denote the hybrid where for all $z \prec x$, CT_z has been changed from $\mathsf{FE.Enc}(\mathsf{PK}_{|z|+1}, (z, S_z, K_z, 0^{\lambda_1}, 0); K'_z)$ to $\mathsf{FE.Enc}(\mathsf{PK}_{|z|+1}, (z, 0^{\lambda_2}, 0^{\lambda_2}, sk, 1); r_z)$. We prove for any two adjacent strings x, x' where $x' \prec x$ in ordered P_1 that $\mathsf{Hyb}_{0,2,x}$ is indistinguishable to $\mathsf{Hyb}_{0,2,x'}$. Since $|P_1| \leq \kappa$, we get $\mathsf{Hyb}_{0,2}$ is indistinguishable to $\mathsf{Hyb}_{0,3}$ through a series a κ hybrids.

---

[3] By iterated decryption procedure on input z we mean decrypting CT_ϕ under $\mathsf{FSK}_1, \cdots, \mathsf{FSK}_{|z|}$ using the bits of z.

* $\mathsf{Hyb}_{0,2,x',1}$: In this hybrid we change CT_x to $\mathsf{FE.Enc}(\mathsf{PK}_{|x|+1}, (x, S_x, K_x,$ $0^{\lambda_1}, 0); r_x)$ where r_x is chosen uniformly at random. Notice that for all strings y that are prefixes of x, CT_y has already been changed to $\mathsf{FE.Enc}(\mathsf{PK}_{|y|+1}, (y, 0^{\lambda_2}, 0^{\lambda_2}, sk, 1); r_y)$ because $y \prec x$ by our ordering. For every y that is a prefix of x, K_y is not needed to simulate $\mathsf{Hyb}_{0,2,x'}$ and $\mathsf{Hyb}_{0,2,x',1}$. It follows from the pseudorandomness at prefix punctured point property of PRF key K_ϕ we have $\mathsf{Hyb}_{0,2,x'}$ is indistinguishable to $\mathsf{Hyb}_{0,2,x',1}$. Illustration for this hybrid change is given in Fig. 7.

* $\mathsf{Hyb}_{0,2,x',2}$: In this hybrid we change CT_x to $\mathsf{FE.Enc}(\mathsf{PK}_{|x|+1}, (x, 0^{\lambda_2}, 0^{\lambda_2},$ $sk, 1); r_x)$. Notice that decrypting $\mathsf{FE.Enc}(\mathsf{PK}_{|x|+1}, (x, 0^{\lambda_2}, 0^{\lambda_2}, sk, 1); r_x))$ and $\mathsf{FE.Enc}(\mathsf{PK}_{|x|+1}, (x, S_x, K_x, 0^{\lambda_2}, 0))$ under the secret key $\mathsf{FSK}_{|x|+1}$ has the same output due to the choice of $\Psi^*_{|x|+1}$. Also, the choice of the two messages and the functionality for which the secret key is obtained do not depend on the public parameters. Hence, it follows from the weakly selective security of FE scheme under $\mathsf{PK}_{|x|+1}$ that $\mathsf{Hyb}_{0,2,x',1}$ and $\mathsf{Hyb}_{0,2,x',2}$ are indistinguishable.

  Notice that $\mathsf{Hyb}_{0,2,x',2}$ is distributed identically to $\mathsf{Hyb}_{0,2,x}$.

- Hyb_2: In this hybrid we are going to change Val_{pk^*} encrypted in Ψ^*_1. Notice that in Hyb_2, Val_{pk^*} is set to be an public key encryption of S_{pk^*} under the public key pk^* (using pseudorandomly generated coins). In this hybrid we are going to change Val_{pk^*} to be an public key encryption of all zeroes string (0^λ) under pk^*.

  • $\mathsf{Hyb}_{1,1}$: In this hybrid we generate the randomness used for encrypting S_{pk^*} under the public key pk^* uniformly instead of generating it pseudorandomly using the key K_{pk^*}. Notice that K_z for every z that is a prefix of pk^* is not needed to simulate either Hyb_1 or $\mathsf{Hyb}_{1,1}$. Therefore, from the pseudorandomness at prefix punctured point property of PRF under key K_ϕ, Hyb_1 is indistinguishable from $\mathsf{Hyb}_{1,1}$.

  • $\mathsf{Hyb}_{1,2}$: In this hybrid we change Val_{pk^*} to be an encryption of 0^κ under pk^*. Indistinguishability of $\mathsf{Hyb}_{1,1}$ and $\mathsf{Hyb}_{1,2}$ follows from the semantic security of public key encryption.

- Hyb_3: In this hybrid we are going to tunnel through the paths from the root to the leaf pk^* in each function secret key FSK_f that is queried by the adversary. We explain the details for a single function key FSK_f and we can extend to all function secret keys by a standard hybrid argument. The indistinguishability argument for a single function secret key FSK_f is similar to our argument to show indistinguishability between Hyb_0 and Hyb_1.

  Let $P_2 := \mathsf{Prefixes}(pk^*)$ and $Q_2 = \mathsf{Siblings}(pk^*)$. For every $z \in P_2 \cup Q_2$ let CT_z^f be the result of the iterated decryption procedure on the function secret key FSK_f with z as input. Additionally, let $\widetilde{C_f}, \{c_{i,b_i}\}_{i \in [\gamma], b_i \in \{0,1\}}$ be the output of the decryption of CT_{pk^*} under $\mathsf{FSK}_{\kappa+1}^f$. Let

$$\mathsf{str}_i^f = \|_{z \in P_2 \cup Q_2 \wedge |z|=i}(z, \mathsf{CT}_z)$$

$$\mathsf{str}_{\kappa+1}^f = (pk^*, \widetilde{C_f}, \{c_{i,b_i}\}_{i \in [\gamma], b_i \in \{0,1\}})$$

We set $\mathsf{len}_i^f(\kappa)$ to be the maximum length of str_i^f over all choices of f. We pad str_i^f to this size.

- $\mathsf{Hyb}_{2,1}$: In this hybrid we are going to change how Ψ_i^f is generated. Instead of encrypting the all zeroes string of length $\mathsf{len}'_i(\kappa)$ we encrypt str_i^f. Indistinguishability follows from the semantic security of the symmetric key encryption since the key sk^f is not needed to simulate Hyb_2 or $\mathsf{Hyb}_{2,1}$.

- $\mathsf{Hyb}_{2,2}$: In this hybrid we change how CT_ϕ^f is generated. Instead of generating CT_ϕ^f to be $\mathsf{FE.Enc}(\mathsf{PK}_1^f, (\phi, S, K_\phi^f, 0^{\lambda_1}, 0))$ we generate it as $\mathsf{FE.Enc}(\mathsf{PK}_1^f, (\phi, 0^{\lambda_2}, 0^{\lambda_2}, sk^f, 1))$. We now argue that $\mathsf{Hyb}_{2,2}$ is indistinguishable from $\mathsf{Hyb}_{2,1}$. Notice that output of $\mathsf{BitExt}_1[\Psi_f^*, \mathsf{PK}_2^f]$ is same on $(\phi, S, K_\phi^f, 0^{\lambda_1}, 0)$ and $(\phi, 0^{\lambda_2}, 0^{\lambda_2}, sk^f, 1)$. Also, the choice of the two messages and the functionality for which the secret key is obtained do not depend on the public parameters. Hence, it follows from the weakly selective security of FE scheme under PK_1^f that $\mathsf{Hyb}_{2,1}$ and $\mathsf{Hyb}_{2,2}$ are indistinguishable.

- $\mathsf{Hyb}_{2,3}$: In this hybrid we are going to tunnel through the paths from the root to the leaf labeled pk^* in FSK_f. To achieve this we are going to change CT_z that is encrypted in Ψ_i^f for every $z \in P_2$. As before, we don't change the encryption when $z \in Q_2$. In particular, we change $\mathsf{CT}_z^f = \mathsf{FE.Enc}(\mathsf{PK}_{|z|+1}^f, (z, S_z, K_z^f, 0^{\lambda_1}, 0); K'^f_z)$ to $\mathsf{FE.Enc}(\mathsf{PK}_{|z|+1}^f, (z, 0^{\lambda_2}, 0^{\lambda_2}, sk^f, 1); r_z)$ where r_z is chosen uniformly at random. The proof of indistinguishability between $\mathsf{Hyb}_{2,2}$ and $\mathsf{Hyb}_{2,3}$ is exactly same as the one between $\mathsf{Hyb}_{0,2}$ and $\mathsf{Hyb}_{0,3}$.

- Hyb_4: In this hybrid we are going to change S_{pk^*} used to generate the challenge ciphertext to an uniformly chosen random κ-bit string T^*. We observe that for z that is a prefix of pk^*, S_z is not needed to simulate either Hyb_3 or Hyb_4 because we have "tunneled" through from the root to leaf node pk^* in the master public key and in all the function secret keys FSK_f. Hence from the pseudorandomness at prefix punctured point property of the PRF under the key S, Hyb_4 is computationally indistinguishable to Hyb_3. Notice that this also implies (from the property of the pseudorandom function) that $\{L_{i,b_i}\}$ for every $i \in [\gamma]$ and for every $b_i \in \{0,1\}$ can be changed to uniformly chosen random strings. This change is made to challenge ciphertext as well as encryption keys used for generating $\{c_{i,b_i}\}_{i \in [\gamma], b_i \in \{0,1\}}$ in $\Psi_{\kappa+1}^f$ in each functional secret key FSK_f.

- Hyb_5: In this hybrid we are going to change to change the randomness used for generating garbled circuit, the encryptions c_{i,b_i} that are encrypted in $\Psi_{\kappa+1}^f$ and the randomness used for permuting c_{i,b_i} in each of the function secret keys FSK_f to uniformly chosen random strings. Observe that since we have "tunneled" through pk^* in each of the function secret keys it follows from pseudorandomness of prefix punctured point property of the PRF under the key K_ϕ^f, Hyb_5 is computationally indistinguishable to Hyb_4.

- Hyb_6: In this hybrid we are going to change $c_{i,1-(m_b)_i}$ to encrypting all zeroes string instead of encrypting $\mathsf{Inp}_{i,1-(m_b)_i}$. This change is made in $\Psi_{\kappa+1}^f$ in each

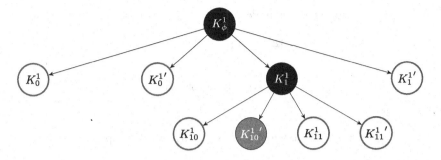

Fig. 7. Illustration for $\mathsf{Hyb}_{0,2,x',1}$ where $x' = 1$ and $x = 10$. The blackened nodes are not needed for simulation.

of the function secret keys FSK_f. Indistinguishablity of Hyb_5 and Hyb_6 follows from the semantic security of secret key encryption under $\mathsf{L}_{i,1-(m_b)_i}$.

- Hyb_7: In this hybrid we are going to change $\{\mathsf{Inp}_{i,(m_b)_i}\}_{i\in[\gamma]}, \widetilde{C_f}$ to be output of the simulator for the garbled circuit. This change is made in $\Psi_{\kappa+1}^f$ in each of the function secret keys FSK_f. More precisely, we set $\{\mathsf{Inp}_{i,(m_b)_i}\}_{i\in[\gamma]}, \widetilde{C_f} \leftarrow \mathsf{Sim}(1^\kappa, C_f, f(m_0))$ (note that $f(m_0) = f(m_b)$). Indistinguishability of Hyb_6 and Hyb_7 follows from the security of garbled circuits.

In Hyb_7, the view of the adversary is independent of the challenge bit b. Hence the advantage that the adversary has in guessing the bit b is 0 in Hyb_7.

6   Efficiency Analysis

In this section we relax the requirement of full compactness from our single-key selectively secure FE scheme to weakly compact ciphertexts. Parts of this section are taken verbatim from Bitansky and Vaikuntanathan [BV15].

Recall that a FE scheme with weakly compact ciphertexts has an encryption circuit whose size grows sub-linearly with the circuit size of functions for which function secret keys are given.

Let $F_1, F_2, \cdots, F_{\kappa+1}$ be the functionalities implemented by the secret keys $\mathsf{FSK}_1^f, \cdots, \mathsf{FSK}_{\kappa+1}^f$.[4] Notice that for any $i = \{1, \cdots, \kappa\}$, F_i implements the encryption circuit E_{i+1} for the functional encryption scheme under PK_{i+1}, symmetric decryption circuit and a prefix puncturing circuit. The size of the functional encryption circuit and the symmetric decryption circuit is bounded by $|E_{i+1}|.\mathsf{poly}(\kappa)$ and the size of the prefix puncturing circuit is bounded by $\mathsf{poly}(\kappa)$. Therefore,

$$|F_i| \leq |E_{i+1}|.\mathsf{poly}(\kappa)$$

---

[4] We restrict our attention to the functional secret keys of our scheme. The analysis of the master public key is exactly the same.

From our assumption that the underlying FE scheme is weakly compact we get:

$$|E_i| \leq |F_i|^{1-\epsilon}.\mathsf{poly}(\kappa)$$

Notice that:

$$|F_{\kappa+1}| \leq |C_f|.\mathsf{poly}(\kappa)$$

Hence we get:

$$|E_i| \leq |F_i|^{1-\epsilon}.\mathsf{poly}(\kappa) \leq |E_{i+1}|^{1-\epsilon}.(\mathsf{poly}(\kappa))^{1-\epsilon}.\mathsf{poly}(\kappa)$$

By recursively enumerating we get:

$$|E_i| \leq |C_f|^{1-\epsilon}.\mathsf{poly}(\kappa). \prod_{j=1}^{\kappa+2-i} \mathsf{poly}(\kappa)^{(1-\epsilon)^j}$$

We observe that:

$$\prod_{j=1}^{\kappa+2-i} \mathsf{poly}(\kappa)^{(1-\epsilon)^j} \leq \prod_{j=0}^{\infty} \mathsf{poly}(\kappa)^{(1-\epsilon)^j} \leq (\mathsf{poly}(\kappa))^{\frac{1}{\epsilon}}$$

Hence, for all $i \in [\kappa+1]$ we get:

$$|E_i| \leq |C_f|^{1-\epsilon}.\mathsf{poly}(\kappa)^{1+\frac{1}{\epsilon}}$$

which implies efficiency of our underlying construction.

Acknowledgements. We would like to thank the anonymous TCC reviewers for useful feedback. Additionally, we thank Divya Gupta, Peihan Miao, Omkant Pandey and Mark Zhandry for insightful discussions.

References

[ABSV15] Ananth, P., Brakerski, Z., Segev, G., Vaikuntanathan, V.: From selective to adaptive security in functional encryption. In: Gennaro, R., Robshaw, M. (eds.) CRYPTO 2015. LNCS, vol. 9216, pp. 657–677. Springer, Heidelberg (2015)

[AJ15] Ananth, P., Jain, A.: Indistinguishability obfuscation from compact functional encryption. In: Gennaro, R., Robshaw, M.J.B. (eds.) CRYPTO 2015. LNCS, vol. 9215, pp. 308–326. Springer, Heidelberg (2015)

[AJS15] Ananth, P., Jain, A., Sahai, A.: Achieving compactness generically: indistinguishability obfuscation from non-compact functional encryption. IACR Cryptology ePrint Archive, 2015:730 (2015)

[AS16] Ananth, P., Sahai, A.: Functional encryption for turing machines. In: Kushilevitz, E., et al. (eds.) TCC 2016-A. LNCS, vol. 9562, pp. 125–153. Springer, Heidelberg (2016). doi:10.1007/978-3-662-49096-9_6

[BF01] Boneh, D., Franklin, M.: Identity-based encryption from the Weil pairing. In: Kilian, J. (ed.) CRYPTO 2001. LNCS, vol. 2139, pp. 213–229. Springer, Heidelberg (2001)

[BGI14]   Boyle, E., Goldwasser, S., Ivan, I.: Functional signatures and pseudoran-
           dom functions. In: Krawczyk, H. (ed.) PKC 2014. LNCS, vol. 8383, pp.
           501–519. Springer, Heidelberg (2014)
[BHR12]   Bellare, M., Hoang, V.T., Rogaway, P.: Foundations of garbled circuits.
           In: Yu, T., Danezis, G., Gligor, V.D. (eds.) ACM CCS 2012, October
           16–18, Raleigh, NC, USA, pp. 784–796. ACM (2012)
[BSW11]   Boneh, D., Sahai, A., Waters, B.: Functional encryption: definitions and
           challenges. In: Ishai, Y. (ed.) TCC 2011. LNCS, vol. 6597, pp. 253–273.
           Springer, Heidelberg (2011)
[BV15]   Bitansky, N., Vaikuntanathan, V.: Indistinguishability obfuscation
           from functional encryption. In: Guruswami, V. (ed.) 56th FOCS,
           October 17–20, Berkeley, CA, USA, pp. 171–190. IEEE Computer Society
           Press (2015)
[BW13]   Boneh, D., Waters, B.: Constrained pseudorandom functions and their
           applications. In: Sako, K., Sarkar, P. (eds.) ASIACRYPT 2013, Part II.
           LNCS, vol. 8270, pp. 280–300. Springer, Heidelberg (2013)
[Coc01]   Cocks, C.: An Identity based encryption scheme based on quadratic
           residues. In: Honary, B. (ed.) Cryptography and Coding 2001. LNCS,
           vol. 2260, pp. 360–363. Springer, Heidelberg (2001)
[GGH+13]   Garg, S., Gentry, C., Halevi, S., Raykova, M., Sahai, A., Waters, B.:
           Candidate indistinguishability obfuscation and functional encryption for
           all circuits. In: 54th FOCS, October 26–29, Berkeley, CA, USA, pp. 40–49.
           IEEE Computer Society Press (2013)
[GGM86]   Goldreich, O., Goldwasser, S., Micali, S.: How to construct random func-
           tions. J. ACM 33(4), 792–807 (1986)
[GKP+13]   Goldwasser, S., Kalai, Y.T., Popa, R.A., Vaikuntanathan, V., Zeldovich,
           N.: Reusable garbled circuits and succinct functional encryption. In: Sym-
           posium on Theory of Computing Conference, STOC 2013, June 1–4,
           Palo Alto, CA, USA, pp. 555–564 (2013)
[GPS15]   Garg, S., Pandey, O., Srinivasan, A.: On the exact cryptographic hard-
           ness of finding a nash equilibrium. Cryptology ePrint Archive, Report
           2015/1078 (2015). http://eprint.iacr.org/2015/1078
[GPSW06]   Goyal, V., Pandey, O., Sahai, A., Waters, B.: Attribute-based encryption
           for fine-grained access control of encrypted data. In: Proceedings of the
           13th ACM Conference on Computer and Communications Security, CCS
           2006, October 30–November 3, Alexandria, VA, USA, pp. 89–98 (2006)
[GPSZ16]   Garg, S., Pandey, O., Srinivasan, A., Zhandry, M.: Breaking the sub-
           exponential barrier in obfustopia. Cryptology ePrint Archive, Report
           2016/102 (2016). http://eprint.iacr.org/2016/102
[GVW13]   Gorbunov, S., Vaikuntanathan, V., Wee, H.: Attribute-based encryption
           for circuits. In: Boneh, D., Roughgarden, T., Feigenbaum, J. (eds.) 45th
           ACM STOC, June 1–4, Palo Alto, CA, USA, pp. 545–554. ACM Press
           (2013)
[GVW15]   Gorbunov, S., Vaikuntanathan, V., Wee, H.: Predicate encryption for cir-
           cuits from LWE. In: Gennaro, R., Robshaw, M. (eds.) CRYPTO 2015.
           LNCS, vol. 9216, pp. 503–523. Springer, Heidelberg (2015)
[HJO+15]   Hemenway, B., Jafargholi, Z., Ostrovsky, R., Scafuro, A., Wichs, D.:
           Adaptively secure garbled circuits from one-way functions. IACR Cryp-
           tology ePrint Archive, 2015:1250 (2015)

[KPTZ13] Kiayias, A., Papadopoulos, S., Triandopoulos, N., Zacharias, T.: Delegatable pseudorandom functions and applications. In: ACM SIGSAC Conference on Computer and Communications Security, CCS 2013, November 4–8, Berlin, Germany, pp. 669–684 (2013)

[KSW08] Katz, J., Sahai, A., Waters, B.: Predicate encryption supporting disjunctions, polynomial equations, and inner products. In: Smart, N.P. (ed.) EUROCRYPT 2008. LNCS, vol. 4965, pp. 146–162. Springer, Heidelberg (2008)

[LM16] Li, B., Micciancio, D.: Compactness vs collusion resistance in functional encryption. Cryptology ePrint Archive, Report 2016/561 (2016). http://eprint.iacr.org/2016/561

[LP09] Lindell, Y., Pinkas, B.: A proof of security of Yao's protocol for two-party computation. J. Cryptol. **22**(2), 161–188 (2009)

[O'N10] O'Neill, A.: Definitional issues in functional encryption. IACR Cryptology ePrint Archive, 2010:556 (2010)

[Sha84] Shamir, A.: Identity-based cryptosystems and signature schemes. In: Blakely, G.R., Chaum, D. (eds.) CRYPTO 1984. LNCS, vol. 196, pp. 47–53. Springer, Heidelberg (1985)

[SS10] Sahai, A., Seyalioglu, H.: Worry-free encryption: functional encryption with public keys. In: Proceedings of the 17th ACM Conference on Computer and Communications Security, CCS 2010, October 4–8, Chicago, Illinois, USA, pp. 463–472 (2010)

[SW05] Sahai, A., Waters, B.: Fuzzy identity-based encryption. In: Cramer, R. (ed.) EUROCRYPT 2005. LNCS, vol. 3494, pp. 457–473. Springer, Heidelberg (2005)

[SW14] Sahai, A., Waters, B.: How to use indistinguishability obfuscation: deniable encryption, and more. In: Shmoys, D.B. (ed.) 46th ACM STOC, May 31–June 3, New York, NY, USA, pp. 475–484. ACM Press (2014)

[Wat15] Waters, B.: A punctured programming approach to adaptively secure functional encryption. In: Gennaro, R., Robshaw, M. (eds.) CRYPTO 2015. LNCS, vol. 9216, pp. 678–697. Springer, Heidelberg (2015)

[Yao86] Yao, A.C.-C.: How to generate and exchange secrets (extended abstract). In: 27th FOCS, October 27–29, Toronto, Ontario, Canada, pp. 162–167. IEEE Computer Society Press (1986)

Compactness vs Collusion Resistance
in Functional Encryption

Baiyu Li[(✉)] and Daniele Micciancio

University of California, San Diego, USA
{baiyu,daniele}@cs.ucsd.edu

Abstract. We present two general constructions that can be used to combine any two functional encryption (FE) schemes (supporting a bounded number of key queries) into a new functional encryption scheme supporting a larger number of key queries. By using these constructions iteratively, we transform any primitive FE scheme supporting a single functional key query (from a sufficiently general class of functions) and has certain weak compactness properties to a collusion-resistant FE scheme with the same or slightly weaker compactness properties. Together with previously known reductions, this shows that the compact, weakly compact, collusion-resistant, and weakly collusion-resistant versions of FE are all equivalent under polynomial time reductions. These are all FE variants known to imply the existence of indistinguishability obfuscation, and were previously thought to offer slightly different avenues toward the realization of obfuscation from general assumptions. Our results show that they are indeed all equivalent, improving our understanding of the minimal assumptions on functional encryption required to instantiate indistinguishability obfuscation.

1 Introduction

Indistinguishability obfuscation ($i\mathcal{O}$), first formalized in [7] and further investigated in [26], is currently one of the most intriguing notions on the cryptographic landscape, and it has attracted a tremendous amount of attention in the last few years. Since Garg et al. [21] put forward a plausible candidate obfuscation algorithm, $i\mathcal{O}$ has been successfully used to solve a wide range of complex cryptographic problems, including functional encryption [21], deniable encryption [32], and much more (e.g., see [8,16].) However, the problem of building an obfuscator with a solid proof of security is still far from being solved. The multilinear-map problems [18–20,23] underlying most known candidate $i\mathcal{O}$ constructions [5,6,11,21,24,31] have recently been subject to attacks [15,17],

This paper was presented jointly with the paper titled "Single-Key to Multi-Key Functional Encryption with Polynomial Loss" by Sanjam Garg and Akshayaram Srinivasan. Research supported in part by the Defense Advanced Research Projects Agency (DARPA) and the U.S. Army Research Office (ARO) under contract numbers W911NF-15-C-0226 and W911NF-15-C-0236, and the National Science Foundation (NSF) under grant CNS-1528068.

© International Association for Cryptologic Research 2016
M. Hirt and A. Smith (Eds.): TCC 2016-B, Part II, LNCS 9986, pp. 443–468, 2016.
DOI: 10.1007/978-3-662-53644-5_17

and basing $i\mathcal{O}$ on a solid, well-understood standard complexity assumption, has rapidly emerged as perhaps the single most important open problem in the area of cryptographic obfuscation.

An alternative path toward the construction of $i\mathcal{O}$ from standard assumptions has recently been opened by Bitansky and Vaikuntanathan [9] and Ananth and Jain [3], who independently showed that $i\mathcal{O}$ can be built from any (subexponentially secure) public key functional encryption scheme satisfying certain compactness requirements. While general constructions of compact functional encryption (for arbitrary functions) are only known using $i\mathcal{O}$, functional encryption is typically considered a weaker primitive than general $i\mathcal{O}$, or, at very least, a more manageable one, closer to what cryptographers know how to build. In fact, several functional encryption schemes (for restricted, but still rather broad classes of functions) are known achieving various notions of security [2,12,14,25,34]. We recall that a (public key) functional encryption scheme [1,10,30,33] is an encryption scheme with a special type of *functional* secret decryption keys sk_f (indexed by functions f) such that encrypting a message m (using the public key) and then decrypting the resulting ciphertext using sk_f produces the output of the function $f(m)$, without revealing any other information about the message. Parameters of interest in the study of functional encryption (in relation to obfuscation) are the time (or circuit) complexity of the encryption function t^{Enc} and the number of functional keys sk_f that can be released without compromising the security of the scheme. (See Sect. 2 for formal definitions and details about security.) Ideally, we would like the encryption time t^{Enc} to depend (polynomially) only on the message size $|m|$ (irrespective of the complexity of the functions f computed during decryption), and the scheme to support an arbitrary polynomial number q of functional decryption keys sk_f. Schemes satisfying these two properties are usually called *compact* (when t^{Enc} is independent of the size $|f|$ of the circuit computing the function) and *collusion-resistant* (when q can be an arbitrary polynomial).

The class of functions f supported by the scheme is also an important parameter, but for simplicity here we will focus on schemes for which f can be any polynomial sized circuit. Interestingly, [25] gives a functional encryption scheme (based on standard lattice assumptions) which supports arbitrary functions f. However, the scheme allows to release only $q = 1$ decryption key (i.e., it is not collusion resistant) and the complexity of encryption depends polynomially on the output size and circuit depth of f (i.e., the scheme is not compact.) It is easy to see that any number q of functional decryption keys can always be supported simply by picking q independent public keys. But this makes the complexity of encryption grow linearly with q. So, technically, the constraint that a scheme is collusion-resistant can be reformulated by requiring that the complexity of encryption t^{Enc} is independent of q. One can also consider weaker versions of both compactness and collusion resistance where the complexity of encryption t^{Enc} is required to be just sublinear in $|f|$ or q.

Using this terminology, the main result of [3,9] states that any (weakly) compact (but not necessarily collusion-resistant) functional encryption scheme can be used to build an $i\mathcal{O}$ obfuscator.[1] In an effort to further reduce (or better understand) the minimal assumptions on functional encryption required to imply obfuscation, the full version of [9] also gives a polynomial reduction from weakly compact functional encryption to (non-compact) weakly collusion-resistant functional encryption. A similar polynomial reduction from compact functional encryption to (non-compact) collusion-resistant functional encryption is also given in [4], where it is suggested that non-compact functional encryption may be easier to achieve, and the reduction is presented as a further step toward basing obfuscation on standard assumptions. In summary, the relation between these four variants of functional encryption (all known to imply $i\mathcal{O}$ by the results of [3,9]) is summarized by the solid arrows in the following diagram:

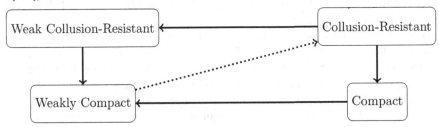

where the horizontal implications are trivial (from stronger to weaker constraints on the t^{Enc}) and the vertical implications are from [3,9].

1.1   Our Results and Techniques

In this paper we further investigate the relation between these four variants of functional encryption, and prove (among other things) the following result:

Theorem 1 (Informal). *There is a polynomial time reduction from collusion-resistant functional encryption to weakly compact functional encryption.*

This adds precisely the (dotted) diagonal arrow to the previous diagram, showing (by transitivity) that all four variants are equivalent under polynomial time reductions. Technically, proving the above theorem requires showing that any single key ($q = 1$) functional encryption scheme satisfying some weak compactness requirement can be turned into a scheme supporting an arbitrary large polynomial number Q of functional key queries. We do so in a modular way, analyzing two general constructions that can be used to combine two arbitrary functional encryption schemes, which we call the SUM construction and the PRODUCT construction.

– The SUM construction takes two functional encryption schemes FE_1, FE_2 supporting q_1 and q_2 functional key queries, and combines them into a new scheme $FE_1 + FE_2$ supporting $q_1 + q_2$ key queries.

---

[1] The reduction incurs a loss in security that is exponential in the input size, which can be accounted for by assuming the functional encryption scheme is exponentially hard to break.

- The PRODUCT construction takes two functional encryption schemes FE_1, FE_2 supporting q_1 and q_2 functional key queries, and combines them into a new scheme $FE_1 \times FE_2$ supporting $q_1 \cdot q_2$ key queries.

The two constructions can be recursively combined in a number of different ways, exhibiting various efficiency/security tradeoffs. For example, Theorem 1 corresponds to starting from a scheme FE_1 supporting a single key ($q_1 = 1$), using the SUM construction $FE_2 = (FE_1 + FE_1)$ to support $q_2 = 2$ keys, and then iterating the PRODUCT construction ($FE_2 \times \cdots \times FE_2$) precisely $\log(Q)$ times, where Q is the desired number of key queries in the final scheme. (Here for simplicity Q is chosen in advance, but our operations are flexible enough to design a scheme where Q is chosen dynamically by the adversary, and the public key does not depend on Q.)

Another possible instantiation is given by repeatedly squaring the scheme FE_2, i.e., defining $FE_4 = FE_2 \times FE_2$, $FE_{16} = FE_4 \times FE_4$, etc. The squaring operation is repeated $\log(\log((Q))$ times, to yield a scheme supporting Q queries. (Again, we are assuming Q is fixed in advance for simplicity, and our results are easily extended to dynamically chosen Q.) Interestingly (and perhaps surprisingly) this produces a different scheme than the iterated product described before, offering different trade-offs. Specifically, the iterated squaring scheme is no longer compact, and the complexity of encryption now depends on Q. However, the dependence is pretty mild, just double-logarithmic $\log(\log(Q))$, as opposed to linear $O(Q)$ as in the trivial construction. This mild dependence on Q results in different ciphertext lengths: while the iterated product construction produces a ciphertext of length twice as long as that of the underlying single-key FE scheme, the iterated squaring construction produces a ciphertext that is a single encryption of a slightly longer message using the underlying FE scheme.

The methods used by the SUM and PRODUCT constructions are relatively standard: the SUM construction is essentially a formalization and generalization of the trivial "repetition" construction to turn a single-key scheme into one supporting q key queries by picking q public keys. The PRODUCT construction is based on the same type of "chaining" techniques used in many bootstrapping theorems before this work. The main technical novelty of this work is the general modular framework to combine the operations, and the detailed analysis of the efficiency and security of the SUM and PRODUCT construction. We remark that, even for the trivial construction, a detailed analysis is needed in order to evaluate the parameters growth when the constructions are applied iteratively an arbitrary (non-constant) number of times. The details of our SUM and PRODUCT constructions are also particularly simple: both constructions combine the component FE schemes making a simple use of just a length doubling pseudorandom generator. Similar constructions in the literature typically make use of more complex building blocks, like puncturable pseudorandom function. We consider the simplicity of the constructions in this work as a positive feature.

1.2   Other Related Work

In a concurrent and independent work, Garg and Srinivasan [22] present a generic transformation from a polynomial-hard single-key weakly-compact FE scheme to a collusion-resistant compact FE scheme. While their work and ours have similar complexity theoretic implications about the existence of functional encryption schemes, they are quite different at the technical level, and produce very different constructions of (collusion resistant) functional encryption, which may be of independent interest. At a high level, the construction of [22] is based on the use of single-key FE schemes to simulate the role of the *obfuscation* in the *iO*-to-FE transformations of [3,9], together with a *prefix puncturable pseudorandom function*. Our constructions are much more direct, and make use of just a simple *pseudorandom generator*. An easy way to see how this leads to very different schemes is that our FE public keys are just the same as the public keys of the underlying (single-key) FE scheme, while the *public keys* of [22] consist of polynomially many *functinal decryption keys* from the underlying (single-key) FE scheme. So, one advantage of our work over [22] is simplicity and, potentially, efficiency.[2] On the other hand the construction of [22] produces a *compact* FE scheme even while starting from a *weakly compact* one, while our construction only preserves the compactness properties, from weakly compact to weakly compact or from fully compact to fully compact.

Our definition of a SUM and PRODUCT construction, and their combined use to build different schemes exhibiting a variety of efficiency/security tradeoffs is somehow similar to the work [28], where sum and product constructions are used to build forward secure signature schemes supporting an arbitrary number of updates, starting from regular signatures (i.e., supporting no updates) and hash functions. However, beside this high level similarity, we deal with completely different cryptographic primitives. The chaining technique used in our product construction has been used many times before in previous bootstrapping theorems for functional encryption, but it is most closely related to the work of [13] where chaining is used in a tree fashion to achieve a hierarchical functional encryption scheme. Our composition approach can be easily adapted to that setting to make the construction and analysis of [13] more modular.

2   Background

We first set up the notation and terminology used in our work.

2.1   Functional Encryption

For notational simplicity, we assume that all randomized algorithms (e.g., key generation and encryption procedure of a cryptosystem) all use precisely κ bits

---

[2] Since the single-key FE schemes underlying [22] and our work lack a known instantiation from standard assumptions, it is hard to make a concrete efficiency comparison between the two schemes. However, based on the asymptotics of the two constructions, one can easily estimate the public keys of [22] to be larger than our public keys at least by a factor κ, proportional to the security parameter.

of randomness, where κ is the security parameter. This is without loss of generality, as κ bits of randomness can be used to generate polynomially many pseudorandom bits using a pseudorandom generator.

We consider only public key functional encryption schemes in our work, so from now on we omit "public key" and just say functional encryption. We use the following syntax for functional encryption schemes, where $R = \{0,1\}^\kappa$.

Definition 1. *A Functional Encryption scheme is specified by four sets* M, R, I, F *(the* message, randomness, index *and* function *spaces) and four algorithms* (Setup, Enc, Dec, KeyGen) *where*

- Setup(sk) = pk *is a public key generation algorithm that on input a random secret key* $sk \in R$, *produces a corresponding public key pk.*
- Enc($pk, m; r$) = c *is an encryption algorithm that on input a public key pk, message* $m \in M$ *and randomness* $r \in R$, *produces a ciphertext c.*
- KeyGen(sk, f, i) = fk *is a functional key derivation algorithm that on input a secret key sk, a function* $f \in F$, *and an index* $i \in I$, *produces a functional decryption key fk associated to* f.
- Dec(fk, c) = m' *is a decryption algorithm that on input a functional decryption key fk and ciphertext c, outputs a plaintext message* m'.

The scheme is correct if with overwhelming probability (over the choice of $sk, r \in R$*), for any message* $m \in M$, *function* $f \in F$ *and index* $i \in I$, *it holds that*

$$\mathsf{Dec}(\mathsf{KeyGen}(sk, f, i), \mathsf{Enc}(\mathsf{Setup}(sk), m; r)) = f(m).$$

Our syntax for functional encryption schemes slightly differs from the standard one in two respects. First, we identify the randomness used by the public key generation procedure Setup with the master secret key of the scheme. This is without loss of generality, but provides a more convenient syntax for our constructions. The other is that the functional key derivation algorithm KeyGen takes an index i as an additional parameter. The only requirement on this index is that different calls to KeyGen($sk, \cdot, i$) use different values of i. The role of i is simply to put a bound on the number of calls to the functional key derivation algorithm. (In particular, the indexes $i \in I$ can be used in any order, and the key derivation algorithm does not need to keep any state.) For example, a functional encryption scheme supporting the release of a single functional key fk will have an index space $I = \{1\}$ of size 1.

Remark 1. We remark that the standard definition of *bounded collusion* resistant functional encryption typically allows an arbitrary number of calls to KeyGen, and imposes a bound only on the number of functional decryption keys that are released to the adversary. This always requires the set I to have exponential (or, at least, superpolynomial) size in the security parameter. So, our definition of I-bounded FE is somehow more restrictive than $|I|$-bounded collusion resistance. When the set I has superpolynomial size (e.g., as obtained in the main results of this paper,) it is easy to make the KeyGen completely stateless and match the standard FE security definition with only a negligible loss in security, e.g., by letting KeyGen pick $i \in I$ at random, or setting $i = H(f)$ for some collision resistant hash function $H: F \to I$.

Security. Since our work is primarily motivated by the application of FE to indistinguishability obfuscation [3,9], we will use an indistinguishability security definition for FE, which is the most relevant one in this context. We follow the indistinguishability security notions as defined in [10], expressed in the functional/equational style of [27,29]. Security for functional encryption is defined by a game between a challenger and an adversary. Both the challenger and the adversary are reactive programs, modeled by monotone functions: the challenger is a function $\mathcal{H}^{\mathsf{FE}}((m_0, m_1), \{f_i\}_{i \in I}) = (pk, c, \{fk^i\}_{i \in I})$ that receives as input a pair of message $(m_0, m_1) \in M^2$ and collection of function queries $f_i \in F$, and outputs a public key pk, ciphertext c and collection of functional keys fk^i for $i \in I$. The adversary is a function $\mathcal{A}(pk, c, \{fk^i\}_{i \in I}) = ((m_0, m_1), \{f_i\}_{i \in I}, b')$ that on input a public key pk, ciphertext c and functional keys $\{fk^i\}_{i \in I}$ outputs a pair of messages (m_0, m_1), function queries $\{f_i\}_{i \in I}$ and decision bit b'. We recall that, as reactive programs, $\mathcal{H}$ and $\mathcal{A}$ can produce some outputs before receiving all the inputs. (Formally, each of the input or output variable can take a special *undefined* value $\bot$, subject to the natural monotonicity requirements. See [29] for details.)

Security for an FE scheme FE is defined using the following challenger $\mathcal{H}_b^{\mathsf{FE}}$, parameterized by a bit $b \in \{0, 1\}$:

$$\mathcal{H}_b^{\mathsf{FE}}((m_0, m_1), \{f_i\}_{i \in I}) = (pk, c, \{fk^i\}_{i \in I})$$
$$\text{where } sk \leftarrow R, \, r \leftarrow R$$
$$pk = \mathsf{Setup}(sk)$$
$$c = \mathsf{Enc}(pk, m_b; r)$$
$$\text{For all } i \in I:$$
$$fk^i = \mathbf{if} \ (f_i(m_0) = f_i(m_1) \neq \bot) \ \mathbf{then} \ \mathsf{KeyGen}(sk, f_i, i) \ \mathbf{else} \ \bot$$

By the notation $x \leftarrow R$ we mean the operation of selecting an element uniformly at random from R. Note that, if $f_i = \bot$ or $m_j = \bot$, then $f_i(m_j) = \bot$. So, this challenger corresponds to a *non-adaptive* security definition where the adversary cannot get any functional key before choosing the challenge messages (m_0, m_1). On the other hand, the public key pk is computed (and given to the adversary) right away, so the (distribution of the) messages (m_0, m_1) may depend on the value of the public key. Alternative definitions can be obtained by setting

- $pk = \mathbf{if} \ ((m_0, m_1) \neq \bot) \ \mathbf{then} \ \mathsf{Setup}(sk) \ \mathbf{else} \ \bot$, which corresponds to the *selective* (i.e., fully non-adaptive) attack where the adversary has to choose the messages before seeing the public key.
- $fk^i = \mathsf{KeyGen}(sk, f_i, i)$ and $c = \mathbf{if} \ (\forall i.f_i(m_0) = f_i(m_1)) \ \mathbf{then} \ \mathsf{Enc}(pk, m_b; r)$ $\mathbf{else} \ \bot$, which corresponds to allowing function queries (only) before choosing the messages (m_0, m_1).

All our results and constructions are easily adapted to all these different definitional variants, as well as *fully adaptive* settings where message and function queries can be specified in any order, subject to the natural non-triviality requirements.

A FE game $\mathsf{Exp}_{FE}[\mathcal{H}^{\mathsf{FE}}_{(\cdot)}, \mathcal{A}]$ is defined by the following system of equations:

$$\mathsf{Exp}_{FE}[\mathcal{H}^{\mathsf{FE}}_{(\cdot)}, \mathcal{A}] = (b \overset{?}{=} b')$$
$$\text{where } b \leftarrow \{0, 1\}$$
$$(pk, c, \{fk^i\}_{i \in I}) = \mathcal{H}^{\mathsf{FE}}_b((m_0, m_1), \{f_i\}_{i \in I})$$
$$((m_0, m_1), \{f_i\}_{i \in I}, b') = \mathcal{A}(pk, c, \{fk^i\}_{i \in I})$$

The output of the game can be obtained by finding the least fixed point of $[\mathcal{H}^{\mathsf{FE}}_b, \mathcal{A}]$, which describes the output when the computation stabilizes. We say that the adversary $\mathcal{A}$ wins the game $\mathsf{Exp}_{FE}[\mathcal{H}^{\mathsf{FE}}_{(\cdot)}, \mathcal{A}]$ if the game outputs $\top$, and we define the advantage of $\mathcal{A}$ in breaking the FE scheme FE as

$$\mathsf{Adv}_{\mathsf{FE}}[\mathcal{A}] = \left| 2 \Pr\{\mathsf{Exp}_{FE}[\mathcal{H}^{\mathsf{FE}}_{(\cdot)}, \mathcal{A}] = \top\} - 1 \right|.$$

Alternatively, we can let the FE game be parameterized by b and output a bit b':

$$[\mathcal{H}^{\mathsf{FE}}_{(b)}, \mathcal{A}] = b'$$
$$\text{where } (pk, c, \{fk^i\}_{i \in I}) = \mathcal{H}^{\mathsf{FE}}_b((m_0, m_1), \{f_i\}_{i \in I}, b')$$
$$((m_0, m_1), \{f_i\}_{i \in I}, b') = \mathcal{A}(pk, c, \{fk^i\}_{i \in I})$$

Then the advantage of $\mathcal{A}$ in breaking the FE scheme FE can be defined as

$$\mathsf{Adv}_{\mathsf{FE}}[\mathcal{A}] = \left| \Pr\{[\mathcal{H}^{\mathsf{FE}}_0, \mathcal{A}] = 1\} - \Pr\{[\mathcal{H}^{\mathsf{FE}}_1, \mathcal{A}] = 1\} \right|.$$

The two formulations are easily seen to be perfectly equivalent.

Definition 2. *A functional encryption scheme* FE *is* (q, ϵ)-*non-adaptively (or selectively/adaptively) secure if* $|I| = q$ *and for any efficient adversary* $\mathcal{A}$ *there exists a negligible function* $\delta(\kappa)$ *such that* $\delta(\kappa) < \epsilon(\kappa)^{\Omega(1)}$ *and the advantage of* $\mathcal{A}$ *in the non-adaptive (or selective/adaptive) FE game is bounded by* $\mathsf{Adv}_{\mathsf{FE}}[\mathcal{A}] \leq \delta(\kappa)$.

When $\epsilon(\kappa)$ is negligible, for simplicity we sometimes omit it and just say a FE scheme is q-secure, where $q = |I|$ as in the definition above.

Efficiency. For a FE scheme to be useful in the real world applications or in building other cryptographic constructs, we need to measure its efficiency. Several notions have been considered in the literature, and here we mention those that are used in our work. Let FE be a FE scheme with security parameter κ, and let n be the length of messages to be encrypted. Then we say

- FE is *compact*[3] if the running time t^{Enc} of the encryption procedure Enc is polynomial in n and κ, and it is independent of other parameters.
- FE is *weakly compact*[4] if t^{Enc} is sub-linear in $|I|$ and the maximal circuit size s of functions in F, and it is polynomial in n and κ.

---

[3] Also known as *fully (circuit) succinct* in [9].
[4] Also known as *weakly (circuit) succinct* in [9].

- FE is *ciphertext-succinct* or simply *succinct* if t^{Enc} is polynomials in n, κ, and the maximal circuit depth d of functions in F.
- FE is *weakly ciphertext-succinct* or simply *weakly succinct* if t^{Enc} is sub-linear in $|I|$ but is polynomials in n, κ, and d.

The notion of compact FE has been considered in [3,4], and also in [9] under the name *fully circuit succinct*. Here we choose the name "compact" to distinguish other variants of succinctness notions. It was shown in [3,9] that a 1-secure compact FE with sub-exponential security for all circuits implies an indistinguishability obfuscation for all circuits.

Succinct FE scheme, a weaker notion, was considered in [25], where their definition was based on ciphertext length. They constructed a succinct FE scheme based on standard sub-exponential lattice assumptions. We note that, although our definition is stronger due to using the complexity of encryption, the [25] FE scheme is still ciphertext-succinct with our definition.

Furthermore, one may naturally require a FE scheme to be secure even when a large number of functional keys are released. We say a FE scheme is *collusion-resistant* if it is secure when $|I|$ is any polynomial in κ. When we also allow sub-linear dependence on $|I|$, the FE scheme is called *weakly collusion-resistant*.

2.2   Pseudorandom Generators

Our construction assumes the existence of pseudorandom generators that can stretch a short random seed to a polynomially long pseudorandom bit-string. In the following we give its definition and some conventions in using it.

Definition 3. *Let* $\mathsf{G} : R \to S$ *be a deterministic function that can be computed in polynomial time. We say that* G *is a* $\mu(\kappa)$*-secure pseudorandom generator of stretch* $\ell(\kappa)$ *if for all* $x \in R$ *we have* $|\mathsf{G}(x)| = \ell(|x|)$, *where* $\ell(\kappa)$ *is a polynominal in* κ, *and for any efficient adversary* $\mathcal{A}$ *we have*

$$\mathsf{Adv}_{\mathsf{G}}[\mathcal{A}] = \left| \Pr_{s \leftarrow S}\{\mathcal{A}(s) = 1\} - \Pr_{r \leftarrow R}\{\mathcal{A}(\mathsf{G}(r)) = 1\} \right| \leq \mu(\kappa).$$

The quantity $\mathsf{Adv}_{\mathsf{G}}[\mathcal{A}]$ *is the advantage of* $\mathcal{A}$ *in breaking the PRG* G.

We write $\mathsf{G}(r)$ to denote the output of a pseudorandom generator on input a (randomly chosen) seed r, with the domain and range of G usually defined implicitly by the context. We write $\mathsf{G}_i(r)$ to denote a specific part of the output, i.e., $\mathsf{G}(r) = \mathsf{G}_0(r)\mathsf{G}_1(r)\ldots\mathsf{G}_k(r)$, where the blocks $\mathsf{G}_i(r)$ usually have all the same length. The assumption is that $\mathsf{G}(r)$ is computationally indistinguishable from a random string of length $|\mathsf{G}(r)|$, i.e., G is μ-secure for some negligible function $\mu(\kappa)$.

3   The SUM Construction

We describe a simple method to combine two functional encryption schemes FE_0 and FE_1 with index spaces I_0 and I_1, into a new scheme $\mathsf{FE} = \mathsf{FE}_0 + \mathsf{FE}_1$ with index space $I = I_0 + I_1 = \{(b,i) \mid b \in \{0,1\}, i \in I_b\}$ given by the disjoint union of I_0 and I_1. Let $\mathsf{FE}_b = (\mathsf{Setup}_b, \mathsf{Enc}_b, \mathsf{Dec}_b, \mathsf{KeyGen}_b)$ for $b \in \{0,1\}$. Then, $\mathsf{FE} = (\mathsf{Setup}, \mathsf{Enc}, \mathsf{Dec}, \mathsf{KeyGen})$ is defined as

- $\mathsf{Setup}(sk) = (\mathsf{Setup}_0(\mathsf{G}_0(sk)), \mathsf{Setup}_1(\mathsf{G}_1(sk)))$
- $\mathsf{Enc}((pk_0, pk_1), m; r) = (\mathsf{Enc}_0(pk_0, m; \mathsf{G}_0(r)), \mathsf{Enc}_1(pk_1, m; \mathsf{G}_1(r)))$
- $\mathsf{Dec}((b, fk), (c_0, c_1)) = \mathsf{Dec}_b(fk, c_b)$
- $\mathsf{KeyGen}(sk, f, (b,i)) = (b, \mathsf{KeyGen}_b(\mathsf{G}_b(sk), f, i))$

for all $sk, r \in R$, $m \in M$, $b \in \{0,1\}$ and $i \in I_b$. Informally, the *SUM* scheme works by generating two public keys (one for each component scheme FE_b), and encrypting each message under both public keys. When applied to two copies of the same scheme $\mathsf{FE}_0 = \mathsf{FE}_1$, this doubles the size of the index space $|I| = 2|I_b|$ (allowing twice as many functional decryption keys,) but at the cost of doubling also the public key and ciphertext size. The complexity of decryption and functional key generation stays essentially the same as that of the component schemes (no doubling, only a small additive increase for multiplexing), as only one of the two ciphertexts gets decrypted.

The correctness of the scheme is easily verified by substitution. Security (proved in the next theorem) is not entirely trivial, as it requires a careful use of the pseudorandom generator, but it still follows by a fairly standard hybrid argument. The construction preserves the non-adaptive/selective/adaptive security properties. We prove the non-adaptive version, which can be easily adapted to the other models.

Theorem 2 (SUM construction). *If FE_i for $i \in \{0,1\}$ is a succinct (q_i, ϵ_i)-non-adaptively secure FE scheme for functions in the class F, with public key size ℓ_i^k and ciphertext length ℓ_i^c, and if G is a μ-secure pseudorandom generator, then $\mathsf{FE} = \mathsf{FE}_0 + \mathsf{FE}_1$ is a succinct $(q_0 + q_1, \epsilon_0 + \epsilon_1 + 4\mu)$-non-adaptively secure FE scheme for F with public-key size $\ell_0^k + \ell_1^k$ and ciphertext length $\ell_0^c + \ell_1^c$.*

Moreover, if the algorithms $\mathsf{Setup}_i, \mathsf{Dec}_i, \mathsf{KeyGen}_i$ and Enc_i of FE_i run in time $t_i^{\mathsf{Setup}}, t_i^{\mathsf{Dec}}, t_i^{\mathsf{KeyGen}}$ and $t_i^{\mathsf{Enc}}(n, \kappa, d_i)$, respectively, where d_i is the maximum depth of functions in F, and if G runs in time t^{G}, then the running times of the algorithms in $\mathsf{FE} = \mathsf{FE}_0 + \mathsf{FE}_1$ are:

- *Setup: $t_0^{\mathsf{Setup}} + t_1^{\mathsf{Setup}} + t^{\mathsf{G}}$*
- *Enc : $t_0^{\mathsf{Enc}} + t_1^{\mathsf{Enc}} + t^{\mathsf{G}}$*
- *Dec : $\max\{t_0^{\mathsf{Dec}}, t_1^{\mathsf{Dec}}\}$*
- *KeyGen : $\max\{t_0^{\mathsf{KeyGen}}, t_1^{\mathsf{KeyGen}}\} + t^{\mathsf{G}}$*

Proof. We build 6 hybrids to reduce the security of the SUM construction $\mathsf{FE}_0 + \mathsf{FE}_1$ to the security of the PRG G and the security of the FE schemes FE_0 and FE_1. We denote a hybrid by $\mathcal{H}_b^{(j)}$ for $b \in \{0,1\}$ and an index j. Like the challenger in a FE game, a hybrid is a monotone function $\mathcal{H}_b^{(j)}((m_0, m_1), \{f_{(h,i)}\}_{(h,i) \in I}) = (pk, c, \{fk^{(h,i)}\}_{(h,i) \in I})$, where $I = I_0 + I_1$. Proofs of lemmas can be found in Appendix A.

$\mathcal{H}_b^{(0)}$: This hybrid is the same as the original challenger $\mathcal{H}_b^{\mathsf{FE}}$ in the FE game for the FE scheme $\mathsf{FE}_0 + \mathsf{FE}_1$. For a fixed $b \in \{0,1\}$, by expanding the SUM construction, we get the following definition of $\mathcal{H}_b^{(0)}$:

$\mathcal{H}_b^{(0)}((m_0, m_1), \{f_{(h,i)}\}) = (pk, c, \{fk^{(h,i)}\})$
    where
        $sk \leftarrow R,\ r \leftarrow R$
        $sk_0 = \mathsf{G}_0(sk),\ sk_1 = \mathsf{G}_1(sk)$
        $pk_0 = \mathsf{Setup}_0(sk_0),\ pk_1 = \mathsf{Setup}_1(sk_1),\ pk = (pk_0, pk_1)$
        $c_0 = \mathsf{Enc}_0(pk_0, m_b; \mathsf{G}_0(r)),\ c_1 = \mathsf{Enc}_1(pk_1, m_b; \mathsf{G}_1(r)),\ c = (c_0, c_1)$
        For all $(h, i) \in I_0 + I_1$:
            $fk^{(h,i)} = $ **if** $(f_{(h,i)}(m_0) = f_{(h,i)}(m_1) \neq \bot)$ **then** $(h, \mathsf{KeyGen}_h(sk_h, f; i))$

$\mathcal{H}_b^{(1)}$: In this hybrid we replace the PRG outputs by truly random strings. So sk and r are no longer needed and hence we remove them from the hybrid.

$\mathcal{H}_b^{(1)}((m_0, m_1), \{f_{(h,i)}\}) = (pk, c, \{fk^{(h,i)}\})$
    where
        $sk_0 \leftarrow R,\ sk_1 \leftarrow R,\ r_0, r_1 \leftarrow R$
        $pk_0 = \mathsf{Setup}_0(sk_0),\ pk_1 = \mathsf{Setup}_1(sk_1),\ pk = (pk_0, pk_1)$
        $c_0 = \mathsf{Enc}_0(pk_0, m_b; r_0),\ c_1 = \mathsf{Enc}_1(pk_1, m_b; r_1),\ c = (c_0, c_1)$
        For all $(h, i) \in I_0 + I_1$:
            $fk^{(h,i)} = $ **if** $(f_{(h,i)}(m_0) = f_{(h,i)}(m_1) \neq \bot)$ **then** $(h, \mathsf{KeyGen}_h(sk_h, f; i))$

Lemma 1. *If* G *is a* μ-*secure pseudorandom generator, then for any* $b \in \{0,1\}$ *and adversary* $\mathcal{A}$ *we have* $|\Pr\{[\mathcal{H}_b^{(0)}, \mathcal{A}] = 1\} - \Pr\{[\mathcal{H}_b^{(1)}, \mathcal{A}] = 1\}| \leq 2\mu(\kappa)$.

$\mathcal{H}_b^{(2)}$: In this hybrid the ciphertext c encrypts both m_0 and m_1:

$\mathcal{H}_b^{(2)}((m_0, m_1), \{f_{(h,i)}\}) = (pk, c, \{fk^{(h,i)}\})$
    where
        $sk_0 \leftarrow R,\ sk_1 \leftarrow R,\ r_0, r_1 \leftarrow R,$
        $pk_0 = \mathsf{Setup}_0(sk_0),\ pk_1 = \mathsf{Setup}_1(sk_1),\ pk = (pk_0, pk_1)$
        $c_0 = \mathsf{Enc}_0(pk_0, m_0; r_0),\ c_1 = \mathsf{Enc}_1(pk_1, m_1; r_1),\ c = (c_0, c_1)$
        For all $(h, i) \in I_0 + I_1$:
            $fk^{(h,i)} = $ **if** $(f_{(h,i)}(m_0) = f_{(h,i)}(m_1) \neq \bot)$ **then** $(h, \mathsf{KeyGen}_h(sk_h, f; i))$

Lemma 2. *If* FE_1 *is a* (q_1, ϵ_1)-*non-adaptively secure FE scheme, then for any adversary* $\mathcal{A}$ *we have* $|\Pr\{[\mathcal{H}_0^{(1)}, \mathcal{A}] = 1\} - \Pr\{[\mathcal{H}_0^{(2)}, \mathcal{A}] = 1\}| \leq \epsilon_1(\kappa)$.

By symmetric argument, we can also obtain the following lemma.

Lemma 3. *If* FE_0 *is a* (q_0, ϵ_0)-*non-adaptively secure FE scheme, then for any adversary* $\mathcal{A}$ *we have* $|\Pr\{[\mathcal{H}_1^{(1)}, \mathcal{A}] = 1\} - \Pr\{[\mathcal{H}_1^{(2)}, \mathcal{A}] = 1\}| \leq \epsilon_0(\kappa)$.

Finally, we observe that the last hybrid $\mathcal{H}_b^{(2)}$ does not depend on the bit b, and therefore $\Pr\{[\mathcal{H}_0^{(2)}, \mathcal{A}] = 1\} = \Pr\{[\mathcal{H}_1^{(2)}, \mathcal{A}] = 1\}$. It follows by triangle inequality that the advantage of adversary $\mathcal{A}$ in breaking the SUM FE scheme is at most $\mathsf{Adv}_{\mathsf{FE}}[\mathcal{A}] = |\Pr\{[\mathcal{H}_0^{(0)}, \mathcal{A}] = 1\} - \Pr\{[\mathcal{H}_1^{(0)}, \mathcal{A}] = 1\}| \le 2\mu + \epsilon_1 + 0 + \epsilon_0 + 2\mu = 4\mu + \epsilon_0 + \epsilon_1$. $\qquad\square$

4   The PRODUCT Construction

We now define a different method to combine FE_0 and FE_1 into a new scheme $\mathsf{FE} = \mathsf{FE}_0 \times \mathsf{FE}_1$ with index space $I_0 \times I_1$ equal to the cartesian product of the index spaces I_0, I_1 of FE_0 and FE_1. Let $\mathsf{FE}_b = (\mathsf{Setup}_b, \mathsf{Enc}_b, \mathsf{Dec}_b, \mathsf{KeyGen}_b)$ for $b \in \{0, 1\}$. First, for each $i \in I_0$, we define a "re-encryption" function $e_i[c, pk]$: $M \times R \to M$, parameterized by $c \in M$ and $pk \in K$:

$$e_i[c, pk](m, \tilde{r}) = \begin{cases} \mathsf{G}_i(\tilde{r}) \oplus c & \text{if } m = \perp \\ \mathsf{Enc}_1(pk, m; \mathsf{G}_i(\tilde{r})) & \text{otherwise} \end{cases}$$

Then, $\mathsf{FE} = (\mathsf{Setup}, \mathsf{Enc}, \mathsf{Dec}, \mathsf{KeyGen})$ is defined as follows:

- $\mathsf{Setup}(sk) = \mathsf{Setup}_0(\mathsf{G}_0(sk))$
- $\mathsf{Enc}(pk, m; r) = \mathsf{Enc}_0(pk, (m, \mathsf{G}_0(r)); \mathsf{G}_1(r))$
- $\mathsf{Dec}((fk_0, fk_1), c) = \mathsf{Dec}_1(fk_1, \mathsf{Dec}_0(fk_0, c))$
- $\mathsf{KeyGen}(sk, f, (i, j)) = (fk_0^i, fk_1^{i,j})$ where

$$sk_0 = \mathsf{G}_0(sk)$$
$$sk_1^i = \mathsf{G}_i(\mathsf{G}_1(sk))$$
$$pk_1^i = \mathsf{Setup}_1(sk_1^i)$$
$$c_i = \mathsf{G}_i(\mathsf{G}_2(sk))$$
$$fk_0^i = \mathsf{KeyGen}_0(sk_0, e_i[c_i, pk_1^i], i)$$
$$fk_1^{i,j} = \mathsf{KeyGen}_1(sk_1^i, f, j)$$

The re-encryption function can work in two modes: in the regular mode where a message m is given, it computes the FE_1 ciphertext of m under a hard-wired public key pk with pseudo-randomness supplied by a random seed from input; in the special mode where m is not given (denoted by the special symbol $\perp$), it pads a hard-wired ciphertext c with pseudo-randomness derived from the random seed from input. Note that the special mode is never invoked in a real world execution of the scheme, but it is only used in security proofs.

Let $\mathbb{RE}_{\mathsf{FE}}$ be the class of functions that include $e_i[c_i, pk_1^i](\cdot, \cdot)$ defined using Enc of the FE scheme FE. Then we state the security of our PRODUCT construction as follows. Again, the analysis can be easily adapted to other (e.g., selective/adaptive) models.

Theorem 3 (PRODUCT construction). *Assume* FE_0 *and* FE_1 *are succinct public-key FE which are* (q_0, ϵ_0)*- and* (q_1, ϵ_1)*-non-adaptively secure for functions in the classes* $\mathbb{RE}_{\mathsf{FE}_0}$ *and* F *respectively, whose key sizes are* ℓ_0^k *and* ℓ_1^k*, ciphertext lengths* $\ell_0^c(n, \kappa, d_0)$ *and* $\ell_1^c(n, \kappa, d_1)$*, where* n *is the message length and* d_0, d_1 *are the maximum depth of functions in* $\mathbb{RE}_{\mathsf{FE}_0}, F$*, respectively. Also assume* G *is a* μ*-secure pseudorandom generator. Then* $\mathsf{FE}_0 \times \mathsf{FE}_1$ *is a* $(q_0 q_1, q_0 \epsilon_1 + 2\epsilon_0 + 12\mu)$*-non-adaptively secure succinct public-key FE scheme for* F *with public-key sizes* ℓ_0^k *and ciphertext length* $\ell_0^c(n + \kappa, \kappa, d_0)$*.*

Moreover, for $i \in \{0, 1\}$*, let* $t_i^{\mathsf{Setup}}, t_i^{\mathsf{Enc}}, t_i^{\mathsf{Dec}}, t_i^{\mathsf{KeyGen}}$ *be the running times of algorithms* $\mathsf{Setup}_i, \mathsf{Enc}_i, \mathsf{Dec}_i, \mathsf{KeyGen}_i$ *of* FE_i*, where* $t_i^{\mathsf{Enc}} = t_i^{\mathsf{Enc}}(n, \kappa, d_i)$*, and let* t^{G} *be the running time of* G*. Then the running times of* FE *are:*

- Setup: $t_0^{\mathsf{Setup}} + t^{\mathsf{G}}$
- Enc : $t_1^{\mathsf{Enc}}(n + \kappa, \kappa, d_0) + t^{\mathsf{G}}$
- Dec : $t_0^{\mathsf{Dec}} + t_1^{\mathsf{Dec}}$
- KeyGen : $t_1^{\mathsf{Setup}} + t_0^{\mathsf{KeyGen}} + t_1^{\mathsf{KeyGen}} + 3t^{\mathsf{G}}$

Proof. We build a series of hybrids to reduce the security of $\mathsf{FE}_0 \times \mathsf{FE}_1$ to the security of the PRG and the security of FE schemes FE_0 and FE_1. We denote our hybrids by $\mathcal{H}_b^{(h)}$ for $b \in \{0, 1\}$ and h an index. Let $I = I_0 \times I_1$. A hybrid is a monotone function $\mathcal{H}_b^{(h)}((m_0, m_1), \{f_i\}_{i \in I}) = (pk, c, \{fk^i\}_{i \in I})$. An adversary $\mathcal{A}$ wins the game against $\mathcal{H}_b^{(h)}$ if $b' = [\mathcal{H}_b^{(h)}, \mathcal{A}] = 1$, and its advantage over $\mathcal{H}_b^{(h)}$ is $\mathsf{Adv}[\mathcal{A}]_b^{(h)} = \Pr\{[\mathcal{H}_b^{(h)}, \mathcal{A}] = 1\}$. Again, proofs of lemmas can be found in Appendix A.

$\mathcal{H}_b^{(0)}$: This is the same as the original challenger $\mathcal{H}_b^{\mathsf{FE}_0 \times \mathsf{FE}_1}$ in the FE game for the scheme $\mathsf{FE}_0 \times \mathsf{FE}_1$. By expanding the PRODUCT construction, we get the following definition of $\mathcal{H}_b^{(0)}$:

$$\mathcal{H}_b^{(0)}((m_0, m_1), \{f_{(i,j)}\}_{(i,j) \in I}) = (pk, c, \{fk^{(i,j)}\}_{(i,j) \in I})$$
$\quad$ **where**
$\quad\quad sk \leftarrow K, r \leftarrow R$
$\quad\quad sk_0 = \mathsf{G}_0(sk), pk = \mathsf{Setup}_0(sk_0)$
$\quad\quad c = \mathsf{Enc}_0(pk, (m_b, \mathsf{G}_0(r)); \mathsf{G}_1(r))$
$\quad\quad$ For all $i \in I_0, j \in I_1$:
$\quad\quad\quad fk^{i,j} = $ **if** $(f_{i,j}(m_0) = f_{i,j}(m_1) \neq \perp)$ **then** $(fk_0^i, fk_1^{i,j})$
$\quad\quad\quad\quad$ **where** $sk_1^i = \mathsf{G}_i(\mathsf{G}_1(sk)), pk_1^i = \mathsf{Setup}_1(sk_1^i), c_i = \mathsf{G}_i(\mathsf{G}_2(sk))$
$\quad\quad\quad\quad\quad fk_0^i = \mathsf{KeyGen}_0(sk_0, e_i[c_i, pk_1^i], i)$
$\quad\quad\quad\quad\quad fk_1^{i,j} = \mathsf{KeyGen}_1(sk_1^i, f_{i,j}, j)$

$\mathcal{H}_b^{(1)}$: In this hybrid some uses of the PRG G are replaced by truly random strings. In addition, sk is no longer needed so we remove it from the hybrid.

$$\mathcal{H}_b^{(1)}((m_0, m_1), \{f_{i,j}\}_{(i,j) \in I}) = (pk, c, \{fk^{i,j}\}_{(i,j) \in I})$$
 where $sk_0 \leftarrow K$, $r \leftarrow R$
  $pk = \mathsf{Setup}_0(sk_0)$
  $r' \leftarrow K$, $r'' \leftarrow K$, $c = \mathsf{Enc}_0(pk, (m_b, r'); r'')$
  For all $i \in I_0, j \in I_1$:
   $fk^{i,j} = \mathbf{if}\ (f_{i,j}(m_0) = f_{i,j}(m_1) \neq \perp)\ \mathbf{then}\ (fk_0^i, fk_1^{i,j})$
    where $sk_1^i \leftarrow K$, $pk_1^i = \mathsf{Setup}_1(sk_1^i)$, $c_i \leftarrow K$
     $fk_0^i = \mathsf{KeyGen}_0(sk_0, e_i[c_i, pk_1^i], i)$
     $fk_1^{i,j} = \mathsf{KeyGen}_1(sk_1^i, f_{i,j}, j)$

Lemma 4. *If* G *is a* μ-*secure pseudorandom generator, then for any* $b \in \{0, 1\}$ *and any efficient adversary* $\mathcal{A}$, *we have* $|\mathsf{Adv}[\mathcal{A}]_b^{(0)} - \mathsf{Adv}[\mathcal{A}]_b^{(1)}| \leq 4\mu(\kappa)$.

$\mathcal{H}_b^{(2)}$: In this hybrid we slightly modify how c_i is generated without changing its distribution.

$$\mathcal{H}_b^{(1)}((m_0, m_1), \{f_{i,j}\}_{(i,j) \in I}) = (pk, c, \{fk^{i,j}\}_{(i,j) \in I})$$
 where $sk_0 \leftarrow K$, $r \leftarrow R$
  $pk = \mathsf{Setup}_0(sk_0)$
  $r' \leftarrow K$, $r'' \leftarrow K$, $c = \mathsf{Enc}_0(pk, (m_b, r'); r'')$
  For all $i \in I_0, j \in I_1$:
   $fk^{i,j} = \mathbf{if}\ (f_{i,j}(m_0) = f_{i,j}(m_1) \neq \perp)\ \mathbf{then}\ (fk_0^i, fk_1^{i,j})$
    where $sk_1^i \leftarrow K$, $pk_1^i = \mathsf{Setup}_1(sk_1^i)$
     $s_i \leftarrow K$, $\tilde{c}_1^i = \mathsf{Enc}_1(pk_1^i, m_b; \mathsf{G}_i(r'))$, $c_i = s_i \oplus \tilde{c}_1^i$
     $fk_0^i = \mathsf{KeyGen}_0(sk_0, e_i[c_i, pk_1^i], i)$
     $fk_1^{i,j} = \mathsf{KeyGen}_1(sk_1^i, f_{i,j}, j)$

Lemma 5. *For any* $b \in \{0, 1\}$ *and adversary* $\mathcal{A}$, *we have* $\mathsf{Adv}[\mathcal{A}]_b^{(1)} = \mathsf{Adv}[\mathcal{A}]_b^{(2)}$.

$\mathcal{H}_b^{(3)}$: In this hybrid we replace the truly random s_i with a pseudorandom string.

$$\mathcal{H}_b^{(3)}((m_0, m_1), \{f_{i,j}\}_{(i,j) \in I}) = (pk, c, \{fk^{i,j}\}_{(i,j) \in I})$$
 where $sk_0 \leftarrow K$, $r \leftarrow R$, $s \leftarrow K$
  $pk = \mathsf{Setup}_0(sk_0)$
  $r' \leftarrow K$, $r'' \leftarrow K$, $c = \mathsf{Enc}(pk, m_b; r) = \mathsf{Enc}_0(pk, (m_b, r'); r'')$
  For all $i \in I_0, j \in I_1$:
   $fk^{i,j} = \mathbf{if}\ (f_{i,j}(m_0) = f_{i,j}(m_1) \neq \perp)\ \mathbf{then}\ (fk_0^i, fk_1^{i,j})$
    where $sk_1^i \leftarrow K$, $pk_1^i = \mathsf{Setup}_1(sk_1^i)$
     $s_i = \mathsf{G}_i(s)$, $\tilde{c}_1^i = \mathsf{Enc}_1(pk_1^i, m_b; \mathsf{G}_i(r'))$, $c_i = s_i \oplus \tilde{c}_1^i$
     $fk_0^i = \mathsf{KeyGen}_0(sk_0, e_i[c_i, pk_1^i], i)$
     $fk_1^{i,j} = \mathsf{KeyGen}_1(sk_1^i, f_{i,j}, j)$

Lemma 6. *If* G *is a* μ-*secure pseudorandom generator, then for any* $b \in \{0, 1\}$ *and adversary* $\mathcal{A}$, *we have* $|\mathsf{Adv}[\mathcal{A}]_b^{(2)} - \mathsf{Adv}[\mathcal{A}]_b^{(3)}| \leq \mu(\kappa)$.

$\mathcal{H}_b^{(4)}$: In this hybrid we modify c to encrypt $(\bot, s)$ instead of (m_b, r).

$\mathcal{H}_b^{(4)}((m_0, m_1), \{f_{i,j}\}_{(i,j) \in I}) = (pk, c, \{fk^{i,j}\}_{(i,j) \in I})$
    where $sk_0 \leftarrow K$, $r \leftarrow R$, $s \leftarrow K$
        $pk = \mathsf{Setup}_0(sk_0)$
        $r' \leftarrow K$, $r'' \leftarrow K$, $c = \mathsf{Enc}_0(pk, (\bot, s); r'')$
        For all $i \in I_0, j \in I_1$:
          $fk^{i,j} = $ **if** $(f_{i,j}(m_0) = f_{i,j}(m_1) \neq \bot)$ **then** $(fk_0^i, fk_1^{i,j})$
               where $sk_1^i \leftarrow K$, $pk_1^i = \mathsf{Setup}_1(sk_1^i)$
                    $s_i = \mathsf{G}_i(s)$, $\tilde{c}_1^i = \mathsf{Enc}_1(pk_1^i, m_b; \mathsf{G}_i(r'))$, $c_i = s_i \oplus \tilde{c}_1^i$
                    $fk_0^i = \mathsf{KeyGen}_0(sk_0, e_i[c_i, pk_1^i], i)$
                    $fk_1^{i,j} = \mathsf{KeyGen}_1(sk_1^i, f_{i,j}, j)$

Lemma 7. *If* FE_0 *is a* (q_0, ϵ_0)-*non-adaptive secure FE scheme for functions in the class* $\mathbb{RE}_{\mathsf{FE}_0}$, *then for any* $b \in \{0, 1\}$ *and any efficient adversary* $\mathcal{A}$, *we have* $|\mathsf{Adv}[\mathcal{A}]_b^{(3)} - \mathsf{Adv}[\mathcal{A}]_b^{(4)}| \leq \epsilon_0(\kappa)$.

$\mathcal{H}_b^{(5)}$: Now we use fresh randomness to generate $\tilde{c}_i$ instead of sharing a pseudorandom string.

$\mathcal{H}_b^{(5)}((m_0, m_1), \{f_{i,j}\}_{(i,j) \in I}) = (pk, c, \{fk^{i,j}\}_{(i,j) \in I})$
    where
        $sk_0 \leftarrow K$, $r \leftarrow R$, $s \leftarrow K$
        $pk = \mathsf{Setup}_0(sk_0)$
        $r'' \leftarrow K$, $c = \mathsf{Enc}_0(pk, (\bot, s); r'')$
        For all $f_{i,j}$ **where** $i \in I_0, j \in I_1$:
          $fk^{i,j} = $ **if** $(f_{i,j}(m_0) = f_{i,j}(m_1) \neq \bot)$ **then** $(fk_0^i, fk_1^{i,j})$
               where $sk_1^i \leftarrow K$
                    $pk_1^i = \mathsf{Setup}_1(sk_1^i)$
                    $s_i = \mathsf{G}_i(s)$, $r_i \leftarrow K$, $\tilde{c}_1^i = \mathsf{Enc}_1(pk_1^i, m_b; r_i)$, $c_i = s_i \oplus \tilde{c}_1^i$
                    $fk_0^i = \mathsf{KeyGen}_0(sk_0, e_i[c_i, pk_1^i], i)$
                    $fk_1^{i,j} = \mathsf{KeyGen}_1(sk_1^i, f_{i,j}, j)$

Lemma 8. *If* G *is a* μ-*secure pseudorandom generator, then for any* $b \in \{0, 1\}$ *and any adversary* $\mathcal{A}$ *we have* $|\mathsf{Adv}[\mathcal{A}]_b^{(4)} - \mathsf{Adv}[\mathcal{A}]_b^{(5)}| \leq \mu(\kappa)$.

Lemma 9. *If* FE_1 *is a* (q_1, ϵ_1)-*non-adaptive secure FE scheme, then for any efficient adversary* $\mathcal{A}$ *we have* $|\mathsf{Adv}[\mathcal{A}]_0^{(5)} - \mathsf{Adv}[\mathcal{A}]_1^{(5)}| \leq q_0 \cdot \epsilon_1(\kappa)$.

Finally, by applying previous lemmas, we see that the advantage of any adversary $\mathcal{A}$ to the PRODUCT scheme FE can be bounded by $\mathsf{Adv}_{\mathsf{FE}}[\mathcal{A}] = |\Pr\{[\mathcal{H}_0^{(0)}, \mathcal{A}] = 1\} - \Pr\{[\mathcal{H}_1^{(0)}, \mathcal{A}] = 1\}| \leq 2(4\mu + 0 + \mu + \epsilon_0 + \mu) + q_0\epsilon_1 = q_0\epsilon_1 + 2\epsilon_0 + 12\mu$. $\qquad \square$

5   Compositions Using SUM and PRODUCT Constructions

SUM and PRODUCT constructions provide ways to build new FE schemes with larger function spaces. They also have nice efficiency preserving properties. Using them as building blocks, we propose two composition methods to define transformations from a FE scheme supporting only one functional key query to a new FE scheme that supports any polynomially many functional key queries without losing much security and efficiency guarantees.

Throughout this section, we assume FE_0 is a $(1, \epsilon_0)$-secure FE scheme, where $\epsilon_0(\kappa)$ is negligible, for functions in a class F with some minimal efficiency guarantees, for example, succinct. FE_0 can be either selective-, non-adaptive-, or adaptive-secure, and our transformations preserve these security notions. We also assume G is a μ-secure PRG, for negligible $\mu(\kappa)$. Let $t_0^{\mathsf{Setup}}, t_0^{\mathsf{Enc}}, t_0^{\mathsf{Dec}}, t_0^{\mathsf{KeyGen}}$ be the running times of the four algorithms in FE_0, and let ℓ_0^k, ℓ_0^c, ℓ_0^{fk} be the lengths of public key, ciphertext, and functional keys of FE_0. Since FE_0 is succinct, $t_0^{\mathsf{Enc}} = t_0^{\mathsf{Enc}}(n, \kappa, d)$ and $\ell_0^c = \ell_0^c(n, \kappa, d)$ are both polynomials in the message length n, security parameter κ, and the maximal depth d of functions in F. Let t^G be the running time of the PRG G. Our main results are two reductions from collusion-resistant (weakly) compact FE schemes for F to FE_0 assuming F meets some requirements (more details later).

5.1   Iterated Squaring Composition

Our first transformation can be obtained by repeatedly squaring the previously composed FE scheme. At the beginning, we use the SUM construction to obtain FE schemes supporting 2 functional key queries. Then PRODUCT construction is used on the FE schemes of the previous iteration.

Formally, we can define the *iterated squaring composition* method by:

$$FE_1 = FE_0 + FE_0, \text{ and for } p \geq 1, FE_{p+1} = FE_p \times FE_p. \tag{1}$$

So FE_1 supports 2 functional queries, and for $p \geq 1$, the FE scheme FE_{p+1} supports 2^{2^p} functional queries. For any polynomial $Q(\kappa)$, when $p \geq \log \log Q$, the FE scheme FE_{p+1} supports $Q(\kappa)$ functional queries, and its security and performance can be characterized as follows.

Security: The advantage of FE_{p+1} over any efficient adversary $\mathcal{A}$ is

$$\mathsf{Adv}_{FE_{p+1}}[\mathcal{A}] = 2^{2^p} \epsilon_0 + 2^{2^p} \mu = Q \cdot \epsilon_0 + Q \cdot \mu. \tag{2}$$

Running times and output lengths:

- Setup: $2t_0^{\mathsf{Setup}} + (p+1)t^G = 2t_0^{\mathsf{Setup}} + \log \log Q \cdot t^G$
- Enc : $2t_0^{\mathsf{Enc}}(n + p\kappa, \kappa, d) + (p+1)t^G = 2t_0^{\mathsf{Enc}}(n + \kappa \log \log Q, \kappa, d) + \log \log Q \cdot t^G$
- Dec : $2^p t_0^{\mathsf{Dec}} = \log Q \cdot t_0^{\mathsf{Dec}}$

- KeyGen : $2(2^p - 1)t_0^{\mathsf{Setup}} + 2^p t_0^{\mathsf{KeyGen}} + (\sum_{i=0}^{p}(p + 2 - i)2^i + 2^{p+1} - 1)t^{\mathsf{G}} = 2\log Q \cdot t_0^{\mathsf{Setup}} + \log Q \cdot t_0^{\mathsf{KeyGen}} + 6\log Q \cdot t^{\mathsf{G}}$

- $\ell_{p+1}^k = 2\ell_0^k$
- $\ell_{p+1}^c = \ell_0^c(n + p\kappa, \kappa, d) = \ell_0^c(n + \kappa\log\log Q, \kappa, d)$
- $\ell_{p+1}^{fk} = 2^p \ell_0^{fk} = \log Q \cdot \ell_0^{fk}$

Clearly FE_{p+1} is a secure FE scheme, and the transformation incurs only linear (in terms of Q) security loss. Since FE_0 is succinct, t_0^{Enc} is a polynomial in n, κ, and d. So t_{p+1}^{Enc} can be bounded by $\mathrm{poly}(\log Q, n, \kappa, d)$ for some fixed polynomial poly, and hence FE_{p+1} is weakly succinct.

Besides, for the iterated squaring composition to be viable, we must be careful about the function classes supported in each iteration of the composition. Let F_h be the class of functions supported by KeyGen_h of the FE scheme FE_h, for $h \geq 0$. First we have $F_1 = F_0$. In the steps using PRODUCT construction on FE_p to derive FE_{p+1}, a functional key $fk = (fk_0, fk_1)$ for any function f consists of two keys under FE_p: fk_0 is for a "re-encryption" function $e_i^{(p)}[c, pk](\cdot, \cdot)$, and fk_1 is for f. Hence for the composition to go through, FE_p must be capable of generating functional keys for these two classes of functions, namely

$$F_{p+1} \cup \{e_i^{(p)}[c, pk] \mid c \in M, pk \in R\} \subseteq F_p.$$

Recall from Sect. 4 that $\mathbb{RE}_{\mathsf{FE}_p}$ is the class containing $e_i^{(p)}[c, pk]$ for all $c \in M, pk \in R$. Let $\mathbb{RE}_{\mathsf{FE}_0}^p = \cup_{h=1}^p \mathbb{RE}_{\mathsf{FE}_h}$. By expanding the above recursion, we see that to support function class F_{p+1} the FE scheme FE_0 must be capable of functional keys for the functions in $F_{p+1} \cup \mathbb{RE}_{\mathsf{FE}_0}^p$ and the PRG G.

Theorem 4. *Fix any polynomial $Q(\kappa)$, and let $p(\kappa) = \Omega(\log\log Q(\kappa))$. Assume FE_0 is a succinct $(1, \epsilon_0)$-non-adaptive (or selective/adaptive) secure FE scheme for the function class F such that $\mathbb{RE}_{\mathsf{FE}_0}^p \subseteq F$ and $\mathsf{G} \in F$, where $\epsilon_0(\kappa)$ is some negligible function; and assume G is a secure PRG. Then FE_{p+1} defined in Eq. 1 is a weakly succinct (Q, ϵ)-non-adaptive (or selective/adaptive, respectively) secure FE scheme for F, where $\epsilon(\kappa)$ is some negligible function.*

5.2 Iterated Linear Composition

A drawback of the iterated squaring composition is that the base scheme FE_0 must be capable of generating functional keys for the re-encryption functions of all iteration steps. It is usually hard to check if this condition holds for a concrete FE scheme. We now present another composition method that only requires the base scheme is capable of functionals keys for its own encryption function.

The *iterated linear composition* is defined recursively by

$$\mathsf{FE}_1 = \mathsf{FE}_0 + \mathsf{FE}_0, \text{ and for } p \geq 1, \mathsf{FE}_{p+1} = \mathsf{FE}_1 \times \mathsf{FE}_p. \tag{3}$$

Under this composition, FE_1 supports 2 functional keys, and for $p \geq 1$, FE_{p+1} supports 2^{p+1} functional keys. For FE_p to achieve $Q(\kappa)$ functional keys, we need $p \geq \log Q$. Then it is straightforward to get the following characteristics of FE_p:

Security: The advantage of FE_p over any efficient adversary is

$$\mathsf{Adv}_{\mathsf{FE}_p}[\mathcal{A}] = (3 \cdot 2^p)\epsilon_0 + (12 \cdot 2^p)\mu = Q\epsilon_0 + Q\mu. \tag{4}$$

Running times and output lengths:

- Setup: $2t_0^{\mathsf{Setup}} + 2t^{\mathsf{G}}$
- Enc : $2t_0^{\mathsf{Enc}}(n + \kappa, \kappa, d) + 2t^{\mathsf{G}}$
- Dec : $pt_0^{\mathsf{Dec}} = \log Q \cdot t_0^{\mathsf{Dec}}$
- KeyGen : $pt_0^{\mathsf{KeyGen}} + 2(p-1)t_0^{\mathsf{Setup}} + (6p-5)t^{\mathsf{G}} = 2\log Q \cdot t_0^{\mathsf{Setup}} + \log Q \cdot t_0^{\mathsf{KeyGen}} + 6\log Q \cdot t^{\mathsf{G}}$

- $\ell_p^k = 2\ell_0^k$
- $\ell_p^c = \ell_0^c(n + \kappa, \kappa, d) = 2\ell_0^c(n + \kappa, \kappa, d)$
- $\ell_p^{fk} = p\ell_0^{fk} = \log Q \cdot \ell_0^{fk}$

The FE scheme FE_p is also secure, and this transformation too incurs linear (in terms of Q) security loss. This time, the running time of the encryption procedure no longer depends on Q, so FE_p is fully succinct.

Again, for this composition method to be viable, we should consider the functions can be handled at each iteration. Let F_h denote the function class supported by FE_h, for $h \geq 0$. As in the squaring composition, we have $F_1 = F_0$. For $h \geq 1$, to derive a functional key for any function f in FE_{h+1}, the scheme FE_1 must generate functional keys for the re-encryption function $e_i[pk, c]$, and FE_h must be capable of generating functional keys of f. This implies that

$$F_p \cup \{e_i[pk, c] \mid pk \in R, c \in M\} \subseteq F_0.$$

Since $e_i[pk, c](\cdot, \cdot)$ can be easily built using basic operations on $\mathsf{Enc}_1(pk, \cdot; \cdot)$ and $\mathsf{G}(\cdot)$, it is sufficient to require that FE_0 can generate functional keys for these two classes of functions.

Theorem 5. *Assume FE_0 is a succinct $(1, \epsilon_0)$-non-adaptive (or selective/adaptive) secure FE scheme for the class F of functions such that $\mathsf{Enc}_0(pk, \cdot; \cdot), \mathsf{G}(\cdot) \in F$ for any $pk \in R$, where $\epsilon_0(\kappa)$ is some negligible function, and assume G is a secure PRG. Then, for any polynomial $Q(\kappa)$, the FE scheme FE_p defined in Eq. 3 for $p = \Omega(\log Q)$ is a succinct (Q, ϵ)-non-adaptive (or selective/adaptive, respectively) secure FE scheme for F, for some negligible function $\epsilon(\kappa)$.*

Comparing with the iterated squaring composition to support Q functional key queries, one can see that the security loss, the running times, and key lengths of Setup and KeyGen are about the same, and the iterated linear composition gives better encryption performance: Enc runs slightly faster. The trade-off is in the ciphertext length: our linear composition simply doubles the ciphertext length of the underlying 1-secure FE scheme, while the iterated squaring composition produces a ciphertext that encrypts a slightly longer message in the 1-secure FE scheme.

5.3   On the Implications of Our Reductions

So far we have obtained two transformations from a 1-secure succinct FE scheme to a (weakly) succinct FE scheme that supports polynomially many functional key queries. In this subsection we explore the implications of our reductions.

A (Q, ϵ)-secure FE scheme for F is called *weakly collusion-succinct* if t^{Enc} grows sub-linearly in Q but polynomially in n, κ, and the maximum circuit size of functions in F. If the sub-linear dependence on Q is removed, then the FE scheme is called *collusion-succinct*. For succinct FE_0, let us consider the following two cases about the encryption time t^{Enc}_{p+1} of FE_{p+1} obtained by our transformations on FE_0:

1. If FE_{p+1} is as in the iterated squaring composition, then $p = \Omega(\log \log Q)$ and $t^{\mathsf{Enc}}_{p+1} = t^{\mathsf{Enc}}_0(n + \kappa \cdot \log \log Q, \kappa, d) + \log \log Q \cdot t^{\mathsf{G}}(\kappa)$. Clearly t^{Enc}_{p+1} is sub-linear in Q, and thus FE_{p+1} is weakly collusion-succinct.
2. If FE_{p+1} is as in the iterated linear composition, then $p = \Omega(\log Q)$ and $t^{\mathsf{Enc}}_{p+1} = 2t^{\mathsf{Enc}}_0(n + \kappa, \kappa, d) + 2t^{\mathsf{G}}(\kappa)$, which is independent of Q. So FE_{p+1} is succinct (hence collusion-succinct).

Remark 2. Security in Theorems 4 and 5 degrades linearly in Q. So it may appear that setting the size of the index space to be superpolynomial results in a superpolynomial security loss. However, a careful analysis shows that Theorems 4 and 5 hold with Q equal to the number of key queries made by any efficient adversary, where the index space is just an upper bound on Q. As long as the adversary runs in polynomial time, the security loss is only polynomial, even when setting p accordingly to achieve superpolynomial-sized index space.

As we have mentioned in Remark 1, when the index space of FE_{p+1} has superpolynomial size, we can eliminate i from the interface of KeyGen to make it completely stateless. To achieve this, we may set $p = \omega(\log \log \kappa)$ in our first transformation, and we may set $p = \omega(\log \kappa)$ in our second transformation. Such conversions incur only a negligible security loss, and they do not affect the security properties of FE_{p+1} in either transformation. Moreover, FE_{p+1} is secure with any polynomial number of functional key queries, so it is collusion-resistant. We can state our transformations in terms of standard FE definition:

Theorem 6. *1. If FE_0 is a succinct 1-secure FE scheme for a class F of functions such that $\mathbb{RE}^p_{\mathsf{FE}_0} \subseteq F$ for $p = O(\log \kappa)$ and that $\mathsf{G} \in F$, then for some $p = \omega(\log \log \kappa)$, FE_{p+1} as in the iterated squaring composition is a weakly collusion-succinct and collusion-resistant FE scheme for F;*
2. *If FE_0 is a succinct 1-secure FE scheme for a class F of functions such that its encryption function Enc_0 satisfies $\mathsf{Enc}_0(pk, \cdot; \cdot) \in F$ for any $pk \in R$ and that $\mathsf{G} \in F$, then for some $p = \omega(\log \kappa)$, FE_{p+1} as in the iterated linear composition is a succinct and collusion-resistant FE scheme for F.*

Bitansky and Vaikuntanathan [9] described a reduction from any (weakly) compact Q-secure FE scheme to a (weakly) collusion-succinct Q-secure FE scheme for the same class of functions. We note that, although in [9] the notion of collusion-succinct was defined in terms of ciphertext length, their reduction still holds with our encryption time based definition. By applying their reduction together with our transformations, we get the following new reductions:

Theorem 7. *1. If there exists a succinct 1-secure FE scheme FE_0 for a class F of functions such that $\mathbb{RE}_{FE_0}^p \subseteq F$ for $p = O(\log \kappa)$ and that $G \in F$, then there exists a weakly compact and collusion-resistant FE scheme for F;*

2. If there exists a succinct 1-secure FE scheme FE_0 for a class F of functions such that its encryption function Enc_0 satisfies $Enc_0(pk, \cdot; \cdot) \in F$ for any $pk \in R$ and that $G \in F$, then there exists a compact and collusion-resistant FE scheme for F.

Notice that a (weakly) compact FE scheme is necessarily (weakly) succinct. Our results show that weakly compact (non-collusion-resistant) FE schemes (supporting a sufficiently general class of functions,) imply collusion-resistant FE schemes. As shown in [3,9], (non-compact) collusion-resistant FE schemes imply compact FE schemes. So now we can see these variants as equivalent notions under polynomial time reductions.

One may attempt to instantiate a compact collusion-resistant FE scheme using our transformations on a succinct 1-secure FE scheme. Based on sub-exponential lattice assumption, Goldwasser et al. [25] showed that, for any polynomial $d(n)$, there exists a succinct 1-secure FE scheme for the class of functions with 1-bit output and depth d circuits. However, it is not clear how to efficiently "upgrade" this FE scheme to be capable of generating a functional key of its own encryption function so that the assumptions of our transformations can be met. This is not surprising because any instantiation would immediately give an indistinguishability obfuscator. We find it very interesting to answer such question and we leave it for future work.

Acknowledgement. We would like to thank Fuyuki Kitagawa for pointing out a mistake in an earlier version of this paper, and we thank anonymous TCC reviewers for useful comments.

A   Proofs of Lemmas

Let $G : R \to S$ be a $\mu(\kappa)$-secure PRG. Recall the following well-known facts:

- The function $G'(r_1 \cdots r_m) = G(r_1) \cdots G(r_m)$ defined by concatenating m pseudorandom strings generated by G on $r_1, \ldots, r_m \in R$ is a $m\mu(\kappa)$-secure pseudorandom generator.
- The function $G''(r) = G(G_i(r))$, where $|G_i(r)| = |r| = n$, is a $2\mu(\kappa)$-secure pseudorandom generator.

We will use them to shorten our security proofs.

First we prove lemmas in Sect. 3 that are used to establish security of the SUM constructions.

Lemma 1. *If G is a μ-secure pseudorandom generator, then for any $b \in \{0,1\}$ and adversary $\mathcal{A}$ we have $|\Pr\{[\mathcal{H}_b^{(0)}, \mathcal{A}] = 1\} - \Pr\{[\mathcal{H}_b^{(1)}, \mathcal{A}] = 1\}| \leq 2\mu(\kappa)$.*

Proof. We define the following adversary $\mathcal{B}$ using $\mathcal{A}$ as an oracle to attack the PRG G, where $\mathcal{H}_b^{(1)}[sk_0, sk_1, r_0, r_1]$ is the hybrid obtained by replacing sk_0, sk_1, r_0, r_1 of $\mathcal{H}_b^{(1)}$ by the given values. By the notation $sk_0 \| sk_1 \| r_0 \| r_1 = x$ we mean to parse x as a concatenation of four bit-strings sk_0, sk_1, r_0, r_1 of appropriate lengths.

$\mathcal{B}(x) = b'$
  where $sk_0 \| sk_1 \| r_0 \| r_1 = x$
    $(pk, c, \{fk^{(h,i)}\}_{i \in I}) = \mathcal{H}_b^{(1)}[sk_0, sk_1, r_0, r_1]((m_0, m_1), \{f_{(h,i)}\}_{i \in I})$
    $((m_0, m_1), \{f_{(h,i)}\}_{i \in I}, b') = \mathcal{A}(pk, c, \{fk^{(h,i)}\}_{i \in I})$

Notice that if x is generated by the PRG G then $\mathcal{B}$ is running the system $[\mathcal{H}_b^{(0)}, \mathcal{A}]$, and if x is uniformly random then $\mathcal{B}$ is running $[\mathcal{H}_b^{(1)}, \mathcal{A}]$. Since in $\mathcal{H}_b^{(1)}$ we replaced two calls to G with truly random seeds, we have $|\mathsf{Adv}[\mathcal{A}]_b^{(0)} - \mathsf{Adv}[\mathcal{A}]_b^{(1)}| = \mathsf{Adv}_G[\mathcal{B}^\mathcal{A}] \leq 2\mu(\kappa)$. □

Lemma 2. *If FE_1 is a (q_1, ϵ_1)-non-adaptively secure FE scheme, then for any adversary $\mathcal{A}$ we have $|\Pr\{[\mathcal{H}_0^{(1)}, \mathcal{A}] = 1\} - \Pr\{[\mathcal{H}_0^{(2)}, \mathcal{A}] = 1\}| \leq \epsilon_1(\kappa)$.*

Proof. We define the following adversary $\mathcal{B}$ using $\mathcal{A}$ as an oracle to attack the FE scheme FE_1.

$\mathcal{B}(pk_1, c_1, \{fk_1^{(1,i)}\}_{i \in I_1}) = ((m_0, m_1), \{f_{(1,i)}\}_{i \in I_1})$
  where
    $(pk, c, \{fk^{(h,i)}\}_{(h,i) \in I}) = \mathcal{H}_0^{(2)}[pk_1, c_1, \{fk_1^{(1,i)}\}_{i \in I_1}]((m_0, m_1), \{f_{(h,i)}\}_{(h,i) \in I})$
    $((m_0, m_1), \{f_{(h,i)}\}_{(h,i) \in I}, b') = \mathcal{A}(pk, c, \{fk^{(h,i)}\}_{(h,i) \in I})$

Since $\mathcal{A}$ is a valid adversary to $\mathsf{FE}_0 + \mathsf{FE}_1$, we must have $f_{(1,i)}(m_0) = f_{(1,i)}(m_1)$ for all $i \in I_1$; and hence $\mathcal{B}$ is valid for FE_1. Notice that if the input c_1 to $\mathcal{B}$ is an encryption of m_0, i.e., $c_1 = \mathsf{Enc}_1(pk_1, m_0; r_1)$ for some random string $r_1 \in R$, then $\mathcal{B}$ is running $[\mathcal{H}_0^{(1)}, \mathcal{A}]$; if $c_1 = \mathsf{Enc}_1(pk_1, m_1; r_1)$ for some $r_1 \in R$, then $\mathcal{B}$ is running $[\mathcal{H}_0^{(2)}, \mathcal{A}]$. Hence the advantage of $\mathcal{B}$ in winning the FE game for the scheme FE_1 is $\mathsf{Adv}_{\mathsf{FE}_1}[\mathcal{B}] = |\mathsf{Adv}[\mathcal{A}]_0^{(1)} - \mathsf{Adv}[\mathcal{A}]_0^{(2)}| \leq \epsilon_1(\kappa)$. □

Next we prove lemmas that are used to establish security of the PRODUCT constructions. From now on, hybrids refer to those defined in Sect. 4.

Lemma 4. *If G is a μ-secure pseudorandom generator, then for any $b \in \{0,1\}$ and any efficient adversary $\mathcal{A}$, we have $|\mathsf{Adv}[\mathcal{A}]_b^{(0)} - \mathsf{Adv}[\mathcal{A}]_b^{(1)}| \leq 4\mu(\kappa)$.*

Proof. We build an adversary $\mathcal{B}$ using $\mathcal{A}$ as an oracle to attack the PRG G. As in previous proofs, by $\mathcal{H}_b^{(1)}[sk_0, r', r'', sk_1^1, \ldots, sk_1^{q_0}, c_1, \ldots, c_{q_0}]$ we mean the hybrid obtained by substituting $sk_0, r', r'', sk_1^1, \ldots, sk_1^{q_0}, c_1, \ldots, c_{q_0}$ with the given values. The adversary $\mathcal{B}$ is defined as:

$$\mathcal{B}(x) = b'$$
where
$$sk_0\|r'\|r''\|sk_1^1\|\cdots\|sk_1^{q_0}\|c_1\|\cdots\|c_{q_0} = x$$
$$(pk, c, \{fk^{i,j}\}_I) = \mathcal{H}_b^{(1)}[sk_0, r', r'', \{sk_1^i\}_{i \in I_0}, \{c_i\}_{i \in I_0}]((m_0, m_1), \{f_{i,j}\}_I)$$
$$((m_0, m_1), \{f_{i,j}\}_{(i,j) \in I}, b') = \mathcal{A}(pk, c, \{fk^{i,j}\}_{(i,j) \in I})$$

Notice that if x is generated by four calls to G then $\mathcal{B}$ is running $[\mathcal{H}_b^{(0)}, \mathcal{A}]$, and if x is truly random then $\mathcal{B}$ is running $[\mathcal{H}_b^{(1)}, \mathcal{A}]$. Since G is a μ-secure pseudorandom generator, we have $|\mathsf{Adv}[\mathcal{A}]_b^{(0)} - \mathsf{Adv}[\mathcal{A}]_b^{(1)}| \le 4\mu(\kappa)$.   □

Lemma 6. *If G is a μ-secure pseudorandom generator, then for any $b \in \{0,1\}$ and adversary $\mathcal{A}$, we have $|\mathsf{Adv}[\mathcal{A}]_b^{(2)} - \mathsf{Adv}[\mathcal{A}]_b^{(3)}| \le \mu(\kappa)$.*

Proof. We build an adversary $\mathcal{B}$ using $\mathcal{A}$ as an oracle to attack G:

$$\mathcal{B}(x) = b'$$
$$\textbf{where } s_1\|\cdots\|s_{q_0} = x$$
$$(pk, c, \{fk^{i,j}\}_{(i,j) \in I}) = \mathcal{H}_b^{(1)}[s_1, \ldots, s_{q_0}]((m_0, m_1), \{f_{i,j}\}_{(i,j) \in I})$$
$$((m_0, m_1), \{f_{i,j}\}_{(i,j) \in I}, b') = \mathcal{A}(pk, c, \{fk^{i,j}\}_{(i,j) \in I})$$

Notice that if x is chosen uniformly random then $\mathcal{B}$ is running $[\mathcal{H}_b^{(2)}, \mathcal{A}]$, and if x is generated by G then $\mathcal{B}$ is running $[\mathcal{H}_b^{(3)}, \mathcal{A}]$. Thus we have $|\mathsf{Adv}[\mathcal{A}]_b^{(2)} - \mathsf{Adv}[\mathcal{A}]_b^{(3)}| \le \mu(\kappa)$.   □

Lemma 7. *If FE_0 is a (q_0, ϵ_0)-non-adaptive secure FE scheme for functions in the class $\mathbb{RE}_{\mathsf{FE}_0}$, then for any $b \in \{0,1\}$ and any efficient adversary $\mathcal{A}$, we have $|\mathsf{Adv}[\mathcal{A}]_b^{(3)} - \mathsf{Adv}[\mathcal{A}]_b^{(4)}| \le \epsilon_0(\kappa)$.*

Proof. We build an adversary $\mathcal{B}$ using $\mathcal{A}$ as an oracle to attack FE_0. For $b \in \{0,1\}$, we define $\mathcal{B}$ as follows:

$$\mathcal{B}(pk_0, c_0, \{fk_0^i\}_{i \in I_0}) = ((x_0, x_1), \{e_i[c_i, pk_1^i]\}_{i \in I_0}, b')$$
$$\textbf{where } pk = pk_0, c = c_0$$
$$r \leftarrow R, r' \leftarrow K, s \leftarrow K$$
$$x_0 = (m_b, r'), x_1 = (\bot, s)$$
$$\text{For all } i \in I_0, j \in I_1:$$
$$fk^{i,j} = \textbf{if } (f_{i,j}(m_0) = f_{i,j}(m_1) \ne \bot) \textbf{ then } (fk_0^i, fk_1^{i,j})$$
$$\textbf{where } sk_1^i \leftarrow K$$
$$pk_1^i = \mathsf{Setup}_1(sk_1^i)$$
$$s_i = \mathsf{G}_i(s), \tilde{c}_1^i = \mathsf{Enc}_1(pk_1^i, m_b; \mathsf{G}_i(r')), c_i = s_i \oplus \tilde{c}_1^i$$
$$fk_1^{i,j} = \mathsf{KeyGen}_1(sk_1^i, f_{i,j}, j)$$
$$((m_0, m_1), \{f_{i,j}\}_{(i,j) \in I}, b') = \mathcal{A}(pk, c, \{fk^{i,j}\}_{(i,j) \in I})$$

We show that $\mathcal{B}$ is a valid adversary for the FE game, that is, the functions $e_i[c_i, pk_1^i]$ appear in $\mathcal{B}$'s queries satisfy $e_i[c_i, pk_1^i](x_0) = e_i[c_i, pk_1^i](x_1)$ for all $i \in I_0$. Since $x_0 = (m_b, r')$ and $x_1 = (\bot, s)$, by definition of $e_i[c_i, pk_1^i]$ we have

$$e_i[c_i, pk_1^i](x_0) = e_i[c_i, pk_1^i](m_b, r') = \mathsf{Enc}_1(pk_1^i, m_b; \mathsf{G}_i(r')),$$
$$e_i[c_i, pk_1^i](x_1) = e_i[c_i, pk_1^i](\perp, s) = \mathsf{G}_i(s) \oplus c_i = \mathsf{Enc}_1(pk_1^i, m_b; \mathsf{G}_i(r')).$$

So indeed $e_i[c_i, pk_1^i](x_0) = e_i[c_i, pk_1^i](x_1)$.

Notice that if the input ciphertext c_0 is an encryption of m_0, i.e., $c_0 = \mathsf{Enc}_0(pk_0, (m_b, r'); r'')$ for some random string r'', then $\mathcal{B}$ is running $[\mathcal{H}_b^{(3)}, \mathcal{A}]$, and if $c_0 = \mathsf{Enc}_0(pk_0, (\perp, s); r'')$ then $\mathcal{B}$ is running $[\mathcal{H}_b^{(4)}, \mathcal{A}]$. Thus the advantage of $\mathcal{B}$ in the FE game is

$$|2\Pr\{b_0' = b_0\} - 1| = |\Pr\{b_0' = 0 \mid b_0 = 0\} + \Pr\{b_0' = 1 \mid b_0 = 1\} - 1|$$
$$= |\Pr\{b' = 1 \mid b_0 = 0\} - \Pr\{b' = 1 \mid b_0 = 1\}|,$$

where $\Pr\{b' = 1 \mid b_0 = 0\} = \mathsf{Adv}[\mathcal{A}]_b^{(3)}$ and $\Pr\{b' = 1 \mid b_0 = 1\} = \mathsf{Adv}[\mathcal{A}]_b^{(4)}$. Since FE_0 is (q_0, ϵ_0)-non-adaptively secure, we have that $|\mathsf{Adv}[\mathcal{A}]_b^{(3)} - \mathsf{Adv}[\mathcal{A}]_b^{(4)}| \le \epsilon_0(\kappa)$. $\square$

Lemma 8. *If* G *is a* μ-*secure pseudorandom generator, then for any* $b \in \{0, 1\}$ *and any adversary* $\mathcal{A}$ *we have* $|\mathsf{Adv}[\mathcal{A}]_b^{(4)} - \mathsf{Adv}[\mathcal{A}]_b^{(5)}| \le \mu(\kappa)$.

Proof. We build an adversary $\mathcal{B}$ to attack G using $\mathcal{A}$ as an oracle.

$$\mathcal{B}(x) = b'$$
$$\textbf{where } r_1 \| \cdots \| r_{q_0} = x$$
$$(pk, c, \{fk^{i,j}\}_{(i,j)\in I}) = \mathcal{H}_b^{(1)}[r_1, \ldots, r_{q_0}]((m_0, m_1), \{f_{i,j}\}_{(i,j)\in I})$$
$$((m_0, m_1), \{f_{i,j}\}_{(i,j)\in I}, b') = \mathcal{A}(pk, c, \{fk^{i,j}\}_{(i,j)\in I})$$

Notice that if x is truly random then $\mathcal{B}$ is running $[\mathcal{H}_b^{(4)}, \mathcal{A}]$, and if x is generated by G then $\mathcal{B}$ is running $[\mathcal{H}_b^{(5)}, \mathcal{A}]$. So we have

$$\Pr\{\mathcal{B}(x) = 1 \mid x \leftarrow R\} - \Pr\{\mathcal{B}(x) = 1 \mid \exists y. x = \mathsf{G}(y)\} = |\mathsf{Adv}[\mathcal{A}]_b^{(4)} - \mathsf{Adv}[\mathcal{A}]_b^{(5)}|.$$

Since G is μ-secure, $|\mathsf{Adv}[\mathcal{A}]_b^{(4)} - \mathsf{Adv}[\mathcal{A}]_b^{(5)}| \le \mu(\kappa)$. $\square$

Lemma 9. *If* FE_1 *is a* (q_1, ϵ_1)-*non-adaptive secure FE scheme, then for any efficient adversary* $\mathcal{A}$ *we have* $|\mathsf{Adv}[\mathcal{A}]_0^{(5)} - \mathsf{Adv}[\mathcal{A}]_1^{(5)}| \le q_0 \cdot \epsilon_1(\kappa)$.

Proof. To close the gap between $\mathcal{H}_0^{(5)}$ and $\mathcal{H}_1^{(5)}$, we build a sequence of q_0 hybrids $\mathcal{H}_0^{(5.h)}$ for $h \in I_0$. Suppose $I_0 = \{1, 2, \ldots, q_0\}$. Let $\mathcal{H}_0^{(5.0)} = \mathcal{H}_0^{(5)}$, and for each $h \in I_0$, we obtain $\mathcal{H}_0^{(5.h)}$ from $\mathcal{H}_0^{(5.(h-1))}$ by changing $\tilde{c}_1^h$ from encrypting m_0 to encrypting m_1. Notice that $\mathcal{H}_0^{(5.q_0)}$ is same as $\mathcal{H}_1^{(5)}$.

For each $h \in I_0$, we can build an adversary $\mathcal{B}$ to attack the FE scheme FE_1 using $\mathcal{A}$ as an oracle.

$$\mathcal{B}(pk_1, c_1, \{fk_1^{h,j}\}_{j \in I_1}) = ((m_0, m_1), \{f_{h,j}\}_{j \in I_1}, b')$$
where
$$pk_1^h = pk_1, \tilde{c}_1^h = c_1$$
$$(pk, c, \{fk^{i,j}\}_{(i,j) \in I}) = \mathcal{H}_b^{(5)}[pk_1^h, \tilde{c}_1^h, \{fk_1^{h,j}\}_{j \in I_1}]((m_0, m_1), \{f_{i,j}\}_{(i,j) \in I})$$
$$((m_0, m_1), \{f_{i,j}\}_{(i,j) \in I}, b') = \mathcal{A}(pk, c, \{fk^{i,j}\}_{(i,j) \in I})$$

If $c_1 = \mathsf{Enc}_1(pk_1, m_0; \tilde{r})$ for some randomness $\tilde{r}$ then $\mathcal{B}$ is running $[\mathcal{H}_0^{(5.(h-1))}, \mathcal{A}]$, and if $c_1 = \mathsf{Enc}_1(pk_1, m_1; \tilde{r})$ then $\mathcal{B}$ is running $[\mathcal{H}_0^{(5.h)}, \mathcal{A}]$. So the advantage of $\mathcal{B}$ in winning the FE game for the FE_1 scheme is $|\mathsf{Adv}[\mathcal{A}]_0^{(5.(h-1))} - \mathsf{Adv}[\mathcal{A}]_0^{(5.h)}| \le \epsilon_1(\kappa)$. Since $\mathcal{H}_0^{(5.0)}$ is same as $\mathcal{H}_0^{(5)}$ and $\mathcal{H}_0^{(5.q_0)}$ is same as $\mathcal{H}_1^{(5)}$, we get

$$|\mathsf{Adv}[\mathcal{A}]_0^{(5)} - \mathsf{Adv}[\mathcal{A}]_1^{(5)}| \le \sum_{h=1}^{q_0} |\mathsf{Adv}[\mathcal{A}]_0^{(5.(h-1))} - \mathsf{Adv}[\mathcal{A}]_0^{(5.h)}|$$
$$= q_0 \cdot \epsilon_1(\kappa).$$

$\square$

References

1. Agrawal, S., Gorbunov, S., Vaikuntanathan, V., Wee, H.: Functional encryption: new perspectives and lower bounds. In: Canetti, R., Garay, J.A. (eds.) CRYPTO 2013, Part II. LNCS, vol. 8043, pp. 500–518. Springer, Heidelberg (2013)
2. Ananth, P., Brakerski, Z., Segev, G., Vaikuntanathan, V.: From selective to adaptive security in functional encryption. In: Gennaro, R., Robshaw, M. (eds.) CRYPTO 2015. LNCS, vol. 9216, pp. 657–677. Springer, Heidelberg (2015)
3. Ananth, P., Jain, A.: Indistinguishability obfuscation from compact functional encryption. In: Gennaro, R., Robshaw, M. (eds.) CRYPTO 2015. LNCS, vol. 9215, pp. 308–326. Springer, Heidelberg (2015)
4. Ananth, P., Jain, A., Sahai, A.: Indistinguishability obfuscation from functional encryption for simple functions. IACR Cryptology ePrint Archive 2015, 730 (2015). http://eprint.iacr.org/2015/730
5. Applebaum, B., Brakerski, Z.: Obfuscating circuits via composite-order graded encoding. In: Dodis, Y., Nielsen, J.B. (eds.) TCC 2015, Part II. LNCS, vol. 9015, pp. 528–556. Springer, Heidelberg (2015)
6. Barak, B., Garg, S., Kalai, Y.T., Paneth, O., Sahai, A.: Protecting obfuscation against algebraic attacks. In: Nguyen, P.Q., Oswald, E. (eds.) EUROCRYPT 2014. LNCS, vol. 8441, pp. 221–238. Springer, Heidelberg (2014)
7. Barak, B., Goldreich, O., Impagliazzo, R., Rudich, S., Sahai, A., Vadhan, S.P., Yang, K.: On the (im)possibility of obfuscating programs. J. ACM **59**(2), 6 (2012). (In: Kilian, J. (ed.) CRYPTO 2001. LNCS, vol. 2139, pp. 1–18. Springer, Heidelberg (2001))
8. Bitansky, N., Goldwasser, S., Jain, A., Paneth, O., Vaikuntanathan, V., Waters, B.: Time-lock puzzles from randomized encodings. In: Innovations in Theoretical Computer Science, pp. 345–356 (2016)
9. Bitansky, N., Vaikuntanathan, V.: Indistinguishability obfuscation from functional encryption. In: Foundations of Computer Science, FOCS, pp. 171–190 (2015)

10. Boneh, D., Sahai, A., Waters, B.: Functional encryption: a new vision for public-key cryptography. Commun. ACM **55**(11), 56–64 (2012)
11. Brakerski, Z., Rothblum, G.N.: Virtual black-box obfuscation for all circuits via generic graded encoding. In: Lindell, Y. (ed.) TCC 2014. LNCS, vol. 8349, pp. 1–25. Springer, Heidelberg (2014)
12. Brakerski, Z., Segev, G.: Function-private functional encryption in the private-key setting. In: Dodis, Y., Nielsen, J.B. (eds.) TCC 2015, Part II. LNCS, vol. 9015, pp. 306–324. Springer, Heidelberg (2015)
13. Brakerski, Z., Segev, G.: Hierarchical functional encryption. IACR Cryptology ePrint Archive 2015, 1011 (2015). http://eprint.iacr.org/2015/1011
14. Chandran, N., Chase, M., Vaikuntanathan, V.: Functional re-encryption and collusion-resistant obfuscation. In: Cramer, R. (ed.) TCC 2012. LNCS, vol. 7194, pp. 404–421. Springer, Heidelberg (2012)
15. Cheon, J.H., Han, K., Lee, C., Ryu, H., Stehlé, D.: Cryptanalysis of the multilinear map over the integers. In: Oswald, E., Fischlin, M. (eds.) EUROCRYPT 2015. LNCS, vol. 9056, pp. 3–12. Springer, Heidelberg (2015)
16. Chung, K.-M., Lin, H., Pass, R.: Constant-round concurrent zero-knowledge from indistinguishability obfuscation. In: Gennaro, R., Robshaw, M. (eds.) CRYPTO 2015. LNCS, vol. 9215, pp. 287–307. Springer, Heidelberg (2015)
17. Coron, J., Gentry, C., Halevi, S., Lepoint, T., Maji, H.K., Miles, E., Raykova, M., Sahai, A., Tibouchi, M.: Zeroizing without low-level zeroes: new MMAP attacks and their limitations. In: Gennaro, R., Robshaw, M. (eds.) CRYPTO 2015. LNCS, vol. 9215, pp. 247–266. Springer, Heidelberg (2015)
18. Coron, J.-S., Lepoint, T., Tibouchi, M.: Practical multilinear maps over the integers. In: Canetti, R., Garay, J.A. (eds.) CRYPTO 2013, Part I. LNCS, vol. 8042, pp. 476–493. Springer, Heidelberg (2013)
19. Coron, J., Lepoint, T., Tibouchi, M.: New multilinear maps over the integers. In: Gennaro, R., Robshaw, M. (eds.) CRYPTO 2015. LNCS, vol. 9215, pp. 267–286. Springer, Heidelberg (2015)
20. Garg, S., Gentry, C., Halevi, S.: Candidate multilinear maps from ideal lattices. In: Nguyen, P.Q., Johansson, T. (eds.) EUROCRYPT 2013. LNCS, vol. 7881, pp. 1–17. Springer, Heidelberg (2013)
21. Garg, S., Gentry, C., Halevi, S., Raykova, M., Sahai, A., Waters, B.: Candidate indistinguishability obfuscation and functional encryption for all circuits. In: Foundations of Computer Science, FOCS, pp. 40–49 (2013)
22. Garg, S., Srinivasan, A.: Single-key to multi-key functional encryption with polynomial loss. In: Hirt, M., Smith, A. (eds.) TCC 2016-B, Part II. LNCS, vol. 9986, pp. 419–442. Springer, Heidelberg (2016). http://eprint.iacr.org/2016/524
23. Gentry, C., Gorbunov, S., Halevi, S.: Graph-induced multilinear maps from lattices. In: Dodis, Y., Nielsen, J.B. (eds.) TCC 2015, Part II. LNCS, vol. 9015, pp. 498–527. Springer, Heidelberg (2015)
24. Gentry, C., Lewko, A.B., Sahai, A., Waters, B.: Indistinguishability obfuscation from the multilinear subgroup elimination assumption. In: Foundations of Computer Science, FOCS, pp. 151–170 (2015)
25. Goldwasser, S., Kalai, Y.T., Popa, R.A., Vaikuntanathan, V., Zeldovich, N.: Reusable garbled circuits and succinct functional encryption. In: Symposium on Theory of Computing Conference, STOC, pp. 555–564 (2013)
26. Goldwasser, S., Rothblum, G.N.: On best-possible obfuscation. J. Cryptol. **27**(3), 480–505 (2014). (In: Vadhan, S.P. (ed.) TCC 2007. LNCS, vol. 4392, pp. 194–213. Springer, Heidelberg (2007))

27. Li, B., Micciancio, D.: Equational security proofs of oblivious transfer protocols. IACR Cryptology ePrint Archive 2016, 624 (2016). http://eprint.iacr.org/2016/624

28. Malkin, T., Micciancio, D., Miner, S.K.: Efficient generic forward-secure signatures with an unbounded number of time periods. In: Knudsen, L.R. (ed.) EUROCRYPT 2002. LNCS, vol. 2332, pp. 400–417. Springer, Heidelberg (2002)

29. Micciancio, D., Tessaro, S.: An equational approach to secure multi-party computation. In: Innovations in Theoretical Computer Science, ITCS 2013, pp. 355–372. ACM, New York (2013)

30. O'Neill, A.: Definitional issues in functional encryption. IACR Cryptology ePrint Archive 2010, 556 (2010). http://eprint.iacr.org/2010/556

31. Pass, R., Seth, K., Telang, S.: Indistinguishability obfuscation from semantically-secure multilinear encodings. In: Garay, J.A., Gennaro, R. (eds.) CRYPTO 2014, Part I. LNCS, vol. 8616, pp. 500–517. Springer, Heidelberg (2014)

32. Sahai, A., Waters, B.: How to use indistinguishability obfuscation: deniable encryption, and more. In: Symposium on Theory of Computing, STOC, pp. 475–484 (2014)

33. Waters, B.: Functional encryption: origins and recent developments. In: Kurosawa, K., Hanaoka, G. (eds.) PKC 2013. LNCS, vol. 7778, pp. 51–54. Springer, Heidelberg (2013)

34. Waters, B.: A punctured programming approach to adaptively secure functional encryption. In: Gennaro, R., Robshaw, M. (eds.) CRYPTO 2015. LNCS, vol. 9216, pp. 678–697. Springer, Heidelberg (2015)

Secret Sharing

Threshold Secret Sharing Requires a Linear Size Alphabet

Andrej Bogdanov[1(✉)], Siyao Guo[2], and Ilan Komargodski[3]

[1] Chinese University of Hong Kong, Hong Kong, China
andrejb@cse.cuhk.edu.hk
[2] New York University, New York, USA
sguo@cims.nyu.edu
[3] Weizmann Institute of Science, Rehovot, Israel
ilan.komargodski@weizmann.ac.il

Abstract. We prove that for every n and $1 < t < n$ any t-out-of-n threshold secret sharing scheme for one-bit secrets requires share size $\log(t + 1)$. Our bound is tight when $t = n - 1$ and n is a prime power. In 1990 Kilian and Nisan proved the incomparable bound $\log(n - t + 2)$. Taken together, the two bounds imply that the share size of Shamir's secret sharing scheme (Comm. ACM '79) is optimal up to an additive constant even for one-bit secrets for the whole range of parameters $1 < t < n$.

More generally, we show that for all $1 < s < r < n$, any ramp secret sharing scheme with secrecy threshold s and reconstruction threshold r requires share size $\log((r + 1)/(r - s))$.

As part of our analysis we formulate a simple game-theoretic relaxation of secret sharing for arbitrary access structures. We prove the optimality of our analysis for threshold secret sharing with respect to this method and point out a general limitation.

1 Introduction

In 1979, Shamir [30] and Blakley [11] presented a method for sharing a piece of secret information among n parties such that any $1 < t < n$ parties can recover the secret while any $t - 1$ parties learn *nothing* about the secret. These methods are called (t, n)-threshold secret sharing schemes. This sharp threshold between secrecy and reconstruction is fundamental in applications where a group of mutually suspicious individuals with conflicting interests must cooperate. Indeed, threshold secret sharing schemes have found many applications in

A. Bogdanov—Supported by RGC GRF grants CUHK410113 and CUHK14208215.
S. Guo—Part of the work done in the Chinese University of Hong Kong supported by RGC GRF grants CUHK410112 and CUHK410113.
I. Komargodski—Part of this work done while visiting CUHK, supported by RGC GRF grant CUHK410113. Supported in part by a Levzion fellowship, by a grant from the I-CORE Program of the Planning and Budgeting Committee, the Israel Science Foundation, BSF and the Israeli Ministry of Science and Technology.

© International Association for Cryptologic Research 2016
M. Hirt and A. Smith (Eds.): TCC 2016-B, Part II, LNCS 9986, pp. 471–484, 2016.
DOI: 10.1007/978-3-662-53644-5_18

cryptography and distributed computing; see the extensive survey of Beimel [3] and the recent book of Cramer et al. [17].

Threshold secret sharing was generalized by Ito et al. [23] to allow more general structures of subsets to learn the secret, while keeping the secret perfectly hidden from all other subsets. The collection of qualified subsets is called an access structure.

A significant goal in secret sharing is to minimize the share size, namely, the amount of information distributed to the parties. Despite the long history of the subject, there are significant gaps between lower and upper bounds both for general access structures and for the special case of threshold structures.

Threshold Access Structures. For (t, n)-threshold access structures (denoted by THR_t^n) and a 1-bit secret, Shamir [30] gave a very elegant and efficient scheme: the dealer picks a random polynomial of degree $t - 1$ conditioned on setting the free coefficient to be the secret, and gives the i-th party the evaluation of the polynomial at the point i. The computation is done over a field $\mathbb{F}$ of size $q > n$.

The correctness follows because one can recover the unique polynomial from any t points (and thus recover the secret). Security follows by a counting argument showing that given less than t points, all possibilities for the free coefficient are equally likely. The share of each party is an element in the field $\mathbb{F}$ that can be represented using $\log q \approx \log n$ bits (all our logarithms are base 2). The efficiency of this scheme makes it very attractive for applications.

A natural question to ask is whether $\log n$-bit shares are necessary for sharing a 1-bit secret for threshold access structures. Kilian and Nisan [25][1] showed that $\log n$ bits are necessary when t is not too large. Specifically, they showed a $\log(n - t + 2)$ lower bound on share size for (t, n)-threshold schemes. For large values of t, especially those close to n, their bound does not rule out schemes with shares much shorter than $\log n$ bits. Their bound leaves open the possibility that, in particular, $(n - 1, n)$-threshold schemes with two-bit shares exist.

Ramp schemes are a generalization of threshold schemes that allow for a gap between the secrecy and reconstruction parameters. In an (s, r, n)-ramp scheme, we require that any subset of at least r parties can recover the secret, while any subset of size at most s cannot learn anything about the secret.[2] When $r = s+1$, an (s, r, n)-ramp scheme is exactly an (r, n)-threshold scheme. Ramp schemes, defined by Blakley and Meadows [10], are useful for various applications (see e.g. [15,27,31]) since if $r - s$ is large, they can sometimes be realized with *shorter* shares than standard threshold schemes (especially in the case of long secret).

---

[1] Their result is unpublished and independently obtained (and generalized in various ways) by [14]. The original argument of Kilian an Nisan appears in [14, Appendix A] and was referenced earlier in [2,4,5].

[2] Another common definition (See [20, Definition 2.7] and [21, Example 2.11] for examples) for a ramp scheme is where the information about the secret increases with the size of the set. We focus only on the definition in which sets of size below a certain threshold have no information about the secret, while sets of size larger than some threshold can recover it.

Generalizing the lower bound of Kilian and Nisan, Cascudo et al. [14] showed that $\log((n - s + 1)/(r - s))$-bit shares are necessary to realize an (s, r, n)-ramp scheme. When $s = n - O(1)$, however, their share size bound is a constant independent of n. Paterson and Stinson [29] showed that this bound is tight for specific small values of s.

General Access Structures. For most access structures, the best known secret sharing schemes require shares of size $2^{O(n)}$ for sharing a 1-bit secret. Specifically, viewing the access structure as a Boolean indicator function for qualified subsets, the schemes of [9, 23, 24] result with shares of size proportional to the DNF/CNF size, monotone formula size, or monotone span program size of the function, respectively. Thus, even for many access structures that can be described by a small monotone uniform circuit, the best schemes have exponential size shares.[3] On the other hand, the best known lower bound on share size for sharing an ℓ-bit secret is $\ell \cdot n / \log n$ bits, by Csirmaz [19] (improving on [13]).

Bridging the exponential gap between upper and lower bounds is the major open problems in the study of secret sharing schemes. While it is widely believed that the lower bound should be exponential (see e.g. [2,3]), no major progress has been obtained in the last two decades. Moreover, a non-explicit linear lower bound is not known, that is, whether there *exists* an access structure that requires linear size shares.[4]

1.1 Our Results

Share Size Lower Bound. We close the gap in share size for threshold secret sharing up to a small additive constant. We assume for simplicity that all parties are given equally long shares.

Theorem 1. *For every $n \in \mathbb{N}$ and $1 < t < n$, any (t, n)-threshold secret sharing scheme for a 1-bit secret requires shares of at least $\log(t + 1)$ bits.*

The assumption $1 < t < n$ is necessary, as $(1, n)$-threshold and (n, n)-threshold secret sharing schemes with share size 1 do exist.

Our bound is tight when $t = n - 1$ and n is the power of a prime; see Appendix A. By combining Theorem 1 with the lower bound of Kilian and Nisan, we determine the share size of threshold schemes up to a small additive constant. That is, we get that any such scheme requires shares of size

$$\max\{\log(n - t + 2), \log(t + 1)\} \geq \log \frac{n + 3}{2}. \tag{1}$$

Theorem 1 is a special case of the following theorem, which applies more generally to ramp schemes.

---

[3] One such notable example is the *directed connectivity* access structure: the parties correspond to edge slots in the complete *directed* graph and the qualified subsets are those edges that connect two distinguished nodes s and t.

[4] The usual counting arguments do not work here since one needs to enumerate over the sharing and reconstruction algorithms whose complexity may be larger than the share size.

Theorem 2. *For every $n \in \mathbb{N}$ and $1 \leq s < r < n$, any (s,r,n)-ramp secret sharing scheme for a 1-bit secret requires shares of at least $\log((r+1)/(r-s))$ bits.*

By combining Theorem 2 with the lower bound of [14], we get that any (s,r,n)-ramp secret sharing scheme must have share size at least

$$\max\left\{\log \frac{n-s+1}{r-s}, \log \frac{r+1}{r-s}\right\} \geq \log \frac{n+r-s+2}{2 \cdot (r-s)}. \tag{2}$$

Proof Technique and Limitations. We prove our lower bounds by analyzing a new game-theoretic relaxation of secret sharing. Here, we focus on threshold schemes, although our argument also applies to ramp schemes.

Given an access structure $\mathcal{A}$ and a real-valued parameter $\theta > 0$ we consider the following zero-sum game $G(\mathcal{A}, \theta)$: Alice and Bob pick sets A and B in the access structure $\mathcal{A}$, respectively, and the payoff is $(-\theta)^{|A \setminus B|}$, where $A \setminus B$ denotes set difference. We say Alice wins if she has a strategy with non-negative expected payoff, and Bob wins otherwise.

We show (in Lemma 2) that if Bob wins in the game $G(\mathcal{A}, 1/(q-1))$, then no secret sharing scheme with share size $\log q$ exists. We prove Theorem 2 by constructing such a strategy for Bob.

On the negative side, we show that our analysis is optimal for threshold access structures, so the lower bound in Theorem 1 is tight with respect to this method:

Theorem 3. *For all $1 < t < n$ and $0 < \theta \leq 1/t$, Alice wins in the game $G(\mathsf{THR}_t^n, \theta)$.*

We also show that, for any total access structure $\mathcal{A}$, this method cannot prove a lower bound exceeding $\log|\min \mathcal{A}| \leq \log \binom{n}{\lfloor n/2 \rfloor} = n - \Omega(\log n)$, where $\min \mathcal{A} = \{A \in \mathcal{A} : \forall B \in \mathcal{A}, B \not\subseteq A\}$ is the set of min-terms in $\mathcal{A}$.

Theorem 4. *For every access structure $\mathcal{A}$ and every $0 < \theta \leq 1/(|\min \mathcal{A}| - 1)$ Alice wins in the game $G(\mathcal{A}, \theta)$.*

1.2 Related Work

Known Frameworks for Proving Lower Bounds. The method of Csirmaz [19] is one of the only previously known general frameworks for proving lower bounds on share size in various access structures.[5] Csirmaz's framework is a linear programming relaxation whose variables are the entropies of the joint distributions of the shares, one for each subset of the parties. Using several Shannon information inequalities, Csirmaz was able to prove an $n/\log n$ lower bound on the entropy of shares (in a specific access structure) which, in turn, imply the same lower bound on share size (for a 1-bit secret).

---

[5] Some lower bounds were proven using other methods such as counting arguments and other tools from information theory.

We note that Csirmaz's framework does not give any non-trivial lower bounds on share size for sharing a 1-bit secret for the threshold access structure. Indeed, Csirmaz's method gives a lower bound on the information ratio of an access structure,[6] namely on the ratio between the size of the shares and the size of the secret, and for threshold schemes this ratio is 1 (using Shamir's scheme for a long enough secret; see Claim 5). Kilian and Nisan's [25] proof is the only known argument for threshold schemes and it does not seem to be useful for any other access structure, including the (t, n)-threshold access structures with t being close to n.

Csirmaz [19] showed that his framework cannot be used to show a super-linear lower bound on share size for any access structure. This claim was strengthened by Beimel and Orlov [8] who showed that certain additional "non-Shannon type" information inequalities cannot bypass the linear share size barrier (see [28] for a follow-up).

Linear Schemes. A secret sharing scheme is *linear* if the reconstruction procedure is a linear function of the shares (over some abelian group). Most previously known schemes are linear (see [7,12,26] for exceptions) and super-polynomial lower bounds for linear schemes were given in [1,6,22] via its equivalence to monotone span programs [24]. In a very recent work, Cook et al. [16] gave the first exponential lower bound for linear secret sharing schemes by giving an exponential lower bound for monotone span programs.

For *linear* $(2, n)$-threshold secret sharing schemes for a 1-bit secret, a $\log n$ lower bound on share size was proven by Karchmer and Wigderson [24]. This was generalized by Cramer et al. [18] (via a duality argument) to get a lower bound as in Equation (1). For *linear* (s, r, n)-ramp secret sharing schemes, Cramer et al. obtained a lower bound as in Eq. (2). We emphasize that our lower bounds match the lower bounds of [18] but are not restricted to linear (ramp) secret sharing schemes.

2 Access Structures and Secret Sharing

Let $\mathcal{P} \triangleq \{1, \ldots, n\}$ be a set of n parties. A collection of subsets $\mathcal{A} \subseteq 2^{\mathcal{P}}$ is *monotone* (upward-closed) if for every $B \in \mathcal{A}$ and $B \subseteq C$ it holds that $C \in \mathcal{A}$. The collection is *anti-monotone* if for every $B \in \mathcal{A}$ and $C \subseteq B$ it holds that $C \in \mathcal{A}$.

Definition 1. *A (partial) access structure $\mathcal{A} = (\mathcal{S}, \mathcal{R})$ is a pair of non-empty disjoint collections of subsets $\mathcal{R}$ and $\mathcal{S}$ of $2^{\mathcal{P}}$ such that $\mathcal{R}$ is monotone and $\mathcal{S}$ is anti-monotone. Subsets in $\mathcal{R}$ are called* qualified *and subsets in $\mathcal{S}$ are called* unqualified.

The access structure is *total* if $\mathcal{R}$ and $\mathcal{S}$ form a partition of $2^{\mathcal{P}}$. If $\mathcal{A} = (\mathcal{S}, \mathcal{R})$ is total we write $R \in \mathcal{A}$ for $R \in \mathcal{R}$ and $S \notin \mathcal{A}$ for $S \in \mathcal{S}$. Our work is mostly about the following two types of access structures:

---

[6] We thank a reviewer for pointing this out.

– The *threshold access structure* THR_t^n is a total access structure over n parties in which any t parties can reconstruct and secrecy is guaranteed against any subset of $t - 1$ parties:

$$\mathcal{S} = \{S\colon |S| \le t - 1\} \qquad \mathcal{R} = \{R\colon |R| \ge t\}.$$

– More generally, in the *ramp access structure* $\mathsf{RAMP}_{s,r}^n$, any r parties can reconstruct and secrecy is guaranteed against any s parties:

$$\mathcal{S} = \{S\colon |S| \le s\} \qquad \mathcal{R} = \{R\colon |R| \ge r\}.$$

A secret sharing scheme involves a dealer who has a secret, a set of n parties, and a partial access structure $\mathcal{A} = (\mathcal{S}, \mathcal{R})$. A secret sharing scheme for $\mathcal{A} = (\mathcal{S}, \mathcal{R})$ is a method by which the dealer distributes shares to the parties such that any subset in $\mathcal{R}$ can reconstruct the secret from its shares, while any subset in $\mathcal{S}$ cannot reveal any information on the secret. We restrict our definition to 1-bit secrets.

Definition 2 (Secret sharing). *A secret sharing scheme of a 1-bit secret for a partial access structure $\mathcal{A} = (\mathcal{S}, \mathcal{R})$ over n parties over share alphabet Σ is a pair of probability distributions p_0 and p_1 over Σ^n with the following properties:*

Reconstruction: *For every $R \in \mathcal{R}$ the marginal distributions[7] of p_0 and p_1 on the set R are disjoint.*

Secrecy: *For every $S \in \mathcal{S}$ the marginal distributions of p_0 and p_1 on the set S are identical.*

An implementation of a secret sharing scheme consists of a sharing algorithm that samples the shares from the probability distribution p_0 or p_1 depending on the value of the secret and of a reconstruction algorithm that recovers the secret from the joint values of the shares of any qualified subsets of parties. The disjointness requirement ensures that recovery by qualified subsets of parties is possible with probability 1. The secrecy requirement ensures that unqualified subsets of parties can extract no information about the secret. Thus, our definition is equivalent to the ones given, for example, in [2, Definition 3.6] and in [3, Definitions 2 and 3].

An Alternative Formulation of Secret Sharing. Here is an equivalent formulation of secret sharing. For $x \in \mathbb{Z}_q^n$, we use $[x]$ to denote the set of non-zero entries of x, namely $[x] = \{i\colon x_i \ne 0\}$, and $[x]^{\complement}$ for the complementary set of zero entries. In this notation, $[x - y]$ is the set of coordinates that x and y differ on and $[x - y]^{\complement}$ is the set of coordinates that they agree on. A function $\phi_S\colon \mathbb{Z}_q^n \to \mathbb{C}$ is an *S-junta* if the value $\phi_S(x_1, \ldots, x_n)$ is determined by the inputs $x_i\colon i \in S$.

---

[7] Given two random variables X and Y whose joint distribution is known, the marginal distribution of X is the probability distribution of X averaging over all possible values of Y. Namely, it is $\mathbf{Pr}[X = x] = \sum_y \mathbf{Pr}[X = x, Y = y]$.

Lemma 1. *A secret sharing scheme of a 1-bit secret for a partial access struc-ture $\mathcal{A} = (\mathcal{S}, \mathcal{R})$ over share alphabet $\mathbb{Z}_q$ exists if and only if there exists a function $f \colon \mathbb{Z}_q^n \to \mathbb{R}$ that is not identically zero satisfying the following properties:*

Reconstruction: *For all $x, y \in \mathbb{Z}_q^n$ such that $[x - y]^{\complement} \in \mathcal{R}$, $f(x) \cdot f(y) \geq 0$.*
Secrecy: *For every $S \in \mathcal{S}$ and every S-junta $\phi_S \colon \mathbb{Z}_q^n \to \mathbb{C}$, $\mathbf{E}[f(x)\phi_S(x)] = 0$, where the expectation is over the uniform probability distribution of $x \in \mathbb{Z}_q^n$.*

Proof. For a secret sharing scheme p_0, p_1, we set $f(x) = p_0(x) - p_1(x)$. The functions p_0 and p_1 have disjoint support (otherwise even reconstruction by all parties is impossible) so f cannot be identically zero. The reconstruction implies that if $[x - y]^{\complement} \in \mathcal{R}$, then at least one of p_0 and p_1 must assign zero probability to both x and y, so $f(x) \cdot f(y)$ equals either $p_0(x) \cdot p_0(y)$ or $(-p_1(x)) \cdot (-p_1(y))$. In either case $f(x) \cdot f(y) \geq 0$. For secrecy, since p_0 and p_1 have the same marginals on $S \in \mathcal{S}$, $\mathbf{E}[p_0(x)\phi_S(x)] = \mathbf{E}[p_1(x)\phi_S(x)]$ so $\mathbf{E}[f(x)\phi_S(x)] = 0$.

In the other direction, let $p_0(x) = C \cdot \max\{f(x), 0\}$ and let $p_1(x) = C \cdot \max\{-f(x), 0\}$ for a suitable scaling constant $C > 0$ that makes p_0 and p_1 be valid probability distributions (it exists since f is nonzero). We show reconstruction by contrapositive: If p_0 and p_1 did not have disjoint support on some set $R \in \mathcal{R}$, there would exist $x, y \in \mathbb{Z}_q^n$ such that $p_0(x) > 0$, $p_1(y) > 0$, and $[x - y]^{\complement} = R$, implying $f(x) > 0$, $f(y) < 0$, and therefore $f(x) \cdot f(y) < 0$. For secrecy, by construction we have $f = (p_0 - p_1)/C$, so $\mathbf{E}[p_0(x)\phi_S(x)] = \mathbf{E}[p_1(x)\phi_S(x)]$ for every test function ϕ_S that only depends on coordinates in $S \in \mathcal{S}$. Since no ϕ_S can distinguish between p_0 and p_1 on S, the statistical distance between the marginal distribution of p_0 and p_1 on S is zero, so the two are identical.

3   A Zero-Sum Game and Proof of Theorem 2

Given a partial access structure $\mathcal{A} = (\mathcal{S}, \mathcal{R})$ and a real parameter $\theta > 0$ we define the following zero-sum game $G(\mathcal{A}, \theta)$ between Alice and Bob. The actions are a set $A \notin \mathcal{S}$ for Alice and a set $B \in \mathcal{R}$ for Bob. The payoff of the game is $(-\theta)^{|A \setminus B|}$. We say Alice wins if she has a strategy with non-negative expected payoff and we say Bob wins if he has a strategy with negative expected payoff (the expectations are over the randomness of Alice and Bob, respecively). By von Neumann's minimax theorem the game has a unique winner.

Lemma 2. *If there exists a secret sharing scheme for $\mathcal{A}$ with alphabet size $q \in \mathbb{N}$, then Alice wins in the game $G(\mathcal{A}, 1/(q - 1))$.*

Our proof of Lemma 2 uses Fourier analysis, which we briefly recall here. The characters of the group $\mathbb{Z}_q^n$ are the complex-valued functions $\chi_a \colon \mathbb{Z}_q^n \to \mathbb{C}$, where a ranges over $\mathbb{Z}_q^n$, defined as $\chi_a(x) = \omega^{\langle a, x \rangle}$, $\omega = e^{2\pi i/q}$. The characters are an orthonormal basis with respect to the inner product $\langle f, g \rangle = \mathbf{E}_x[f(x) \cdot \overline{g(x)}]$ with x chosen uniformly from $\mathbb{Z}_q^n$. The characters inherit the group structure:

$\chi_a \cdot \chi_b = \chi_{a+b}$ and $\chi_a^{-1} = \overline{\chi_a} = \chi_{-a}$. Every function $f \colon \mathbb{Z}_q^n \to \mathbb{C}$ can then be uniquely written as a linear combination $f = \sum_{a \in \mathbb{Z}_q^n} \hat{f}(a) \cdot \chi_a$ with the Fourier coefficients $\hat{f}(a)$ given by $\hat{f}(a) = \langle f, \chi_a \rangle = \mathbf{E}_x[f(x) \cdot \overline{\chi_a(x)}]$.

Proof of Lemma 2. We show that Alice has a winning strategy. That is, we show that Alice has a strategy such that for every possible action of Bob, the expected payoff of the game is non-negative.

We identify the alphabet with the elements of the group $\mathbb{Z}_q$. Let $f \colon \mathbb{Z}_q^n \to \mathbb{R}$ be the function $f(x) = p_0(x) - p_1(x)$. Alice plays set A with probability proportional to $\sum_{a \colon [a] = A} |\hat{f}(a)|^2$. By the secrecy part of Lemma 1, $\mathbf{E}[f(x) \cdot \overline{\chi_a(x)}] = 0$ whenever $[a] \in \mathcal{S}$, so Alice's strategy is indeed supported on sets outside $\mathcal{S}$.

Now let B be an arbitrary set in $\mathcal{R}$. By the reconstruction part of Lemma 1 and the fact that f is real-valued, for every $x \in \mathbb{Z}_n^q$ and every $z \in \mathbb{Z}_n^q$ such that $[z]^{\complement} = B$, we have that

$$f(x) \cdot \overline{f(x - z)} = f(x) \cdot f(x - z) \geq 0. \tag{3}$$

Let x be uniform in $\mathbb{Z}_q^n$ and z be uniform in $\mathbb{Z}_q^n$ conditioned on $[z]^{\complement} = B$. Averaging over this distribution, we have

$$\mathbf{E}_{x,z}[f(x) \cdot \overline{f(x - z)}] = \sum_{a,b \in \mathbb{Z}_q^n} \hat{f}(a) \cdot \overline{\hat{f}(b)} \cdot \mathbf{E}_{x,z}[\chi_a(x) \cdot \overline{\chi_b(x - z)}]$$

$$= \sum_a |\hat{f}(a)|^2 \cdot \mathbf{E}_z[\chi_a(z)]$$

$$= \sum_a |\hat{f}(a)|^2 \cdot \prod_{i \in [a]} \mathbf{E}_z[\omega^{a_i z_i}],$$

where the first equality follows by writing $f(x)$ and $\overline{f(x-z)}$ using their Fourier representation and using linearity of expectation, the second equality follows since x and z are independent and since $\mathbf{E}_x[\chi_a(x) \cdot \overline{\chi_b(x)}] = 0$ for $a \neq b$, and the last equality follows since z is chosen from a product distribution.

The expression $\mathbf{E}[\omega^{a_i z_i}]$ evaluates to one when i is in B (since z_i is fixed to zero). Otherwise, z_i is uniformly distributed over the set $\mathbb{Z}_q \setminus \{0\}$ and

$$\mathbf{E}_z[\omega^{a_i z_i}] = \frac{1}{q-1} \sum_{z_i \in \mathbb{Z}_q \setminus \{0\}} \omega^{a_i z_i} = \frac{1}{q-1} \left(\sum_{z_i \in \mathbb{Z}_q} \omega^{a_i z_i} - 1 \right) = -\frac{1}{q-1}.$$

Therefore, $\prod_{i \in [a]} \mathbf{E}_z[\omega^{a_i z_i}] = (-1/(q-1))^{|[a] \setminus B|}$, and by Eq. (3)

$$\sum_a |\hat{f}(a)|^2 \cdot \left(\frac{-1}{q-1} \right)^{|[a] \setminus B|} \geq 0.$$

Grouping all a's for which $[a] = A$, we get that

$$\sum_A \left(\sum_{a \colon [a] = A} |\hat{f}(a)|^2 \right) \cdot \left(-\frac{1}{q-1} \right)^{|A \setminus B|} \geq 0 \qquad \text{for all } B \in \mathcal{R}.$$

Therefore, Alice's strategy has non-negative expected payoff with respect to every possible action of Bob. ∎

Proof of Theorem 2. It is sufficient to prove Theorem 2 in the case $n = r + 1$: If a secret sharing scheme for $\mathsf{RAMP}^n_{s,r}$ existed, then a secret sharing for $\mathsf{RAMP}^{r+1}_{s,r}$ over the same alphabet can be obtained by discarding the remaining $n - r - 1$ parties and their shares.

We now give a winning strategy for Bob in the game $G(\mathsf{RAMP}^{r+1}_{s,r}, \theta)$ for any $\theta > (r - s)/(s + 1)$. By Lemma 2 it then follows that no secret sharing scheme over an alphabet of size $(r + 1)/(r - s)$ exists.

Bob's strategy is to uniformly choose a set B of size r (which is in $\mathcal{R}$). Then for every set $A \notin \mathcal{S}$, either $A \subseteq B$ and then $|A \setminus B| = 0$, or $A \not\subseteq B$ and then $|A \setminus B| = 1$ (since B includes all parties except one). Thus, for every $A \notin \mathcal{S}$, the expected payoff is

$$\mathbf{E}_B \left[(-\theta)^{|A \setminus B|} \right] = 1 \cdot \mathbf{Pr}_B[A \subseteq B] - \theta \cdot \mathbf{Pr}_B[A \not\subseteq B]$$

$$= 1 \cdot \frac{r + 1 - |A|}{r + 1} - \theta \cdot \frac{|A|}{r + 1}$$

$$\leq \frac{r - s}{r + 1} - \theta \cdot \frac{s + 1}{r + 1}, \tag{4}$$

where the inequality follows since $|A| \geq s+1$. If $\theta > (r-s)/(s+1)$ this expression is less than zero, i.e., Bob wins. ∎

It is also possible to deduce Theorem 2 directly from Lemma 2 by showing the existence of a winning strategy for Bob in the game $G(\mathsf{RAMP}^n_{s,r}, \theta)$ whenever $\theta > (r-s)/(s+1)$ (rather than for $G(\mathsf{RAMP}^{r+1}_{s,r}, \theta)$, as we did above). Let R be a random subset of $r + 1$ parties. Bob's strategy has the form $B = B_0 \cup B_1$, where B_0 is a uniformly random subset of R of size r and B_1 is a random subset of $R^{\complement}$ obtained by including each element independently with probability $p = \theta/(1+\theta)$. The value of p is chosen so that a random variable that equals 1 with probability p and $-\theta$ with probability $1 - p$ is unbiased.

Let A, where $|A| \geq s + 1$, be any action of Alice. For a fixed choice of R, if $A \setminus R$ is nonempty, by our choice of probability p the expected payoff is zero. Otherwise, A is a subset of R, and by Eq. (4) the expected payoff is at most $-(s + 1) \cdot \theta + (r - s) < 0$. Since the event $A \subseteq R$ has positive probability the expected payoff is negative and Bob wins.

4   Limitations of the Game Relaxation

In the case of threshold access structures Theorem 2 shows that Bob has a winning strategy in the game $G(\mathsf{THR}^n_t, \theta)$ whenever $\theta > 1/t$. We now prove Theorem 3, which states that our analysis is optimal: There exists a winning strategy for Alice when $\theta \leq 1/t$.

We also prove Theorem 4: For every total access structure $\mathcal{A}$ over n parties, Alice has a winning strategy in $G(\mathcal{A}, \theta)$ for every $\theta \leq 1/(|\mathcal{A}| - 1)$. As the proof of Theorem 4 is simpler we present that one first. We remark Theorem 4 can be generalized to any partial access structure $(\mathcal{S}, \mathcal{R})$ by replacing $\mathcal{A}$ by $\mathcal{R}$ in the proof.

Proof of Theorem 4. Alice's strategy is uniformly random over all minterms $A \in \min \mathcal{A}$. Then, for every $B \in \mathcal{A}$ and $\theta < 1$, it holds that

$$
\begin{aligned}
\mathbf{E}_A[(-\theta)^{|A \backslash B|}] ={}& \mathbf{E}_A[(-\theta)^{|A \backslash B|} \mid A \subseteq B] \cdot \mathbf{Pr}_A[A \subseteq B] + \\
& \mathbf{E}_A[(-\theta)^{|A \backslash B|} \mid A \not\subseteq B] \cdot \mathbf{Pr}_A[A \not\subseteq B] \\
\geq{}& 1 \cdot \mathbf{Pr}_A[A \subseteq B] - \theta \cdot \mathbf{Pr}_A[A \not\subseteq B] \\
={}& (1 + \theta) \cdot \mathbf{Pr}_A[A \subseteq B] - \theta \\
\geq{}& (1 + \theta) \cdot \frac{1}{|\min \mathcal{A}|} - \theta.
\end{aligned}
$$

This is non-negative when $\theta \leq 1/(|\min \mathcal{A}| - 1)$. ∎

Proof of Theorem 3. Let $a_0, \ldots, a_n$ be the following sequence of integers:

$$
a_0 = \cdots = a_{t-1} = 0, \quad a_t = 1, \quad a_s = k_t \cdot a_{s-1} + \cdots + k_0 \cdot a_{s-t-1}
$$

for $t + 1 \leq s \leq n$, where k_j is the coefficient of x^j in the formal expansion of $(x + 1)^t \cdot (1/\theta - x)$. By expanding this expression according to the Binomial formula, we see that the numbers $k_0, \ldots, k_t$ are non-negative when $\theta \leq 1/t$ because

$$
k_j = \binom{t}{j} \left(\frac{1}{\theta} - \frac{j}{t - j + 1} \right) \geq 0
$$

for all $0 \leq j \leq t$. Therefore a_s is also non-negative for all s.

Alice plays set A with probability proportional to the number $a_{|A|}$. We will prove that this is a winning strategy for Alice. When $B = \{1, \ldots, n\}$, then $\mathbf{E}_A[(-\theta)^{|A \backslash B|}] = 1$ and Alice wins. Now let $B \subseteq \{1, \ldots, n\}$ be any set such that $t \leq |B| < n$. Let

$$
\theta_j = \begin{cases} 1, & \text{if } j \in B, \\ -\theta, & \text{if } j \notin B. \end{cases}
$$

Then,

$$
\mathbf{E}_A[(-\theta)^{|A \backslash B|}] \propto \sum_A a_{|A|} \prod_{j \in A} \theta_j = \sum_{s=0}^{n} a_s w_s \quad \text{where} \quad w_s = \sum_{A:\, |A| = s} \prod_{j \in A} \theta_j.
$$

The number w_s can be represented as the coefficient of z^s in the formal expansion of $g_0(z) = \prod_{j=1}^{n}(1 + \theta_j z)$. Since exactly $|B|$ of the θ_j's equal 1 and the other $n - |B|$ equal $-\theta$, it follows that

$$
g_0(z) = (1 + z)^{|B|} \cdot (1 - \theta z)^{n - |B|}. \tag{5}
$$

The numbers $a_0, \ldots, a_n$ (as defined in the beginning of the proof) are defined by an order t homogeneous linear degree relation with constant coefficients whose characteristic equation is $(x+1)^t \cdot (1/\theta - x) = 0$. This equation has roots -1 (with multiplicity t) and $1/\theta$ (with multiplicity 1). Therefore,

$$a_s = C \cdot \theta^{-s} + \sum_{i=0}^{t-1} c_i \cdot s^i \cdot (-1)^s$$

where $c_0, \ldots, c_{t-1}$ and C are constants determined by the initial conditions on $a_0, \ldots, a_t$. We can now write

$$\sum_{s=0}^{n} a_s \cdot w_s = C \cdot \sum_{s=0}^{n} w_s \cdot \theta^{-s} + \sum_{i=0}^{t-1} c_i \cdot \sum_{s=0}^{n} w_s \cdot s^i \cdot (-1)^s.$$

Recall that g_0 is the generating function of w_s which means that $g_0(z) = \sum_{s=0}^{n} w_s \cdot z^s$. So, the term $\sum_{s=0}^{n} w_s \cdot \theta^{-s}$ equals $g_0(1/\theta) = 0$. To finish the proof, we show that $\sum_{s=0}^{n} w_s \cdot s^i \cdot (-1)^s = 0$ for all $i \leq t - 1$ (this implies that Alice's strategy has a 0 payoff, which means that she wins the game). Let $g_i(z) = z \cdot g'_{i-1}(z)$ for $1 \leq i \leq t - 1$ where g'_{i-1} is the derivative of g_{i-1}. On the one hand, since -1 is a root of g_0 of multiplicity t, $g_i(-1) = 0$ for all $i \leq t - 1$. On the other hand, $g_i(z)$ has the formal expansion $\sum_{s=0}^{n} w_s \cdot s^i \cdot z^s$. Therefore, $\sum_{s=0}^{n} w_s \cdot s^i \cdot (-1)^s$ must equal zero. ∎

5   Concluding Remarks

Theorem 1 requires that the shares given to all parties have the same length. Its proof extends easily to yield the following generalization: For every n, every $1 < t < n$, and every (t, n)-threshold secret sharing scheme in which party i receives a $\log q_i$-bit share and $q_1 \leq q_2 \leq \cdots \leq q_n$ it must hold that

$$\frac{1}{q_1} + \cdots + \frac{1}{q_{t+1}} \leq 1. \tag{6}$$

In particular, inequality (6) implies that the *average* share size must be at least $\log(t+1)$. We sketch the Proof in Appendix B. Kilian and Nisan [25] prove the same for $(n-t+1, n)$-threshold access structures.

By Theorem 3 our analysis of threshold secret sharing is tight within the game-theoretic relaxation that we introduce here. As the lower bound of Kilian and Nisan [25] is incomparable with ours, their analysis cannot be cast in terms of a winning strategy in our game. It is, however, possible to capture both our analysis and that of Kilian and Nisan by a single *linear program*. We performed computer experiments to investigate the feasibility of one such family of linear programs, but were unable to obtain better lower bounds on share size.

We do not know what is the best possible lower bound on share size that our method can give among all access structures on n parties. Theorem 1 shows a lower bound of $\log(n-1)$ is attainable, while Theorem 4 shows that a lower bound of $\log\binom{n}{\lfloor n/2 \rfloor}$ cannot be proved. The best possible bound is the logarithm of

$$b_n = \min_{\mathcal{A}} \max \{q : \text{Bob wins in } G(\mathcal{A}, 1/(q-1))\},$$

where the minimum is taken over all access structures $\mathcal{A}$ on n parties. We can prove that if the payoff function is replaced by $(-\theta)^{|A\triangle B|}$, where $\triangle$ is symmetric set difference, then the quantity analogous to b_n is upper bounded by $O(n^2)$.

Acknowledgments. We thank Moni Naor for telling us about the work of Kilian and Nisan. We thank the anonymous reviewers for their useful advice.

A   On the Tightness of Theorem 2

We show that Theorem 2 is tight when $t = n - 1$ and n is the power of a prime. This result is known (see e.g. [17, Theorem 11.13]) and we give it here for completeness.

Claim 5. *For every power of a prime n there exists a $(n-1)$-out-of-n secret sharing scheme for $\log n$-bit secrets with $\log n$-bit shares.*

Claim 5 follows by a small optimization of Shamir's secret sharing scheme. We give the construction and sketch the correctness proof.

To share a secret $s \in \mathbb{F}_n$, let $p(x) = sx^{n-2} + r(x)$, where r is a random polynomial of degree $n - 3$ and all algebra is over the finite field $\mathbb{F}_n$. The shares are the n values $p(x)$ as x ranges over $\mathbb{F}_q$. Reconstruction is immediate as the polynomial p can be interpolated from any $n - 1$ of its values.

For secrecy, we show for any $s \in \mathbb{F}_n$ and distinct elements $x_1, \ldots, x_{n-2} \in \mathbb{F}_n$, the vector $(p(x_1), \ldots, p(x_{n-2}))$ is uniformly random in $\mathbb{F}_n^{n-2}$. Since $p(x) = sx^{n-2} + r(x)$ it suffices to show that $(r(x_1), \ldots, r(x_{n-2}))$ is uniformly random. This is true because the evaluation map that takes the coefficients of r into its values $r(x_1), \ldots, r(x_{n-2})$ is a full-rank Vandermonde matrix.

B   Proof Sketch of Inequality (6)

The proof of inequality (6) is a direct extension of the proof of Theorem 1. We describe the differences. The payoff function in the game G in Lemma 2 becomes $\prod_{i\in A\setminus B} -1/(q_i - 1)$. The generalized lemma can be proved via Fourier analysis over the product group $\mathbb{Z}_{q_1} \times \cdots \times \mathbb{Z}_{q_n}$.

As in the proof of Theorem 1, it is sufficient to establish inequality (6) in the special case $t = n - 1$. Bob then plays set $B = \{1, \ldots, n\} \setminus \{i\}$ with probability proportional to $1 - 1/q_i$. It can be verified that when $\sum_{i=1}^n 1/q_i > 1$ this is a winning strategy for Bob.

References

1. Babai, L., Gál, A., Wigderson, A.: Superpolynomial lower bounds for monotone span programs. Combinatorica **19**(3), 301–319 (1999)
2. Beimel, A.: Secure schemes for secret sharing and key distribution. Ph.D. thesis, Technion - Israel Institute of Technology (1996)

3. Beimel, A.: Secret-sharing schemes: a survey. In: Chee, Y.M., Guo, Z., Ling, S., Shao, F., Tang, Y., Wang, H., Xing, C. (eds.) IWCC 2011. LNCS, vol. 6639, pp. 11–46. Springer, Heidelberg (2011)

4. Beimel, A., Chor, B.: Universally ideal secret-sharing schemes. IEEE Trans. Inf. Theor. **40**(3), 786–794 (1994)

5. Beimel, A., Franklin, M.K.: Weakly-private secret sharing schemes. In: Vadhan, S.P. (ed.) TCC 2007. LNCS, vol. 4392, pp. 253–272. Springer, Heidelberg (2007)

6. Beimel, A., Gál, A., Paterson, M.: Lower bounds for monotone span programs. Comput. Complex. **6**(1), 29–45 (1997)

7. Beimel, A., Ishai, Y.: On the power of nonlinear secrect-sharing. In: 16th Annual IEEE Conference on Computational Complexity, CCC, pp. 188–202 (2001)

8. Beimel, A., Orlov, I.: Secret sharing and non-shannon information inequalities. IEEE Trans. Inf. Theor. **57**(9), 5634–5649 (2011)

9. Benaloh, J.C., Leichter, J.: Generalized secret sharing and monotone functions. In: Goldwasser, S. (ed.) CRYPTO 1988. LNCS, vol. 403, pp. 27–35. Springer, Heidelberg (1990)

10. Blakley, G.R., Meadows, C.: Security of ramp schemes. In: Blakely, G.R., Chaum, D. (eds.) CRYPTO 1984. LNCS, vol. 196, pp. 242–268. Springer, Heidelberg (1985)

11. Blakley, G.R.: Safeguarding cryptographic keys. Proc. AFIPS Natl. Comput. Conf. **22**, 313–317 (1979)

12. Bogdanov, A., Ishai, Y., Viola, E., Williamson, C.: Bounded indistinguishability and the complexity of recovering secrets. In: Robshaw, M., Katz, J. (eds.) CRYPTO 2016. LNCS, vol. 9816, pp. 593–618. Springer, Heidelberg (2016). doi:10.1007/978-3-662-53015-3_21

13. Capocelli, R.M., Santis, A.D., Gargano, L., Vaccaro, U.: On the size of shares for secret sharing schemes. J. Cryptol. **6**(3), 157–167 (1993)

14. Cascudo Pueyo, I., Cramer, R., Xing, C.: Bounds on the threshold gap in secret sharing and its applications. IEEE Trans. Inf. Theor. **59**(9), 5600–5612 (2013)

15. Chen, H., Cramer, R.: Algebraic geometric secret sharing schemes and secure multiparty computations over small fields. In: Dwork, C. (ed.) CRYPTO 2006. LNCS, vol. 4117, pp. 521–536. Springer, Heidelberg (2006)

16. Cook, S.A., Pitassi, T., Robere, R., Rossman, B.: Exponential lower bounds for monotone span programs. Electron. Colloq. Comput. Complex. **23**, 64 (2016)

17. Cramer, R., Damgård, I., Nielsen, J.B.: Secure Multiparty Computation and Secret Sharing. Cambridge University Press, Cambridge (2015)

18. Cramer, R., Fehr, S., Stam, M.: Black-box secret sharing from primitive sets in algebraic number fields. In: Shoup, V. (ed.) CRYPTO 2005. LNCS, vol. 3621, pp. 344–360. Springer, Heidelberg (2005)

19. Csirmaz, L.: The size of a share must be large. J. Cryptol. **10**(4), 223–231 (1997)

20. Farràs, O., Hansen, T., Kaced, T., Padró, C.: Optimal non-perfect uniform secret sharing schemes. In: Garay, J.A., Gennaro, R. (eds.) CRYPTO 2014, Part II. LNCS, vol. 8617, pp. 217–234. Springer, Heidelberg (2014)

21. Farràs, O., Molleví, S.M., Padró, C.: A note on non-perfect secret sharing. IACR Cryptology ePrint Archive, p. 348 (2016)

22. Gál, A.: A characterization of span program size and improved lower bounds for monotone span programs. Comput. Complex. **10**(4), 277–296 (2001)

23. Ito, M., Saito, A., Nishizeki, T.: Multiple assignment scheme for sharing secret. J. Cryptol. **6**(1), 15–20 (1993)

24. Karchmer, M., Wigderson, A.: On span programs. In: 8th Annual Structure in Complexity Theory Conference, pp. 102–111 (1993)

25. Kilian, J., Nisan, N.: Unpublished (1990). Referenced in [4,2,5,14]
26. Komargodski, I., Naor, M., Yogev, E.: How to share a secret, infinitely. IACR Cryptology ePrint Archive 2016, 194 (2016)
27. Martin, K.M., Paterson, M.B., Stinson, D.R.: Error decodable secret sharing and one-round perfectly secure message transmission for general adversary structures. Crypt. Commun. **3**, 65–86 (2011)
28. Mollevi, S.M., Padró, C., Yang, A.: Secret sharing, rank inequalities, and information inequalities. IEEE Trans. Inf. Theor. **62**(1), 599–609 (2016)
29. Paterson, M.B., Stinson, D.R.: A simple combinatorial treatment of constructions and threshold gaps of ramp schemes. Crypt. Commun. **5**, 229–240 (2013)
30. Shamir, A.: How to share a secret. Commun. ACM **22**(11), 612–613 (1979)
31. Stinson, D.R., Wei, R.: An application of ramp schemes to broadcast encryption. Inf. Process. Lett. **69**, 131–135 (1999)

How to Share a Secret, Infinitely

Ilan Komargodski$^{(\boxtimes)}$, Moni Naor, and Eylon Yogev

Weizmann Institute of Science, Rehovot, Israel
{ilan.komargodski,moni.naor,eylon.yogev}@weizmann.ac.il

Abstract. Secret sharing schemes allow a dealer to distribute a secret piece of information among several parties such that only qualified subsets of parties can reconstruct the secret. The collection of qualified subsets is called an **access structure**. The best known example is the k-threshold access structure, where the qualified subsets are those of size at least k. When $k = 2$ and there are n parties, there are schemes where the size of the share each party gets is roughly $\log n$ bits, and this is tight even for secrets of 1 bit. In these schemes, the number of parties n must be given in advance to the dealer.

In this work we consider the case where the set of parties is not known in advance and could potentially be infinite. Our goal is to give the tth party arriving the smallest possible share as a function of t. Our main result is such a scheme for the k-threshold access structure where the share size of party t is $(k - 1) \cdot \log t + \mathsf{poly}(k) \cdot o(\log t)$. For $k = 2$ we observe an *equivalence* to prefix codes and present matching upper and lower bounds of the form $\log t + \log \log t + \log \log \log t + O(1)$. Finally, we show that for any access structure there exists such a secret sharing scheme with shares of size 2^{t-1}.

1 Introduction

> 640K ought to be enough for anybody
>
> *Misattributed to Bill Gates, 1981*

Engineering scalable systems is a delicate business: important decisions have to be made regarding balancing scalability and efficiency when fixing system parameters (such as the representation size of a date, the number of clients the system can serve simultaneously, security parameters and more). This inherent tradeoff between scalability and efficiency has had devastating consequences. There are many Y2K [34] style horror stories such as losing contact with the NASA spacecraft "Deep Impact" when its internal clock overflowed, triggering

I. Komargodski, et al.—Research supported in part by grants from the Israel Science Foundation grant no. 1255/12, BSF and from the I-CORE Program of the Planning and Budgeting Committee and the Israel Science Foundation (grant no. 4/11). Moni Naor is the incumbent of the Judith Kleeman Professorial Chair. Ilan Komargodski is supported in part by a Levzion fellowship.

© International Association for Cryptologic Research 2016
M. Hirt and A. Smith (Eds.): TCC 2016-B, Part II, LNCS 9986, pp. 485–514, 2016.
DOI: 10.1007/978-3-662-53644-5_19

an endless series of computer reboots [20], and the IPv4 address exhaustion problems caused by the limited allocation size for numeric Internet addresses [33]. Can we design scalable systems without suffering a great deal of efficiency costs? In this work we investigate methods that do not assume a fixed upper bound on the number of participants in the area of secret sharing.

Secret sharing is a method by which a secret piece of information can be distributed among n parties so that any qualified subset of parties can reconstruct the secret, while every unqualified subset of parties learns nothing about the secret. The collection of qualified subsets is called an **access structure**. Secret sharing schemes are a basic primitive and have found applications in cryptography and distributed computing; see the extensive survey of Beimel [2]. A significant goal in secret sharing is to minimize the share size, namely, the amount of information distributed to the parties.

Secret sharing schemes were introduced in the late 1970s by Shamir [31] and Blakley [8] for the k-out-of-n threshold access structures that includes all subsets of cardinality at least k for $1 \leq k \leq n$. Their constructions are fairly efficient both in the size of the shares and in the computation required for sharing and reconstruction. Ito et al. [22] showed the existence of a secret sharing scheme for every (monotone) access structure. In their scheme the size of the shares is proportional to the depth 2 complexity of the access structure when viewed as a Boolean function (and hence shares are exponential for most structures). Benaloh and Leichter [5] gave a scheme with share size polynomial in the monotone *formula* complexity of the access structure. Karchmer and Wigderson [24] generalized this construction so that the size is polynomial in the monotone span program complexity.

All of these schemes require that an upper bound on the number of participants is known in advance. However, in many scenarios this is either unrealistic or prone to disaster. Moreover, even if a crude upper bound n is known in advance, it is preferable to have shares as small as possible if the eventual number of participants is much smaller than this bound on n.

In this work we consider the well motivated, yet almost unexplored[1], case where the set of parties is *not* known in advanced and could potentially be infinite. Our goal is to give the tth party arriving the smallest possible share as a function of t. We require that in each round, as a new party arrives, there is no communication to the parties that have already received shares, i.e. the dealer distributes a share only to the new party. We call such access structures evolving: the parties arrive one by one and, in the most general case, a qualified subset is revealed to the dealer only when all parties in that subset are present (in special cases the dealer knows the access structure to begin with, just does not have an upper bound on the number of parties). For this to make sense, we assume that the changes to the access structure are monotone, namely, parties are only added and qualified sets remain qualified.

Our first result is a construction of a secret sharing scheme for *any* evolving access structure.

---

[1] But see the work of Csirmaz and Tardos [15] discussed below.

Theorem 1. *For every evolving access structure there is a secret sharing scheme for a 1-bit secret where the share size of the tth party is 2^{t-1}.*

Then, we construct more efficient schemes for specific access structures. We focus on the evolving k-threshold access structure for $k \in \mathbb{N}$, where at any point in time any k parties can reconstruct the secret but no $k - 1$ parties can learn anything about the secret.

Theorem 2 (Informal). *There is a secret sharing scheme for the evolving k-threshold access structure and a 1-bit secret in which the share size of the tth party is $(k - 1) \cdot \log t + \mathsf{poly}(k) \cdot o(\log t)$.*

For $k = 2$, we present a construction for the evolving 2-threshold access structure with slightly better low order terms. In this scheme the share size of the tth party is $\log t + \log \log t + 2 \log \log \log t + O(1)$.[2] To complement this construction, we prove a matching lower bound showing that our scheme is tight.

Theorem 3. *For any constant $c \in \mathbb{N}$, there is no secret sharing scheme for the evolving 2-threshold access structure and a 1-bit secret in which the share size of the tth party is at most $\log t + \log \log t + c$.*

Finally, we present a tight connection to prefix codes for the integers. A prefix code is a code in which no codeword is a prefix of any other codeword. These codes are widely used, for example in country calling codes, the UTF-8 system for encoding Unicode characters, and more.

Theorem 4. *Let $\sigma \colon \mathbb{N} \to \mathbb{N}$. A prefix code for the integers in which the length of the tth codeword is $\sigma(t)$ exists if and only if a secret sharing scheme for the evolving 2-threshold access structure and 1-bit secret in which the share size of the tth party is $\sigma(t)$.*

1.1   Discussion

Schemes for General Access Structures. In the classical setting of secret sharing many schemes are known for general access structures, depending on their representation [5,22,24]. All of these schemes result with shares of exponential size for general access structures. One of the most important open problems in the area of secret sharing is to prove the necessity of long shares, namely, find an access structure (even a non-explicit one) that requires exponential size shares.

Our scheme for general evolving access structures also results with exponential size shares. Since any access structure can be made evolving, we cannot hope to obtain anything better than exponential in general (unless we have a major breakthrough in the classical setting).

Threshold Schemes. In the classical setting there are several different schemes for the threshold access structure. One of the best such schemes (in terms of the

---

[2] See Sects. 4 and 5 for efficient generalizations that support larger domains of secrets.

computation needed for sharing and reconstruction and in terms of the share size) is due to Shamir [31]. In this scheme, to share a 1-bit secret among n parties, roughly $\log n$ bits have to be distributed to each party. It is known that $\log n$ bits are essentially required, so Shamir's scheme is optimal (see [12] for the original proof of Kilian and Nisan [25], an improvement, and a discussion of the history; see also [9]).

Let us review Shamir's scheme for the k-out-of-n threshold access structure. The dealer holding a secret bit s, samples a random polynomial $p(\cdot)$ of degree $k - 1$ with coefficients over $\mathsf{GF}(q)$, where the free coefficient is fixed to be s, and gives party $i \in [n]$ the field element $p(i)$. q is chosen to be the smallest prime (or a power of a prime) larger than n. Correctness of the scheme follows by the fact that k points on a polynomial of degree $k - 1$ completely define the polynomial and allow for computing $p(0) = s$. Security follows by a counting argument showing that given less than k points, both possibilities for the free coefficient are equally likely. The share of each party is an element in the field $\mathsf{GF}(q)$ that can be represented using $\log q \approx \log n$ bits. Notice that the share size is independent of k.

As a first attempt one might try to adapt this procedure to the *evolving* setting. But since n is not fixed, what q should we choose? A natural idea is to use an extension field. Roughly, we would simulate the dealer for Shamir's scheme, sample a random polynomial of degree $k - 1$ and increase the field size from which we compute shares as more parties arrive. Ideally, for the share of the tth party we will use a field of size $O(t)$. This implies that the share size of party t would be $\log(O(t)) \ll \log t + \log \log t$ for large enough t. The lower bound in Theorem 3 means that *no such solution can work!*

We take a different path for obtaining efficient schemes. For example, for $k = 2$, our scheme results with essentially optimal share size for the tth party: the first two high order terms are $\log t + \log \log t$ (without hidden constant factors) and there is an additional lower order term of $2 \log \log \log t + 6$. See the simplified scheme in Sect. 4.

Linearity of Our Schemes. In a linear scheme the secret is viewed as an element of a finite field, and the shares are obtained by applying a linear mapping to the secret and several independent random field elements. Equivalently, a linear scheme is defined by requiring that each qualified set reconstructs the secret by applying a linear function to its shares [1, Sect. 4.1]. Most of the known schemes are linear (see [3] for an exception). Linear schemes are very useful for updating and manipulating secret shares (cf. proactive secret sharing [21]) and have many applications, most notably for secure multi-party computation [4,14]. Our schemes from Theorems 1 and 2 are linear (see Sect. 5.5 for details), whereas the scheme based on prefix codes from Theorem 4 is non-linear.

1.2 Related Work

Most similar to our setting is the notion of *on-line* secret sharing of Csirmaz and Tardos [15]. Csirmaz and Tardos present a scheme for any access structure in which every party participates in at most d qualified sets, where d is an upper bound known in advance. The share size of every party in this scheme is linear in d. In addition, Csirmaz and Tardos presented a scheme for the evolving 2-threshold access structure in which the share size of party t is linear in t. Our Theorem 2 is an exponential improvement on the latter.

There are numerous areas where systems are designed to work without any fixed upper bound on the size or the duration they will be used. A few examples include prefix codes of the integers (a.k.a. prefix-free encodings), such as the Elias code [17] or the online encoding of Dodis et al. [16], labeling nodes for testing adjacency in possibly infinite graphs [23], forward-secure signatures with an unbounded number of time periods [29], and data structures for approximate set membership (Bloom filters) for sets of unknown size [30].

1.3 Overview of Our Constructions and Techniques

First, we overview our construction for general evolving access structures. Then, we describe our construction for the evolving 2-threshold access structure. This serves as a warm-up for our more general construction for k-threshold access structures. Lastly, we discuss the connection with prefix codes.

General Evolving Access Structures. Let $\mathcal{A} = \mathcal{A}_1, \mathcal{A}_2, \ldots$ be any evolving access structure with corresponding monotone characteristic functions $f_1, f_2, \ldots$, where $f_t \colon \{0,1\}^t \to \{0,1\}$. Note that the dealer does not know $\mathcal{A}$ in advance but is only given $\mathcal{A}_t$ when the tth party arrives. Let $s \in \{0,1\}$ be the secret to be shared. The share of party $t \in \mathbb{N}$ consists of 2^{t-1} bits, each denoted by $w_{(b_1,\ldots,b_{t-1},1)}$, where $b_1, \ldots, b_{t-1} \in \{0,1\}$. The $w_{(b_1,\ldots,b_{t-1},1)}$'s are generated as follows: if party t "completes" a minimal qualified set whose indicator vector is $(b_1, \ldots, b_{t-1}, 1)$, then the dealer gives party t the bit $w_{(b_1,\ldots,b_{t-1},1)} = w_{(b_1,\ldots,b_{t-1})} \oplus \cdots \oplus w_{(b_1)} \oplus s$ (where $w_{(b_1,\ldots,b_i,0)} = 0$), so XORing the appropriate shares will recover s. Otherwise, if $(b_1, \ldots, b_{t-1}, 1)$ is unqualified, then the dealer sets $w_{(b_1,\ldots,b_{t-1},1)} \leftarrow \{0,1\}$ to be a uniformly random bit. See Sect. 3 for the exact details.

Evolving 2-Threshold Access Structure. The approach of [15] for the evolving 2-threshold access structure is to give party t a random bit b_t and all bits $s \oplus b_1, \ldots, s \oplus b_{t-1}$. This clearly allows for each pair of parties to reconstruct the secret and ensures that for every single party the secret remains hidden. The share size of the tth party in this scheme is t. (Essentially the same scheme also follows from our general construction in Sect. 3 with a simple efficiency improvement described towards the end of that section.) Generalizing this idea to larger values of k results with shares of size roughly t^{k-1}.

Whereas the above approach is somewhat naive (and very inefficient in terms of share size), our construction is more subtle and results with exponentially shorter shares. Our main building block is a *domain reduction* technique which allows us to start with a naive solution and apply it only on a small number of parties to get an overall improved construction. Details follow.

We assign each party a generation, where the gth generation consists of 2^g parties (i.e. the generations are of geometrically increasing size). Within each generation we execute a standard secret sharing scheme for 2-threshold. Notice that here we know exactly how many parties are in the same generation: party t is part of generation $g = \lfloor \log t \rfloor$ and the size of that generation is $\text{SIZE}(g) \leq t$. A standard secret sharing scheme for 2-out-of-t costs roughly $\log t$ bits (using Shamir's scheme; see Claim 5). This solves the case in which both parties come from the same generation.

To handle the case where the two parties come from different generations we use a (possibly naive) scheme for the *evolving* 2-threshold access structure. For each generation we generate *one* share for the evolving scheme and give it to each party in that generation. Thus, if two parties from different generations come together they hold two different shares for the evolving scheme that allow them to reconstruct the secret. Since we generate one share of the evolving scheme per generation, party t holds the share of the $(g = \log t)^{th}$ party of the evolving scheme!

Summing up, if we start with a scheme in which the share size of the tth party is $\sigma(t)$, then we end up with a scheme with share size roughly $\sigma'(t) = \log t + \sigma(\log t)$. To get our result we start with a scheme in which $\sigma(t) = t$ (described above) and iteratively apply this argument to get better and better schemes.

Evolving k-Threshold Access Structure. There are several ideas underlying the generalization of the 2-threshold scheme to work for any k. As before, we assign each party a generation, but now the gth generation is roughly of size $2^{(k-1) \cdot g}$. This means that party t is in generation $g = \lfloor (\log t)/(k-1) \rfloor$ that includes $\text{SIZE}(g) = t \cdot 2^{k-1}$ parties. Again, within a generation we apply a standard k-out-of-$\text{SIZE}(g)$ secret sharing scheme. This costs us $\log(\text{SIZE}(g)) \leq \log t + k$ bits using Shamir's scheme. This solves the problem if k parties come from the same generation.

We are left with the case where the k parties come from at least two different generations. For this we use a (possibly naive) scheme for the *evolving* k-threshold access structure. For each generation we generate $k - 1$ shares $s_1, \ldots, s_{k-1}$ for the evolving scheme and share each s_i using a standard i-out-of-$\text{SIZE}(g)$ secret sharing scheme. Thus, if $\ell \leq k - 1$ parties from some generation come together, they can reconstruct $s_1, \ldots, s_\ell$ which are ℓ shares for the evolving scheme. Therefore, any k parties (that come from at least two generations) can reconstruct k shares for the evolving k-threshold scheme that enable them to reconstruct the secret. Since we generate $k - 1$ shares of the evolving scheme per generation, party t holds (roughly) the share of the $(\log t + k)^{th}$ party of the evolving scheme.

The share size needed to share each s_i is $\max\{\log(\text{SIZE}(g)), |s_i|\} \leq \max\{\log t + k, \sigma(\log t + k)\}$ (using Shamir's scheme; see Claim 5). Summing up, if we start with a scheme in which the share size of the tth party is $\sigma(t)$, then we end up with a scheme with share size roughly $\sigma'(t) = \log t + (k - 1) \cdot$

$\max\{\log t + k, \sigma(\log t + k)\}$. A small optimization is that sharing s_1 costs just $|s_1|$, as we can give s_1 to each party (similarly to what we did in the $k = 2$ case).

We want to iteratively apply this domain reduction procedure. For this we have to specify the initial scheme. If we start with the scheme that results from the construction in Theorem 1 which has share size roughly 2^t (or roughly t^k with the optimization described above), then the resulting scheme will have a factor that depends *exponentially* on k. This makes the scheme impractical even for small values of k.

A Formula for the Future. To get around this we present a tailor-made construction for the evolving k-threshold in which the share size of party t has *almost linear* dependence on t and k. Specifically, the share size in this scheme is $kt \cdot \log(kt)$. For this, we construct, at least intuitively, a Boolean monotone formula for k-threshold that counts to k.[3] For this counting to make sense in the evolving setting we notice that counting to k can be done by summing up the number of 1's so far with the number of 1's that will come in the future. Since we are counting to k, both of these numbers can be bounded by k, so we have to prepare only k possibilities for the unknown future. To make this construction efficient, we combine it with a generation-like mechanism. See Sect. 5 for the full details.

Prefix Codes and Evolving 2-Threshold. There are several clues that point to a connection with prefix codes: the construction with the repeated domain reduction is reminiscent of the Elias code construction; the lower bound on schemes for the evolving 2-threshold access structure in Theorem 3 uses what is identical to a Kraft inequality, which is a characterization of prefix codes. We are able to formalize this tight relationship:

- Given any prefix code in which the length of the tth codeword is $\sigma(t)$, we construct a secret sharing scheme for the evolving 2-threshold access structure in which the share size of the tth party is $\sigma(t)$. Using the best prefix code constructions we get a scheme in which the share size is the same as in our direct construction described above (but it is less efficient for sharing secrets longer than 1 bit). See Sect. 7 for the transformation.
- On the other hand, any secret sharing scheme for the evolving 2-threshold access structure in which the share size of the tth party is $\sigma(t)$, implies the existence of a prefix code in which the length of the tth codeword is $\sigma(t)$. This comes from the fact that the sufficient condition of Kraft's inequality yields prefix codes.

---

[3] Even though we can make our construction a monotone formula, our final construction is not phrased as a formula since we want to optimize share size. To exemplify this gap notice that the secret sharing scheme that results from the best formula for k-threshold on n parties has share size $\mathsf{poly}(k) \cdot \log n$ [10,19], while the scheme of Shamir has size roughly $\log n$, independently of k.

2   Model and Definitions

For an integer $n \in \mathbb{N}$ we denote by $[n]$ the set $\{1, \ldots, n\}$. We denote by log the base 2 logarithm and assume that $\log 0 = 0$. For a set $\mathcal{X}$ we denote by $x \leftarrow \mathcal{X}$ the process of sampling a value x from the uniform distribution over $\mathcal{X}$

We start by briefly recalling the standard setting of (perfect) secret sharing. Let $\mathcal{P}_n = \{1, \ldots, n\}$ be a set of n parties. A collection of subsets $\mathcal{A} \subseteq 2^{\mathcal{P}_n}$ is monotone if for every $B \in \mathcal{A}$, and $B \subseteq C$ it holds that $C \in \mathcal{A}$.

Definition 1 (Access structure). *An access structure $\mathcal{A} \subseteq 2^{\mathcal{P}_n}$ is a monotone collection of subsets. Subsets in $\mathcal{A}$ are called* **qualified** *and subsets not in $\mathcal{A}$ are called* **unqualified**.

Definition 2 (Threshold access structure). *For every $n \in \mathbb{N}$ and $1 \leq k \leq n$, let (k, n)-THR be the* **threshold** *access structure over n parties which contains all subsets of size at least k.*

A (standard) secret sharing scheme involves a dealer who has a secret, a set of n parties, and an access structure $\mathcal{A}$. A secret sharing scheme for $\mathcal{A}$ is a method by which the dealer distributes shares to the parties such that any subset in $\mathcal{A}$ can reconstruct the secret from its shares, while any subset not in $\mathcal{A}$ cannot reveal any information on the secret.

More precisely, a secret sharing scheme for an access structure $\mathcal{A}$ consists of a pair of probabilistic algorithms (SHARE, RECON). SHARE gets as input a secret s (from a domain of secrets S) and a number n, and generates n shares $\Pi_1^{(s)}, \ldots, \Pi_n^{(s)}$. RECON gets as input the shares of a subset B and outputs a string. The requirements are:

1. For every secret $s \in S$ and every qualified set $B \in \mathcal{A}$, it holds that $\Pr[\mathsf{RECON}(\{\Pi_i^{(s)}\}_{i \in B}, B) = s] = 1$.
2. For every unqualified set $B \notin \mathcal{A}$ and every two different secrets $s_1, s_2 \in S$, it holds that the distributions $(\{\Pi_i^{(s_1)}\}_{i \in B})$ and $(\{\Pi_i^{(s_2)}\}_{i \in B})$ are identical.

The share size of a scheme is the maximum number of bits each party holds in the worst case over all parties and all secrets.

The well known scheme of Shamir [31] for the (k, n)-THR access structure (based on polynomial interpolation) satisfies the following.

Claim 5 ([31]). *For every $n \in \mathbb{N}$ and $1 \leq k \leq n$, there is a secret sharing scheme for secrets of length m and the (k, n)-THR access structure in which the share size is ℓ, where $\ell \geq \max\{m, \log q\}$ and $q > n$ is a prime number (or a power of a prime). Moreover, if $k = 1$ or $k = n$, then $\ell = m$.[4]*

---

[4] Schemes in which the share size is equal to the secret size are known as *ideal* secret sharing schemes.

2.1   Secret Sharing for Evolving Access Structures

We proceed with the definition of an evolving access structure. Roughly speaking, the parties arrive one by one and, in the most general case, a qualified subset is revealed only when all parties in that subset are present (in special cases the access structure is known to begin with, but there is no upper bound on the number of parties). To make sense of sharing a secret with respect to such a sequence of access structures, we require that the changes to the access structure are monotone, namely, parties are only added and qualified sets remain qualified.

To define evolving access structures we need to define a restriction.

Definition 3 (Restriction). *Let $\mathcal{A}$ be an access structure on n parties and let $0 < m < n$. We denote by $\mathcal{A}|_m$ the restriction of $\mathcal{A}$ to the first m parties. That is,*

$$\mathcal{A}|_m = \{X \in \mathcal{A} \mid \{m+1, \ldots, n\} \cap X = \emptyset\}.$$

Due to monotonicity of the access structure, we have the following claim.

Claim 6. *If $\mathcal{A}$ is an access structure on n parties, then $\mathcal{A}|_m$ is an access structure over m parties.*

Proof. By definition of $\mathcal{A}|_m$, it contains only parties from the set $\{1, \ldots, m\}$. Thus, to prove the claim it is enough to show that $\mathcal{A}|_m$ is a monotone set, namely, that if $B \in \mathcal{A}|_m$ then for any $B \subseteq C \subseteq \mathcal{P}_m$. Indeed, since $\mathcal{A}$ is an access structure, for $B \in \mathcal{A}|_m$ and $B \subseteq C \subseteq \mathcal{P}_m \subseteq \mathcal{P}_n$, we have that $B, C \in \mathcal{A}$. By definition of $\mathcal{A}|_m$, it holds that $C \in \mathcal{A}|_m$.

Definition 4 (Evolving access structure). *A (possibly infinite) sequence of access structures $\{\mathcal{A}_t\}_{t \in \mathbb{N}}$ is called* **evolving** *if the following conditions hold:*

1. *For every $t \in \mathbb{N}$, it holds that $\mathcal{A}_t$ is an access structure over t parties.*
2. *For every $t \in \mathbb{N}$, it holds that $\mathcal{A}_t|_{t-1}$ is equal to $\mathcal{A}_{t-1}$.*[5]

This definition naturally gives rise to an evolving variant of threshold access structures (see Definition 2). Here, we think of k as fixed, namely, independent of the number of parties.

Definition 5 (Evolving threshold access structure). *For every $k \in \mathbb{N}$, let* evolving *k-THR be the* **evolving** *threshold access structure which contains for any access structure in the sequence all subsets of size at least k.*

We generalize the definition of a standard secret sharing scheme to apply for evolving access structures. Intuitively, in this setting, at any point $t \in \mathbb{N}$ in time, there is an access structure $\mathcal{A}_t$ which defines the qualifies and unqualified subsets of parties.

Definition 6 (Secret sharing for evolving access structures). *Let $\mathcal{A} = \{\mathcal{A}_t\}_{t \in \mathbb{N}}$ be an evolving access structure. Let S be a domain of secrets, where $|S| \geq 2$. A secret sharing scheme for $\mathcal{A}$ and S consists of a pair of algorithms* (SHARE, RECON). *The sharing procedure* SHARE *and the reconstruction procedure* RECON *satisfy the following requirements:*

---

[5] Recall the definition of a *restriction* from Definition 3.

1. SHARE$(s, \{\Pi_1^{(s)}, \ldots, \Pi_{t-1}^{(s)}\})$ *gets as input a secret* $s \in S$ *and the secret shares of parties* $1, \ldots, t-1$. *It outputs a share for the* t^{th} *party. For* $t \in \mathbb{N}$ *and secret shares* $\Pi_1^{(s)}, \ldots, \Pi_{t-1}^{(s)}$ *generated for parties* $\{1, \ldots, t-1\}$, *respectively, we let*

$$\Pi_t^{(s)} \leftarrow \text{SHARE}(s, \{\Pi_1^{(s)}, \ldots, \Pi_{t-1}^{(s)}\})$$

be the secret share of party t.
We abuse notation and sometimes denote by $\Pi_t^{(s)}$ *the random variable that corresponds to the secret share of party* t *generated as above.*

2. **Correctness:** *For every secret* $s \in S$ *and every* $t \in \mathbb{N}$, *every qualified subset in* $\mathcal{A}_t$ *can reconstruct the secret. That is, for* $s \in S$, $t \in \mathbb{N}$, *and* $B \in \mathcal{A}_t$, *it holds that*

$$\Pr\left[\text{RECON}(\{\Pi_i^{(s)}\}_{i \in B}, B) = s\right] = 1,$$

where the probability is over the randomness of the sharing and reconstruction procedures.

3. **Secrecy:** *For every* $t \in \mathbb{N}$, *every unqualified subset* $B \notin \mathcal{A}_t$, *and every two secret* $s_1, s_2 \in S$, *the distribution of the secret shares of parties in* B *generated with secret* s_1 *and the distribution of the shares of parties in* B *generated with secret* s_2 *are identical. Namely, the distributions* $(\{\Pi_i^{(s_1)}\}_{i \in B})$ *and* $(\{\Pi_i^{(s_2)}\}_{i \in B})$ *are identical.*

The share size of the tth party in a scheme for an evolving access structure is $\max |\Pi_t|$, namely the number of bits party t holds in the worst case over all secrets and previous assignments.[6]

On Choosing the Access Structure Adaptively. One can also consider a stronger definition in which $\mathcal{A}_t$ is chosen at time t (rather than ahead of time) as long as the sequence of access structures $\mathcal{A} = \{\mathcal{A}_1, \ldots, \mathcal{A}_t\}$ is evolving. In this variant, the RECON procedure gets the access structure $\mathcal{A}_t$ as an additional parameter. Our construction of a secret sharing scheme for general evolving access structures in Sect. 3 works for this notion as well.

On the Domain of Secrets. Unless otherwise stated, we usually assume that the secret is a single bit (either 0 or 1). One can generalize any such scheme to support longer secrets by secret sharing every bit of the secret independently, suffering a multiplicative factor in share size that depends on the length of the secret. When we generalize our schemes to support long secrets, this naive generalization will be our benchmark.

2.2   Warm-Up: Undirected s-t-Connectivity

We start with a simple warm-up scheme. We show that the standard scheme for the st-connectivity access structure can be easily adapted to the evolving

---

[6] This means that the share size is bounded, which is almost always the case. An exception is the scheme (for rational secret sharing) of Kol and Naor [26] in which the share size does not have a fixed upper bound.

setting. In this access structure parties correspond to edges of an *undirected* graph $G = (V, E)$. There are two fixed vertices in the graph called s and t (where $s, t \in V$). A set of parties (i.e. edges) is qualified if and only if they include a path from s to t. Around 1989 Benaloh and Rudich [6] (see also [2, Sect. 3.2]) constructed a (standard) secret sharing for this access structure. The dealer, given a secret $s \in \{0, 1\}$, assigns with each vertex $v \in V$ a label. For $v = s$ the label is $w_s = s$, for $v = t$ the label is $w_t = 0$ and for the rest of the vertices the label is chosen independently uniformly at random $w_v \leftarrow \{0, 1\}$. The share of a party $e = (u, v) \in E$ is $w_u \oplus w_v$.

Consider a set of parties that include a path $s = v_1 v_2 \ldots v_k = t$ from s to t. To reconstruct the secret, the parties XOR their shares to get

$$(w_{v_1} \oplus w_{v_2}) \oplus (w_{v_2} \oplus w_{v_3}) \oplus \cdots \oplus (w_{v_{k-1}} \oplus w_{v_k}) = w_{v_1} \oplus w_{v_k} = s.$$

One can observe that this access structure and scheme naturally generalize to the evolving setting. In this setting, we consider an evolving (possibly infinite) graph, where the set of nodes and edges are unbounded. At any point in time an arbitrary set of vertices and edges can be added to the graph. An addition of an edge corresponds to a new party added to the scheme. The special vertices s and t are fixed ahead of time and cannot change (this is to ensure the access structure is *evolving*).

Initially, the dealer assigns labels for the special vertices s and t, as before (i.e. it sets $w_s = s$ and $w_t = 0$). For the rest of the vertices the dealer assigns (uniformly random) labels only on demand: When a new edge $e = (u, v)$ is added to the graph (which corresponds to a new party), the dealer gives the party corresponding to the edge e the XOR of the labels of the vertices u and v. Correctness and security of this scheme follow similarly to the correctness and security of the standard scheme. One can see that the share size of each party is exactly the size of the secret.

3 A Scheme for General Evolving Access Structures

We give a construction of a secret sharing scheme for every evolving access structure. We emphasize that our construction also works in the scenario in which the access structure is chosen adaptively; see remark after Definition 6. We focus on the case where the secret is a single bit.

Theorem 7 (Theorem 1 restated). *For every evolving access structure there is a secret sharing scheme where the share size of the tth party is at most 2^{t-1}.*

The fact that our construction results with shares of exponential size should come as no surprise, as the best constructions known for standard secret sharing schemes for general access structures have shares of exponential size (in the number of parties). Proving that shares of exponential size are necessary to realize some *evolving* access structure is a very interesting open problem.

Proof of Theorem 7. Let $\mathcal{A} = \{\mathcal{A}_t\}_{t \in \mathbb{N}}$ be an evolving access structure.[7] Let $\{f_t\}_{t \in \mathbb{N}}$ be the sequence of functions, where $f_i : \{0,1\}^i \to \{0,1\}$ is the (monotone) characteristic function of $\mathcal{A}_i$.

Let $s \in \{0,1\}$ be the secret to be shared. We describe what the dealer stores and how it prepares a share for an arriving party. At time t (before party t arrives) the dealer maintains a set of bits we denote by $w_{(b_1,\ldots,b_i)}$ for all $i \in [t-1]$ and $b_1, \ldots, b_i \in \{0,1\}$. These bits are defined iteratively. First, the dealer sets $w_{(1)} = s$ if $f_1(1) = 1$ and it is a uniformly random bit otherwise. Moreover, for every $i \geq 1$, the dealer sets $w_{(b_1,\ldots,b_{i-1},0)} = 0$. The rest of the bits are defined as follows.

1. If $f_t(b_1,\ldots,b_{t-1},1) = 1$ and $f_{t-1}(b_1,\ldots,b_{t-1}) = 0$, then the dealer sets

$$w_{(b_1,\ldots,b_{t-1},1)} = w_{(b_1,\ldots,b_{t-1})} \oplus \cdots \oplus w_{(b_1)} \oplus s.$$

2. If $f_t(b_1,\ldots,b_{t-1},1) = 1$ and $f_{t-1}(b_1,\ldots,b_{t-1}) = 1$, then the dealer sets

$$w_{(b_1,\ldots,b_{t-1},1)} = 0.$$

3. If $f_t(b_1,\ldots,b_{t-1},1) = 0$, then the dealer sets

$$w_{(b_1,\ldots,b_{t-1},1)} \leftarrow \{0,1\}$$

to be a uniformly random bit.

The share of party $t \in \mathbb{N}$ consists of 2^{t-1} bits $w_{(b_1,\ldots,b_{t-1},1)}$ for all $b_1, \ldots, b_{t-1} \in \{0,1\}$.

Correctness and Security. We argue correctness and security at time $t \in \mathbb{N}$. Let $\boldsymbol{b} = (b_1,\ldots,b_t) \in \{0,1\}^t$ be an indicator vector of a minimal qualified set of parties at time t. For every $i \in [t-1]$ such that $b_i = 1$, party i holds the bit $w_{(b_1,\ldots,b_i)}$. Party t, by construction, holds the bit $w_{(b_1,\ldots,b_{t-1})} \oplus \cdots \oplus w_{(b_1)} \oplus s$, where $w_{(b_1,\ldots,b_i,0)} = 0$ for $0 \leq i \leq t-2$. Therefore, by XOR-ing all the shares, namely, computing

$$\bigoplus_{i=1}^{t} w_{(b_1,\ldots,b_i)},$$

the parties present can compute s.

For security it is instructive to give a simple example that illustrates how the scheme works and why it is secure. Consider the access structure at time $t = 4$ that consists of the following qualified sets $\{\{1,2\},\{1,3\},\{1,4\},\{2,4\}\}$ and we will argue security for the set $\{3,4\}$. Party 1 is unqualified so its share is $w_{(1)}$ is a uniformly random bit. Party 2 completes a qualified set with party 1 and so it's share consists of two bits $w_{(0,1)}, w_{(1,1)}$, where $w_{(0,1)}$ is a uniformly random bit and $w_{(1,1)} = w_{(1)} \oplus s$. Similarly, the share of party 3 consists of four bits

---

[7] As mentioned, our construction actually works in the setting where $\mathcal{A}_t$ itself is chosen at time t (and it is not known at any time $t' < t$).

$w_{(0,0,1)}, w_{(0,1,1)}, w_{(1,0,1)}, w_{(1,1,1)}$, where $w_{(0,0,1)}$ and $w_{(0,1,1)}$, uniformly random, $w_{(1,1,1)} = 0$ since $\{1, 2\}$ is qualified as well, and $w_{(1,0,1)} = w_{(1)} \oplus s$. Finally, the share of party 4 consists of 8 bits most of which are either 0 or uniformly random, and the interesting ones are $w_{(1,0,0,1)} = w_{(1)} \oplus s$ and $w_{(0,1,0,1)} = w_{(0,1)} \oplus s$. Let us assume that the shares given to parties $\{1, 2, 3\}$ do not reveal s and show that the shares of party 4 do not reveal it as well. Indeed, all its uniformly random bits and the zero bits do not help, so we focus on $w_{(1,0,0,1)}$ and $w_{(0,1,0,1)}$. We observe that since parties 4 and 3 both complete party 1 to be a qualified set, they both have the same share $w_{(1,0,0,1)} = w_{(1,0,1)} = w_{(1)} \oplus s$, so we can ignore $w_{(1,0,0,1)}$ as well and be left with $w_{(0,1,0,1)} = w_{(0,1)} \oplus s$. Now, the point is that since party 3 does not complete party 2 to get a qualified set, the element $w_{(0,1)}$ completely masks the secret. More generally, the formal vector space generated by the share $w_{(0,1,0,1)}$ is linearly independent of all other shares.

We sketch security in the general case by induction. For $t = 1$ it is immediate and assume that the scheme is secure for $t - 1$. For every $(b_1, \ldots, b_t) \in \{0, 1\}^t$, party t receives a bit $w_{(b_1, \ldots, b_{t-1}, 1)}$ which is either a uniformly random bit or the bit $w_{(b_1, \ldots, b_{t-1})} \oplus \cdots \oplus w_{(b_1)} \oplus s$, depending on the value of $f_t(b_1, \ldots, b_{t-1}, 1)$. Let $b = (b_1, \ldots, b_t) \in \{0, 1\}^t$ be an indicator vector of an unqualified set of parties at time t. Assume that $b_t = 1$, as the other case follows immediately from the induction hypothesis. For every $w_{(b_1, \ldots, b_{t-1}, 1)}$, the uniformly random bits given to party t do not give an unqualified set any additional information about the secret as they are independent of everything else this set posses, so we can ignore them. Let us consider all the bit of the form $w_{(b'_1, \ldots, b'_i)} \oplus \cdots \oplus w_{(b'_1)} \oplus s$ held by parties in b. If there are two parties i, j such that $b_i = b_j = 1$ that complete the same set, then they posses the *same* bit so we can ignore one of them. We are left with the case in which all parties complete different subsets. In this case, one can see that all the shares are linearly independent and thus the secret remains hidden. Security follows by the hypothesis.

Share Size. The share size of party t is 2^{t-1} bits. ∎

3.1   Efficiency Improvements

In some cases, depending on the access structure, it is possible to reduce the share size by slightly optimizing the above scheme. The 0 bits that occur due to Theorem 3, do not have to be remembered as they can be inferred from the access structure.

At time t, the shares of party t will consists of:

1. A bit for each unqualified subset of $[t]$ that party t participates in. For the case when the access structure is known ahead of time, the only unqualified sets to consider our those that can be expanded to a qualified subset using future parties.
2. A bit for each qualified subset of $[t]$ that party t completes (i.e. is the last one).

This optimization is useful for access structures in which the number of unqualified sets is small. For example, for the evolving 2-THR access structure, the fact that there are only t unqualified sets, implies a scheme in which the share size of the tth party is exactly t (we use this fact in Sect. 4). More generally, for the evolving k-THR access structure, there are $\sum_{i=0}^{k-2} \binom{t-1}{i}$ unqualified sets and $\binom{t-1}{k-1}$ qualified sets which t completes, implying a scheme with share size roughly t^{k-1}.

4   An Efficient Scheme for Evolving 2-Threshold

We now describe the efficient construction for a secret sharing scheme for the evolving 2-THR access structure. Recall that evolving 2-THR is the sequence of access structures $(2,1)$-THR, $(2,2)$-THR, $(2,3)$-THR, ... which allow, at any point in time, for every pair of parties to learn the secret while disallowing singletons to learn anything about it. We first focus on the case where the secret is a single bit and discuss the more general case in Sect. 4.1.

Theorem 8. *There is a secret sharing scheme for the evolving 2-THR access structure in which the share size of the tth party is bounded by*

$$\log t + \log \log t + 2 \log \log \log t + 6.$$

Recall that in the classical setting of secret sharing, where an upper bound on the number of parties is known, there is a very efficient scheme for $(2,n)$-THR in which the share size of each party is roughly $\log n$ (see Claim 5). In Sect. 6 we show that in the evolving setting, for any $c \in \mathbb{N}$, a scheme in which the share size of the tth party is $\log t + \log \log t + c$ cannot exist. Thus, up to an additive $\log \log \log t$ term, our scheme is optimal.

Our main technical claim used to prove Theorem 8 is given in the following lemma.

Lemma 1. *Assume that there exists a secret sharing scheme for the evolving 2-THR access structure in which the share size of the tth party is $\sigma(t)$. Then, there exists a secret sharing scheme for the evolving 2-THR access structure in which the share size of the tth party is*

$$\log t + \sigma(\log t + 1).$$

Proof of Theorem 8 Assuming Lemma 1. Recall that in Sect. 3 we constructed a secret sharing scheme for any evolving access structure that results with shares of size 2^{t-1}. However, using the efficiency improvements described in Sect. 3.1,[8] we get a scheme in which the share size of the tth party is

$$\sigma^{(0)}(t) = t.$$

Using Lemma 1 this gives rise to a scheme $\Pi^{(1)}$ in which the share size of the tth party is

$$\sigma^{(1)}(t) = \log t + \sigma^{(0)}(\log t + 1)$$
$$= 2 \log t + 1.$$

---

[8] Alternatively, we can use the construction of [15] (see Sect. 1.3) which gives the tth party a share of size t.

Applying Lemma 1 again we get a scheme $\Pi^{(2)}$ in which the share size of the tth party is

$$\begin{aligned} \sigma^{(2)}(t) &= \log t + \sigma^{(1)}(\log t + 1) \\ &\leq \log t + 2\log(\log t + 1) + 1 \\ &\leq \log t + 2\log\log t + 3. \end{aligned}$$

Applying Lemma 1 one last time we get a scheme $\Pi^{(3)}$ in which the share size of the tth party is

$$\begin{aligned} \sigma^{(3)}(t) &= \log t + \sigma^{(2)}(\log t + 1) \\ &\leq \log t + \log(\log t + 1) + 2\log\log(\log t + 1) + 3 \\ &\leq \log t + \log\log t + 2\log\log\log t + 6. \end{aligned}$$

This proves the theorem.

We note that this bound is tight according to the lower bound in Theorem 3 up to the low-order term $\log\log\log t$.

We note that by applying Lemma 1 i times we can improve the share size for large enough t. This will match the lower bound up to a low order term of $\log^{(i)}(t)$ (See Remark 1). We choose to stop after three applications of Lemma 1 due to aesthetic reasons (but see Sect. 7). ∎

We are left to prove Lemma 1.

Proof of Lemma 1. Let Π be a construction of a secret sharing scheme for evolving 2-THR in which the share size of the tth party is $\sigma(t)$. We construct a scheme Π' for the same access structure in which the share size is $\log t + \sigma(\log t + 1)$. We proceed with the description of the scheme.

Let $s \in \{0,1\}$ be the secret to be shared. Each party, when it arrives, is assigned to a generation. The generations are growing in size: For $g = 0, 1, 2\ldots$ the gth generation begins when the 2^g-th party arrives. Therefore, the size of the gth generation, namely, the number of parties that are part of this generation, is $\text{SIZE}(g) = 2^g$ and party $t \in \mathbb{N}$ is part of generation $g = \lfloor \log t \rfloor$.

When a generation begins the dealer prepares shares for all parties that are part of that generation. Let us focus on the beginning of the gth generation and describe the dealer's procedure:

1. Split s using a secret sharing scheme for $(2, \text{SIZE}(g))$-THR. Denote the resulting shares by $u_1^{(g)}, \ldots, u_{\text{SIZE}(g)}^{(g)}$.
2. Generate one share using the secret sharing scheme Π given the secret s and previous shares $\{v^{(i)}\}_{i \in \{0,\ldots,g-1\}}$. Denote the resulting share by $v^{(g)}$.
3. Set the secret share of the jth party in the gth generation (i.e. $j \in [\text{SIZE}(g)]$) to be

$$\left(u_j^{(g)}, v^{(g)} \right).$$

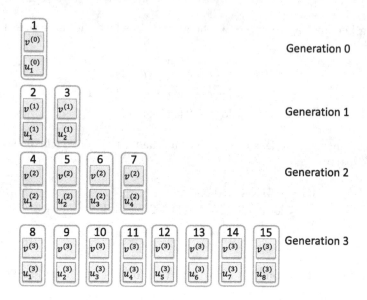

Fig. 1. The shares of parties $1, \ldots, 15$ from generations $0, \ldots, 3$.

The output of the scheme is depicted in Fig. 1.

Correctness and Security. Let $t_1, t_2 \in \mathbb{N}$ be any two different parties. We show that the secret s can be computed from their shares. If t_1 and t_2 are from the same generation g (i.e. if $g = \lfloor \log t_1 \rfloor = \lfloor \log t_2 \rfloor$), then they can reconstruct the secret s using the reconstruction procedure of the $(2, \text{SIZE}(g))$-THR scheme using the corresponding $u^{(g)}$ shares. If they are from different generations $g_1 \neq g_2$, then the parties can compute s using the reconstruction procedure of the evolving 2-THR scheme and the two shares $v^{(g_1)}$ and $v^{(g_2)}$.

For security consider any single party $t \in \mathbb{N}$ from generation g. By the security of the $(2, \text{SIZE}(g))$-THR scheme, the security of the evolving 2-THR scheme, and the fact that both parts of the share are generated independently, the shares cannot be used to learn anything about the secret.

Share Size Analysis. We analyze the share size of parties in the scheme Π'. Denote by $\sigma(t)$ the share size of party t in the scheme Π. We bound the size of each component in the share of party t. The share of party t that is the jth party of generation $g = \lfloor \log t \rfloor$ is $(u_j^{(g)}, v^{(g)})$.

1. $u_j^{(g)}$ – generated by secret sharing s using a scheme for $(2, \text{SIZE}(g))$-THR. Since $\text{SIZE}(g) = 2^g$ and using Claim 5 we get that

$$|u_j^{(g)}| \leq \log(\text{SIZE}(g)) \leq \lfloor \log t \rfloor.$$

2. $v^{(g)}$ – generated by generating one share of a secret sharing scheme Π for evolving 2-THR. Recall that g shares were generated for previous generations.

Therefore,

$$|v^{(g)}| = \sigma(g+1) = \sigma(\lfloor \log t \rfloor + 1).$$

Thus, the total share size in the scheme Π' is bounded by

$$\log t + \sigma(\log t + 1).$$

■

4.1 Generalization to Larger Domains of Secrets

This scheme can be generalized to larger domains of secrets in an efficient way (in particular, better than sharing each bit independently). Roughly speaking, this follows since Shamir's threshold scheme can be used to share a secret longer than 1 bit without increasing the share size; see Claim 5. More generally, sharing a secret of ℓ-bits long, requires shares of size roughly $\max\{\log n, \ell\}$, where n is the number of parties in the scheme.

Let the secret be a string of length ℓ. Using the above feature of Shamir's scheme, a slight variant of Lemma 1 still holds (following the same proof). Namely, given any secret sharing scheme for the evolving 2-THR access structure and ℓ-bit secrets in which the share size of the tth party is $\sigma(t)$. Then, there exists a secret sharing scheme for the evolving 2-THR access structure and ℓ-bit secrets in which the share size of the tth party is

$$\max\{\log t, \ell\} + \sigma(\log t + 1).$$

We have to specify the initial scheme that supports ℓ-bit secrets to start the recursive composition with. We use our scheme for Theorem 8 by secret sharing every bit independently. The share size will be $\sigma^{(0)}(t) \leq \ell \cdot (\log t + \log \log t + 2 \log \log \log t + 6)$. For large enough t it holds that $\sigma^{(0)}(t) \leq t$ and $\max\{\log t, \ell\} = \log t$. Thus, one can follow the same outline of the proof of Theorem 8 and obtain the *same* share size as in Theorem 8 for large enough t. (For smaller values of t one can follow the analysis and obtain a bound as a function of t and ℓ).

5 A Scheme for Evolving k-Threshold

In this section we give a construction for a secret sharing scheme for the evolving k-THR access structure for general k. As in Sect. 4, we first focus on the case where the secret is a single bit and discuss the more general case in Sect. 5.4.

Theorem 9 (Theorem 2 restated). *There is a secret sharing scheme for the evolving k-THR access structure in which the share size of the tth party is at most*

$$(k-1) \cdot \log t + 6k^3 \cdot \log \log t \cdot \log \log \log t + 7k^4 \cdot \log k.$$

As in the case of $k = 2$ (see the discussion after Theorem 8), the best one could hope to obtain is a scheme in which the share of the tth party is close to $\log t$.[9] Our construction has a linear dependence on k and we leave open the question whether this can be improved.

We note that the bound in Theorem 9 applies for any $t \in \mathbb{N}$ and $k \geq 2$. For specific values of t and k it is possible to follow the analysis and obtain a better bound.

Our approach is to start with some basic scheme that has good dependency on k but high dependency on t and use a domain reduction technique in order to obtain better dependency on t.

Our main technical lemma used to prove Theorem 9 is a general transformation where we take any scheme for the evolving k-THR access structure (possibly with large share size), and convert it into a different scheme with smaller share size. Formally we prove following lemma.

Lemma 2. *Let $k \in \mathbb{N}$. Assume that there exists a secret sharing scheme for the evolving k-THR access structure in which the share size of the tth party is $\sigma(t)$. Then, there exists a secret sharing scheme for the evolving k-THR access structure in which the share size of the tth party is at most*

$$(k-1) \cdot \log t + k \cdot \sigma(\log t + k) + k^2.$$

The proof of Theorem 9 is done via repeated applications of Lemma 2, somewhat similarly to the proof of Theorem 8. However, naively the resulting parameters are not very good. Specifically, if we start with the scheme for the evolving k-THR access structure in which the share size is exponential in t or k (which is what we get using the scheme from Theorem 7; see Sect. 3.1), then by applying Lemma 2, the share size will eventually depend *exponentially* on k.

To overcome this, we first present a tailor-made construction for the evolving k-THR access structure in which the share size of party t has *almost linear* dependence on t and k. Using this scheme as a basic building block, we repeatedly apply Lemma 2 to obtain Theorem 9. The proof of the latter can be found in Sect. 5.3. The tailor-made construction for the evolving k-THR access structure appears next in Sect. 5.1. Finally, the proof of Lemma 2 appears in Sect. 5.2.

5.1 The Basic Scheme for Evolving k-Threshold

The main result of this subsection is a construction of a secret sharing scheme for the evolving k-THR access structure and 1-bit secrets in which the share size of party t is almost linear in t and k. This scheme will be used later as the basic building block in our final scheme for evolving k-THR satisfying Theorem 9.

Lemma 3. *There is a secret sharing scheme for the evolving k-THR access structure in which the share size of the tth party is bounded by $kt \cdot \log(kt)$.*

---

[9] Shamir's scheme for (k, n)-THR results with shares of size roughly $\log n$. In particular, independent of k.

In the construction used to prove Lemma 3 we will employ two secret sharing schemes: (1) Shamir's threshold scheme and (2) a secret sharing scheme for a new access structure. The latter access structure C_ℓ over $2k$ parties, where $\ell \leq k$, is defined via its characteristic monotone function that we denote by C_ℓ as well. Let $(x, y) \in \{0, 1\}^k \times \{0, 1\}^k$ be an inputs to $C_\ell \colon \{0, 1\}^k \times \{0, 1\}^k \to \{0, 1\}$, where we think of x and y as unary encoding of two numbers in $\{0, \ldots, k\}$. Jumping ahead, the variable x will represent the number of parties present so far and y will represent the number of parties to come. The access structure contains all pairs whose sum is at least ℓ. Formally, we define $C_\ell(x, y) = 1$ if and only if at least one of the following conditions hold:

1. $\exists i, j \in [\ell - 1]$ such that $x_i = 1$, $y_j = 1$, and $i + j = \ell$.
2. $y_\ell = 1$ or $x_\ell = 1$.

Claim 10. *Let $\ell, k \in \mathbb{N}$ such that $\ell \leq k$. There exists a secret sharing scheme for the access structure C_ℓ in which the share size of each party is exactly the size of the shared secret.*

Proof. The following monotone formula computes C_ℓ:

$$C_\ell(x, y) = \bigvee_{i=1}^{\ell-1} (x_i \wedge y_{\ell-1}) \vee (x_\ell \vee y_\ell).$$

Notice that this formula is a DNF and every input variable appears exactly once. This formula gives rise to a simple secret sharing scheme for the access structure C_ℓ using the method of [5]. Since each variable appears at most once in the formula ($x_1, \ldots, x_\ell$ and $y_1, \ldots, y_\ell$ appear once, but $x_{\ell+1} \ldots, x_k$ and $y_{\ell+1}, \ldots, y_k$ do not appear), the share of each party is bounded by the length of the secret. The theorem follows by padding all shares to be of the same length.

Intuition for the Construction. Our goal is to allow any combination of k parties to learn s. The main idea is not to consider all possible combinations of k parties, but to group parties into generations, ignore the identities of the parties within a generation, and only focus on their *quantity*. For simplicity, let us focus on the first and second generation. How many quantities should we consider? Exactly k, since the presence of $i \leq k$ parties from the first generation requires the presence of $k - i$ parties from the second generation. Therefore, the idea is to generate $2k$ strings $x_1, \ldots, x_k$ and $y_1, \ldots, y_k$, such that only a proper combination of x_i and y_{k-i} will recover the secret s (for this we use the scheme for the access structure C_k). These $2k$ strings are generated when the first generation begins and the x's (the values corresponding to the "present") are shared among the parties of that generation in a way that allows any i parties to learn x_i. The y's (the values corresponding to the "future") will be shared among the parties of the second generation in a similar way allowing any $k - i$ parties to learn y_{k-i}. Together, they will be able to recover s.

To formalize the above intuition and extend it to more generation we need some notation. For a generation $g \geq 0$, we denote by $[k]^g = \{1, \ldots, k\}^g$ the set

$\{1, \ldots, k\} \times \ldots \times \{1, \ldots, k\}$. We will be using vectors of the form $\mathbf{z} = (i_1, \ldots, i_g)$
$\underbrace{\phantom{\{1, \ldots, k\} \times \ldots \times \{1, \ldots, k\}}}_{g \text{ times}}$
$\in [k]^g$ in our notation. For such a vector $\mathbf{z}$ and $i_{g+1} \in [k]$, we denote by $(\mathbf{z}, i_{g+1})$
the vector $(i_1, \ldots, i_g, i_{g+1}) \in [k]^{g+1}$.

Proof of Lemma 3. Let $s \in \{0, 1\}$ be the secret to be shared. Each party, when
it arrives, is assigned to a generation. Party $t \in \mathbb{N}$ is assigned to generation
$g = \lfloor \log_k t \rfloor$. The generations are growing in size: For $g = 0, 1, 2 \ldots$ the gth
generation begins when the k^g-th party arrives. Therefore, the size of the gth
generation (i.e. the number of parties that are members of this generation), is

$$\text{SIZE}(g) = k^{g+1} - k^g = (k - 1) \cdot k^g.$$

When a generation g begins the dealer prepares shares for all parties that
are members of that generation, and in addition, it generates k^{g+1} strings
$\{y_{\mathbf{z}}^{(g+1)}\}_{\mathbf{z} \in [k]^{g+1}}$ which it remembers for the next generation. Initially, the dealer
sets $y_{\emptyset}^{(0)} = s$. Let us focus on the beginning of the gth generation and describe
the dealer's procedure (for consistency of notation we define $[k]^0 = \emptyset$):

1. (a) If $g = 0$: Split the string $y_{\emptyset}^{(0)} = s$ using the secret sharing scheme for C_k of
    Claim 10. Denote the resulting $2k$ shares by $x_{(1)}^{(0)}, \ldots, x_{(k)}^{(0)}, y_{(1)}^{(1)}, \ldots, y_{(k)}^{(1)}$.
   (b) If $g \geq 1$: For all $\mathbf{z} = (i_1, \ldots, i_g) \in [k]^g$ split the string $y_{\mathbf{z}}^{(g)}$ using the
    secret sharing scheme for C_{i_g} of Claim 10. Denote the resulting $2k$ shares
    by $x_{(\mathbf{z},1)}^{(g)}, \ldots, x_{(\mathbf{z},k)}^{(g)}, y_{(\mathbf{z},1)}^{(g+1)}, \ldots, y_{(\mathbf{z},k)}^{(g+1)}$.
    The x's will be shared amongst the parties in the current (g^{th}) genera-
    tion, whereas the y's will be used to generate shares for parties in the next
    $((g+1)^{th})$ generation.
2. For all $\mathbf{z} = (i_1, \ldots, i_{g+1}) \in [k]^{g+1}$ secret share $x_{\mathbf{z}}^{(g)}$ using a scheme for
   $(i_{g+1}, \text{SIZE}(g))$-THR. Denote the resulting $\text{SIZE}(g)$ shares by $u_{\mathbf{z},1}^{(g)}, \ldots, u_{\mathbf{z},\text{SIZE}(g)}^{(g)}$.
3. The secret share of the jth party in the gth generation (that is, the tth party
   where $t = k^g + j - 1$) is composed of all the strings $u_{\mathbf{z},j}^{(g)}$ for any possible $\mathbf{z}$.
   Namely, it is the sequence of strings

$$\{u_{\mathbf{z},j}^{(g)}\}_{\mathbf{z} \in [k]^{g+1}}.$$

Correctness and Security

Claim 11. *Any $c \leq k$ parties from generation g can compute $\{x_{(\mathbf{z},i)}^{(g)}\}_{\mathbf{z} \in [k]^g, i \in [c]}$.*

Proof. Let $j_1, \ldots, j_c \in [\text{SIZE}(g)]$ be the indices of parties present from that generation. Thus, the parties can compute

$$\{u_{\mathbf{z},j_1}^{(g)}, \ldots, u_{\mathbf{z},j_c}^{(g)}\}_{\mathbf{z} \in [k]^{g+1}}.$$

Therefore, all the x values that were shared via a threshold scheme in
which the threshold was at most c can be reconstructed. Namely, the values
$\{x_{(\mathbf{z},i)}^{(g)}\}_{\mathbf{z} \in [k]^g, i \in [c]}$.

Claim 12. *Fix a generation $g \geq 0$, two numbers $c_1, c_2 \in [k]$ and $z = (i_1, \ldots, i_g) \in [k]^g$. Then, given $x^{(g)}_{(z,c_1)}$ and $y^{(g+1)}_{(z,c_2)}$ such that $c_1 + c_2 \geq i_g$, one can compute $y^{(g)}_z$. Moreover, given $x^{(g)}_{(z,c_1)}$ such that $c_1 \geq i_g$, one can compute $y^{(g)}_z$.*

Proof. Follows from the correctness of the secret sharing scheme for C_ℓ.

Now, let us assume that k parties come together and try to reconstruct s. Assume that c_0 parties come from generation 0, c_1 come from generation 1 and so on. That is, for some generation g it holds that $\sum_{i=0}^{g} c_i = k$ and without loss of generality $c_g > 0$. We show that these parties can learn $y^{(0)}_{(k)} = s$, as required. This is done by applying Claims 11 and 12 iteratively. Details follow.

By Claim 11, using the shares of the c_i parties in generation $i \in \{0, \ldots, g\}$ we can compute

$$x^{(0)}_{(c_0)}, \quad \{x^{(1)}_{(z,c_1)}\}_{z \in [k]^1}, \quad \cdots, \quad \{x^{(g)}_{(z,c_g)}\}_{z \in [k]^g}.$$

By the second part of Claim 12, using $\{x^{(g)}_{(z,c_g)}\}_{z \in [k]^g}$ we can reconstruct

$$\{y^{(g)}_z\}_{z \in [k]^{g-1} \times \{c_g\}}.$$

By the first part of Claim 12, using $\{x^{(g-1)}_{(z,c_{g-1})}\}_{z \in [k]^{g-1}}$ with $\{y^{(g)}_z\}_{z \in [k]^{g-1} \times \{c_g\}}$ we can reconstruct

$$\{y^{(g-1)}_z\}_{z \in [k]^{g-2} \times \{c_g + c_{g-1}\}}.$$

Using the first part of Claim 12 iteratively as above, one can eventually compute $y^{(1)}_{(\sum_{i=1}^{g} c_i)}$. Combining with $x^{(0)}_{(c_0)}$, one can compute $y^{(0)}_\emptyset = s$, as required.

To argue security, fix any set of parties as above where $\sum_{i=0}^{g} c_i < k$. We claim that these parties cannot learn the value $y^{(0)}_\emptyset = s$. From the security of the scheme for C_ℓ, it is enough to show that they cannot learn any value in $y^{(1)}_{(k-c_0)}$. Applying this logic once again, it is enough to show that they cannot learn any value in $\{y^{(2)}_{(z,k-c_0-c_1)}\}_{z \in [k]}$. Applying this argument g times, we get that s cannot be learned if and only if $\{y^{(g+1)}_{(z,k-\sum_{i=0}^{g} c_i)}\}_{z \in [k]^g}$ cannot be learned. Indeed, these values are independent of the shares of parties up to generation g.

Share Size Analysis. We analyze the share size of parties in the scheme Π_k described above. The share of party t from generation g is composed of k^{g+1} shares generated via standard threshold schemes over $\text{SIZE}(g)$ parties. Thus, in total, the share size of party t is bounded by $k^{g+1} \cdot \log(\text{SIZE}(g))$. Recall that $g = \lfloor \log_k t \rfloor$ and $\text{SIZE}(g) = (k-1) \cdot k^g$. Therefore, the share size is bounded by

$$k \cdot t \cdot \log((k-1) \cdot t) \leq kt \cdot \log(kt).$$

∎

5.2   Recursive Composition: Proof of Lemma 2

Let Π_k be a construction of a secret sharing scheme for evolving k-THR in which the share size of the tth party is $\sigma_k(t)$. We construct a scheme Π'_k for the same access structure in which the share size is $\sigma'_k(t) = \log t + (k-1) + \sigma(\log t + (k-1)) + (k-2) \cdot \max\{\log t + (k-1), \sigma(\log t + (k-1))\}$.

Let $s \in \{0,1\}$ be the secret to be shared. Each party is assigned to a generation. The generations are growing in size: For $g = 0,1,2\ldots$ the gth generation begins when the $2^{(k-1)\cdot g}$-th party arrives. Thus, party $t \in \mathbb{N}$ is part of generation $g = \lfloor (\log t)/(k-1) \rfloor$, and the number of parties that are part of generation g, is

$$\text{SIZE}(g) = 2^{(k-1)\cdot(g+1)} - 2^{(k-1)\cdot g} = 2^{(k-1)\cdot g} \cdot (2^{k-1} - 1) \le t \cdot 2^{k-1}.$$

As in Sect. 5.1, when a generation begins the dealer prepares shares for all parties that are members of that generation. We focus on the beginning of generation g and describe the dealer's procedure:

1. Split s using a secret sharing scheme for $(k, \text{SIZE}(g))$-THR. Denote the resulting shares by $u_1^{(g)}, \ldots, u_{\text{SIZE}(g)}^{(g)}$.
2. Generate $k-1$ shares using the secret sharing scheme Π_k given the secret s and previous shares $\{v_j^{(i)}\}_{i\in[g-1],j\in[k-1]}$. Denote the resulting shares by $v_1^{(g)}, \ldots, v_{k-1}^{(g)}$.
3. For $i \in [k-1]$, split $v_i^{(g)}$ using a secret sharing scheme for $(i, \text{SIZE}(g))$-THR. Denote the resulting shares by $\{w_{i,1}^{(g)}, \ldots, w_{i,\text{SIZE}(g)}^{(g)}\}_{i\in[k-1]}$.
4. Set the secret share of the jth party in the gth generation (i.e. $j \in [\text{SIZE}(g)]$) to be

$$\left(u_j^{(g)}, w_{1,j}^{(g)}, \ldots, w_{k-1,j}^{(g)}\right).$$

Correctness and Security. We show that any k parties can learn the secret. If all the parties come from the same generation g, then they can use their $u^{(g)}$ in order to run the reconstruction procedure of the $(k, \text{SIZE}(g))$-THR scheme and learn s. For k parties that come from at least two generations we show that they can jointly learn k shares for the evolving k-THR scheme Π_k. By correctness of Π_k, using these shares they can reconstruct s. Indeed, assume that c_0 parties come from generation 0, c_1 come from generation 1 and so on, where there is some generation g where $\sum_{i=0}^{g} c_i = k$ and for all i it holds that $c_i \le k-1$.

Claim 13. *Any $c \in [\text{SIZE}(g)]$ parties from generation g can compute $v_c^{(g)}$.*

Proof. The c parties hold c shares for $(1, \text{SIZE}(g))$-THR scheme that give $v_1^{(g)}$, c shares for the $(2, \text{SIZE}(g))$-THR scheme that give $v_2^{(g)}$ and so on.

Using this claim we get that the k parties can learn $\sum_{i=0}^{g} c_i = k$ shares of the evolving k-THR scheme, as required.

For security consider any set of $k-1$ parties. First, the u shares of the $(k, \text{SIZE}(g))$-THR scheme are independent of the secret. Thus, to complete the proof we need to show that the parties cannot learn any k shares of the evolving k-THR scheme Π_k. Indeed, any c parties from generation g cannot learn more than c shares $v_1^{(g)}, \ldots, v_c^{(g)}$; this follows from the security of the schemes $(c+1, \text{SIZE}(g))$-THR, $\ldots, (k-1, \text{SIZE}(g))$-THR. Therefore, in total, the parties can learn at most $\sum_{i=0}^{g} c_i < k$ shares.

Share Size Analysis. We bound the size of each component in the share of party t in the scheme Π_k'. The share of party t that is the jth party of generation $g = \lfloor (\log t)/(k-1) \rfloor$ is composed of $u_j^{(g)}$ and $w_{1,j}^{(g)}, \ldots, w_{k-1,j}^{(g)}$:

1. $u_j^{(g)}$ – generated by secret sharing s using a scheme for $(k, \text{SIZE}(g))$-THR. By Claim 5 it holds that

$$|u_j^{(g)}| \le \log(\text{SIZE}(g)) \le \log t + (k-1)$$

2. $w_{i,j}^{(g)}$ – generated by secret sharing $v_i^{(g)}$ using a scheme for $(i, \text{SIZE}(g))$-THR. By Claim 5 for $1 < i \le k-1$ it holds that

$$|w_{i,j}^{(g)}| \le \max\{\log(\text{SIZE}(g)), |v_i^{(g)}|\} \le \max\{\log t + (k-1), |v_i^{(g)}|\}$$

and for $i = 1$ it holds that

$$|w_{1,j}^{(g)}| = |v_i^{(g)}|.$$

– $v_i^{(g)}$ – generated by generating a share of a sharing scheme Π_k for evolving k-THR. Recall that $g \cdot (k-1) \le \log t + (k-1)$ shares were generated for previous g generations. Therefore, for all $i \in [k-1]$

$$|v_i^{(g)}| \le \sigma(\log t + (k-1)).$$

Therefore, for $1 < i \le k-1$

$$|w_{i,j}^{(g)}| \le \max\{\log t + (k-1), \sigma(\log t + (k-1))\}$$

and for $i = 1$

$$|w_{1,j}^{(g)}| \le \sigma(\log t + (k-1)).$$

Thus, the total share size in the scheme Π_k' is bounded by:

$$\log t + (k-1) + \sigma(\log t + (k-1)) + (k-2) \cdot \max\{\log t + (k-1), \sigma(\log t + (k-1))\} \tag{1}$$

$$\le \log t + (k-1) + \sigma(\log t + (k-1)) + (k-2)(\log t + (k-1) + \sigma(\log t + (k-1)))$$

$$\le (k-1)\log t + k \cdot \sigma(\log t + k) + k^2.$$

5.3   Proof of Theorem 9 Assuming Lemma 2

Let $k \in \mathbb{N}$ be such that $k \geq 2$. We use the scheme for evolving k-THR constructed in Sect. 5.1 in which the share size of the tth party is $\sigma_k^{(0)}(t) = kt \cdot \log(kt)$. Using Lemma 2 this gives rise to a scheme $\Pi_k^{(1)}$ for evolving k-THR in which the share size of the tth party is:

$$\sigma_k^{(1)}(t) = (k-1) \cdot \log t + k \cdot \sigma_k^{(0)}(\log t + k) + k^2. \tag{2}$$

We bound $\sigma_k^{(0)}(\log t + k)$. If $k > \log t$, then

$$\sigma_k^{(0)}(\log t + k) \leq \sigma_k^{(0)}(2k) \leq 2k^2 \cdot \log(2k^2) \leq 4k^2 \cdot \log(2k)$$

If $k \leq \log t$ then

$$\begin{aligned}
\sigma_k^{(0)}(\log t + k) &\leq \sigma_k^{(0)}(2\log t) \\
&\leq k \cdot 2\log t \cdot \log(k \cdot 2\log t) \\
&\leq 2k \cdot \log t \cdot \log\log t + 2k \cdot \log t \cdot \log(2k) \\
&\leq 4k \cdot \log t \cdot \log\log t + 4k^2 \cdot \log(2k),
\end{aligned}$$

where the last inequality follows since $2k \cdot \log t \cdot \log(2k) \leq 2k \cdot \log t \cdot \log\log t + 4k^2 \cdot \log(2k)$. Together we get that

$$\sigma_k^{(0)}(\log t + k) \leq \max\{\sigma_k^{(0)}(2\log t), \sigma_k^{(0)}(2k)\} \leq 4k \cdot \log t \cdot \log\log t + 4k^2 \cdot \log(2k).$$

Plugging this in Eq. (2), we get that

$$\begin{aligned}
\sigma_k^{(1)}(t) &= (k-1) \cdot \log t + k \cdot \sigma_k^{(0)}(\log t + k) + k^2 \\
&\leq (k-1) \cdot \log t + 4k^2 \cdot \log t \cdot \log\log t + 4k^3 \cdot \log(2k) + k^2 \\
&\leq 5k^2 \cdot \log t \cdot \log\log t + 5k^3 \cdot \log k.
\end{aligned}$$

Using Lemma 2 again, we get a scheme $\Pi_k^{(2)}$ in which the share size of the tth party is

$$\sigma_k^{(2)}(t) = (k-1) \cdot \log t + k \cdot \sigma_k^{(1)}(\log t + k) + k^2. \tag{3}$$

We bound $\sigma_k^{(1)}(\log t + k)$ as follows.

$$\begin{aligned}
\sigma_k^{(1)}(\log t + k) &\leq \max\{\sigma_k^{(1)}(2\log t), \sigma_k^{(1)}(2k)\} \\
&\leq 5k^2 \cdot \log(2\log t) \cdot \log\log(2\log t) + 5k^2 \cdot \log(2k) \cdot \log\log(2k) \\
&\quad + 5k^3 \cdot \log k \\
&\leq 6k^2 \cdot \log\log t \cdot \log\log\log t + 6k^3 \cdot \log k.
\end{aligned}$$

Plugging this back in Eq. (3), we get that

$$
\begin{aligned}
\sigma_k^{(2)}(t) &= (k-1) \cdot \log t + k \cdot \sigma_k^{(1)}(\log t + k) + k^2 \\
&\leq (k-1) \cdot \log t + 6k^3 \cdot \log\log t \cdot \log\log\log t + 6k^4 \cdot \log k + k^2 \\
&\leq (k-1) \cdot \log t + 6k^3 \cdot \log\log t \cdot \log\log\log t + 7k^4 \cdot \log k.
\end{aligned}
$$

Remark. As in the proof of Theorem 8, one can iteratively apply Lemma 2 again and again to decrease the dependence on $\log\log t \cdot \log\log\log t$. However, the dependence on $\log t$ cannot be improved using this method.

5.4 Generalization to Larger Domains of Secrets

Similarly to the generalization of the scheme from Sect. 4 to support larger domains of secrets (see Sect. 5.4), we generalize the above scheme. Let the secret be of length ℓ. Following the proof of Lemma 2, we obtain that given a scheme for the evolving k-THR access structure that supports secrets of length ℓ in which the share size of the tth party is $\sigma(t)$, there exists a scheme for the same access structure and same length of secrets in which the share size of the tth party is bounded by (cf. Eq. (1))

$$
\begin{aligned}
\max\{\log t + (k-1), \ell\} &+ \sigma(\log t + (k-1)) + (k-2) \cdot \\
&\max\{\log t + (k-1), \sigma(\log t + (k-1))\}
\end{aligned}
$$

Notice that for large enough $t \in \mathbb{N}$ the above bound is the *same* as the bound we had in Eq. (1). For the recursive composition step (cf. Sect. 5.2) we start with the naive generalization of the scheme from Theorem 9 to support several input bits (i.e. bit by bit). This gives a scheme in which the share size is $\sigma^{(0)}(t) \leq \ell \cdot ((k-1) \cdot \log t + k \cdot \sigma(\log t + k) + k^2)$. For large enough t it holds that $\sigma^{(0)}(t) \leq kt \cdot \log(kt)$. Thus, one can follow the same outline of the proof of Theorem 9 (see Sect. 5.3) and obtain the *same* share size as in Theorem 9 for large enough t. (For smaller values of t one can follow the analysis and obtain a bound as a function of t and ℓ).

5.5 Linearity of the Scheme

The scheme from Theorem 7 is linear over $\mathsf{GF}(2)$. In the scheme from this section the shares are composed of several different parts each being an element coming from a different scheme. Consider the scheme from Sect. 5.1 (denoted by $\Pi_k^{(0)}$ in Sect. 5.3). Each share there is a composition of several linear schemes (the threshold scheme of Shamir and the scheme of Benaloh and Leichter). Since composition of linear schemes results with a linear scheme, the scheme is linear. Next, for the basic construction $\Pi_k^{(1)}$ in Sect. 5.3, each share is composed of several parts each being either a share of a linear scheme (Shamir's scheme) or a composition of linear schemes (Shamir's scheme and the scheme $\Pi_k^{(0)}$), resulting with a linear scheme. The same argument applies for the recursive composition which eventually gives that the final construction is linear.

6 A Lower Bound

For general access structures the best standard secret sharing schemes require exponential-size shares. Instantiating our scheme for n parties, results with the nth party holding a share of size 2^{n-1}. Thus, any improvement in the share size on our scheme for general access structures, will imply a non-trivial improvement for general access structures in the standard setting.

In the case of k-threshold access structures, we do not know if our scheme is tight. Specifically, for $k > 2$, using our scheme to implement a standard secret sharing scheme for k-out-of-n is not tight. Indeed, the most significant term in the share size in our scheme depends linearly on $k - 1$, while the best schemes in the standard setting are independent of k (see Claim 5).

Thus, one may ask whether there exists a secret sharing scheme for the evolving k-THR in which the share size of the tth party is roughly $\log t$. We show that such a scheme cannot exist.

Theorem 14 (Theorem 3 restated). *For any constant $c \in \mathbb{N}$, there is no secret sharing scheme for the evolving 2-THR access structure in which the share size of the tth party is at most*

$$\log t + \log \log t + c.$$

Proof. Assume (towards contradiction) that there is a secret sharing scheme for the evolving 2-THR access structure in which the share size of the tth party is at most $\log t + \log \log t + c$ for a constant $c \in \mathbb{N}$. We can use this scheme to implement a standard secret sharing scheme for $(2, n)$-THR in which the share size of party $t \in [n]$ is $m_t \le \log t + \log \log t + c$.

We use the following claim that underlies the lower bound of Kilian and Nisan. This inequality is the same as Kraft's (see [13, Chapter 5.2]), a fact that we use in Sect. 7.

Claim 15. *([25] and [12, Appendix A]).* *For any $n \in \mathbb{N}$, in any secret sharing scheme for $(2, n)$-THR, it holds that*

$$\sum_{t=1}^{n} \frac{1}{2^{m_t}} \le 1,$$

where m_t is the share size of the tth party.

Using this claim we get that

$$1 \ge \sum_{t=1}^{n} \frac{1}{2^{m_t}} \ge \sum_{t=2}^{n} \frac{1}{2^{\log t + \log \log t + c}} = \frac{1}{2^c} \cdot \sum_{t=2}^{n} \frac{1}{t \cdot \log t}.$$

To get a contradiction we need to show that $\sum_{t=2}^{n} \frac{1}{t \cdot \log t} > 2^c$ for large enough n. Indeed, letting $n \to \infty$, we have that

$$\sum_{t=2}^{\infty} \frac{1}{t \cdot \log t} \ge \int_{2}^{\infty} \frac{1}{t \cdot \log t} dt = \log \log t \Big|_{2}^{\infty} \to \infty.$$

This completes the proof.

Remark 1 (A stronger lower bound). We note that the lower bound can be strengthened to show that even schemes in which the share size is $\sum_{t=1}^{\ell} \log^{(i)}(t) + c$ cannot exist for any $\ell \in \mathbb{N}$ and where $\log^{(i)}(t)$ is the i-times repeated log of t (letting $\log^{(0)}(t) = t$). This follows similarly to the above argument noting that for every $\ell \in \mathbb{N}_0$ it holds that $\int_1^{\infty} \frac{1}{\prod_{i=0}^{\ell} \log^{(i)}(t)} dt = \log^{(\ell+1)} t$ and using that $\log^{(\ell+1)} t \geq 2^c$ for any constant $c \in \mathbb{N}$ and large enough t.

This is reminiscent of bounds in the literature on prefix codes [7,18]. This is not surprising given the equivalence (in terms of complexity) between prefix codes and secret sharing for the evolving 2-THR access structures developed in Sect. 7.

7   The Equivalence Between Evolving 2-Threshold and Prefix Codes

We now show the very tight connection between schemes for the evolving 2-THR access structure and prefix codes.

Theorem 16 (Theorem 4 restated). *Let $\sigma \colon \mathbb{N} \to \mathbb{N}$. A prefix code for the integers in which the length of the tth codeword is $\sigma(t)$ exists if and only if a secret sharing scheme for the evolving 2-threshold access structure and 1-bit secret in which the share size of the tth party is $\sigma(t)$.*

Proof of the "if" Part of Theorem 16. Kraft's inequality (see [13, Theorem 5.2.2]) gives a *necessary and sufficient* condition for the existence of a prefix code for a given set of codeword lengths. The proof of the sufficient direction is constructive: given the collection of lengths of codewords it is possible to construct the code. Furthermore, we do not need to know the collection of lengths in advance, i.e. we can create the code on the fly, as long as the demand ($\sum_t \frac{1}{2^{m_t}}$) does not exceed 1. This inequality is the same as the one given in Claim 15 that must be satisfied by any secret sharing scheme for the evolving 2-THR access structure. Thus, any secret sharing scheme for the evolving 2-THR access structure in which the share size of the tth party is $\sigma(t)$, implies the existence of a prefix code in which the length of the tth codeword is $\sigma(t)$.   ∎

Proof of the "only if" Part of Theorem 16. Let $\Sigma \colon \mathbb{N} \to \{0,1\}^*$ be a prefix code for the integers. That is, for any $t_1, t_2 \in \mathbb{N}$ such that $t_1 \neq t_2$, it holds that $\Sigma(t_1)$ is not a prefix of $\Sigma(t_2)$. For $t \in \mathbb{N}$ denote by $\sigma(t)$ the length of the codeword $\Sigma(t)$.

The Scheme. Let $s \in \{0,1\}$ be the secret to be shared. Let w be an infinite random binary string. The dealer generates the string as needed: at time $t \in \mathbb{N}$ the dealer holds the prefix of length $\sigma(t)$ of the string w, denoted by $w_{[\sigma(t)]}$ (for simplicity we assume that $\sigma(t)$ is monotonically increasing, but this is not necessary). The share of party t is a string u_t such that:

1. If $s = 0$, then $u_t = w_{[\sigma(t)]}$.
2. If $s = 1$, then $u_t = \Sigma(t) \oplus w_{[\sigma(t)]}$ (bit-wise XOR).

Reconstruction. Any two different parties t_1 and t_2, holding shares u_1 and u_2, respectively, where $|u_1| \leq |u_2|$, should check if u_1 is a prefix of u_2. If it is a prefix, then they output $s = 0$ and otherwise, they output $s = 1$.

Correctness and Security. If $s = 0$, then since u_1 and u_2 are both prefixes of the same string w it holds that u_1 is a prefix of u_2. On the other hand, if $s = 1$ then $u_1 = \Sigma(t_1) \oplus w_{[\sigma(t_1)]}$ and $u_2 = \Sigma(t_2) \oplus w_{[\sigma(t_2)]}$, where $w_{[\sigma(t_1)]}$ is a prefix of $w_{[\sigma(t_2)]}$. Since Σ is a prefix code, $\Sigma(t_1)$ is *not* a prefix of $\Sigma(t_2)$, and thus u_1 is not a prefix of u_2.

Security follows since for both $s = 0$ and for $s = 1$ each single party t holds a single string u_t which is uniformly distributed over $\{0,1\}^{\sigma(t)}$. In case $s = 0$ this is true by construction, and in case $s = 1$ this is true since all the party sees is the codeword $\Sigma(t)$ XORed with $w_{[\sigma(t)]}$ which is uniform.

Share Size. The share size of the tth player in this scheme is $\sigma(t)$. Using the best constructions of prefix codes [7,18], we get the share size of Theorem 8.

Generalization to Larger Domains of Secrets. One can support sharing of longer secrets by sharing every bit independently. Our direct construction presented in Sect. 4 is more efficient for sharing longer secrets (see Sect. 4.1 for more details). ∎

Efficiency Preservation. Note that the transformation from the prefix code to secret sharing preserves the efficiency of the code, i.e. dealing a share to party t is as easy as computing $\Sigma(t)$. However, the other direction, with the construction based on Kraft's inequality, does not preserve the efficiency. That is, we cannot say that encoding the number t, i.e. computing $\Sigma(t)$, is as easy as dealing a secret to party t.

8   Further Work and Open Problems

This work suggests several research directions. The most evident one is to investigate the necessity of the linear dependence on k in the most significant term in our scheme for the evolving k-THR access structure. In particular, are more algebraic-oriented constructions possible?

There are several interesting access structures for which we do not have efficient constructions. For example, a very natural evolving access structure is the one in which qualified subsets are the ones which form a *majority* of the present parties at *some* point in time. The only scheme that realized this access structure we are aware of stems from our construction for general access structures from Sect. 3 which results with very long shares.

When $k = 2$, we show a tight connection between evolving secret sharing and prefix codes (see Sect. 7). Is there a generalization of prefix codes that is related to the evolving k-THR access structure for $k > 2$?

Secret sharing has had many applications in cryptography and distributed computing. One of the most notable examples is *multi-party computation* (MPC). Can secret sharing for evolving access structures be useful for MPC?

We focused on schemes in which correctness and security are *perfect*. One can relax correctness to work with high probability and to allow small statistical error in security. Can these relaxations be used to obtain more interesting and efficient schemes? Another variant of secret sharing schemes is the *computational* one. In these schemes security is required only against computationally bounded adversaries. Efficient computational schemes for much richer classes of access structures are known [27,28,32,35]. Is there a meaningful way to define computationally secure secret sharing schemes for evolving access structures? Can this be used to obtain efficient schemes for more classes of evolving access structures? Cachin [11] studied a similar question in a model in which there is a large public bulletin board.

Other natural variants of secret sharing can be adapted to the evolving setting. For example, verifiable, robust and visual secret sharing. We leave these as interesting directions for future exploration.

References

1. Beimel, A.: Secure Schemes for Secret Sharing and Key Distribution. Ph.D. thesis, Technion - Israel Institute of Technology (1996). http://www.cs.bgu.ac.il/beimel/Papers/thesis.ps
2. Beimel, A.: Secret-sharing schemes: a survey. In: Chee, Y.M., Guo, Z., Ling, S., Shao, F., Tang, Y., Wang, H., Xing, C. (eds.) IWCC 2011. LNCS, vol. 6639, pp. 11–46. Springer, Heidelberg (2011)
3. Beimel, A., Ishai, Y.: On the power of nonlinear secrect-sharing. In: 16th Annual IEEE Conference on Computational Complexity, CCC, pp. 188–202 (2001)
4. Ben-Or, M., Goldwasser, S., Wigderson, A.: Completeness theorems for non-cryptographic fault-tolerant distributed computation (extended abstract). In: STOC, pp. 1–10 (1988)
5. Benaloh, J.C., Leichter, J.: Generalized secret sharing and monotone functions. In: Goldwasser, S. (ed.) CRYPTO 1988. LNCS, vol. 403, pp. 27–35. Springer, Heidelberg (1990)
6. Benaloh, J.C., Rudich, S.: Unpublished, private Communication with Steven Rudich. (1989)
7. Bentley, J.L., Yao, A.C.: An almost optimal algorithm for unbounded searching. Inf. Process. Lett. **5**(3), 82–87 (1976)
8. Blakley, G.R.: Safeguarding cryptographic keys. In: Proceedings of the AFIPS National Computer Conference, vol. 22, pp. 313–317 (1979)
9. Bogdanov, A., Guo, S., Komargodski, I.: Threshold secret sharing requires a linear size alphabet. Electronic Colloquium on Computational Complexity (ECCC) 23, 131 (2016). http://eccc.hpi-web.de/report/2016/131, to appear in TCC 2016B
10. Boppana, R.B.: Threshold functions and bounded depth monotone circuits. J. Comput. Syst. Sci. **32**(2), 222–229 (1986)
11. Cachin, C.: On-line secret sharing. In: Boyd, C. (ed.) Cryptography and Coding 1995. LNCS, vol. 1025, pp. 190–198. Springer, Heidelberg (1995)
12. Pueyo, C.I., Cramer, R., Xing, C.: Bounds on the threshold gap in secret sharing and its applications. IEEE Trans. Inf. Theory **59**(9), 5600–5612 (2013)
13. Cover, T.M., Thomas, J.A.: Elements of Information Theory, 2nd edn. Wiley, New York (2006)

14. Cramer, R., Damgård, I.B., Maurer, U.M.: General secure multi-party computation from any linear secret-sharing scheme. In: Preneel, B. (ed.) EUROCRYPT 2000. LNCS, vol. 1807, pp. 316–334. Springer, Heidelberg (2000)
15. Csirmaz, L., Tardos, G.: On-line secret sharing. Des. Codes Crypt. **63**(1), 127–147 (2012)
16. Dodis, Y., Patrascu, M., Thorup, M.: Changing base without losing space. In: STOC, pp. 593–602 (2010)
17. Elias, P.: Universal codeword sets and representations of the integers. IEEE Trans. Inf. Theory **21**(2), 194–203 (1975)
18. Even, S., Rodeh, M.: Economical encoding of commas between strings. Commun. ACM **21**(4), 315–317 (1978)
19. Friedman, J.: Constructing $O(n \log n)$ size monotone formulae for the k-th threshold function of n boolean variables. SIAM J. Comput. **15**(3), 641–654 (1986)
20. Geographic, N.: NASA declares end to deep impact comet mission. http://news.nationalgeographic.com/news/2013/09/130920-deep-impact-ends-comet-mission-nasa-jpl/. Acccessed 07 Feb 2016
21. Herzberg, A., Jarecki, S., Krawczyk, H., Yung, M.: Proactive secret sharing or: how to cope with perpetual leakage. In: Coppersmith, D. (ed.) CRYPTO 1995. LNCS, vol. 963, pp. 339–352. Springer, Heidelberg (1995)
22. Ito, M., Saito, A., Nishizeki, T.: Multiple assignment scheme for sharing secret. J. Cryptol. **6**(1), 15–20 (1993)
23. Kannan, S., Naor, M., Rudich, S.: Implicit representation of graphs. SIAM J. Discrete Math. **5**(4), 596–603 (1992)
24. Karchmer, M., Wigderson, A.: On span programs. In: 8th Annual Structure in Complexity Theory Conference, pp. 102–111 (1993)
25. Kilian, J., Nisan, N.: Unpublished (1990). see [12]
26. Kol, G., Naor, M.: Games for exchanging information. In: STOC, pp. 423–432 (2008)
27. Komargodski, I., Naor, M., Yogev, E.: Secret-sharing for NP. In: Sarkar, P., Iwata, T. (eds.) ASIACRYPT 2014, Part II. LNCS, vol. 8874, pp. 254–273. Springer, Heidelberg (2014)
28. Krawczyk, H.: Secret sharing made short. In: Stinson, D.R. (ed.) CRYPTO 1993. LNCS, vol. 773, pp. 136–146. Springer, Heidelberg (1994)
29. Malkin, T., Micciancio, D., Miner, S.K.: Efficient generic forward-secure signatures with an unbounded number of time periods. In: Knudsen, L.R. (ed.) EUROCRYPT 2002. LNCS, vol. 2332, pp. 400–417. Springer, Heidelberg (2002)
30. Pagh, R., Segev, G., Wieder, U.: How to approximate a set without knowing its size in advance. In: 54th Annual IEEE Symposium on Foundations of Computer Science, FOCS, pp. 80–89 (2013)
31. Shamir, A.: How to share a secret. Commun. ACM **22**(11), 612–613 (1979)
32. Vinod, V., Narayanan, A., Srinathan, K., Pandu Rangan, C., Kim, K.: On the power of computational secret sharing. In: Johansson, T., Maitra, S. (eds.) INDOCRYPT 2003. LNCS, vol. 2904, pp. 162–176. Springer, Heidelberg (2003)
33. Wikipedia: IPv4 address exhaustion. https://en.wikipedia.org/wiki/IPv4_address_exhaustion. Acccessed 07 Feb 2016
34. Wikipedia: Year 2000 problem. https://en.wikipedia.org/wiki/Year_2000_problem. Acccessed 07 Feb 2016
35. Yao, A.C.: Unpublished, mentioned in [2]. See also [32]

New Models

Designing Proof of Human-Work Puzzles
for Cryptocurrency and Beyond

Jeremiah Blocki[1]([⊠]) and Hong-Sheng Zhou[2]

[1] Purdue University, West Lafayette, USA
jblocki@purdue.edu
[2] Virginia Commonwealth University, Richmond, USA
hszhou@vcu.edu

Abstract. We introduce the novel notion of a Proof of Human-work
(PoH) and present the first distributed consensus protocol from hard Artificial Intelligence problems. As the name suggests, a PoH is a proof that
a *human* invested a moderate amount of effort to solve some challenge. A
PoH puzzle should be moderately hard for a human to solve. However, a
PoH puzzle must be hard for a computer to solve, including the computer
that generated the puzzle, without sufficient assistance from a human. By
contrast, CAPTCHAs are only difficult for other computers to solve —
not for the computer that generated the puzzle. We also require that a
PoH be publicly verifiable by a computer without any human assistance
and without ever interacting with the agent who generated the proof of
human-work. We show how to construct PoH puzzles from indistinguishability obfuscation and from CAPTCHAs. We motivate our ideas with two
applications: HumanCoin and passwords. We use PoH puzzles to construct HumanCoin, the first cryptocurrency system with human miners.
Second, we use proofs of human work to develop a password authentication scheme which provably protects users against offline attacks.

1 Introduction

The emergence of decentralized cryptocurrencies like Bitcoin [45] has the potential to significantly reshape the future of distributed interaction. These recent
cryptocurrencies offer several advantages over traditional currencies, which rely
on a centralized authority. At the heart of Bitcoin-like cryptocurrencies is an
efficient distributed consensus protocol that allows for all users to agree on the
same public ledger. When combined with other cryptographic tools like digital signatures the distributed consensus protocol prevents users from engaging
in dishonest behavior like "double spending" their money or spending another
user's money. Fundamentally, the applications of a tamper-proof blockchain like
the one in Bitcoin are not limited to cryptocurrency. For example, a tamper
proof blockchain could help us construct secure and fair multiparty computation protocols [1,7,36,38], develop smart contracts [38,53], and build distributed
autonomous agents, to name a few applications. In this paper we propose a fundamentally new technique, Proofs of Human-work (PoH), for constructing a
secure blockchain, and we show that our techniques have several other valuable
applications like password protection and non-interactive bot detection.

© International Association for Cryptologic Research 2016
M. Hirt and A. Smith (Eds.): TCC 2016-B, Part II, LNCS 9986, pp. 517–546, 2016.
DOI: 10.1007/978-3-662-53644-5_20

At its core, Bitcoin's distributed consensus protocol is based on moderately hard Proofs of Work (PoW) [23]. In Bitcoin the Hashcash [3] PoW puzzles are used to extend the blockchain, a cryptographic data-structure in which the public ledger is recorded. A PoW puzzle should be moderately hard for a computer to solve, but the PoW solution should be easy for a computer to verify. Cryptocurrencies like Bitcoin require that this hardness parameter of PoW puzzles be tunable. An adversary would need to control 51 % of the computational power in the Bitcoin network to be able to alter the blockchain and prevent users from reaching the correct consensus[1]. While Bitcoin cleverly avoids the Sybil attack by using PoW puzzles, there are still many undesirable features of this distributed consensus protocol. For example, constructing the proofs of work is energy intensive making the mining process in this distributed consensus protocol environmentally unfriendly. Furthermore, the mining process is dominated by a smaller number of professional miners with customized hardware making it unprofitable for others to join — this raises the natural concern that a few professional miners might collude to alter the public ledger [46]. Indeed, the mining pool GHash.io[2] recently exceeded 50 % of the computational power in Bitcoin. While other techniques like Proofs of Space [25,47] or Proofs of Stake [8] have been proposed to build the blockchain in a distributed consensus protocol each of these techniques has its own drawbacks. It is clearly desirable to find new techniques for reaching a stable distributed consensus. In this paper we ask the following question:

Is it possible to design proof of human-work puzzles that are suitable for a decentralized cryptocurrency?

We believe that a cryptocurrency based on Proof of Human-work might offer many advantages over other approaches. First, the mining process would be eco-friendly. Second, instead of wasting 'human cycles,' it might be possible to base the proofs of human work on activities that are fun [34], educational [33] or even beneficial to society [35,56]. Third, proofs of human work are fair by nature in the sense that two individuals will generally perform a comparable amount of work to produce a proof of human work. Thus, professional or rich miners would not have an significant advantage over regular users. By contrast, in Bitcoin the cost of computing the SHA256 hash function on customized hardware is dramatically less than the cost of computing SHA256 on personal computing[3]. Finally, we believe that the cryptocurrency would be less-vulnerable to 51 % attacks by nation states or by a few professional miners. However, we stress that our purpose is not to enumerate all of the possible social consequences of

---

[1] Technically byzantine agreement is only possible when the adversary has less than 50 % of the hashing power and the network has high synchronicity — otherwise we need to ensure that the adversary has at most 33.3 % of the hashing power [29].

[2] See http://arstechnica.com/security/2014/06/bitcoin-security-guarantee-shattered-by-anonymous-miner-with-51-network-power/.

[3] See https://bitcoinmagazine.liberty.me/bitmain-announces-launch-of-next-generation-antminer-s7-bitcoin-miner/ (Retrieved 5/4/2016).

a cryptocurrency based on Proofs of Human-work. As with any new technology HumanCoin could potentially be used for good or for evil. See the full version [12] for additional discussions.

1.1 Cryptocurrencies Meet AI: Proof of Human-Work Puzzles

In this work we introduce the novel notion of Proofs of Human-work (PoH) which would be suitable for cryptocurrencies. Proofs of Human-work are fundamentally different from standard Proofs of Work. Informally, a PoH puzzle should be moderately hard for a human to solve meaning that it should require modest effort for a human to produce a valid proof of human work — again we require that this hardness parameter should be tunable. Furthermore, the puzzles should be easy for a computer to generate, but they need to be difficult for a computer to solve without sufficient human assistance — even for the computer that generated the puzzle. Finally, the puzzles need to be publicly verifiable meaning that it should be easy for a computer to verify the solution to the puzzle without any human assistance — even if the computer did not generate the puzzle. We stress that there is no interaction during the puzzle generation or during the puzzle verification process, and there is no trusted server in our distributed setting. Thus, a computer will need to validate proofs of human-work that were generated and solved by agents with whom it has never interacted.

Our description of a PoH puzzle might remind the reader of a CAPTCHA (Completely Automated Public Turing-Test to tell Computers and Humans Apart) [55]. CAPTCHAs have been widely deployed on the Internet to fight spam and protect against sybil attacks. Informally, a CAPTCHA is a puzzle that is easy for a human to solve, but difficult for a computer. CAPTCHAs are based on the assumption that some underlying artificial intelligence (AI) problem is hard for computers, but easy for humans (e.g., reading distorted letters).

While we do use CAPTCHAs to construct proofs of human work, we stress that a CAPTCHA itself *cannot* achieve our notion of proofs of human-work. Let (Z, σ) be a CAPTCHA puzzle-solution pair. Verifiers who receive the pair (Z, σ) would not necessarily be able to check that σ is the correct solution without interacting with a human. More importantly, the computer that generates the puzzle Z could produce the solution σ *without any human effort* because CAPTCHA generation algorithms start by randomly selecting a target solution σ and then outputting a randomly generated puzzle Z with the solution σ. Thus, a pair (Z, σ) does not constitute a proof of human work. The PoH verifier would need to ensure, without interacting with any other human agent or any other computer agent, that the challenge generator did not already have the answer σ to the puzzle Z.

We believe that our Proof of Human-work puzzles could also have applications in many other contexts. For example, to limit spam or prevent phishing attacks it might useful to verify that some human effort went into producing a message. When a human user is busy it would be convenient if the computer could validate this proof of human effort automatically without needing to interact with the sender who may no longer be available when the message is received. Similarly, proofs of human-work might be a useful tool for honest preference elicitation — a challenging problem in mechanism design. A human could demonstrate that a particular issue or outcome is truly important to him by producing a proof of human-work.

1.2   AI Meets Obfuscation: Constructing Proof of Human-Work Puzzles

It is not immediately clear how to construct PoH puzzles. CAPTCHAs allow a computer to generate puzzles that other computers cannot solve, but how could a computer generate a puzzle that is meaningful to a human without learning the answer itself? Even if this were possible how could a puzzle verifier be convinced that the puzzle(s) was generated honestly (e.g., in a way that does not reveal the answer) without any interaction? How could the verifier be convinced that the answer is correct without help from a human? Building PoH puzzles is a challenging problem.

To address these issues, we need to have a way to generate CAPTCHAs *obliviously* in the sense that a computer is able to generate a well-formed puzzle instance Z without learning the corresponding solution σ. This is feasible by leveraging recent breakthroughs in indistinguishability obfuscation [30]. At an intuitive level, we can have a CAPTCHA puzzle Z generated inside an obfuscator, and now the corresponding answer σ remains hidden inside the obfuscated program. We note that the puzzle solution verification can also take place inside an obfuscated program, even without having human effort involved.

Once we have the idea of generating a CAPTCHA puzzle obliviously as mentioned above, we then can mimic the steps of constructing Proof of Work puzzle in Bitcoin to get a PoH scheme. In PoW, a prover/miner is given a puzzle instance x. The prover will compute the cryptographic hash $H(x, s)$ for many distinct witness s until the value $H(x, s)$ is smaller than a target value. In PoH, the miner uses (x, s) as the input for an obfuscated program, and inside the obfuscated program, a pseudorandom string r is generated from the input (x, s), and this r will be used for generating the solution σ and the puzzle instance Z. The miner obtains Z but has no access to the internal state r and σ.

A human miner is now able to obtain the solution σ from the puzzle Z. As in PoW, the miner will repeat this process until he finds a witness s so that $H(x, s, \sigma, Z)$ is smaller than a target value. We note that, once a successful miner publishes a valid tuple (x, s, σ, Z), any verifier is able to verify it without interaction with human: The verifier can reproduce Z inside the obfuscated program along with a verification tag, *tag*. While the verification tag allows the verifier to check whether a given solution σ is correct this value will not expose the solution σ (e.g., *tag* might be an obfuscated point function which outputs 1 on input $x = \sigma$ and 0 on all other inputs).

Our PoH scheme maintains many of the same desirable properties as a PoW. For example, we can tune the hardness of our PoH puzzle generator by having the verifier reject a valid triple (x, s, σ, Z) with probability $1 - 2^\omega$ so that a human would need to generate and solve 2^ω on average to produce a valid proof of human-work. Thus, the hardness of the PoH puzzles could be tuned by adjusting ω.

While the conceptual understanding of our PoH construction is quite simple, the security analysis is a bit tricky. In the PoW, we sample from a uniform distribution via random oracle, here we need to sample from a more sophisticated distribution. We rely on a newly developed tool *universal samplers* by Hofheinz et al. [32], which is based on the existence of indistinguishability obfuscation and one-way functions in the random oracle model. As discussed in [32], we stress that the random oracle is only used outside of obfuscated programs. There has been tidal wave of new cryptographic constructions using indistinguishability obfuscation since the roundbreaking results of Garg et al. [30]. However, to the best of our knowledge we are the first rigorous paper to explore the connection between AI and program obfuscation[4]. We believe that obfuscation is a powerful new tool that has the potential to fundamentally shape the nature of human-computer interaction. Could program obfuscation allow for a human to interact with a computer in fundamentally new ways? We view our work as a first step towards answering this question.

Remark 1. We view our Proof of Human-work construction as a novel proof of concept that is not yet practical due to the use of indistinguishability obfuscation. Since the work of Garg et al. [30] several other candidate indistinguishability obfuscation schemes have been proposed, but a practical obfuscation scheme would still be a major breakthrough. We note that PoH puzzles do not necessarily require general purpose indistinguishability obfuscation. It would be sufficient to obfuscate a few very simple programs (e.g., a CAPTCHA puzzle generator and a pseudorandom function). Constructing PoH puzzles without obfuscation (or without general purpose obfuscation) is an interesting open problem.

Other Applications. The applications of our techniques are not limited to cryptocurrency. In Sect. 5 we use our ideas to build a password authentication scheme that provably resists offline attacks even if the adversary breaches the authentication server. The basic idea to to require a proof of human-work during the authentication process so that it is not economically feasible for the adversary to check millions of password guesses. We also show how to develop a non-interactive bot detection protocol which allows Alice to send a message m to Bob along with a proof of human-work. Bob is able to verify that human-effort was used in the production/transmission of the message m without ever interacting with Alice.

---

[4] Several existing altcoins (e.g., Bytecent, CaptchaCoin) do involve CAPTCHAs, but they rely on a trusted third party to generate the CAPTCHAs. There has also been informal discussion on the Bitcoin research chatroom about using obfuscation to base cryptocurrency on proofs of human labor. For example, see https://download.wpsoftware.net/bitcoin/wizards/2014-05-29.html or http://vitalik.ca/files/problems.pdf.

1.3   Related Work

While there are many variations of CAPTCHAs [55], they are all based on the fundamental assumption that some underlying AI problem is hard (e.g., reading garbled text [56], voice recognition with distorted audio [52], image recognition [26] or even motion recognition). While several CAPTCHAs have been broken (e.g., [16,43,54]) there is still a clear gap between human intelligence and artificial intelligence. We conjecture that in the foreseeable future we will continue to have viable CAPTCHA candidates suitable for proofs of human work. CAPTCHAs have many applications in security: fighting spam [55], mitigating Sybil attacks [20], preventing denial of service attacks [57] and even preventing fully automated man-in-the-middle attackers [24]. As we noted earlier CAPTCHAs alone are not suitable as PoH puzzles. Kumarasubramanian et al. [39] introduced the notion of human-extractable CAPTCHAs, and used them to construct concurrent non-malleable zero-knowledge protocols.

Canneti et al. [18] proposed a slight modification of the notion of CAPTCHAs that they called HOSPs (Human Only Solvable Puzzles) as a defense against offline attacks on passwords. HOSPs are similar to PoHs in that the puzzles must be difficult even for the computer that generates them, but HOSP puzzles are not publicly verifiable by a computer and their construction assumes the existence of a large centralized storage server filled with unsolved CAPTCHA challenges. This makes their protocol vulnerable to pre-computation attacks[5]. By contrast, in Sect. 5 we present a protocol for password storage that provably protects users against offline attacks, does not require a large centralized storage server and is not vulnerable to pre-computation attacks. Blocki et al. [9] introduced GOTCHAs (Generating panOptic Turing Tests to Tell Computers and Humans Apart) as a defense against offline dictionary attacks on passwords. However, GOTCHAs have a high usability cost and are not suitable for cryptocurrency because the puzzle generation protocol requires interaction with a human and the solutions are not publicly verifiable by a computer. We refer an interested reader to the full version [12] for more details about CAPTCHAs and HOSPs.

The problem of designing distributed consensus protocols that work in the presence of an adversarial (Byzantine) parties has been around for decades [2, 22,40]. Typically distributed consensus requires that 2/3 of the parties are honest [40]. On the Internet this assumption is typically not valid because it is often possible for a malicious user to register for multiple fake accounts — a Sybil attack [20]. However, amazing ideas have been proposed in the original Bitcoin white paper [45] under a pseudo identity 'Nakamoto'. At its core Bitcoin is based on an elegant distributed consensus protocol which in turn is based on Proof of Work puzzles [23] to allow users to agree on a common blockchain. Bitcoin uses the Hashcash Proof of Work algorithm due to Back [3]. Very recently, the underlying consensus protocol in the Bitcoin system have been rigorously analyzed in

---

[5] In particular, the adversary might pay to solve every CAPTCHA challenge on the server. While expensive, this one-time cost would amortize over the number of users being attacked.

the cryptographic setting [29,48]; intensive analysis has also been given in the rational setting (e.g. [27,51]).

Since the breakthrough result of Garg et al. [30], demonstrating the first candidate of indistinguishability obfuscation for all circuits, a myriad of uses for indistinguishability obfuscation in cryptography have been found. Among these results, the puncturing methodology by Sahai and Waters [50] has been found very useful. Hofheinz et al. explored the puncturing technique further introducing and constructing universal samplers in the random oracle model [32]. Their universal sampler is one of the key building blocks in our construction of proof of human-work puzzles. We remark that our work is distinct from previous applications in that we are using obfuscation to develop a new way for *humans* to interact with computers.

2   Preliminaries

We adopt the following notational conventions: Given a randomized algorithm $\mathcal{A}$ we use $y \leftarrow \mathcal{A}(x)$ to denote a random sample from the distribution induced by an input x. If we fix the random bits r then we will use $y := \mathcal{A}(x; r)$ to denote the deterministic result.

We will consider two types of users: machine-only users and human-machine users. A machine-only user is a probabilistic polynomial time (PPT) algorithm who does not interact with a human. In general, when we say "human" user we mean a "human user equipped with a PPT machine." Accordingly, we also consider two types of adversaries: a machine-only adversary $\mathcal{A}$, and a human-machine adversary $\mathcal{B}^{\mathcal{H}}$. The machine-only adversary is a PPT algorithm that does not get to query a human. The human-machine adversary $\mathcal{B}^{\mathcal{H}}$ is a PPT algorithm that gets to interact with a human oracle $\mathcal{H}$ which could, for example, solve CAPTCHA puzzles. We typically restrict the total number of queries that human-machine adversary can make to the human oracle. We say that an human-machine adversary $\mathcal{B}^{\mathcal{H}}$ has m human-work units if it is allowed to query $\mathcal{H}$ at most m times. We intentionally under-specify the behavior of the human oracle $\mathcal{H}$. At minimum we assume $\mathcal{H}$ is capable of solving a CAPTCHA puzzle for one human-work unit (one query to the oracle). However, the human-machine adversary may use his queries to ask the human oracle to perform arbitrary tasks $\mathcal{H}$ (e.g., solve basic arithmetic problems, write poetry) so long as each task takes (approximately) the same amount of human-effort as a single CAPTCHA puzzle.

2.1   CAPTCHAs

CAPTCHAs are a fundamental building block in our construction of Proof of Human-work puzzles. Traditionally, a CAPTCHA generator G is defined as a randomized PPT algorithm that outputs a puzzle Z and a solution σ. In every CAPTCHA generator that we are aware of the program G first generates a random target solution σ and then produces a random puzzle Z with solution

σ (e.g., by distorting the string σ). Given public parameters PP for CAPTCHA puzzle generation we adopt the syntax $(Z, tag) \leftarrow \mathtt{G}(\mathrm{PP}, \sigma)$ to emphasize that the target puzzle Z is generated with complete knowledge of the CAPTCHA solution. In traditional CAPTCHA applications it is desirable for the agent who generates a puzzle Z to have knowledge of the corresponding answer σ so that he can verify another agent's response to the challenge Z. However, in our setting this property is problematic since the agent who generates the puzzle Z is trying to produce a convincing proof of human-work. Thus, we will need additional tools to obtain proof of human-work puzzles from CATPCHAs. Formally, a CAPTCHA puzzle-system is defined as follows.

Definition 1 (CAPTCHA). *A CAPTCHA puzzle system consists of a tuple of algorithms* $(\mathtt{Setup}, \mathtt{W}, \mathtt{G}, \mathtt{C}^{\mathcal{H}}, \mathtt{Verify})$, *where*

- $\mathtt{Setup}$ *is a randomized system setup algorithm that takes as input* 1^{λ} *(λ is the security parameter), and outputs a system public parameter* $\mathrm{PP} \leftarrow \mathtt{Setup}(1^{\lambda})$, *which includes a puzzle size parameter* $\ell = \mathrm{poly}(\lambda)$;
- $\mathtt{W}$ *is a randomized sampling algorithm that takes as input the public parameter* PP *and outputs a target solution* $\sigma \leftarrow \mathtt{W}(\mathrm{PP})$ *(e.g., a witness) of length* ℓ;
- $\mathtt{G}$ *is a randomized puzzle generation algorithm that takes as input the public parameter* PP *and a solution* σ, *and outputs* $(Z, tag) \leftarrow \mathtt{G}(\mathrm{PP}, \sigma)$ *where* Z *is a CAPTCHA puzzle and tag is a string that may be used to help verify a solution to* Z;
- $\mathtt{Verify}$ *is a verification algorithm that takes as input the public parameters* PP, *a puzzle* Z *along with the associated tag and a proposed solution* σ' *outputs a bit* $b := \mathtt{Verify}(\mathrm{PP}, Z, tag, \sigma')$. *We require that* $b = 1$ *whenever* $(Z, tag) \leftarrow \mathtt{G}(\mathrm{PP}, \sigma)$ *and* $\sigma' = \sigma$;
- $\mathtt{C}^{\mathcal{H}}$ *is a solution finding algorithm (i.e., human-machine solver) that takes as input the public parameter* PP *and a puzzle* Z, *and outputs a value* $a \leftarrow \mathtt{C}^{\mathcal{H}(\cdot)}(\mathrm{PP}, Z)$ *as the solution to the puzzle* Z. *Here,* $\mathcal{H}(\cdot)$ *denotes the human oracle which takes intermediate human-efficient objects (such as images) as inputs, and returns machine-efficient values as outputs.*

We typically require that $\mathtt{Setup}$, $\mathtt{W}$, $\mathtt{G}$ *are probabilistic polynomial time algorithms, and* $\mathtt{Verify}$ *a deterministic polynomial time algorithm.* $\mathtt{C}$ *should be a probabilistic polynomial time oracle machine.*

For example, if we are defining a text based CAPTCHA puzzle-system the public parameters PP might specify the set of characters Σ, the set of fonts and a set of font sizes/colors. The public parameters PP would also describe the length $\ell = |\sigma|$ of the target solution (e.g., the number of characters in the CAPTCHA). In general, larger security parameters λ would imply longer puzzles. $\mathtt{W}$ is a randomized algorithm that outputs a random string $\sigma \in \Sigma^*$ (the target solution), and $\mathtt{G}$ is the randomized algorithm that produces a puzzle Z along with a *tag* which may be used for public verification of a potential solution σ'. We view the solution function $\mathtt{C}^{\mathcal{H}}$ as a human equipped with a PPT computer. Typically the computer would just be used to display the challenge to the user, but it could also apply a more sophisticated algorithm to post-process the user's answer.

Fixing the security parameter λ we define one human work unit to be the amount of time/energy that it takes a human to solve one honestly generated CAPTCHA puzzle $Z \leftarrow \texttt{G}(\text{PP}, \sigma)$. Any CAPTCHA puzzle-system should be human usable, meaning that a typical human can consistently solve randomly generated CAPTCHA puzzles. While we recognize that solving a CAPTCHA puzzle may require more effort for some people than for others we will use the term human-work unit to denote the amount of human effort necessary to solve one CAPTCHA puzzle with security parameter λ.[6]

Definition 2 (Honest Human Solvability). *We say that a human-machine solver* $\texttt{C}^{\mathcal{H}}$ *controls* m *human-work units if the machine* $\texttt{C}$ *can query the human oracle* $\mathcal{H}(\cdot)$ *at least* m *times. We say a CAPTCHA puzzle-system* (Setup, $\texttt{W}$, $\texttt{G}$, $\texttt{C}^{\mathcal{H}}$, Verify) *is honest human solvable if for every polynomial* $m = m(\lambda)$ *and for any human* $\texttt{C}^{\mathcal{H}}$ *who controls* m *human-work units, it holds that*

$$\Pr \left[\begin{array}{c} \text{PP} \leftarrow \texttt{Setup}(1^\lambda); \forall i \in [m] \big(\sigma_i^* \leftarrow \texttt{W}(\text{PP})\big); \\ \forall i \in [m] \big((Z_i^*, tag_i^*) \leftarrow \texttt{G}(\text{PP}, \sigma_i^*)\big) \quad : \\ (\sigma_1^*, \ldots, \sigma_m^*) \leftarrow \texttt{C}^{\mathcal{H}(\cdot)}(\text{PP}, Z_1^*, \ldots, Z_m^*) \end{array} \right] \geq 1 - \mathsf{negl}(\lambda)$$

Finally, we require that CAPTCHAs are hard for computers to invert. More concretely, no (known) PPT adversarial machine should be able to find the solutions to $m + 1$ honestly-generated puzzles given only m-human work units. We introduce two similar notions of computer uncrackable CAPTCHAs. The first version states that an adversary with m human-work units cannot find the solution to $m + 1$ CAPTCHAs with non-negligible probability when he is only given the puzzles $Z_1^*, \ldots, Z_n^*$ ($n > m$).

Philosophical Remark. There are two philosophical positions that one could take regarding CAPTCHA puzzles, the human oracle $\mathcal{H}$ and Artificial Intelligence in general. The first view is that for any class of problems that a human oracle $\mathcal{H}$ can solve there exists a (possibly unknown to mankind) PPT computer algorithm to solve the same class of problems. The second philosophical view is that there are some tasks that humans can solve that computers will never be able to solve (i.e., no PPT computer algorithm can consistently/accurately solve the task).

We will implicitly follow view 1 in our CAPTCHA security definitions. However, we do not advocate for either view and we stress that our construction would also work under view 2. Under this second view the class of PPT machine-human hybrid adversaries is *strictly more powerful* than the class of PPT adversaries. Thus, one would need to make the assumption that the cryptographic

---

[6] In the same way some computers (ASICs) are much faster at evaluating the SHA256 hash function than others. However, we expect this difference to be less extreme for human users.

primitives used in our construction (e.g., $i\mathcal{O}$, OWF) are secure against machine-human hybrids. This assumption is highly plausible[7], but also non-standard.

Following view 1 we can avoid such non-standard assumptions about cryptographic primitives. In particular, we assume that the behavior of the human oracle $\mathcal{H}$ is fully described by some (unknown) PPT algorithm. We note that because there exists a PPT algorithm specifying the behavior of $\mathcal{H}$ the class PPT$^{\mathcal{H}}$ (the class of PPT algorithms with oracle access to $\mathcal{H}$) is no more powerful than the class PPT. Thus, we do not need to rely on non-standard cryptographic assumptions (e.g., $i\mathcal{O}$ is secure against adversaries in PPT$^{\mathcal{H}}$). How can a CAPTCHA scheme be secure if there exists some PPT algorithm that accurately solves challenges without human assistance? We will use the set Discoverable to denote a subset containing all known turing machines and all turing machines that mankind might plausibly discover in the near future (e.g., 10–20 years). More specifically, Discoverable$_X = \{M \mid M$ is a turing machine that mankind will build within the next X years $\}$. The security of a CAPTCHA scheme relies on the assumption that no PPT algorithm in Discoverable$_X$ will be able to accurately solve CAPTCHA puzzles for some reasonably large value of X (e.g., 10–20 years).

Stating that no PPT algorithm $A \in$ Discoverable$_X$ breaks CAPTCHAs is a statement about human ignorance. While the meaning of this statement is clear at an intuitive level it is vague in a formal mathematical sense. As Rogaway observed the same issue arises in the definition of (keyless) collision resistant hash functions [49]. There is an efficient algorithm to find collisions, but it is not known to mankind and the hope is that no such algorithm will be known to mankind for a long time in the future. We will follow the same approach taken by Rogaway [49] when making security statements about constructions (e.g., PoH) that rely on CAPTCHAs. For example, we prove that there is an explicit PPT blackbox reduction (blackbox-constructive form [49]) transforming an adversary who breaks Proof of Human-work security to an adversary who breaks CAPTCHAs.

Definition 3 (CAPTCHA Break v1). *We say that a PPT adversary $\mathcal{A}$ who has at most m human-work units breaks security of a CAPTCHA puzzle-system* (Setup, W, G, C$^{\mathcal{H}}$, Verify) *if if for some polynomials $m = m(\lambda)$, $n = poly(\lambda)$ and $\mu(\lambda)$ when $\mathcal{A}$ controls at most m human-work units, it holds that*

$$\Pr \left[\begin{array}{l} \text{PP} \leftarrow \text{Setup}(1^{\lambda}); \ \forall i \in [n] \big(\sigma_i^* \leftarrow \text{W(PP)} \big); \\ \forall i \in [n] \big((Z_i^*, tag_i^*) \leftarrow \text{G(PP}, \sigma_i^*) \big); \\ S \leftarrow \mathcal{A}^{\mathcal{H}(\cdot)}(\text{PP}, Z_1^*, \ldots, Z_n^*); \\ \forall i \in [n] \big(b_i \leftarrow \max_{\sigma \in S} \text{Verify(PP}, Z_i^*, tag_i^*, \sigma) \big) \quad : \\ \hfill \sum_{i \in [n]} b_i \geq m + 1 \end{array} \right] \geq \frac{1}{\mu(\lambda)}$$

---

[7] If a cryptographic primitives like $i\mathcal{O}$ or one-way functions were not secure against machine-human hybrids then these primitives would have to be considered broken in practice.

We say that the CAPTCHA puzzle-system is *computer uncrackable* for the next X years if for any PPT adversary $\mathcal{A} \in$ Discoverable$_X$, $\mathcal{A}$ does not break security of the CAPTCHA puzzle system.

Our second formulation of CAPTCHA security is slightly non-standard due to the fact that the adversary is given a tag tag_i along with each challenge Z_i. In particular, the value tag_i allows the adversary to run Verify(PP, Z_i, tag_i, σ'_i) to test different candidate CAPTCHA solutions. While this formulation is non-standard we argue that we would expect that any CAPTCHA that is secure under Definition 3 can be transformed into a CAPTCHA that is secure under Definition 4. For example, tag_i might be the cryptographic hash of the solution σ_i or we might set $tag_i = \mathrm{i}\mathcal{O}(I_{Z_i,\sigma_i})$ to be the indistinguishability obfuscation of a point function $I_{Z_i,\sigma_i}(x) = 1$ if $x = (Z_i, \sigma_i)$; otherwise $I_{Z_i,\sigma_i}(x) = 0^8$.

It is reasonable to believe that G could produces a tag tag_i, which allows us verify whether or not a solution σ' is correct without revealing σ_i. For example, we might set $tag_i = \mathrm{i}\mathcal{O}(I_{Z_i,\sigma_i})$ to be the indistinguishability obfuscation of a point function $I_{Z_i,\sigma_i}(x) = 1$ if $x = (Z_i, \sigma_i)$; otherwise $I_{Z_i,\sigma_i}(x) = 0$. In this case Verify(PP, Z_i, tag_i, σ') would simply output $tag_i(Z_i, \sigma')$.

Definition 4 (CAPTCHA Break v2). *We say that a PPT adversary $\mathcal{A}$ units breaks security of a CAPTCHA puzzle-system* (Setup, W, G, C$^{\mathcal{H}}$, Verify) *if for some polynomials* $m = m(\lambda)$, $n = poly(\lambda)$ *and* $\mu(\lambda)$ *when* $\mathcal{A}$ *controls at most m human-work units, it holds that*

$$\Pr\left[\begin{array}{l} \text{PP} \leftarrow \text{Setup}(1^\lambda); \; \forall i \in [n]\big(\sigma_i^* \leftarrow \text{W}(\text{PP})\big); \\ \forall i \in [n]\big((Z_i^*, tag_i^*) \leftarrow \text{G}(\text{PP}, \sigma_i^*)\big); \\ S \leftarrow \mathcal{A}^{\mathcal{H}(\cdot)}\big(\text{PP}, (Z_1^*, tag_1^*), \ldots, (Z_n^*, tag_n^*)\big); \\ \forall i \in [n]\big(b_i \leftarrow \max_{\sigma \in S} \text{Verify}(\text{PP}, Z_i^*, tag_i^*, \sigma)\big) \quad : \\ \qquad\qquad\qquad\qquad\qquad \sum_{i \in [n]} b_i \geq m+1 \end{array}\right] \geq \frac{1}{\mu(\lambda)}$$

We say that the CAPTCHA puzzle-system is *computer uncrackable* for the next X years if for any PPT adversary $\mathcal{A} \in$ Discoverable$_X$ $\mathcal{A}$ does not break security of the CAPTCHA puzzle-system.

We will require λ to be large enough that a computer cannot reasonably find a solution by brute force. As Von Ahn et al. [55] observed we can always increase λ by composing CAPTCHA puzzles. Of course this will increase the amount of time that it would take to solve a puzzle. Bursztein et al. [17] conducted a large scale experiment on Amazon's Mechanical Turk to evaluate human performance on a variety of different CAPTCHAs. Based on these results we estimate that, if we define one human work unit to be about two minutes of human effort, it is plausible to believe that security could be amplified to the extent that

---

[8] Indistinguishability obfuscation provides 'best case' obfuscation [31] so it would be highly surprising if an adversary could use tag_i to extract σ_i as this would immediately imply that a host of alternative cryptographic techniques (e.g., one way functions, collision resistance hash functions) fail to hide σ_i. A recent result of Barak et al. [4] provides evidence that evasive circuit families (e.g., point functions) can be obfuscated.

that adversary's odds of solving the long CAPTCHA challenge correctly (and without human assistance) is negligible (e.g., 2^{-100})[9]. For traditional CAPTCHA applications like bot detection this would make the solution impracticable due to the high usability costs. However, for our applications such a delay can be acceptable (e.g., in Bitcoin the parameters are tuned so that a new block is mined every 10 min).

While some spammers have paid human workers to solve CAPTCHAs in bulk [44] we do not consider this an attack on our definition because human effort was involved to find the solution. A HumanCoin miner could pay users to solve CAPTCHAs for him, but human users would have incentive to mine their own HumanCoins if compensation was unfair.

2.2 Universal Samplers

In [32], Hofheinz et al. introduce the notion of universal samplers. The essential property of a universal sampler scheme is that given the sampler parameters U, and given any program d that generates samples from randomness, it should be possible for any party to use the sampler parameters U and the description of d to obtain induced samples that look like the samples that d would have generated given uniform and independent randomness.

Definition 5. *A universal sampler scheme consists of algorithms* (Setup, Sample) *where*

- $U \leftarrow$ Setup(1^λ) *is a randomized algorithm which takes as input a security parameter* 1^λ *and outputs sampler parameters* U.
- $p_d \leftarrow$ Sample(U, d) *takes as input sampler parameters* U *and a circuit* d *of size at most* $\ell = \mathrm{poly}(\lambda)$, *and outputs induced samples* p_d.

In our construction in the next section, we will use a slightly extended version of universal sampler scheme which allows an additional input. Note that in the basic version of universal sampler scheme in Definition 5 above, the algorithm Sample(U, d) receives as input a program d which specifies certain distribution. In our application the program d will be fixed ahead of time, and Sample takes an additional input β where β is an index for specifying randomness for the program to generate a CAPTCHA puzzle Z with tag. Thus, for the slightly extended version of universal sampler scheme with an additional input, we will use the

---

[9] Some CAPTCHA candidates have already been "broken" by PPT algorithms (e.g. [16,19,43,54]). For example, [16] was able to solve reCAPTCHA with accuracy 33.34 %. There are solid guidelines about generating CAPTCHAs that are harder for a computer to crack (e.g., see [58]). Furthermore, we stress that even apparently "broken" CAPTCHAs may still be useful in our proof of human work context because it is acceptable to use CAPTCHA puzzles that take a long time (e.g., 2 min) for a human to solve. By contrast, most deployed CAPTCHAs (e.g., reCAPTCHA) are meant to be solvable in a few seconds. As long as there is some gap between human intelligence and artificial intelligence we can use standard hardness amplification techniques (e.g., parallel repetition) to obtain stronger CAPTCHAs [55].

notation $\mathtt{Sample}(U, d, \beta)$ instead of $\mathtt{Sample}(U, d)$. This allows us to provide alternative and flexible description for a circuit d without changing its functionality. We note that this slightly extended version has been explored in [32], and it is straightforward to extend $\mathtt{Sample}$, without requiring a new construction or security analysis.

The formal security definition of adaptive security for the slightly extended universal samplers with additional inputs can be found in Appendix A. We briefly overview the notion of adaptive security here. Intuitively, adaptive security guarantees that induced samples are indistinguishable from honestly generated samples to an arbitrary interactive system of adversarial and honest parties. In a universal sampler with additional inputs, the program d is fixed, and when an additional input β is provided, the induced sample can be computed as $p_\beta \leftarrow \mathtt{Sample}(U, d, \beta)$.

We first consider an "ideal world," where a trusted party with a fixed program description d, on input β, simply outputs $d(r_\beta)$ where r_β is independently chosen true randomness, chosen once and for all for each given β. In other words, if F is a truly random function, then the trusted party outputs $d(F(\beta))$. In this way, if any party asks for samples corresponding to a specific value of β, they are all provided with the same honestly generated value.

In the real world, however, all parties would only have access to the trusted sampler parameters. Parties would use the sampler parameters to derive induced samples $d(r_\beta)$ for any specific inputs β. Now r_β is a pseudo random value corresponding to the randomness index β. We will require that for every real-world adversary $\mathcal{A}$, there exists a simulator $\mathcal{S}$ that can provide simulated sampler parameters U to the adversary such that these simulated sampler parameters U actually induce the completely honestly generated samples $d(F(\beta))$ created by the trusted party: in other words, that $\mathtt{Sample}(U, d, \beta) = d(F(\beta))$. Note that since honest parties are instructed to simply compute induced samples, this ensures that honest parties in the ideal world would obtain these completely honestly generated samples $d(F(\beta))$.

3   Proof of Human-Work Puzzles

In this section, we first define the syntax and security for proof of human-work puzzles; then we demonstrate a construction using universal samplers and CAPTCHAs.

3.1   Definitions

In a proof of work (PoW) puzzle, a party (i.e., prover) is allowed to prove to a bunch of verifiers that he completed some amount of computation/work. In general, those parities are machines. A typical PoW puzzle scheme consists of several algorithms: setup algorithm $\mathtt{Setup}()$ for generating the global system parameters and policies, puzzle instance generation algorithm $\mathtt{G}()$, puzzle solution finding algorithm $\mathtt{C}()$, and solution verification algorithm $\mathtt{V}()$. To enable a consensus protocol, the PoW puzzle has to meet the following requirements: (i) it has to be moderately hard to compute (for machines), and no prover can create a proof of work in no time; (ii) it has to be easy to verify (for machines), and all verifiers can efficiently check if a proof is valid; (iii) the difficulty needed in order to solve the proof has to be adjustable in a linear way; and (iv) it has to be possible to ensure that proofs of work cannot be reused multiple times, and the proofs of work should be linked to some public information, e.g., the hash of the block header in a consensus protocol.

Proof of human-work puzzles are very similar to PoW puzzles, except that we intend to have the human in the loop for finding the solution. The key difference is that the prover (problem solver) should not be machine-only. In the above listed requirements, we therefore expect the PoH puzzle to be *moderately hard to compute for humans, and infeasible to compute for machines*. On the other hand, as in PoW, we expect the verification to be easy for machines[10]. The syntax is as follows:

Definition 6 (Proof of Human-work Puzzle). *A proof of human-work puzzle-system consists of a tuple of algorithms* $(\texttt{Setup}, \texttt{G}, \texttt{C}^{\mathcal{H}}, \texttt{V})$, *where*

- $\texttt{Setup}$ *is a randomized system setup algorithm that takes as input* 1^λ *(λ is the security parameter) and* 1^ω *(ω is the difficulty parameter), and outputs a system public parameter* $\text{PP} \leftarrow \texttt{Setup}(1^\lambda, 1^\omega)$;
- $\texttt{G}$ *is a randomized puzzle generation algorithm that takes as input the public parameter* PP, *and outputs puzzle* $x \leftarrow \texttt{G}(\text{PP})$;
- $\texttt{C}^{\mathcal{H}}$ *is a solution finding algorithm (i.e., human-machine solver) that takes as input the public parameter* PP *and a puzzle* x, *and outputs value* $a \leftarrow \texttt{C}^{\mathcal{H}(\cdot)}(\text{PP}, x)$ *as the solution to the puzzle* x. *Here,* $\mathcal{H}(\cdot)$ *denotes the human oracle which takes intermediate human-efficient objects (such as images) as inputs, and returns machine-efficient values as outputs.*
- $\texttt{V}$ *is a deterministic puzzle-solution verification algorithm that takes as input the public parameter* PP *and a puzzle-solution pair* (x, a), *and outputs bit* $b := \texttt{V}(\text{PP}, x, a)$ *where* $b = 1$ *if* a *is a valid solution to the puzzle* x, *and* $b = 0$ *otherwise.*

Following notation of Miller et al. [42] we will let $\zeta(m, \omega) \doteq 1 - (1 - 2^{-\omega})^m$. Intuitively, $\zeta(m, \omega)$ denotes the probability of finding a valid solution with m queries to the human-oracle.

Definition 7 (Honest Human Solvability). *A PoH puzzle system* $(\texttt{Setup}, \texttt{G}, \texttt{C}^{\mathcal{H}}, \texttt{V})$ *is honest human solvable if for every polynomial* $m = m(\lambda)$, *and for any honest human-machine solver* $\texttt{C}^{\mathcal{H}(\cdot)}$ *who controls* m *human-work units, it holds that*

$$\Pr\left[\begin{array}{ll} \text{PP} \leftarrow \texttt{Setup}(1^\lambda, 1^\omega); & x^* \leftarrow \texttt{G}(\text{PP}); \\ a^* \leftarrow \texttt{C}^{\mathcal{H}(\cdot)}(\text{PP}, x^*) & : \quad \texttt{V}(\text{PP}, x^*, a^*) = 1 \end{array} \right] \geq \zeta(m, \omega) - \mathsf{negl}(\lambda)$$

---

[10] We remark that, it might also be interesting to consider the variant in which verification is easy for human but not for machine-verifiers.

Definition 8 (Adversarial Human Unsolvability). *We say that a PPT algorithm $\mathcal{B}$ breaks security of the a PoH puzzle system ($\mathtt{Setup}, \mathsf{G}, \mathsf{C}^{\mathcal{H}}, \mathsf{V}$) if for some polynomials $m = m(\lambda)$ and $\mu(\lambda)$ when $\mathcal{B}$ controls at most m human-work units, it holds that*

$$\Pr \left[\begin{array}{l} \mathtt{PP} \leftarrow \mathtt{Setup}(1^\lambda, 1^\omega); \ x^* \leftarrow \mathsf{G}(\mathtt{PP}); \\ a^* \leftarrow \mathcal{B}^{\mathcal{H}(\cdot)}(\mathtt{PP}, x^*) \quad : \quad \mathsf{V}(\mathtt{PP}, x^*, a^*) = 1 \end{array} \right] \geq \zeta(m+1, \omega) + \frac{1}{\mu(\lambda)}$$

If no PPT human-machine adversary $\mathcal{B}^{\mathcal{H}(\cdot)} \in \mathtt{Discoverable}_X$ breaks security we say that the PoH puzzle system is *adversarial human unsolvable* for the next X years.

Remark 2. We remark that the above definition can be strengthened by providing the adversarial $\mathcal{B}$ additional access to a polynomial number of (x_i, a_i) pairs, where $x_i \leftarrow \mathsf{G}(\mathtt{PP})$ and $\mathsf{V}(\mathtt{PP}, x_i, a_i) = 1$. The definition can be further strengthened further by providing the adversarial $\mathcal{B}$ multiple puzzle instances $x_1^*, \ldots, x_k^*$, and asking $\mathcal{B}$ to output a valid a_j^* for any $j \in [k]$. Our construction in next section can achieve these strengthened notions. For simplicity, we focus on the above simplified notion in this paper.

3.2   Construction

In this subsection, we show how to construct PoH puzzles for cryptocurrency. In Bitcoin each PoW puzzle instance is specified by the public ledger x. A motivated miner (i.e., the PoW prover) will produce a PoW by repeatedly querying a random oracle RO (e.g., the SHA256 hash function) to sample uniformly random elements in an attempt to produce a "small" output. More concretely, the miner computes random elements $y_i = \mathtt{RO}(x, s_i)$ for different strings s_i's. If there exist i so that $y_i < T_\omega$, then the corresponding s_i can be viewed as the PoW solution. Given a random oracle $\mathtt{RO} : \{0,1\}^* \rightarrow \{0,1\}^n$ we will use the notation $T_\omega \doteq 2^{n-\omega}$. Intuitively, this ensures that $\mathtt{RO}(x, s_i) < T_\omega$ with probability $2^{-\omega}$.

To have human in the loop, we need to first sample CAPTCHA instances for human solvers. Those instances are not in uniform distribution, and it is unclear if we can use a random oracle RO to generate such instances. We here use a cryptographic tool called "universal sampler" recently developed by Hofheinz et al. [32] to generate such CAPTCHA instances. Universal sampler can be viewed as an extended version of RO, which can generate elements in any efficiently samplable distributions. More concretely, we fix d to be a circuit for computing the CAPTCHA generation function CAPT.G. Thus, $d(r)$ generates a CAPTCHA puzzle Z_r and a tag tag_r from randomness r. Now, the miner begins by computing $(Z_i, tag_i) = \mathtt{Sample}(U, d, \beta = (x, s_i))$; then the miner solves Z_i via human effort to get the corresponding CAPTCHA solution σ_i; at this moment, we can adapt the strategy in the original PoW by computing $y_i = \mathtt{RO}(x, s_i, \sigma_i)$ and if $y_i < T_\omega$ and if the CAPTCHA solution σ_i is correct, then the corresponding pair (s_i, σ_i) can be viewed as the PoH solution. We can verify that the solution is correct by re-sampling $(Z_i, tag_i) \leftarrow \mathtt{Sample}(U, d, \beta = (x, s_i))$ and checking that $\mathtt{Verify}(Z_i, tag_i, \sigma_i) = 1$ and that $\mathtt{RO}(x, s_i, \sigma_i) < T_\omega$.

Construction Details. In our proof of human-work puzzle construction, we use a universal sampler scheme UNI = UNI.{Setup, Sample}, a CAPTCHA scheme CAPT = CAPT.{Setup, W, G, $C^\mathcal{H}$, Verify}, and a hash function **G**. We will treat **G** as a random oracle in our analysis. The constructed PoH puzzle scheme consists of algorithms POH.{Setup, G, $C^\mathcal{H}$, V}. Note that $\mathcal{H}$ denotes a human oracle.

- The setup algorithm PP $\leftarrow$ POH.Setup($1^\lambda, 1^\omega$): Compute $\tilde{PP}$ $\leftarrow$ CAPT.Setup(1^λ); Compute $U \leftarrow$ UNI.Setup(1^λ); Define a program d as follows: On input randomness $r = (r_1, r_2)$, compute $\sigma := $ CAPT.W($\tilde{PP}; r_1$), $(Z, tag) := $ CAPT.G($\tilde{PP}, \sigma; r_2$), and output (Z, tag). Set PP $:= (U, d, \tilde{PP}, T = T_\omega, \text{PARAM})$ where PARAM denotes the instructions of using the system.
- The puzzle generation algorithm $x \leftarrow$ POH.G(PP): Parse PP into $(U, d, \tilde{PP}, T, \text{PARAM})$; Based on the description of PARAM, sample x.
- The solution function $a \leftarrow$ POH.$C^\mathcal{H}$(PP, x): Upon receiving puzzle instance x, parse PP into $(U, d, \tilde{PP}, T, \text{PARAM})$; Randomly choose $s \leftarrow \{0,1\}^\lambda$; Compute CAPTCHA puzzle instance $(Z, tag) \leftarrow$ UNI.Sample($U, d, \beta = (x, s)$) ; Use the human oracle $\mathcal{H}$ to find a solution to CAPTCHA puzzle instance Z, i.e., $\sigma \leftarrow$ CAPT.$C^\mathcal{H}$($\tilde{PP}, Z$); If **G**(x, s, σ) < T, then set $a := (s, \sigma)$. Otherwise set $a := \perp$.
- The puzzle verification algorithm $b :=$ POH.V(PP, x, a): Parse a into (s, σ); Parse PP into $(U, d, \tilde{PP}, T, \text{PARAM})$; Compute $(Z, tag) \leftarrow$ UNI.Sample($U, d, \beta = (x, s)$); If CAPT.Verify($\tilde{PP}, Z, tag, \sigma$) = 1 and **G**($x, s, \sigma$) < T, then set $b := 1$. Otherwise set $b := 0$.

It is easy to verify that the PoH scheme is honest human solvable if the underlying universal sampler is correct and the CAPTCHA scheme is honest-human solvable. Next we state a theorem for the security of our PoH scheme, and the proof can be found in the full version [12]. In our proof we give an explicit PPT reduction R from CAPTCHA to PoH security. Intuitively, this means that if mankind finds a PPT algorithm $\mathcal{B} \in \text{Discoverable}_X$ attacking PoH security then mankind will quickly find a PPT algorithm $\mathcal{A}$ breaking CAPTCHA security. We take this to mean that if $\mathcal{B} \in \text{Discoverable}_X$ then $\mathcal{A} \in \text{Discoverable}_{X+\epsilon}$, where ϵ is the time necessary to apply a known, efficient blackbox reduction[11].

---

[11] In this (informal) line of reasoning we implicitly assume if there is an PPT algorithm breaking PoH is discovered then the reduction R will be quickly implemented by someone because it is publicly known and leads to a very useful result (breaking CAPTCHAs). We stress that we are not claiming that $R(\text{Discoverable}_X) \subset \text{Discoverable}_{X+\epsilon}$ for every known, explicit reduction R so that the set $\text{Discoverable}_{X+\epsilon}$ contains the result of applying every explicit, known reduction to every machine in the set Discoverable_X. If this was the case then we could claim that (at minimum) the set Discoverable_X contains all strings of length X/ϵ since the reductions $R_1 = $ "append 1" and $R_0 = $ "append 0" are explicit, known reductions. In our case we are assuming that the known, explicit reduction R from CAPTCHAs to PoHs would be implemented *because* it is known that the reduction leads to a useful result when we have an algorithm to break PoH security (breaking CAPTCHAs). By contrast, the reduction "append 1" is unlikely to lead to useful results when applied to most Turing Machines.

Thus, our POH construction is essentially as secure as the underlying construction. CAPT is a computer uncrackable CAPTCHA for the next $X + \epsilon$ years (Definition 4), then the above proof of human-work scheme POH is adversarial human unsolvable for the next X years. We use ϵ to denote the time necessary (e.g., 1 day) to implement the reduction and build the resulting PPT CAPTCHA solver.

Theorem 1. *If* UNI *is an adaptively secure universal sampler then given any* PPT *algorithm* $\mathcal{B}$ *that breaks* POH *security (Definition 8) there is an explicit* PPT *blackbox reduction producing a* PPT *algorithm* $\mathcal{A}$ *that breaks CAPTCHA security (Definition 4).*

Proof (idea). The security of our PoH relies on the security of underlying building blocks, the universal sampler scheme UNI, and the CAPTCHA scheme CAPT. We start from the real security game. Based on the security of the universal sampler scheme UNI, we can modify the real security game into a hybrid world where CAPTCHA puzzle instances are generated independently and based on uniform randomness. Then we can use the security of CAPT to argue about the security of PoH. That is, we can construct a CAPT attacker $\mathcal{A}_{\mathrm{CAPT}}$ based on a PoH attacker $\mathcal{A}_{\mathrm{POH}}$. The CAPT attacker $\mathcal{A}_{\mathrm{CAPT}}$ can simulate an internal copy of $\mathcal{A}_{\mathrm{POH}}$, and embed his challenge into a simulated hybrid for $\mathcal{A}_{\mathrm{POH}}$. If $\mathcal{A}_{\mathrm{POH}}$ wins with more than specified probability (i.e., $\zeta(m + 1, \omega)$) plus non-negligible probability, then $\mathcal{A}_{\mathrm{CAPT}}$ can also win the computer-unbreakable game with non-negligible probability.

4   Application 1: HumanCoin

In this section we outline how a new cryptocurrency called *HumanCoin* could be built using Proofs of Human-work. At a high level HumanCoin closely follows the Bitcoin protocol, except that we use PoH puzzles to extend the blockchain instead of PoW puzzles. We will not attempt to describe HumanCoin in complete detail. Instead we will focus on the key modifications that would need to be made to an existing cryptocurrency like Bitcoin to use Proof of Human work puzzles. In our discussion we will use lowercase bitcoin (resp. humancoin) to denote the base unit of currency in the Bitcoin (resp. HumanCoin) protocol.

4.1   Bitcoin Background

We begin by highlighting several of the key features of Bitcoin. Our overview follows the systemization of knowledge paper by Bonneau et al. [14]. However, our discussion of Bitcoin is overly simplified and this choice is intentional. For example, we will completely ignore the use of Merkle Trees [41] in Bitcoin to compress the blockchain even though it is quite useful in practice. We make this choice so that we can focus on the key differences of HumanCoin (the use of Merkle Trees [41] in HumanCoin and Bitcoin would be identical). We do include additional discussion of Bitcoin in the full version, but even this discussion is not intended to be complete. We refer interested readers to the excellent lectures by Narayanan et al. [46] for more details about Bitcoin or the original paper published under the pseudonym Nakamoto [45].

Blockchain. In Bitcoin all transactions (e.g., "Alice sends Bob 50 bitcoins") are published on a public ledger. This public ledger is stored on a cryptographic data structure called a blockchain $b = B_0, \ldots, B_t$. A blockchain b is valid if and only if all of the blocks B_i ($i \leq t$) are valid and an individual block $B_i = (tx_i, s_i, h_{i-1})$ is valid if and only if three key conditions are satisfied. First, all of the transactions recorded in the transcript tx_i must be valid (e.g., each transaction is signed by the sender and the spender has sufficient funds). Second, the block B_i must contain the cryptographic hash $h_{i-1} = hash\,(B_{i-1})$ of the previous block B_{i-1}[12]. Finally, the block B_i should contain a nonce s_i which ensures that cryptographic hash $hash\,(B_i)$ begins with at least ω leading zeros, where ω is a hardness parameter that we will discuss later. Finding such a nonce s constitutes a proof of work in the Hashcash [3] puzzle system. The first property ensures that users cannot spend money they don't have and that they cannot spend someone else's money. The second property ensures that it is impossible to tamper with blocks B_i in the middle of the blockchain without creating an entirely new blockchain $b' = B_0, \ldots, B_{i-1}, B'_i, B'_{i+1}, \ldots, B'_t$. Finally, the third property ensures that it is moderately difficult to add new blocks to a blockchain. To incentivize miners to help validate transactions (i.e. extend the blockchain by finding a valid nonce s) the miner is allowed to add a special transaction (e.g., "I create 25 new bitcoins and give them to myself") to the new block as a reward.

Distributed Consensus Protocol. Bitcoin's distributed consensus protocol is simple, yet elegant. An agent should accept a transaction if and only if it is recorded on a block B_i of a valid blockchain $b = B_0, \ldots, B_t$ and b is the longest valid that the agent has seen and $i \leq t - 6$. Unless a miner controls at least 25% of the hash power in the network the rational mining strategy is always to extend the longest blockchain because nobody will accept the Bitcoins they try to mine in a shorter blockchain (e.g., the special transaction in which a miner claims 25 bitcoins' would only be recorded on a shorter blockchain which nobody accepts) [27]. Assuming that the network has high synchronicity [29] and that a malicious user controls at most 49% of the computational mining power he will never be able to tamper with any of the transactions in a block B_i from the middle of the blockchain because he would need to eventually produce a new blockchain $b' = B_0, \ldots, B_{i-1}, B'_i, B'_{i+1}, \ldots, B'_t$ that is at least as long as the true blockchain b and he will fail to accomplish this goal with high probability [45].

4.2   HumanCoin

Similar to Bitcoin all HumanCoin transactions (e.g., "Alice sends Bob 50 human-coins") are recorded inside a blockchain $b = B_0, \ldots, B_t$, where each block

---

[12] Bitcoin uses the cryptographic hash function $hash = $ SHA256. The function $hash$ is typically treated as a random oracle in security analysis of Bitcoin.

$B_i = (tx_i, a_i, h_{i-1})$ contains three components: a list of transactions tx_i, a hash $h_{i-1} = hash(B_{i-1})$ of the previous block, and a Proof of Human-work which is encoded by a_i. As before all of the transactions in tx_i must be valid and the block must contain the hash $h_{i-1} = hash(B_{i-1})$ of the previous block. We additionally require that the PoH verifier accepts the Proof of Human-Work solution a_i. More formally, suppose that we are given a PoH puzzle system $(\text{Setup}, \text{G}, \text{C}^{\mathcal{H}}, \text{V})$ and that we have already run $\text{Setup}(1^\lambda, 1^\omega)$ to obtain public parameters PP which are available to every miner. A valid block B_i must contain a value a_i such that the public verifier $\text{V}(\text{PP}, x_i, a_i)$ outputs 1, where $x_i = \text{G}(\text{PP}; r = hash(tx_i, h_{i-1}))$. Given a valid blockchain $b = B_0, \ldots, B_t$ a miner can earn HumanCoins by finding a valid block $B_{t+1} = (tx_{t+1}, a_{t+1}, x_{t+1}, h_t)$ extending b. To find such a block the human-computer miner would first set $r = hash(tx_{t+1}, h_t)$ and then sample $x \leftarrow \text{G}(\text{PP}; r)$. Finally, the human-computer miner can run $\text{C}^{\mathcal{H}}(\text{PP}, x)$ to obtain a potential solution a. If $a = \bot$ then the miner will need to try again. Otherwise, the miner has found a valid proof of human-work and he can produce a valid new block $B_{t+1} = (tx_{t+1}, a, h_t)$ by adding inserting the PoH solution a into the block B_{t+1}. As before the miner is allowed to insert a special transaction into the new block (e.g., "I create 25 humancoins and give them to myself") as a reward for extending the blockchain.

Parameter Selection. In Bitcoin ω is a public parameter is tuned to ensure that, on average, miners will add one new block to the blockchain every 10 min [46] — on average we need 2^ω hash evaluations to create one new block. The Bitcoin protocol would most likely work just fine with a shorter delay (e.g., 5 min) or a slightly longer delay (e.g., 20 min) between consecutive blocks — there is nothing magical about the specific target value of 10 min. However, it is clear that there needs to be some delay to promote stability. If multiple miners find a new block at the same time then we could end up with competing blockchains resulting in temporary confusion. Note that if the value of ω remains fixed then the average time to create one new block would begin to decrease as more miners join Bitcoin, or as existing miners upgrade their computational resources. Thus, the value of ω must be adjusted periodically. In Bitcoin the value of ω is adjusted every $2,016$ blocks, which works out to two weeks on average (2 weeks = 2016×10 min), using the formula $\omega = \omega_{old} - \log\left(\frac{t_{elapsed}}{2016 \times 10 \text{ min}}\right)$, where $t_{elapsed}$ denotes the time span that it actually took to generate the last $2,016$ blocks [46].

In HumanCoin we adjust ω in exactly the same way. Note that the PoH hardness parameter $T_\omega = 2^{n-\omega}$ in our PoH construction is a public parameter PP and can easily be modified as it is not embedded into any of the obfuscated programs. In HumanCoin we will need to select an initial value of ω that is *much* smaller than in Bitcoin if we want ensure that new block are discovered every 10 min. This is because computers can evaluate a hash function $hash$ much faster than a human can solve a long CAPTCHA puzzle. However, we could still use the same basic formula to tune the hardness parameter ω of our proof of work puzzles in the event that many miners join/leave.

5   Application 2: Password Protection

An adversary who breaches an authentication server is able to mount an automated brute-force attack by comparing the cryptographic hash of each user's password with the cryptographic hashes of likely password guesses. These offline attacks have become increasingly prevalent and dangerous as password cracking resources has improved. In particular, the cost of computing a hash function H like SHA256 or MD5 on an Application Specific Integrated Circuit (ASIC) is orders of magnitude smaller than the cost of computing H on traditional hardware [21, 46]. Similarly, data from previous breaches allow adversaries to improve their guessing strategies. Recent security breaches (e.g., Ashley Madison, LastPass, RockYou, LinkedIn and eBay to name a few[13]), which have affected millions of users, highlight the importance of this problem.

Canneti et al. [18] had a clever idea to deter an offline attacker that they called Human Only Solvable Puzzles. They proposed filling a hard drive with a dataset of unsolved CAPTCHA puzzles. When a user authenticates he will be challenged with a pseudorandom CAPTCHA puzzle from the dataset, and the server will append the solution to the user's password before computing the hash value. The choice of the pseudorandom CAPTCHA puzzle becomes deterministic once the user's password and username are fixed. Thus, if the user types in the same password he will receive the exact same CAPTCHA puzzle as a challenge. If the underlying CAPTCHA system is human usable, then the user will always be able to authenticate successfully provided that he can remember his password. If an offline advesary wants to verify a password guess he will need to find and solve the corresponding CAPTCHA puzzle. The key point is that each time the adversary tries a new guess he will need to solve a different CAPTCHA challenge.

Unfortunately, the Human Only Solvable Puzzles solution of [18] has one critical drawback. There are a finite number of CAPTCHAs on the hard drive, and the defense will break down once the adversary manages to solve all (or most) of them. Blocki et al. [9] estimated that it would cost about 10^6 to solve all of the CAPTCHAs on an 8 TB hard drive. While this is certainly an expensive start-up cost it may not be sufficient to deter the adversary because these costs would amortize over all user accounts. Many password breaches affect millions of users, and each cracked password has significant value on the black market (e.g., $4–$30). Blocki et al. [9] introduced their own scheme called GOTCHA based on inkblot images, but their protocol had higher usability costs and was based on newer untested AI assumptions.

In this section we introduce a provably secure password authentication scheme in the Random Oracle model using CAPTCHAs and program obfuscation. Unlike Blocki et al. [9] our solution can be based on standard CAPTCHA

---

[13] See http://www.privacyrights.org/data-breach/ (Retrieved 9/1/2015).

assumptions. Unlike Canneti et al. [18] our solution is not vulnerable to pre-computation attacks[14].

5.1 Password Authentication Scheme

We first formalize the notion of a password authentication scheme. Definition 9 formalizes the account creation and authentication algorithms from the perspective of an authentication server. We note that the server is allowed to interact with the human user $\mathcal{H}$ during the account creation and authentication protocols.

Definition 9. *A password authentication scheme consists of a tuple of algorithms* (Setup, CreateAccount$^{\mathcal{H}}$, Authenticate$^{\mathcal{H}}$) *and a random oracle* **G**, *where*

- Setup *is a randomized system setup algorithm that takes as input* 1^{λ} *(λ is the security parameter) and outputs a system public parameter* PP $\leftarrow$ Setup(1^{λ});
- CreateAccount$^{\mathcal{H}}$ *is an account creation algorithm that takes as input the public parameter* PP, *a username* u *and a password* pwd *and outputs a tuple* (h, s). *Here,* s *is typically a random bit string (salt) and* h *is a hash value produced by the random oracle. We note that* CreateAccount$^{\mathcal{H}}$ *is a human-machine algorithm and thus the hash value* h *may include the solution to CAPTCHAs that the human solves as well as the password* pwd *and salt* s;
- Authenticate$^{\mathcal{H}}$ *is the algorithm that is invoked when a user wants to authenticate. The algorithm takes as input the public parameter* PP, *a username* u, *a password* pwd, *a hash* h *and a salt value* s *and outputs a bit* $b \in \{0, 1\}$ *indicating whether or not the authentication attempt was successful. We note that* Authenticate$^{\mathcal{H}}$ *is a human-machine algorithm and thus the human* $\mathcal{H}$ *may be asked to solve CAPTCHAs as part of the authentication procedure.*

Our next definition says what it means for a password authentication scheme to be costly to crack. The game mimics an offline adversary who has breached the authentication server and stolen the record (u, h, s) indicating that user u has an account with salt value s and the salted hash of the user's password needs to match h. In our definition we let $\mathcal{P}$ denote a distribution over the passwords $\{pwd_1, \ldots, pwd_n\}$ that the user u might select and let $p_i = \Pr_{\mathcal{P}}[pwd_i]$ denote the probability that the user selects password pwd_i. We assume that p_i and pwd_i are known to the adversary for all i and for convenience we assume that the passwords are ordered such that $p_1 \geq p_2 \geq \ldots \geq p_n$. Informally, our definition states that an adversary with B units of human-work will succeed in cracking the user's password with probability at most $p_1 + \ldots + p_B + \mathsf{negl}(\lambda)$.

Definition 10 (Costly to Crack). *We say a* PPT *adversary* $\mathcal{A}$ *breaks security of a password authentication scheme* {Setup, CreateAccount$^{\mathcal{H}}$,

---

[14] Of course the main downside to our approach is the dependence on indistinguishability obfuscation, which does not have practical solutions at this time.

$\mathbf{Authenticate}^{\mathcal{H}}, \mathbf{G}\}$ *if for some polynomials* $B = B(\lambda)$ *and* $\mu(\lambda)$ *and user* u, *whenever* $\mathcal{A}$ *has* B *human-work units it holds that*

$$
\Pr \left[\begin{array}{l} \text{PP} \leftarrow \mathbf{Setup}(1^\lambda); \ pwd \leftarrow \mathcal{P}; \\ (h, s) \leftarrow \mathbf{CreateAccount}^{\mathcal{H}}(\text{PP}, u, pwd); \\ pwd' \leftarrow \mathcal{A}^{\mathcal{H}(\cdot)}(\text{PP}, h, s) \quad : \\ \mathbf{Authenticate}^{\mathcal{H}}(\text{PP}, u, pwd', h, s) = 1 \end{array} \right] \geq p_1 + \ldots + p_B + \frac{1}{\mu(\lambda)}
$$

We say that the password authentication scheme is *costly to crack* for the next X years if for any PPT adversary $\mathcal{A} \in \mathbf{Discoverable}_X$ $\mathcal{A}$ does not break security. We remark that we do not require the adversary's success probability to be negligibly small. Indeed, if the user selects passwords from a distribution with low entropy (and many users do [13]) then the adversary may have a good success rate. Thus, the problem is unavoidable as long as users are allowed to select low-entropy passwords[15]. We do not focus on helping users to select strong passwords [10,15], although this is indeed an important direction of research. Our goal is to provide the best possible protection for the passwords that users actually select.

The next definition quantifies human usability. Informally, the password authentication scheme is usable if an honest human user will always be able to authenticate if he remembers his password. We stress that our definition does not say anything about how easy it will be to remember the password. While this is certainly an important consideration it is orthogonal to our work. We are not focused on how to get users to choose stronger passwords, but rather how to more effectively protect the passwords that users actually choose. Our definition merely says that an honest user won't be locked out of his account as long as he remembers his password (e.g., because he cannot solve the CAPTCHAs).

Definition 11 (Human Usable). *We say that a password authentication scheme* $\{\mathbf{Setup}, \mathbf{CreateAccount}^{\mathcal{H}}, \mathbf{Authenticate}^{\mathcal{H}}, \mathbf{G}\}$ *is human usable if for every human user* $\mathcal{H}$ *who controls* 1 *human-work unit during authentication and* 1 *human work unit during account creation, it holds that*

$$
\Pr \left[\begin{array}{l} \text{PP} \leftarrow \mathbf{Setup}(1^\lambda); \ pwd \leftarrow \mathcal{P}; \\ (h, s) \leftarrow \mathbf{CreateAccount}^{\mathcal{H}}(\text{PP}, u, pwd) \quad : \\ \mathbf{Authenticate}^{\mathcal{H}}(\text{PP}, u, pwd, h, s) = 1 \end{array} \right] \geq 1 - \mathsf{negl}(\lambda)
$$

5.2 Construction

Construction Details. In our construction we use a universal sampler scheme UNI = UNI.$\{\mathbf{Setup}, \mathbf{Sample}\}$, a CAPTCHA scheme CAPT = CAPT.$\{\mathbf{Setup}, \mathbf{W}, \mathbf{G}, \mathbf{C}^{\mathcal{H}}, \mathbf{Verify}\}$, and a hash function $\mathbf{G}$. We will treat $\mathbf{G}$ as a random oracle in our analysis. The constructed password authentication scheme consists of algorithms Password.$\{\mathbf{Setup}, \mathbf{CreateAccount}^{\mathcal{H}}, \mathbf{Authenticate}\}$. Note that $\mathcal{H}$ denotes a human oracle.

---

[15] In addition to their high usability costs [28], policies aimed at forcing users to chose stronger passwords (e.g., requiring numbers and capital letters) can have the opposite affect on password strength [11,37].

- The setup algorithm PP ← `Password.Setup`(1^λ): Compute $\tilde{\text{PP}}$ ← CAPT. `Setup`(1^λ); Compute U ← UNI.`Setup`(1^λ); Define a program d as follows: On input randomness $r = (r_1, r_2)$, compute $\sigma := $ CAPT.$\mathsf{W}(\tilde{\text{PP}}; r_1)$, $(Z, tag) := $ CAPT.$\mathsf{G}(\tilde{\text{PP}}, \sigma; r_2)$, and output Z. Set PP $:= (U, d, \tilde{\text{PP}}, $ PARAM$)$ where PARAM denotes the instructions of using the system.
- The account creation algorithm (h, s) ← `Password.CreateAccount`$^{\mathcal{H}}$ (PP, u, pwd): Parse PP into $(U, d, \tilde{\text{PP}}, $ PARAM$)$; randomly choose s ← $\{0, 1\}^\lambda$. Set $\beta = (u, pwd, s)$ and compute CAPTCHA puzzle instance Z ← UNI.`Sample`$(U, d, \beta = (u, pwd, s))$; Use the human oracle $\mathcal{H}$ to find a solution to CAPTCHA puzzle instance Z, i.e., σ ← CAPT.$\mathsf{C}^{\mathcal{H}}(\tilde{\text{PP}}, Z)$; Compute h ← $\mathbf{G}(pwd|\sigma|s)$ and output (h, s).
- The authentication algorithm b ← `Password.Authenticate`$^{\mathcal{H}}$(PP, u, pwd, h, s): Parse PP into $(U, d, \tilde{\text{PP}}, $ PARAM$)$, set $\beta = (u, pwd, s)$ and compute CAPTCHA puzzle instance Z ← UNI.`Sample`(U, d, β); Use the human oracle $\mathcal{H}$ to find a solution to CAPTCHA puzzle instance Z, i.e., σ ← CAPT.$\mathsf{C}^{\mathcal{H}}(\tilde{\text{PP}}, Z)$; Compute h' ← $\mathbf{G}(pwd|\sigma|s)$. If $h' = h$ then output $b = 1$; otherwise output 0.

It is easy to verify that `Password` is human usable if the underlying CAPTCHA scheme CAPT is honest human solvable. At an philosophical level we can interpret Theorem 2, our main technical result in this section, to say that the above password authentication scheme `Password.{Setup, CreateAccount`$^{\mathcal{H}}$, `Authenticate`$^{\mathcal{H}}$} is costly to crack for the next X years as long as the underlying CAPTCHA scheme is computer uncrackable for the next $X + \epsilon$ years (Definition 10). Here, ϵ denotes the time it takes to implement an explicit (blackbox) PPT reduction from CAPT to the password scheme (e.g., one day). We stress that we only need to assume that the underling CAPTCHA scheme is computer uncrackable in the more traditional sense of Definition 3 (e.g., the adversary is only given the puzzles $Z_1, \ldots, Z_n$ and not the associated verification tags).

Theorem 2. *If* UNI *is an adaptively secure universal sampler then given a* PPT *algorithm* $\mathcal{B}$ *that breaks security of our password authentication scheme there is a* PPT, *blackbox reduction which produces a* PPT *algorithm* $\mathcal{A}$ *to break CAPTCHA security (Definition 3).*

Proof (Idea). At a high level we show that we can construct an adversary that breaks CAPTCHAs (under Definition 3) from an adversary that breaks the password authentication scheme. To do this we embed challenge CAPTCHAs $Z_1, \ldots, Z_n$ inside the UniversalSampler UNI (we can do this by the security of the Universal Sampler scheme). Intuitively, in order to check that a password guess pwd_i is correct the adversary will need to query the random oracle $\mathbf{G}$ with the value $\beta_i = (pwd_i|\sigma_i|u)$, where σ_i is the correct solution to CAPTCHA Z_i. If the adversary queries $\mathbf{G}$ with $B + 1$ unique solutions then we can win the CAPTCHA challenge (Definition 3) by simply outputting these $B + 1$ solutions. If the adversary queries $\mathbf{G}$ with at most B unique solutions then we can show that his success rate is at most $p_1 + \ldots + p_B + \mathsf{negl}(\lambda)$. A formal proof of Theorem 2 can be found in the full version [12].

Discussion. We believe that the construction of our secure password authentication scheme might lead to many other useful applications. For example, the scheme might allow us to use human memorable (i.e., lower entropy) secrets to secure highly confidential data like secret keys. Let pwd_i be the user's password and let σ_i denote the solution to the corresponding CAPTCHA challenge. The random oracle value $R_i = \mathbf{G}(pwd_i, \sigma_i, s, 1|i)$ is completely uncorrelated with any information that the adversary can obtain without discovering the user's password. The random values $R_1, R_2, \ldots$ could be used as a one-time pad to efficiently encrypt/decrypt information on a hard drive or to (re)derive private keys for a signature scheme.

We also note that our authentication scheme could potentially be modified to make the proof of human work *safely exportable* and that the amount of human work during authentication can easily be tuned. For example, suppose that Bob wants to protect his passwords, but that he is too busy to solve CAPTCHAs. During authentication, after Bob enters his password and receives the CAPTCHA challenge, Bob might like to pay other human(s) to solve the CAPTCHA challenge for him. However, he wants to make sure that his password is not exposed if these contracted workers are malicious. For example, we might replace the hash of the password with an obfuscation of two program $P_{K,pwd}$ and G_K. Here, $G_K(x, pwd)$ generates a CAPTCHA puzzle Z using randomness $(r_1, r_2) = \mathrm{PRF}_K(x, pwd)$ in the procedures CAPT.W and CAPT.G respectively. $P_{K,pwd}(pwd', \sigma, x)$ outputs 1 if and only if $pwd' = pwd$ and σ is the correct solution to the puzzle Z output by $G_K(x, pwd)$. During authentication we can obtain the puzzle Z by running $G_K(x, pwd')$ with a uniformly random string $x \in \{0,1\}^\lambda$ which should be discarded immediately after the authentication session finishes. As long as Bob keeps the value x secret he can safely share the puzzle Z with other users. However, this modified authentication protocol is merely a heuristic as we do not have any formal security proof that it is hard for a computer to solve CAPTCHA puzzles generated by G_K when given the obfuscated source code $i\mathcal{O}(G_K)$.

As another application we could use the same general framework as a way to detect bots *without* interaction! Suppose that we rename the algorithms CreateAccount$^{\mathcal{H}}$ and Authenticate$^{\mathcal{H}}$ to algorithms GenerateVerified Message$^{\mathcal{H}}$ and VerifyMessage$^{\mathcal{H}}$. The algorithms have essentially the same functionality except for a few minor modifications: 1) the password field pwd is renamed to denote a message m that a user Alice wishes to send to Bob, 2) we replace the username u with a pair (u_1, u_2) where u_1 denotes the sender and u_2 denotes the intended receiver, and we fix the salt value $s = \mathbf{G}(u_1, u_2, m)$ for a given message m that a user u_1 wishes to send to u_2. To send the message m to Bob Alice would first execute GenerateVerifiedMessage$^{\mathcal{H}}$(PP, $(Alice, Bob), m$) and solve the corresponding CAPTCHA to obtain a tuple (h, s). Now Alice sends the tuple $(Alice, Bob, m, h, s)$ to Bob. At this point Alice is finished with the protocol. Bob runs VerifyMessage$^{\mathcal{H}}$(PP, $(Alice, Bob), m, h, s$) and solves the corresponding CAPTCHA to obtain a bit b. If $b = 1$ then Bob accepts that a human (possibly Alice) spent time and energy to send the him the message m^{16}. If $b = 0$ then Bob may dismiss the message as potentially being produced by a bot.

6 Future Challenges

While we believe that Proofs of Human Work could have many benefits, we see three primary challenges for future research. First, because our construction of PoH puzzles is based on $i\mathcal{O}$ HumanCoin is not practical without a large breakthrough in the design of practical $i\mathcal{O}$ schemes. Could we design efficient targeted obfuscation schemes for specific programs like our PoH algorithms? Second, because our PoH puzzles rely on the assumption that some underlying AI problem is hard it is possible that a cryptocurrency like HumanCoin might have a shorter shelf life (e.g., if it takes 15 years for AI researchers to break the underlying CAPTCHA then HumanCoin would expire in at most 15 years). Would it possible for HumanCoin participants to reach a consensus to change the underlying CAPTCHA in the event of an AI breakthrough? Finally, our Proof of Human Work construction, and by extension HumanCoin, requires an initial trusted setup phase for the Proof of Human Work construction. If the Proof of Human Work system is generated by a malicious party then that party might be able to insert a trapdoor which would allow him to mine HumanCoins without any human effort. We note that this concern is not unique to HumanCoin. Other cryptocurrencies like Zerocash [5] also require an initial trusted setup phase[17]. Ben-Sasson et al. [6] proposed to run this trusted setup phase using secure multiparty computation. As long as at least one of the parties in this computation are honest it would be impossible for a malicious adversary to insert a backdoor. Similar techniques could also be used to minimize risks during the HumanCoin setup phase.

In addition to cryptocurrency we also showed that our PoH techniques could be applied to protect passwords and to detect bots without interaction. What other applications are possible?

Acknowledgments. The authors thank paper shepherd Peter Gaži for his very constructive feedback which helped us to improve the quality of the paper. In particular, we are thankful for his suggestions about formalizing security statements involving hard AI problems.

The authors also thank Andrew Miller, and the PC of ITCS 2016 and TCC 2016B for their helpful comments.

---

[16] If Bob wanted to additionally verify that Alice was the human that sent the message Alice and Bob would need to use other cryptographic tools like digital signatures.

[17] Arguably, even Bitcoin does require some trust assumptions during setup. For example, we need to trust that the cryptographic hash function $h = \text{SHA256}$, which is modeled as a random oracle in the Bitcoin protocol, does not have any secret backdoors. A malicious miner with a secret backdoor could easily reverse old transactions.

A   Universal Samplers: Security Definition

Definition 12. *Consider efficient algorithms* (Setup, Sample) *where* $U \leftarrow$ Setup$^{\text{RO}}(1^\lambda)$, d *is the fixed program supporting additional input, and* $p_\beta \leftarrow$ Sample$^{\text{RO}}(U, d, \beta)$. *We say* (Setup, Sample) *is an adaptively-secure ·universal sampler scheme for a circuit* d, *if there exist efficient interactive Turing Machines* SimSetup, SimRO *such that for every efficient admissible adversary* $\mathcal{A}$, *there exists a negligible function* negl() *such that the following two conditions hold:*

$$\Pr[\mathbf{Real}(1^\lambda) = 1] - \Pr[\mathbf{Ideal}(1^\lambda) = 1] = \mathsf{negl}() \text{ and } \Pr[\mathbf{Ideal}(1^\lambda) = aborts] < \mathsf{negl}()$$

where admissible adversaries, the experiments **Real** *and* **Ideal** *and the notion of the* **Ideal** *experiment aborting, are described below*

- *An admissible adversary* $\mathcal{A}$ *is an efficient interactive Turing Machine that outputs one bit, with the following input/output behavior:*
  - $\mathcal{A}$ *initially takes input security parameter* 1^λ *and sampler parameters* U, *as well as the program* d.
  - $\mathcal{A}$ *can send a message* (RO, x) *corresponding to a random oracle query. In response,* $\mathcal{A}$ *expects to receive the output of the random oracle on input* x.
  - $\mathcal{A}$ *can send a message* (sample, β). *The adversary does not expect any response to this message. Instead, upon sending this message,* $\mathcal{A}$ *is required to honestly compute* $p_\beta = $ Sample(U, d, β), *making use of any additional RO queries, and* $\mathcal{A}$ *appends* (β, p_β) *to an auxiliary tape.*

    Remark. *Intuitively,* (sample, β) *messages correspond to an honest party seeking a sample generated by the fixed program* d *on input* β. *Recall that* $\mathcal{A}$ *is meant to internalize the behavior of honest parties.*
- *The experiment* **Real**(1^λ) *is as follows:*
  - *Throughout this experiment, a random oracle* RO *is implemented by assigning random outputs to each unique query made to* RO.
  - $U \leftarrow$ Setup$^{\text{RO}}(1^\lambda)$.
  - $\mathcal{A}(1^\lambda, U, d)$ *is executed; when* $\mathcal{A}$ *sends every message of the form* (RO, x), *it receives the response* RO(x).
  - *The output of the experiment is the final output of the execution of* $\mathcal{A}$ *(which is a bit* $b \in \{0, 1\}$).
- *The experiment* **Ideal**(1^λ) *is as follows:*
  - *Throughout this experiment, a Samples Oracle* $\mathcal{O}$ *is implemented as follows: On input* β, $\mathcal{O}$ *outputs* $d(F(\beta))$, *where* F *is a truly random function.*
  - $(U, \tau) \leftarrow$ SimSetup(1^λ). *Here,* SimSetup *can make arbitrary queries to the Samples Oracle* $\mathcal{O}$.
  - $\mathcal{A}(1^\lambda, U, d)$ *and* SimRO(τ) *begin simultaneous execution. Messages for* $\mathcal{A}$ *or* SimRO *are handled as:*
    1. *Whenever* $\mathcal{A}$ *sends a message of the form* (RO, x), *this is forwarded to* SimRO, *which produces a response to be sent back to* $\mathcal{A}$.

2. SimRO *can make any number of queries to the Samples Oracle* $\mathcal{O}$.
3. *In addition, after* $\mathcal{A}$ *sends messages of the form* (sample, β), *the auxiliary tape of* $\mathcal{A}$ *is examined until* $\mathcal{A}$ *adds entries of the form* (β, p_β) *to it. At this point, if* $p_\beta \neq d(F(\beta))$, *the experiment aborts and we say that an "Honest Sample Violation" has occurred. Note that this is the only way that the experiment* **Ideal** *can abort. In this case, if the adversary itself "aborts", we consider this to be an output of zero by the adversary, not an abort of the experiment itself.*

- *The output of the experiment is the final output of the execution of* $\mathcal{A}$ *(which is a bit* $b \in \{0, 1\}$).

References

1. Andrychowicz, M., Dziembowski, S., Malinowski, D., Mazurek, L.: Secure multiparty computations on Bitcoin. In: 2014 IEEE Symposium on Security and Privacy, pp. 443–458. IEEE Computer Society Press, May 2014
2. Aspnes, J., Jackson, C., Krishnamurthy, A.: Exposing computationally-challenged Byzantine impostors. Technical report YALEU/DCS/TR-1332, Yale University Department of Computer Science, July 2005
3. Back, A.: Hashcash – a denial of service counter-measure (2002). http://hashcash.org/papers/hashcash.pdf
4. Barak, B., Bitansky, N., Canetti, R., Kalai, Y.T., Paneth, O., Sahai, A.: Obfuscation for evasive functions. In: Lindell, Y. (ed.) TCC 2014. LNCS, vol. 8349, pp. 26–51. Springer, Heidelberg (2014)
5. Ben-Sasson, E., Chiesa, A., Garman, C., Green, M., Miers, I., Tromer, E., Virza, M., Zerocash: decentralized anonymous payments from Bitcoin. In: 2014 IEEE Symposium on Security and Privacy, pp. 459–474. IEEE Computer Society Press, May 2014
6. Ben-Sasson, E., Chiesa, A., Green, M., Tromer, E., Virza, M.: Secure sampling of public parameters for succinct zero knowledge proofs. In: 2015 IEEE Symposium on Security and Privacy, pp. 287–304. IEEE Computer Society Press, May 2015
7. Bentov, I., Kumaresan, R.: How to use Bitcoin to design fair protocols. In: Garay, J.A., Gennaro, R. (eds.) CRYPTO 2014, Part II. LNCS, vol. 8617, pp. 421–439. Springer, Heidelberg (2014)
8. Bentov, I., Lee, C., Mizrahi, A., Rosenfeld, M.: Proof of activity: extending Bitcoins proof of work via proof of stake. In: Proceedings of the ACM SIGMETRICS 2014 Workshop on Economics of Networked Systems, NetEcon (2014)
9. Blocki, J., Blum, M., Datta, A.: GOTCHA password hackers! In: AISec 2013, Proceedings of the 2013 ACM Workshop on Artificial Intelligence and Security, pp. 25–34 (2013). http://www.cs.cmu.edu/jblocki/papers/aisec2013-fullversion.pdf
10. Blocki, J., Komanduri, S., Cranor, L.F., Datta, A.: Spaced repetition and mnemonics enable recall of multiple strong passwords. In: NDSS 2015. The Internet Society, February 2015
11. Blocki, J., Komanduri, S., Procaccia, A., Sheffet, O.: Optimizing password composition policies. In: Proceedings of the Fourteenth ACM Conference on Electronic Commerce, pp. 105–122. ACM (2013)
12. Blocki, J., Zhou, H.-S.: Designing proof of human-work puzzles for cryptocurrency and beyond. In: IACR Cryptology ePrint Archive 2016/145 (2016). http://eprint.iacr.org/2016/145

13. Bonneau, J.: The science of guessing: analyzing an anonymized corpus of 70 million passwords. In: 2012 IEEE Symposium on Security and Privacy, pp. 538–552. IEEE Computer Society Press, May 2012

14. Bonneau, J., Miller, A., Clark, J., Narayanan, A., Kroll, J.A., Felten, E.W.: SoK: research perspectives and challenges for Bitcoin and cryptocurrencies. In: 2015 IEEE Symposium on Security and Privacy, pp. 104–121. IEEE Computer Society Press, May 2015

15. Bonneau, J., Schechter, S.: Toward reliable storage of 56-bit keys in human memory. In: Proceedings of the 23rd USENIX Security Symposium, August 2014

16. Bursztein, E., Aigrain, J., Moscicki, A., Mitchell, J.C.: The end is nigh: generic solving of text-based captchas. In: 8th USENIX Workshop on Offensive Technologies (WOOT 2014), San Diego, CA, August 2014. USENIX Association (2014)

17. Bursztein, E., Bethard, S., Fabry, C., Mitchell, J.C., Jurafsky, D.: How good are humans at solving CAPTCHAs? A large scale evaluation. In: 2010 IEEE Symposium on Security and Privacy, pp. 399–413. IEEE Computer Society Press, May 2010

18. Canetti, R., Halevi, S., Steiner, M.: Mitigating dictionary attacks on password-protected local storage. In: Dwork, C. (ed.) CRYPTO 2006. LNCS, vol. 4117, pp. 160–179. Springer, Heidelberg (2006)

19. Chellapilla, K., Simard, P.Y.: Using machine learning to break visual human interaction proofs (HIPs). In: Neural Information Processing Systems (NIPS), pp. 265–272 (2004). https://papers.nips.cc/paper/2571-using-machine-learning-to-break-visual-human-interaction-proofs-hips.pdf

20. Douceur, J.R.: The Sybil attack. In: Druschel, P., Kaashoek, F., Rowstron, A. (eds.) IPTPS 2002. LNCS, vol. 2429, pp. 251–260. Springer, Heidelberg (2002)

21. Dwork, C., Goldberg, A.V., Naor, M.: On memory-bound functions for fighting spam. In: Boneh, D. (ed.) CRYPTO 2003. LNCS, vol. 2729, pp. 426–444. Springer, Heidelberg (2003)

22. Dwork, C., Halpern, J.Y., Waarts, O.: Performing work efficiently in the presence of faults. SIAM J. Comput. 27(5), 1457–1491 (1998)

23. Dwork, C., Naor, M.: Pricing via processing or combatting junk mail. In: Brickell, E.F. (ed.) CRYPTO 1992. LNCS, vol. 740, pp. 139–147. Springer, Heidelberg (1993)

24. Dziembowski, S.: How to pair with a human. In: Garay, J.A., De Prisco, R. (eds.) SCN 2010. LNCS, vol. 6280, pp. 200–218. Springer, Heidelberg (2010)

25. Dziembowski, S., Faust, S., Kolmogorov, V., Pietrzak, K.: Proofs of space. In: Gennaro, R., Robshaw, M. (eds.) CRYPTO 2015. LNCS, vol. 9216, pp. 585–605. Springer, Heidelberg (2015)

26. Elson, J., Douceur, J.R., Howell, J., Saul, J.: Asirra: a CAPTCHA that exploits interest-aligned manual image categorization. In: Ning, P., di Vimercati, S.D.C. Syverson, P.F. (eds.) ACM CCS 2007, pp. 366–374. ACM Press, October 2007

27. Eyal, I., Sirer, E.G.: Majority is not enough: Bitcoin mining is vulnerable. In: Christin, N., Safavi-Naini, R. (eds.) FC 2014. LNCS, vol. 8437, pp. 431–449. Springer, Heidelberg (2014)

28. Florêncio, D., Herley, C.: Where do security policies come from. In: Proceedings of SOUPS, p. 10 (2010)

29. Garay, J., Kiayias, A., Leonardos, N.: The Bitcoin backbone protocol: analysis and applications. In: Oswald, E., Fischlin, M. (eds.) EUROCRYPT 2015. LNCS, vol. 9057, pp. 281–310. Springer, Heidelberg (2015)

30. Garg, S., Gentry, C., Halevi, S., Raykova, M., Sahai, A., Waters, B.: Candidate indistinguishability obfuscation and functional encryption for all circuits. In: 54th FOCS, pp. 40–49. IEEE Computer Society Press, October 2013

31. Goldwasser, S., Rothblum, G.N.: On best-possible obfuscation. In: Vadhan, S.P. (ed.) TCC 2007. LNCS, vol. 4392, pp. 194–213. Springer, Heidelberg (2007)

32. Hofheinz, D., Jager, T., Khurana, D., Sahai, A., Waters, B., Zhandry, M.: How to generate and use universal samplers. Cryptology ePrint Archive, Report 2014/507 (2014). http://eprint.iacr.org/2014/507

33. Hwang, K.-F., Huang, C.-C., You, G.-N.: A spelling based CAPTCHA system by using click. In: 2012 International Symposium on Biometrics and Security Technologies (ISBAST), pp. 1–8, March 2012

34. Kani, J., Nishigaki, M.: Gamified CAPTCHA. In: Marinos, L., Askoxylakis, I. (eds.) HAS 2013. LNCS, vol. 8030, pp. 39–48. Springer, Heidelberg (2013)

35. Khot, R.A., Srinathan, K.: iCAPTCHA: image tagging for free. In: Proceedings of Conference on Usable Software and Interface Design (2009)

36. Kiayias, A., Zhou, H.-S., Zikas, V.: Fair and robust multi-party computation using a global transaction ledger. In: Fischlin, M., Coron, J.-S. (eds.) EUROCRYPT 2016. LNCS, vol. 9666, pp. 705–734. Springer, Heidelberg (2016). doi:10.1007/978-3-662-49896-5_25

37. Komanduri, S., Shay, R., Kelley, P., Mazurek, M., Bauer, L., Christin, N., Cranor, L., Egelman, S.: Of passwords, people: measuring the effect of password-composition policies. In: Proceedings of the Annual Conference on Human Factors in Computing Systems, pp. 2595–2604. ACM (2011)

38. Kosba, A., Miller, A., Shi, E., Wen, Z., Papamanthou, C.: Hawk: the blockchain model of cryptography and privacy-preserving smart contracts. In: IEEE Symposium on Security and Privacy (2016)

39. Kumarasubramanian, A., Ostrovsky, R., Pandey, O., Wadia, A.: Cryptography using captcha puzzles. In: Kurosawa, K., Hanaoka, G. (eds.) PKC 2013. LNCS, vol. 7778, pp. 89–106. Springer, Heidelberg (2013)

40. Lamport, L., Shostak, R., Pease, M.: The byzantine generals problem. ACM Trans. Program. Lang. Syst. (TOPLAS) 4(3), 382–401 (1982)

41. Merkle, R.C.: A digital signature based on a conventional encryption function. In: Pomerance, C. (ed.) CRYPTO 1987. LNCS, vol. 293, pp. 369–378. Springer, Heidelberg (1988)

42. Miller, A., Kosba, A.E., Katz, J., Shi, E.: Nonoutsourceable scratch-off puzzles to discourage bitcoin mining coalitions. In: Ray, I., Li, N., Kruegel, C. (eds.) ACM CCS 15, pp. 680–691. ACM Press, October 2015

43. Mori, G., Malik, J.: Recognizing objects in adversarial clutter: breaking a visual CAPTCHA. In: IEEE Computer Society Conference on Computer Vision and Pattern Recognition (CVPR), pp. 134–144 (2003)

44. Motoyama, M., Levchenko, K., Kanich, C., McCoy, D., Voelker, G.M., Savage, S.: Re: CAPTCHAs-understanding CAPTCHA-solving services in an economic context. In: 19th USENIX Security Symposium, Washington, DC, USA, 11–13 August 2010, Proceedings, pp. 435–462 (2010)

45. Nakamoto, S.: Bitcoin: a peer-to-peer electronic cash system (2008). https://bitcoin.org/bitcoin.pdf

46. Narayanan, A., Bonneau, J., Felten, E., Miller, A.: Bitcoin and Cryptocurrency Technology (online course) (2015). https://piazza.com/princeton/spring2015/btctech/resources

47. Park, S., Pietrzak, K., Kwon, A., Alwen, J., Fuchsbauer, G., Gaži, P.: Spacemint: a cryptocurrency based on proofs of space. Cryptology ePrint Archive, Report 2015/528 (2015). http://eprint.iacr.org/2015/528
48. Pass, R., Seeman, L.: abhi shelat. Analysis of the blockchain protocol in asynchronous networks. In: Cryptology ePrint Archive, Report 2016/454 (2016). http://eprint.iacr.org/2016/454
49. Rogaway, P.: Formalizing human ignorance. In: Nguyên, P.Q. (ed.) VIETCRYPT 2006. LNCS, vol. 4341, pp. 211–228. Springer, Heidelberg (2006)
50. Sahai, A., Waters, B.: How to use indistinguishability obfuscation: deniable encryption, and more. In: Shmoys, D.B. (ed.) 46th ACM STOC, pp. 475–484. ACM Press, May/June 2014
51. Sapirshtein, A., Sompolinsky, Y., Zohar, A.: Optimal selfish mining strategies in bitcoin. In: FC (2016). http://arxiv.org/abs/1507.06183
52. Sauer, G., Hochheiser, H., Feng, J., Lazar, J.: Towards a universally usable CAPTCHA. In: Proceedings of the 4th Symposium on Usable Privacy and Security (2008)
53. Szabo, N.: Formalizing and securing relationships on public networks. In: First Monday (1997). http://firstmonday.org/ojs/index.php/fm/article/view/548/469
54. Tam, J., Simsa, J., Hyde, S., Von Ahn, L.: Breaking audio captchas. Advan. Neural Inf. Process. Syst. 1(4), 1625–1632 (2008)
55. Ahn, L., Blum, M., Hopper, N.J., Langford, J.: CAPTCHA: using hard AI problems for security. In: Biham, E. (ed.) EUROCRYPT 2003. LNCS, vol. 2656, pp. 294–311. Springer, Heidelberg (2003)
56. Von Ahn, L., Maurer, B., McMillen, C., Abraham, D., Blum, M.: reCAPTCHA: human-based character recognition via web security measures. Science 321(5895), 1465–1468 (2008)
57. Waters, B., Juels, A., Halderman, J.A., Felten, E.W.: New client puzzle outsourcing techniques for DoS resistance. In: Atluri, V., Pfitzmann, B., Mc-Daniel, P. (eds.) ACM CCS 2004, pp. 246–256. ACM Press, October (2004)
58. Wilkins, J.: Strong CAPTCHA guidelines v1.2. (2009). http://bitland.net/captcha.pdf

Access Control Encryption: Enforcing Information Flow with Cryptography

Ivan Damgård, Helene Haagh[(✉)], and Claudio Orlandi

Aarhus University, Aarhus, Denmark
{ivan,haagh,orlandi}@cs.au.dk

Abstract. We initiate the study of *Access Control Encryption (ACE)*, a novel cryptographic primitive that allows fine-grained access control, by giving different rights to different users not only in terms of which messages they are allowed to *receive*, but also which messages they are allowed to *send*.

Classical examples of security policies for information flow are the well known Bell-Lapadula [BL73] or Biba [Bib75] model: in a nutshell, the Bell-Lapadula model assigns roles to every user in the system (e.g., *public*, *secret* and *top-secret*). A users' role specifies which messages the user is allowed to receive (i.e., the *no read-up* rule, meaning that users with *public* clearance should not be able to read messages marked as *secret* or *top-secret*) but also which messages the user is allowed to send (i.e., the *no write-down* rule, meaning that a malicious user with *top-secret* clearance should not be able to write messages marked as *secret* or *public*). To the best of our knowledge, no existing cryptographic primitive allows for even this simple form of access control, since no existing cryptographic primitive enforces any restriction on what kind of messages one should be able to encrypt. Our contributions are:

- Introducing and formally defining access control encryption (ACE);
- A construction of ACE with complexity linear in the number of the roles based on classic number theoretic assumptions (DDH, Paillier);
- A construction of ACE with complexity polylogarithmic in the number of roles based on recent results on cryptographic obfuscation;

1 Introduction

Traditionally, cryptography has been about providing secure communication over insecure channels. We want to protect honest parties from external adversaries: only the party who has the decryption key can access the message. More recently, more complicated situations have been considered, where we do not want to trust everybody with the same information: depending on who you are and which keys you have, you can access different parts of the information sent (this can be done using, e.g., functional encryption [BSW11]).

However, practitioners who build secure systems in real life are often interested in achieving different and stronger properties: one wants to control the

Full version: [DHO16].

M. Hirt and A. Smith (Eds.): TCC 2016-B, Part II, LNCS 9986, pp. 547–576, 2016.
DOI: 10.1007/978-3-662-53644-5_21

information flow in the system, and this is not just about what you can receive, but also about what you can send. As an example, one may think of the first security policy model ever proposed, the one by Bell and Lapadula [BL73]. Slightly simplified, this model classifies users of a system in a number of levels, from "public" in the bottom to "top-secret" on top. Then two rules are defined: (1) "no read-up" – a user is not allowed to receive data from higher levels and (2) "no write-down" – a user is not allowed to send data to lower levels. The idea is of course to ensure confidentiality: data can flow from the bottom towards the top, but not in the other direction. Clearly, both rules are necessary, in particular we need no write-down, since a party on top-secret level may try to send information she should not, either by mistake or because her machine has been infected by a virus.

In this paper we study the question of whether cryptography can help in enforcing such security policies. A first thing to realize is that this problem cannot be solved without some assumptions about physical, i.e., non-cryptographic security: if the communication lines cannot be controlled, we cannot prevent a malicious user from sending information to the wrong place. We therefore must introduce a party that controls the communication, which we will call the *sanitizer* San. We assume that all outgoing communication must pass through this party. San can then be instructed to do some specific processing on the messages it gets.

Of course, with this assumption the problem can be solved: San is told what the security policy is and simply blocks all messages that should not be sent according to the policy. This is actually a (simplified) model of how existing systems work, where San is implemented by the operating system and various physical security measures.

However, such a solution is problematic for several reasons: users must securely identify themselves to San so that he can take the correct decisions, this also means that when new users join the system San must be informed about this, directly or indirectly. A side effect of this is that San necessarily knows who sends to whom, and must of course know the security policy. This means that a company cannot outsource the function of San to another party without disclosing information on internal activities of the company.

Therefore, a better version of our basic question is the following: *can we use cryptography to simplify the job of* San *as much as is possible, and also ensure that he learns minimal information?*

To make the goal more precise, note that it is clear that San must process each message that is sent i.e., we cannot allow a message violating the policy to pass through unchanged. But we can hope that the processing to be done does not depend on the security policy, and also not on the identities of the sender and therefore of the allowed receivers. This way we get rid of the need for users to identify themselves to San. It is also clear that San must at least learn when a message was sent and its length, but we can hope to ensure he learns nothing more. This way, one can outsource the function of running San to a party that is only trusted to execute correctly.

Our goal in this paper is therefore to come up with a cryptographic notion and a construction that reduces the sanitizer's job to the minimum we just described. To the best of our knowledge, this problem has not been studied before in the cryptographic literature, and it is easy to see that existing constructions only solve "half the problem": we can easily control which users you can *receive from* by selecting the key material we give out (assuming that the sender is honest). This is exactly what attribute based [GPSW06] or functional encryption [BSW11] can do. But any such scheme of course allows a malicious sender to encrypt what he wants for any receiver he wants.

Our Contribution. In this paper we propose a solution based on a new notion called Access Control Encryption (ACE). In a nutshell ACE works as follows: an ACE scheme has a key generation algorithm that produces a set of sender keys, a set of receiver keys and a sanitizer key. An honest sender S encrypts message m under his sender key and sends it for processing by San using the sanitizer key. San does not need to know the security policy, nor who sends a message or where it is going (so a sender does not have to identify himself): San simply executes a specific randomised algorithm on the incoming ciphertext and passes the result on to a broadcast medium, e.g., a disk from where all receivers can read. So, as desired, San only knows when a message was sent and its length.[1] An honest receiver R who is allowed to receive from S is able to recover m using his key and the output from San. On the other hand, consider a corrupt sender S who is not allowed to send to R. ACE ensures that no matter what S sends, what R receives (after being processed by San) looks like a random encryption of a random message. In fact we achieve security against collusions: considering a subset S of senders and a subset R of receivers, if none of these senders are allowed to send to any of the receivers, then S cannot transfer any information to R, even if players in each set work together. We propose two constructions of ACE: one based on standard number theoretic assumptions (DDH, Pailler) which achieves complexity linear in the number of roles, and one based on recent results in cryptographic obfuscation, which achieves complexity polylogarithmic in the number of roles.

Example. A company is working on a top-secret military project for the government. To protect the secrets the company sets up an access policy that determines which employees are allowed to communicate (e.g., a researcher with top-secret clearance should not be allowed to send classified information to the intern, who is making the coffee and only has public clearance). To implement the access policy, the company sets up a special server that sanitizes every message sent on the internal network before publishing it on a bulletin board or broadcasting it. Using ACE this can be done without requiring users to log into the sanitizer.

---

[1] Note that the sanitizer has to send the ciphertext to all receivers – both those who are allowed to decrypt and those who are not. A sanitizer who could decide whether a particular receiver is allowed to receive a particular ciphertext would trivially be able to distinguish between different senders with different writing rights.

Furthermore, if corrupted parities (either inside or outside the company) want to intercept the communication they will get no information from the sanitizer server, since it does not know the senders identities and the messages sent over the network.

In the following sections, we describe ACE in more detail and take a closer look at our technical contributions.

1.1   Access Control Encryption: The Problem it Solves

Senders and Receivers. We have n (types of) senders $S_1, \ldots, S_n$ and n (types of) receivers $R_1, \ldots, R_n$.[2] There is some predicate $P : [n] \times [n] \rightarrow \{0, 1\}$, where $P(i, j) = 1$ means that S_i is allowed to send to R_j, while $P(i, j) = 0$ means that S_i is not allowed to send to R_j.

Network Model. We assume that senders are connected to all receivers via a public channel i.e., a sender cannot send a message only to a specific receiver and any receiver can see all traffic from all senders (also from those senders they are not allowed to communicate with).

Requirements. Informally we want the following properties[3]

1. *Correctness:* When an honest sender S_i sends a message m, all receivers R_j with $P(i, j) = 1$ learn m;
2. *No-Read Rule:* At the same time all receivers R_j with $P(i, j) = 0$ learn no information about m;
3. *No-Write Rule:* No (corrupt) sender S_i should be able to communicate any information to any (possibly corrupt) receiver R_j if $P(i, j) = 0$

Note that the *no-read rule* on its own is a simple *confidentiality* requirement, which can be enforced using standard encryption schemes. On the other hand standard cryptographic tools do not seem to help in satisfying the *no-write rule*. In particular the *no-write* rule is very different from the standard *authenticity* requirement and e.g., signature schemes cannot help here: had we asked for a different property such as "*a corrupt sender S_i should not be allowed to communicate with an honest receiver R_j if $P(i, j) = 0$*" then the problem could be solved by having R_j verify the identity of the sender (using a signature scheme) and ignore messages from any sender i with $P(i, j) = 0$. Instead, we are trying to block communication even between corrupt senders and corrupt receivers.

The problem as currently stated is impossible to solve, since a corrupt sender can broadcast m in the clear to all receivers (the corrupt sender might not care that other receivers also see the message). As mentioned above, we therefore enhance the model by adding a special party, which we call the *sanitizer* San. The sanitizer receives messages from senders, performs some computation on them, and then forwards them to all receivers. In other words, we allow the

---

[2] The number of senders equals the number of receivers only for the sake of exposition.

[3] The security model, formalized in Definitions 2 and 3, is more general than this.

public channel to perform some computation before delivering the messages to the receivers. Hence, the output of the sanitizer is visible to all receivers (i.e., the sanitizer cannot give different outputs to different receivers). We therefore add the following requirement to our *no-read rule*:

2b. The sanitizer should not learn anything about the communication it routes. In particular, the sanitizer should not learn any information about the message m which is being transmitted nor the identity of the sender i;

In Sect. 2 we formalize properties 2 and 2b as a single one (i.e., no set of corrupt receivers, even colluding with the sanitizer, should be able to break the *no-read* rule). When considering property 3, we assume the sanitizer not to collude with the corrupt senders and receivers: after all, since the sanitizer controls the communication channel, there is no way of preventing a corrupt sanitizer from forwarding messages from corrupt senders to the corrupt receivers.[4]

We stress that previous work is not sufficient to achieve property 3: Even encryption schemes with fine-grained decryption capabilities (such as predicate- and attribute based- encryption [GPSW06, KSW13]) do not offer security guarantees against colluding senders and receivers.

1.2 Technical Overview

Linear ACE. The main idea behind our construction of ACE with linear complexity (described in Sect. 3) is the following: we start with an ACE for a single identity i.e., where $n = 1$ and $P(1,1) = 1$. First we need to make sure that even a corrupt sender with encryption rights (i.e., $i = 1$) cannot communicate with a corrupt receiver with no decrypting right (i.e., with a special identity $j = 0$). To prevent this, since the receiver cannot decrypt the ciphertext, it is enough to use a randomizable public key encryption and let the sanitizer refresh the ciphertext. This ensures that the outgoing ciphertext is distributed exactly as a fresh encryption.

The more challenging task is to ensure that a corrupt sender with no rights (i.e., with a special identity $i = 0$) cannot transfer any information to a corrupt receiver with decrypting rights (i.e., $j = 1$), since in this case the receiver knows the decryption key. Thus, we cannot use the security of the underlying encryption scheme. We solve the problem using any encryption scheme which is homomorphic both in the message and in the randomness (such as ElGamal or Pailler). The main idea is to let the encryption key ek as well as the randomizer

---

[4] Note that it is possible to reduce the trust on the sanitizer in different ways: in a black-box way, one could imagine several parties emulating the work of the sanitizer using MPC. In a more concrete way, it is possible to have a *chain* of sanitizers, where the senders send their encryptions to sanitizer 1, the receivers receive ciphertexts from sanitizer n, and sanitizer $i + 1$ further sanitizes the output of sanitizer i. We note that all definitions and constructions in this paper can be easily generalized to this scenario but, to keep the presentation as simple as possible, we do not discuss this solution further and stick to the case of a single sanitizer.

key rk be some secret value α, and an encryption of a message m being a tuple $(c_0, c_1) = (E(ek), E(m))$. On input such a tuple the sanitizer picks a random s and outputs c', a fresh encryption of $(ek - rk) \cdot s + m$ (which can be computed thanks to the homomorphic properties of E): note that sanitization does not interfere with honestly generated encryptions (since $ek = rk = \alpha$), while the sanitized version of a ciphertext produced by anyone who does not know α is indistinguishable from a random encryption of a random value.

We then turn this into a scheme for any predicate $P : [n] \times [n] \to \{0, 1\}$ by generating n copies of the single identity ACE scheme. Each receiver j is given the decryption key for one of the schemes, and each sender i is given the encryption key for all instances j such that $P(i, j) = 1$. The resulting scheme has linear complexity in n, the number of the roles in the system, which makes our scheme impractical for large predicates.

Polylogarithmic ACE. At first it might seem easy to construct an ACE scheme with compact ciphertexts using standard tools (such as non-interactive zero-knowledge proofs). In Sect. 4 we discuss why this is not the case before presenting our construction of an ACE with complexity polylogarithmic in n. To construct the scheme we first introduce the notion of a *sanitizable functional encryption* (sFE) scheme which is a functional encryption (FE) scheme enhanced with a sanitization algorithm. Informally we require that given any two ciphertexts c_0, c_1 that decrypt to the same message and a sanitized ciphertext c', no one (even with access to the master secret key), should be able to tell whether c' is a sanitized version of c_0 or c_1.[5] We are able to construct such a scheme by modifying the FE based on indistinguishability obfuscation of Garg et al. [GGH+13]: in their scheme ciphertexts consist of two encryptions and a *simulation statistically-sound* NIZK proof that they contain the same message. We instantiate their construction with a sanitizable encryption scheme[6], and we instruct the sanitizer to sanitize the two encryptions, drop the original proof and append a *proof of a proof* instead (that is, a proof of the fact that the sanitizer saw a proof who would make the original verifier accept). This preserves the functionality of the original FE scheme while making the sanitized ciphertext independent of the randomness used by the sender. We formally define sFE in Sect. 4.1 and present a construction in Sect. 4.2.

Finally, armed with such a sFE scheme, we construct a polylog ACE scheme in Sect. 4.3 in the following way: ciphertexts are generated by encrypting tuples of the form (m, i, y) with $y = F_{ek_i}(m)$ for a PRF F (where ek_i is the the encryption key of the sender S_i), using the sFE scheme. Decryption keys are sFE secret keys for the function that outputs m only if $P(i, j) = 1$ (and ignores y). The sanitizer key is a sFE secret key which outputs 1 only if y is a valid MAC on

---

[5] We note that this is a relaxation of re-randomizability for FE, in the sense that we do not require sanitized ciphertexts to be indistinguishable from fresh encryptions, but only independent of the randomness used in the original encryption. However, to the best of our knowledge, no re-randomizable FE scheme for all circuits exist.

[6] Similar to a re-randomizable encryption scheme, where we do not require sanitized ciphertexts to look indistinguishable from fresh encryptions.

m for the identity i (note that this can be checked by a compact circuit by e.g., generating all the keys ek_i pseudorandomly using another PRF). This key allows the sanitizer to check if an encryption contains a valid MAC or not, but without learning anything about the message nor the identity. Now the sanitizer drops invalid encryptions (or replaces them with random encryptions of random values for a special, undecryptable identity $i = 0$) and forwards valid encryptions (after having refreshed them).

Open Questions. We identify two major opens questions: the first one is to construct practically interesting ACE from noisy, post-quantum assumptions such as LWE – the challenge here is that it always seems possible for a malicious sender to encrypt with just enough noise that any further manipulation by the sanitizer makes the decryption fail. This can be addressed using "bootstrapping" techniques, but this is not likely to lead to schemes with efficiency comparable to the ones based on DDH or Pailler described above. The second open question is to design sublinear ACE scheme with practical efficiency even for limited classes of interesting predicates such as e.g., $P(i,j) = 1 \Leftrightarrow i \geq j$.

1.3   Related Work

One of the main challenges in our setting is to prevent corrupt senders to communicate to corrupt receivers using subliminal channels (e.g., by producing the encryptions with maliciously generated randomness). In some sense we are trying to prevent steganography [HLA02]. Recent work on cryptographic firewalls [MS15, DMS15] also deals with this problem, but in the context of preventing malicious software implementations to leak information via steganographic techniques. Raykova et al. [RZB12] presented solutions to the problem of access control on outsourced data, with focus on hiding the access patterns from the cloud (this is not a concern in our application since all receivers receive all ciphertexts) and in preventing malicious writers from updating files they are not allowed to update. However they only guarantee that malicious writers are caught if they do so, while we want to prevent any communication between corrupt senders and receivers. Backes and Pfitzmann introduced the notion of probabilistic non-interference which allows to relate cryptography to the notion of information flow for both transitive [BP03] and intransitive policies [BP04]. Halevi et al. [HKN05] address the problem of enforcing confinement in the T10 OSD protocol, in the presence of a fully trusted manager (which has a role similar to the sanitizer in our model). Fehr and Fischlin [FF15] study the case of sanitizable signatures in the context of an intermediate party that sanitizes messages and signatures send over the channel. The special party learns as little as possible about the messages and signatures. However, they do not prevent corrupt senders from sending information to corrupt receivers. Finally, the problem of hiding policies and credentials in the context of attribute based encryption has been studied by Frikken et al. [FAL06], Kapadia et al. [KTS07], Müller and Katzenbeisser [MK11], and

Ferrara et al. [FFLW15]. However, they do not consider the case of preventing corrupt sender from communicating with corrupt receivers (e.g. by sending the message unencrypted over the channel).

2 Defining ACE

ACE Notation. An *access control encryption* (ACE) scheme is defined by the following algorithms:

Setup: The Setup algorithm on input the security parameter κ and a policy $P : [n] \times [n] \rightarrow \{0,1\}$ outputs a master secret key msk and public parameters pp, which include the message space $\mathcal{M}$ and ciphertext spaces $\mathcal{C}, \mathcal{C}'$.[7]

Key Generation: The Gen algorithm on input the master secret key msk, an identity $i \in \{0, \dots, n+1\}$,[8] and a type $t \in \{\mathsf{sen}, \mathsf{rec}, \mathsf{san}\}$ outputs a key k. We use the following notation for the three kind of keys in the system:

- $ek_i \leftarrow \mathsf{Gen}(msk, i, \mathsf{sen})$ and call it an *encryption key for* $i \in [n]$
- $dk_j \leftarrow \mathsf{Gen}(msk, j, \mathsf{rec})$ and call it a *decryption key for* $j \in [n]$
- $ek_0 = dk_0 = pp$;
- $rk \leftarrow \mathsf{Gen}(msk, n+1, \mathsf{san})$ and call it the *sanitizer key*;

Encrypt: The Enc algorithm on input an encryption key ek_i and a message m outputs a ciphertext c.

Sanitizer: San transforms an incoming ciphertext $c \in \mathcal{C}$ into a sanitized ciphertext $c' \in \mathcal{C}'$ using the sanitizer key rk;

Decryption: Dec recovers a message $m' \in \mathcal{M} \cup \{\bot\}$ from a ciphertext $c' \in \mathcal{C}'$ using a decryption key dk_j.

ACE Requirements. We formalize Properties 1–3 from the introduction in the following way:

Definition 1 (Correctness). *For all* $m \in \mathcal{M}$, $i,j \in [n]$ *such that* $P(i,j) = 1$:

$$\Pr\left[\mathsf{Dec}\left(dk_j, \mathsf{San}\left(rk, \mathsf{Enc}\left(ek_i, m\right)\right)\right) \neq m\right] \leq \mathsf{negl}\left(\kappa\right)$$

with $(pp, msk) \leftarrow \mathsf{Setup}(1^\kappa, P)$, $ek_i \leftarrow \mathsf{Gen}(msk, i, \mathsf{sen})$, $dk_j \leftarrow \mathsf{Gen}(msk, j, \mathsf{rec})$, *and* $rk \leftarrow \mathsf{Gen}(msk, n+1, \mathsf{san})$, *and the probabilities are taken over the random coins of all algorithms.*

---

[7] We use the convention that all other algorithms take pp as input even if not specified. Formally, one can think of the pp as being part of msk and all other keys ek, dk, sk.

[8] To make notation more compact we define two special identities: $i = 0$ representing a sender or receiver with no rights such that $P(0,j) = 0 = P(i,0)$ for all $i,j \in [n]$; $i = n+1$ to be the sanitizer identity, which cannot receive from anyone but can send to all i.e., $P(n+1,j) = 1 \ \forall j \in [n]$ and $P(i,n+1) = 0 \ \forall i \in [n]$

Definition 2 (No-Read Rule). *Consider the following game between a challenger C and a stateful adversary A:*

No-Read Rule	
Game Definition	**Oracle Definition**
1. $(pp, msk) \leftarrow \mathsf{Setup}(1^\kappa, P)$; 2. $(m_0, m_1, i_0, i_1) \leftarrow A^{\mathcal{O}_G(\cdot), \mathcal{O}_E(\cdot)}(pp)$; 3. $b \leftarrow \{0, 1\}$; 4. $c \leftarrow \mathsf{Enc}(\mathsf{Gen}(msk, i_b, \mathsf{sen}), m_b)$; 5. $b' \leftarrow A^{\mathcal{O}_G(\cdot), \mathcal{O}_E(\cdot)}(c)$;	$\mathcal{O}_G(j, t)$: 1. Output $k \leftarrow \mathsf{Gen}(msk, j, t)$; $\mathcal{O}_E(i, m)$: 1. $ek_i \leftarrow \mathsf{Gen}(msk, i, \mathsf{sen})$; 2. Output $c \leftarrow \mathsf{Enc}(ek_i, m)$;

We say that A wins the No-Read game if $b = b'$, $|m_0| = |m_1|$, $i_0, i_1 \in \{0, \ldots, n\}$ and one of the following holds:

Payload Privacy: *For all queries q to $\mathcal{O}_G$ with $q = (j, \mathsf{rec})$ it holds that*

$$P(i_0, j) = P(i_1, j) = 0$$

Sender Anonymity: *For all queries q to $\mathcal{O}_G$ with $q = (j, \mathsf{rec})$ it holds that*

$$P(i_0, j) = P(i_1, j) \text{ and } m_0 = m_1$$

We say an ACE scheme satisfies the No-Read rule if for all PPT A

$$\mathsf{adv}^A = 2 \cdot \left| \Pr[A \, wins \, the \, No-Read \, game] - \frac{1}{2} \right| \leq \mathsf{negl}(\kappa)$$

Definition 2 captures the requirement that only intended receivers should be able to learn anything about the message (payload privacy) and that no one (even intended receivers) should learn anything about the identity of the sender (sender anonymity). Note that the ciphertext c sent by the challenger to the adversary has *not* been sanitized and that the adversary is allowed to query for the sanitizer key rk. This implies that even the sanitizer (even with help of any number of senders and unintended receivers) should not learn anything. Note additionally that the adversary is allowed to query for the encryption keys ek_{i_0}, ek_{i_1} corresponding to the challenge identities i_0, i_1, which implies that the ability to *encrypt* to a particular identity does not automatically grant the right to *decrypt* ciphertexts created with that identity (e.g., a user might be able to *write* top-secret documents but not to *read* them). Note that if $i_b = 0$ for some $b \in \{0, 1\}$, then the definition implies that it is possible to create "good looking" ciphertexts even without having access to any of the senders' keys. This is explicitly used in our solution with linear complexity. Furthermore note that if there exist multiple keys for a single identity (e.g., the output of $\mathsf{Gen}(msk, i, \mathsf{sen})$

is randomized), then our definition does not guarantee that the adversary can ask the oracle $\mathcal{O}_G$ for the encryption key used to generate the challenge ciphertext. The definition can be easily amended to grant the adversary this power but (since in all our constructions ek_i is a deterministic function of msk and i) we prefer to present the simpler definition. Finally, the encryption oracle $\mathcal{O}_E$ models the situation that the adversary is allowed to see encrypted messages under identities for which he does not have the encryption key.

Definition 3 (No-Write Rule). *Consider the following game between a challenger C and a stateful adversary A:*

No-Write Rule	
Game Definition	**Oracle Definition**

1. $(pp, msk) \leftarrow \mathsf{Setup}(1^\kappa, P)$;
 2. $(c, i') \leftarrow A^{\mathcal{O}_E(\cdot), \mathcal{O}_S(\cdot)}(pp)$;
 3. $ek_{i'} \leftarrow \mathsf{Gen}(msk, i', \mathsf{sen})$;
 4. $rk \leftarrow \mathsf{Gen}(msk, n+1, \mathsf{san})$;
 5. $r \leftarrow \mathcal{M}$;
 6. $b \leftarrow \{0, 1\}$,
 	
   – If $b = 0$, $c' \leftarrow \mathsf{San}(rk, \mathsf{Enc}(ek_{i'}, r))$;
   – If $b = 1$, $c' \leftarrow \mathsf{San}(rk, c)$;
 	
 7. $b' \leftarrow A^{\mathcal{O}_E(\cdot), \mathcal{O}_R(\cdot)}(c')$;
	$\mathcal{O}_S(j, t)$:
	 1. Output $k \leftarrow \mathsf{Gen}(msk, j, t)$;
	
	 $\mathcal{O}_R(j, t)$:
	 1. Output $k \leftarrow \mathsf{Gen}(msk, j, t)$;
	
	 $\mathcal{O}_E(i, m)$:
	 1. $ek_i \leftarrow \mathsf{Gen}(msk, i, \mathsf{sen})$;
	 2. $c \leftarrow \mathsf{Enc}(ek_i, m)$;
	 3. Output $c' \leftarrow \mathsf{San}(rk, c)$;

Let Q_S (resp. Q) be the set of all queries $q = (j, t)$ that A issues to $\mathcal{O}_S$ (resp. both $\mathcal{O}_S$ and $\mathcal{O}_R$). Let I_S be the set of all $i \in [n]$ such that $(i, \mathsf{sen}) \in Q_S$ and let J be the set of all $j \in [n]$ such that $(j, \mathsf{rec}) \in Q$. Then we say that A wins the No-Write game if $b' = b$ and all of the following hold:

1. $(n+1, \mathsf{san}) \notin Q$;
2. $i' \in I_S \cup \{0\}$;
3. $\forall i \in I_S, j \in J, P(i, j) = 0$;

We say an ACE scheme satisfies the No-Write rule if for all PPT A

$$\mathsf{adv}^A = 2 \cdot \left| \Pr[A \, wins \, the \, No - Write \, game] - \frac{1}{2} \right| \leq \mathsf{negl}(\kappa)$$

Definition 3 captures the property that any set of (corrupt) senders $\{S_i\}_{i \in I}$ cannot transfer any information to any set of (corrupt) receivers $\{R_j\}_{j \in J}$ unless at least one of the senders in I is allowed communication to at least one of the receivers in J (Condition 3)[9]. This is modelled by saying that in the eyes of the

---

[9] Note that the adversary is allowed to ask for any senders' key in the post-challenge queries.

receivers, the sanitized version of a ciphertext coming from this set of senders looks like the sanitized version of a fresh encryption of a random value produced by one of these senders (Condition 2). Note that if the adversary does not ask for any encryption key (i.e., $I_S = \emptyset$), then the only valid choice for i' is 0: this implies that (as described for the no-read rule) there must be a way of constructing "good looking" ciphertexts using the public parameters only[10] and this property is used crucially in the construction of our linear scheme. Furthermore, we require that the adversary does not corrupt the sanitizer (Condition 1) which is, as discussed in the introduction, an unavoidable condition. Finally, the encryption oracle $\mathcal{O}_E$ again models the situation that the adversary is allowed to see encrypted messages under identities for which he do not have the encryption key.

3   Linear ACE from Standard Assumptions

The roadmap of this section is the following: we construct an ACE scheme for a single identity (i.e., $n = 1$ and $P(1,1) = 1$) from standard number theoretic assumptions, and then we construct an ACE scheme for any predicate $P : [n] \times [n] \to \{0,1\}$ using a *repetition scheme*. The complexity of the final scheme (in terms of public-key and ciphertext size) is n times the complexity of the single-identity scheme.

3.1   ACE for a Single Identity

We propose two constructions of ACE for a single identity (or 1-ACE for short). The first is based on the DDH assumption and is presented in this section, while the second is based on the security of Pailler's cryptosystem and is deferred to the full version [DHO16]. Both schemes share the same basic idea: the encryption key ek is some secret value α, and an encryption of a message m is a pair of encryptions $(c_0, c_1) = (E(\alpha), E(m))$. The sanitizer key is also the value α, and a sanitized ciphertext is computed as $c' = c_1 \cdot (c_0 \cdot E(-\alpha))^s$ which (thanks to the homomorphic properties of both ElGamal and Pailler) is an encryption of a uniformly random value unless c_0 is an encryption of α, in which case it is an encryption of the original message m. The decryption key is simply the decryption key for the original encryption scheme, which allows to retrieve m from c'. Note that even knowing the decryption key is not enough to construct ciphertexts which "resist" the sanitization, since the receiver never learns the value α.

1-ACE from DDH: Our first instantiation is based on the ElGamal public-key encryption scheme [Gam85]. The construction looks similar to other *double-strand* versions of ElGamal encryption which have been used before in the literature to achieve different goals (e.g., by Golle et al. [GJJS04] in the context of universal re-encryption and by Prabhakaran and Rosulek [PR07] in the context of rerandomizable CCA security).

---

[10] Recall that we defined $ek_0 = pp$.

Construction 1. *Let* EGACE $= ($Setup, Gen, Enc, San, Dec$)$ *be a 1-ACE scheme defined by the following algorithms:*

Setup: Let (G, q, g) be the description of a cyclic group of prime order q generated by g. Let $(\alpha, x) \leftarrow \mathbb{Z}_q \times \mathbb{Z}_q$ be uniform random elements, and compute $h = g^x$. Output the public parameter $pp = (G, q, g, h)$ and the master secret key $msk = (\alpha, x)$. The message space is $\mathcal{M} = G$ and the ciphertext spaces are $\mathcal{C} = G^4$ and $\mathcal{C}' = G \times G$.

Key Generation: Given the master secret key msk, the encryption, decryption and sanitizer key are computed as follows:
- $ek = \alpha$;
- $dk = -x$;
- $rk = -\alpha$;

Encryption: Given the message m and an encryption key ek, sample random $r_1, r_2 \in \mathbb{Z}_q$ and output:

$$(c_0, c_1, c_2, c_3) = (g^{r_1}, g^{ek}h^{r_1}, g^{r_2}, mh^{r_2})$$

(and encryptions for the identity 0 are random tuples from G^4).

Sanitize: Given a ciphertext $c = (c_0, c_1, c_2, c_3) \in \mathcal{C}$ and a sanitizer key rk, sample uniform random $s_1, s_2 \in \mathbb{Z}_q$ and output:

$$(c_0', c_1') = (c_2 c_0^{s_1} g^{s_2}, c_3 (g^{rk} c_1)^{s_1} h^{s_2})$$

Decryption: Given a ciphertext $c' = (c_0', c_1') \in \mathcal{C}'$ and a decryption key dk, output:

$$m' = c_1'(c_0')^{dk}$$

Lemma 1. *Construction 1 is a correct 1-ACE scheme that satisfies the No-Read Rule and the the No-Write Rule assuming that the DDH assumption holds in G.*

Proof. Correctness: Let $c = (c_0, c_1, c_2, c_3)$ be an honestly generated ciphertext, and let $c' = (c_0', c_1')$ be a sanitized version of c. We check that (c_0', c_1') is still an encryption of the original message m:

$$\begin{aligned}
c_1'(c_0')^{dk} &= c_3(g^{rk}c_1)^{s_1}h^{s_2}(c_2 c_0^{s_1} g^{s_2})^{dk} \\
&= mh^{r_2}(g^{-\alpha}g^{\alpha}h^{r_1})^{s_1}h^{s_2}(g^{r_2}g^{r_1 s_1 + s_2})^{-x} \\
&= mh^{r_2 + r_1 s_1 + s_2}g^{-x(r_2 + r_1 s_1 + s_2)} = m
\end{aligned}$$

Thus, the sanitization of a valid ciphertext produces a new valid ciphertext under the same identity and of the same message.

No-Read Rule: There are three possible cases, depending on which identities the adversary queries during the game: the case $(i_0, i_1) = (0, 0)$ is trivial as both

$\mathsf{Enc}(ek_0, m_b)$ for $b \in \{0, 1\}$ are random ciphertexts; the case $(i_0, i_1) = (1, 1)$ is trivial if the adversary asks for the decryption key dk, since in this case it must be that $m_0 = m_1$. The case where the adversary does not ask for the decryption key and $i_0 \neq i_1$ implies the case where the adversary does not ask for the decryption key and $(i_0, i_1) = (1, 1)$ using standard hybrid arguments (i.e., if $\mathsf{Enc}(ek, m)$ is indistinguishable from a random ciphertext $c \leftarrow \mathcal{C}$ for all m, then $\mathsf{Enc}(ek, m_0)$ is indistinguishable from $\mathsf{Enc}(ek, m_1)$ for all m_0, m_1). So we are only left to prove that honest encryptions are indistinguishable from a random element in $\mathcal{C} = G^4$, which follows in a straightforward way from the DDH assumption. In particular, since (g, h, g^{r_2}, h^{r_2}) is indistinguishable from (g, h, g^{r_2}, h^{r_3}) for random r_2, r_3 we can replace c_3 with a uniformly random element (independent of m). Notice that neither α (the encryption and sanitizer key) nor encryptions from oracle $\mathcal{O}_E$ will help the adversary distinguish. Thus, we can conclude that the adversary's advantage is negligible, since he cannot distinguish in all three cases.

No-Write Rule: We only need to consider two cases, depending on which keys the adversary asks for *before* producing the challenge ciphertext c and identity i': (1) the adversary asks for ek before issuing his challenge (c, i') with $i' \in \{0, 1\}$ (and receives no more keys during the distinguishing phase) and (2) the adversary asks for dk before issuing his challenge $(c, 0)$ and then asks for ek during the distinguishing phase. Case (1) follows directly from the DDH assumption: without access to the decryption key the output of the sanitizer is indistinguishable from a random ciphertext thanks to the choice of the random s_2, in particular since (g, h, g^{s_2}, h^{s_2}) is indistinguishable from (g, h, g^{s_2}, h^{s_3}) for random s_2, s_3 we can replace (c_0', c_1') with uniformly random elements in G. Case (2) instead has to hold unconditionally, since the adversary has the decryption key. We argue that the distribution of $\mathsf{San}(rk, (c_0, c_1, c_2, c_3))$ is independent of its input. In particular, given any (adversarially chosen) $(c_0, c_1, c_2, c_3) \in G^4$ we can write:

$$(c_0, c_1, c_2, c_3) = (g^{\delta_0}, g^{\delta_1}, g^{\delta_2}, g^{\delta_3})$$

Then the output $c' \leftarrow \mathsf{San}(rk, c)$ is

$$(c_0', c_1') = (c_2 c_0^{s_1} g^{s_2}, c_3 (g^{rk} c_1)^{s_1} h^{s_2})$$
$$= (g^{\delta_2 + s_1 \delta_0 + s_2}, g^{\delta_3 + s_1(\delta_1 - \alpha) + s_2 x})$$

Which is distributed exactly as a uniformly random ciphertext $(g^{\gamma_0}, g^{\gamma_1})$ with $(\gamma_0, \gamma_1) \in \mathbb{Z}_q \times \mathbb{Z}_q$ since for all (γ_0, γ_1) there exists (s_0, s_1) such that:

$$\gamma_0 = \delta_2 + s_1 \delta_0 + s_2 \quad \text{and} \quad \gamma_1 = \delta_3 + s_1(\delta_1 - \alpha) + s_2 x$$

This is guaranteed unless the two equations are linearly dependent i.e., unless $\alpha = (\delta_1 - x\delta_0)$ which happens only with negligible probability thanks to the principle of deferred decisions. The adversary is allowed to see sanitized ciphertext from the encryption oracle. However, this does not help him distinguish, since the output of the San algorithm is distributed exactly as a uniform ciphertext. Thus, we can conclude that the adversary's advantage is negligible, since he cannot distinguish in both cases. $\qquad\square$

3.2   Construction of an ACE Scheme for Multiple Identities

In this section we present a construction of an ACE scheme for multiple identities, which is based on the 1-ACE scheme in a black-box manner. In a nutshell, the idea is the following: we run n copies of the 1-ACE scheme and we give to each receiver j the decryption key dk_j for the j-th copy of the scheme. An encryption key for identity i is given by the set of encryption keys ek_j of the 1-ACE scheme such that $P(i, j) = 1$. To encrypt, a sender encrypts the same message m under all its encryption keys ek_j and puts random ciphertexts in the positions for which he does not know an encryption key. The sanitizer key contains all the sanitizer keys for the 1-ACE scheme: this allows the sanitizer to sanitize each component independently, in such a way that for all the positions for which the sender knows the encryption key, the message "survives" the sanitization, whereas in the other positions the output is uniformly random.

Example. We conclude this informal introduction of our repetition scheme by giving a concrete example of an ACE scheme for the Bell-LaPadula access control policy with three levels of access: *Level 1: top-secret, level 2: secret, level 3: public.* The predicate of this access control is defined as $P(i, j) = 1 \Leftrightarrow i \geq j$. This predicate ensures the property of *no write down* and *no read up* as discussed in the introduction. Table 1 shows the structure of the keys and the ciphertext for the different levels of access.

Table 1. Access Control Encryption Scheme for Bell-LaPadula access control policy. c_2', c_3', c_3'' are random ciphertexts from $\mathcal{C}$.

i	ek_i	dk_i			c
1	$\{ek_1\}$	dk_1			$(\mathsf{Enc}(ek_1, m), c_2', c_3')$
2	$\{ek_1, ek_2\}$		dk_2		$(\mathsf{Enc}(ek_1, m), \mathsf{Enc}(ek_2, m), c_3'')$
3	$\{ek_1, ek_2, ek_3\}$			dk_3	$(\mathsf{Enc}(ek_1, m), \mathsf{Enc}(ek_2, m), \mathsf{Enc}(ek_3, m))$

Construction 2. *Let* $1\mathsf{ACE} = (\mathsf{Setup}, \mathsf{Gen}, \mathsf{Enc}, \mathsf{San}, \mathsf{Dec})$ *be a 1-ACE scheme. Then we can construct an ACE scheme* $\mathsf{ACE} = (\mathsf{Setup}, \mathsf{Gen}, \mathsf{Enc}, \mathsf{San}, \mathsf{Dec})$ *defined by the following algorithms:*

Setup: Let n be the number of senders/receivers specified by the policy P. Then run n copies of the 1ACE setup algorithm

$$(pp_i^{1\mathsf{ACE}}, msk_i^{1\mathsf{ACE}}) \leftarrow 1\mathsf{ACE}.\mathsf{Setup}(1^\kappa) \quad \text{for } i = 1, \dots, n$$

For each of the 1ACE master secret keys run the 1ACE key generation algorithm on each of the three modes. For $i \in [n]$ do the following

$$ek_i^{1\mathsf{ACE}} \leftarrow 1\mathsf{ACE}.\mathsf{Gen}(msk_i^{1\mathsf{ACE}}, \mathsf{sen})$$

$$dk_i^{1\mathsf{ACE}} \leftarrow 1\mathsf{ACE}.\mathsf{Gen}(msk_i^{1\mathsf{ACE}}, \mathsf{rec})$$

$$rk_i^{1\mathsf{ACE}} \leftarrow 1\mathsf{ACE}.\mathsf{Gen}(msk_i^{1\mathsf{ACE}}, \mathsf{san})$$

Output the public parameter $pp = \{pp_i^{1\mathsf{ACE}}\}_{i\in[n]}$ and the master secret key $msk := \{ek_i^{1\mathsf{ACE}}, dk_i^{1\mathsf{ACE}}, rk_i^{1\mathsf{ACE}}\}_{i\in[n]}$.[11]

Key Generation: On input an identity $i \in \{0,\ldots,n+1\}$, a mode $\{\mathsf{sen, rec, san}\}$ and the master secret key msk, output a key depending on the mode

- $ek_i := \{ek_j^{1\mathsf{ACE}}\}_{j\in S}$, where $S \subseteq [n]$ is the subset s.t. $j \in S$ iff $P(i,j) = 1$;
- $dk_i := dk_i^{1\mathsf{ACE}}$;
- $rk := \{rk_j^{1\mathsf{ACE}}\}_{j\in[n]}$;

Encrypt: On input an encryption key ek_i and a message m encrypt the message under each of the 1ACE encryption keys in ek_i and sample uniform random ciphertext for each public key not in the encryption key. Thus, for $j = 1,\ldots,n$ do the following

- If $ek_j^{1\mathsf{ACE}} \in ek_i$ then compute $c_j^{1\mathsf{ACE}} \leftarrow 1\mathsf{ACE}.\mathsf{Enc}(ek_j^{1\mathsf{ACE}}, m)$.
- If $ek_j^{1\mathsf{ACE}} \notin ek_i$ then sample $c_j^{1\mathsf{ACE}} \leftarrow_\$ C_j^{1\mathsf{ACE}}$.[12]

Output the ciphertext $c := \left(c_1^{1\mathsf{ACE}},\ldots,c_n^{1\mathsf{ACE}}\right)$.

Sanitizer: On input a ciphertext c and a sanitizer key rk, sanitize each of the n 1ACE ciphertexts as follows

$$c_i'^{1\mathsf{ACE}} \leftarrow 1\mathsf{ACE}.\mathsf{San}(rk_i^{1\mathsf{ACE}}, c_i^{1\mathsf{ACE}}) \quad \text{for } i = 1,\ldots,n$$

Output the sanitized ciphertext $c' := (c_1'^{1\mathsf{ACE}},\ldots,c_n'^{1\mathsf{ACE}})$.

Decryption: On input a ciphertext c and a decryption key dk_i decrypt the i'th 1ACE ciphertext

$$m' \leftarrow 1\mathsf{ACE}.\mathsf{Dec}(dk_i^{1\mathsf{ACE}}, c_i^{1\mathsf{ACE}})$$

We can prove that the scheme presented above satisfies *correctness* as well as the *no-read* and the *no-write* rule, by reducing the properties of the repetition scheme to properties of the scheme with a single identity using hybrid arguments. The formal proofs are deferred to the full version [DHO16].

4 Polylogarithmic ACE from iO

In this section, we present our construction of ACE with polylogarithmic complexity in the number of roles n.

At first it might seem that it is easy to construct an ACE scheme with short ciphertexts by using NIZK and re-randomizable encryption: the sender would send to the sanitizer a ciphertext and a NIZK proving that the ciphertext is a well-formed encryption of some message using a public key that the sender is allowed to send to (for instance, each sender could have a signature on their

---

[11] There exists some encoding function that takes a message m from the message space of the ACE scheme and encodes it into a message of each of the 1-ACE message spaces. The ciphertext spaces of the ACE scheme are the crossproduct of all the 1-ACE ciphertext spaces, thus $C = C_1^{1\mathsf{ACE}} \times \cdots \times C_n^{1\mathsf{ACE}}$ and $C' = C_1'^{1\mathsf{ACE}} \times \cdots \times C_n'^{1\mathsf{ACE}}$.

[12] Here $c_j^{1\mathsf{ACE}} \leftarrow_\$ C_j^{1\mathsf{ACE}}$ is a shorthand for $c_j^{1\mathsf{ACE}} \leftarrow 1\mathsf{ACE}.\mathsf{Enc}(pp_j^{1\mathsf{ACE}}, \perp)$.

identity to be able to prove this statement). Now the sanitizer drops the NIZK and passes on the re-randomized ciphertext. However, the problem is that the sanitizer would need to know the public key of the intended receiver to be able to re-randomize (and we do not want to reveal who the receiver is).

As described in the introduction, we build our ACE scheme on top of a FE scheme which is *sanitizable*, which roughly means that given a ciphertext it is possible to produce a new encryption of the same message which is independent of the randomness used in the original encryption (this is a relaxation of the well-known *re-randomizability* property, in the sense that we do not require sanitized ciphertexts to look indistinguishable from fresh encryptions e.g., they can be syntactically different). We construct such an FE scheme by modifying the FE scheme of Garg et al. [GGH+13], and therefore our construction relies on the assumption that indistinguishability obfuscation exists. We define and construct sFE in Sect. 4.1 and then construct ACE based on sFE (and a regular PRF) in Sect. 4.3.

4.1 Sanitizable Functional Encryption Scheme – Definition

A *sanitizable functional encryption* (sFE) scheme is defined by the following algorithms:

Setup: The Setup algorithm on input the security parameter κ outputs a master secret key msk and public parameters pp, which include the message space $\mathcal{M}$ and ciphertext spaces $\mathcal{C}, \mathcal{C}'$.

Key Generation: The Gen algorithm on input the master secret key msk and a function f, outputs a corresponding secret key SK_f.

Encrypt: The Enc algorithm on input the public parameters pp and a message m, outputs a ciphertext $c \in \mathcal{C}$

Sanitizer: The San algorithm on input the public parameters pp and a ciphertext $c \in \mathcal{C}$, transforms the incoming ciphertext into a sanitized ciphertext $c' \in \mathcal{C}'$

Decryption: The Dec algorithm on input a secret key SK_f and a sanitized ciphertext $c' \in \mathcal{C}'$ that encrypts message m, outputs $f(m)$.

For the sake of exposition we also define a *master decryption algorithm* that on input $c \leftarrow \mathsf{Enc}(pp, m)$, returns $m \leftarrow \mathsf{MDec}(msk, c)$.[13] We formally define *correctness* and *IND-CPA* security for an sFE scheme (which are essentially the same as for regular FE), and then we define the new *sanitizable* property which, as described above, is a relaxed notion of the re-randomization property.

Definition 4 (Correctness for sFE). *Given a function family $\mathcal{F}$. For all $f \in \mathcal{F}$ and all messages $m \in \mathcal{M}$:*

$$\Pr\left[\mathsf{Dec}(\mathsf{Gen}(msk, f), \mathsf{San}(pp, \mathsf{Enc}(pp, m))) \neq f(m)\right] \leq \mathsf{negl}(\kappa)$$

---

[13] Formally MDec is a shortcut for $\mathsf{Dec}(\mathsf{Gen}(msk, f_{id}), \mathsf{San}(pp, c))$, where f_{id} is the identity function.

where $(pp, msk) \leftarrow \mathsf{Setup}(1^\kappa)$ and the probabilities are taken over the random coins of all algorithms.

Definition 5 (IND-CPA Security for sFE). *Consider the following game between a challenger C and a stateful adversary A:*

IND-CPA Security	
Game Definition	*Oracle Definition*
1. $(pp, msk) \leftarrow \mathsf{Setup}(1^\kappa)$; 3. $(m_0, m_1) \leftarrow A^{\mathcal{O}(\cdot)}(pp)$; 4. $b \leftarrow \{0, 1\}$; 5. $c^* \leftarrow \mathsf{Enc}(pp, m_b)$ 6. $b' \leftarrow A^{\mathcal{O}(\cdot)}(c^*)$;	$\mathcal{O}(f_i)$: 1. Output $SK_{f_i} \leftarrow \mathsf{Gen}(msk, f_i)$;

We say that A wins the IND-CPA game if $b = b'$, $|m_0| = |m_1|$, and that $f_i(m_0) = f_i(m_1)$ for all oracle queries f_i. We say a sFE scheme satisfies the IND-CPA security property if for all PPT A

$$\mathsf{adv}^A = 2 \cdot \left| \Pr[A \text{ wins the } IND-CPA \text{ game}] - \frac{1}{2} \right| \leq \mathsf{negl}(\kappa)$$

Definition 6 (Sanitization for sFE). *Consider the following game between a challenger C and a stateful adversary A:*

Sanitization
Game Definition

1. $(pp, msk) \leftarrow \mathsf{Setup}(1^\kappa)$;
 2. $c \leftarrow A(pp, msk)$;
 3. $b \leftarrow \{0, 1\}$,
 	
   – If $b = 0$, $c^* \leftarrow \mathsf{San}(pp, c)$;
   – If $b = 1$, $c^* \leftarrow \mathsf{San}(pp, \mathsf{Enc}(pp, \mathsf{MDec}(msk, c)))$;
 	
 4. $b' \leftarrow A(c^*)$;

We say that A wins the sanitizer game if $b = b'$. We say a sFE scheme is sanitizable if for all PPT A

$$\mathsf{adv}^A = 2 \cdot \left| \Pr[A \text{ wins the sanitizer game}] - \frac{1}{2} \right| \leq \mathsf{negl}(\kappa)$$

Note that in Definition 6 the adversary has access to the master secret key.

4.2 Sanitizable Functional Encryption Scheme – Construction

We now present a construction of a sFE scheme based on iO. The construction is based on the functional encryption construction by Garg et. al [GGH+13]. In their scheme a ciphertext contains two encryptions of the same message and a NIZK of this statement, thus an adversary can leak information via the randomness in the encryptions or the randomness in the NIZK. In a nutshell we make their construct sanitizable by:

1. Replacing the PKE scheme with a *sanitizable PKE* (as formalized in Definition 7).
2. Letting the sanitizer drop the original NIZK, and append a *proof of a proof* instead (i.e., a proof that the sanitizer knows a proof that would make the original verifier accept). Thanks to the ZK property the new NIZK does not contain any information about the randomness used to generate the original NIZK.
3. Changing the decryption keys (obfuscated programs) to check the new proof instead.

Building Blocks. We formalize here the definition of sanitization for a PKE scheme. Any re-randomizable scheme (such as Paillier and ElGamal) satisfies perfect PKE sanitization, but it might be possible that more schemes fit the definition as well.

Definition 7 (Perfect PKE Sanitization). *Let $\mathcal{M}$ be the message space and $\mathcal{R}$ be the space from which the randomness for the encryption and sanitization is taken. Then for every message $m \in \mathcal{M}$ and for all $r, s, r' \in \mathcal{R}$ there exists $s' \in \mathcal{R}$ such that*

$$\mathsf{San}(pk, \mathsf{Enc}(pk, m; r); s) = \mathsf{San}(pk, \mathsf{Enc}(pk, m; r'); s')$$

Our constructions also uses (by now standard) tools such as *pseudo-random functions (PRF)*, *indistinguishability obfuscation (iO)* and *statistical simulation-sound non-interactive zero-knowledge (SSS-NIZK)*, which are defined for completeness in Appendix A.

Constructing sFE. We are now ready to present our construction of sFE.

Construction 3. *Let* $\mathsf{sPKE} = (\mathsf{Setup}, \mathsf{Enc}, \mathsf{San}, \mathsf{Dec})$ *be a perfect sanitizable public key encryption scheme. Let* $\mathsf{NIZK} = (\mathsf{Setup}, \mathsf{Prove}, \mathsf{Verify})$ *be a statistical simulation-sound NIZK. Let* iO *be an indistinguishability obfuscator. We construct a sanitizable functional encryption scheme* $\mathsf{sFE} = (\mathsf{Setup}, \mathsf{Gen}, \mathsf{Enc}, \mathsf{San}, \mathsf{Dec})$ *as follows:*

Setup: On input the security parameter κ the setup algorithm compute the following

1. $(pk_1, sk_1) \leftarrow \mathsf{sPKE.Setup}(1^\kappa)$;
2. $(pk_2, sk_2) \leftarrow \mathsf{sPKE.Setup}(1^\kappa)$;
3. $crs_E \leftarrow \mathsf{NIZK.Setup}(1^\kappa, R_E)$;
4. $crs_S \leftarrow \mathsf{NIZK.Setup}(1^\kappa, R_S)$;
5. Output $pp = (crs_E, crs_S, pk_1, pk_2)$ and $msk = sk_1$;

The relations R_E and R_S are defined as follows: Let $x_E = (c_1, c_2)$ be a statement and $w_E = (m, r_1, r_2)$ a witness, then R_E is defined as

$$R_E = \{(x_E, w_E) \mid c_1 = \mathsf{sPKE.Enc}(pk_1, m; r_1) \wedge c_2 = \mathsf{sPKE.Enc}(pk_2, m; r_2)\}$$

Let $x_S = (c'_1, c'_2)$ be a statement and $w_S = (c_1, c_2, s_1, s_2, \pi_E)$ a witness, then R_S is defined as

$$R_S = \left\{ (x_S, w_S) \middle| \begin{array}{l} c'_1 = \mathsf{sPKE.San}(pk_1, c_1; s_1) \wedge c'_2 = \mathsf{sPKE.Enc}(pk_2, c_2; s_2) \\ \wedge \mathsf{NIZK.Verify}(crs_E, (c_1, c_2), \pi_E) = 1 \end{array} \right\}$$

Key Generation: On input the master secret key msk and a function f output the secret key $SK_f = iO(P)$ as the obfuscation of the following program

Program P
Input: c'_1, c'_2, π_S; Const: crs_S, f, sk_1; 1. If $\mathsf{NIZK.Verify}(crs_S, (c'_1, c'_2), \pi_S) = 1$;   output $f(\mathsf{sPKE.Dec}(sk_1, c'_1))$; 2. else output fail;

Encrypt: On input the public parameters pp and a message m compute two PKE encryptions of the message

$$c_1 \leftarrow \mathsf{sPKE.Enc}(pk_1, m; r_1)$$
$$c_2 \leftarrow \mathsf{sPKE.Enc}(pk_2, m; r_2)$$

with randomness (r_1, r_2). Then create a proof π_E that $(x_E, w_E) \in R_E$ with $x_E = (c_1, c_2)$ and witness $w_E = (m, r_1, r_2)$

$$\pi_E \leftarrow \mathsf{NIZK.Prove}(crs_E, x_E, w_E; t_E)$$

using randomness t_E. Output the triple $c = (c_1, c_2, \pi_E)$ as the ciphertext.

Sanitizer: On input the public parameter pp and a ciphertext $c = (c_1, c_2, \pi_E) \in \mathcal{C}$ compute the following

1. If $\mathsf{NIZK.Verify}(crs_E, x_E, \pi_E) = 1$ then
$$c'_1 \leftarrow \mathsf{sPKE.San}(pk_1, c_1; s_1)$$
$$c'_2 \leftarrow \mathsf{sPKE.San}(pk_2, c_2; s_2)$$
$$\pi_2 \leftarrow \mathsf{NIZK.Prove}(crs_S, x_S, w_S; t_S)$$
Output $c' = (c'_1, c'_2, \pi_2)$

2. Else

Output $c' \leftarrow \mathsf{sFE.San}(pp, \mathsf{sFE.Enc}(pp, \bot))$

with randomness (s_1, s_2) and t_S in the PKE and NIZK respectively. The generated NIZK is a proof that $(x_S, w_S) \in R_S$ with $x_S = (c'_1, c'_2)$ and $w_S = (c_1, c_2, s_1, s_2, \pi_E)$.

Decryption: On input a secret key SK_f and a ciphertext $c' = (c'_1, c'_2, \pi_S) \in \mathcal{C}'$, run the obfuscated program $SK_f(c'_1, c'_2, \pi_S)$ and output the answer.

Lemma 2. *Construction 3 is a correct functional encryption scheme.*

Proof. Correctness follows from the correctness of the iO, PKE, and SSS-NIZK schemes, and from inspection of the algorithms. □

Lemma 3. *For any adversary A that breaks the IND-CPA security property of Construction 3, there exists an adversary B for the computational zero-knowledge property of the NIZK scheme, an adversary C for the IND-CPA security of the PKE scheme, and an adversary D for iO such that the advantage of adversary A is*

$$\mathsf{adv}^{\mathsf{sFE},A} \leq 4|\mathcal{M}| \left(\mathsf{adv}^{\mathsf{NIZK},B} + \mathsf{adv}^{\mathsf{sPKE},C} + q \cdot \mathsf{adv}^{iO,C}(1 - 2p_{sss}) \right)$$

where q is the number of secret key queries adversary A makes during the game, and p_{sss} is the negligible soundness error of the SSS-NIZK scheme.

Proof. This proof follows closely the selective IND-CPA security proof of the FE construction presented by Garg et. al. [GGH+13]. See the full version [DHO16] for the complete proof. □

Lemma 4. *For any adversary A that breaks the sanitizer property of Construction 3, there exists an adversary B for the computational zero-knowledge property of the NIZK scheme such that the advantage of adversary A is*

$$\mathsf{adv}^{\mathsf{sFE},A} \leq 2|\mathcal{M}|\mathsf{adv}^{\mathsf{NIZK},B}$$

Proof. This lemma is proven via a series of indistinguishable hybrid games between the challenger and the adversary. For the proof to go through we notice that the challenger needs to simulate the NIZK proof. At a first look it might seem that the reduction needs to guess the entire ciphertext before setting up the system parameter, but in fact we show that it is enough to guess the *message* beforehand! Thus, we can use a complexity leveraging technique to get the above advantage. See the full version [DHO16] for the complete proof. □

4.3 Polylog ACE Scheme

In this section, we present a construction of an ACE scheme for multiple identities based on sanitizable functional encryption. The idea of the construction is the following: an encryption of a message m is a sFE encryption of the message together with the senders identity i and a MAC of the message based on the identity. Crucially, the encryption keys for all identities are generated in a pseudorandom way from a master key, thus it is possible to check MACs for all identities using a compact circuit. The sanitizer key is a sFE secret key for a special function that checks that the MAC is correct for the claimed identity. Then the sanitization consists of sanitizing the sFE ciphertext, and then using the sanitizer key to check the MAC. The decryption key for identity j is a sFE secret key for a function that checks that identity i in the ciphertext and identity j are allowed to communicate (and ignores the MAC). The function then outputs the message iff the check goes through.

Construction 4. *Let* sFE $=$ (Setup, Gen, Enc, San, Dec) *be a sanitizable functional encryption scheme. Let* F_1, F_2 *be pseudorandom functions. Then we can construct an ACE scheme* ACE $=$ (Setup, Gen, Enc, San, Dec) *defined by the following algorithms:*

Setup: Let $K \leftarrow \{0,1\}^{\kappa}$ be a key for the pseudorandom function F_1. Run $(pp^{\mathsf{sFE}}, msk^{\mathsf{sFE}}) \leftarrow \mathsf{sFE.Setup}(1^{\kappa})$. Output the public parameter $pp = pp^{\mathsf{sFE}}$ and the master secret key $msk = (msk^{\mathsf{sFE}}, K)$

Key Generation: Given the master secret key msk and an identity i, the encryption, decryption and sanitizer key are computed as follows:
  - $ek_i \leftarrow F_1(K, i)$
  - $dk_i \leftarrow \mathsf{sFE.Gen}(msk^{\mathsf{sFE}}, f_i)$
  - $rk \leftarrow \mathsf{sFE.Gen}(msk^{\mathsf{sFE}}, f_{rk})$

where the functions f_i and f_{rk} are defined as follows

Decryption function	Sanitizer function
$f_i(m, j, t)$: 1. If $P(j, i) = 1$: output m; 2. Else output $\bot$;	$f_{rk}(m, j, t)$: 1. $ek_j = F_1(K, j)$; 2. If $t = F_2(ek_j, m)$: output 1; 3. Else output 0;

Encryption: On input a message m and an encryption key ek_i, compute $t = F_2(ek_i, m)$ and output

$$c = \mathsf{sFE.Enc}(pp^{\mathsf{sFE}}, (m, i, t))$$

Sanitizer: Given a ciphertext c and the sanitizer key $rk = SK_{rk}$ check the MAC and output a sanitized FE ciphertext
  1. $c' = \mathsf{sFE.San}(pp^{\mathsf{sFE}}, c)$
  2. If $\mathsf{sFE.Dec}(SK_{rk}, c') = 1$: output c'
  3. Else output $\mathsf{San}(rk, \mathsf{Enc}(ek_0, \bot))$

Decryption: Given a ciphertext c' and a decryption key $dk_j = SK_j$ output

$$m' = \mathsf{sFE.Dec}(SK_j, c')$$

Lemma 5. *Construction 4 is a correct ACE scheme*

Proof. Let $P(i, j) = 1$ for some i, j. Let c' be a honest sanitization of a honest generated encryption of message m under identity i:

$$c' = \mathsf{San}(rk, \mathsf{Enc}(ek_i, m)) = \mathsf{sFE.San}(pp^{\mathsf{sFE}}, \mathsf{sFE.Enc}(pp^{\mathsf{sFE}}, (m, i, F_2(ek_i, m))))$$

Given the decryption key $dk_j = SK_j \leftarrow \mathsf{sFE.Gen}(msk, f_j)$. Then the correctness property of the sFE scheme gives

$$\Pr\left[\mathsf{Dec}(dk_j, c') = m\right] = \Pr\left[\mathsf{sFE.Dec}(SK_j, c') = m\right] \leq \mathsf{negl}(\kappa)$$

□

Theorem 1. *For any adversary A that breaks the No-Read Rule of Construction 4, there exists an adversary B for the IND-CPA security of the sanitizable functional encryption scheme, such that the advantage of A is*

$$\mathsf{adv}^{\mathsf{ACE}, A} \leq \mathsf{adv}^{\mathsf{sFE}, B}$$

Proof. Assume that any adversary wins the IND-CPA security game of the sanitizable functional encryption (sFE) scheme with advantage at most ϵ. Assume for contradiction that there is an adversary A that wins the ACE no-read game with advantage greater than ϵ, then we can construct an adversary B that wins the IND-CPA security game for the sFE scheme with advantage greater than ϵ.

B starts by generating $K \leftarrow \{0, 1\}^\kappa$ for some pseudorandom function F_1. Then B receives pp^{sFE} from the challenger and forwards it as the ACE public parameter to the adversary A. Adversary A then performs some oracle queries to $\mathcal{O}_G$ and $\mathcal{O}_E$ to which B replies as follows:

- B receives (j, sen), then he sends $ek_j \leftarrow F_1(K, j)$ to A.
- B receives (j, rec), then he makes an oracle query $\mathcal{O}(f_j)$ to the challenger and gets back SK_j. B sends $dk_j = SK_j$ to A.
- B receives (j, san), then he makes an oracle query $\mathcal{O}(f_{rk})$ to the challenger and gets back SK_{rk}. B sends $rk = SK_{rk}$ to A.
- B receives (i, m), then he computes $ek_i \leftarrow F_1(K, i)$ and sends to A

$$c \leftarrow \mathsf{sFE.Enc}(pp^{\mathsf{sFE}}, (m, i, F_2(ek_i, m)))$$

After the oracle queries B receives messages m_0, m_1 and identities i_0, i_1 from adversary A. Then B computes $ek_{i_l} \leftarrow F_1(K, i_l)$ for $l \in \{0, 1\}$ and sends m_0^{sFE} and m_1^{sFE} to the challenger, where $m_l^{\mathsf{sFE}} = (m_l, i_l, F_2(ek_{i_l}, m_l))$ for $l \in \{0, 1\}$. Then the sFE challenger sends a ciphertext c', which B forwards to A as the ACE ciphertext. This is followed by a new round of oracle queries.

If the sFE challenger is in case $b = 0$, then c' is generated as an sFE encryption of message m_0^{sFE}, and we are in the case $b = 0$ in the no-read game. Similar, if the sFE challenger is in case $b = 1$, then we are in the case $b = 1$ in the no-read game. Note that our adversary respects the rules of the IND-CPA game, since $f_{rk}(m_0^{\mathsf{sFE}}) = f_{rk}(m_1^{\mathsf{sFE}}) = 1$ and $f_j(m_0^{\mathsf{sFE}}) = f_j(m_1^{\mathsf{sFE}})$ for all j such that SK_j was queried. This follows directly from the payload privacy (the function outputs $\perp$) and sender anonymity ($m_0^{\mathsf{sFE}} = m_1^{\mathsf{sFE}}$) properties of the no-read rule. Thus, we can conclude that if A wins the no-read game with non-negligible probability, then B wins the IND-CPA security game for the sFE scheme. $\square$

Theorem 2. *For any adversary A that breaks the No-Write Rule of Construction 4, there exists an adversary B for the PRF security, an adversary C for the sanitizer property of the sFE scheme, and an adversary D for the IND-CPA security of the sFE scheme, such that the advantage of A is*

$$\mathsf{adv}^{\mathsf{ACE},A} \leq 3 \cdot \mathsf{adv}^{\mathsf{PRF},B} + \mathsf{adv}^{\mathsf{sFE},C} + \mathsf{adv}^{\mathsf{sFE},D} + 2^{-\kappa}$$

Proof. This theorem is proven by presenting a series of hybrid games.

Hybrid 0. The no-write game for $b = 1$

Hybrid 1. As Hybrid 0, except that when the challenger receives a oracle request (i, sen) he saves the identity: $I_S = I_S \cup i$, and the encryption key $ek_i \leftarrow F_1(K, i)$. When the challenger receives the challenge (c, i') he uses the sFE master decryption to get

$$(m^*, i^*, t^*) \leftarrow \mathsf{sFE.MDec}(msk^{\mathsf{sFE}}, c)$$

If $i^* \notin I_S$, then the challenger generates ek_{i^*} honestly. Next, he checks that $t^* = F_2(ek_{i^*}, m^*)$. If the check goes through he computes the challenge response as $c^* \leftarrow \mathsf{sFE.San}(pp^{\mathsf{sFE}}, c)$, otherwise $c^* \leftarrow \mathsf{San}(rk, \mathsf{Enc}(ek_0, \perp))$.

Hybrid 2. As Hybrid 1, except that the encryption keys are chosen uniformly at random: $ek_i \leftarrow_{\$} \{0, 1\}^\kappa$ for all i, (note that ek_{i^*} is also chosen at random).

Hybrid 3. As Hybrid 2, except that after receiving and master decrypting the challenge, the challenger check whether $i^* \in I_S$. If this is the case the challenger checks the MAC t^* as above, otherwise he compute the response as $c^* \leftarrow \mathsf{San}(rk, \mathsf{Enc}(ek_0, \perp))$.

Hybrid 4. As Hybrid 3, except that if the checks $i^* \in I_S$ and $t^* = F_2(ek_{i^*}, m^*)$ go through, then the challenger computes the response as

$$c^* \leftarrow \mathsf{sFE.San}(pp^{\mathsf{sFE}}, \mathsf{sFE.Enc}(pp^{\mathsf{sFE}}, (m^*, i^*, t^*)))$$

Hybrid 5. As Hybrid 4, except that the challenge response is computed as

$$c^* = \mathsf{San}(rk, \mathsf{Enc}(ek_{i'}, r))$$

where $r \leftarrow_{\$} \mathcal{M}$ and $rk \leftarrow \mathsf{Gen}(msk, n+1, \mathsf{san})$.

Hybrid 6. As Hybrid 5, except that the encryption keys are generated honestly: $ek_i \leftarrow F_1(K, i)$ for all i. Observe, this is the no-write game for $b = 0$.

Now we show that each sequential pair of the hybrids are indistinguishable.

Claim 1. Hybrids 0 and 1 are identical.

Proof. This follows directly from the definition of the sanitization and sanitizer key rk. □

Claim 2. For any adversary A that can distinguish Hybrid 1 and Hybrid 2, there exists an adversary B for the security of PRF F_1 such that the advantage of A is $\mathsf{adv}^A \leq \mathsf{adv}^{\mathsf{PRF},B}$.

Proof. Assume that any adversary can break the PRF security with advantage ϵ, and assume for contradiction that we can distinguish the hybrids with advantage greater than ϵ. Then we can construct an adversary B that breaks the PRF security with advantage greater than ϵ.

B starts by creating the public parameters honestly and sends it to the adversary. All the adversary oracle queries are answered as follows: whenever B receives (i, sen) from the adversary, he sends i to the PRF challenger, receives back y_i, set $ek_i := y_i$, and sends ek_i to the adversary. When B receives the challenge (i, m) he ask the challenger for the encryption key (as before), and encrypts m. The rest of adversary's queries are answered honestly by using the algorithms of the construction. When B receives (c, i') from the adversary, he master decrypts the ciphertext to get (m^*, i^*, t^*). If $i^* \notin I_S$, then B creates ek_{i^*} by sending i^* to the challenger. B concludes the game by forwarding the adversary's guess b' to the challenger.

Observe that the if $y_i \leftarrow F_1(K, i)$ then we are in Hybrid 1, and if y_i is uniform random, then we are in Hybrid 2. Thus, if adversary A can distinguish between the hybrids, then B can break the constraint PRF property. □

Claim 3. For any adversary A that can distinguish Hybrid 2 and Hybrid 3, there exists an adversary B' for the security of PRF F_2 such that the advantage of A is $\mathsf{adv}^A \leq \mathsf{adv}^{\mathsf{PRF},B'} + 2^{-\kappa}$.

Proof. Assume that any adversary can break the PRF security with advantage $\epsilon - 2^{-\kappa}$, and assume for contradiction that we can distinguish the hybrids with advantage greater than ϵ. Then we can construct an adversary B' that breaks the PRF security with advantage greater than $\epsilon - 2^{-\kappa}$.

B' starts by creating the public parameters and sending them to the adversary. The adversary's oracle queries are answered honestly by using the algorithms of the construction. When C receives the challenge (c, i') he master decrypts the ciphertext to get (m^*, i^*, t^*). Then he sends m^* to the challenger and receives back t'. If $t' = t^*$ then B' guess that the challenger is using the pseudorandom function F_2, otherwise B' guess that the challenger is using a random function.

We evaluate now the advantage of B' in the PRF game: Observe, if t' is generated using F_2, then B' outputs "PRF" with probability exactly ϵ. In the case when t' is generated using a random function, then it does not matter how t^* was created, and the probability that $t' = t^*$ is $2^{-\kappa}$. Thus, the advantage of adversary B' is greater than $\epsilon - 2^{\kappa}$. $\square$

Claim 4. For any adversary A that can distinguish Hybrid 3 and Hybrid 4, there exists an adversary C for the sanitizer property of the sFE scheme such that the advantage of A is $\mathsf{adv}^A \leq \mathsf{adv}^{\mathsf{sFE},C}$.

Proof. Assume that any adversary wins the sanitizer game for the sFE scheme with advantage ϵ, and assume for contradiction that we can distinguish the hybrids with advantage greater than ϵ. Then we can construct an adversary C that wins the sanitizer game with advantage greater than ϵ.

C starts by receiving the sFE system parameters from the challenger, and he forwards the public parameters as the ACE public parameters to the adversary. The adversary's oracle queries are answered honestly by using the algorithms of the construction, since C receives the sFE master secret key from the challenger. When C receives the challenge (c, i') he master decrypts the ciphertext to get (m^*, i^*, t^*). Then he checks that $i^* \in I_S$ and $t^* = F_2(ek_{i^*}, m^*)$. If the check goes through he sends c to the challenger and receives back a sFE sanitized ciphertext c'. Thus, the challenge response is $c^* = c'$. C concludes the game by forwarding the adversary's guess b' to the challenger.

Observe, if $c' = \mathsf{sFE.San}(pp^{\mathsf{sFE}}, c)$, then we are in Hybrid 3. On the other hand, we are in Hybrid 4 if

$$c' = \mathsf{sFE.San}(pp^{\mathsf{sFE}}, \mathsf{sFE.Enc}(pp^{\mathsf{sFE}}, \mathsf{sFE.MDec}(msk^{\mathsf{sFE}}, c)))$$

Thus, if adversary A can distinguish between the hybrids, then C can break the sFE sanitizer property. $\square$

Claim 5. For any adversary A that can distinguish Hybrid 4 and Hybrid 5, there exists an adversary D for the IND-CPA security of the sFE scheme such that the advantage of A is $\mathsf{adv}^A \leq \mathsf{adv}^{\mathsf{sFE},D}$.

Proof. Assume that any adversary wins the IND-CPA game for the sFE scheme with advantage ϵ, and assume for contradiction that we can distinguish the hybrids with advantage greater than ϵ. Then we can construct an adversary D that wins the IND-CPA game with advantage greater than ϵ.

D start by receiving the sFE public parameters from the challenger and forwards it to the challenger. The adversary's oracle queries are answered by sending secret key queries to the challenger, and otherwise using the algorithms of the construction (see the proof of Theorem 1 for more details). When D receives the challenge (c, i') he master decrypts the ciphertext to get (m^*, i^*, t^*). Then he checks that $i^* \in I_S$ and $t^* = F_2(ek_{i^*}, m^*)$. If the check goes through he set $m_0 = (m^*, i^*, t^*)$, otherwise he sets $m_0 = (\bot, 0, \bot)$. Then he creates $m_1 = (r, i', F_2(ek_{i'}, r))$, sends m_0 and m_1 to the challenger, and receives back an sFE encryption c'. Next, D creates the response $c^* = \mathsf{sFE.San}(pp^{\mathsf{sFE}}, c')$. D concludes the game by forwarding the adversary's guess b' to the challenger.

If c' is an encryption of the message m_0, then we are in Hybrid 4, and if it is an encryption of m_1, then we are in Hybrid 5. Thus, if adversary A can distinguish between the hybrids, then D can break the sFE IND-CPA security.   □

Claim 6. For any adversary A that can distinguish Hybrid 5 and Hybrid 6, there exists an adversary B for the security of PRF F_1 such that the advantage of A is $\mathsf{adv}^A \le \mathsf{adv}^{\mathsf{PRF},B}$.

The proof follow the same structure as the proof for Claim 2.

From these claims we can conclude that for any adversary A that can distinguish Hybrid 0 and Hybrid 6, there exists an adversary B for the PRF security, an adversary C for the sanitizer property of the sFE scheme, and an adversary D for the IND-CPA security of the sFE scheme, such that the advantage of A is

$$\mathsf{adv}^{\mathsf{ACE},A} \le 3 \cdot \mathsf{adv}^{\mathsf{PRF},B} + \mathsf{adv}^{\mathsf{sFE},C} + \mathsf{adv}^{\mathsf{sFE},D} + 2^{-\kappa}$$

□

A   Standard Building Blocks

A.1   Pseudorandom Function

Definition 8 (PRF). *We say $F : \{0,1\}^\kappa \times \{0,1\}^* \to \{0,1\}^\kappa$ is a pseudorandom function if for all PPT A*

$$\mathsf{adv}^A = 2 \cdot |\Pr[A^{\mathcal{O}_b(\cdot)}(1^\kappa) = b] - 1/2| < \mathsf{negl}(\kappa)$$

with $\mathcal{O}_0$ a uniform random function and $\mathcal{O}_1 = F_K$.

A.2 Statistical Simulation-Sound Non-Interactive Zero-Knowledge Proofs

The content of this subsection is taken almost verbatim from [GGH+13]. Let L be a language and R a relation such that $x \in L$ if and only if there exists a witness w such that $(x, w) \in R$. A non-interactive proof system [BFM88] for a relation R is defined by the following PPT algorithms

Setup: The Setup algorithm takes as input the security parameter κ and outputs common reference string crs.

Prove: The Prove algorithm takes as input the common reference string crs, a statement x, and a witness w, and outputs a proof π.

Verify: The Verify algorithm takes as input the common reference string crs, a statement x, and a proof π. It outputs 1 if it accepts the proof, and 0 otherwise.

The non-interactive proof system must be complete, meaning that if $R(x, w) = 1$ and $crs \leftarrow \mathsf{Setup}(1^\kappa)$ then

$$\mathsf{Verify}(crs, x, \mathsf{Prove}(crs, x, w)) = 1$$

Furthermore, the proof system must be statistical sound, meaning that no (unbounded) adversary can convince a honest verifier of a false statement. Moreover, we define the following additional properties of a non-interactive proof system.

Definition 9 (Computational Zero-Knowledge). *A non-interactive proof* $\mathsf{NIZK} = (\mathsf{Setup}, \mathsf{Prove}, \mathsf{Verify})$ *is computational zero-knowledge if there exists a polynomial time simulator* $\mathsf{Sim} = (\mathsf{Sim}_1, \mathsf{Sim}_2)$ *such that for all non-uniform polynomial time adversaries* A *we have for all* $x \in L$ *that*

$$\Pr\left[crs \leftarrow \mathsf{Setup}(1^\kappa); \pi \leftarrow \mathsf{Prove}(crs, x, w) : A(crs, x, \pi) = 1\right]$$
$$\approx$$
$$\Pr\left[(crs, \tau) \leftarrow \mathsf{Sim}_1(1^\kappa, x); \pi \leftarrow \mathsf{Sim}_2(crs, \tau, x) : A(crs, x, \pi) = 1\right]$$

where crs *is the common reference string,* x *is the statement,* w *is the witness,* π *is the proof, and* τ *is the trapdoor.*

Thus, the definition states that the proof do not reveal any information about the witness to any bounded adversary. In the definition this is formalized by the existence of two simulators, where Sim_1 returns a simulated common reference string together with a trapdoor that enables Sim_2 to simulate proofs without access to the witness.

Definition 10 (Statistical Simulation-Soundness). *A non-interactive proof* NIZK = (Setup, Prove, Verify) *is statistical simulation-sound (SSS) if for all statements* x *and all (unbounded) adversaries* A *we have that*

$$\Pr\left[\begin{array}{l}(crs,\tau) \leftarrow \mathsf{Sim}_1(1^{\kappa}, x); \pi \leftarrow \mathsf{Sim}_2(crs,\tau,x): \\ \exists(x',\pi') : x' \neq x : \mathsf{Verify}(crs,x',\pi') = 1 : x' \notin L\end{array}\right] \leq p_{sss}$$

where $p_{sss} = \mathsf{negl}(\kappa)$ *is negligible in the security parameter.*

Thus, the definition states that it is not possible to convince a honest verifier of a false statement even if the adversary is given a simulated proof.

Remark 1. If a proof system is statistical simulation-sound then it is also statistical sound. Thus, we can upper bound the negligible probability of statistical soundness by the negligible probability of the statistical simulation-soundness.

A.3   Indistinguishability Obfuscation

We use an *indistinguishability obfuscator* like the one proposed in [GGH+13] such that $\bar{C} \leftarrow iO(C)$ which takes any polynomial size circuit C and outputs an obfuscated version $\tilde{C}$ that satisfies the following property.

Definition 11 (Indistinguishability Obfuscation). *We say* iO *is an* indistinguishability obfuscator *for a circuit class* $\mathcal{C}$ *if for all* $C_0, C_1 \in \mathcal{C}$ *such that* $\forall x : C_0(x) = C_1(x)$ *and* $|C_0| = |C_1|$ *it holds that:*

1. $\forall C \in \mathcal{C}, \forall x \in \{0,1\}^n, iO(C)(x) = C(x);$
2. $|iO(C)| = \mathsf{poly}(\lambda|C|)$
3. *for all PPT* $\mathcal{A}$*:*

$$\mathsf{adv}^{\mathcal{A}} = 2 \cdot |\Pr[\mathcal{A}(iO(C_0)) = 1] - \Pr[\mathcal{A}(iO(C_1)) = 1]| < \mathsf{negl}(\lambda)$$

References

[BFM88] Blum, M., Feldman, P., Micali, S.: Non-interactive zero-knowledge and its applications (extended abstract). In: Proceedings of the 20th Annual ACM Symposium on Theory of Computing, May 2–4, 1988, Chicago, Illinois, USA, pp. 103–112 (1988)

[Bib75] Biba, K.J.: Integrity considerations for secure computer systems. No. MTR-3153-REV-1. MITRE Corp., Bedford, MA (1975)

[BL73] Bell, D.E., LaPadula, L.J.: Secure computer systems: Mathematical foundations. Draft MTR, The MITRE Corporation, 2 (1973)

[BP03] Backes, M., Pfitzmann, B.: Intransitive non-interference for cryptographic purpose. In: 2003 IEEE Symposium on Security and Privacy (S&P 2003), 11–14 May 2003, Berkeley, CA, USA, p. 140 (2003)

[BP04] Backes, M., Pfitzmann, B.: Computational probabilistic noninterference. Int. J. Inf. Sec. **3**(1), 42–60 (2004)

[BSW11] Boneh, D., Sahai, A., Waters, B.: Functional encryption: definitions and challenges. In: Ishai, Y. (ed.) TCC 2011. LNCS, vol. 6597, pp. 253–273. Springer, Heidelberg (2011)

[DHO16] Damgård, I., Haagh, H., Orlandi, C.: Access control encryption: enforcing information flow with cryptography. Cryptology ePrint Archive, Report 2016/106 (2016). http://eprint.iacr.org/2016/106

[DMS15] Dodis, Y., Mironov, I., Stephens-Davidowitz, N.: Message transmission with reverse firewalls—secure communication on corrupted machines. In: Robshaw, M., Katz, J. (eds.) CRYPTO 2016. LNCS, vol. 9814, pp. 341–372. Springer, Heidelberg (2016). doi:10.1007/978-3-662-53018-4_13

[FAL06] Frikken, K., Atallah, M., Li, J.: Attribute-based access control with hidden policies and hidden credentials. IEEE Trans. Comput. **55**(10), 1259–1270 (2006)

[FF15] Fehr, V., Fischlin, M.: Sanitizable signcryption: Sanitization over encrypted data (full version). IACR Cryptology ePrint Archive, 2015:765 (2015)

[FFLW15] Ferrara, A.L., Fuchsbauer, G., Liu, B., Warinschi, B.: Policy privacy in cryptographic access control. In: IEEE 28th Computer Security Foundations Symposium, CSF 2015, Verona, Italy, 13–17 July, 2015, pp. 46–60 (2015)

[Gam85] El Gamal, T.: A public key cryptosystem and a signature scheme based on discrete logarithms. IEEE Trans. Inform. Theory **31**(4), 469–472 (1985)

[GGH+13] Garg, S., Gentry, C., Halevi, S., Raykova, M., Sahai, A., Waters, B.: Candidate indistinguishability obfuscation and functional encryption for all circuits. In: 54th Annual IEEE Symposium on Foundations of Computer Science, FOCS 2013, 26–29 October 2013, Berkeley, CA, USA, pp. 40–49 (2013)

[GJJS04] Golle, P., Jakobsson, M., Juels, A., Syverson, P.: Universal re-encryption for mixnets. In: Okamoto, T. (ed.) CT-RSA 2004. LNCS, vol. 2964, pp. 163–178. Springer, Heidelberg (2004). doi:10.1007/978-3-540-24660-2_14

[GPSW06] Goyal, V., Pandey, O., Sahai, A., Waters, B.: Attribute-based encryption for fine-grained access control of encrypted data. In: Proceedings of the 13th ACM Conference on Computer and Communications Security, CCS 2006, Alexandria, VA, USA, October 30 - November 3, 2006, pp. 89–98 (2006)

[HKN05] Halevi, S., Karger, P.A., Naor, D.: Enforcing confinement in distributed storage and a cryptographic model for access control. IACR Cryptology ePrint Archive 2005:169 (2005)

[HLA02] Hopper, N.J., Langford, J., Ahn, L.: Provably secure steganography. In: Yung, M. (ed.) CRYPTO 2002. LNCS, vol. 2442, pp. 77–92. Springer, Heidelberg (2002). doi:10.1007/3-540-45708-9_6

[KSW13] Katz, J., Sahai, A., Waters, B.: Predicate encryption supporting disjunctions, polynomial equations, and inner products. J. Cryptology **26**(2), 191–224 (2013)

[KTS07] Kapadia, A., Tsang, P.P., Smith, S.W.: Attribute-based publishing with hidden credentials and hidden policies. In: Proceedings of the Network and Distributed System Security Symposium, NDSS 2007, San Diego, California, USA, 28th February - 2nd March 2007 (2007)

[MK11] Müller, S., Katzenbeisser, S.: Hiding the policy in cryptographic access control. In: Meadows, C., Fernandez-Gago, C. (eds.) STM 2011. LNCS, vol. 7170, pp. 90–105. Springer, Heidelberg (2012). doi:10.1007/978-3-642-29963-6_8

[MS15] Mironov, I., Stephens-Davidowitz, N.: Cryptographic reverse firewalls. In: Oswald, E., Fischlin, M. (eds.) EUROCRYPT 2015. LNCS, vol. 9057, pp. 657–686. Springer, Heidelberg (2015). doi:10.1007/978-3-662-46803-6_22

[PR07] Prabhakaran, M., Rosulek, M.: Rerandomizable RCCA encryption. In: Menezes, A. (ed.) CRYPTO 2007. LNCS, vol. 4622, pp. 517–534. Springer, Heidelberg (2007). doi:10.1007/978-3-540-74143-5_29

[RZB12] Raykova, M., Zhao, H., Bellovin, S.M.: Privacy enhanced access control for outsourced data sharing. In: Keromytis, A.D. (ed.) FC 2012. LNCS, vol. 7397, pp. 223–238. Springer, Heidelberg (2012). doi:10.1007/978-3-642-32946-3_17

Author Index

Printed in the United States
By Bookmasters